T0251697

COMPUTER METHODS IN BIOMECHANICS & BIOMEDICAL ENGINEERING – 2

SYMPOSIUM ORGANISERS

J. MIDDLETON
University of Wales Swansea
Swansea, UK

M. L. JONES
University of Wales College of Medicine
Cardiff, UK

G. N. PANDE
University of Wales Swansea
Swansea, UK

TECHNICAL ADVISORY PANEL

J. Bonet
Barcelona, Spain
J. M. Crolet
Besancon, France
J. H. Heegaard
Stanford, USA
S. J. Hollister
Michigan, USA
R. Huiskes
Nijmegen, The Netherlands
I. Knets
Riga, Latvia
A. McCulloch
San Diego, USA
A. N. Natali
Padova, Italy
C. Oomens
Eindhoven, The Netherlands
T. M. Peters
Montreal, Canada
B. R. Simon
Arizona, USA
N. G. Shrive
Calgary, Canada
R. L. Spilker
Renssalaer, USA
G. Steven
Sydney, Australia
R. Summers
London, UK
K. Tanne
Hiroshima, Japan
D. Taylor
Dublin, Ireland
A. Toni
Bologna, Italy

SPONSORS

3M Unitek (Division of 3M Health Care Ltd.)
Gordon & Breach Publishers
Welsh Development Agency, Wales, UK
Wilde & Partners (FE Division)

COMPUTER METHODS IN BIOMECHANICS & BIOMEDICAL ENGINEERING – 2

Edited by

J. Middleton

*University of Wales Swansea,
Swansea, UK*

M. L. Jones

*University of Wales College of Medicine,
Cardiff, UK*

G. N. Pande

*University of Wales Swansea,
Swansea, UK*

Gordon and Breach Science Publishers
Australia • Canada • China • France • Germany • India
Japan • Luxembourg • Malaysia • The Netherlands
Russia • Singapore • Switzerland

Amsteldijk 166
1st Floor
1079 LH Amsterdam
The Netherlands

British Library Cataloguing in Publication Data

A catalogue record for this book is available from the British Library

ISBN: 90-5699-206-6

CONTENTS

3. BONE ADAPTATION, STRUCTURAL MODELS AND ARCHITECTURE

PREFACE

The first two symposia on *Computer Methods in Biomechanics and Biomedical Engineering* were held in Swansea, May 1992 and September 1994, and provided a focus for the utilisation of computer methods for solving the many complex problems encountered in the field of biomechanics. These symposia, we believe, have been successful in bringing together scientists, clinicians and analysts involved in this multidisciplinary subject and based on the previous format the third symposium in the series was held in Barcelona, Spain in May 1997. This book contains over one hundred papers that were presented at the meeting from contributors throughout the world.

It is clear, from the interest shown in this subject area, that the use of computer-based models in biomedical engineering is still rapidly expanding and new and novel solutions are being developed many of these being for previously intractable problems. The selection of invited and contributed papers which form this proceedings reflect these new and recent advances and with notable presentations on topics such as virtual reality techniques and reconstructive surgery providing a glimpse of exciting trends for the future. Many new advances in areas such as bone structures and image analysis, spine biomechanics, mixture theories for soft tissue and biofluid computations were presented. Also important contributions in multibody systems, 3D modelling and non-linear computations were given. These have been subdivided into the following headings:

1. MULTIBODY SYSTEMS AND JOINT MODELS
2. HIP REPLACEMENTS: PROSTHESIS/CEMENT/BONE ANALYSIS
3. BONE ADAPTATION, STRUCTURAL MODELS AND ARCHITECTURE
4. SPINE AND VERTEBRA MECHANICS
5. RECONSTRUCTIVE SURGERY, VIRTUAL REALITY AND IMPLANT ANALYSIS
6. SOFT TISSUE STRUCTURES, CONTACT AND BIOFLUID MECHANICS
7. DENTAL MATERIALS, BEHAVIOUR AND BIOMECHANICS
8. CRANIO-FACIAL MECHANICS AND DIAGNOSTIC METHODS

These presentations and the optimism shown during the meeting provide further evidence that computer modelling is making remarkable progress and is indeed becoming an essential toolkit in the field of biomechanics and biomedical engineering. We believe that the contents of this book provides evidence of this and furthermore gives a comprehensive review of the theoretical developments and application of modelling techniques in biomechanics.

We wish to thank the invited speakers, authors, delegates and session chairs for the many stimulating presentations and lively discussions that contributed to the success of the meeting. Finally we would like to thank the Technical Advisory Panel for their support, the organisations who endorsed and sponsored the event and the City of Barcelona Convention Bureau for providing a delightful venue.

J. Middleton
M. L. Jones
G. N. Pande

CONTRIBUTORS

B. S. Acar	*Loughborough University, UK*
G. Acquisti	*Rizzoli Institute, Bologna, Italy*
S. Albalat	*Universidad Politecnica Valencia, Spain*
M. Alcañiz	*Universidad Politecnica Valencia, Spain*
P. Allard	*University of Montréal, PQ, Canada*
F. C. Anderson	*University of Texas at Austin, USA*
J. r. Arlaud	*Centre Borely-Mermoz, Marseille, France*
E. Astoin	*Laboratory of Biomechanics, ENSAM, Paris, France*
A. L. Baldwin	*University of Arizona, Tucson, USA*
M. Baleani	*Rizzoli Institute, Bologna, Italy*
C. Baréa	*Inserm U305, Centre Hospitalier Hôtel-Dieu, Toulouse, France*
D. S.Barker	*Repatriation General Hospital, Daw Park, S.A. Australia*
E. Barnes	*University of Teesside, Middlesbrough, UK*
F. Baruffaldi	*Istituti Ortopedici Rizzoli, Bologna, Italy*
A. D. Bassanta	*Sao Paulo University, Brazil*
J. F. Benvenuti	*Orthopaedic Hospital and Federal Institute of Technology, Lausanne, Switzerland*
G. Bessonnet	*bessonne@l.m.suniv-poitiers.fr*
C. Birnie	*University of Cape Town, South Africa*
M. I. G. Bloor	*University of Leeds, UK*
N. Börlin	*Umeå University, Sweden*
P. H. M. Bovendeerd	*Eindhoven University of Technology, The Netherlands*
J. C. Brignola	*Société Euros, La Ciotat, France*
J. M. Brown	*University of Leeds, UK*
R. Brown	*University of Wales Institute Cardiff, UK*
S. Brumby	*Royal Adelaide Hospital, S.A., Australia*
B. Buck	*Technical University Darmstadt, Germany*
M. Bürgi	*Sulzer Orthopedics, Winterthur, Switzerland*
C. Capozzolo	*University di Napoli 'Federico II', Italy*
A. Cappello	*Universita di Bologna, Italy*
F. Carli	*University of Pavia, Italy*
F. Catini	*Istituti Ortopedici Rizzoli, Bologna, Italy*
E. Cendte	*Laboratoire de Contrôle Non Destructif, INSA, Villeurbanne, France*
F. Chinesta	*Universidad Politecnica Valencia, Spain*
K. Chinzei	*Mechanical Engineering Laboratory, Ibaraki, Japan*
Y. S. Choo	*CAD/CAM/CAE Centre, National University of Singapore*
R. Cimdins	*Riga Technical University, Latvia*
S. E. Clift	*University of Bath, UK*
S. Conforto	*University of Rome 'La Sapienza', Italy and University of Rome III*

R. Contro	*Politecnico di Milano, Milan, Italy*
E. Cordier	*INRIA Rhône-Alpes, Montbonnot St. Martin, France*
O. Còussi	*coussi@lms.univ-poitiers.fr*
B. Couteau	*Inserm U305, Centre Hospitalier Hôtel Dieu, Toulouse, France*
L. Cristofolini	*University of Bologna, Italy*
J. M. Crolet	*Université de Franche-Compté, Besançon, France*
T. D'Alessio	*University of Rome III, Italy*
F. Danes	*danes@lms.univ-poitiers.fr*
C. Danias	*University of Wales College of Medicine, Cardiff, UK*
J. Dansereau	*École Polytechnique de Montréal, PQ, Canada*
R. Darmana	*Centre Hospitalier Hôtel Dieu Toulose, France*
B. R. Davis	*Pennsylvania State University, Hershey PA, USA*
M. De Cooman	*Katholieke Universiteit Leuven, Heverlee, Belgium*
P. Deforge	*Laboratoire Vision Robotique, Bourges, France*
M. R. de Leval	*Great Ormond Street Hospital for Children NHS Trust, London, UK*
G. Delòge	*Stewal Implants, Ronse, Belgium*
M. del Pilar	*Federal University of Rio de Janeiro, Brazil*
F. M. Denaro	*Second Università di Napoli, Aversa, Italy*
R. A. Dickson	*St. James' University Hospital, Leeds, UK*
M. Dobelis	*Riga Technical University, Latvia*
C. P. Doube	*Nulite System International Pty Ltd, Hornsby, NSW, Autralia*
H. Druyts	*Katholieke Universitèit Leuven, Heverlee, Belgium*
G. Dubini	*Politecnico di Milano, Italy*
M. Duhaime	*Shriners Hospital, Montréal, PQ, Canada*
N. A. Duncan	*University of California at San Francisco, USA*
J. Duyck	*Katholieke Universiteit Leuven, Heverlee, Belgium*
B. Espiau	*INRIA, Montbonnot St. Martin, France*
F. X. Espiau	*INRIA, Montbonnot St. Martin, France*
É. Estivalèzes	*Inserm U305, Centre Hospitalier Hôtel Dieu, Toulouse, France*
G. Fabry	*Katholieke Universiteit Leuven, Pellenberg, Belgium*
F. Fedele	*Università 'La Sapienza', Rome, Italy*
J. D. Feikes	*Oxford Orthopaedic Engineering Centre and Nuffield Orthopaedic Centre, University of Oxford, UK*
C. B. Frank	*University of Calgary, Alberta, Canada*
P. Fridez	*Federal Institute of Technology, Lausanne, Switzerland*
A. J. H. Frijns	*Eindhoven University of Technology and University of Maastricht, The Netherlands*
R. Fumero	*Politecnico di Milano, Italy*
D. P. Fyhrie	*Henry Ford Hospital, Detroit MI USA*
T. N. Gardner	*University of Oxford, UK*
B. A. Garner	*University of Texas at Austin, USA*

E. Genda	*Rosai Rehabilitation Center, Nagoya, Japan*
C. R. Gentle	*Nottingham Trent University, UK*
A. W. J. Gielen	*Eindhoven University of Technology, The Netherlands*
H. S. Gill	*Nuffield Orthopaedic Centre, University of Oxford, UK*
M. Giuntini	*Pontíficia Universidade, Católica do Rio de Janeiro, Brazil*
R. Gobin	*Katholieke Universiteit Leuven Heverlee, Belgium*
C. Götz	*Universität Tübingen, Germany*
C. Gomes	*Federal University of Rio de Janeiro, Brazil*
A. Goswami	*INRIA Rhône-Alpes, Montbonnot St. Martìn, France*
R. M. Grassmann	*University of Calgary, Alberta, Canada*
V. Grau	*Universidad Politecnica Valencia, Spain*
S. L. Grilli	*Loughborough, University, UK*
P. Gross	*University of Strathclyde, Glasgow, UK*
B. Haex	*Katholieke Universiteit Leuven, Heverlee, Belgium*
N. Haibara	*Tokoyo Medical and Dental University, Japan*
T. Hara	*Niigata University, Japan*
S. L. Hartle	*Nottingham Trent University, UK*
G. W. Hastings	*IMRE and BIOMAT, National University of Singapore*
M. B. Hecke	*Federal University of Paraná, Brazìl*
J. H. Heegaard	*Stanford University, USA*
F. Heitaplatz	*Nottingham Trent University, UK*
R. Herrera	*Laboratoire d'Orthèses et de Prothèses Medicus, Montréal, PQ, Canada*
J. Hickman	*University of Wales College of Medicine, Cardiff, UK*
M. C. Hobatho	*Inserm U305, Centre Hospitalier Hôtel Dieu, Toulouse, France*
R. J. Hollands	*University of Sheffield, UK*
E. Hòrii	*Nagoya University, Japan*
D. R. Hose	*University of Sheffield, UK*
I. C. Howard	*University of Sheffield, UK*
D. Howie	*Royal Adelaide Hospital, S.A., Australia*
J. M. Huyghe	*Eindhoven University of Technology, The Netherlands*
A. Iafrati	*CIRA, Capua, Italy*
A. Imràn	*Orthopaedic Engineering Centre, University of Oxford, UK*
D. Inglis	*McMaster University, Hamilton, Ont., Canada*
M. Ishida	*Hiroshima University School of Dentistry, Japan*
T. Ishida	*Tokoyo Medical and Dental University, Japan*
C. R. Jacobs	*Pennsylvania State University, Hershey PA, USA*
J. D. Janssen	*Eindhoven University of Technology, The Netherlands*
J. Jíra	*Czech Technical University, Prague, Czech Republic*
J. Jirová	*Institute of Theoretical and Applied Mechanics, Prague, Czech Republic*
M. L. Jones	*University of Wales College of Medicine Cardiff, UK*
T. Kasahara	*Rosai Rehabilitation Engineering Center, Nagoya, Japan*

M. V. Kaufmann	*The Hewlett-Packard Company, Palo Alto, CA, USA*
K. Kędzior	*Warsaw University of Technology, Poland*
K. Khálaf	*Ohio State University, Columbus, OH, USA*
P. J. Kilner	*Royal Brompton Hospital, London, UK*
I. Knets	*Riga Technical University, Latvia*
M. Knothe-Tate	*Federal Institute of Technology, Zürich, Switzerland*
J. Knox	*University of Wales College of Medicine, Cardiff, UK*
B. Kralj	*University of Wales College of Medicine and University of Wales, Swansea, UK*
G. Kullmer	*University of Paderborn, Germany*
L. Labey	*Katholieke Universiteit Leuven, Heverlee, Belgium*
A. Laib	*University and ETH Zürich, Switzerland*
J. Laizans	*Riga Technical University, Latvia*
F. Lavaste	*Laboratory of Biomechanics, ENSAM, Paris, France*
F. Lbath	*Laboratoire de Mècanique des Solides, INSA, Villeurbanne, France*
A. Leardini	*Department of Engineering Science, University of Oxford and Oxford Orthopaedic Engineering Centre, Oxford UK*
M. Lee	*University of Sydney, NSW, Australia*
V. S. P. Lee	*University of Strathclyde, Glasgow, UK*
M. Lengsfeld	*Philipps-University Marburg, Germany*
A. B. Lennon	*University College Dublin, Ireland*
P. F. Leyvraz	*Orthopaedic Hospital, Lausanne, Switzerland*
W. Li	*University of Sydney, NSW, Australia*
S. Lievens	*Katholieke Universiteit Leuven, Heverlee, Belgium*
G. Limbert	*Inserm U305, Centre Hospitalier Hôtel Dieu, Toulouse, France*
T. Lindh	*Umeå University, Sweden*
J. Liu	*University of Arizona, Tucson, USA*
J. C. Lotz	*University of California at San Francisco, USA*
T. W. Lu	*Orthopaedic Engineering Centre, University of Oxford, UK*
R. M. Manfredi	*Università 'La Sapienza', Rome, Italy*
M. Mansat	*CHU Purpan, Service d'Orthopédie Traumatologie, Toulouse, France*
S. Mantero	*Politecnico di Milano, Milan, Italy*
P. Marché	*Laboratoire Vision Robotique, Bourges, France*
L. Marks	*Desktop Engineering, Long Hanborough, UK*
M. Martelli	*University of Trento, Italy*
S. Matsubara	*Nara Medical University, Japan*
A. D. McCarthy	*University of Sheffield, UK*
B. A. O. McCormack	*University College Dublin, Ireland*
L. L. Menegaldo	*Escola Politécnica da Universidade de São Paulo, Brazil*
R. Mericske-Stern	*University of Berne, Switzerland*
E. A. Meroi	*Università dì Padova, Italy*

B. Merz	*Institut StraumannAG, Waldenburg, Switzerland*
M. Micka	*Institute of Theoretical and Applied Mechanics, Prague, Czech Republic*
J. Middleton	*University of Wales, Swansea, UK*
B. Migeon	*Laboratoire Vision Robotique, Bourges, France*
F. Migliavacca	*Politecnico di Milano, Italy*
K. Miller	*University of Western Australia, Perth, Australia*
P. A. Millner	*St. James's University Hospital, Leeds, UK*
D. Mitton	*Laboratoire de Mècanique des Solides, INSA, Villeurbanne, France*
C. Monserrat	*Universidad Politecnica Valencia, Spain*
M. Mulier	*Catholic University of Leuven, Pellenberg, Belgium*
I. Naert	*Katholieke Universiteit Leuven, Heverlee, Belgium*
A. Nakatsuka	*Hiroshima University School of Dentistry, Japan*
Y. Nakayama	*Toray Research Centre, Shiga, Japan*
A. N. Natali	*Università di Padova, Italy*
N. Nawaña	*Howmedica, Herouville-Saint-Clair Cedex, France*
A. M. R. New	*Robert Jones and Agnes Hunt Orthopaedic Hospital, Oswestry, Shropshire and Queen Mary and Westfield College, London, UK*
M. D. Northmore-Ball	*Robert Jones and Agnes Hunt Orthopaedic Hospital Oswestry, Shropshire, UK*
J. J. O' Connor	*Oxford Orthopaedic Engineering Centre and Department of Engineering Science, University of Oxford, UK*
C. W. Oomens	*Eindhoven University of Technology, The Netherlands*
D. Pamplona	*Pontifícia Universidade Cátolica do Rio de Janeiro, Brazil*
G. N. Pande	*University of Wales, Swansea, UK*
M. G. Pandy	*University of Texas at Austin, USA*
M. Parnianpour	*Ohio State University, Columbus, OH, USA*
E. M. Paul	*Pennsylvania State University, Hershey PA, USA*
J. P. Paul	*University of Strathclyde, Glasgow, UK*
M. Pearcy	*Queensland University of Technology Brisbane, Queensland, Australia*
G. Pennati	*Politecnico di Milano, Italy*
J. M. T. Penrose	*University of Sheffield, UK*
R. Perucchio	*University of Rochester, NY, USA*
L. Pierotti	*Policlinico S. Orsola-Malpighi, Bologna, Italy*
R. Pietrabissa	*Politecnico di Milano, Milan, Italy*
S. Pietruszczak	*McMaster University, Hamilton, Ont., Canada*
D. P. Pioletti	*Orthopaedic Hospital and Federal Institute of Technology, Lausanne, Switzerland*
S. Piszczatowski	*Bialystok University of Technology, Poland*
H. L. Ploeg	*Queen's University, Kingston, ON, Canada and Sulzer Orthopedics Ltd, Winterthur, Switzerland*

A. L. Possobom   *Federal University of Paraná, Brazil*
P. J.' Prendergast   *Trinity College Dublin, Ireland*
Ch. Provatidis   *National Technical University of Athens, Greece*
R. Puers   *Katholieke Universiteit Leuven, Heverlee, Belgium*

V. Quaglini   *Politecnico di Milano, Milan, Italy*
M. Raìmondi   *Politecnico di Milano, Milan, Italy*
L. Rakotomanana   *Orthopaedic Hospital and Swiss Federal Institute of*
                        Technology, Lausanne, Switzerland
J. S. Rees   *University of Wales College of Medicine, Cardiff, UK*
G. Riccardi   *Dipartimento di Ingegneria Aerospaziale, Aversa, Italy*
H. A. Richard   *University of Paderborn, Germany*
C. J. Richardson   *University of Sheffield, UK*
A. M. Rodriguez   *Federal University of Rio de Janeiro, Brazil*
P. Rüegsegger   *University and ETH Zürich, Switzerland*
C. Rumelhart   *Laboratoire de Mècanique des Solides, INSA,*
                        Villeurbanne, France
M. Runza   *IRCCS Ospedale Maggiore di Milano, Milan, Italy*
A. M. Saad   *Henry Ford Hospital, Detroit MI, USA*
M. Sakamoto   *Niigata College of Technology, Japan*
F. Sarghini   *University of Maryland at College Park, MD, USA*
A. Sasaki   *Hiroshima University School of Dentistry, Japan*
J. Schmitt   *Philipps-University, Marburg, Germany*
R. Schreiner   *Rizzoli Institute, Bologna, Italy*
K. B. Shelburne   *University of Texas at Austin, USA*
N. G. Shrive   *University of Calgary, Alberta, Canada*
B. R. Simon   *University of Arizona, Tucson, USA*
M. Simondi   *Laboratory of Biomechanics, ENSAM, Paris, France*
K. Skalski   *Warsaw University of Technology, Poland*
G. Slugocki   *Warsaw University of Technology, Poland*
S. E. Solomonidis   *University of Strathclyde Glasgow, UK*
K. Soma   *Tokyo Medical and Dental University, Japan*
P. Sparto   *Ohio State University, Columbus, OH, USA*
E. L. N. Spelier   *CMG Den Haag B.V., Den Haag, The Netherlands*
W. D. Spence   *University of Strathclyde, Glasgow, UK*
R. Srinivasan   *University of Rochester, NY, USA*
V. Srinivasan   *University of Rochester, NY, USA*
G. R. Starke   *University of CapeTown, South Africa*
G. P. Steven   *University of Sydney, NSW, Australia*
T. Stoll   *Federal Institute of Technology, Zürich, Switzerland*
J. Subke   *Universität Tübingén, Germany*
Y. Sùzuki   *Rosai Rehabilitation Engineering Center, Nagoya, Japan*
B. Swaelens   *Materialise NV, Heverlee, Belgium*
W. Swięszkowski   *Warsaw University of Technology, Poland*

L. Taber	*University of Rochester, NY, USA*
Z. Taha	*University of Wales Institute Cardiff, UK*
C. T. Tan	*Singapore General Hospital*
J. S. Tan	*BIOMAT, National University of Singapore*
Y. Tanaka	*Rosai Rehabilitation Engineering Center, Nagoya, Japan*
K. Tanne	*Hiroshima University School of Dentistry, Japan*
K. E. Tanner	*Queen Mary and Westfield Collge, London, UK*
D. Taylor	*Trinity College Dublin, Ireland*
S. J. G. Taylor	*University College London, UK*
W. R. Taylor	*University of Bath, UK*
E. C. Teo	*Nanyang Technological University, Singapore*
S. H. Teoh	*IMRE and BIOMAT, National University of Singapore*
A. Terrier	*Orthopaedic Hospital, Lausanne, Switzerland*
A. Tòni	*University of Bologna, Italy*
I. Tonković	*University Hospital Rebro, Zagreb, Croatia*
S. Tonković	*Faculty of Electrical Engineering and Computing, Zagreb, Croatia*
G. Tonti	*Servizio di Cardiologia con UTIC Pescara, Italy*
H. Trebacz	*Università di Padova, Italy*
E. A. Trowbridge	*University of Sheffield, UK*
F. Trudeau	*University of Montréal, PQ, Canada*
N. L. Ulbrich	*Federal University of Paraná, Brazil*
D. Ulrich	*University and ETH Zürich, Switzerland*
R. Van Audekercke	*Katholieke Universiteit Leuven, Heverlee, Belgium*
K. Van Brussel	*Katholieke Universiteit Leuven, Heverlee, Belgium*
R. Van Noort	*University of Sheffield, UK*
H. Van Oosterwyck	*Katholieke Universiteit Leuven, Heverlee, Belgium*
B. Van Rietbergen	*University and ETH Zürich, Switzerland*
L. Vanden Berghe	*Katholieke Universiteit Leuven, Pellenberg, Belgium*
P. A. Van den Blink	*University of Cape Town, South Africa*
J. Vander Sloten	*Katholieke Universiteit Leuven, Heverlee, Belgium*
G. Van der Perre	*Katholieke Universiteit Leuven, Heverlee, Belgium*
M. Viceconti	*Rizzoli Institute, Bologna, Italy*
V. Vitins	*Riga Technical University, Latvia*
D. Voloder	*Faculty of Electrical Engineering and Computing, Zagreb, Croatia*
C. Volp	*University of Wales College of Medicine, Cardiff, UK*
K. Wakasa	*Hiroshima University School of Dentistry, Japan*
A. Wakulicz	*Institute of Mathematics, Warsaw, Poland*
P. S. Walker	*University College London, UK*
H. Walter	*Laboratoire de Mécanique des Solides, INSA, Villeurbanne, France*
A. Wang	*QE2 Medical Centre, Nedlands, WA., Australia*
H. l. Weber	*Pontificia Universidadé Católica de Rio de Janeiro, Brazil*

H. -D. Wehner	*Universität Tübingen, Germany*
J. Weiser	*University of Paderborn, Germany*
H. W. Wevers	*Queen's University, Kingston, ON, Canada*
N. W. Williams	*Northern General Hospital NHS Trust, Sheffield, UK*
D. R. Wilson	*Beth Israel Hospital, Boston, USA*
M. J. Wilson	*University of Leeds, UK*
H. P. Woelfel	*Technical University Darmstadt, Germany*
D. Wright	*University of Wales Institute, Cardiff, UK*
U. P. Wyss	*Queen's University, Kingston, ON, Canada and Sulzer Orthopedics Ltd, Winterthur, Switzerland*
M. Yamaki	*Hiroshima University School of Dentistry, Japan*
M. F. Yeo	*University of Adelaide, S.A., Australia*
Y. Yoshida	*Hiroshima University School of Dentistry, Japan*
C. Zannoni	*Università di Bologna, Italy*

1. MULTIBODY SYSTEMS AND JOINT MODELS

ANATOMICAL MODELS OF DIARTHRODIAL JOINTS: RIGID MULTIBODY SYSTEMS AND DEFORMABLE STRUCTURES

J.H. Heegaard[1]

1. ABSTRACT

Computer models of diarthrodial joint are commonly represented by a set of constraints limiting the possible motion between limb segments. The nature of these constraints determines the joint motion and the forces and stresses acting across the joint. This paper presents a brief review of commonly used mathematical methods to model diarthrodial joints. Merits and limitations of each method are also presented and possible trends for future research in computational joint biomechanics are briefly discussed.

2. INTRODUCTION

Diarthrodial joints provide mobility to the skeletal system. At the same time they must also provide stability to the skeletal assemblage. For example, mobility at the knee is required to ensure clearance during the swing phase of the leg, while stability at the stance leg knee ensures proper support of the body. Mobility and stability of the skeletal system are antagonistic features which are realized by subjecting diarthrodial joints to high loading. Most joints in the leg bear several times body weight during daily activities (1).

Computer models of diarthrodial joints have been recognized as effective tools to better understand the relationship existing between joint anatomy, joint kinematics and loading at the joint(2, 3). Each limb segment can be modeled as a rigid body resulting in a finite dimensional description of its kinematics. If stresses need to be evaluated, the limb segments must be modeled as deformable continua, requiring numerical methods, such as the finite element method, to discretize the segment's kinematics into a finite dimensional space.

From a mathematical point of view, joints can be viewed as constraint equations limiting the range of motion of the connected limb segments. The most common types of joint used in computer models include hinges, linkages, and contact between anatomical articular surfaces. The objective of this paper is to briefly review these different joint models, and to discuss their capabilities and limitations.

In Section 3 we model limb segments as rigid bodies and express their dynamics using Lagrange's equations of motion. We further derive the constraint equations used to model the most commonly used types of joints. In Section 4 we model limb segments as deformable continua and discuss the constraint equations used to represent deformable anatomical joints. The features of these joint models are illustrated with a 3D model of the knee used to calculate the motion of the patella and the stresses in the adjacent tissues during knee flexion. We conclude in Section 5 with a few remarks concerning computer models for diarthrodial joints and possible improvements to current models.

Keywords: Diarthrodial joint, Contact, Dynamics, Continuum mechanics
[1]Professor, Mechanical Engineering Department, Stanford University, Stanford, CA 94305-4040

3. RIGID BODY MODELS

3.1. Equations of motion

In this Section we treat each limb segment as a rigid body. The kinematics of the system can therefore be expressed with a finite number N of *generalized coordinates* $q_1, ..., q_N$. The Lagrangian L of the system is defined as

$$L = T - V \tag{1}$$

where T and V represent respectively the kinetic energy and the potential energy. The N Lagrange's equations of motion are given by

$$\frac{d}{dt}\frac{\partial L}{\partial \dot{q}_r} - \frac{\partial L}{\partial q_r} = Q_r \tag{2}$$

where Q_r is the rth. generalized force

$$Q_r = \sum_\beta \boldsymbol{F}_\beta \frac{\partial \boldsymbol{r}_\beta}{\partial q_r} \tag{3}$$

corresponding to those forces $\boldsymbol{F}_\beta$ not already included in V.

3.2. Motion constraints

Additional constraints may limit the possible motion of each segment. For instance a joint between two segments B_1 and B_2 may prevent some configurations to occur.

An important motivation for using Lagrange's equations is the possibility to eliminate constraint equations by choosing a proper set of generalized coordinates. In some circumstances however, it may be helpful not to eliminate the constraint equations. Such may be the case when handling difficult to express constraints, or when constraint forces need to be evaluated. In those cases, additional constraint equations $f_s(\boldsymbol{q}, t) = 0$; $s = 1, ..., C$ must be introduced to ensure consistency between the system configuration and the motion constraints.

The equations of motion take then the usual form

$$\frac{d}{dt}\frac{\partial L}{\partial \dot{q}_r} - \frac{\partial L}{\partial q_r} + \sum_{s=1}^{C} \lambda_s \frac{\partial f_s}{\partial q_r} = Q_r \tag{4}$$

where the λ_s are Lagrange multiplier associated to the generalized constraint forces necessary to enforce the constraints. In the remaining subsections we briefly review the constraint equations associated with hinge joints, and anatomical joints.

3.3. Hinge joint

Hinge joint models provide only one DOF between the linked segments in the form of a simple rotation about the hinge's axis. This type of joint is commonly used to model the hip, knee and ankle in human locomotion models (4, 5, 6, 7).

For planar models, the constraint equation simply expresses the collocation of two points on each limb segment. For instance by denoting $\boldsymbol{r}^F$ the distal extremity of the femur and $\boldsymbol{r}^T$ the proximal extremity of the tibia, the constraint equations representing the knee joint expresses simply as

$$\boldsymbol{r}^F - \boldsymbol{r}^T = \boldsymbol{0} \tag{5}$$

For 3D models, the collocation conditions are completed by two additional conditions ensuring rotations about the hinge axis.

To illustrate the use of hinge joints in musculoskeletal models, we consider the ballistic gait model originally proposed by Mochon and MacMahon (4) (Fig. 1-A). The knee $K1$

of the stance leg is assumed to remain fully extended throughout the swing phase of the contralateral leg. Therefore, extensor and flexor moments must be applied to $K1$ to ensure full extension. Denoting the stance leg flexion angle by q_4, the following simple constraint equation is introduced

$$f(q) := q_4 = 0 \qquad (6)$$

The corresponding Lagrange multiplier λ_4 represents the generalized forces associated with the moment applied to $K1$ to keep it extended (Fig. 1-B). We further notice qualitative agreement with experimental data (8).

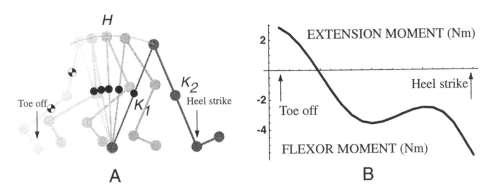

Fig. 1. Simple 1 DOF hinge joint. A. Computed gait kinematics. B. Moments acting on the stance leg knee.

Overconstraints of the normal range of motion is a major drawback of hinge joints, and may generate high constraint forces at the hinge. Nonetheless the extreme simplicity of these joint models is advantageously used in large musculoskeletal models including several body segments actuated by complex controls..

3.4. Anatomical joints

More realistic simulations are obtained by modeling diarthrodial joints with contacting anatomical surfaces. However, the contact conditions ensuring non-penetrability of the articular surfaces lead to more complex constraint equations. The contact problem can be efficiently solved by introducing a parametrization of the configuration space leading to trivial expressions for the contact conditions(9), avoiding thus costly contact detection computations.

Consider two interacting planar rigid limb segments Ω_1 and Ω_2. Let $\partial\Omega_\alpha$ ($\alpha = 1, 2$) denote their boundaries defined in parametric form by $\partial\Omega_\alpha = c_\alpha(\xi)$ The system has 5 DOF when frictionless contact occurs. There is always a unique pair of points ($\mathscr{P}_1, \mathscr{P}_2$) belonging respectively to $\partial\Omega_1$ and $\partial\Omega_2$ representing the closest points between Ω_1 and Ω_2. Non penetrability of the rigid bodies expresses as

$$d_n \geq 0 \qquad (7)$$

where d_n is the gap between Ω_1 and Ω_2.

Without loss of generality, we consider body Ω_1 to be fixed. The usual parametrization for Ω_2 includes translations (q_1, q_2) of a reference point in Ω_2 and rotation q_3 of Ω_2 about axis $\hat{e}_3$.

The contact condition [7] can be expressed in a simple form by introducing a new parametrization. To this end consider three new parameters $q_1^* = \xi_1$, $q_2^* = \xi_2$, $q_3^* = d_n$. It follows that

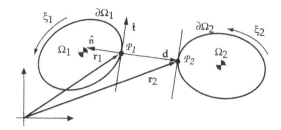

Fig. 2. Rigid multibody anatomical joint

the orientation of Ω_2 can be defined by the parametric coordinates of points $\mathscr{P}_1$ and $\mathscr{P}_2$. The last generalized coordinates q_3^* describes the translation of Ω_2 along the direction normal to $\partial\Omega_1$. Using this parametrization the contact constraint simply expresses as $q_3^* = 0$

The efficiency of the method is illustrated on a three segment planar model of the knee joint (Fig. 4). When compared to more conventional parametrizations the present method

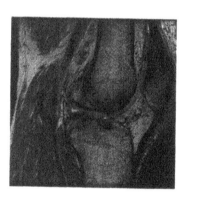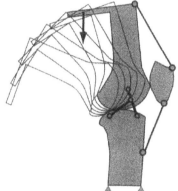

Fig. 3. Anatomical multi-body knee model

provides simpler constraint equations and faster solution procedure as indicated summarized in Table 1. This new parametrization transforms the nonlinear contact constraints

Numerical values		Algorithm performance		
Applied force:	500 N	Contact	#iteration	CPU [s]
Ligament stiffness:	150 N/mm	Node–Node:	1687	120
Quadriceps lengthening:	1 mm/s	Node-Facet:	2734	200
		Smooth:	–	40

Table 1. CPU times to solve multibody knee model

into simple linear ones avoiding to search for the pair of closest points on the contacting surfaces. In its current form this parametrization is however limited to planar convex articular surfaces.

4. DEFORMABLE MODELS

4.1. Motion constraints

When the limb segments are assumed deformable the equations of motion express in weak form as

$$\int_\Omega \nabla w : P\, dV - \int_{\partial\Omega} w \cdot \bar{p}\, dA = 0 \tag{8}$$

where w is an arbitrary virtual displacement field compatible with the essential boundary conditions (displacements), u is the displacement field, P is the first Piola-Kirchhoff stress tensor, and $\bar{p}$ the natural boundary conditions (force). In the absence of inertia and body forces (quasi-static problem) these equations characterize an equilibrium configuration with stationary energy. The problem is therefore equivalent to the minimization of the total energy π

$$\min_u \pi(u) \tag{9}$$

with respect to the unknown displacement field u

4.2. Motion constraints

Additional constraints must be introduced to ensure non-penetrability of the contacting limb segments, leading to a constrained minimization problem

$$\min_u \pi(u) \tag{10}$$

$$d_n \geq 0 \tag{11}$$

where d_n represents the gap between the contacting segments. This optimization problem can be solved using either a penalty method in which the strict impenetrability condition is relaxed by introducing stiff springs to model the contacting surface (10, 11) or by using Lagrangian based methods (12, 13).

Several alternatives exist to express the contact conditions $d_n \geq 0$. The simplest one involves small relative motion between the contacting limb segments; contact constraints can then easily be expressed on a node-to-node basis (Fig. 4-A) as

$$d_n := (r^1 - r^2) \cdot \hat{n} \geq 0 \tag{12}$$

where r^α denotes the closest nodes on the articular surfaces and $\hat{n}$ is the inward normal at r^1. When the relative motion between the limb segments becomes more important,

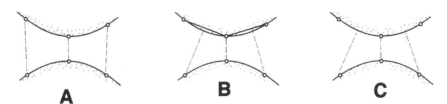

Fig. 4. Common types of contact elements. A. node-to-node, B. node-to-facet, C. note-to-surface geometry.

node-to-facet (14) (Fig. 4-B) or node-to-surface gap elements (15) (Fig. 4-C) must be used to handle the larger relative slip. The gap distance d_n is then expressed implicitly as

$$d_n := (r_1 - r_2) \cdot \hat{n}$$

where r_α are mutual projections of points on the respective articular surfaces.

4.3. Anatomical joint models

The two most challenging problems arising during the development of a quasi-static deformable joint model are the expression of the articular contact constraints, and the possible existence of rigid body modes.

Joint contact constraints include the expression of geometrical and kinematical conditions describing the gap distance d_n. Furthermore the contact constitutive behavior is characterized by nonsmooth relations between gap distance, contact force, slip velocity and friction force.

Rigid body motion arise because of the possibility of limb segments to slide on one another. The essential boundary conditions on the displacements, instead of being well defined, depend on the contact solution. The tangent matrices are thus only semi-positive definite, due to rigid body modes (inertia is discarded in quasi-static models).

Current contact models can handle large relative slip between the contacting segments(15) and are therefore well suited to model deformable joints undergoing physiological motion. The capabilities of these joint models are illustrated in a 3D model of the patello-femoral joint (12) used to compute patellar tracking (Fig. 5) during knee flexion.

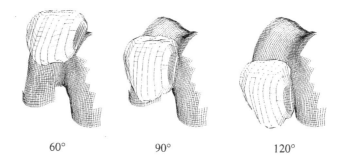

60°   90°   120°

Fig. 5. Computed patellar tracking during knee flexion

The model can also compute the joint contact pressure distribution during knee flexion (Fig. 6).

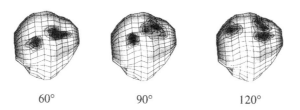

60°   90°   120°

Fig. 6. Patello-femoral contact pressure during knee flexion

Finally patellar stresses are also calculated during knee flexion (Fig. 7).

Validation against experimental results (16) have demonstrated the excellent accuracy that could be achieved with a detailed anatomical model of the joint. This type of model is currently being used to assess the biomechanical effects of orthopaedic corrective procedures (17).

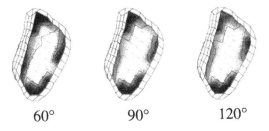

60° 90° 120°

Fig. 7. Compressive hydrostatic pressure in the patella during knee flexion

5. CONCLUSION

Rigid multibody models produce very fast solution procedures, allowing real time animation and feedback. These models lack however the ability to compute contact pressure and tissue stress. On the other hand deformable models are well suited to compute joint contact pressure and tissue stress but lead to significantly slower algorithms, precluding their use in real time application such as virtual reality. The computational cost associated with deformable models arises from the nonlinearities associated with large displacements and constitutive models. While constitutive models can be linearized to obtain fast approximation of stress distribution, large displacement are inherently present during joint motion. One approach to linearize the displacements without compromising the ability to handle finite rotations is to decouple the motion into rigid body finite rotations and small linear deformations. The nonlinear part of the resulting problem has only a few dimensions (3 DOF per segment in 3D). Such approaches have already been successfully implemented in structural problems involving rods, beams and shells (*e.g.*, the corotational method (18)) but need still be developed for general 3D solids.

The contact models discussed in this paper did not account for viscoelastic or multiphasic behavior of the contacting cartilage. Additional kinematical conditions for the fluid phase of cartilage at the contact must then be included. Analytical solutions have been proposed for simple geometries (19) and finite element implementation are currently being developed (20) to handle more complex geometries.

6. REFERENCES

1. Morrison, J. Bioengineering analysis of force actions transmitted transmitted by the knee joint. *Biomed Engng 3* (1968), 164–170.
2. Hefzy, M. S., and Grood, E. S. Review of knee models. *Appl. Mech. Rev. 48* (1988), 1–13.
3. Huiskes, R. Mathematical modeling of the knee. In *Biology and biomechanics of the traumatized synovial joint: the knee as a model*, G. A. M. Finerman and F. R. Noyes, Eds. American Academy of Orthopedic Surgeons, Rosemont, IL, 1992, pp. 419–439.
4. Mochon, S., and McMahon, T. A. Ballistic walking. *J Biomechanics 13* (1980), 49–57.
5. Pandy, M., Zajac, F., Sim, E., and Levine, W. An optimal control model for maximum-height human jumping. *J Biomechanics 23* (1990), 1185–1198.
6. Zajac, F. E., and Winters, J. M. Modeling musculoskeletalmovement systems: joint and body segmental dynamics, musculoskeletal actuation, and neuromuscular control. In *Multiple muscle systems*, J. C. Winters and S. L. Woo, Eds. Springer Verlag, New York, 1990, pp. 359–411.
7. McGeer, T. Passive dynamic walking. *Int J Robot Res 9* (1990), 62–82.
8. Inman, V. T., Todd, F., and Ralston, H. J. *Human walking*. Williams & Wilkins, Baltimore, MD, 1981.
9. Heegaard, J. Parametrization of the configuration space for multibody systems with

frictional contact. In *Euromech 351* (1996).

10. Essinger, J., Leyvraz, P. F., Heegaard, J. H., and Robertson, D. D. A mathematical model for the evaluation of the behaviour during flexion of condylar type knee prostheses. *J Biomechanics 22* (1989), 1229–1241.

11. Blankevoort, L., and Huiskes, R. Ligament bone interaction in a three-dimensional model of the knee. *J Biomech Engng 113* (1991), 263–269.

12. Heegaard, J., Leyvraz, P., Curnier, A., Rakotomanana, L., and Huiskes, R. Biomechanics of the human patella during passive knee flexion. *J Biomechanics (ESB-95 Best Paper Award) 28* (1995), 1265–1279.

13. Donzelli, P., and Spilker, R. Iterative contact detection algorithm for a mixed-penalty biphasic finite element. In *Proc. BED ASME* (1995), vol. 29, pp. 169–170.

14. Heegaard, J. H., and Curnier, A. An augmented Lagrangian method for discrete large slip contact problems. *Int. J. Num. Meth. Engng. 36* (1993), 569–593.

15. Heegaard, J. H., and Curnier, A. Geometric properties of 2d and 3d unilateral large slip contact operators. *Comp Meth Appl Mech Engng 131* (1996), 263–286.

16. Heegaard, J., Leyvraz, P., van Kampen, A., Rakotomanana, L., Rubin, P., and Blankevoort, L. Influence of soft structures on patellar 3d tracking. *Clin Orthop Rel Res 299* (1994), 235–243.

17. Heegaard, J., and Leyvraz, P. Computer aided surgery: application to the maquet procedure. In *ASME Bioengineering Summer Meeting* (1995), pp. 221–222.

18. Criesfield, M. *Non-linear finite element analysis of solids and structure.* Wiley, Chichester, NY, 1991.

19. Ateshian, G., and Wang, H. A theoretical solution for the frictionless rolling contact of cylindrical biphasic articular cartilage layers. *J Biomechanics 28* (1995), 1341–1355.

20. Donzelli, P., and Spilker, R. A contact finite element formulation for biological soft hydrated tissues. *Comp Meth Appl Mech Engng To appear* (1997).

ACKNOWLEDGMENTS

This work was supported by a Young Investigator's award from the Swiss National Fund, the Department of Veterans Affairs, the San Diego Supercomputer Center and the Powell Foundation. The author would like to thank Alain Curnier, Lalao Rakotomanana, Marc Levenston and Chris Jacobs for their valuable comments.

A DYNAMIC OPTIMIZATION SOLUTION FOR JUMPING IN THREE DIMENSIONS

Frank C. Anderson[1] and Marcus G. Pandy[2]

1. ABSTRACT

A mathematical model of the human body comprising 23 degrees of freedom and 54 muscles is used to solve an optimal control problem for jumping in three dimensions. The optimal control solution was computed on an IBM SP2 parallel supercomputer located at NASA Ames Research Center in California. The computational performance of the SP2 is far better than that of other parallel machines such as the Connection Machine CM-5. The predicted ground-reaction forces, body-segmental motions, and muscle activation patterns agree closely with measurements of the same variables obtained for maximum-height human jumping.

2. INTRODUCTION

The emergence of fast, parallel supercomputers has enabled more detailed simulations of human movement to be undertaken. Anderson et al. [1] developed a parallel computational algorithm which reduced computation times by factors in excess of 100, depending on the type of supercomputer used. This study demonstrated that it is now feasible to solve optimal control problems for fully three-dimensional activities, using musculoskeletal models comparable to those used in static optimization studies [2].

The purpose of this paper is three-fold. First, a detailed musculoskeletal model of the body is described. The model incorporates 23 degrees of freedom, allowing full three-dimensional motion of the body segments. The model is also actuated by 54 muscles, and is able to break and establish contact with the ground. Second, the feasibility of obtaining an optimal control solution using a musculoskeltal model of this complexity is demonstrated. All computations are performed on an IBM SP2 parallel machine. Third, the accuracy of the model is evaluated by comparing the optimal control solution for a maximum-height jump with experimental data obtained from human subjects.

Keywords: Dynamic Optimization, Musculoskeletal Model, Maximum-height Jumping

[1] Department of Mechanical Engineering, University of Texas at Austin, Austin, Texas 78712, USA
[2] Department of Kinesiology and Department of Mechanical Engineering, University of Texas at Austin, Austin, Texas 78712, USA

3. METHODS

3.1 Model of the Skeleton

The model comprises 10 segments and 23 degrees of freedom (DOF) (Fig. 1). The reference or base segment of the model is the pelvis, whose global position is described by 6 generalized coordinates: three coordinates to specify the translation of the pelvis in the global reference frame; and three to specify the orientation of the pelvis about the body-fixed x, y, and z axes. The remaining 9 segments branch in an open chain from the pelvis. These are the head-arms-torso (HAT) segment and the left and right femur, tibia, hindfoot, and toes. The HAT segment is articulated with the pelvis by a 3 DOF back joint located at the approximate level of the third lumbar vertebrae. The exact location of the back joint was determined by averaging *in-vivo* kinematic data obtained from five human subjects (§ 3.5). The back joint is included to decouple the motion of the pelvis from that of the massive HAT segment. The right and left femora are articulated with the pelvis by 3 DOF hip joints. The right and left tibiae are articulated with their respective femora by 1 DOF knee joints. Although this neglects the action of the patella and the sliding of the femoral condyles on the tibial plateaux, for computational reasons, it is unrealistic to incorporate more detailed models of the knee into a three-dimensional, musculoskeletal model of the body. Estimates of the moment arms for vasti and rectus femoris were obtained from Spoor et al. [3]. The right and left feet are each divided into two segments: a hindfoot and the toes. The right and left hindfoot segments articulate with their respective tibiae via 2-DOF ankle/subtalar joints. Each of these joints is represented by a u-joint, whose two axes are skewed with respect to the inertial axes. Directions for the ankle and subtalar axes were based on data reported by Inman [4]. The right and left toes segments articulate with their respective hindfoot segments via 1 DOF revolute joints, whose axes are also skewed with respect to the inertial axes.

3.2 Foot-Ground Interaction

The interaction of the foot with the ground is modeled by a set of springs and dampers distributed on the sole of the hindfoot and toes segments. Specifically, four three-dimensional, spring-and-damper units are placed near the corners of each hindfoot, and one three-dimensional spring and damper is placed toward the distal end of the toes. A three-dimensional spring means that there are three separate springs acting at each spring location: one in each of the $\hat{x}$, $\hat{y}$, and $\hat{z}$ directions. The force producing characteristics of each spring are represented by an exponential function in the vertical ($\hat{y}$) direction and by two simple linear springs in the horizontal ($\hat{x}$ and $\hat{z}$) directions.

3.3 Model of the Muscles

Each musculotendinous unit is modeled as a Hill-type contractile element in series with tendon. Driving each musculotendinous unit is a first-order differential equation which couples neural excitation to muscle activation [5]. The values of maximum isometric strength, optimal fiber length, pinnation angle, and tendon slack length for each musculotendon actuator are based on data reported by Delp [6]. In cases where several muscles have been combined into one, the final values of muscle strength are based on a strength-weighted average of the parameters for the separate muscles reported in [6]. The maximum isometric strength of each muscle was then scaled so that the summed muscle torques about each joint matched torque-angle curves measured for humans [7].

Where possible, the action of a muscle group is represented by a single line of force. Many of the smaller muscles which originate on the tibia and insert on the foot or toes were grouped into one of four lines of action (e.g., PFIN, PFEV, DFIN, and DFEV; Fig. 1). Gluteus maximus and gluteus medius/minimus, which have fan-like origins, were each represented by two lines of force. The path of a muscle from origin to insertion is specified by a series of viapoints or a combination of viapoints and

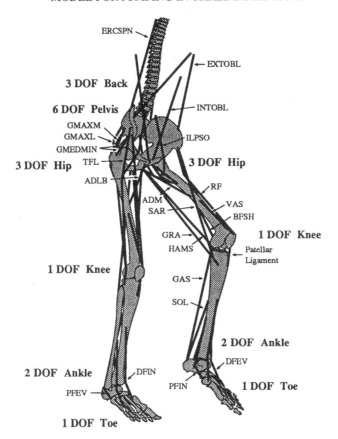

Fig. 1: Three-dimensional, 54-muscle, 23-degree-of-freedom musculoskeletal model. The degrees of freedom of the model enable the simulation of unconstrained three-dimensional jumping motion. The muscles in the model are erector spinae (ERCSPN), external obliques (EXTOBL), internal obliques (INTOBL), medial portion of gluteus maximus (GMAXM), lateral portion of gluteus maximus (GMAXL), anterior and posterior portion of gluteus medius and minimus (GMEDMIN), iliopsoas (ILPSO), adductor magnus (ADM), adductor longus brevis (ADLB), gracilis (GRA), sartorius (SAR), tensor fasciae latae (TFL), hamstrings (HAMS), rectus femoris (RF), a combination of vastus medialis, intermedius, and lateralis (VAS), biceps femoris short head (BFSH), gastrocnemius (GAS), soleus (SOL), a combination of tibialis posterior, flexor digitorum longus, flexor hallucis longus (PFIN), a combination of peroneus brevis and longus (PFEV), a combination of tibialis anterior and extensor hallucis longus (DFIN), a combination of peroneus tertius and extensor digitorum longus (DFEV), periformis (not shown), pectinius (not shown), and a toes extensor and flexor (not shown).

viacylinders. A viacylinder is a cylinder of specified radius and three-dimensional orientation around which a muscle may wrap and slide frictionlessly.

3.4 Optimal Control Problem

The equations of motion for the model were integrated using a variable-step, Runge-Kutta-Feldberg 5-6 integrator. A parallel computational algorithm developed by Anderson et al. [1] was used to solve the optimal control problem for maximum-height jumping. To reduce the number of controls, jumping was assumed to be a bilaterally symmetric activity. Specifically, it was assumed that the optimal muscle excitation histories for the left and right sides of the body are the same during the jump. While this assumption reduces the number of muscles by a factor of two, from 54 to 27, it does not preclude three-dimensional motion of the body segments. For jumping, the optimal control problem was to maximize the height reached by the center of mass of the body. The initial joint angles for the model were taken as the average of the subject joint angles at the beginning of the jump. Both the model and the subjects began from static equilibrium.

3.5 Human Experiments

Five healthy, adult males (age 26 ± 3 yr, height 177 ± 3 cm, and mass 70.1 ± 7.8 kg) were asked to perform five maximum-height squat jumps. Detailed anthropometric measurements were taken for each subject as described in McConville et al. [8]. Model anthropometry was set to the average of subject measures. Each subject began from a static, squatting position, with the depth of the squat fixed so that the shoulders were at 80% of their height at standing. Subjects determined when they were in the proper starting position by viewing themselves on a large-screen video monitor. Kinematic, force-plate, and EMG data were recorded simultaneously during each jump. EMG activity was monitored by placing surface electrodes over the following muscles: tibalis anterior, soleus, lateral gastrocnemius, vastus lateralis, rectus femoris, hamstrings, adductor magnus, gluteus maximus, gluteus medius, erector spinae, and the external obliques.

4. RESULTS AND DISCUSSION

The optimal control solution was computed using an IBM SP2 located at NASA Ames Research Center in California. Excellent single processor speed was achieved on this machine. Relative to past computations performed on a Connection Machine CM–5 [1], each SP2 processor ran approximately 15 times faster than each CM–5 processor. With one processor, one iteration of the optimal control algorithm required about 4 hours of CPU time. However, each time the number of processors was doubled, CPU time was nearly halved. Using 128 processors, one iteration of the algorithm was completed in just over 3 minutes of CPU time.

There is close agreement between model and experiment. Peak vertical ground-reaction forces for the model and the subjects ranged from 1500 to 2000 N. The duration of force exertion was approximately 0.3 seconds for both the model and the subjects (Fig. 2A). The rapid decrease in vertical force as the body leaves the ground is shifted back in time for the model relative to the subjects. This difference may be due to a greater extension of the ankle joint in the model than in the subjects.

There is also close agreement between the vertical trajectories of the center of mass of the model and the subjects. Peak vertical accelerations of the body's center of mass are approximately 15 m/s^2, while vertical velocities at liftoff ranged from 2.0 to 2.5 m/s. Both the model and the subjects leave the ground with a vertical displacement of the center-of-mass between 10 and 15 cm above standing.

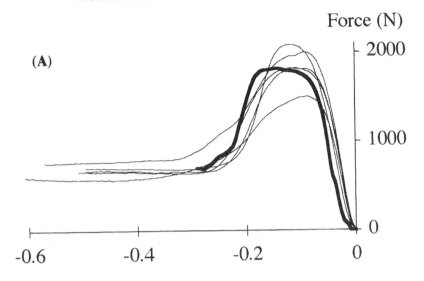

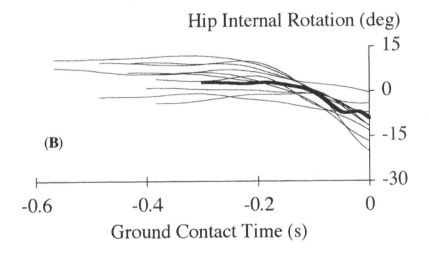

Fig. 2: (A) Vertical ground reaction force measured for the subjects (thin lines) and predicted by the model (thick line). There is close agreement between model and subjects in terms of the duration, peak, and shape of the vertical ground reaction force.

(B) Hip internal rotation measured for the subjects (thin lines) and predicted by the model (thick line). The model externally rotated at the hips by about 10 degrees. Subjects demonstrated a similar but slightly larger external rotation.

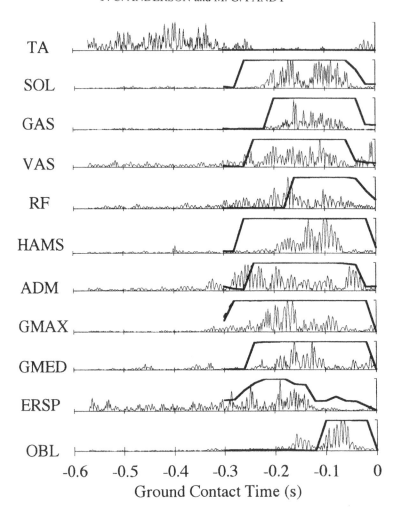

Fig. 3:   EMG activity for one subject during a maximum-height jump (thin lines) and muscle excitation histories predicted by the model (thick lines). The records have been shifted in time so that zero seconds corresponds to the time at which the subject and the model left the ground. There is good agreement between model and subject.

There are small differences between the model and subject kinematics during the jump. Because the model is perfectly symmetric, the model's pelvis neither abducts nor internally rotates; subjects, however, abduct and internally rotate their pelvis' by about 5°. Pelvic extension is also slightly greater in the model than in the subjects. Furthermore, ankle plantarflexion is about 15° greater in the model than in the experiments, and the model plantarflexes the ankles later in the jump. This discrepancy may be due in part to the model having greater range of motion at the ankle than the subjects.

Nonetheless, the major movements of the body segments are reproduced in the model. The model and the subjects show roughly the same amount of hip extension, and the knee is brought to full extension in both the model and the experiments. Hip abduction and hip internal rotation are also nearly the same for the model and the subjects (Fig. 2B). The subjects and the model also invert both feet just prior to liftoff.

There is excellent agreement between the calculated and measured muscle activation patterns (Fig. 3). A minor difference was found between the model and one subject in the activation of the oblique muscles: the model activated the oblique muscles at the end of the jump, whereas one subject activated these muscles at the very beginning. All of the other subjects used an activation pattern for the obliques that is similar to the result predicted by the model.

5. REFERENCES

[1]   Anderson F.C., Ziegler J.M., Pandy M.G., and Whalen R.T. Application of high-performance computing to numerical simulation of human movement. ASME Journal of Biomechanical Engineering, 1995, Vol. 117: 155-157.

[2]   Hardt, D.E. Determining muscle forces in the leg during human walking: an application and evaluation of optimization methods. ASME Journal of Biomechanical Engineering, 1978, Vol. 100: 88-92.

[3]   Spoor C.W. and van Leeuwen J.L. Knee muscle moment arms from MRI and from tendon travel. Journal of Biomechanics, 1992, Vol. 25: 201-206.

[4]   Inman V.T. The Joints of the Ankle. The Williams & Wilkins Company. Baltimore, 1976.

[5]   Pandy M.G., Zajac F.E., Sim E., Levine W.S. An optimal control model for maximum-height human jumping. Journal of Biomechanics, 1990, Vol. 23: 1185-1198.

[6]   Delp S.L. Surgery simulation: A computer graphics system to analyze and design musculoskeletal reconstructions of the lower limb. PhD Dissertation. Stanford University, Stanford, California, 1990.

[7]   Shelburne, K.B. Modeling the mechanics of the normal and reconstructed knee joint. PhD Dissertation. University of Texas at Austin, Austin, Texas, 1996.

[8]   McConville J., Clauser C., Churchill T., Cuzzi J., Kaleps I. Anthropometirc relationships of body and body segment moments of inertia. Technical Report AFAMRL-TR-80-119. Air Force Aerospace Medical Research Laboratory, Wright-Patterson AFB, Ohio, 1980.

A METHOD OF ESTIMATING CONTROL FORCES TO ACHIEVE A GIVEN SWING PHASE TRAJECTORY DURING NORMAL GAIT

H.S. Gill [1]

1. ABSTRACT

A method, which does not require dynamic optimisation, for predicting the control forces required for a given swing phase trajectory is presented. This method is a Predictor-Corrector method, and will be termed the P-C method. The method has the advantage of being less computationally expensive than Dynamic Optimisation (D-O) methods and requires fewer parameters and is less sensitive to the parameters than D-O methods. The P-C method appears to be well suited for non-maximal activities, D-O methods are probably more suitable for maximal activities.

2. INTRODUCTION

The strategy used by the human locomotor system to select control forces to achieve a given motion remains elusive. A large amount of effort has been made to determine the control forces in various human activities by many workers. For the lower limb, these activities have included maximal jumping, maximal kicking and swing phase during gait. A review of the literature indicates that most authors consider the use of dynamic optimisation techniques, involving the use of models that consider muscle properties and limitations, essential (Hatze [1]). This type of approach gives rise to a model formulation which has to consider many state variables to produce an optimal solution; leading to a very large scale problem requiring significant computational cost to solve. In addition these formulations are very sensitive to

KEYWORDS

Gait, Simulation, Swing Phase, Control

---
[1]Research Engineer, Oxford Orthopaedic Engineering Centre, University of Oxford, Nuffield Orthopaedic Centre, Headington, Oxford, OX3 7LD, UK

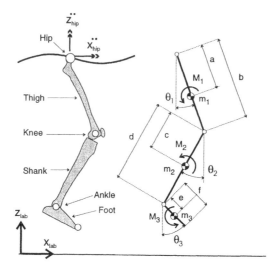

Figure 1: Three degree of freedom sagittal plane model of the swing leg. Motion of the hip is assumed to be prescribed. All angles and moments are positive in the anti-clockwise direction.

parameter variation as indicated by Davy and Audu 1987, [2].

Whilst the limitations of muscle force generation may be important for maximal tasks (with a well defined cost function), this approach may produce unnecessary complications for non-maximal tasks such as swing during gait. A new, simpler, method for estimating control effort during swing phase has been developed.

The study reported here aimed to reveal the differences in control effort between two distinct patterns of swing phase identified in normal subjects by previous work (Gill 1996, [3]). Pattern 1 gave rise to higher impulsive loads at heelstrike than Pattern 2, the differences between the patterns were very subtle and involved only the kinematics of the thigh, shank and foot; for these two patterns no significant differences existed in the trajectories of the hip joint.

3. METHOD

3.1 Simulation Model

A 2D sagittal plane forward dynamic model of the human lower limb was developed, considering the thigh, shank and foot as linked segments and using joint torques at the hip, knee and ankle as control forces, see Figure 1.

The simulation model was described by equations of motion derived using the Lagrange formulation. The system was considered to have three degrees of freedom, described by the following generalised coordinates:

θ_1 The angle between the vertical and the thigh long axis.

θ_2 The angle between the vertical and the shank long axis.

θ_3 The angle between the vertical and the foot long axis.

The general equations of motion were expressed in a convenient matrix form as shown below:

$$[A1] \left\{ \begin{array}{c} \ddot{\theta}_1 \\ \ddot{\theta}_2 \\ \ddot{\theta}_3 \end{array} \right\} + [A2] \left\{ \begin{array}{c} \dot{\theta}_1{}^2 \\ \dot{\theta}_2{}^2 \\ \dot{\theta}_3{}^2 \end{array} \right\} + [A3] \left\{ \begin{array}{c} \sin(\theta_1) \\ \sin(\theta_2) \\ \sin(\theta_3) \end{array} \right\} + [A4] \left\{ \begin{array}{c} \ddot{x}_{hip} \\ \ddot{z}_{hip} \end{array} \right\} = \left\{ \begin{array}{c} M_1 \\ M_2 \\ M_3 \end{array} \right\} \quad (1)$$

where:

$$[A1] = \begin{bmatrix} \begin{array}{c} I_1 + a^2.m_1 \\ +b^2.(m_2+m_3) \end{array} & \begin{array}{c} (c.m_2+d.m_3). \\ b.\cos(\theta_1-\theta_2) \end{array} & \begin{array}{c} b.e.m_3. \\ \cos(\theta_1-\theta_3) \end{array} \\ \begin{array}{c} (c.m_2+d.m_3). \\ b.\cos(\theta_2-\theta_1) \end{array} & \begin{array}{c} I_2 + (c^2).m_2+ \\ (d^2).m_3 \end{array} & \begin{array}{c} d.e.m_3. \\ \cos(\theta_2-\theta_3) \end{array} \\ \begin{array}{c} b.e.m_3. \\ \cos(\theta_3-\theta_1) \end{array} & \begin{array}{c} d.e.m_3. \\ \cos(\theta_3-\theta_2) \end{array} & \begin{array}{c} I_3+ \\ (e^2).m_3 \end{array} \end{bmatrix} \quad (2)$$

$$[A2] = \begin{bmatrix} 0 & \begin{array}{c} (c.m_2+d.m_3). \\ \sin(\theta_1-\theta_2).b \end{array} & \begin{array}{c} b.e.m_3. \\ \sin(\theta_1-\theta_3) \end{array} \\ \begin{array}{c} (c.m_2+d.m_3). \\ \sin(\theta_2-\theta_1).b \end{array} & 0 & \begin{array}{c} d.e.m_3. \\ \sin(\theta_2-\theta_3) \end{array} \\ \begin{array}{c} b.e.m_3. \\ \sin(\theta_3-\theta_1) \end{array} & \begin{array}{c} d.e.m_3. \\ \sin(\theta_3-\theta_2) \end{array} & 0 \end{bmatrix} \quad (3)$$

$$[A3] = \begin{bmatrix} \begin{array}{c} (a.m_1+b.m_2 \\ +b.m_3).g \end{array} & 0 & 0 \\ 0 & \begin{array}{c} (c.m_2+ \\ d.m_3).g \end{array} & 0 \\ 0 & 0 & e.g.m_3 \end{bmatrix} \quad (4)$$

$$[A4] = \begin{bmatrix} \begin{array}{c} (a.m_1+b.m_2+ \\ b.m_3).\cos\theta_1 \end{array} & \begin{array}{c} (a.m_1+b.m_2+ \\ b.m_3).\sin\theta_1 \end{array} \\ (c.m_2+d.m_3).\cos\theta_2 & (c.m_2+d.m_3).\sin\theta_2 \\ e.m_3.\cos\theta_3 & e.m_3.\sin\theta_3 \end{bmatrix} \quad (5)$$

These equations were re-arranged to express the acceleration terms as functions of the velocity and displacement terms, and the non-conservative generalised forces.

$$\left\{ \begin{array}{c} \ddot{\theta}_1 \\ \ddot{\theta}_2 \\ \ddot{\theta}_3 \end{array} \right\} = [A1]^{-1} \left\{ \begin{array}{c} M_1 \\ M_2 \\ M_3 \end{array} \right\} - [A6] \left\{ \begin{array}{c} \dot{\theta}_1{}^2 \\ \dot{\theta}_2{}^2 \\ \dot{\theta}_3{}^2 \end{array} \right\} - [A7] \left\{ \begin{array}{c} \sin(\theta_1) \\ \sin(\theta_2) \\ \sin(\theta_3) \end{array} \right\} - [A8] \left\{ \begin{array}{c} \ddot{x}_{hip} \\ \ddot{z}_{hip} \end{array} \right\} \quad (6)$$

where:

$$\begin{array}{rcl} [A6] & = & [A1]^{-1}[A2] \\ [A7] & = & [A1]^{-1}[A3] \\ [A8] & = & [A1]^{-1}[A4] \end{array} \quad (7)$$

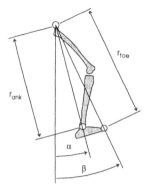

Figure 2: Definition of spherical coordinates used for describing positions of the ankle and toe relative to the hip joint centre, note that r_{ank} and r_{toe} are normalised by leg length.

3.2 Swing Phase Description

Swing phase trajectories were represented in radial coordinates centred on the hip joint, see Figure 2. The length values of the radial coordinates were normalised by the subject's leg length. With the description of relative ankle and toe movement in spherical coordinates, the absolute trajectories of the ankle and toe could then be generated for any given hip trajectory and leg length.

3.3 Control Effort Estimation

The numerical method for estimating the control effort required to achieve a given swing phase is described in this section.

The spherical coordinates used to describe ankle (α, r_a) and toe trajectories (β, r_t) in the sagittal plane relative to the hip were used to establish desired displacement profiles for the ankle and toe for any given hip trajectory and leg length. The velocity profiles of the ankle and toe were then obtained by numerical differentiation of fifth order splines fitted to the x and z components of the displacement profiles.

Considering the ankle, its velocity could be expressed in terms of the hip motion and segment angles:

$$\left\{ \begin{array}{c} \dot{x}_{ank} \\ \dot{z}_{ank} \end{array} \right\} = \left\{ \begin{array}{c} \dot{x}_{hip} \\ \dot{z}_{hip} \end{array} \right\} + b.\dot{\theta}_1. \left\{ \begin{array}{c} \cos\theta_1 \\ \sin\theta_1 \end{array} \right\} + d.\dot{\theta}_2. \left\{ \begin{array}{c} \cos\theta_2 \\ \sin\theta_2 \end{array} \right\} \qquad (8)$$

Thus the velocity of the ankle relative to the hip is given by:

$$\left\{ \begin{array}{c} \dot{x}_{ank}r \\ \dot{z}_{ank}r \end{array} \right\} = b.\dot{\theta}_1. \left\{ \begin{array}{c} \cos\theta_1 \\ \sin\theta_1 \end{array} \right\} + d.\dot{\theta}_2. \left\{ \begin{array}{c} \cos\theta_2 \\ \sin\theta_2 \end{array} \right\} \qquad (9)$$

Small increments in the segment angular velocities, $\delta\dot{\theta}_1$ and $\delta\dot{\theta}_2$, give rise to small increments in relative ankle velocity:

$$\left\{ \begin{array}{c} \dot{x}_{ank}r + \delta\dot{x}_{ank}r \\ \dot{z}_{ank}r + \delta\dot{z}_{ank}r \end{array} \right\} = b.(\dot{\theta}_1 + \delta\dot{\theta}_1). \left\{ \begin{array}{c} \cos\theta_1 \\ \sin\theta_1 \end{array} \right\} + d.(\dot{\theta}_2 + \delta\dot{\theta}_2). \left\{ \begin{array}{c} \cos\theta_2 \\ \sin\theta_2 \end{array} \right\} \qquad (10)$$

the accompanying small increments in segment angle are not significant, since:

$$\begin{aligned}
\sin(\theta + \delta\theta) &= \sin\theta.\cos\delta\theta + \cos\theta.\sin\delta\theta \\
\cos(\theta + \delta\theta) &= \cos\theta.\cos\delta\theta - \sin\theta.\sin\delta\theta
\end{aligned} \tag{11}$$

if $\delta\theta$ is small $\sin\delta\theta \approx \delta\theta \approx 0$ and $\cos\delta\theta \approx 1$. Thus, the increments in relative ankle velocity can be expressed as:

$$\left\{ \begin{array}{c} \delta\dot{x}_{ank}r \\ \delta\dot{z}_{ank}r \end{array} \right\} = b.\delta\dot{\theta}_1.\left\{ \begin{array}{c} \cos\theta_1 \\ \sin\theta_1 \end{array} \right\} + d.\delta\dot{\theta}_2.\left\{ \begin{array}{c} \cos\theta_2 \\ \sin\theta_2 \end{array} \right\} \tag{12}$$

Similarly the velocity of the toe relative to the hip can be expressed in terms of the segment angular velocities:

$$\left\{ \begin{array}{c} \delta\dot{x}_{toe}r \\ \delta\dot{z}_{toe}r \end{array} \right\} = \left\{ \begin{array}{c} \delta\dot{x}_{ank}r \\ \delta\dot{z}_{ank}r \end{array} \right\} + f.\delta\dot{\theta}_3.\left\{ \begin{array}{c} \cos\theta_3 \\ \sin\theta_3 \end{array} \right\} \tag{13}$$

Since the desired velocity profiles for the ankle and toe, as well as the hip velocities, were known, the desired relative ankle and toe velocities could be calculated for any instant in the swing phase. The swing phase was divided into n equal time increments of δt. The conditions at the start of the swing phase, i.e. at the start of the first increment, were known and the desired ankle and toe relative velocities at the end of the first time increment were also known. Thus the changes required in ankle and toe relative velocities during this time increment could be calculated. Generalising to the ith time increment:

$$\begin{aligned}
\delta\dot{x}_{ank}r &= \dot{x}_{ank}r_i^* - \dot{x}_{ank}r_{(i-1)} \\
\delta\dot{z}_{ank}r &= \dot{z}_{ank}r_i^* - \dot{z}_{ank}r_{(i-1)} \\
\delta\dot{x}_{toe}r &= \dot{x}_{toe}r_i^* - \dot{x}_{toe}r_{(i-1)} \\
\delta\dot{z}_{toe}r &= \dot{z}_{toe}r_i^* - \dot{z}_{toe}r_{(i-1)}
\end{aligned} \tag{14}$$

where the $*$ denotes the desired relative velocities. Thus, for each time increment, the desired increments in angular segment velocities could be determined from application of Equations 12 and 13. From these equations it can be seen that the four target velocities, composed of the cartesian components of the desired ankle and toe velocities, can be achieved by increments in the three angular velocities. Thus there are four equations in three unknowns, constituting an overconstrained system. These equations were rewritten in the below form:

$$\begin{aligned}
\delta\dot{x}_{ank}r - (b.\delta\dot{\theta}_1.\cos\theta_1 + d.\delta\dot{\theta}_2.\cos\theta_2) &= 0 \\
\delta\dot{z}_{ank}r - (b.\delta\dot{\theta}_1.\sin\theta_1 + d.\delta\dot{\theta}_2.\sin\theta_2) &= 0 \\
\delta\dot{x}_{toe}r - (b.\delta\dot{\theta}_1.\cos\theta_1 + d.\delta\dot{\theta}_2.\cos\theta_2 + f.\delta\dot{\theta}_3.\cos\theta_3) &= 0 \\
\delta\dot{z}_{toe}r - (b.\delta\dot{\theta}_1.\sin\theta_1 + d.\delta\dot{\theta}_2.\sin\theta_2 + f.\delta\dot{\theta}_3.\sin\theta_3) &= 0
\end{aligned} \tag{15}$$

An iterative numerical method was used to solve Equations 15 for $\delta\dot{\theta}_1$, $\delta\dot{\theta}_2$ and $\delta\dot{\theta}_3$. Using an initial guess for the increments in angular velocity the residuals from each equation were determined. A least squares optimisation method was then used to obtain values of $\delta\dot{\theta}_1$, $\delta\dot{\theta}_2$ and $\delta\dot{\theta}_3$ which minimised these residuals; this set of increments in angular velocity thus minimised the deviation from the desired ankle and toe velocity profiles. Once the changes required in angular velocity

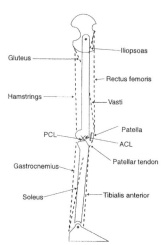

Figure 3: Two dimensional whole leg model, incorporating four-bar linkage tibiofemoral model and finite patellofemoral model. The muscles considered by the model are indicated.

were determined the necessary segment moments for each time increment could be estimated from:

$$\begin{Bmatrix} M_1 \\ M_2 \\ M_3 \end{Bmatrix}_i = [A1] \begin{Bmatrix} \delta\dot{\theta}_1/\delta t \\ \delta\dot{\theta}_2/\delta t \\ \delta\dot{\theta}_3/\delta t \end{Bmatrix} + [A2] \begin{Bmatrix} (\dot{\theta}_1 + \delta\dot{\theta}_1)^2 \\ (\dot{\theta}_2 + \delta\dot{\theta}_2)^2 \\ (\dot{\theta}_3 + \delta\dot{\theta}_3)^2 \end{Bmatrix} + [A3] \begin{Bmatrix} \sin(\theta_1) \\ \sin(\theta_2) \\ \sin(\theta_3) \end{Bmatrix} + [A4] \begin{Bmatrix} \ddot{x}_{hip} \\ \ddot{z}_{hip} \end{Bmatrix} \quad (16)$$

Applying these moments, the simulation equations (Equations 6) were integrated over the time increment and the actual position of the segments at the end of the increment were determined. The values of the segment angles θ_1, θ_2 and θ_3 were then updated, and the same procedure was then repeated for the next time increment. In this way it was possible to determine the time histories of the segment moments required to achieve any given desired ankle and toe trajectory.

3.4 Muscle Force Estimation

A previously developed 2D inverse dynamics model of the human lower limb, see Figure 3. (Collins 1995, [4]), which considered 8 muscles, was then used to analyse the two trajectory patterns using the *DDOSC* approach [4].

DDOSC stands for *D*ynamically *D*eterminate *O*ne-sided *C*onstraint. The method solves problems involving indeterminate systems by reducing the number of unknowns to the number of available equations, thereby generating multiple sets of solutions for any given condition, each solution only containing values for a reduced set of variables. This method produces all the solutions at the vertices of the solution space. One-sided constraints are then used to reduce the number of acceptable solutions. In the case of the swing leg the one sided constraints used were:

- Ligaments, tendons and muscles can only transmit tensile loads.

- Compressive forces can only be transmitted by bones.

4. RESULTS AND DISCUSSION

The forward dynamics P-C formulation was able to produce joint torques which generated the desired trajectory. However, the method proved to be very sensitive to step size and required significant computation time (30 minutes CPU on a single processor DEC AXP21064 (150MHz) workstation). A step size of 4 milliseconds was found be suitable, larger step sizes caused the soultion procedure to become unstable. A sensitivity analysis revealed that changes in the parameters of the thigh segment were not significant, however the solutions were sensitive to shank and foot length, and foot mass.

The *DDOSC* method was able to generate muscle solutions for 80% of the swing phase. This approach only considers the vertices of the possible solution space, this limitation may explain the inability of the method to solve for all of the swing phase. The solutions generated were consistent with EMG records given by Perry 1992 [5].

The method presented generates estimates of control effort which are dependent upon significantly fewer parameters than the dynamic optimisation approaches. The method also requires significantly less computational expense. However, the shortcomings of the *DDOSC* approach need to be addressed.

5. ACKNOWLEDGEMENTS

This study was supported by the Arithritis and Rheumatism Council (UK).

6. REFERENCES

[1] H. Hatze. The Complete Optimization of a Human Motion. *Math. Biosci.*,28:99-135, 1976.

[2] D.T. Davy and M.L. Audu. A dynamic optimization technique for predicting muscle forces in the swing phase of gait. *J. Biomech*, 20:187-201, 1987.

[3] H.S. Gill. *The Mechanics of Heelstrike*. PhD thesis, University of Oxford, 1996.

[4] J.J. Collins. The redundant nature of locomotor optimization laws. *J. Biomech*, 28:251-67, 1995.

[5] J. Perry. *Gait Analysis: Normal and Pathological Function*. McGraw-Hill, New York, 1992.

A NEW MATHEMATICAL MODEL FOR THE HUMAN ANKLE JOINT

A.Leardini [1-2], J.J. O'Connor[1], F.Catani[2]

1. ABSTRACT

The objective of the present study is to develop a preliminary two-dimensional mathematical model that could predict the sagittal kinematics of the human ankle/subtalar joint complex in unloaded conditions. This is a necessary preliminary step to the study of ankle joint stability in response to applied loads. From the results of a preliminary experimental investigation, the calcaneofibular and the tibiocalcaneal ligaments were selected as the ligament links of a four-bar linkage model of the ankle as they showed quasi-isometric patterns of rotation during passive ankle motion. The model predictions of calcaneus movements, ligament orientations, instantaneous centres of rotation, and conjugate talus surface profile compared well with the experimental measurements from one specimen and therefore support the assumptions underlying the formulation of the geometric model. For the first time, many features of rearfoot kinematics were explained by the linkage model. The model is intended to be applied to contact and ligament force predictions during activities, and thus, ultimately to contribute to the improvement of knowledge of these joints, to the enhancement of arthritis treatment and to the development of ankle prosthesis design and techniques of ligament reconstruction.

2. INTRODUCTION

The rearfoot plays a fundamental role in human locomotion. A better understanding of normal behaviour of the ankle joint complex still remains a crucial issue in the prevention of joint degeneration, treatment of bone fractures, surgical techniques for ligament reconstruction but above all to improve the disappointing results of total ankle replacement arthroplasty for the surgical treatment of severe

Keywords: Ankle, Ligaments, Model, Four-bar linkage, Degrees of freedom

[1]University of Oxford, Department of Engineering Science, and Oxford Orthopaedic Engineering Centre, Windmill Road, Oxford OX3 7LD, England

[2]Movement Analysis Laboratory, Istituti Ortopedici Rizzoli, Via di Barbiano 1/10, 40136 Bologna, Italy

joint degeneration [1, 2, 3]. Although the early model of the ankle as a pure hinge joint has been recently questioned [4, 5], no models have been proposed that could explain and predict experimental findings such as the change in both position and orientation of ankle axis of rotation [6, 7, 8] and the shifting contact area on the tibial mortise (tibiotalar articulation) during ankle flexion [9].

Many factors affect human joint kinematics and mechanics such as the shapes of the articular surfaces, the geometric arrangement of ligaments, the mechanical properties of the bones and soft tissues, the muscle forces and patterns of muscle activation and any external loads to which the joint is subjected during activities. To distinguish between the factors which control mobility and stability, a sequential approach can be used. The geometry of the joint in all possible unloaded configurations is investigated first. The geometric model can predict the lines of action of muscle, ligament and contact forces in the unloaded state. The model is intended to be an essential preliminary step to the study of the displacements from these positions due to the application of load and deformation of the soft tissues.

The preliminary results [10, 11] from the experimental part of the investigation are first summarized. They have shown that the ankle/subtalar complex may behave as a single degree of freedom (DOF) mechanism. These results were used as the basis for the formulation of a preliminary two-dimensional (2D) computer-based mathematical model of the ankle/subtalar joint complex. The validation of the model was obtained by comparing its predictions with experimental measurements from one specimen.

3. RESULTS FROM EXPERIMENTS

In the experimental part of the investigation, six skeleto-ligamentous preparations including the intact tibia, fibula, talus and calcaneus bones were tested. A rig able to move the ankle/subtalar complex through its entire range of flexion while allowing unconstrained joint motion was built. The movement of the 4 bones, constrained only by the articular surfaces and the ligaments, were tracked with a stereophotogrammetric system (Elite, BTS, Milan) using a cluster of four reflective markers mounted on a balsawood plate on each bone. Passive motion of the complex through the entire range of flexion was collected. In deviations trials at a number of flexion angles, the calcaneal pin was displaced manually along the horizontal arm of the rig in both directions until significant resistance was felt, and was then released. The 'anatomical landmark calibration' procedure [12] defined anatomically based segment coordinate systems embedded on each bone. Origin and insertion points of each of the main ligamentous structures of the joint complex were also tracked during motion.

In all specimens, the calcaneus movement described in the tibial frame followed virtually the same path in dorsiflexion as it did in plantarflexion (the neutral path). Ligaments located anteriorly slackened during dorsiflexion and posterior ligaments slackened during plantarflexion. Only the calcaneofibular (CaFiL) and tibiocalcaneal (TiCaL) ligaments remained more or less isometric, with maximum strains over the whole range of flexion, averaged over all the specimens, of 6 and 8% respectively. During passive motion, most of the rotation occurred at the ankle, whereas during deviation trials additional rotation occurred mainly at the subtalar joint. Any displacement from the neutral path was resisted, as

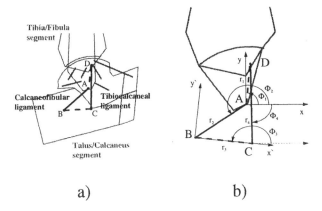

a)   b)

Fig. 1: Geometry of the four-bar linkage model in ankle neutral position. a) AB (CaFiL) and CD (TiCaL) are the ligament links (solid bold), AD (tibio/fibula) and BC (talus/calcaneus) the segment links (dashed bold). b) Reference symbols for the description of the geometry of the four-bar linkage

the calcaneus sprang back to its equilibrium position on the neutral path when the external disturbance was removed. These two observations imply that the ankle joint has one degree of unresisted freedom (DoUF) and that the subtalar joint has no DoUF. We also observed that the 3D position and orientation of the instantaneous helical axis (IHA) changed during passive motion. The bundle of axes tended to run as the line drawn through the tips of the two malleoli but a large angular and linear dispersion of the bundle was detectable.

The experiments demonstrated that the ankle/subtalar behaves as a single DOF system during passive motion, with a variable axis of rotation, with the bulk of the movement occurring at the level of the ankle, with fibres within the CaFiL and TiCaL ligaments remaining approximately isometric (neutral fibres), more anterior fibres being tight only at the limit of plantarflexion, most posterior fibres being tight only at the limit of dorsiflexion. These observations suggest that the subtalar joint behaves as a flexible structure but that the ankle/subtalar complex behaves as a single DOF mechanism with ligaments and articular surfaces acting together as constraints to motion.

4. MODEL DESCRIPTION

As a first approximation, a four-bar linkage (4BL) model was formulated to describe motion in the sagittal plane. The tibia/fibula and the talus/calcaneus were taken to be the two rigid segments of the linkage. The linkage ABCD is formed by the line segments AB and CD (solid-bold lines in figure 1a) representing the ligaments CaFiL and TiCaL respectively, and by line segments (dashed-bold lines) which join their attachments on the tibia/fibula and talus/calcaneus segments and which are rigidly fixed in the bones. Figure 1b shows the mathematical descriptors for the geometry of the linkage. The motion of the coordinate system (x', y'), rigid with the talus/calcaneus segment, with respect to the reference frame (x, y), rigid with the tibia/fibula segment, is guided only by the rotation of the two lig-

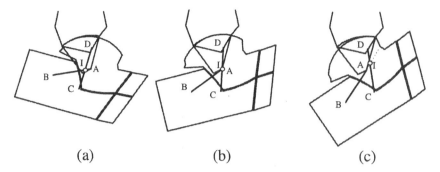

Fig. 2: Geometry of the model in three different joint positions: a) at 30° plan-
     tarflexion, b) at neutral, c) at 20° dorsiflexion.

ament links and described by the single parameter ϕ_3, which can also represent
dorsi/plantarflexion (DP) of the ankle. The orientation angles of the two ligaments (ϕ_2 and ϕ_4) change with flexion (ϕ_3) and the relationships between them
are governed only by the geometry of the 4BL model (the lengths of the four links
r_1, r_2, r_3, r_4) [13].

An important property of the linkage is that the point at which the two ligaments
cross (point I in figure 2) is the instant centre of the joint, i.e. the instant point of
zero velocity of the talus/calcaneus rotation relative to the tibia/fibula segment.
The path of the instantaneous positions of the centre of rotation is the so-called
'centrode' of the linkage. The flexion axis of the joint passes through I.

The shapes of the articular surfaces of the bones must be 'compatible' with the ligaments, i.e. they move in contact with one another while maintaining the neutral
fibres of the ligaments at constant length. If the shape of one articular surface is
given, the shape of the complementary surface of the other bone can be deduced.
In order to avoid interpenetration or separation of the bones, the construction of
the complementary surface depends on the principle that the common normal to
the articular surfaces at their point of contact must pass through I. Using this
principle, the shape of the talus surface complementary to a concave tibial mortise and the position of the contact point during motion on both the shapes were
deduced.

5. MODEL GENERAL PREDICTIONS

Figure 2 shows the predicted motion of the talus/calcaneus segment from a typical
geometry of the two ligament links, in three different joint positions: at 30° of
plantarflexion (a), at 0° (neutral) position (b) and at 20° of dorsiflexion (c). The
orientation of the two ligament links AB and CD and the position of the instant
centre I change relative to both bony segments during DP.

The contact area moves from the posterior part of the tibial mortise in maximal
plantarflexion to the anterior part in maximal dorsiflexion. Because of the movement of the instant centre, the talus rolls forwards on the tibial mortise during
dorsiflexion, backwards during plantarflexion.

The deduced shape of the complementary surface of the talus, compatible with
the mortise shape as an arc of a circumference is a polycentric and polyradial
curve. The position of the centre of curvature and radius of curvature vary for

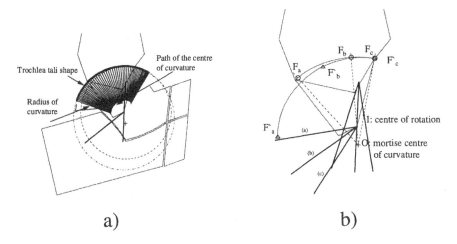

a)   b)

Fig. 3: Shape of the articular surfaces and contact points: a) the calculated polycentric and polyradial shape (bold curve) of the trochlea tali; the best-fit
circumferences for the posterior 75% (dash-dotted) and the anterior 25%
(dashed) of the curve are indicated. b) Positions of the contact point on
both surfaces at the three ankle positions of figure 2.

different points along the articular surface. In a typical configuration shown in
figure 2, with the mortise shape assumed to be an arc of circumference of 6 cm
radius, the complementary shape of the talus was deduced by the model (bold
curve in figure 3a). The calculated radius of curvature of the trochlea tali ranged
from 2.7 cm in the anterior aspect to 3.5 cm at the posterior aspect.

The positions of the contact point in the tibial mortise in the three ankle positions of figure 2 are indicated in figure 3b with small circles (F_a, F_b and F_c). The
corresponding contact points F'_a, F'_b and F'_c on the trochlea are indicated with
small triangles on the surface in 20° of dorsiflexion. The dashed lines are the corresponding common normals, passing through the instant centre I and through
the centre of curvature of the mortise O. The larger distance between successive
contact points on the trochlea compared with corresponding points on the tibial
mortise demonstrates that the talus must slide forward on the tibia while rolling
backward during plantarflexion and must slide backward while rolling forward
during dorsiflexion. The proportion of sliding and rolling depends on the shapes
of the articular surfaces and on the flexion angle. This proportion may be well
expressed with the 'slip ratio', that is the ratio of the length of the talar curve
between two successive contact points and the length of the tibial curve between
the corresponding contact points. The values found in this geometric configuration vary with ankle position from 2.2 to 1.6.

6. MODEL VALIDATION

Using a set of anatomical parameters obtained from one specimen, the predictions from this specimen-specific model were analyzed. Figure 4 shows the model
predictions for both anatomical landmark trajectories and ligament link rotations
and the corresponding experimental results obtained from the relevant specimen.

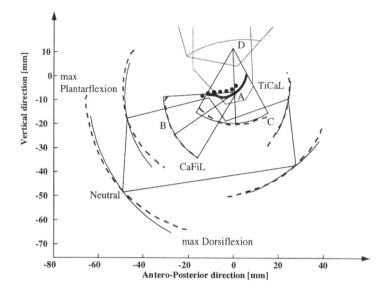

Fig. 4: Model predictions (solid lines) and experimental results (dashed lines). The trajectories of the points of insertion of the CaFiL and TiCaL ligaments and of the 4 calcaneus landmarks are shown. The centrode of the instant centre (bold curve) is compared with the intersection of the IHA's with the mid-sagittal plane (dotted line).

The trajectories of four calcaneus anatomical landmarks and the points of insertion of the CaFiL and TiCaL ligaments as predicted from the 4BL model (solid lines) are shown together with the projections in the mid-sagittal plane of the relevant recorded 3D trajectories (dashed lines). The predicted instant centre of the joint (solid bold) together with the intersection of the calculated IHA's with the mid-sagittal plane (dotted line) are shown in figure 4. The centrode goes more or less from the apex of the lateral malleolus to the apex of the medial malleolus. The trajectory of the instant centre of rotation over the entire range of passive motion, averaged over several trials, was calculated. The mean and the standard deviation over trials of the antero-posterior and proximal-distal translation of such trajectory was found to be 18.3 (0.2 S.D.) and 9.4 (0.3 S.D.) mm respectively. The trajectories obtained from the experiments did not compare well with any hinge-like model prediction.

7. DISCUSSION

Results from the experimental part of the investigation suggested that the human ankle exhibits one DoUF, and the subtalar joint zero DoUF. The observed moving axis of rotation, and the almost isometric pattern of rotation of the CaFiL and TiCaL ligaments led us to model the ankle/subtalar complex as a single DOF mechanism. The preliminary 2D model predictions based on a set of anatomical parameters from one of the specimen tested compare well with the experimental measurements. The match between specimen-specific predictions of calcaneus landmark movements, ligament orientations, instantaneous centres of rotation, and conjugate talus surface profile support the assumptions underlying the formulation of the geometric model.

The 4BL model is a purely kinematic construct that can provide an analytical description of the ankle/subtalar complex motion starting from the definition of the geometry of the joint, without reference to the muscle or external forces needed to produce such movements. The model can therefore be used to demonstrate the relationship between the geometry of the ankle ligaments and the shapes of the articular surfaces.

The calculated anterior translation of the calcaneus during a complete range of plantarflexion is in reasonable agreement with the measurements reported by Siegler *et al.* [8]. According to Barnett and Napier [14] and Raimann [15] it was confirmed that the shape of the posterior part of the trochlea tali is described as an arc of a circle having a bigger radius than the anterior part. The model predicted also the anterior motion of the contact point on the tibial mortise as the ankle dorsiflexes, as recently found in an ankle contact area study [9].

Many are the clinical implications of the results of this investigation. Study of the geometric and mechanical interactions between ligaments and articular surfaces emphasises their complementary roles in guiding the joint motion. Any change to the original geometry of the intact joint, such as bone fracture, erosion of the articular surfaces due to osteoarthritis, ligament injury and reconstruction, or total joint arthroplasty, will alter the original number of DOF and lead to abnormal kinematics of the joint. Consideration of the factors which govern the mobility of a joint has immediate application to the geometric aspects of prosthesis design and is central therefore to the treatment of arthritis as well to the planning of ligament reconstruction. The appreciable antero-posterior and proximal translation of the centre of rotation observed in the experiments and predicted from the model suggests some rolling motion at the ankle. This may call for new ankle prosthesis designs, moving forward from the current practice of converting the ankle into a kinematic hinge. All this evidence also suggests that careful reconstruction of the ligaments, restoring the original origins and insertions, may improve the outcome of both total ankle replacement and ligament reconstruction.

The model based on the set of parameters taken from one specimen successfully predicted what was observed in experiments. However, each specimen has shown different geometry and different range and pattern of motion, pointing out the importance of subject-specific geometric configuration of the model. Having observed most of the ankle/subtalar complex motion during passive rotations in the sagittal plane, much has been learned by developing a preliminary 2D model of the ankle. However, future development of a 3D analogue of the 4BL model to study the interaction of ligament and articular constraints in the ankle joint motion is needed.

Acknowledgements
This work was supported by the Italian National Institute of Health Care. The authors would also thank Dr. Tung-Wu Lu for help with the first version of the model.

References

[1] Bentley G, Shearer J. The foot and ankle. In *Mercer's Orthopaedic Surgery*,

edited by R.B.Duthie, G.Bentley. Arnold, 1996; 1193–1253

[2] Johnson K. Ankle replacement arthroplasty. In *Joint replacement arthro-plasty*, edited by BF Morrey. Churchill Livingstone, 1991; 1173–1182

[3] Kitaoka H, Patzer G. Clinical results of the Mayo total ankle arthroplasty. *J Bone Joint Surg* 1996;78-A(11):1658–1664

[4] Sangeorzan B, Sidles J. Hinge like motion of the ankle and subtalar articulations. *Orthop Trans* 1995;19(2):331–332

[5] Sands A, Early J, Sidles J, Sangeorzan B. Uniaxial description of hindfoot angular motion before and after calcaneocuboid fusion. *Orthop Trans* 1995; 19(4):936

[6] Lundberg A, Goldie I, Kalin B, Selvik G. Kinematics of the ankle-foot complex: Plantarflexion and dorsiflexion. *Foot and Ankle* 1989;9(4):194–200

[7] Lundberg A, Svensson O, Nemeth G, Selvik G. The axes of rotation of the ankle joint. *J Bone Jt Surg Br* 1989;71-B:94–99

[8] Siegler S, Chen J, Schneck C. The three-dimensional kinematics and flexibility characteristics of the human ankle and subtalar joints. Part 1: kinematics. *J Biomech Engng* 1988;110:364–373

[9] Bertsch C, Rosenbaum D, Claes L. Effects of various foot positions on intra-articular and plantar pressure distribution in the intact ankle and chopart joints. In *Transactions of the 7th Conference of the European Orthopaedic Research Society*, Barcellona (Spain), 22-23 April 1997. EORS, 1997; 71

[10] Leardini A, Catani F, O'Connor J. The one degree of freedom nature of the human ankle complex. In *Proceedings of the Autumn 1996 Meeting*, Brighton (UK), September 1996. British Orthopaedic Research Society, 1996; 41

[11] O'Connor J, Leardini A, Catani F, Giannini S. A four bar linkage model for the human ankle joint. In *Transactions of the 7th Conference of the European Orthopaedic Research Society*, Barcellona (Spain), 22-23 April 1997. EORS, 1997; 49

[12] Cappozzo A, Catani F, Della Croce U, Leardini A. Position and orientation in space of bones during movement: anatomical frame definition and determination. *Clinical Biomechanics* 1995;10(4):171–8

[13] Zavatsky A, O'Connor J. A model of human knee ligaments in the sagittal plane: Part I. Response to passive flexion. *Proc Inst Mech Eng Part H J Engng Med* 1992;206:125–34

[14] Barnett C, Napier J. The axis of rotation at the ankle joint in man. its influence upon the form of the talus and mobility of the fibula. *J Anatomy* 1952;86:1–9

[15] Reimann R, Anderhuber F, Gerold J. Geometrical shape of the human trochlea tali. *Acta Anatomica* 1986;127(4):271–278

GEOMETRIC MODELING OF THE HUMAN UPPER EXTREMITY BASED ON RECONSTRUCTED MEDICAL IMAGES

B. A. Garner[1] and M. G. Pandy[2]

1. ABSTRACT

Dynamic simulation is a valuable tool for studying the biomechanics of human movement. Fundamental to this approach is an accurate model of human anatomy and physiology. In this paper, an innovative approach is presented by which the geometric parameters for a dynamic model of the human arm are defined. The parameters are derived based on medical cross-section images obtained from the male data set of the Visible Human Project. The model includes joint motions at the shoulder girdle, elbow, and wrist. Methods for determining joint characteristics and for defining local reference frames are described.

2. INTRODUCTION

The knowledge gained through the study of human biomechanics has application to many fields including athletics, performing arts, medical diagnosis and treatment, ergonomic design, military training, and space exploration. Despite years of research, the principles behind the coordination of human movement remain poorly understood. High degrees of coupling and nonlinearity, high system dimension, and a high degree of over-determinacy contribute to the complexity of multijoint movement. In recent years, musculoskeletal modeling has become increasingly valuable as a tool for studying human movement. The purpose of this work is to clearly present a geometric model of the human upper extremity, comprising the shoulder, elbow, and wrist, which may be used to study the biomechanics of arm movement.

2.1 Previous Work

Due to the many degrees-of-freedom, the many broad, wrapping muscles, the large ranges of motion, and the complex motion constraints, the shoulder is perhaps the most difficult region in the body to model. Early attempts at modeling the shoulder were restricted to a single plane [1]. A few recent efforts have attempted to model

Keywords: Mathematical model, Upper-extremity, Medical images

[1] Department of Mechanical Engineering, University of Texas, Austin, Texas 78712
[2] Departments of Kinesiology and Mechanical Engineering, University of Texas, Austin, Texas 78712

three-dimensional motion of the shoulder [2]. Perhaps the most detailed model of the shoulder mechanism presented to date is that due to Van der Helm et. al. [3]. This model, based on cadaver measurements, includes ligaments, wrapping muscles, and scapulothoracic constraints. It was used for inverse kinematic studies of humeral abduction/adduction and fast, goal-directed movements. However, the coordinate system in which their work is presented makes reproducing the model and applying it to other biomechanical problems difficult [4]. Mathematical models of the elbow [5] and wrist [6] are more common than those of the shoulder.

To our knowledge, only Siereg and Arvikar [7] have attempted to combine the motions of the shoulder, elbow, and wrist into a single, comprehensive model. Their model is based on scaled diagrams from a reference book, and is relatively simple: only glenohumeral motion is modeled at the shoulder, with the clavicle and scapula considered fixed to the thorax; and straight-line muscle representations are used to study muscle and joint forces in the arm during static tasks.

2.2 Overview of Present Work

We have developed a dynamic model of the upper extremity, which includes the relative movements of the entire shoulder girdle, the elbow, and the wrist. The model allows 13 degrees-of-freedom between the thorax, clavicle, scapula, humerus, radius, ulna, and hand. In addition, two constraints are applied to enforce contact between the thorax and scapula. The model is actuated by 41 muscle segments, each represented as an elastic band joining the attachment sites of the muscle tendons. Obstacles which divert the paths of certain muscle segments are represented by simple geometric shapes. Parameters defining the geometry of the model are based on reconstructed surface data of bones and muscles from the two-dimensional image set made available through the Visible Human Project [8].

In this paper, we describe the method used to derive the geometric parameters of the model. The method used to reconstruct the bone and muscle surfaces is summarized. The properties of the modeled joints and the location and orientation of the local reference frames assigned to each modeled body are then defined. Properties and parameters defining the muscle segments and their corresponding obstacles will be presented in a subsequent paper.

3. RECONSTRUCTION OF BONE AND MUSCLE GEOMETRY

The development of previous biomechanical models has typically relied on geometric data obtained from cadaver measurements. This approach can be difficult and time-consuming, and can lead to a number of inaccuracies. First, it is difficult to measure accurately and with high resolution the three-dimensional coordinates of points on the body. Second, the measuring device or cadaver may be subject to inadvertent perturbation from its reference position and location. Third, and perhaps most importantly, measuring the location of one anatomical structure may require the removal of neighboring structures.

Recently the National Library of Medicine has made available, through the Visible Human Project, a large collection of digital, medical, cross-sectional images obtained from frozen male and female cadavers. These images represent a fixed, high-resolution "snap shot" of the intact human body, and provide a new way of obtaining accurate, three-dimensional coordinate data for anatomical structures. Through a variety of methods the images can be reconstructed into three-dimensional data sets representing surface or volume geometry of structures.

Reconstructed surface data of the bones and muscles of the upper extremity of the male image set were used as the basis for defining the geometric parameters of the present model. Bone surfaces were used to define joint centers, segment lengths, muscle origin and insertion sites, and muscle path obstacles. Muscle surfaces were used to define muscle centroid line-of-action paths, origin and insertion sites, and path obstacles. The surface of each bone and muscle was reconstructed using a thresholding formula similar to that given by Lorensen and Cline [9]. The reconstruction method comprises four steps: 1) manually identifying surface boundaries within each image, 2) reconstructing surface boundaries from neighboring images into a three-dimensional triangle mesh, 3) smoothing the reconstructed surface, and 4) decimating the triangle mesh.

The high-resolution of the images and the nature of the reconstruction algorithm produced a very large number of vertices and triangles for each bone and muscle. Prior to decimation, the longest edge of any triangle in the bone surfaces was less than 2.0 millimeters; and in the muscle surfaces, less than 4.0 millimeters. By eliminating redundant triangles, the decimation process typically achieved a 90% reduction in mesh density (Fig. 1).

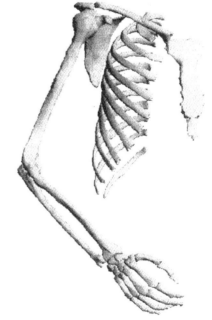

Figure 1: Computer-generated rendering of the surfaces of the upper-extremity bones represented in the model. Surfaces were reconstructed from medical images obtained from the Visible Human Project male data set.

4. DETERMINATION OF JOINT PARAMETERS

The reconstructed bone surfaces were used to define geometric parameters of the upper-extermity model. To facilitate location of landmarks and shapes from the surface data, an interactive, computer-graphics utility was developed: it displays the reconstructed three-dimensional surfaces, allows the user to move a three-dimensional pointer, and allows the user to select and isolate individual triangles. All body segments were assumed to be rigid with fixed centers of mass; all joints were modeled with fixed axes or centers of rotation [10].

4.1 Joints of the Shoulder Girdle

Each joint of the shoulder girdle was modeled as a three-degree-of-freedom spherical joint (Fig. 2). The sternoclavicular joint was assumed to lie medial to the clavicular attachment of the costoclavicular ligament, and was found by locating this point on the reconstructed clavicle using the three-dimensional pointer. The acromioclavicular joint was assumed to lie in the gap between the lateral end of the clavicle and the medial face of the acromion process; its location was found by computing the center of a number of points (identified with the three-dimensional pointer) used to define the boundary of this gap. The glenohumeral joint was assumed to lie at the center of the humeral head. This point was computed as the center of a sphere fitted to triangles isolated from the spherically-shaped portion of the reconstructed humeral head.

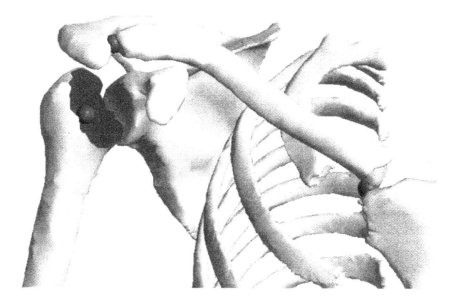

Figure 2: The locations of the sternoclavicular, acromioclavicular, and glenohumeral joint centers in the model. Each joint is modeled as a three degree-of-freedom, spherical joint.

4.2 Scapulothoracic Joint

The scapulothoracic joint was modeled by forcing two scapula-fixed points to remain in contact with an ellipsoid, which represented the shape of the rib cage. The ellipsoid was calculated by best fitting those triangles isolated from the dorsal surface of ribs one through nine. A gap of about 30mm, formed by the thickness of subscapular muscles, was observed between the medial edge of the scapula and the surface of the ribs. To account for this gap, two points of the ellipsoid, lying closest to the medial border of the scapula, were computed and constrained. One point lies just beneath the medial end of the scapular spine; the other, just beneath the inferior angle.

4.3 Joints of the Lower Arm

The joints of the lower arm were each represented by a fixed-axis hinge (Fig. 3). The humeroulnar joint axis was assumed to pass through the center of the trochlea and capitulum, and was found by fitting a variable-radius cylinder to triangles isolated from this region of the reconstructed humerus. The radioulnar joint axis was assumed to join the center of the capitulum (intersecting the humeroulnar axis) and the center of the distal head of the ulna. Each of these joint centers was found by fitting a sphere to triangles isolated from the respective regions of the humerus and ulna.

The radiocarpal joint was represented by *two* non-intersecting, perpendicular axes. These axes, which permit wrist flexion-extension and radio-ulnar deviation, respectively, pass through the proximal end of the capitate bone, but miss each other at the closest point by 5mm [11]. The flexion-extension axis was defined by fitting a variable-radius cylinder to triangles representing the cylindrically-shaped concavity at the distal end of the radius. The radio-ulnar deviation axis was defined to be perpendicular to both the flexion-extension axis and the axis of the third metacarpal (found by fitting a variable-radius cylinder to triangles isolated from the metacarpal shaft), and was located 5mm distal to the flexion-extension axis along the metacarpal shaft axis.

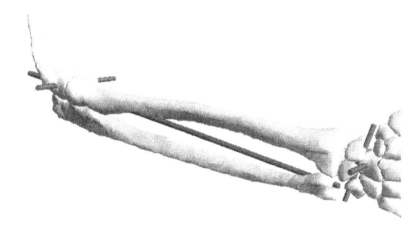

Figure 3: The hinge axes of the humeroulnar, radioulnar, and radiocarpal joints. The radiocarpal joint is formed by two hinge axes representing wrist flexion-extension and radial-ulnar deviation, respectively.

5. LOCAL REFERENCE FRAMES

To facilitate dynamic modeling of the upper extremity, body-fixed, local reference frames were defined for the thorax and each independently moving bone in the model. General conventions for these frames are such that, in the anatomical position, the x, y, and z axes point in the lateral, anterior, and superior directions, respectively. Where appropriate, one axis if each frame is aligned with the shaft of the bone. In all cases, two axes were defined based on anatomical landmarks, and the third axis was computed as the cross product of the first two.

5.1 Thorax

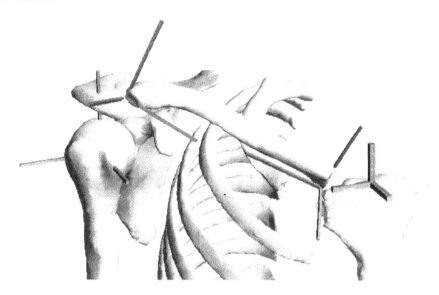

Figure 4: Local reference frames assigned to the thorax and bones of the shoulder girdle. The origin of each moving frame coincides with the proximal joint center.

The thorax (ground) frame was aligned with anatomical axes (Fig. 4). The sagittal plane was delimited by four readily identifiable points: the jugular notch (IJ); the inferior center of the xyphoid process (XP); and the spinous processes of the seventh cervical (C7) and eighth thoracic (T8) vertebrae [3]. The x-axis of the thorax frame was assumed to lie normal to the sagittal plane, pointing laterally to the right. The z-axis was assumed to lie along the line joining the XP-T8 midpoint to the IJ-C7 midpoint. The jugular notch (IJ) was chosen as the origin of the thorax frame.

5.2 Frames of the Upper Arm

The sternoclavicular joint was chosen as the origin of the clavicle frame. The x-axis of the clavicle frame, assumed to lie along the line joining the sternoclavicular and acromio-clavicular joint centers, points laterally in the model. To locate the y-axis, a plane was fitted to triangles which represented the flat region of the superior surface of the medial end of the clavicle. The y-axis was chosen to lie in this plane, perpendicular to the x-axis, and pointing anteriorly (Fig. 4).

The acromioclavicular joint was chosen as the origin of the scapula frame. The z-axis of the scapula frame was assumed to lie along the line joining the glenohumeral and acromioclavicular joint centers. The x-axis was chosen to be perpendicular to the z-axis, passing through the medial border of the scapula, and pointing laterally (Fig. 4).

The glenohumeral joint was chosen as the origin of the humerus frame. The z-axis of the humerus frame was defined to be collinear with the axis of the humeral shaft, pointing superiorly; the axis of the humeral shaft was found by fitting a variable-radius cylinder to triangles isolated from this region of the reconstructed humerus. The y-axis of the humerus frame, assumed to be perpendicular to both the z-axis and the humeroulnar joint axis, points anteriorly (Fig. 4).

5.3 Frames of the Lower Arm

The point on the humeroulnar joint axis closest to the olecranon process was chosen as the origin of the ulna frame. The z-axis of the ulna frame was found by fitting a variable-radius cylinder to triangles isolated from the shaft of the reconstructed ulna. The y-axis was defined to be perpendicular to both the z-axis and the humeroulnar joint axis (Fig. 5). Interestingly, the resulting x-axis of the ulna frame was found to make a 10° angle with the humeroulnar joint axis, contributing to the controversial "carrying angle" [12].

The intersection of the humeroulnar and radioulnar joint axes at the center of the capitulum was chosen as the origin of the radius frame. The z-axis of the radius frame was assumed to point proximally along the shaft of the radius, described by a line joining the distal tip of the radial styloid process and the local origin. The y-axis was chosen to be perpendicular to both the z-axis and the flexion-extension axis of the radiocarpal joint (Fig. 5).

The hand frame was defined so that the x-axis aligns with the wrist flexion-extension axis, and the y-axis aligns with the wrist radio-ulnar deviation axis. The point on the radio-ulnar deviation axis nearest the flexion-extension axis was chosen as the origin of the hand frame (Fig. 5).

6. SUMMARY

A geometric model of the upper extremity, which includes the relative movements of the bones at the shoulder girdle, elbow, and wrist, is presented. Local reference frames are defined for the thorax, clavicle, scapula, humerus, ulna, radius, and hand.

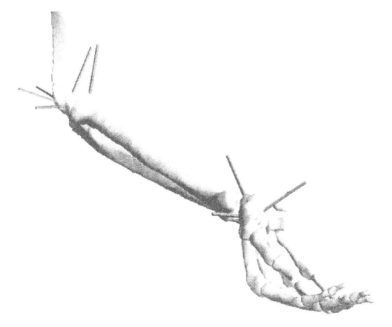

Figure 5: Local reference frames assigned to the bones of the lower arm in the model. The hand was modeled as a single, rigid body with its own reference frame.

Geometric parameters are based on reconstructed surface data obtained from medical images, which is a new approach for defining the kinematics of movement.

7. REFERENCES

[1] DeLuca, C.J., Forrest, W.J., "Force Analysis of Individual Muscles Acting Simultaneously on the Shoulder Joint During Isometric Abduction", *Journal of Biomechanics*, vol. 6, pp 385-393, 1973.

[2] Karlsson, D., Peterson, B., "Towards a Model for Force Predictions in Human Shoulder", *Journal of Biomechanics*, vol. 25, no. 2, pp. 189-199, 1992.

[3] Van der Helm, F.C.T., "Analysis of the Kinematic and Dynamic Behavior of the Shoulder Mechanism", *Journal of Biomechanics*, vol. 27, no. 5, pp. 527-550, 1994.

[4] Johnson, G.R., Spalding, D., Nowitzke, A., and Bogduk, N., "Modelling the Muscles of the Scapula: Morphometric and Coordinate Data and Functional Implications", *Journal of Biomechanics*, vol. 29, no. 8, pp. 1039-1051, 1996.

[5] Hutchins, E.L., Gonzalez, R.V., Barr, R.E., "Comparison of Experimental and Analytical Torque Angle Relationships of the Human Elbow Joint Complex", *Biomedical Scientific Instrumentation*, vol. 29, pp. 17-24, 1993.

[6] Buchanan, T.S., and Shreeve, D. A., "An Evaluation of Optimization Techniques for the Prediction of Muscle Activation Patterns During Isometric Tasks", *Journal of Biomechanical Engineering*, vol. 118, pp. 565-574, November 1996.

[7] Seireg, A., Arvikar, R., "Modeling of the Musculoskeletal System for the Upper and Lower Extremities", *Biomechanical Analysis of the Musculoskeletal Structure*, 1989

[8] Visible Human Project, National Library of Medicine, 1997. http://www.nlm.nih.gov/research/visible/visible_human.html.

[9] Lorensen, W. E. and Cline, H. E., "Marching Cubes: A High Resolution 3D Surface Construction Algorithm," *Computer Graphics*, vol. 21, no. 3, pp. 163-169, July 1987.

[10] Raikova, R., "A General Approach for Modelling and Mathematical Investigation of the Human Upper Limb", *Journal of Biomechanics*, vol. 25, no. 8, pp. 857-867, 1992.

[11] Sommer, H. J. and Miller, N. R., "A Technique for Kinematic Modeling of Anatomical Joints," *Journal of Biomechanical Engineering*, vol. 102, p. 311-317, 1980.

[12] Morrey, M.D., and Chao, Y.S., "Passive Motion of the Elbow Joint", The *Journal of Bone and Joint Surgery*, vol. 58-A, no. 4, June 1976.

3-D SPECIMEN-SPECIFIC GEOMETRIC MODELLING OF THE KNEE

J.D. Feikes [1], D.R.Wilson [2] and J.J. O'Connor [3]

1. ABSTRACT

This paper describes the practical need for and the development of a three-dimensional mathematical model of the human knee capable of predicting the spatial displacement of the femur relative to the tibia in the unloaded state. The knee joint is modelled as a single degree-of-freedom parallel spatial mechanism, the three-dimensional analogue to the four-bar linkage model of the knee. The femur and tibia are linked by five ligament and articular contact connections. By solving for the unique displacement path of the mechanism from a starting position, the model predicts the relative locations of the bones and as a result the positions and orientations of the contact normals and ligaments throughout a range of unresisted motion. Model parameters can be derived from the measured geometry of a single knee including the tibial and femoral articular surface shapes and the attachment sites of nearly isometric fascicles in the ACL, PCL and MCL. The model successfully predicts a physiological range of unresisted passive flexion in which both internal tibial rotation and ab/adduction are coupled to the flexion angle. The model allows the distinction between the kinematic geometry of the knee and its mechanics to be made; in other words, the distinction between the interactions between the articular surfaces and ligaments of the joint which constrain its mobilty and those which contribute to its stability. The development of such a geometric model is the first step in producing a practical three-dimensional mechanical model of the knee.

2. INTRODUCTION

2.1 Joint Kinematic Geometry and Mechanics

The musculoskeletal system is highly redundant in the sense that there are many more load-bearing structures than are necessary for dynamic equilibrium. With no agreed basis for the calculation of the forces transmitted by the individual muscles or ligaments or by the articular surfaces at the joint, the most modern gait analysis systems provide only the *resultant* force and moment transmitted at

---

[0] Keywords: Knee, Model, Geometry, Kinematics,Mechanism

[1] D.Phil. Candidate, Oxford Orthopaedic Engineering Centre, Nuffield Orthopaedic Centre, Windmill Road, Headington, Oxford, OX3 7LD, U.K.

[2] Postdoctoral Fellow, Orthopaedic Biomechanics Laboratory, Department of Orthopaedic Surgery, Beth Israel Hospital, 330 Brookline Ave. RN 115, Boston, MA 02215, U.S.A.

[3] Professor and Director, Oxford Orthopaedic Engineering Centre, see [1]

each joint. Further effort is required to develop practical methods of calculating levels of these forces in daily activities.

Geometric factors are prime determinants of the levels of force transmitted by the joint structures. The joint *geometry* controls the lines of action of muscle tendons, ligaments and contact forces at the joint. Understanding the *kinematic geometry* of the joint is therefore key to understanding load transmission through the musculoskeletal system.

The joints of the lower leg vary with regard to kinematic geometry. Both the range of unresisted movement they allow and the details of movement within that range differ. The differences in mobility do not stem from differences in the mechanical properties of the tissues but from differences in the geometry of the articular surfaces and from differences in the geometric arrangements of the ligaments and capsule.

Joints allow mobility but they must also transmit load. Muscles are required to stabilise the skeleton by suppressing movement at the joints in the presence of external loads. During activity, muscle forces also initiate and maintain movement. The passive joint structures contribute to load transfer and joint stabilisation through the transmission of compressive stress between the articular surfaces and the transmission of tensile stress along the fibres of the ligaments and capsules.

It is important to distinguish between that which we can learn from the *kinematic geometry* of a joint and that which we can learn from studying its *mechanics*. The features which control mobility can be deduced mainly from kinematic geometric analysis while the study of stability and load transfer obviously also requires mechanics.

2.2 Mathematical Models of the Knee

Many mathematical models of the knee [1, 3, 18] choose to ignore the study of kinematic geometry. Tackling the study of knee mechanics directly, these models aim to predict the forces transmitted by various joint structures and the resulting movement of the joint under specified external loads. This approach requires the simultaneous solution of the equations describing mechanical equilibrium and those describing geometric constraints [1, 18] or the minimization of system energy [3]. The models can be computationally costly [3] and they are not capable of predicting the motion of the unloaded joint, *i.e.* joint mobility. More importantly, they do not bring us any closer to achieving the goal of developing *practical* methods of calculating the forces in individual joint structures during activity.

2.2.1 The Sequential Approach

By contrast, examining the kinematic geometry of the joint *first* not only allows the study of the influence of joint geometry on knee mobility, but it also provides a highly practical and efficient starting point for subsequent mechanical analysis. The geometric and mechanical analyses are undertaken *sequentially* versus *simultaneously*.

Geometric Analysis

The development of an appropriate geometric model of the joint constitutes the necessary first step in the sequential approach. Within the range of unresisted movement of some joints, the articulating bones and the ligaments which hold the bones together can be modelled as a mechanical linkage or mechanism. For the model to be predictive it must possess a single degree-of-freedom (dof) so that the position of all links in the mechanism can be specified fully by one quantity alone. The displacement path of the links from a known starting position is therefore

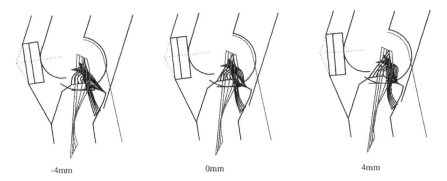

-4mm   0mm   4mm

Figure 1: The two-dimensional knee model with continuous arrays of extensible ligament fibres (ACL, PCL, MCL and LCL) in the neutral configuration (0 mm) as determined by the four-bar linkage and with ±4 mm anterior-posterior tibial translation. The changing muscle and tendon lines of action and a patello-femoral model are also shown (from Lu and O'Connor [9]).

unique. In the sagittal plane, the knee joint has been previously modelled as a one dof kinematic pair, a hinge joint [11], and more accurately as a single dof 2-D mechanism, a four-bar linkage, based on the cruciate mechanism [4, 7, 6]. To model the complete spatial behaviour of the knee, however, a single dof 3-D mechanism is required.

Geometric models predict how the unloaded joint moves with flexion. They provide the knee's *neutral* configuration, *i.e.* the relative positions and orientations of the bones, and as a result the positions of the origins and insertions of muscle tendons and ligaments and the positions of articular contact points at any given flexion angle. Such a model also provides the location of the instantaneous centre of rotation (2-D) or the instantaneous screw axis (3-D) throughout the range of motion. A sagittal-plane four-bar linkage model of the knee has, for example, in this manner, predicted and provided an explanation for the posterior movement of the instantaneous centre of rotation and the combination of rolling and sliding of the femur on the tibia during flexion [4, 12].

Mechanical Analysis

At any flexion angle, the knee's neutral configuration may be used as a starting point for subsequent mechanical analyses. A geometric model thereby reduces the number of unknowns in the problem of force and moment distribution at the joint. In the manner of previous two-dimensional work based on the four-bar linkage model, it may be assumed, as a first approximation, that the directions of the structures do not change under load and that the ligaments are adequately represented by single fibres [2, 10].

A better approximation, however, which incorporates continuous arrays of extensible ligament fibres has already proved successful in two-dimensions [19, 20, 21]. Figure 1 shows a two-dimensional knee model whose neutral path is determined by the four-bar linkage made up of isometric fibres in the ACL and PCL, the tibia and the femur. The recruitment of fibres in the ACL, PCL, MCL and LCL, their relative stretching and slackening with tibial displacement, is a function of geometry alone. The external forces which have to be applied to balance the ligament and contact forces resulting from any assumed tibial translation can be calculated by incorporating the appropriate stress-strain relationships for the ligament fibres [19, 20]. The compressive loading and subsequent deformation of the cartilage due to the tension forces developed in the ligaments must also be taken into account.

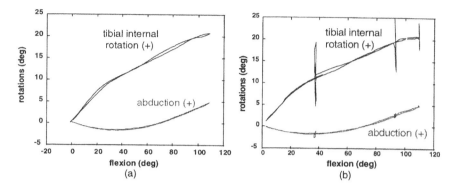

Figure 2: Joint rotations *in vitro*. (a) The unique path of motion of a representative knee specimen flexed and extended in the virtually unloaded state. (b) Deviations from the neutral path.

The resulting non-linear relationships between applied force and tibial displacement vary with flexion.

Taking tissue deformation into account, mechanical analysis of the joint with the addition of muscle forces may proceed by considering incremental loading. An iterative numerical solution of the aforementioned non-linear equations is required at each increment, with sequential geometric and mechanical analysis during each iteration [21]. The two-dimensional knee model with inextensible [10] and extensible [13] ligament fibres has been successfully incorporated into a two-dimensional locomotor system model which has been used in conjunction with kinematic and external force gait data to calculate individual muscle, ligament and joint contact forces at the knee during various activities. A three-dimensional geometric model is required to account for the knee's behaviour in the frontal and transverse planes, to allow the complete three-dimensional analysis of knee mechanics and for incorporation into a three-dimensional locomotor system model [8] [1].

2.3 Objective of the Study

This paper describes both the experimental and theoretical work involved in the development of an appropriate three-dimensional geometric model of the human knee - the analogue of the two-dimensional four-bar linkage model. The current objective was to demonstrate the usefulness of such a model as the first important step in a sequential approach to the study of three-dimensional knee mechanics.

3. PREVIOUS EXPERIMENTAL WORK

A geometric model can successfully predict the passive displacement of a joint which behaves as a one dof mechanism in the unloaded state. *In vitro* passive flexion/extension experiments have been undertaken to assess the mobility of the intact knee [15, 16].

Figure 2 (a) shows the curves of internal/external tibial rotation and ab/adduction plotted versus flexion angle for a post-mortem knee specimen which was flexed and extended repeatedly under very light loads (the weight of about 15 cm of distal femur, about 5 N). The plot demonstrates that the knee follows virtually the same path of movement in flexion as in extension. All components of the movement (including the three translations of a reference point not shown here) are uniquely

---

[1] Citation for the paper in the current proceedings by Lu *et al.*

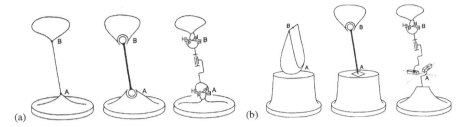

Figure 3: Kinematically equivalent mechanical connections: (a) isometric ligament fibre (b) articular contact (after Hunt [5]).

coupled to the flexion angle. Figure 2 (b) shows the path of movement for the same knee stopped at three flexion angles, displaced from its preferred path of movement and then quickly released. Further internal/external rotation or ab/adduction were induced only by applying load. Upon removal of the displacing force, the knee returned to its preferred position on the original track. It then continued to follow this track when flexion or extension was resumed.

By specifying one of the six components of movement, such as the flexion angle, the relative positions of the bones can be specified completely. The passive flexion path represents a series of neutral equilibrium positions. Displacement of the bones along the neutral path is not resisted or involves minimal tissue deformation, whereas beyond this path lies a region where elastic deformation of the joint tissues accompanies any further displacement of the bones. Deviations from the neutral path in this region are fully recoverable. The experiments demonstrate that the intact unloaded knee has a single degree of unresisted freedom [15, 16].

4. THEORETICAL WORK

4.1 Model Formulation

The experimental observations described above justify modelling the unloaded knee joint geometrically as a three-dimensional one dof mechanism. In formulating the model, the joint structures which 'guide' the knee's passive motion were considered. These structures act as constraints to the relative movement between the tibia and femur, and must reduce the number of dof between the bones from the maximum of six to one.

It is assumed that there exists a fibre in each of the anterior cruciate (ACL), posterior cruciate (PCL) and medial collateral (MCL) ligaments which remains isometric in passive flexion and extension, each of which is represented by a rigid rod connecting a spherical pair at its femoral origin to a spherical pair at its tibial insertion (Figure 3 (a)). Each of the ligament connections reduces the number of dof between the two bones by one. It is further assumed that the tibia and femur are rigid and that contact is continuous in both the medial and lateral compartments throughout the range of unresisted motion. As a first approximation, the femoral condyles are assumed to be spherical and the tibial condyles planar. The articular contacts in the medial and lateral compartments are each represented by a spherical pair at the centre of the corresponding femoral condyle connected by a rod to a planar pair at the contact point on the tibial condyle (Figure 3 (b)). The two sphere-plane contact connections allow rolling and sliding at the contact and supply the remaining two constraints, leaving the knee with a single dof.

The resulting mechanism resembles a platform-type manipulator similar to those used in aircraft cockpit simulators. In each of the five ligament and contact connections a spherical pair may be replaced by three revolute pairs in series (with non-coplanar axes) and each planar pair by two prismatic pairs and a revolute pair in series (with non-coplanar axes). Figure 3 shows the ligament and contact connections represented by chains of one dof pairs with the redundancy in axial rotational freedom removed. The spatial mechanism is then made up of 22 links (including the tibia and femur) and 25 pairs where each pair has 1 dof. Grübler's criterion for mobility [5]: $M = 6(n - m - 1) + \sum_{i=1}^{m} f_i$ is used to verify that the mechanism does indeed have a single dof or mobility one where n is the number of links, m is the number of pairs and each pair i possesses f_i dof ($M = 6(22 - 25 - 1) + 25 = 1$).

4.2 Model Parameters

Model parameters are derived from the geometry of a single cadaver knee measured *in vitro* [17]. Using a least-squares criterion, spheres are fitted to measured points on the femoral condyles, planes are fitted to measured points on the tibial condyles. For each of the ACL, PCL and MCL, individual fibres are defined by lines joining all possible combinations of measured femoral and tibial attachment points in full extension. The lengths of the lines are tracked in a simulation of the knee's measured passive flexion and extension. In each ligament, the fibre found to experience the least change in overall length was chosen as the isometric fibre. The positions of the femoral sphere centres, the positions and orientations of the tibial planes and the femoral and tibial attachments of the isometric ligament fibres at full extension constitute the only input to the geometric model. Each tibial contact point is defined as the point where the line normal to the tibial plane passing through the femoral sphere centre intersects the plane. The configuration of the model in full extension is depicted in Figure 4 (a).

4.3 Model Displacement

With the formulation of the three-dimensional model as described in the previous sections, the next step involves calculating the displacement of the mechanism model from a known starting position. Whereas the displacement of the two-dimensional four-bar linkage mechanism can be calculated analytically, there exists no closed-form solution for that of a three-dimensional mechanism such as the one proposed.

A numerical method developed by Uicker *et al.* [14] was therefore implemented [17]. The mechanism is divided into four independent closed-loops between the tibia and femur: ACL-PCL, ACL-MCL, ACL-Medial Contact and ACL-Lateral Contact. The iterative solution method is based on the four *loop* equations which together ensure that the ligament and contact constraints are satisfied. The calculation of the spatial displacement of the knee model through over 100 degrees of flexion takes less than 5 seconds computation time on an Indy workstation (Silicon Graphics, Mountain View, CA, U.S.A.).

4.4 Model Predictions

The model predicts not only the specific *range* of unresisted movement, but also the three-dimensional *pattern* of movement within that range for a specifc knee geometry. The femoral link follows a unique track of motion relative to the tibia as observed in the unloaded knee in experiment. Beyond the range of unresisted movement the mechanism locks and the femur can only reverse along the calculated unique path.

A model based on parameters derived from a single cadaver knee predicts joint rotations which compare well with the measured motion of that same knee[17].

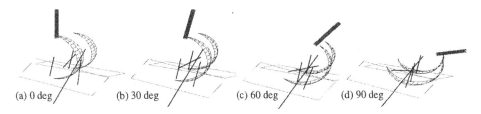

(a) 0 deg   (b) 30 deg   (c) 60 deg   (d) 90 deg

Figure 4: The three-dimensional specimen-specific geometric model in the calculated configurations at (a) 0°, (b) 30°, (c) 60° and (d) 90°. Solid lines: ligament and contact connections. Spherical sections: femoral condyles. Planes: tibial condyles. Cylinder: the femoral shaft.

Both internal tibial rotation and ab/adduction are coupled to knee flexion. The model also successfully predicts the posterior translations of the tibial contact points in both compartments with flexion. The magnitudes of the translations are still larger than previously reported translations of the menisci [7], but incorporating more anatomical surface shapes for the tibial condyles may provide more physiological values for the contact point excursions. Figure 4 shows the model in the calculated position at 0°, 30°, 60° and 90° flexion.

5. DISCUSSION

Force analysis of the musculoskeletal system requires, first, a geometrical model of the system in the unloaded state, preferably customized through choice of parameters to individual subjects, animated with data from a gait analysis system. The neutral configuration of joint structures provided by the geometric models of the hip, knee and ankle is the starting point for mechanical analysis. It reduces the number of unknowns in the mechanically indeterminate locomotor system.

Whereas the hip is easily represented by a spherical pair, the knee's spatial kinematic behaviour is more complex and can only be represented by a three-dimensional mechanism. By predicting the displacement of the knee in the unloaded state, the current spatial mechanism model of the knee provides the joint's neutral configuration at any given flexion angle. On the assumption that the directions of the lines of action of the muscles and ligaments do not change under load, such a three-dimensional geometric model based on preliminary parameters (derived from the literature) has already been incorporated into a three-dimensional locomotor system model [8] for use in gait studies. The knee model, however, has now been customized and will be developed further in the manner of previous two-dimensional modelling. Arrays of extensible ligament and capsule fibres will be included. A three-dimensional knee model capable of taking into account tissue deformation will then be incorporated with muscles into the three-dimensional locomotor system model.

6. CONCLUSIONS

A three-dimensional *geometric* model such as the one described in this paper is a highly practical tool in the study of knee mechanics in activity. In the context of calculating muscle, ligament and contact forces in activity, such a model constitutes the first step in a less computationally intensive, sequential approach to the problem of distributing forces among the load-bearing structures at the joint.

7. Acknowledgements

This work was supported by CAMARC-II, the Arthritis and Rheumatism Council

(U.K.), the Fonds FCAR (Québec, Canada), the Univeristy of Oxford and the Overseas Research Studentship Scheme (U.K.).

References

[1] Blankevoort, L., Kuiper, J.H., Huiskes, R. and Grootenboer, H.J., Articular contact in a three-dimensional model of the knee, *J Biomech*, 1991, 24(11):1019–1031.

[2] Collins, J.J. and O'Connor, J.J., Muscle-ligament interactions at the knee during walking, *Proc Inst Mech Eng Part H, J Engng Med*, 1991, 205:11–18.

[3] Essinger, J.R., Leyvraz, P.F., Heegard, J.H. and Robertson, D.D., A mathematical model for the evaluation of the behaviour during flexion of condylar-type knee prostheses, *J Biomech*, 1989, 22:1229–1241.

[4] Goodfellow, J.W. and O'Connor, J.J., The mechanics of the knee and prosthesis design, *J Bone Jt Surg [Br]*, 1978, 60-B:358–369.

[5] Hunt, K.H., *Kinematic Geometry of Mechanisms*, 1978, Oxford University Press, Oxford.

[6] Huson, A., Biomechanische Probleme des Kniegelenks, *Orthpade*, 1974, 3:119–126.

[7] Kapandji, I.A., *The Physiology of the Joints, Vol. 2 The Lower Limb*, 1987, Churchill Livingstone, London, 5th edition. (Translated by L.H. Honore).

[8] Lu, T. W., O'Connor, J. J., Taylor, S.J.G. and Walker, P.S., Comparison of femoral forces during double and single leg stance: application to knee replacement, In *Proceedings of the 7th Conference of European Orthopaedic Research Society*, 1997, page 68, Barcelona, Spain.

[9] Lu, T. W. and O'Connor, J.J., Fibre recruitment and shape changes of knee ligaments during motion: as revealed by a computer graphics-based model, *Proc Inst Mech Eng Part H, J Engng Med*, 1996, 210:71–9.

[10] Lu, T.W., O'Connor, J.J., Taylor, S.J.G. and Walker, P.S., Validation of a lower limb model with *in vivo* femoral forces telemetred from two subjects, *J Biomech*, 1997, in press.

[11] Morrison, J.B., Bioengineering analysis of force actions transmitted by the knee joint, *Bio-Med Eng*, April 1968, 90:164–70.

[12] O'Connor, J.J., Shercliff, T., FitzPatrick, D., Bradley, J., Daniel, D., Biden, E., and Goodfellow, J., Geometry of the knee, In D.M. Daniel, W.H. Akeson, and J.J. O'Connor, editors, *Knee Ligaments: Structure, Function, Injury, and Repair*, 1990, chapter 10, pages 163–200. Raven Press, New York.

[13] Toutoungi, D.E., *The mechanics of rehabilitation of the knee joint*, PhD thesis, University of Oxford, 1996, (in preparation).

[14] Uicker, J.J., Denavit, J. and Hartenberg, R.S., An iterative method for the displacement analysis of spatial mechanisms, *Journal of Applied Mechanics, Trans ASME*, 1964, pages 309–14, June.

[15] Wilson, D.R., Feikes, J., Zavatsky, A.B., Bayona, F., and O'Connor, J.J., The human knee: a one degree-of-freedom joint?, In *British Orthopaedic Research Society, Oswestry, U.K.*, 1996.

[16] Wilson, D.R., Feikes, J., Zavatsky, A.B., Bayona, F. and O'Connor, J.J., The one degree-of-freedom nature of the human knee joint - basis for a kinematic model, In *Canadian Society of Biomechanics, Vancouver, Canada*, August 1996, pages 194–5.

[17] Wilson, D.R., Feikes, J. and O'Connor, J.J., Kinematic geometry of the knee: A theory of screw-home motion. under review, 1997.

[18] Wismans, J., Veldpaus, F. and Janssen, J., A three-dimensional mathematical model of the knee-joint, *J Biomech*, 1980, 13:677–685.

[19] Zavatsky, A.B. and O'Connor, J.J., A model of human knee ligaments in the sagittal plane: Part I. Response to passive flexion, *Proc Inst Mech Eng Part H, J Engng Med*, 1992, 206:125–34.

[20] Zavatsky, A.B. and O'Connor, J.J., A model of human knee ligaments in the sagittal plane: Part II. Fibre recruitment under load, *Proc Inst Mech Eng Part H, J Engng Med*, 1992, 206:135–45.

[21] Zavatsky, A.B. and O'Connor, J.J., Muscle-ligament interaction at the knee during isometric quadriceps contractions, *Proc Inst Mech Eng Part H, J Engng Med*, 1993, 207:7–18.

MUSCULOSKELETAL MODEL OF THE KNEE FOR STUDYING LIGAMENT FUNCTION DURING ACTIVITY

Marcus G. Pandy[1] and Kevin B. Shelburne[2]

1. ABSTRACT

A sagittal-plane model of the knee is used to determine the relationships between the forces developed by the muscles and the forces induced in the cruciate ligaments during activity. The geometry of the model bones is adapted from cadaver data. Eleven elastic elements describe the geometric and mechanical properties of the knee ligaments. The model is actuated by eleven musculotendinous units, each unit represented as a three-element muscle in series with tendon. For isometric contractions of the quadriceps, ACL force increases as quadriceps force increases; for isotonic contractions, ACL force decreases monotonically as knee-flexion angle increases. The relationships between ACL force, quadriceps force, and knee-flexion angle are explained by the geometry of the knee-extensor mechanism and by the changing orientation of the ACL in the sagittal plane. For isometric contractions of the hamstrings, PCL force increases as hamstrings force increases; for isotonic contractions, PCL force increases monotonically with increasing flexion. The relationships between PCL force, hamstrings force, and knee-flexion angle are explained by the geometry of the hamstrings and by the changing orientation of the PCL in the sagittal plane. Hamstrings co-contraction is an effective means of reducing ACL force at all flexion angles, except near full extension.

2. INTRODUCTION

The relationships between the forces developed by the muscles, the external loads applied to the leg, and the forces produced in the knee ligaments are unknown. The reason is that measurements of muscle and ligament forces are difficult to obtain *in vivo*. Numerous *in vitro* studies have determined the effect of muscle forces on cruciate-ligament loading, but the forces applied in these experiments are well below the forces developed by the muscles during activity [1, 2]. We have developed a sagittal-plane model of the knee to study the load sharing between the muscles, ligaments, and bones during activity. Calculations obtained from the model were analyzed previously to explain the pattern of cruciate-

Keywords: Cruciate-ligament forces, Muscle activity, Knee model.

[1] Department of Kinesiology and Department of Mechanical Engineering, University of Texas at Austin, Austin, Texas 78712, U.S.A.
[2] Department of Mechanical Engineering, University of Texas at Austin, Austin, Texas 78712, U.S.A.

ligament loading during maximum contractions of the extensor and flexor muscles [3]. In this paper, the model is used to evaluate the effects of isolated and simultaneous contractions of the knee-extensor and knee-flexor muscles on the forces induced in the cruciate ligaments during isometric exercise.

3. METHODS

The model has been described in detail by Shelburne and Pandy [3]. The geometry of the model bones is based on the shapes of parasagittal sections taken from 23 cadaver knees [4]. Two-dimensional profiles for the lateral femoral condyle and the femoral groove were obtained by fitting splines to the parasagittal sections of the distal femur. The tibial plateau is assumed to be flat, and it slopes 8° posteriorly in the sagittal plane. The position of the tibia relative to the femur is defined by three variables: anterior-posterior tibiofemoral translation, proximal-distal tibiofemoral distraction, and the angle between the long axis of the tibia and the long axis of the femur (i.e., the knee angle). One holonomic constraint is used to prevent interpenetration of the bones at the tibiofemoral contact point. The model patella is massless, and its shape, rectangular in the sagittal plane. The patellar tendon is assumed to be inextensible. Under these conditions, three forces act to equilibrate the patella: the force from the quadriceps tendon, the force from the patellar tendon, and the patellofemoral-contact force. Given the configuration of the tibiofemoral joint, the position of the patella is found by solving the equations for the patellofemoral model iteratively. Details of this procedure are provided by Shelburne [5].

Eleven separate bundles describe the geometric and mechanical properties of the cruciate ligaments, the collateral ligaments, and the posterior capsule (Fig. 1, inset). The ACL and PCL are each represented by two bundles: an anterior bundle and a posterior bundle. The MCL is separated into two portions: the superficial fibers, represented by an anterior bundle, an intermediate bundle, and a posterior bundle; and the deep-lying fibers, represented by an anterior bundle and a posterior bundle. The LCL and the posterior capsule are each represented by one bundle. Each bundle is assumed to be elastic, and its mechanical properties are described by a force-length curve. The path of each bundle in the model is approximated as a straight line, joining the insertion sites reported by Garg and Walker [4]. The parameters for the model ligaments are reported in [3].

The model knee is actuated by eleven musculotendinous units (Fig. 1). Each unit is modeled as a three-element muscle in series with tendon. Parameters defining the nominal properties of each actuator were adapted from data reported by Friederich and Brand [6] and Delp [7]. Values reported for peak isometric muscle forces and tendon rest lengths were adjusted until the maximum, isometric torque-angle curves for the model matched data obtained from humans (see Figs 8 and 9 in [3]). The musculoskeletal geometry of the model is based on data reported by Delp [7]. The three-dimensional path of each actuator is approximated as a straight line, except where it contacts and wraps around bone and other muscles. Under these conditions, the path of the actuator is modeled using via-cylinders, the radii of which were obtained by scaling the model to MR images of a cadaver knee.

The model is used to simulate a variety of isometric exercises: isolated contractions of the quadriceps, isolated contractions of the hamstrings and gastrocnemius, and simultaneous contractions of the quadriceps, hamstrings, and gastrocnemius. Only the knee-flexion angle is varied in the model; hip flexion is fixed at 60° and the ankle is held in the neutral (standing) position. An external restraining force is applied perpendicular to the long axis of the tibia, at the ankle. Given the activation of each muscle in the model, the conditions for static equilibrium produce five equations with five unknowns: anterior-posterior

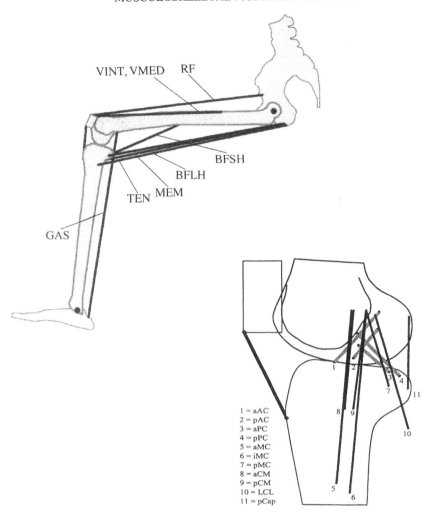

Fig. 1: Musculoskeletal model of the knee. Muscles included in the model are rectus femoris (RF), vastus medialis (VMED), vastus intermedius and vastus lateralis lumped together (VINT), biceps femoris long head (BFLH), biceps femoris short head (BFSH), semimembranosus (MEM), semitendinosus (TEN), sartorius (SAR), gracilis (GRA), tensor fascia latae (TFL), and gastrocnemius (GAS). TFL, SAR, and GRA are not shown. The inset shows the location of the ligament bundles in the model: aAC, anteromedial bundle of the ACL; pAC, posterolateral bundle of the ACL; aPC, anterolateral bundle of the PCL; pPC, posteromedial bundle of the PCL; aMC, anterior bundle of the superficial fibers of the MCL; iMC, intermediate bundle of the superficial fibers of the MCL; pMC, posterior bundle of the superficial fibers of the MCL; aCM, anterior bundle of the deep fibers of the MCL; pCM, posterior bundle of the deep fibers of the MCL; LCL, lateral collateral ligament; pCap, posterior capsule. The LCL inserts on the fibula.

tibiofemoral translation, proximal-distal tibiofemoral distraction, the tibiofemoral-contact force, the patellar-tendon force, and the magnitude of the restraining force. At each knee-flexion angle, the values of these unknown variables are found by solving the model equations iteratively [5].

4. RESULTS

For isolated, isometric contractions of the quadriceps muscles, ACL force increases as quadriceps activation increases. As quadriceps force is held constant, ACL force decreases monotonically as knee-flexion angle increases (Fig. 2A, ACL). For isometric contractions of the hamstrings and gastrocnemius muscles, PCL force increases as hamstrings activation increases. As hamstrings force is held constant, PCL force decreases monotonically as flexion angle decreases (Fig. 2B, PCL).

Hamstrings co-contraction decreases the magnitude of ACL force and the range of knee flexion over which the ACL is loaded. There is a concomitant increase in the magnitude of PCL force and the range of flexion over which the PCL is loaded (cf. 0 and 100% for ACL and PCL in Fig. 3A). Except near full extension, hamstrings co-contraction is an effective means of reducing ACL force (cf. 0 and 15° in Fig. 3B).

5. DISCUSSION

5.1 Limitations of the model

Since the model is two-dimensional, it cannot account for axial rotation of the bones which accompanies flexion-extension movements of the real knee. During passive extension, for example, the tibia rotates externally relative to the femur. Since external rotation is known to decrease the resultant force in the ACL, the model results represent an upper-bound for ACL forces in the intact knee. The model cannot account also for the load-sharing between the menisci, cartilage, and the knee ligaments. Because the model bones are rigid, contact between the articulating surfaces of the femur and tibia occurs at a single point. In life, the tibiofemoral-contact force, created by the action of the muscles, causes the menisci and cartilage to compress, so that a finite area of contact is created at the joint. Including the behavior of the menisci and cartilage in the model is not likely to alter the estimates of the forces transmitted to the bones; however, the forces in the model ligaments would be lower had the effect of soft-tissue compression been taken into account. Compression of the menisci and cartilage causes a decrease not only in the distance between the ligament attachment sites, but also in the a-p translation of the bones at the tibiofemoral joint. Because the model is unable to reproduce this behavior, the analysis may overestimate the forces in the cruciate ligaments.

5.2 Comparison with reported experimental data

Although experimental data are not available for the forces produced in the knee ligaments during activity, qualitative comparisons may be drawn between the model results and measurements of tibiofemoral displacements and ligament strains obtained *in vivo*. Beynnon et al. [8] measured the strains produced in the anteromedial bundle of the ACL as subjects contracted their quadriceps muscles during isometric knee extension. The results showed that ACL strains are significantly higher between 0 and 30° of flexion for isometric extension than for passive extension of the knee; ACL strain was zero beyond 60° of flexion. By comparison, the model ACL is loaded between 0 and 80° of flexion during maximum, isolated contractions of the quadriceps (Fig. 2A, ACL). This difference

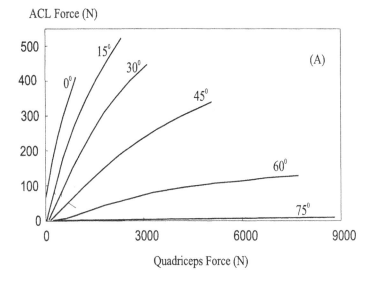

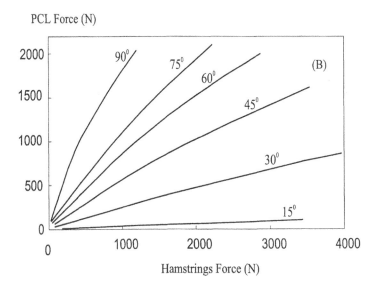

Fig. 2: (A) Relationship between quadriceps force and the resultant ACL force at various knee flexion angles during isolated contractions of the quadriceps. At each knee-flexion angle, quadriceps activation varies from 0 to 100%.

(B) Relationship between hamstrings force and the resultant PCL force at various flexion angles during isolated contractions of the hamstrings and gastrocnemius. At each flexion angle, hamstrings and gastrocnemius activation varies from 0 to 100%.

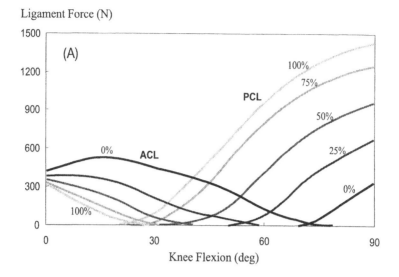

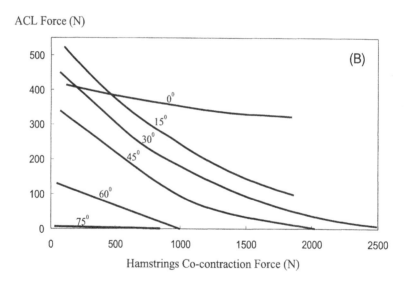

Fig. 3: (A) Relationship between ligament forces, hamstrings activation, and knee-flexion angle during simultaneous contractions of the extensor and flexor muscles. At each flexion angle, the quadriceps are fully activated and the activation of each flexor muscle is varied from 0 to 100%.

(B) Relationship between hamstrings co-contraction force and the resultant ACL force at various knee-flexion angles during simultaneous contractions of the extensor and flexor muscles.

between model and experiment may be explained by the fact that maximum contractions of the quadriceps were not elicited during the experiments.

Beynnon et al. [8] also measured ACL strains during isolated contractions of the hamstrings and during simultaneous contractions of the quadriceps and hamstrings. For isometric contractions of the hamstrings, ACL strain was minimal at $15°$ of flexion and zero beyond $30°$. By comparison, the model ACL is loaded only between 0 and $10°$ of flexion during maximum, isolated contractions of the hamstrings and gastrocnemius muscles (Fig. 2B, ACL). The ACL may be loaded over a larger range of knee flexion in the experiments because the peak knee-flexor torque developed by the subjects is much lower than that calculated in the model.

For maximum, simultaneous contractions of the quadriceps and hamstrings, the ACL was strained from 0 to $60°$ of flexion in the experiments, compared with a range of 0 to $30°$ in the model (Fig. 3A, ACL at 100%). In the experiments, simultaneous contractions of the quadriceps and hamstrings were performed with no restraining force applied to the leg; in the model, a small value of restraining force is needed to equilibrate the leg during maximum contractions of the extensor and flexor muscles. The fact that some resisting force is needed to equilibrate the leg in the model suggests that the subjects may not have contracted all of their muscles maximally during the experiments.

5.3 Relationships between muscle and ligament forces

For isotonic contractions of the quadriceps, the resultant force in the ACL decreases monotonically as flexion angle increases (cf. ACL forces for 3000 N of quadriceps force in Fig. 2A). Two mechanisms contribute to this result: first, the shear force applied by the patellar tendon to the lower leg decreases as knee-flexion angle increases; and second, the orientation of the ACL relative to the tibial plateau decreases as knee flexion increases.

As knee flexion increases, the angle between the patellar tendon and the long axis of the tibia decreases in the model and in the real knee. For a constant value of quadriceps force, as the patellar-tendon angle decreases, so too does the patellar-tendon shear force. ACL force decreases in proportion to a decrease in the patellar-tendon shear force because the ACL is the primary restraint to anterior drawer. The decrease in patellar-tendon shear force with increasing flexion is caused by the geometry of the knee-extensor mechanism, which in turn is determined by the shapes of the articulating surfaces of bones and the insertion sites of the patellar tendon. An increase in knee-flexion angle results also in a decrease in the angle between the tibial plateau and the resultant line of action of the ACL. ACL force decreases as the angle between the ACL and the tibial plateau decreases because the ligament is then better oriented to resist an anteriorly-directed shear force. The contours of constant flexion angle in Fig. 2A are explained, therefore, by the interaction between the geometry of the knee-extensor mechanism and the changing orientation of the ACL in the sagittal plane.

Two mechanisms also explain why, for isotonic contractions of the hamstrings, the resultant PCL force increases monotonically as knee-flexion angle increases: first, the shear force applied by the hamstrings muscles increases as flexion angle increases; and second, the orientation of the PCL relative to the tibial plateau increases with increasing flexion. As knee-flexion angle increases, the angle between the line of action of the hamstrings and the long axis of the tibia increases. For a constant value of hamstrings force, the shear force applied by the hamstrings increases, and this increase requires a greater PCL force. An increase in knee-flexion angle results also in an increase in the angle between the PCL and the tibial plateau; PCL force increases because the ligament is then more poorly oriented to resist a posteriorly-directed shear force. Thus, the contours of constant flexion angle in Fig. 2B are explained by the interaction between the geometry of the hamstrings muscles and the changing orientation of the PCL in the sagittal plane.

The relationships between muscle forces and cruciate-ligament forces are nonlinear and asymptotic at all angles of knee flexion (Fig. 2A and B). This behavior is consistent with that noted by Zavatsky and O'Connor [9] and is due to the effects of ligament elasticity in the model. Had the cruciate ligaments in the model been inelastic, the contours of constant flexion angle would be linear: inelastic knee ligaments would prevent any additional translation of the bones as muscle force is increased. If tibiofemoral translation was independent of muscle force during an isometric contraction, the orientation of the ACL relative to the tibial plateau also would not change. Cruciate-ligament force would therefore depend linearly on muscle force, and any departure from such a linear relationship must then reflect the effects of ligament stretch. For example, as the ACL stretches during an isometric contraction of the quadriceps, the angle between the patellar tendon and the long axis of the tibia decreases slightly; the shear force applied by the patellar tendon to the lower leg therefore decreases; the inclination of the ACL to the tibial plateau also decreases; and the resultant force in the ACL is lower.

6. REFERENCES

[1] Markolf, K.L., Gorek, J.F., Kabo, J.M., and Shapiro, M.S.. Direct measurement of resultant forces in the anterior cruciate ligament: An in vitro study performed with a new experimental technique. J. Bone Jt. Surg., 1990, Vol. 72-A: 557-567.

[2] Hirokawa, S., Solomonow, M., Luo, Z., Lu, Y., and D'Ambrosia, R.. Muscular co-contraction and control of knee stability. J. Elect. and Kinesiol., 1991, Vol. 1: 199-208.

[3] Shelburne, K.B. and Pandy, M.G. A musculoskeletal model of the knee for evaluating ligament forces during isometric contractions. J. Biomechanics, 1997, Vol. 30: 163-176.

[4] Garg, A. and Walker, P.S. Prediction of total knee motion using a three-dimensional computer-graphics model. J. Biomechanics, 1990, Vol. 23: 45-58.

[5] Shelburne, K.B. Modeling the mechanics of the normal and reconstructed knee. Ph.D. thesis, Department of Mechanical Engineering, The University of Texas at Austin, 1996.

[6] Friederich, J. and Brand, R.A. Muscle fiber architecture in the human lower limb. J. Biomechanics, 1990, Vol. 23: 91-95.

[7] Delp, S.L. Surgery simulation: A computer-graphics system to analyze and design musculoskeletal reconstructions of the lower limb. Ph.D. dissertation, Department of Mechanical Engineering, Stanford University, California, 1990.

[8] Beynnon, B.D., Fleming, B.C., Johnson, R.J., Nichols, C.E., Renstrom, P.A., and Pope, M.H. Anterior cruciate ligament strain behavior during rehabilitation exercises in vivo. American Journal of Sports Medicine, 1995, Vol. 23: 24-34.

[9] Zavatsky, A.B. and O'Connor, J.J. Ligament forces at the knee during isometric quadriceps contractions. Proceedings of the Institute of Mechanical Engineers, Part H, 1993, Vol. 207: 7-18.

THREE DIMENSIONAL RIGID BODY SPRING MODELING AND ITS APPLICATION FOR HUMAN JOINTS

E. Genda[1], E. Horii[2], Y.Suzuki[1], T. Kasahara[1], Y. Tanaka[1]

1. ABSTRACT

Three dimensional Rigid Body Spring Modeling was applied for stress analysis of multi-articular joints such as the wrist, elbow and ankle joint. Geometric bone models were constructed using normal CT data with a 1 to 2 mm slice interval. Articular surface mesh models were constructed automatically by calculating the mid-points between two articulating bone surfaces. In the wrist radio-carpal joint under neutral loading condition, stress was concentrated on the volar side of the joint. The force distribution of radio-ulno-carpal joint was consistent with previous experimental results. In the elbow joint, ulnar-radial loading ratio of seven to three seemed reasonable in order to get equal stress distribution in both radio-humeral and ulno-humeral joints.

This model has much promise in solving multi-articular joint problems. An easy modeling procedure, simple formulation, and short calculation time enable this model to be applied to very complex structures—structures for which even FEA cannot easily be applied. However, further experimental studies will need to be performed in order to achieve complete validation of this method.

key words: Rigid Body Spring Model, wrist, elbow, ankle

[1]Rosai Rihabilitation Engineering Center, 1-10-5 Koumei, Minato District, Nagoya, Japan. [2] Nagoya Univ. Dept. of Orthopaedic Surgery, 65 Tsurumai, Shyouwa District, Nagoya, Japan.

2. INTRODUCTION

Using computer simulation techniques to estimate joint pressure distribution is a powerful tool with many applications.   These include understanding the pathogenesis of degenerative arthritis, the mechanics of an injury, the design of joint prostheses and preoperative planning of a surgical procedure.   Finite Element Aanalysis (FEA) has been an effective method for analyzing human structures such as bone.   However, there are many limitations regarding its use in analyzing contact problems in human joints, especially in multiple-body contact-force analysis such as in the wrist, elbow and ankle joint.   Additionally,   FEA is sensitive to the values of material properties—many of which are difficult to obtain in biological materials.

If the purpose of the analysis is not to find the actual stress inside the cartilage and bone, but only the contact stress distribution on the articulating surface, a simpler approach can be performed[1]. The cartilage interposed between the bones and ligaments connects bones to each other, and this is where deformations mainly occur.   In this situation,   it could be assumed that bones are rigid bodies and that the cartilage and ligaments are linear springs interposed between the rigid bodies (Rigid Body Spring Modeling: RBSM, another name is Discrete Element Analysis: DEA).

3. MATERIAL AND METHODS

RBSM is a type of finite element analysis originally developed and validated within the field of civil engineering by Kawai and Takeuchi[2] to solve tension crack problems. When load is applied to the joint, deformations of the cartilage and ligaments are relatively larger than the deformation of bones,   therefore bones can be considered to be rigid bodies, while cartilage (including the subchondral bone and ligaments) can be considered to be a matrix of linear springs interposed between the two rigid bodies (Figure 1). Cartilage does not resist tensile force and ligaments do not resist compressive force, therefore cartilage should be represented as compressive-resistive springs and ligaments as tensile-resistive springs.   This non-linear problem can be solved by an iterative process in this method.

A set of equilibrium equations is formulated between the spring forces and external forces.   If the load is applied through the centroid of the rigid body,   some springs may be in compression, while others may be in tension.   Compressive springs carrying tension and tensile springs carrying compression are eliminated from the overall equilibrium equation and the new equations are formulated.   This calculation is carried out in an iterative manner until all existing springs bear reasonable values.   The pressure distribution is obtained by calculating each spring force according to its deformation.

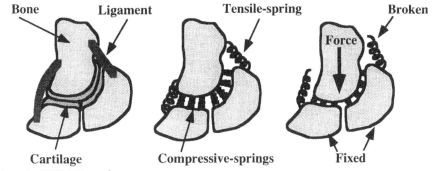

Figure 1   RBSM illustration

Three-dimensional RBSM was used in the stress analysis of following multi-articular joints: wrist, elbow and ankle.   Geometric bone models were constructed using normal CT data with a 1 to 2 mm slice interval.   Contour lines were digitized manually and the line was interpolated using the B-spline method.   From the wire frame models, shaded models were constructed using original software.   Figure 2 shows constructed shaded models, wrist and elbow joint.   We assumed a perfecty congruent joint to simplify the problems.   There is initially no gap between cartilage surfaces.   Articulating surfaces with distances less than the thickness of the cartilage were assumed to be in contact.   Contact surface models were constructed as a matrix of small quadrilateral mesh elements which were equidistant from articulating bone surfaces.

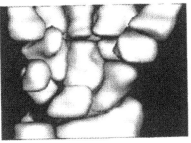

Shaded joint models are depicted in the antero-posterior view and the lateral view.   The three dimensional location of ligaments' attachment points were determined on the computer screen using two orthogonal projections.   The locations and stiffness of the ligaments were obtained from previous reports.

Compressive springs were set on the each mesh element in the normal direction, and ligaments were expressed as line segments connecting two bones.

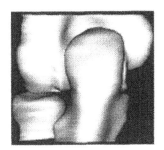

Figure 2

We assumed the same stiffness of springs along the whole joint surface to simplify the problems.   Ligaments' stiffness were obtained from previous reports.   The calculations were performed on a Silicon Graphics workstation, taking less than one minute to complete.

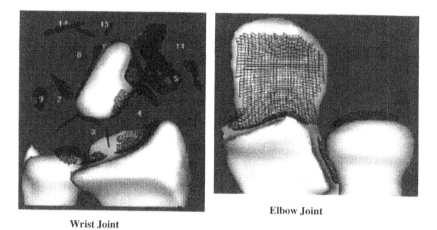

Wrist Joint   **Elbow Joint**

Figure 3   Contact surface mesh models: Wrist and Elbow.   Some bones are elminated to show meshed surface.

4. RESULTS

Wrist joint model consisted of 15 bones, 27 contact surfaces, and 47 ligaments.   A total load of 140N was applied through five metacarpal bones.   The right slide shows the stress distribution in the entire wrist joint, and ligament tension using color gradients. In radio-ulno-carpal joint, 46% of load was transmitted through radio-scaphoid joint,

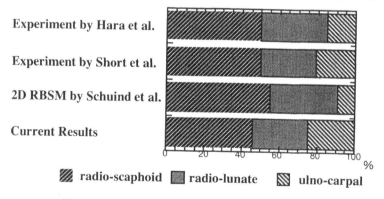

Figure 4   Comparison with previous reports

30% through radio-lunate joint and 24% through ulno-carpal joint.   This result was favorably consistent with previous experimental and simulation results [3,4,5](Figure 4). Loading direction was inclined from 10 degrees dorsal to 10 degrees palmar and the change of stress distribution in radio-ulna-carpal joint was investigated.   Under neutral loading condition, stress was concentrated on the palmar side of joint.   As load direction inclined dorsally, the stress became concentrated on the palmar side.   On the other hand, as the load inclined in the palmar direction, the stress was distrubuted more evenly,   becoming   concentrated   on   the   dorsal   side   of   joint   (Figure   5).

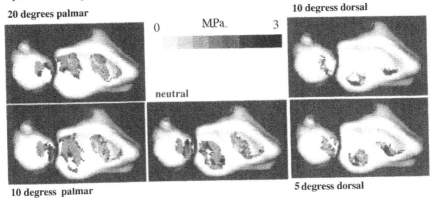

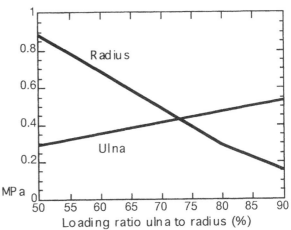

Figure 5 The change of stress distribution on radi-ulno-carpal joint

In   the   elbow   joint, humerus was fixed and a total of 200N was applied through ulna and radius, parallel to their axes. Ulnar-radial loading ratio was changed from 5:5 to 9:1 and the peak pressure change of each joint was   investigated.   Ulnar-radial loading ratio of seven to three seemed reasonable for achieving an equivalent stress distribution in both radio-humeral and ulno-humeral joints (Figure 6).

Figure 6

Foot and ankle model consisted of 14 bones from the tibia to metatarsal bones, 22 contact surfaces, and 87 ligaments.   A total load of 320N was applied through tibia and fibula to simulate standing.   The calcaneus and five metatarsal bones were in contact with floor (Figure 7).

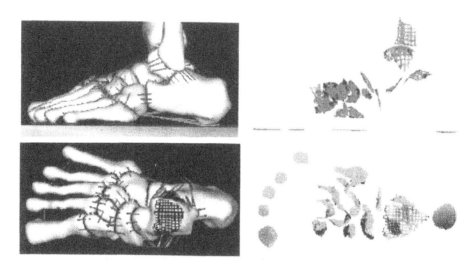

Figure 7 Ankle and foot model (left), Stress distribution of foot joints and floor (right)

5. DISCUSSION

There are both advantages and limitations to this method. Several simplifying assumptions enable solving multiple contact problems.   In this discrete element model, element nodes are considered to be at the centroid of rigid bodies, and consequently the concept of superposition of element nodes cannot be applied. However, the use of elements with arbitrary shapes is possible. This concept allows us to use a more anatomical contact surface geometry.   Easy modeling procedure and short calculation time are also advantages of this method. For example, in wrist joint model, only 90 simultaneous equations must be solved.

There were several limitations derived from this simplication. Articulating joint surfaces were assumed to be perfectly congruent—there was initially no gap between articulating surfaces. There was no laxity of ligaments. It could not analyze the stress inside the cartilage and neglected visco-elastic nature of the cartilage. Therefore, the results of these calculations are approximations of joint pressure distributions.

6. CONCLUSION

An easy modeling procedure, simple formulation, and short calculation time enable this model to be applied to very complex structures. However, the results should be carefully examined considering the limitation of this method and further experimental studies will need to be performed in order to achieve complete validation.

REFERENCES

1) An, K. N., Himeno, S., Tsumura, T., Kawai, T., & Chao, E. Y. S. Pressure Distribution on Articular Surfaces: Application of Joint Stability Evaluation. J. Biomechanics.,1990, 23(10):1013-1020.

2)Kawai, T., and Takeuchi, N. A discrete method of limit analysis with simplified element. American Society Civil Engineers, International Conference on Computing in Civil Engineering, New York 1981.

3) Hara, T., Horii, E., An, K.N. Cooney, W.P., Linschcid, R.L. and Chao, E.Y.S. Force distribution across wrist joint: Application of pressure-sensitive conductive rubber. J. Hand Surg.,1992, 17A:339-47.

4) Schuind, F., Cooney , W. P., Linscheid, R. L., An, K. N. and Chao, E. Y. S. Force and pressure transmission through the normal wrist. A theoretical two-dimensional study in the posteoanterior plane. Journal of Biomechanics., 1995, 28: 587-601.

5) Short, W.H., Palmer, A.K., Werner, F.W. and Murphy, D.J. A biomechanical study of distal radial frauctures. J. Hand Surg.,1987,12A:529-34.

DYNAMIC SIMULATION OF HUMAN FLEXION-EXTENSION MOVEMENT USING OPTIMAL CONTROL THEORY.

O. Coussi [1,2] , F. Danes [3,2] and G. Bessonnet [4,2]

1. ABSTRACT

The inverse dynamic problem of a human flexion-extension movement is solved in order to obtain the joint intersegmental forces, the actuating torques and the energy consumption. Afterwards, a similar movement with the same boundary conditions is simulated using optimal control theory by implementing Pontryagin's Maximum Principle. Real and simulated movements are compared in terms of energy consumption.

KEYWORDS : HUMAN MOVEMENT, FLEXION-EXTENSION, OPTIMIZATION, DYNAMIC ANALYSIS

2. INTRODUCTION

In biomechanics the human body is commonly modelled as an equivalent multibody system articulated as a treelike kinematic structure. Chow [1] assumed that human locomotion involves the optimization of a performance criterion. He introduced the control theory to simulate the human gait by using a multiarc programing problem transformed in a two point boundary value problem. Chow simulated a walking step with an optimal energy consumption along some specified trajectory. Chao [2] used optimization principles for solving the inverse dynamic problem in order to compute the joint actuating torques which generate and control the human gait. Yen [3] specified a movement of a five-link human model and computed the boundary conditions and total time of a walking step in such a way the energy consumption is minimized. Zefran [4] used the calculus of variations for solving an optimal planning problem. He introduced a new performance criterion based upon the minimization

---

[1]Researcher, `coussi@lms.univ-poitiers.fr`

[2]Laboratoire de Mécanique des Solides - UMR 6610 CNRS, Université de Poitiers, Faculté des Sciences, Bat. SP2MI, Bd. 3, Téléport 2, BP 179, 86960 Futuroscope Cedex, France

[3]Researcher, `danes@lms.univ-poitiers.fr`

[4]Professor, `bessonne@lms.univ-poitiers.fr`

of the integral norm of the actuator input derivatives. The comparison between genuine human movement and results obtained through simulated movement is made. The objective of the present study is to compare a real human flexion-extension movement with that of a simulated one which satisfies the same initial and final kinematic values.

3. KINEMATIC AND ANTHROPOMETRIC MODEL

We consider a planar kinematic model which can be reduced to three links : L_1 representing the two shanks, L_2 representing the two thighs and L_3 representing the H.A.T. (=head+arm+trunk). This simple multibody model includes 3 joints : the hip, with axis $(O_3, \mathbf{z}_0)$, the knee, with axis $(O_2, \mathbf{z}_0)$ and the ankle, with axis $(O_1, \mathbf{z}_0)$ ($\mathbf{z}_0 = \mathbf{x}_0 \times \mathbf{y}_0$). A set of relative coordinates is introduced as indicated in figure 1. The time-dependent generalized coordinates q_i are constructed from the experimental kinematic data obtained by an expert vision device. They are filtered using a wavelet transform technique [5] before differentiations yielding the time derivatives $\dot{q}_i$ and $\ddot{q}_i$. The figure 2 represents the time chart of q_i's and $\dot{q}_i$'s.

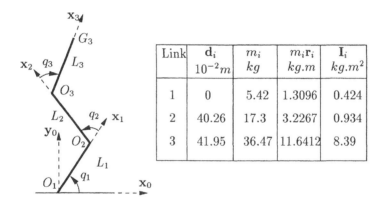

Link	$\mathbf{d}_i$ $10^{-2}m$	m_i kg	$m_i\mathbf{r}_i$ $kg.m$	$\mathbf{I}_i$ $kg.m^2$
1	0	5.42	1.3096	0.424
2	40.26	17.3	3.2267	0.934
3	41.95	36.47	11.6412	8.39

Figure 1: Kinematic model and anthropometric data

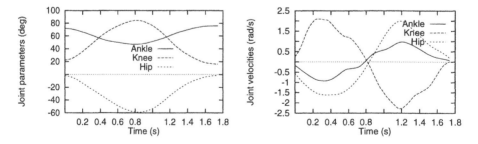

Figure 2: Data of real human movement : joint parameters q_i and joint velocities $\dot{q}_i$

Estimations of mass m_i, center of mass position $\mathbf{r}_i$ and moment of inertia $\mathbf{I}_i$ with respect to $(O_i, \mathbf{z}_0)$ joint axis of each link L_i have been calculated using Winter's anthropometric table [6].

4. INVERSE DYNAMIC ANALYSIS

This section is aimed at computing all intersegmental and contact forces and moments involved in the moving kinematic chain. The Newton-Euler formalism is quite relevant for solving this problem [7]. The corresponding dynamic model is quite easely constructed using the Luh, Walker and Paul [8] double recursive technique. First, a forward recursive formulation of joint velocities and accelerations is achieved from the proximal segment to the distal one. Next, Newton-Euler dynamic equations are formulated following a backward step. In the backward recursion the motion equations of each link L_i can easily be organized under the form of the backward recursive vector equations (for $i = 3, 2, 1$) :

$$\begin{cases} \mathbf{F}_{i-1} &=& \mathbf{F}_i + m_i(\mathbf{g} - \boldsymbol{\gamma}_{G_i}) \\ \mathbf{M}_{i-1} &=& \mathbf{M}_i + \mathbf{d}_i \times \mathbf{F}_i + m_i\mathbf{r}_i \times \mathbf{g} - \boldsymbol{\delta}_i \end{cases} \qquad (1)$$

where $\mathbf{g}$ is the gravity acceleration, $\mathbf{d}_i = \mathbf{O}_i\mathbf{O}_{i+1}$, $\mathbf{F}_i$ is the force exerted by L_{i+1} on L_i, $\mathbf{M}_i$ is the moment at O_{i+1} exerted by L_{i+1} on L_i, $\boldsymbol{\gamma}_i$ (resp. $\boldsymbol{\gamma}_{G_i}$) is the acceleration of O_i (resp. G_i) in the inertial frame, $\boldsymbol{\omega}_i = \dot{q}_i\mathbf{z}_i$ is the vector-angular velocity at joint i, $\boldsymbol{\delta}_i$ stands for the dynamic moment of L_i at point O_i and is defined as

$$\boldsymbol{\delta}_i = \mathbf{I}_i\dot{\boldsymbol{\omega}}_i + \boldsymbol{\omega}_i \times \mathbf{I}_i\boldsymbol{\omega}_i + m_i\mathbf{r}_i \times \boldsymbol{\gamma}_i \qquad (2)$$

This set of equations establishes a linear recursive relationship between the forces and moments exerted by L_{i+1} on L_i and by L_i on L_{i-1}. For the inverse dynamic problem, the kinetic quantities $m_i\boldsymbol{\gamma}_{G_i}$ and $\boldsymbol{\delta}_i$ in equations (1) are determined independently of forces $\mathbf{F}_i$ and torques $\mathbf{M}_i$. The recursion is initialized by $\mathbf{F}_3 = 0$ and $\mathbf{M}_3 = 0$ and is terminated for $i = 1$ by the determination of $\mathbf{F}_0$ and $\mathbf{M}_0$ which represents respectively the force and moment exerted by the proximal link L_1 on the ground. When measured by a force plate FP placed under the subject, this contact forces, then noted $\mathbf{F}(L_1 \to FP)$ and $\mathbf{M}(O_1, L_1 \to FP)$, must agree with the force and moment $\mathbf{F}_0$, $\mathbf{M}_0$ calculated by inverse dynamic. Thus, we intend to verify the equalities :

$$\begin{cases} \mathbf{F}_0 = \mathbf{F}(L_1 \to FP) \\ \mathbf{M}_0 = \mathbf{M}(O_1, L_1 \to FP) \end{cases} \qquad (3)$$

Therefore, the forceplate appears as an experimental instrument to check the computed values of $\mathbf{F}_0$ and $\mathbf{M}_0$. The results of the computation of intersegmental and contact forces and torques are shown in figures 3.

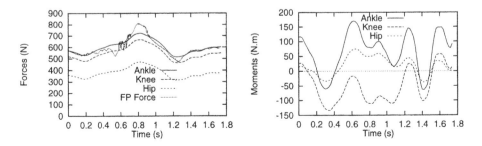

Figure 3: Results of Inverse Dynamic Analysis : Forces $\mathbf{F}_i$ and Moments $\mathbf{M}_i$

As there is a good correlation between the computed and measured forces one can think that the computed torques correspond with those produced by the tested subject. The difference between the extremum values of the measured and computed ground reaction forces is principally due to the difference between kinematic and dynamic data sampling (50 Hz and 1000 Hz respectively).

5. DYNAMIC SIMULATION

The dynamic simulation problem is transformed into a classical robotic task : find an optimal motion between an initial and a final configuration minimizing the actuator inputs along an unspecified path. Introducing the Hamiltonian phase variables $x_i = q_i$ and $x_{n+i} = p_i$, with $p_i = \frac{\partial L}{\partial \dot{q}_i}$, where L is the Lagrangian of the mechanical system, one obtains the Hamiltonian vector-equation of motion :

$$t \in [0, T] \, , \, \dot{\mathbf{x}}(t) = \mathbf{G}(\mathbf{x}(t)) + \mathbf{B}\mathbf{u}(t) \equiv \mathcal{F}(\mathbf{x}(t), \mathbf{u}(t)) \tag{4}$$

where $\mathbf{x} = (x_1, \cdots, x_{2n})^T$, $\mathbf{u}$ is the vector of joint actuating torques, $\mathbf{G}$ is a non-linear function in $\mathbf{x}$, and $\mathbf{B}$ is the $[2n \times n]$-matrix ($n = 3$ for this study) :

$$\mathbf{B} = \begin{bmatrix} \mathbf{0}_{n \times n} \\ \mathbf{I}_{n \times n} \end{bmatrix} \tag{5}$$

The total motion time T is specified and the state equation (4) goes together with the boundary conditions :

$$\mathbf{x}(0) = \mathbf{x}^0 \text{ and } \mathbf{x}(T) = \mathbf{x}^T \tag{6}$$

which specify the initial and final state of the mechanical system. Joint actuating torques are submitted to bound such as :

$$|u_i(t)| \leq u_i^{max} \, , \, \forall t \in [0, T] \, , \, i = 1, \cdots, n \tag{7}$$

The performance criterion to be minimized is an integral functionnal of the type :

$$J(\mathbf{u}) = \int_0^T L(\mathbf{u}(t))dt \equiv \int_0^T \sum_{i=1}^n [\frac{u_i(t)}{u_i^{max}}]^2 dt \tag{8}$$

In order to obtain an anthropomorphic optimal movement, a phase constraint limiting q_1's joint motion must be introduced. This constraint is defined by the inequality

$$g(\mathbf{x}(t)) \equiv x_1(t) - x_1^0 \leq 0 \tag{9}$$

Then, the constrained optimization problem can be recast into an unconstrained problem by using an exterior penalty method [9]. Define the penalized criterion :

$$J^*(\mathbf{u}(t), \mathbf{x}(t)) = \int_0^T [L(\mathbf{u}(t)) + r[g^+(\mathbf{x}(t))]^2]dt \tag{10}$$

where r is a penalty multiplier and g^+ is the function :

$$g^+(\mathbf{x}) = \max_{t \in [0,T]} (0, g(\mathbf{x}(t))) \tag{11}$$

Thus J^* minimizes the quadratic positive values of the constraint when it is infringed. The Pontryagin Maximum Principle (PMP) [10] states necessary optimality conditions which allow us to convert the dynamic optimization problem into an ordinary differential problem. For $\mathbf{w} \in \mathbb{R}^{2n}$, define the Hamiltonian :

$$H(\mathbf{x}, \mathbf{u}, \mathbf{w}) = \mathbf{w}^T \mathcal{F}(\mathbf{x}, \mathbf{u}) - L(\mathbf{u}) - r[g^+(\mathbf{x})]^2 \tag{12}$$

The (PMP) states that if $(\mathbf{x}, \mathbf{u})$ is an optimal control process then :

1. It exists a costate-vector function $\mathbf{w}(t)$ which is solution of the adjoint system

$$t \in [0, T] \ , \ \dot{\mathbf{w}}(t) = -(\frac{\partial H(\mathbf{x}(t), \mathbf{u}(t), \mathbf{w}(t)}{\partial t})^T \tag{13}$$

2. The optimal control $\mathbf{u}(t)$ satisfies the Hamiltonian maximality condition :

$$H(\mathbf{x}(t), \mathbf{u}(t), \mathbf{w}(t)) = \max_{\mathbf{v} \in \mathcal{U}} H(\mathbf{x}(t), \mathbf{v}, \mathbf{w}(t)) \tag{14}$$

where $\mathcal{U}$ is the set of admissible control defined by inequality (7).

As in [9], an explicit expression of control variables u_i is derived from (14) under the form :

$$u_i(t) = \text{Sat}[(u_i^{max})^2 w_{n+i}(t)] \ , \ i = 1, \cdots, n \tag{15}$$

where Sat is the saturation function defined as :

$$\text{If } |\mathbf{v}| < \mathbf{v}^{max} \ , \ \text{Sat}(\mathbf{v}) = \mathbf{v} \text{ else } \text{Sat}(\mathbf{v}) = \mathbf{v}^{max}\text{sign}(\mathbf{v}) \tag{16}$$

Substituting the expression (15) of $\mathbf{u}$ in the state equation (4) and the adjoint system (13), the unknown vector functions $\mathbf{x}$ and $\mathbf{w}$ appear as a solution of a $4n$-dimensionnal differential system :

$$\begin{cases} \dot{\mathbf{x}} = \mathcal{F}_1(\mathbf{x}, \mathbf{w}) \\ \dot{\mathbf{w}} = \mathcal{F}_2(\mathbf{x}, \mathbf{w}) \end{cases} \tag{17}$$

which must satisfy the boundary conditions (6). This ordinary two-point boundary value-problem is numerically solved by using the routine D02RAF of the Fortran Nag Library. This routine is initialized with a solution obtained by means of first order gradient method describe in [11].

5. ENERGY CONSUMPTION

The purpose of this section is to compare the different movements in term of criterion J defined by equation (8) and internal work for both real and simulated movements. The obtention of the energy consumption is immediately available with the knowledge of joint actuator torques and velocities. The instantaneous power is defined by the relationship :

$$P_i = \Gamma_i \dot{q}_i \tag{18}$$

where $\Gamma_i = \mathbf{M}_i.\mathbf{z}_i$ for the observed movement and $\Gamma_i = u_i$ for the simulated movements. By integrating these quantities with respect to the running time we compute the internal energy comsumption :

$$W = \int_0^T \sum_{i=1}^n |P_i| dt \qquad (19)$$

Then it is possible to compare the real movement and the simulated one in terms of energy consumption and total amount of actuating torques. This is done in the next section.

6. RESULTS

First results shown in figure 4 concerns an optimal motion computed without introducing kinematic constraint of type (9). One can remark that the simulated movement does not correspond to the real one shown in figure 2 because all joint coordinates exceed the initial and final configurations. To avoid this problem we must add some constraints concerning joint coordinates magnitudes.

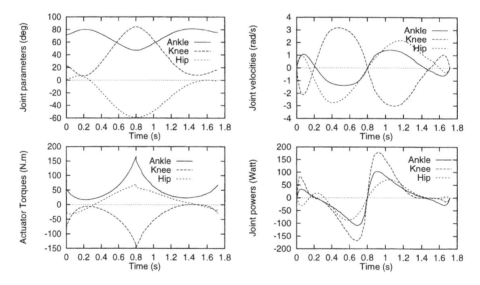

Figure 4: Free movement optimization : q_i, $\dot{q}_i$, u_i and P_i

Indeed, introduction of a bound on the q_1 joint coordinate as defined by (9) has proved to be sufficient in order to obtain a flexion-extension motion which can be compare to the real movement without surpassing its specified end values. Thus, this constrained optimal movement can be compared with the real one. One can see in figure 5 that this simulated movement is slightly different from the movement of the tested subject and corresponds to the assigned task of flexion-extension. Magnitudes of joint velocities are more important for the optimal movement than the real one. Optimal actuating torques are less varying than the real one. This can be explained by the fact that the tested subject remains strained when moving and does not manage his effort and equilibrium at best.

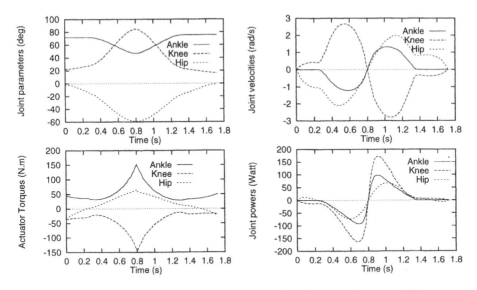

Figure 5: Constrained movement optimization : q_i, $\dot{q}_i$, u_i and P_i

The comparison between observed and simulated movements in terms of internal work and quadratic amount of actuating torques is shown on figure 6. In term of energy the simulated movement is less consumming than the observed one. The constrained simulated movement is also more consumming in term of internal work than the free simulated movement. Results are similar when considering the values of criterion J.

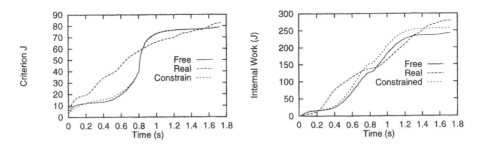

Figure 6: Criterion J and Internal Work W for real and optimized movements

7. CONCLUSION

We have shown that there is no need for a force plateform to compute the joint actuating torques in a human movement as flexion-extension. A comparison between observed and simulated flexion-extension movements was made. It showed that for the tested subject the optimized movement is less energy consuming than the real one. Some constraints must been introduced in order to obtain an optimal movement closer to the real one. We have also demonstrated the efficiency of optimal control theory in simulating a simple human movement. In the future, the approach presented will be developped for studying more complex movements as walking and running.

8. REFERENCES

[1] C.K. CHOW D.H. JACOBSON. Studies of human locomotion via optimal programming. *Mathematical Biosciences*, 10:239–306, 1971.

[2] E.Y. CHAO K. RIM. Application of optimization principles in determining the applied moments in human joints during gait. *Journal of Biomechanics*, 6:497–510, 1973.

[3] V. YEN and L. NAGURKA. Suboptimal trajectory planning of a five-link human locomotion model. In *ASME Winter Annual Meeting on Biomechanics of Normal and Prosthetic gait, Boston*, pages 17–22. ASME, 1987.

[4] M. ZEFRAN V. KUMAR J. DESAI E. HENIS. Two-arm manipulation: What can we learn by studying humans. In *Proc. IEEE International Conference on Intelligent Robots and Systems, IROS'95*, pages 70–75. IEEE, 1995.

[5] O. COUSSI F. BREMAND G. BESSONNET. Wavelet transform and biomechanical data filtering. In *4th International Symposium on 3-D Analysis of Human Movement, Grenoble*, 1996.

[6] D.A. WINTER. *Biomechanics and motor control of human movement*. Wiley Intersciences, 1990. second edition.

[7] O. COUSSI G. BESSONNET L. FROSSARD P. ALLARD. A solution of the inverse dynamic problem for a 2-d human walking biped : single and double support phases. In *IX World Congress on Theory of Machines and Mecanism Milan*, pages 3:2287–2291. IFToMM, 1995.

[8] J.Y.S. LUH M.W. WALKER R.C.P. PAUL. On-line computational scheme for mechanical manipulators. *Transactions of ASME Journal of Dynamic Systems, Measurement and control*, 102(2):69–76, 1980.

[9] G. BESSONNET A.D. JUTARD. Optimal free path planning of robot arms submitted to phase constraints. In *Third International Conference on Robotics and Manufacturing*, pages 51–56, Cancun, Mexico, June 14-16 1995. IASTED.

[10] L. PONTRYAGIN V. BOLTIANSKY A. GAMKRELIDZE E. MISHCHENKO. *The Mathematical Theory of Optimal Process*. Wiley Interscience, 1962.

[11] A.E. BRYSON Y.C HO. *Applied optimal control*. Hemisphere Publishing Corporation, 1969.

9. ACKNOWLEDGMENTS

This work was supported by the Poitou-Charentes region (Grant number 96/RPC-R-163).

2. HIP REPLACEMENTS: PROSTHESIS/CEMENT/BONE ANALYSIS

TORSIONAL STABILITY OF TOTAL HIP ARTHROPLASTY: *IN-VITRO* AND FEM ANALYSIS WITH NEW TRENDS FOR THE FUTURE

Aldo Toni[1], Marco Viceconti[2], Luca Cristofolini[3], Massimiliano Baleani[4], Gianni Acquisti[5] and Robert Schreiner[6]

1. ABSTRACT

Uncemented hip stems are advantageous for the lack of risk related to fatigue failure of the cement mantle. Their main problem is primary stability, which is necessary to guarantee osteointegration. A new fixation concept was developed for a partially cemented stem (Cement Locked Uncemented, CLU), derived from an anatomic uncemented pressfit stem (AncaFit, Cremascoli, Italy). To improve the primary stability, the stem was modified machining two pockets for cement injection in the lateral-proximal area. The CLU stem was found to be significantly more stable than a press-fitted AncaFit stem. Micromotion recorded in the cemented type was comparable with that for the CLU. At the same time, the FEM

Keywords: Hip prosthesis, stress shileding, primary stability, Finite element analysis

---

1 Confirmed Researcher, Orhtopaedics Clinic, University of Bologna, ITALY
2 Researcher, Laboratory for Biomaterials Technology, Rizzoli Institute, Bologna, ITALY
3 Researcher, Dept. of Mechanical Engineering (DIEM), University of Bologna, ITALY
4 Assistant Researcher, Laboratory for Biomaterials Technology, Rizzoli Institute
5 Assistant Researcher, Laboratory for Biomaterials Technology, Rizzoli Institute
6 Visiting Researcher, Laboratory for Biomaterials Technology, Rizzoli Institute

analysis confirmed that the CLU design does not induce excessive stresses in the surrounding cement. The CLU stem thus opens a new way to reach primary fixation for cementless stems.

2. INTRODUCTION

The total hip prostheses available are based on two different means of fixation: cemented and cementless. While both approaches were available since long time, none of them has been proven to be ultimately better. Especially for the femoral components, it is common opinion that both methods of fixation have advantages and disadvantages. Thus most orthopaedic departments, at least in Europe, tend to use both fixation methods, in relation to the patient age, bone stock quality, ethiology, etc. In the present work a wide pre-clinical validation of a prototype femoral stem is presented. This device, hereinafter called Cement-Locked Uncemented (CLU) stem, addressed some specific problems of both types of fixation.

Cemented implants have usually the largest and longest follow-up. Apart from wear, most of the reported failures are related to cement failure [1, 2]. Every new cemented design should thus minimize the stress induced in the cement mantle; the amplitude of the relative micromotion between cement and stem should also be kept as small as possible [3, 4]. Also, in some older cases, the incidence of mechanical failure of the stem was not negligible [5, 6]. On the other hand, cementless stems rarely present mechanical failure [7] and obviously do not have any cement to fail. The main reason for clinical failure of cementless stems is aseptic loosening [8, 9]. Under this definition a complex mechanism related is hidden to the biochemical and biomechanical interaction of the implant with the host bone. Stress shielding and primary stability have been proven to be potential causes of failure [9-11].

There is agreement on the fact that close stem-bone fit would improve most of cementless stems stability problems [12, 10]. However, the great variability of the femoral canal prevents any non-customised device (even when available in many sizes) to perfectly fit every canal. Highly modular systems were developed to address the perfect fit problem [13]; however, their complexity may be another source of problems, and the fit is only mildly improved. The only real way to obtain a perfect fit seems still to be the cement.

The CLU stem was derived from a cementless anatomic stem into which two slots are cut laterally in the anterior and posterior faces of the proximal stem. When the stem is implanted, the two pockets are in communication with the outside through a lateral inlet. Once the femur has been reamed and the stem positioned, the two pockets are filled with PMMA through the aforementioned inlet. The cement is supposed to give a much better proximal fit and thus an improved primary stability. With respect to a cemented stem however, the PMMA should be sensibly less stressed. The CLU concept was developed to obtain anatomic cementless stem stability enhanced by cement without the high cement mantle stresses as in cemented stems. Once the prototype design was devised a pre-clinical experimental and numerical *in vitro* validation was carried out.

3. MATERIALS AND METHODS

3.1 The implants

Three stem types were then chosen:
- anatomic Ti AnCAFit 9.5 mm stem (Cremascoli, Italy), cemented with PMMA (CEM);
- AnCAFit 14 mm stem press-fitted (AFIT);
- Cement Locked Uncemented (CLU) prototype, having an identical shape to the AnCAFit 14 mm, except for the cement pockets.

The stems were press-fitted using a machine controlled insertion protocol that allows to obtain better reproducible preparation [14].

3.2 The composite femur and the cement

Synthetic femur models (Mod. 3103, Pacific Research Labs) were used, as they allow reproducible conditions for comparative studies [15]. A comparative study [16] has shown that they are an excellent substitute to cadaver specimens. The CLU stems were inserted in position and PMMA bone cement was subsequently injected. This cement has a Young modulus of 2721 ± 76 MPa (measurements following the ISO 5833). The CEM stems were inserted after filling the medullary canal with bone cement injected and pressurized with a "cement gun" (Cerim, Cremascoli, Italy).

3.3 The experimental stress shielding procedure

The forces applied to the femur were chosen based on the literature and represented a compromise between reproducibility of the set-up, and physiological simulation of the selected phase of gait (immediately after heel strike, corresponding to the highest peak force) and were based on an extensive methodological investigation [16, 18]. Nine femurs were prepared with uniaxial strain gauges [19], at 10 optimal locations [20]. Repeated measurement were taken disassembling and re-aligning the entire set-up for ten times. The testing protocol showed an overall repeatability on the same femur of better than 0.8% at all strain gauge locations, and a reproducibility between femurs of better than 11%.

3.4 The experimental torsional stability procedure

Primary stability was assessed applying a torque of 18.9 Nm, corresponding to peak during stair climbing [14]. The torque was applied for 1000 cycles, corresponding to about 1 month of patient activity [21]. The femurs were clamped distally. The vertical force of the machine was converted into a torque which was transmitted to the femur by means of a double universal joint (thus avoiding transmission of any other force component (Fig. 1). Relative micromotion was measured transcortically with precision LVDT's (D5/40G8, RDP Electronics, UK; accuracy < 1 micron). The transducers were bonded to the cortical bone at four locations (proximal-medial, proximal-anterior, mid-stem medial, stem-tip medial).
Six femurs were prepared for each of the 3 stem types. The stems were evaluated in terms of elastic (peak to valley) and permanent relative displacement (peak to peak).

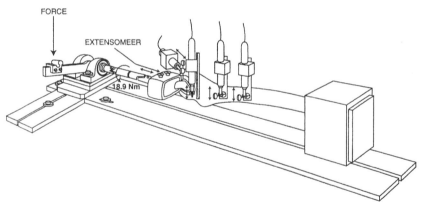

Fig. 1 - Experimental set-up for the measurement of torsional stability. On the left, the loading system is visible, including the double universal joint. Also indicated are the four LVDT's to monitor the rotational micromotion.

3.5 FEM model definition

The geometry of the stem-femur complex was determined from a composite femur prepared by an orthopaedic surgeon. The specimen was sampled using a CT scanner; the CT data set was segmented and converted in a solid model. The solid models of bone, cement and prosthesis were imported into PATRAN (MacNeal-Schwendler Corp.; Los Angeles, USA) via Initial Graphics Exchange Specifications (IGES) neutral format and used to create a mapped mesh using a similar procedure to that described in [22]. The resulting mesh is reported in Fig. 2.

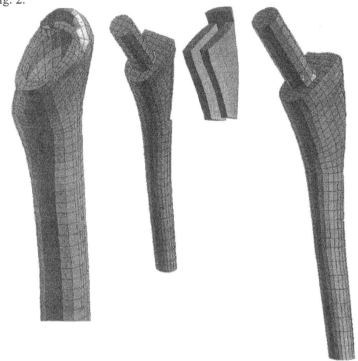

Fig. 2 - FE meshes used: from left to right: operated femur, CLU stem, CLU cement mantle and AFIT stem meshes.

Two load cases were simulated: a heel strike (HS) load case, defined after the stress shielding experimental protocol, and a pure torsion (PT) load case, defined after the primary stability protocol. The first load case (HS) was used (i) to validate the FE model, (ii) to investigate the structural behaviour of the model and (iii) to predict cement stresses under the conditions of peak external load. The torsional load case was used to analyse the accuracy of the interface modelling and to predict cement stresses in the most critical load case for assessing primary stability. In both load cases a boundary condition was imposed such that all nodes on the most distal slice (cortical bone only) were fixed with zero displacement. Material properties were obtained by the manufacturers of the bone analogue and of the implants.

The AFIT model was designed with a gap interface between the metal and the bone tissue. To simulate the initial press-fit, an initial interference of 0.004 mm was imposed to the gap elements. This value was obtained in a preliminary study where the primary stability measurements were used a reference. The CLU model instead was designed with two different interface models. The "GAP" model was designed with a gap interfaces at all the interfaces between metal and the other materials (bone and cement). The "GRASP" model had most of the nodes at the cement-metal interface merged while the stem-bone interface was modelled with gap elements. Only the lateral most nodes of the cement inlet were left separated from the stem surface, as in that area a clear detachment of the stem-cement interface exists. The GAP and the GRASP were comparatively validated with respect to the various experimental measurements. All the models described were solved using the PATRAN/FEA solver (MacNeal-Schwendler Corp.; Los Angeles, USA).

4. RESULTS and DISCUSSION

4.1 Experimental stress shielding results

The stem type was a significant factor only proximally and laterally (Fig. 3; in L1 and L2, ANOVA p<0.05). A Scheffé's F post-hoc test revealed that a significant difference existed between the CLU prototype and the AnCAFit 9.5 mm cemented stem (p=0.002), while scarcely significant was the difference of the uncemented AnCAFit 14 mm versus the cemented stem and the CLU prototype (p=0.04 and p=0.055 respectively).

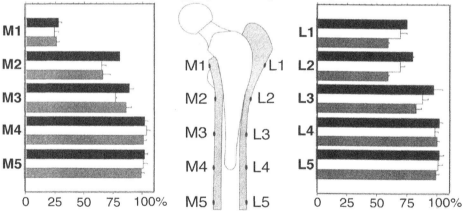

Fig. 3 - Stress shielding for the AnCAFit 9.5 mm cemented (black), AnCAFit 14 mm pressfitted (white) and CLU (grey). Strains in the implanted femur are expressed as percent of strains in the intact (mean ± 95% C.I. for 3 femurs)

The CLU prototype generally behaves like the AnCAFit 14 mm from which it was derived. The increase in stress shielding for the CLU prototype and the AnCAFit 14 mm uncemented stem, with respect to a comparable cemented stem was moderate (less than 15-20% in the worst locations). The only differences observed between the AnCAFit 14 mm and the CLU prototype corresponded to the areas of cement injection. The region of the greater trochanter is normally strained by the abducting forces applied in this area. In the CLU stem, the cement tends to bond the bone to the prosthesis; this seems to stiffen this area.

4.2 Torsional stability results

During the stability test all stems showed no significant permanent displacement after 200-500 cycles, indicating consistent stem sitting (total permanent displacements recorded in the first 500 cycles were equal or lower to 360 microns for the AFIT stem; the CLU stem did not exhibit measurable permanent displacements). Considering the elastic motion measured after 1000 cycles, the CLU stem was significantly more stable than the AFIT stem in the proximal-anterior and stem-tip medial locations. Also axial elastic and permanent motion resulted significantly lower for the CLU prototype than the AFIT stem. Furthermore in the proximal-medial femur the CLU recoverable micromotions resulted lower than 30 μm. If these result are confirmed *in vivo* these small movements should not inhibit bone ingrowth [23]. In any case, since the CLU stem micromovements are one order of magnitude lower than those recorded for the press-fit stem, it is strongly expected higher chances of osteointegration for the CLU implant. Micromotion recorded in the cemented type was comparable with that for the CLU. These results were confirmed by a statistical analysis. These findings suggest that there is an excellent initial torsional stability of the CLU stem.

4.3 FEM models validation

Comparison of measured structural displacements at relevant points with the values computed by the FE models were used to evaluate the correctness of assigned boundary conditions. In the HS load case, the displacement of the calcar was used as reference. Both the AFIT and the CLU models predicted under the HS load case a calcar displacement of 7.4 mm (experimental measurements gave 7.5 ± 0.1 mm). No differences were observed between the GAP and GRASP CLU models. The PT load case was validated by measuring the periosteal rotation of the medial line at the distal-most location where the trans-cortical pins are inserted for the micro-motion test. Measurements gave a rotation of 1.3 ± 0.2° for both implants. The AFIT FEM model predicted a rotation of 1.4°. The CLU GAP model underestimated the rotation at 0.8°, while the GRASP model predicted a rotation of 1.2°.
To validate the local prediction of stresses from the FE models, the predicted surface strains were compared with the experimental strain gauge measurements. The AFIT model showed an overall agreement between the gauge measurements and the FE predicted strains. The Root Mean Squared Error (RMSE) over the 10 strain gauges measurements and the surface

strains predicted at the same nominal locations by the FE model, was 150 με, approximately 5% of the peak measured strain. A similar agreement was found with the CLU GRASP results (RMSE = 81 με or 2% of peak strain). The results from the CLU GAP model were sensibly lower than the measured values (RMSE = 403 με or 12% of peak strain).

To verify the ability of the FE models to accurately represent the non-linear contact of the implant against the host bone, the relative micro-motion between stem and bone, as recorded experimentally by the LVDT transducers, was compared with the similar relative motions computed by the FE models. Again the overall agreement was good for the AFIT model and for the CLU/GRASP model, while the CLU/GAP model overestimated the relative motion. The relative micro-motion results are reported in fig. 4.

4.4 cement stress

Under the PT load case action, most of the cement presented a 4 MPa tensile stress or lower. Zones of higher stresses were found, located in correspondence of some stress raisers; their intensity was always below 10 MPa. In the HS load case the cement is more stressed than under the PT load case. The inlet connection to the cement blades presents sharp angles which act as stress risers. In these points the tensile stress reaches up to 25 MPa. However, the most part of the cement is stressed below 8.5 MPa.

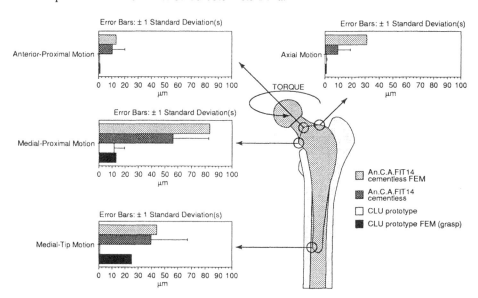

Fig. 4 - Comparison of LVDT-measured and FE-predicted stem-femur relative micromotion for the AFIT and CLU stems. The FEM results for the CLU prototype refer to the GRASP interface model.

5. CONCLUSIONS

Although total hip arthroplasty results are significantly improved in the last 20 years, at least 5-10% implants fail within 10 years from the operation. Thus, the actual implant designs may be still significantly improved. However, with respect to the pioneer years, today it is mandatory to thoroughly investigate any new solution before clinical use. The pre-clinical validation protocol here described, combined *in vitro* biomechanical simulation with finite element analysis, and showed to be an effective an reliable approach to this complex matter.

6. REFERENCES

1. Harris W.H., Will stress shielding limit the longevity of cemented femoral components of total hip replacement?, Clin. Orthop. Rel. Res., 1992, Vol. 274, 120-126.

2. Jasty M., Jiranek W. and Harris W.H., Acrylic fragmentation in total hip replacements and its biological consequences, Clin. Orthop., 1992, Vol. 285, 116-128.

3. Jasty M., Burke D.W. and Harris W.H., Biomechanics of cemented and cementless prostheses, Chir. Organi Mov., 1992, Vol 77, 349-358.

4. Kärrholm J., Malchau H., Snorranson F. and Herberts P., Micromotion of femoral stems in total hip arthroplasty. A randomized study of cemented, hydroxylapatite-coated, and porous-coated stems with rontgen stereophotogrammetric analysis, J. Bone Joint Surg., 1994, Vol. 76A, 1692-1705.

5. Dall D.M., Learmonth I.D., Solomon M.I., Miles A.W. and Davenport J.M., Fracture and loosening of Charnley femoral stems, J. Bone Joint Surg., 1993, Vol. 75B, 259-265.

6. Wroblewsky B.M. and Siney P.D., Charnley low friction arthroplasty of the hip. Long-term results, Clin. Orthop., 1993, Vol. 292, 191-201.

7. Sotereanos N.G., Engh C.A., Glassman A.H., Macalino G.E. and Engh C.A., Cementless femoral components should be made from cobalt chrome, Clin. Orthop., 1995, Vol. 313, 146-153.

8. Wroblewski B.M., The mechanism of fracture of the femoral prosthesis in total hip replacement. Int. Orthop., 1979, Vol. 3, 137-139.

9. Charnley J., Low friction arthorplasty of the hip: theory and practice. Springer-Verlag, New York, 1979, 112-124.

10. Sumner, D.R., Galante, J.O., Determinants of stress shielding: design versus material versus interface, Clin. Orthop. Rel. Res., 1992, Vol. 274, 202-212.

11. Schreurs B.W., Buma P., Huiskes R., Slagter J.L.M.and Sloof T.J.J., Morsellized allograft for fixation of the hip prosthesis femoral component, Acta Orthop. Scand., 1994, Vol. 65, 267-275.

12. Walker, P.S. and Robertson D.D., Design and fabrication of cementless hip stems, Clin. Orthop., 1988, Vol. 235, 25-34.

13. Oh M.D., Whiteside L.A., McCarthy D.S. and White S. E., Torsional Fixation of a modular hip

component, Clin. Orthop., 1993, Vol. 287, 135-141.

14. Harman M.K. Toni A., Cristofolini L. and Viceconti M., Initial stability of uncemented hip stems: an in-vitro protocol to measure torsional interface motion, Medical Engineering and Physics, 1995, Vol. 17(3), 163-171.

15. Viceconti, M., Casali M., Massari B., Cristofolini, L., Bassini S. and Toni, A., The 'Standardized femur program'. Proposal for a reference geometry to be used for the creation of finite element models of the femur, J. Biomech., 1996, Vol. 29, 1241.

16. Cristofolini, L. Viceconti, M. Cappello A. and Toni A., Mechanical validation of whole bone composite femur models, J. Biomech., 1996, Vol. 29(4), 525-535.

17. Cristofolini L., McNamara B.P., Cappello A., Toni A. and Giunti A., A new protocol for stress shielding tests of hip prostheses. In: Blankervoort, L., Kooloos, J.G.M., Eds., Abstracts of the 2nd World Congress of Biomechanics, Nijmegen: Stichting World Biomechanics publ., 1994, Vol.2, 338.

18. Cristofolini L., Viceconti M., Toni A. and Giunti A., Influence of thigh muscles on the axial strains in a proximal femur during early stance in gait, J. Biomech., 1995, Vol. 28(5), 617-624.

19. Cristofolini L. and Viceconti, M., Comparison of uniaxial and triaxial rosette gauges for strain measurement in the femur, Experimental Mechanics, 1997, *In press.*

20. Cristofolini L., McNamara B.P., Freddi A. and Viceconti, M., In-vitro measured strains in the loaded femur: quantification of experimental error, J. of Strain Analysis for Engineering Design, 1997, *In press.*

21. McLeod P.C., Kettelkamp D.B., Srinivasan V. and Henderson O.L., Measurement of repetitive activities of the knee, J. Biomech., 1975, Vol. 8, 369-373.

22. McNamara B.P., Viceconti M., Cristofolini L., Toni A. and Taylor D., Experimental and numerical pre-clinical evaluation relating to total hiparthroplasty. In: Computer Methods in Biomechanics and Biomedical Engineering, edited by Middleton, J., Jones, M.J. and Pande, G.N.Amsterdam: gordon and Breach, 1996, 1-10.

23. Pilliar R.M., Lee J.M. and Maniatopulos C., Observations on the effects of movement on bone ingrowth into porous-surfaced implants, Clin. Orthop., 1986, Vol. 208, 108-113.

STRUCTURAL ANALYSIS OF PHYSICAL MODELS OF TOTAL HIP REPLACEMENTS USING ANALYTICAL AND FINITE ELEMENT METHODS

A. B. Lennon*   B. A. O. McCormack[†]   P. J. Prendergast[‡]

1   ABSTRACT

The authors have developed a physical model of the femoral portion of a hip replacement which allows visualisation of fatigue damage accumulation in the cement mantle. The extent to which the model accurately replicates damage accumulation within such a structure was assessed by calculating cement stresses using two- and three-dimensional finite element analyses. The strains on the surface of the cement were experimentally measured and compared to the results from the finite element models. The results show that the magnitude of the stresses generated in the cement mantle of the experimental model are within the range of those reported in the literature. However the distribution of stress differs from finite element models with more physiological geometries. The experimental model was found to be suitable for quantification of damage and the accuracy of the two-dimensional finite element analysis was found to be sufficient for modelling of the damage accumulation phenomenon. Fatigue crack data from the experimental model can be used to validate finite element predictions of damage.

2   INTRODUCTION

Polymethylmethacrylate 'cement' is still one of the most popular artificial joint fixation techniques in use today. However, the cement mantle is prone to mechanical degradation and this is likely to be a preliminary cause of clinical failure of the joint reconstruction [1]. Breakdown of mechanical integrity most likely occurs in the form of 'damage accumulation', which can be described as the initiation and propagation of cracks and flaws at the microscopic scale [2] .

Damage has been reported in autopsy retrieved specimens [3], in *in vitro* models [4], and has been simulated in finite element studies [5]. However, there remains a lack of studies of this phenomenon in the literature to date. One reaon for this is the difficulty in monito-

Keywords: Damage accumulation, Finite element analysis, Experimental analysis.

---

*Research student, Dept. Mechanical Engineering, University College Dublin, Belfield, Dublin, Ireland.
[†]Lecturer, Dept. Mechanical Engineering, University College Dublin, Belfield, Dublin, Ireland.
[‡]Lecturer, Dept. Mechanical and Manufacturing Engineering, Trinity College Dublin, Dublin 2, Ireland

ring cracks as they initiate and propagate within the structure. Crack monitoring requires sectioning of the joint and this negates the possibility of using the same specimen for further study of damage accumulation. To overcome this difficulty, an experimental model was developed which incorporates the essential features of the femoral portion of a hip reconstruction while allowing cracks within the cement mantle to be viewed [6]. The model consists of a tapered stem with an offset head, with the stem embedded into a sandwich layer of bone and cement. This, in turn, is encased within an aluminium holder which simulates the out-of-plane cortical connection in the real femur. Cut-outs, fitted with transparent perspex windows, in the aluminium holder allow direct viewing of the cement mantle (Figure 1). Under fatigue loading, the amount of damage at any time, the rate of damage accumulation, and the distribution of damage throughout the structure can be monitored. This data can be used to develop realistic damage models of cemented hip reconstructions.

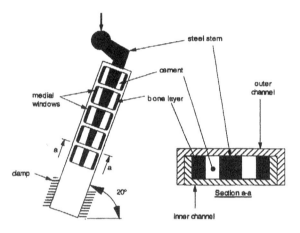

Figure 1: Schematic of experimental specimen, showing the geometry of the layered structure, the clamping position, and the point of load application

Whether or not the damage generated in the cement of the model is realistic depends on how closely the stresses within the cement, and at the interfaces, match those of the physiological structure. The magnitude of the cement stresses predicted by finite element models depend significantly on the geometry and loading conditions of the model. Svesnsson *et al.* [7] reported normal and shear stem-cement interface stresses in the range −5.0 to 2.0 MPa and −5.0 to 1.0 MPa using a two-dimensional side-plate model. A wide range of values for the principal stresses have been reported along the stem-cement interface from 3-D finite element models, ranging from: −13.5 to 1.5 MPa [8], −2.2 to 5.5 MPa [9], and −3.0 to 2.8 MPa [10]. Similarly for the bulk cement, the range of principal stress values varies considerably between studies: −9.6 to 5.4 MPa [11], −6.0 to 2.5 MPa [12], and −7.0 to 5.0 MPa [13]. Principal stresses in the bulk cement of −2.8 to 0.5 MPa were reported by Rohlmann *et al.* [14] for a 3-D finite element model with similar loading to the experimental specimen of Figure 1. An analysis of the above studies shows the distribution of normal stress to be compressive in the proximal medial and distal lateral cement regions and tensile in the proximal lateral and distal medial region [7, 12]. Principal stress distributions are predominantly compressive medially and tensile laterally [14, 8, 9, 10].

The purpose of this study is to investigate the stress distribution in the cement mantle of the experimental specimen. Both two- and three-dimensional finite element models are developed. The results from the finite element models are validated against (i) experimentally measured strains on the surface of the cement mantle, and (ii) the displacement of the prosthesis head as calculated using the principle of virtual work.

3   METHODS

Both two- and three-dimensional analyses were undertaken using ANSYS 5.2. The three-dimensional mesh can be seen on the left of Figure 2 and was developed using eight-noded isoparametric bricks. The two-dimensional mesh (on the right of Figure 2) was built from eight-noded isoparametric plane elements with a thickness option to allow both axial and flexural stiffness to be matched to the physical model. A 'side-plate' was attached to the outer aluminium layers of the mesh, to model the out-of-plane connection of the holder.

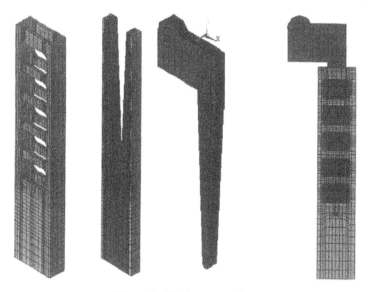

Figure 2: Mesh geometries

All materials were assumed to be linearly isotropic (values for material constants can be seen in Table 1) and all interfaces are fully bonded.

Material	Young's Modulus(GPa)	Poisson's Ratio
Steel stem	215	0.3
Cement	1.7	0.4
Bone	20	0.35
Aluminium	73	0.33
Perspex	2	0.4

Table 1: Material properties used in analyses

The head of the stem was loaded at 20° to the longitudinal axis of the stem with a magnitude of 3 kN. The distal portion of the mesh was constrained to simulate the clamped portion of the experimental model. The effect of inclusion of muscle loads was investigated by applying a load of 1.25 kN and 0.25 kN to the proximal lateral corner of the holder of the two-dimensional model. The first of these was oriented upwards and at 20° to the longitudinal axis of the holder, and simulates an abductor load, while the second was oriented downwards and parallel to the same axis, simulating an iliotibial tract load.

To validate the finite element analysis, eight strain gauges were applied to medial and lateral anterior cement surfaces of the experimental specimen (Figure 1). The four proximal gauges were aligned parallel to the longitudinal axis of the stem while the four gauges placed within the distal two windows were applied perpendicular to this axis.

The deflection of the prosthesis head was calculated using the Unit Load method for determining deflections in structures, which is based on the principle of virtual work [15]. The deflection, δ, is given by equation 1:

$$\delta = \sum_i \frac{\int M_U M_L dx}{EI} \tag{1}$$

where M_U = moment due to the hypothetical unit load and M_L = moment due to actual load applied to the prosthesis head for a given section.

The term 'i' refers to a given section number, 'x' to the length along that section, and EI to the flexural stiffness.

The tapered portion of the stem was modelled as six discrete sections, each of constant cross-section, resulting in a total of nine sections for the entire specimen. The flexural stiffness was calculated for each section and Equation 1 was summed to give the deflection of the entire model at the prosthesis head.

4   RESULTS

For the models excluding the muscle forces, the peak tensile stresses for the two- and three-dimensional models were 4.5 and 5.5 MPa respectively, and were located at the proximal end of the lateral stem-cement interface. Figure 3 shows the maximum and minimum principal cement stresses for the stem-cement interface for both the two- and three-dimensional models. The results show that there is no predominantly lateral tensile or medial compressive stress distribution in the models, as would be expected for the physiological case.

The results for the model including the muscle forces can be seen in Figure 4. Overall, cement stresses are decreased in magnitude apart from the peak compressive stress which increases from -5.7 to -9.0 MPa. There is still no predominant tensile stress distribution on the lateral side—however, the stresses on the medial side are more predominantly compressive, along most of its length, as compared to the 'no muscle' models.

Measured and predicted strains are presented in Table 2. The results show good agreement medially, apart from the most proximal gauge, while laterally, there are substantial

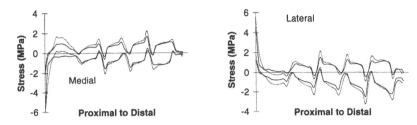

Figure 3: Maximum and minimum principal cement stresses for the stem-cement interface: aluminium holder and no muscle inclusion. Dashed lines = 3-D principal stresses. Solid lines = 2-D principal stresses.

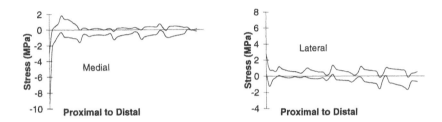

Figure 4: Maximum and minimum principal cement stresses for the stem-cement interface (flexible holder with muscle inclusion)

differences, with the more distal of each of the longitudinal and transverse strains being completely reversed.

Good agreement was obtained for the deflection of the prosthesis head as predicted by the principle of virtual work and by both the finite element models (Table 3).

5   DISCUSSION

There is no precise understanding of the stress distribution in the cement mantle of a hip reconstruction. Given the variety of the different model geometries and loading conditions used in finite element studies it is not realistic to quote an absolute value for the stress in any location throughout the mantle (Figure 5). However, the common trend would appear to consist of medially compressive and laterally tensile principal stresses. On this basis, the model investigated in this study does not show the physiological cement stress distribution. This may be explained by the fact that the model (Figure 1) does not have the typical curvature of the proximal stem and bone. A holder and stem that were curved proximally , to recreate the geometry of a hip reconstruction more accurately, would likely alter the distribution, as would the inclusion of muscle loads (Figure 4). On the other hand, the magnitude of the cement stresses predicted by the finite element models

Gauge	Calculated	Measured	Gauge	Calculated	Measured
2M	−250	−482	2L	90	476
3M	−130	−162	3L	−100	202
4M	140	152	4L	113	103
5M	−125	−112	5L	450	−91

Table 2: Predicted and measured strains. L and M represent lateral and medial respectively. The windows are numbered 1 to 5, from proximal to distal. Window 1 had no gauges applied. Gauges for windows 2 and 3 measure longitudinal strain. Gauges for windows 4 and 5 measure transverse strain

Model	3-D F.E.	2-D F.E.	Unit Load
Deflection (mm)	0.19	0.189	0.187

Table 3: Comparison of the deflection of prosthesis head as predicted by unit load method and the two finite element models.

are within the range of those reported in the literature (Figure 5). Thus, under fatigue loading, the experimental specimen would be expected to generate damage at the same rate as the *in-vivo* cement mantle, but, due to the inaccurate stress distribution, may not accurately model the spatial distribution of damage throughout the cement mantle.

The good agreement for prosthesis head deflection between the three models, indicates that, at least theoretically, the overall structural behaviour of the experimental specimen is accurately recreated. However, the substantial differences between some of the predicted and measured strains indicate that, locally, within specific regions of the cement layer, the finite element models are not accounting for some phenomena that are present in the experimental specimen. The most obvious such phenomenon is that of localised interface debonding. Debonding significantly alters the strain distribution in the region at which it occurs, as suggested by Harrigan et al. [16] and Akay and Aslan [10]. A non-linear analysis could be undertaken to investigate this more fully.

There was close agreement between the two- and three-dimensional finite element models, although the two-dimensional model predicts slightly lower values for some of the peak stresses. These differences could be due to the use of higher order elements in the two-dimensional mesh and/or the presence of some poorly shaped elements in the three-dimensional mesh.

Because of its smaller size, the two-dimensional model possesses substantial advantages, in terms of memory requirements and processor time, when running non-linear simulations. These advantages can be retained, even with the use of higher order elements (which would allow more sophisticated simulation of damage accumulation). Also, out-of-plane stresses, for the three-dimensional model, are negligible compared to those in the mid-frontal plane (as would be expected), since all loading occurs in this plane, and there is no asymmetry about it. Therefore, the stress state is essentially two-dimensional. For these reasons, the two-dimensional finite element model will be used for the development of a predictive damage accumulation model for cemented joint reconstructions.

Cement Interior

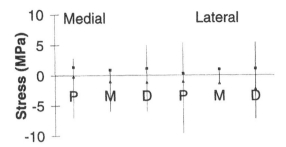

Figure 5: Comparison of present study and literature. P = Proximal, M = Middle, and D = Distal. Lines indicate principal stress range from the literature. Data points indicate corresponding max and min principal stresses from present study.

6 CONCLUSIONS

The model should accurately predict the overall amount and rate of damage accumulation in cemented hip reconstructions but may not predict a similar spatial distribution of damage. Possible improvements to the model include more accurate representation of the geometry and loading conditions of the structure. The two-dimensional finite element model was deemed suitable for development of predictive models to investigate damage accumulation in cemented hip reconstructions.

REFERENCES

[1] Huiskes, R., Mechanical failure in total hip arthoplasty with cement, Current Orthopaedics, 1993, vol. 7, 239–247.

[2] Lemaitre, J. and Chaboche, J. L., Mechanics of Solid Materials, Cambridge University Press, 1994.

[3] Jasty, M., Maloney, W. J., Bragdon, C. R., O'Connor, D. O., Haire, T. and Harris, W. H., The initiation of failure in cemented femoral components of hip arthroplasties, Journal of Bone and Joint Surgery, 1991, vol. 73B, 551–558.

[4] McCormack, B. A. O., Prendergast, P. J. and Gallgher, D. G., An experimental study of damage accumulation in cemented hip prostheses, Clinical Biomechanics, 1996, vol. 11, 214–219.

[5] Verdonschot, N. and Huiskes, R., A combination of continuum damage mechanics and the finite element method to analyze acrylic bone cement cracking around im-

plants, in: *Second International Conference on Computer Methods in Biomechanics and Biomedical Engineering* (Middleton, J., Pande, G. and Jones, M., eds.), Gordon and Breach, 1996, pp. 25–33.

[6] McCormack, B. A. O., Prendergast, P. J., Curley, L. and Lennon, D., Visualisation of fatigue crack growth in a model of a hip reconstruction, in: *EORS Trans. 7th Annual Conference*, vol. 7, 1997 .

[7] Svesnsson, N., Valliappan, S. and Woods, R. D., Stress analysis of human femur with implanted charnley prosthesis, Journal of Biomechanics, 1977, vol. 10, 581–588.

[8] Rohlmann, A., Mössner, U. and Bergmann, G., The use of finite element analysis in assessment of endoprosthetic fixation, in: *Engineering and Clinical Aspects of Endoprosthetic Fixation IMechE Conference*, 1984 .

[9] Prendergast, P. J., Monaghan, J. and Taylor, D., Materials selection in the artificial hip joint using finite element analysis, Clinical Materials, 1989, vol. 4, 361–376.

[10] Akay, M. and Aslan, N., Numerical and experimental stress analysis of a polymeric composite hip joint prosthesis, Journal of Biomedical Materials Research, 1996, vol. 31, 167–182.

[11] Crowninshield, R. D., Brand, R. A., Johnston, R. C. and Pedersen, D. R., An analysis of femoral component stem design in total hip arthroplasty, Journal of Bone and Joint Surgery, 1980, vol. 62-A, 68–78.

[12] Tarr, R. R., Clarke, I. C., Gruen, T. A. and Sarmiento, A., Predictions of cement-bone failure criteria: three-dimensional finite element models versus clinical reality of total hip replacement, John Wiley and Sons, Ltd., 1982 .

[13] Lewis, J. L., Askew, M. J., Wixson, R. L., Kramer, G. M. and Tarr, R. R., The influence of prosthetic stem stiffness and of a calcar collar on stresses in the proximal end of the femur with a cemented femoral component, Journal of Bone and Joint Surgery, 1984, vol. 66-A, 280–286.

[14] Rohlmann, A., Mössner, U. and Bergmann, G., Finite-element-analysis and experimental investigation in a femur with hip endoprosthesis, Journal of Biomechanics, 1983, vol. 16, 727–742.

[15] Gere, J. M. and Timoshenko, S. P., Mechanics of Materials, Chapman and Hall, 1991.

[16] Harrigan, T. P., Kareh, J. A., O'Connor, D. O., Burke, D. W. and Harris, W. H., A finite element study of the initiation of failure of fixation in cemented femoral total hip components, Journal of Orthopaedic Research, 1992, vol. 10, 134–144.

PARAMETERIZED 3D FINITE ELEMENT MODEL OF THE CEMENTED REPLACED FEMUR: NUMERICAL ROLE OF CONTACT ELEMENTS IN THE PROSTHESIS/CEMENT INTERFACE

E.Astoin* ; M.Simondi* ; F.Lavaste*

1. ABSTRACT.

The numerical models, commonly used to evaluate the chances for acrylic cement failure and the mechanical behavior of the bone, have results very dependent of the characteristic of the modelization. So it is for the type of prosthesis/cement interface, often modeled with contact elements to simulate a failure. The present paper proposes to study different kind of cemented stems comparing when modeling the prosthesis/cement interface linearly or with contact elements. This method is first applied with a parameterized model of the cemented femur, which gives global results on the bone. They show large differences in the Von Mises stresses values and in their distribution. It is then applied to a more schematic model of the fixation, which gives more precise and local results on the cement. They point out a large increasing of the compressive stresses, while the shear ones remain almost constant, when introducing contact elements. It is concluded that the type of interface modelization is of large importance, and that a linear modelization may not be well adapted to represent a perfect fixation.

2. INTRODUCTION.

Many numerical models represent the cemented fixation of the femur. They are commonly used to explain the mechanical behavior of the replaced joint. It is therefore important to evaluate the influence of numerical parameters, as it has been done for the material mechanical properties. In such, we propose to study different cemented femoral stems with 2 different types of modelization of the prosthesis/cement interface : linear modelization or modelization with contact element. For this comparison, a complex parameterized model, and a more schematic one of the cemented femur are presented. The different stems tested, characterized by different metaphyseal width in the frontal plane, represent different kind of cementing : regular cement mantle, and discontinue cement mantle with direct prosthesis/bone contact.

Keywords : Femur, F.E.M., Interface modeling, Cement.

---

* Laboratory of Biomechanics - ENSAM Paris - 151, boulevard de l'hôpital - 75013 France

3. MATERIAL & METHOD.

Two different finite elements models were developed, using the ANSYS® package, release 5.2. The first one is a 3D parameterized model of the healthy and replaced femur. The personalized values of the parameters were obtained from frontal and sagittal roentgenograms of the patient. The roentgenograms are selected on the following criteria :
- Perfect AP pelvis view, including at least the upper third of the both femurs and the pelvis. For the frontal roentgenograms, femurs had to be internally rotated in order to bring the femoral neck into the coronal plane. This criteria was supposed to be met, when both medial aspects of the great trochanter were superimposed. For the sagittal plane, the femoral head center had to be on the diaphyseal axis.
- A 1 inch diameter metallic sphere was fixed on the skin of the patient at the level of the greater trochanter, to allow calculation of the magnification. Since the measured values was very close to the rest of the femoral bone, it was assumed that the magnification was the same for the femoral head and the rest of the femoral bone.

The second model is also a 3D model of the cemented femur. This time, the geometrical shapes of the bone, cement and prosthesis are modeled using elementary geometrical elements.

3-1. Definition of the parameterized model.

Geometric measurements. The measurements were performed on the roentgenograms using a digitalising table (Digitizer CX 1000). Internal and external width are measured every 10mm under the lesser trochanter till 130mm, on frontal and sagittal roentgenograms. On the frontal ones, 10 points were picked on the femoral head border, in order to calculate the best fit circle of the femoral head. The center of the circle was taken as the femoral head center, and its diameter as the femoral head diameter.

Parameterization and meshing of the healthy bone. The geometrical parameters of the model are calculated from the measurements on the roentgenograms. To realize the Parameterization, the bone geometrical shape is divided in 3 parts :
- The diaphyseal part, defined as the bone under the level of the lesser trochanter.
- The femoral head part. Going along the neck axis, defined as the line passing throw the center of the femoral head and forming a 135° with the diaphyseal axis of the femur, the femoral head section is the part located above the head center.
- The metaphyseal part is the one completing the 2 precedent parts: it includes the bone above the lesser trochanter, the greater trochanter and the femoral neck.

Diaphyseal part : For each section, the frontal and sagittal values of the internal and external width define the main axes of an ellipse. Internal and external ellipses are then built in each horizontal section, and 24 elements were defined between successive ones.

Femoral head part : The part of the head defined here is approximated by a hemi-sphere. Then, on the coordinates of the center of the femoral head measured on the frontal roentgenogram, 2 concentric half-spheres were constructed and meshed. Their diameters are fixed to the diameter of the femoral head measured on the frontal roentgenogram for the external half-sphere, and 2mm less for the internal one.

Metaphyseal part : Considering that this part of the bone geometry is difficult to model using simple geometrical elements, it was decided to use the method of kriging[1] : The metaphyseal section of an average femur was meshed manually, based on reconstruction upon CT scans. It was used as the reference mesh. In order to adapt this mesh to the diaphyseal and head mesh, the border nodes were imposed to have identical 3D coordinates. The kriging method allow then to deform all the others nodes of the reference mesh, keeping the global shape as constant as possible.

Realization of the replaced model : meshing of the prosthesis and the cement. The prosthesis was meshed semi-automatically from its geometry defined in CAD. From the healthy model, the replaced one was obtained by deleting the elements of the femoral head and of the upper part of the femoral neck. The prosthesis was then inserted in the bone in it theoretical position:
- Diaphyseal axis of the prosthesis on the diaphyseal axis of the femur.
- Horizontal reference section at the level of the lesser trochanter, as defined in the surgical technique of this stem.

In the space between the prosthesis and the bone, 2 sets of cement elements were inserted. The shape of the cement mantle was the consequence of both the cortical bone shape, and the prosthesis shape.

For the 4 following models, the bone geometry used was maintained constant, while different prosthesis were used. The first model was realized with an anatomical prosthesis (called reference prosthesis), setting close prosthesis/bone elements on the internal metaphyseal cortical bone (between 0.2 and 0.5 mm of cement). This was called *model 0*. Three other prosthesis were realized from the reference one by reducing of 10%, 20% and 30% it frontal metaphyseal width. They were called respectively *models 1, 2* and *3*. For all those models, the cement mantle was then continue, and there were no direct contact between the prosthesis and the bone.

Finally, a last model, called *model C*, was realized from the model 0 : The bone geometry was lightly deformed in order to set direct prosthesis/bone contact in the internal metaphyseal zone, where the cement mantle was very thin. The cement mantle was then discontinue. Frontal sections of this 5 models are presented in figure 1.

Mechanical parameters. The material were all modeled as linearly isotropic. Numerical characteristics were the following :

Cortical bone :	E = 20 000 MPa	$v = 0.3$
Cement :	E = 2 000 MPa	$v = 0.3$
Prosthesis :	E = 200 000 MPa	$v = 0.3$

These values were taken as representative of data reported in the literature[2][3].

For each model, the bone/cement interface was modeled linearly. On the other hand, the

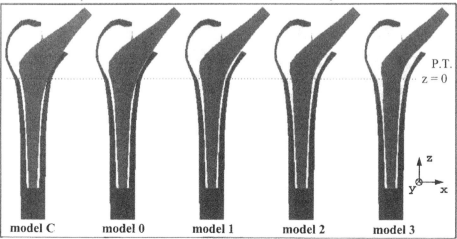

Figure 1 : The 5 geometrical models represented in their frontal section.

prosthesis/cement interface was modeled either linearly too, either with contact elements[5]. The friction coefficient was then settled to 0.2., and virtual springs were used to assure initial stability. The penalty method + Lagrange multiplier was used.

Boundary conditions and loads. This model allow to reproduce physiologic loads, and to study a global behavior of the bone and the cement. So, the calculations were realized by fixing the displacement of all the distal diaphyseal cortical nodes in every direction, and applying 3D physiologic loads both on the prosthesis head ($F_p x = -500$ N ; $F_p y = -153$ N ; $F_p z = -1364$ N) and the greater trochanter ($F_t x = 382$ N ; $F_t y = 117$ N ; $F_t z = 784$ N). This loads are issue from Burke et al.[4].
The Von Mises stresses were then calculated in the nodes of the cortical bone.

3-2. Definition of the simple model.

In the upper section (z=0), the geometry of the bone is modeled by two half-circle separated par a straight portion. So it is for the same for the 4 different model of prosthesis. But the length of their straight portion is variable, so that they allow a variable cement layer in the large axis (x axis) : it is respectively of 0, 2, 4 and 8 mm for the models 1 to 4. This upper section is presented in figure 2.
For the bone, the upper section is identically reproduced along the vertical axis, so that the global shape is a cylinder of 120 mm long. For all the prosthesis, the shape of the upper section is also reproduced, regularly deceased in its dimensions while descending along the vertical axis (z axis), Then, the global shape of each prosthesis is a cone of 100 mm long. One quarter of each 3D model is presented in figure 3.

Mechanical parameters. For each model, the prosthesis/cement interface was modeled successively linearly and with contact elements, as for the first model. Similarly, the numerical parameters modeling the mechanical behavior of the materials and of the prosthesis/cement interface are the same as the one used for the first model.

Boundary conditions and loads : This model allow to get more localized information about the mechanical behavior of the cement. So, the calculations were realized by blocking the displacement of

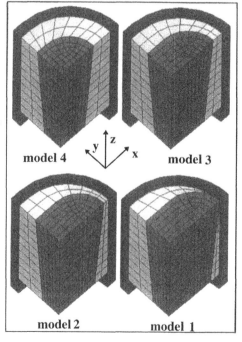

model 4   model 3

model 2   model 1

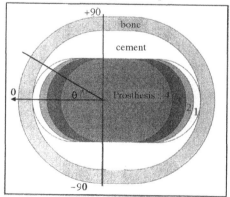

Figure 2 : Upper section of the 4 geometrical models

Figure 3 : Quarter sections of the 4 geometrical models.

all the external node of the cortical bone in every direction, and applying a vertical rotation couple on each of the 11 nodes of the vertical axis of 100 N.mm.

In order to evaluate the stability of the model, the rotations of some nodes located in the upper plane (z=0) and on the large axis (x axis) were measured : node P (external node of the prosthesis, in contact with the cement) and nodes C1, C2 and C3 (nodes of the cement respectively in contact with the prosthesis, middle, and in contact with the bone. The shear stresses ($\sigma_{r\theta}$) and the compressive stresses (σ_r) were also calculated for the nodes of the cement mantle located on one of the two circle (cf. figure 2 : $-90<\theta<90$).

4. RESULTS.

First model : The Von Mises stresses are measured in the cortical bone for each of the model. The results are represented in figures 4a and 4b for a modelization of the prosthesis/cement interface respectively linear or with contact elements.

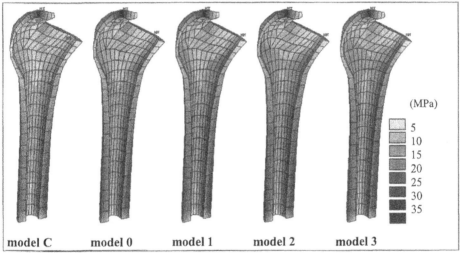

Figure 4a : Von Mises stresses in the cortical bone. Case linear modelization.

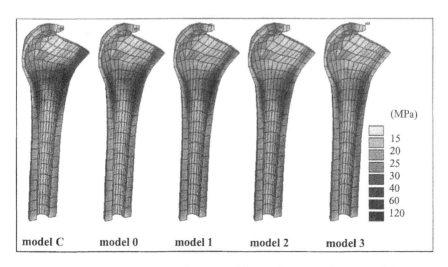

Figure 4b : Von Mises stresses in the cortical bone. Case non-linear modelization.

Second model : The rotation of the nodes P, C1, C2 and C3 are presented on figures 5a and 5b for a modelization of the prosthesis/cement interface respectively linear or with contact elements. In the linear case, the nodes P and C1 are identical. The scale on each figure is different, but numerical values of the rotation are written, so that comparison is possible.

The shear stresses are presented for the internal nodes (the one in contact with the prosthesis) of 2 sections : z=-10 in figures 6a and 6b, and z=-90 in figures 7a and 7b, the both for a modelization of the prosthesis/cement interface respectively linear or with contact elements. The scales were settled identical in order to make comparison easier.

The compressive stresses are represented for the same nodes in figures 8a and 8b for the section z=-10 and in figures 9a and 9b for the section z=-90. The scale are here different for the 2 type of modelization, so that each figure is easier to read.

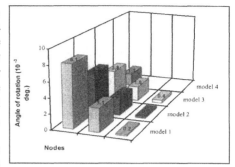

Figure 5a : Rotations in case « linear ».

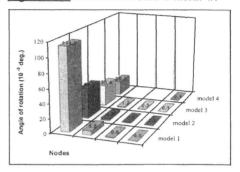

Figure 5b : Rotations in case « contact ».

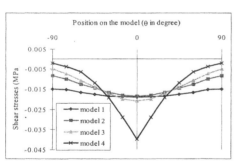

Figure 6a : Shear stresses in the cement internal nodes. Z=-10. Case « linear »

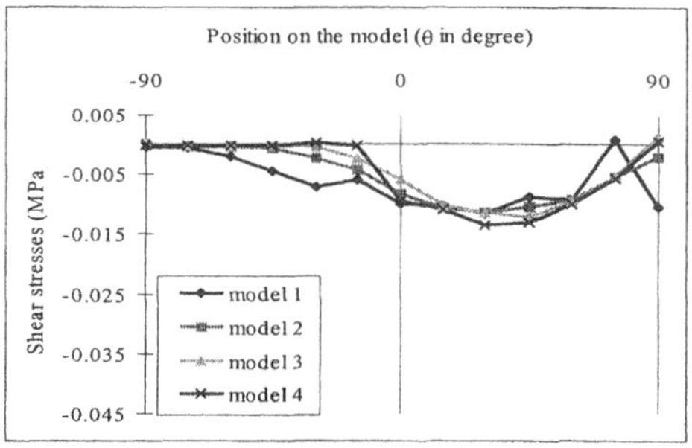

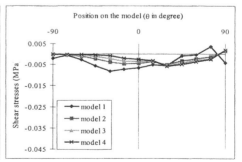

Figure 6b : Shear stresses in the cement internal nodes. Z=-10. Case « contacts »

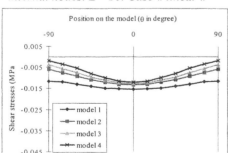

Figure 7a : Shear stresses in the cement internal nodes. Z=-90. Case « linear »

Figure 7b : Shear stresses in the cement internal nodes. Z=-90. Case « contacts »

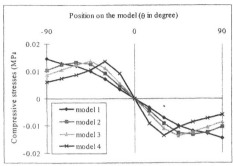

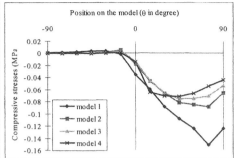

Fig.8a :Compressive stresses in the cement internal nodes. Z=-10. Case « linear »   *Fig.8b :Compressive stresses in the cement internal nodes. Z=-10. Case « contacts »*

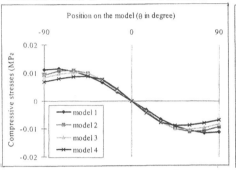

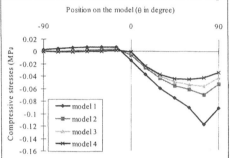

Fig.9a :Compressive stresses in the cement internal nodes. Z=-90. Case « linear »   *Fig9b :Compressive stresses in the cement internal nodes. Z=-90. Case « contacts »*

5. DISCUSSION.

Analyzing the first model, we notice large differences in the Von Mises stresses calculated in the cortical bone when modeling differently the prosthesis/cement interface : first, the calculated values are less important when modeling linearly the interface. Furthermore, differences in the distribution are also important. The maximum Von Mises stresses are in the distal bone when modeling linearly, while it is in the lesser trochanter zone, where the prosthesis is the closest to the bone, when modeling with contact elements. Actually, this maximum of this stresses is more important when the prosthesis is closer to the bone. We particularly notice a very important increase when there is direct prosthesis/bone contact (model C). This result is inversed is the linear case, the Von Mises stresses being more important in the internal metaphyseal bone for the smallest prosthesis.

Actually, contact elements are commonly used to simulate a cement failure in an interface, while a perfect fixation correspond to a linear model. But, analyzing roentgenograms of long term prosthesis having an important metaphyseal filling, we often observed bone growth in the zones where maximum Von Mises stresses are calculated in the modelization with contact element : lower part of the zone where the prosthesis is the larger. This results suggests that the modelization of the prosthesis/cement interface using contact elements may be more representative than a linear modelization. Anyway, it suggests also question about the situation really represented by each type of modelization.

The results of the second model gives precision about the mechanical behavior of the cement mantle. The shear stresses have similar order of values in the 2 cases of prosthesis/cement interface modelization, while the compressive stresses are very different : Calculated values are about 10 times more important when modeling with contact element. This is explained by the examination of the rotational stabilities : Modeling the interface with contact elements allow a large movement between the prosthesis and the cement mantle. Then, the rotation of the node P is much more important than when modeling linearly, while the rotation of the nodes C1, C2 and C3 are less important.

This large mobility between the prosthesis and the cement in the case with non-linear modelization interface can be considered as the possible movement when the stem fixation fails. This seems to invalidate the use of contact elements in the modelization of the prosthesis/cement interface when wanting to represent a perfect fixation. The fact that the non-linear modelization seems to be valid or not in different cases may be the consequence of the major difficulty to model micro-movement. We may then imagine, in the case of cemented stems, that the difference between those 2 modelizations of the prosthesis/cement interface may be related to the difference between cemented stem with coated and smooth surface.

6. CONCLUSION.

We first want to insist about the major influence of the kind of modelization of the prosthesis/cement interface, when using finite element models of the cemented femur to validate prosthesis : On one hand, the calculated values are very different, but on the other hand, the comparison between different stems gives inverse results. Trying to define the situation exactly represented by each type of modelization, we are more reserved : We could imagine that a linear interface represents a perfectly fixed prosthesis having a coated surface, and that a non-linear interface, with contact elements, represents a loose fixation. To model a perfectly fixed prosthesis having a smooth surface, we imagine further kind of modelization. We can imagine a global linear modelization, using contact elements only in some precise location. Or a more complex interface, using contact elements with different static and dynamic friction coefficient. Finally, we may use the improvement of the software, allowing to realize crack modeling. Those different possibilities will be at the origin of future works.

7. REFERENCES.

1. Trochu,F., A contouring program based on dual kriging interpolation, Eng. With Computers, 1993, vol. 9,
2. Huiskes,R. and Boeklagen,R., Mathematical shape optimization of hip prosthesis design, J. Biomechanics, 1989, Vol. 22, N°8/9, 793-804.
3. Harrigan,T.P. and Harris,W.H., A three-dimensional non-linear finite element study of the effect of cement-prosthesis debonding in cemented femoral total hip components, J. Biomechanics, 1991, Vol. 24, N°11, 1047-1058.
4. Burke,D.W., O'Connor,D.O., Zalenski,E.B., Jasty,M. and Harris,W.H., Micromotion of cemented and uncemented femoral component in simulated stance and stair climbing, J. Bone Joint Surg. (Br.), 1990, Vol. 73, 33-37.
5. ANSYS User's Manual, Vol. IV, 1992, 196-204

AN *IN VIVO* AND NUMERICAL ANALYSIS OF THE EFFECT OF FEMORAL STEM SURFACE FINISH ON CORTICAL STRAIN IN CEMENTED HIP REPLACEMENT

D.S. Barker[1], A. Wang[2], M.F. Yeo[3], N. Nawana[4], S. Brumby[5], M. Pearcy[6] and D. Howie[7]

1. ABSTRACT

The aim of this study was to assess the effect of femoral cemented stem surface finish on cortical strain. Eleven sheep underwent hemiarthroplasty of the left hip (study) with either a matt (M) or polished (P) finish, double tapered cemented stem and were subjected to controlled exercise for nine months. The sheep were then killed and the right (control) femur implanted with an identical stem type. The study and control femurs were then subject to strain gauge analysis. A 3D finite element (FE) model was constructed which assumed either a bonded or slipping stem-cement interface. The stem-cement friction coefficient was varied from 0 to 0.22 to 0.5. A viscoelastic cement mantle was included in the model.

The *in vivo* model showed a significant decrease in the proximal strain of the study femurs compared to the controls, however no significant difference was noted between the M and P femurs. The FE model indicated that strain levels in the proximal femur

Keywords: Animal model, hip replacement, finite element, interface mechanics, creep

[1]Research Engineer, Department of Orthopaedics, Repatriation General Hospital, Daw Park, S.A. 5041 Australia
[2]Consultant Orthopaedic Surgeon, Division of Orthopaedics, QE2 Medical Centre, Nedlands, W.A. 6009 Australia
[3]Senior Lecturer, Department of Civil Engineering, University of Adelaide, Adelaide, S.A. 5000 Australia
[4]Biomedical Engineer, Howmedica, 203, Boulevard de la Grande Delle BP8, 14201 Herouville-Saint-Clair Cedex France
[5]Orthopaedic Registrar, Department of Orthopaedics and Trauma, Royal Adelaide Hospital, Adelaide, S.A. 5000 Australia
[6]Professor of Biomedical Engineering, School of Mechanical, Manufacturing and Medical Engineering, Queensland University of Technology, Brisbane, Queensland 4001 Australia
[7]Professor, Department of Orthopaedics and Trauma, Royal Adelaide Hospital, Adelaide, S.A. 5000 Australia

increased significantly as the coefficient of friction decreased. It appeared in the present study that the stem-cement interface remained bonded *in vivo*, for both matt and polished stems. A polished stem surface finish may only be effectual in increasing proximal femur loading if complete stem-cement debonding occurs. FE models which consider stem-cement interface mechanics should include contact elements which disallow Coulomb friction effects until a failure capacity is exceeded.

2. INTRODUCTION

A major controversy in cemented total hip replacement is whether to use a femoral stem with a polished surface or to use a matt surface. Surface finish of the femoral stem is thought to influence the way that load is transferred from the stem to the cement and bone and thereby loosening of the femoral component [1].

Polished femoral stems have been associated with decreased loosening of hip replacements and less osteolysis at 17 years follow-up [2]. The polished surface in conjunction with a collarless double tapered design is thought to be advantageous for the effective transmission of load from the femoral component to the femoral bone by allowing subsidence of the stem within the cement mantle [2]. Recent review of results from a national data base suggest the matt collarless tapered stem of the Exeter design may have higher loosening rates than the polished collarless Exeter stem [3].

Jasty *et al.* [4] suggested aseptic loosening of the cemented femoral stem is initiated by mechanical failure of the stem-cement bond. The opposite findings to those of Jasty *et al.* have been reported by Gardiner and Hozack [5]. The authors noted that when measures were taken to increase the stem-cement bond, failure occurred at the cement-bone interface, the stem-cement interface was solid without evidence of de-bonding.

Studies utilising the FE method have shown that the stem-cement interface condition and cement creep may strongly influence the mechanical behaviour of cemented femoral stems [6,7]. The present study examined the differences in behaviour of a polished and matt surfaced femoral stem in a sheep hip arthroplasty model. A corresponding FE model of the sheep hip hemiarthroplasty was constructed. The primary hypothesis tested was: there is no significant difference in the change in cortical strain pattern of the proximal femur after remodelling around a polished double tapered cemented stem compared to a matt stem of the same geometric design.

3. METHODS

3.1 The animal model

Twenty mature Australian Merino sheep underwent unilateral cemented hemiarthroplasty of the left hip, utilising modern cementing techniques, with either a matt or polished, double tapered stem. The prosthesis was specially designed for insertion into the sheep femur and was based on the Exeter human hip prosthesis (Howmedica International, Staines, UK). Two versions were used in this study. One version had a highly polished surface and was collarless, type (P) (Ra = 0.02 - 0.04 μm), whilst the second version had a matt surface and was collarless, type (M) (Ra = 1 μm). At 4 weeks post operation the sheep were exercised daily to ensure the sheep were walked a distance of 4 km per day.

Nine months later, 11 sheep (7P, 4M with nine sheep lost to complications) underwent hemiarthroplasty of the right hip using the same design of prosthesis. After hardening of the cement, the sheep were killed and the femurs explanted and stripped of all soft tissue to allow assessment of this prosthesis immediately after implantation and compare it to the prosthesis nine months *in situ*. These control and study femurs and were then subjected to strain gauge analysis. For comparison, 4 intact femurs, derived from the same flock, were also tested. Four triaxial rosette strain gauges were placed on each femur. A proximal medial strain gauge was located at approximately the same transverse level as the base of the lesser trochanter, 10 mm distal to the level of the medial cut surface of the femoral neck. The proximal lateral strain gauge was placed at the junction of the proximal femoral metaphysis with the flare out into the greater trochanter. The distal medial and distal lateral strain gauges were positioned approximately 5mm proximal to the tip of the implant. The position of the gauges is shown schematically in Figures 2, 3 and 4.

For each rosette the axial, transverse and oblique strain recording was used to calculate the principal tensile and compressive strain magnitude and direction. Following strain gauge application the femur was prepared for mechanical testing by potting the distal end in methyl methacrylate cement. A loading jig was designed which enabled the implant to be loaded in three orthogonal directions: along the axis of the implant, in the anterior-posterior (A-P) and medio-lateral (M-L). The axial load applied was 0.592 x the body weight (0.592 BW) in Newtons, the AP load was 0.215 BW and the ML load was 0.415 BW. These load components were derived from the *in vivo* study of Lanyon *et al.* [8] who used instrumented prostheses to determine the hip joint reaction force for mature sheep, during normal gait. High contrast radiography, sectional photography and ground section histology were also performed.

3.2 The finite element model

The femoral shell was modelled, including topographical data from a CT scan as a conical section, with a proximally expanding diameter and thinning cortex. The supporting shell was assumed to consist purely of cortical bone. The prosthesis was modelled to be rectangular in section with dimensions 4x4mm at the tip, located 60 mm proximally to the base femoral level, and expanding to 18x7.2 mm at a level 120 mm proximally to the base level. A distal air-gap was modelled by assuming a 4x4x20 mm volume under the prosthesis tip.

The finite element mesh was constructed with enhanced strain 8 node hexahedron elements, available in the LUSAS element library (FEA Pty. Ltd., Surrey, UK). The FE mesh contained 2825 nodes for the bonded stem-cement interface and 3126 nodes for the sliding interfaces. Material properties were considered to be linear, homogenous, isotropic, and elastic except when cement creep was considered. The values used were; a)cortical bone, Young's modulus, $E = 17.0x10^3$ MPa, Poisson's ratio, $v = 0.3$, [9], b) cement, $E = 2.1x10^3$ MPa, $v = 0.3$, [10], c) steel, $E = 200.0x10^3$ MPa, $v = 0.3$ [11].

The bone-cement interface was assumed to be fully bonded. Four separate conditions were considered at the stem-cement interface.
1) a fully bonded stem-cement interface

2) a slipping stem-cement interface with a) coefficient of friction = 0, b) coefficient of friction = 0.22, c) coefficient of friction = 0.5

A power law was used to model the time dependent straining of PMMA. The power law was based on the work of McKellop *et al* [12], who calculated the creep coefficients for commercially available bone cements (Equation 1):

$$\varepsilon_{cr} = \frac{\sigma^{1.55} t^{0.1}}{45997}$$ ....Equation 1

Where,

ε_{cr} - creep strain, σ – uniaxial stress, t - loading time

In vivo loading is cyclic in nature and involves long periods of unloading (ie at night). In order to model the long term effects of creep two different scenarios were considered;

1) the fully bonded and frictionless interfaces. For these stem-cement interface conditions, there is no long-term accumulation of creep. This is because overnight, there is full strain recovery due to unloading. Daytime (short-term) creep was based on the work of Lu and McKellop [7]. One half of the peak load was applied for 8 hours with the remainder of the load applied instantaneously at the end of the 8 hour loading period. This simulation of sinusoidal loading is valid for linear viscoelastic materials. In the present study, a nonlinear viscoelastic law was used, hence the simulations represent an approximation of the theoretical solution. The cement loading curve is shown in Figure 1a), demonstrating the short and long term effects of creep.

2) the frictional stem-cement interfaces. Verdonschot and Huiskes [6] have demonstrated that cemented femoral stems, which are de-bonded from the surrounding cement, will subside in a slip-stick fashion. The authors also pointed out that there are both static and dynamic loading components associated with de-bonded stems. The prosthesis will remain "stuck" in the cement mantle on unloading due to the frictional effects of the stem surface finish. This will leave a residual static stress during periods of unloading, ie at night. In the present study, for the frictional stem-cement interface conditions, the creep loading was simulated by applying the full load and then removing it. The residual static stress which remained in the cement mantle was then applied for nine months. The dynamic component of creep recovered overnight, hence it was not considered over the nine month period. After nine months static creep loading, the dynamic component was added for a loading time of 8 hours. The cement loading curve is shown in Figure 1b), demonstrating the short and long term effects of creep.

After the simulated loading was complete, principal compressive and tensile strains were obtained as output from the FE program at four nodes geometrically similar to the rosette gauges used in the strain gauge study.

4. RESULTS

4.1 The Animal Model

The strain gauge results for the study, control and intact femurs studies are shown in Figures 2 and 3. Although the load components, axial, medio-lateral, and anterior-posterior were applied independently, the resultant principal strains are shown in Figures

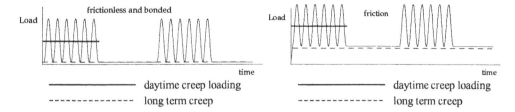

Figure 1: a) There is no long-term accumulation of creep. Daytime loading is modelled by applying half the peak load for 8 hours then the remainder of the load instantaneously b) Long term loading is modelled by applying the peak load and then removing it. A residual static stress is then applied for nine months. After nine months a further day's creep is added, including the dynamic component.

2 and 3. These resultant strains were obtained by combining the strains contributed from the three load components, thus allowing comparison with the FE results.

Principal compressive strains for the matt and polished groups are shown in Figure 2. The insertion of a femoral cemented stem significantly decreased the compressive strain levels at the four rosette gauges, for both matt and polished stems. There was, however, no significant difference in compressive strain levels between the control and study groups. Importantly, there was no significant difference in compressive strain levels after nine months between the matt and polished groups.

Principal tensile strains for the matt and polished groups are shown in Figure 3. As for the compressive strains, the insertion of a femoral cemented stem significantly decreased the tensile strain levels at the four rosette gauges for both matt and polished stems. There was also no significant difference in strain levels between the control and study groups. Similarly, there was no significant difference in tensile strain levels after nine months between the matt and polished groups.

The radiographic, photographic and histological investigations did not reveal the presence of a radiolucent line at the stem-cement interface. The stem-cement interface appeared to be primarily intact with only small (less than 100μm width) and incomplete (less than 5 mm length) gaps.

4.2 The finite element model

Principal tensile and compressive strains obtained from the FE output, after nine months simulated loading, are shown in Figure 4. Results are shown for the four stem-cement interface conditions. The horizontal lines show averaged results from the strain gauge tests for comparison. These averaged results were obtained from the mean of the matt and polished strain gauge tests.

There was no large variation in compressive strains due to stem-cement interface condition at the four node locations, geometrically representing the rosette gauge positions. The compressive strains correlated well with the strain gauge values, except at gauge 4 where bending led to an overestimation of the FE strain level. The FE model

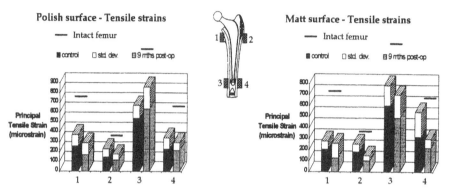

Figure 2: Compressive femoral principal strains for the four rosette gauges for both study and control femurs.

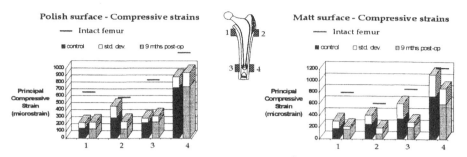

Figure 3: Tensile femoral principal strains for the four rosette gauges for both study and control femurs.

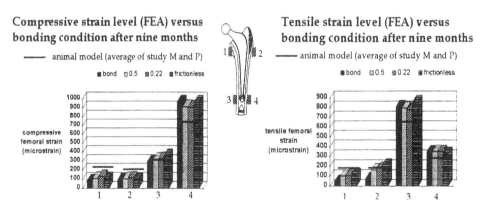

Figure 4: FEA output of principal tensile and compressive strains for the four stem-cement interface conditions after nine months simulated loading.

also showed that at the distal node locations (3 and 4), variation of the stem-cement interface condition did not greatly influence the magnitude of the tensile strains. At the proximal locations (1 and 2) however, the magnitude of the tensile strains significantly increased as the coefficient of friction at the stem-cement interface decreased. The tensile strains correlated well with the strain gauge values, except at gauge 3 where bending led to an overestimation of the FE strain level.

5. DISCUSSION

There was no significant difference in the proximal femur strain levels after *in vivo* remodelling between matt or polished stems in the sheep model The finite element model, however, suggested that a polished stem should induce larger tensile hoop strains in the proximal femur compared to a stem which bonds to the cement. This anomaly may be explained by examining the results of the histological investigations. The stem-cement interface remained intact *in vivo* without de-bonding for both the matt and polished stems. The FE model however assumed that the stem was either completely bonded (scenario 1) or completely debonded with varying coefficients of friction (scenarios 2a), b) and c)). Scenario 2 did not actually occur *in vivo*.

Bundy and Penn [13], and more recently, Davies et al. [14] have demonstrated that there is a strong bond between a highly polished metallic stem and cement. Bundy and Penn suggested that this bond is due to chemical attraction and may be as strong as rough surfaced stems which 'bond' to cement through mechanical integration. A combination of the *in vivo* sheep hip joint loading and the stem shape design may have produced less stem-cement interface stresses than found in humans. This stress decrease, combined with second generation cementing techniques, may have prevented complete de-bonding at the stem-cement interface in the sheep model.

In the present study both matt and polished stems remained 'bonded' to the cement (scenario 1) *in vivo*, which meant dynamic frictional effects were not present between the stem and cement. It is believed that the magnitude of these frictional effects, which are dependent on the coefficient of friction between stem and cement, may alter load transmission to the proximal femur. This study has shown that differences in proximal femur loading, due to stem surface finish, may only be apparent if there is stem-cement de-bonding. This de-bonding, which is noted radiologically with Exeter prostheses in humans within 12 months [2], may explain the difference in results between the present *in vivo* study and a recent long-term clinical review [3].

The present study has also shown that when using FE analysis to model cemented hip replacement, contact elements must be included at the stem-cement interface, which allow slip with friction, only after a specified failure criteria has been reached. The assumption of either complete bonding or complete de-bonding may lead to unrealistic results.

6. REFERENCES

1. Lee, A.J.C., Perkins, R.D. and Ling, R.S.M., Time-Dependent Properties of Polymethylmethacrylate Bone Cement. In: *Implant Bone Interface*, edited by Older, J., London: Springer-Verlag, 1990, 85-90.

2. Fowler, J.L., Gie, G.A., Lee, A.J.C., and Ling, R.S.M., Experience with the Exeter total hip Since 1970, Orthop. Clin. North Am., 1988 19, 477-489.

3. Ahnfelt, L., Herberts, P., Malchau, H., and Andersson, G.B.J., Prognosis of total hip replacement. A Swedish multicenter study of 4,664 revisions. Acta Orthop. Scand. Suppl., 1990, 238.

4. Jasty, M., Maloney, W.J., Bragdon, C.R., O'Connor, D.O., Haire T., and Harris W.H., The initiation of failure in cemented femoral components of hip arthroplasties. J. Bone Jt. Surg., 1991, 73-B, 551-558.

5. Gardiner, R.C. and Hozack, W.J., Failure of the cement-bone interface. A consequence of strengthening the cement-prosthesis interface? J. Bone Jt. Surg., 1994, 76-B, 49-52.

6. Verdonschot, N. and Huiskes, R., Subsidence of THA stems due to acrylic cement creep is extremely sensitive to interface friction, J. Biomechanics, 1996, 29, 1569-1575.

7. Lu, Z. and McKellop, H., Effects of cement creep on stem subsidence and stresses in the cement mantle of a total hip replacement, J. Biomed. Mat. Res., 1997, 34, 221-226.

8. Lanyon, L.E., Paul, I.L., Rubin, C.T., Thrasher, E.L., DeLaura, R., Rose, R.M., and Radin, E.L., *In vivo* strain measurements from bone and prosthesis following total hip replacement. J. Bone Jt Surg., 1981, 63-A, 989-1001.

9. Reilly, D.T. and Burstein, A.H., The mechanical properties of cortical bone. J. Bone Jt. Surg., 1974 56-A, 1001-1022.

10. Saha, S. and Pal, S., Mechanical properties of bone cement: A review, J. Biomed. Mat. Res., 1984, 18, 435-462.

11. Gere, J.M. and Timoshenko, S.P., Mechanics of Materials, UK, Von Nostrand Reinhold, 1987, 744.

12. McKellop, H., Narayan, S., Ebramzadeh, E., and Sarmiento, A., Viscoelastic creep properties of PMMA surgical cement, 3rd World Biomaterials Congress, 1988, p.328.

13. Bundy, K.J. and Penn, R.W., The effect of surface preparation on metal/bone cement interfacial strength. J. Biomed. Mat. Res., 1987, 21, 773-805

14. Davies, J., Anderson, M.J. and Harris, W.H., The cement-metal interface of the Exeter stem during fatigue loading of a simulated THA in-situ, 42nd Trans ORS, 1996, Atlanta, 525.

COMPARATIVE STUDY OF THE RESULTS BETWEEN CUSTOM NON-COATED CEMENTLESS HIP IMPLANTS AND MIRRORED CEMENTLESS HA - COATED HIP IMPLANTS ON THE CONTRA-LATERAL SIDE.

M. Mulier[1] , G. Deloge [2]

Abstract

For 30 patients with a minimum follow-up of 3 years, we compared the results of the contralateral custom-made hip implants. The second prosthesis was coated with hydroxylapathite and based on a mirrored image of the contralateral side.

Introduction

Hydroxylapatite coatings on prostheses are reported to give a very early and reliable stability and show a better osseointegration compared with identical non-coated implants.

Materials & Methods:

Since 1990 we have implanted 43 HAP- coated femoral stems based on a mirrored image of the contra-lateral side stem of the patient which was performed earlier using the I.M.P. (= Intra-operative Manufactured Prosthesis) system.

Keywords: Cementless, Custom-made, I.M.P., Hydroxylapatite

[1] M.D., Orthopaedic department , Catholic University of Leuven, U.Z. Pellenberg, Weligerveld,1, 3212 Pellenberg (Lubbeek), Belgium
[2] Engineer; Stewal Implants; Industriepark, 37 ; 9600 Ronse; Belgium

The IMP system was developed in collaboration with *Stewal Implants* and the KUL university in Belgium, the department of biomechanics (supervised by Prof. Vanderperre) and the department of orthopaedics (supervised by Prof. Mulier). The system aims to achieve the best fit and fill for a every unique patient, in order to diminish the interference with anatomical loading of cortex.

It is also our goal to restore of the anatomical centre of rotation and thus the lever arm of the abductor muscles and also to correct the leg length discrepancies
The system is operational since 1987 and more than 3000 THP have been implanted.
The surgeon makes a femoral cavity according to the available bone stock of that specific patient, in which a custom-made stem is to be fixed. Of this cavity a mould is made.
This mould is then transferred to manufacturing unit near to the hospital where it is measured with in a specially designed 3D laser measuring machine.
The measured data is then converted to an NC millingprogram in the computer.
The software was especially written for this purpose and has different aspects:
* Calculation of antiloosenings to make the prosthesis introducible at all times.
* Interference fit to allow load transfer in preferential zones.
* Immediate feed back of important data to the surgeon, such as:
    - ante/ retroversion.
    - varus / valgus tilt.
    - contact between stem and bone.
    - resistant surfaces against vertical, inclination and torsional loads.
After the surgeon agrees with the resulting design, the prosthesis is manufactured and decontaminated.
The prosthesis is transferred to the operating theatre, sterilised and introduced in the patient.
The complete manufacturing takes about 25 minutes, during which time the surgeon performs the acetabular part of the THR.
Recent follow-up studies have shown us that fit and fill is not enough.
An early fixation of the implant is necessary in order to get good long term results. This early fixation can be obtained by applying an Hydroxylapatite coating to the Ti - alloy substrate.

This study was set-up in order to determine the influence of HA coatings on the performance of the I.M.P. implant.

The data of the first stem, which was gathered during the operation on the first side, was mirrored in order to have the right neck orientation and stemgeometry for the other side.
A custom-made prosthesis was manufactured according to the GMP of the I.M.P. manufacturing unit.
We also provided a custom-made rasp complete with neck in order to be able to perform a trial reduction with the rasp.
The stem was HA coated on the proximal 1/3 of the stem by plasma spraying.

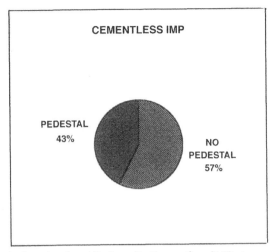

Figure 1: Radiological evaluation (Pedestal): Cementless IMP stem.

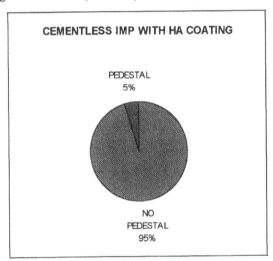

Figure 2: Radiological evaluation (Pedestal): Cementless IMP stem with HA coating proximal.

Of the 43 implanted prosthesis we have 30 patients with a minimum follow-up of 3 years on both sides.(3 - 7 years).

We compared the contralateral sides, both clinically and radiographically.

Results:

All the 43 rasps and respective stems could be introduced without much difficulty. This proves the high grade of similarity of the femoral cavities of both femurs

Radiographically:

On the uncoated side bony pedestal formation below the distal tip was observed in 50 % of the cases.

On the contra-lateral HA coated side this phenomenon only occurred in 5 % of the cases. (see figure 1 and 2)

This pedestal formation may be triggered by excess force transmission from the prosthesis to the bone possibly by late subsidence.

No other significant differences were observed radiographically.

The HA coated implants didn't show any form of osteolysis caused by defragmentation of the HA coating.

Clinically:

Most of the patients have a more secure feeling on the HA coated side during monopodal stance; this appeared insignificant.

It is impossible to compare Harris scores, because both sides influence each other.

Conclusions:

Our system is a good example of a practical application of computers in orthopaedics with continuous evolution and with unlimited possibilities.

The HA coated implants show a much better radiographical result then the non -coated implants.

It is a challenge to coat the implants during surgery. This way the I.M.P. system could become available to all the patients with a difficult and unpredictable femoral geometry (i.e.: revisions, C.D.H.).

EFFECT OF LOADING HISTORY ON SHORT TERM BONE ADAPTATION AFTER TOTAL HIP ARTHROPLASTY.

Terrier A[1,2], Rakotomanana L[1,2], Leyvraz PF[1]

1. ABSTRACT

After an alteration of the physiological stress whithin bone, experimental studies reported some time-delay in bone adaptation. To study this phenomenon, a bone adaptation model including the stress history was developed. The model consisted in a first-order nonlinear integro-differential equation, coupled to the equilibrium partial differential equation. Existence, unicity and stability of the solutions were investigated and established. A finite element code was used to solve the equilibrium equation and a third order accuracy numerical scheme was developed to integrate the evolution law. The unknown parameters of the model were identified by using published clinical data. The model was then applied to analyse adaptation of bone after Total Hip Arthroplasty (THA). Two situations were compared: with and without stress history. In both cases, the density evolution converged approximately to the same values after one year but were notably different during the first two months. During this period, the use of the stress history induced a smooth and delayed bone response, which was more conform to clinical data.

2. INTRODUCTION

After a THA, the degree of stress shielding induced by the femoral stem depends on many parameters: interface properties, material stiffness, shape and size of the implant, etc (*e.g.* [1]). However, it was noticed from X-ray analysis that all designs presented a trend to proximal bone loss and distal cortical hypertrophy [2]. As changes of bone density may affect the quality of the stem fixation and may eventually require a revision of

Keywords: bone adaptation model, stress history, total hip arthroplasty

---

1. Orthopaedic Hospital, Av. Pierre-Decker 4, 1005 Lausanne, Switzerland.
2. Biomedical Engineering Laboratory, Swiss Federal Institute of Technology, 1015 Lausanne, Switzerland.

the implant [1], several numerical simulations were performed to evaluate the effect of the stem on bone adaptation (e.g. [3], [4], [5]). Most of these numerical studies paid only attention to the equilibrium density distribution, occurring after several years after surgery. However, recent adaptation models were developed to obtain more realistic density evolution (e.g. [6], [7]). In parallel, many experimental studies were devoted to the measurement of the transition period and it was observed that some time-delay occured in bone adaptation after a load alteration: resorption rate was maximum after 2-4 weeks of immobilization [8], resorption went on 2-4 weeks after remobilizing [9]. To account for this time-delay, it was necessary to consider not only the present stress state but also the whole stress history. Thus, the goal of this study was first to develop a mathematical model for internal bone adaptation accounting for the stress history and second to evaluate the effects of the stress history on bone adaptation. To perform the second part of this work, a THA with a non-cemented anatomical collarless design was considered.

3. METHODS

3.1. Bone adaptation model

To relate bone adaptation to mechanical environment, internal variables had to be defined and then linked to the stress. Inertial effects being negligible, the static equilibrium equation completed by boundary conditions was used to obtain the stress tensor within bone:

$$\begin{cases} \nabla S(x) + b(x) = 0, & \forall x \in \Omega \\ S(x) = \bar{S}(x), & \forall x \in \partial\Omega_1 \\ u(x) = \bar{u}(x), & \forall x \in \partial\Omega_2 \end{cases} \tag{1}$$

S being the second Piola-Kirchhoff stress tensor, u the displacement vector and b the volumic forces. Bone was assumed to be non-homogeneous and transverse isotropic. Therefore, it might be characterized by two internal variables: the relative density ϕ and the structural tensor M [10]. Bone adaptation to mechanical stimulus was then entirely defined by three constitutive laws:

$$S = D(\phi, M):E \tag{2}$$

$$\frac{\partial}{\partial t}\phi(x, t) = \begin{cases} v_r \cdot [\psi(x, t) - \psi_r] & \psi(x, t) < \psi_r \\ 0 & \psi_r \le \psi(x, t) \le \psi_d \\ v_d \cdot [\psi(x, t) - \psi_d] & \psi(x, t) > \psi_d \end{cases} \tag{3}$$

$$\frac{\partial}{\partial t}M(x, t) = 0 \tag{4}$$

where D was the elastic tensor and E the Green-Lagrange strain tensor (e.g. [11]). The relative density ϕ was related to the mechanical stimulus ψ by a piece-wise linear evolution law including a lazy-zone (bounded by ψ_r and ψ_d) and two different slopes (v_r and v_d were respectively the resorption and deposition rate). M was held constant with respect to time.

To account for the whole stress history, the integral stimulus function ψ was defined by using an integral (Volterra) operator of a present stimulus function Y [6]:

$$\psi(x, t) = \int_{-\infty}^{t} Y(x, s)k(t - s)ds \qquad (5)$$

For the sake of simplicity, the nonlinear dependence on ϕ and S (contained in Y and consequently in ψ) did not appear explicitly in equation (5):

$$Y(x, t) = Y(\phi(x, t), M(x, t), S(x, t)).$$

The present mechanical stimulus Y was taken to be an invariant scalar function of ϕ, M and S [12]. In the present study, Y was set to the anisotropic plastic Hill criterion [13]. By this choice, the dependence on density might be separated from anisotropy and stress dependence:

$$Y(\phi, M, S) = \tilde{Y}(M, S) \cdot \phi^{-4} \qquad (6)$$

Introducing equation (6) in (5) and then in (3), the evolution law could be described by a nonlinear integro-differential equation:

$$\frac{\partial}{\partial t}\phi(x, t) = \begin{cases} v_r \cdot \left\{ \int_{-\infty}^{t} (\tilde{Y}(x, s)\phi(x, s)^{-4} \cdot k(t - s)ds) - \psi_r \right\} & \psi(x, t) < \psi_r \\ 0 & \psi_r \le \psi(x, t) \le \psi_d \\ v_d \cdot \left\{ \int_{-\infty}^{t} (\tilde{Y}(x, s)\phi(x, s)^{-4} \cdot k(t - s)ds) - \psi_d \right\} & \psi(x, t) > \psi_d \end{cases} \qquad (7)$$

Continuity and Lipschitz conditions on $Y(\phi) = \tilde{Y}\phi^{-4}$ being here satisfied, continuity of k implied the existence and unicity of solution to equation (7) [14].

Inspired from the theory of materials with fading memory [15], the kernel function was set to an exponential:

$$k(t) = \frac{1}{\tau}e^{-\frac{t}{\tau}} \qquad (8)$$

where the time constant τ characterized the persistence of the memory. With this latter choice, the integro-differential equation (7) might be transformed into a system of two ordinary differential equations. Indeed, by introducing (8) in (7), changing the integration variables ($s' = t - s$) and integrating by parts, an equivalent system was found:

$$\begin{cases} \frac{\partial}{\partial t}\phi(x, t) = \begin{cases} v_r \cdot [\psi(x, t) - \psi_r] & \psi(x, t) < \psi_r \\ 0 & \psi_r \le \psi(x, t) \le \psi_d \\ v_d \cdot [\psi(x, t) - \psi_d] & \psi(x, t) > \psi_d \end{cases} \\ \frac{\partial}{\partial t}\psi(x, t) = \frac{1}{\tau}[\tilde{Y}(x) \cdot \phi(x, t)^{-4} - \psi(x, t)] \end{cases} \qquad (9)$$

These nonlinear and coupled equations had to be completed with two initial conditions: the initial density distribution $\phi(x, 0) = \phi_0(x)$ and the initial stimulus value $\psi(x, 0)$, describing the stress history before $t = 0$. Assuming that the system was in equilibrium before $t = 0$, we deduced $\psi(x, 0) = \psi_0(x) = \psi_e$ with $\psi_e = (\psi_r + \psi_d)/2$.

3.2. Stability analysis

The mathematical model defined above was made up of a (space) partial differential equation (1) and a (time) differential equation (9), completed with boundary and initial conditions. To analyse the time dependence of this system, we used a uniform axial compressive load applied to a homogeneous anisotropic hollow cylinder. The stress tensor S and the structural tensor M were thus constant (in space and time) within bone volume and the function Y became only a function of ϕ:

$$Y(\phi, M, S) = Y(\phi) = \tilde{Y} \cdot \phi^{-4} \tag{10}$$

Moreover, without loss of generality, the equilibrium zone was set to zero ($\psi_r = \psi_d = \psi_e$) and a single density rate was considered ($v_r = v_d = v$). A stability analysis of this system (9) proved that the (unique) solution possessed a unique equilibrium point (when $t \to \infty$) which was always asymptotically stable. Furthermore, the equilibrium point was independent of the parameter τ. Finally, a condition over the type of stability of the equilibrium has been found. Indeed, if

$$\tau < \tau_{cr} = \frac{1}{16v} \cdot \sqrt[4]{\frac{\tilde{Y}}{\psi_e^5}} \tag{11}$$

the solution converged monotonically towards the equilibrium, otherwise, the solution oscillated near the equilibrium. This dynamical behavior was observed in the classical graph $\{\phi(t), t\}$, but also in the phase space $\{\phi(t), \psi(t)\}$, using different initial conditions (near the equilibrium) and different values of the parameter τ (Fig. 1).

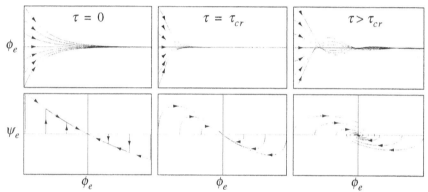

Fig. 1. Evolutions of the perturbed system near the equilibrium point $\{\phi_e, \psi_e\}$. Upper part: density evolutions $\{\phi(t), t\}$. Lower part: orbits in the phase space $\{\phi(t), \psi(t)\}$. Effect of increasing values of the memory parameter τ over the solution behavior and the type of stability.

3.3. Numerical strategies and parameters identification

To investigate complex 3D systems, equations (1) and (2) were solved by FEM-code [16] to get the stress field. The nonlinear evolution equation (9) was then solved at each nodal point of the FEM-mesh by using a (semi-implicit) Crank-Nicholson scheme (*e.g.* [16]). This scheme was completed by an adaptive stepsize control to provide a third-order accuracy method [12].

The mathematical model contained five unknown parameters: the memory parameter (τ), the two limits of the equilibrium zone (ψ_r, ψ_d) and the resorption and densification rates (v_r, v_d) which defined the piece-wise linear evolution law (3). To identify these parameters, the experimental results of Ulivieri *et al.* [8] were used. These results consisted in density measurements of the human tibia during an immobilization period. By a least-squares fitting of these data to the solutions of the differential equations, a value of $\tau = 10$ days was directly obtained for the memory parameter. The ψ_r and ψ_d values were then chosen in order to correspond to 1000 and 2000 microstrains [17]. We also obtained v_r and finally deduced $v_d = v_r / 4$ [18].

3.4. Application to Total Hip Arthroplasty

The 3D geometry and the density distribution of a proximal femur were reconstructed from CT-Scan. Bone anisotropy was taken into account from anatomical observation. The femoral component, made out of Titanium alloys (Ti6 Al4 V), was a press-fit, collar-less anatomical stem filling optimally the proximal femur. Stem was implanted according to usual surgical techniques. Finite element model of the bone-implant system was obtained with a 3D mesh generator [19]. A discontinuous frictional contact between bone and implant was assumed (friction coefficient: 0.6). The loading condition corresponding to the single leg stance phase during the gait cycle including major muscular forces (*gluteus maximus, gluteus medius, psoas*) was used [20]. A post-operative loading schedule was tested with the model. During the first two weeks, 20% of the normal forces were applied. This load was then increased to 50% during the following 2 weeks and finally to 100% thereafter. Two situations were compared: without memory ($\tau = 0$) and with memory ($\tau = 10$ days).

4. RESULTS

Evolution of bone density distribution was calculated for both cases: without and with memory. The initial bone density distribution corresponded to the density measured just before implantation and the final one when no more adaptation occurred (Fig. 2). The final density distributions were almost the same for both situations. In both cases, implants generated resorption in the medial proximal part and hypertrophy of the distal cortical wall. Thus, the cortical wall became thinner in the proximal part and thicker in the distal one. Furthermore, a significant hypertrophy of cancellous bone was also observed in the lateral endosteal cortex, near the insertion point of the abductor muscles.

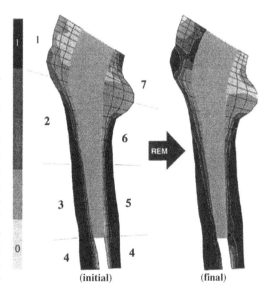

Fig. 2. Initial and final relative density distributions in the frontal plane. Description of the 7 Gruen zones.

To study more precisely the effect of the memory parameter τ, the density evolution was calculated in each Gruen zones (Fig. 3) and for both values of τ. The main difference between the two simulations lies mainly in the short-term (before 2 months). Indeed, without memory, adaptation took place immediately after arthroplasty and each loading alteration (at 2 and 4 weeks) produced an immediate and abrupt response of bone. At the opposite, when the memory parameter was set to 10 days, the evolution curves were smoother and we also observed the time delay of bone adaptation. Indeed, the maximal bone resorption rate occurred only after 3 weeks in the memory case. As a consequence, the non-memory model predicted a 18% density reduction in the medial proximal region (Gruen zone 7) after 2 weeks, whereas the memory model only predicted a 2% resorption. The difference between the two case, which was noticeable during the third week (17%) reduced notably after 1 year (6%) and the final density distributions were quite the same.

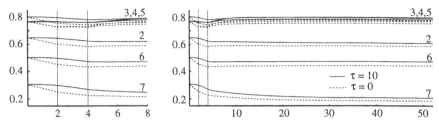

Fig. 3. Evolution of the mean relative density in Gruen zones (as defined in Fig. 2) during the first 8 weeks (left) and until equilibrium (right). Each of the graphs contains both cases: continuous line for the memory case and doted line for the non-memory case. The two vertical lines at 2 and 4 weeks mark the change of loading conditions.

5. DISCUSSION

As far as we know, most bone adaptation models used to predict density changes after THA were focused on the final equilibrium density, occurring after several years. The goal of this work was to develop a bone adaptation model which took into account the stress history. This model was based on a previous model, taking only into account the present stress state [12]. This extension permitted to simulate the delay of the response of bone tissue to stress alteration reported by experimental studies. This delay may be explained by almost two ways. First, the lag in cellular response, due to cells transformation, recruitment, activation [21]. Secondly, the fatigue damage accumulation which was proposed [22] as a factor controlling bone adaptation. The fatigue of bone was related to the apparition of micro-fractures which might be the real process involving bone adaptation. The integral stimulus developed in this study was based on a plastic criterion. Since fatigue in materials corresponds to an accumulation of stress occurring beyond some limit of plasticity, the integral stimulus defined in this paper might be considered in some sense as a measure of the fatigue of bone.

A mathematical analysis of the present model showed that the system possessed a unique solution (bone density evolution) which possessed a unique equilibrium state, independent on the initial density ϕ_o and the memory parameter τ. It was also shown that the final density distribution depended on the initial density distribution and on the memory parameter if and only if the width of the lazy zone did not vanished. To display the stability of the system, orbits in the phase space were obtained for different values of the parameter τ (Fig. 1).

These analytical results were limited to a simple case (homogeneous hollow cylinder), allowing to get rid of the spatial dependency. However, the existence, unicity and stability of the general case were established by a numerical study of complex 3D-systems (mesh dependency, convergence, space-continuity). It was also demonstrated that the use of a lazy-zone introduced a dependency on the initial density and the memory parameter to the final equilibrium density. Indeed, as shown in the cylinder study (no lazy-zone), each density curve converged to the same equilibrium value (Fig. 1), depending neither on the initial density nor on the memory parameter. A the opposite, in the femur case ($\psi_r \neq \psi_d$), the final values of the density were slightly different in each Gruen zone (less than 6%) with two values of the memory parameter (Fig. 3).

The final distributions of femal bone density were quite conform to histomorphometric studies: loss of bone mass in the medial proximal part and distal cortical hypertrophy (stress shielding) (e.g. [1], [2]). The high density spot in the lateral cortex resulted probably from the abductors force distribution, which was concentrated only on a few nodes. The effect of the memory parameter was more obvious in the first 8 weeks. Indeed, the integration of the stress history caused some inertia in bone response. This inertia created some time-delay but also a smoothing of the density curves; at each change of the loading conditions, the reaction is not immediate but staggered over many days. This delayed and smooth adaptation observed seems more conform to clinical studies (e.g. [8]).

In their paper, Levenston et al. [6] used three different values of the memory parameter τ (5, 20 and 100 days) and, from the data of Ulivieri et al. [8], estimated this latter to be on the order of 20 days. Using the same clinical data and the hollow cylinder model, we obtained $\tau = 10$ days with the least-squares method. However, this parameter should certainly vary with species, age, bone location, etc [6].

The critical value τ_{cr} (11) being an increasing function of the stress (through the function Y), it was meaningless to search for a unique value of τ_{cr} within the whole proximal femur. In fact, τ_{cr} varies whithin the femur according to the density and stress field. Therefore, with a fixed value of the memory parameter, each location of the femur may be either in the oscillating or non-oscillating regime. This aspect of the problem was not investigated in the present paper. However, using a equilibrium loading (20 MPa) applied to a cortical hollow cylinder, the value of τ_{cr} was estimated to 15 days. This result gives the order of the τ_{cr} when physiological loading is supported by the diaphyse of long bones (tibia, femur). Thus, if $\tau_{cr} = 15$ and $\tau = 10$, a small variation around the physiological loading would produce a non-oscillating ($\tau < \tau_{cr}$) adaptation of the density.

The time-delay response of bone to stress was observed by many experimental studies but it is not yet precisely measured (e.g. [9], [8]). Due to the lack of data, it was therefore difficult to validate the identification of the model's parameters obtained in this paper. However, the density evolution calculated after THA seems in agreement with measured bone density evolution after THA. Finally, for the sake of completness, an experimental protocol is under development to improve the parameter identification.

6. REFERENCES

1. Engh C. A., and Bobyn J. D., The influence of stem size and extent of porous coating on femoral bone resorption after primary cementless hip arthroplasty, Clin Orthop, 1988, Vol. 231, 7-28.
2. Haddad R., Cook S., and Brinker M., A comparison of three varieties of noncemented porous-coated hip replacement, J Bone Joint Surg, 1990, Vol. 72B:1, 2-8.
3. Huiskes R., Weinans H., and van Rietbergen B., The relationship between stress

shielding and bone resorption around total hip stems and the effects of flexible materials, Clin Orthop, 1992, , 124-134.

4. Weinans H., Huiskes R., and Grootenboer H. J., Effects of fit and bounding characteristics of femoral stems on adaptive bone remodeling, J Biomed Engng, 1994, Vol. 116, 400.

5. Terrier A., Rakotomanana R. L., Ramaniraka N., and Leyvraz P. F., Non-cemented femoral stems fixation: effects of shape design on bone adaptation and on post-remodeling stability, In Trans 43rd ORS, San-Fransisco, 1997, 308.

6. Levenston M., Beaupre G., Jacobs C., and Carter D., The role of loading memory in bone adaptation simulations, Bone, 1994, Vol. 15, 177-186.

7. Fyhrie D., and Schaffler M., The adaptation of bone aparent density to applied load, J Biomech, 1995, Vol. 28, 135-146.

8. Ulivieri F., Bossi E., Azzoni R., Ronzani C., Trevisan C., Montesano A., and Ortolani S., Quantification by dual photonabsorptiometry of local bone loss after fracture, Clin Orthop, 1990, , 291-296.

9. Jaworski Z., and Uhtohoff H., Reversibility of non traumatic disuse osteoporosis during its active phase, Bone, 1986, Vol. 7, 431-439.

10. Cowin S., The relationship between the elasticity tensor and the fabric tensor, Mechanics of Materials, 1985, Vol. 4, 137-147.

11. Gurtin M. E., An Introduction to Continuum Mechanics, Academic Press, New York, 1981.

12. Terrier A., Rakotomanana R., Ramaniraka N, and Leyvraz P., Adaptation models of anisotropic bone, Computer Methods in Biomechanical and Biomedical Engineering, 1997, Vol. 1, 47-59.

13. Rakotomanana R. L., Leyvraz P. F., Curnier A., Heegaard J. H., and Rubin P. J., A finite element model for evaluation of tibial prosthesis-bone interface in total knee replacement, J Biomech, 1992, Vol. 25, 1413-1424.

14. Corduneanu C., Integral equations and applications, Cambridge University Press, 1991.

15. Truesdell C., and Noll W., The non-linear field theories of mechanics-second edition, Springer-Verlag, 1992.

16. Curnier A., Computational methods in solids mechanics, Kluwer Academic Publishers, Dordrecht-Boston-London, 1994.

17. Rubin C. T., and Lanyon L. E., Regulation of bone mass by mechanical strain magnitude, Calc Tissue Int, 1985, Vol. 37, 411-417.

18. Nauenberg T., Bouxsein M. L., Mikic B., and Carter D. R., Using clinical data to improve computational bone remodeling theory, In 39th Meeting Orthopaedic Research Society, San-Fransisco, 1993, .

19. Rubin P. J., Rakotomanana R. L., Leyvraz P. F., Zysset P. K., Curnier A., and Heegaard J. H., Frictional interface micromotions and anisotropic stress distribution in a femoral total hip component, J Biomech, 1993, Vol. 26, 725-739.

20. Ramaniraka N., Leyvraz P., Rakotomanana L, Rubin P., and Zysset P., Micromotion at the bone-stem interface during the gait cycle after cementless total hip replacement: influence of stem design and loading level, Hip International, 1996, Vol. 6, p 51-58.

21. Martin R., and Burr D., Structure, function and adaptation of compact bone, Raven Press, New York, 1989.

22. Burr D. B., Martin R. B., Schaffler M. B., and Radin E. L., Bone remodelling in reponse to in vivo fatigue microdamage, J Biomech, 1985, Vol. 18, 189-200.

LOAD TRANSFER BETWEEN ELASTIC HIP IMPLANT
AND VISCOELASTIC BONE

S. Piszczatowski[1], K. Skalski[2], W.Święszkowski[3]

1. ABSTRACT

One of the most important elements in the process of designing bone - implant structures (in the case of the alloplastics of joints) is the selection of a stiffness for the implant stem which is appropriate to the surrounding bone tissue. This is critical under conditions of load transfer between implant and bone.

The present work attempts to describe the influence of rheological changes in the mechanical properties of bone tissue when the implanted joint is under working conditions, with the assumption of purely elastic properties of the implant. Shown are the results of numerical analysis carried out on the finite elements method of the bone - implant system, taking into account the viscoelasticity and non-homogeneity of bone tissue.

2. INTRODUCTION

During the last decade one can observe much research activity in the field of stress-strain analysis for the need of implant design. Such works has concentrated on the three following problems:
1) stress transfer mechanism from implant to bone,
2) bone functional adaptation,
3) design and selection of the different shape - stiffness implants.

Keywords: Viscoelasticity, Hip endoprosthesis, Stress , Strain

---

[1] Assistant, Mechanical Department, Bialystok University of Technology, Wiejska 45A, 15-351 Bialystok, POLAND
[2] Professor, Department of Manufacturing Technology, Warsaw University of Technology, Warsaw, POLAND
[3] Postgraduate student, Faculty of Power and Aeronautical Engineering, Warsaw University of Technology, Warsaw, POLAND

According to these problems, the concepts of elastic behaviour of the bone - implant system have been usually presented [1]. This means that from the biomechanical point of view the considered system has not changed its physical - mechanical properties in time. Nevertheless, the viscoelastic properties of the bone tissue are well known [2],[3],[4],[5]. This signifies that the load transfer conditions from the implant to the bone and vice-versa are time dependent during the working of the system.

3. MATERIALS AND METHODS

3.1. Mathematical model of the bone - implant system

A characteristic feature of viscoelastic materials is the memory effect. That is, the material response is not only determined by the current state of stress but is also determined by all past history of deformation. This dependence can be seen in the integral constitutive equations:

$$\sigma_{ij}(t) = \int_{-\infty}^{t} G_{ijkl}(t - \tau)\frac{\partial \varepsilon_{kl}(\tau)}{\partial \tau} d\tau , \qquad (1)$$

$$\varepsilon_{ij}(t) = \int_{-\infty}^{t} J_{ijkl}(t - \tau)\frac{\partial \sigma_{kl}(\tau)}{\partial \tau} d\tau . \qquad (2)$$

In the integral forms of the constitutive equations (1,2) t, τ denote current and variable time respectively, the functions $J_{ijkl}(t)$ and $G_{ijkl}(t)$ represent material properties of the material and are termed the creep and the relaxation functions respectively. For isotropic cases fourth order $G_{ijkl}(t)$ tensor is given by the form:

$$G_{ijkl}(t) = \frac{1}{3}\left[G^{II}(t) - G^{I}(t)\right]\delta_{ij}\delta_{kl} + \frac{1}{2}\left[G^{I}(t)\right]\left(\delta_{ik}\delta_{jl} + \delta_{il}\delta_{jk}\right) \qquad (3)$$

where: $G^{I}(t)$ and $G^{II}(t)$ are independent relaxation functions and δ_{ij} is the Kronecker symbol. The $G^{I}(t)$ is the relaxation function appropriate to states of shear, while $G^{II}(t)$ is defined relative to a state of dilatation. The creep function tensor J_{ijkl} can be presented in a similar form.

As a practical approach, the different physical models are applied to describe the viscoelastic material properties. One of them used in the work is the generalized Maxwell model. The relaxation functions G^{I} and G^{II} can be presented as the expotantional function:

$$G^{I,II}(t) = \sum_{i=1}^{N} G_{i}^{I,II} e^{(-t/\lambda_{i})} + G^{I,II}(\infty) \qquad (4)$$

where: λ_{i} denotes the relaxation time, $G^{I,II}(\infty)$ is the value of the relaxation functions for the infinite time.

In the case of the bone-implant system, where bone (V_{I}) and implant (V_{II}) have viscoelastic and elastic properties respectively, the constitutive equation may be written following the Stieltjes convolution:

$$\sigma_{ij}(t,x) = \left(\tilde{G}_{ijkl} * d\varepsilon_{kl}\right)(t,x), \quad \text{where: } i,j = 1 \dots N, \qquad (5)$$

where:

$$\tilde{G}_{ijkl} = \begin{cases} 0, t \le 0 \\ D_{ijkl}(x), \ t > 0, \ x \in V_{II} \\ \tilde{G}_{ijkl}(t,x), t > 0, \ x \in V_I \end{cases} \tag{6}$$

Procedures for the formulation of the quasi-static boundary value problem, the variational formulation and finite element approach for the problem in question is given in paper [7].

3.2. Geometrical model of implant - femur system

Finite element methods has been used in the analysis of femur-implant systems. A three-dimensional geometric model of the femur has been obtained based on The Standardised Femur [8]. The stem shape was taken from OSTEONICS Omnifit implant (size No.8). FEM model has been build up in ANSYS 5.3 (Fig.1). The 20 nodal points isoparametric VISCO89 elements have been used for modelling the structure. The bone has been modelled by 4328 elements and the implant by 3337 elements, the total number of elements being 7665. The total number of nodes in the analysed model is 26223.

a)   *b)*   *c)*

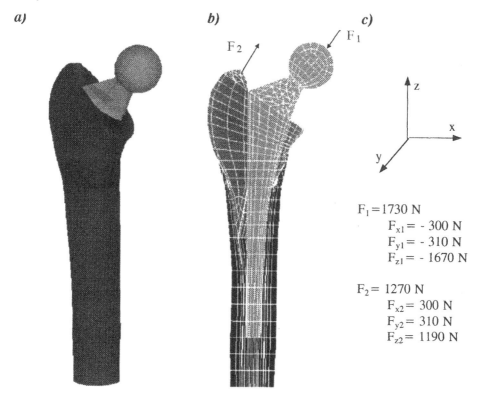

$F_1 = 1730$ N
$F_{x1} = -300$ N
$F_{y1} = -310$ N
$F_{z1} = -1670$ N

$F_2 = 1270$ N
$F_{x2} = 300$ N
$F_{y2} = 310$ N
$F_{z2} = 1190$ N

Fig. 1. Geometrical model of the femur - hip endoprosthesis system:
    a) a view of 3D geometrical model,
    b) cross-section of the meshed FEM model,
    c) load conditions.

3.2. Material properties

The stem of the implant taken into analysis was made of Ti6Al4V alloy. Assumed was that its mechanical properties don't change in time and can be expressed by Young modulus (1.1×10^{11} Pa) and Poisson's ratio (0,3).

The bone tissue has been divided into a cortical and a trabecular part. Both of these have been modelled as a viscoelastic material. For the sake of simplicity, the linear viscoelastic, isotropic model of bone tissue was accepted. With reference to cortical and trabecular bones the same relaxation functions were applied to the shear G^I and bulk G^{II} modulus.

The rheological properties of cortical bone have been modelled on the basis of results obtained by Lakes at.el. [3]. Initial values of shear and bulk modulus have been taken respectively as 7,4 GPa and 22,2 GPa.

Time t [s]	10	10^2	10^3	10^4	10^5
$G^{I,II}(t)/G^{I,II}(0)$	0.973	0.950	0.915	0.853	0.773

Tab.1. Parameters of the rheological model of the cortical bone

For trabecular bone initial values of shear and bulk modulus have been taken respectively as 0,36 GPa and 1,67 GPa. Rheological properties of that type of bone tissue have been modelled on the base of results presented by Deligianni at.el. [2].

Time t [s]	1	2	3	4	5	6	7	10	30	100
$G^{I,II}(t)/G^{I,II}(0)$	0.930	0.900	0.888	0.873	0.865	0.857	0.852	0.834	0.740	0.700

Tab.2. Parameters of the rheological model of the trabecular bone

3.3. Boundary condition

The three-dimensional forces were applied quasi-statically to the implant head (F_1) and the greater trochanter (F_2) (Fig.1). The joint reaction force 1730 N directed through the centre of the implant head and the resultant force imposed by the abductor muscles 1270 N applied on the surface of the greater trochanter simulate a single-legged stance, based the work of Keyak and Skinner [9]. The model of the femur was constrained under the distal end of the stem of the implant. The stem and the femur were rigidly bounded at the proximal part in the region of the stem covering by a hydroxyapatyt layer.

4. RESULTS

The model was analysed under a constant load during a period of 10000 sec. For the sake of significant differences in relaxation time between the trabecular and cortical bones, results will be presented in two cases, after $t_1 = 100$ sec. and $t_2 = 10000$ sec. time of loading. Viscoelastic changes of strain, stress and strain energy density will be presented by a relative quotient of appropriate values at time $t_1 = 100$ sec. to their values at time $t_0 = 0$ sec. (i.e. elasticity state.

The most significant changes in strain and stress were observed during times up to 100 sec. These resulted from the rheological properties of the trabecular bone, which are most evident in this period. Changes in the fields of strain, stress and strain energy density are shown in Fig.2. Visible is an increase in equivalent strain in the bone tissue. In the region of the greater trochanter this reaches more than 10%. The biggest increase in strain appears, however, in the proximo-medial part of the bone, where the value of change reaches 25%. Along with the increase in strain is visible an increase in strain energy density (Fig.2c) and the region of its greatest change matches the area of changes in equivalent strain. It is important to note that level of observed changes in strain is lower than would result from modelled rheological properties of bone tissue under conditions of an unchanging load on the system for the period in question. The explanation for this effect is a decline in the value of stress in the bone, this being the results of the redistribution of stress in the structure (Fig.2b).

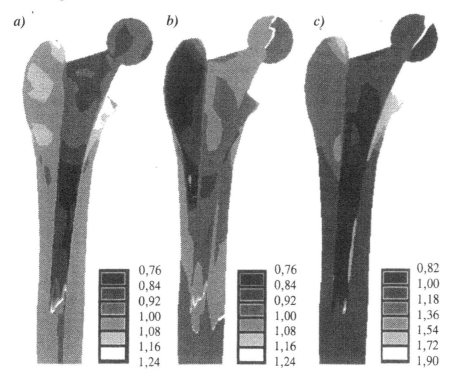

Fig. 2. Rheological changes of strain, stress and strain energy distributions after $t1 = 100$ s of loading;

a) changes of equivalent strain, express by relative quotient $\varepsilon_{eqv}^{t1=100} / \varepsilon_{eqv}^{t0=1}$,

b) changes of equivalent stress, express by relative quotient $\sigma_{eqv}^{t1=100} / \sigma_{eqv}^{t0=1}$,

c) changes of strain energy density (SEND), express by relative quotient $SEND_{eqv}^{t1=100} / SEND_{eqv}^{t0=1}$.

a)

b)

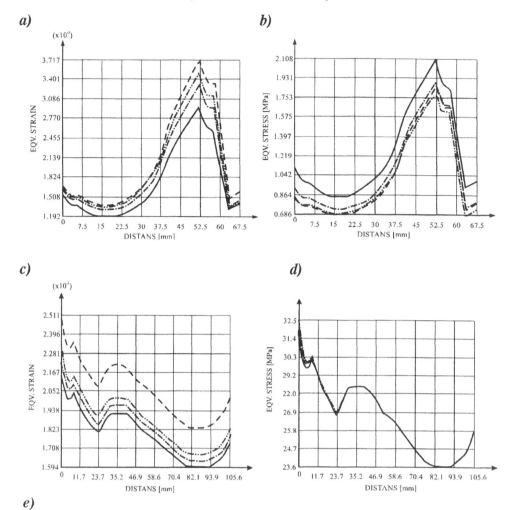

c)

d)

e)

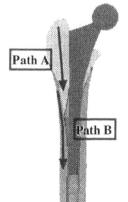

Fig. 3. The stress and strain curves along path in bone tissue at the following moments of time:

——————— $t_0 = 1$ sec.,

—·—·—·—·— $t_1 = 16$ sec.,

—··—··—··— $t_2 = 100$ sec.,

— — — — — $t_3 = 10000$ sec.

a) equivalent strain along path A,
b) equivalent stress along path A,
c) equivalent strain along path B,
d) equivalent stress along path B,
e) the scheme showing location of paths A and B.

Detailed analysis of the correlations between stress and strain in the trabecular bone is possible on the basis of Figs. 3a,b. Shown are graphs of equivalent stress and strain along the length of selected sections through the greater trochanter. Visible is a systematic fall in the value of stress along with a simultaneous increase in strain. Also evident is that the highest rate of change occurs in the field at the beginning of the period of activity and this thereafter systematically declines.

Other types of changes in strain and stress are shown in Fig.3c and d. The graphs show a section through the cortical bon (Path B in Fig.3e) The basis differences here are the domination of changes in strain occurring in the period from 100 to 10,000 sec., as well as the lack of changes in stress. In this case of the cortical bone, rheological effects over a longer period of time are shown. The lack of changes in stress in this part of the bone confirms, however, the thesis earlier formulated that this type of effect under conditions of constant load on the outer system can be the result of the redisribution of stress in the elastic implant - viscoelastic bone system. In the model presented sole assumption was a connection of the prosthesis stem with the proximal part of the implant, leaving the distal end free. In this situation, in section Path B", there is no interaction between the bone and the implant, which is expressed as a lack of change in stress in this region

5. CONCLUSIONS AND DISCUSSION

Bone - implant system analysis taking viscoelastc features of the bone into account indicates that stress and strain distributions change in time. Generally, deformation of the bone increases in time being under constant loading.

Stiffness changes of the bone leading to changes of load transfer conditions and stress redistribution in the implant - bone system can be observed. Finally the deepening of stress shielding effect can appear.

Many simplification have been made in presented model. In particular, taking into consideration anisotropic properties of the bone tissue can be important. More realistic, changing in time, load conditions may allow to observation of some new effects.

Acknowledgements:

Grant of KBN (State Committee of Scientific Research) No.7T07A 008 10 is gratefully acknowledgement.
The Standardized Femur" is the computer geometry of a femoral bone analogue developed by Laboratory for Biomaterials Technology of Istituti Ortopedici Rizzoli.

REFERENCES:

1. Huiskes, R., The various stress patterns of press-fit, ingrown, and cemented femoral stems, Clin.Orthop., 1990, No.261, 27-38.
2. Deligianni, D. D., Maris, A., Missirlis, Y. F., Stress relaxation behaviour of trabecular bone specimens, J.Biomech., 1994, Vol.27, 1469-1476.
3. Lakes, R. S., Katz, J. L., Sternstein, S. S., Viscoelastic properties of wet cortical bone - I - torsional and biaxial studies, J.Biomech., 1979, Vol.12, 657-678.

4. Knets, I., Viscoelastic properties of compact bone tissue, Lecture at the CISM Course on bone mechanics, Udine 1987.
5. Sasaki, N., Nakayama, Y., Yoshikawa, M., Enyo, A., Stress relaxation function of bone and bone collagen, J.Biomech., 1993, Vol.26, 1369-1376.
6. Christensen, R. M., Theory of viscoelasticity, An introduction, 1971, Academic Press, New York,
7. Piszczatowski, S., Skalski, K., S³ugocki, G., Wakulicz, A., Finite element formulation for the interactions between various elastic - viscoelastic structure in biomechanical model, Third International Symposium on Computer Method in Biomechanics & Biomedical Engineering, Barcelona 1997 (this proceding).
8. Viceconti, M., Casali, M., Massari, B., Cristofolini, L., Bassini, S., Toni, A., The Standardised Femur Program" proposal for a reference geometry to be used for the creation of finite element models of the femur, J.Biomech., 1996, Vol.29, 1241.
9. Keyak, J. H., Skinner, H. B., Three - dimensional finite element modelling of bone: effects of element size, J.Biomed.Eng., 1992, Vol.14, 483-489.

EVALUATION OF ACETABULAR WEAR IN HIP JOINT PROSTHESES

R. Pietrabissa [1], M. Raimondi [2], V. Quaglini [3] and R. Contro [4]

1. ABSTRACT

Polymeric wear debris generated from total hip arthroplasties (THA) has been suggested as a major cause of osteolysis, which may lead to loosening of implants. As the polyethylene wear depends on several parameters a parametric mathematical model has been developed to preview wear volume and localisation. In this paper we focus on an experimental work that has been carried out to measure, on retrieved cups, the wear volume. Experimental results will help in setting the mathematical model.

Several ultra high molecular weight polyethylene (UHMWPE) cups coupled to the relevant femoral heads were obtained from hip revisions. The surface roughness was measured on each femoral head. Two linear voltage displacement transducers (LVDT) were mounted on a mobile support to allow planar geometric measurements inside the cups while 10 degree step rotations of the cup allowed to gain 36 equi-spaced transpolar tracks of the worn surface. The measurements enabled 3D reconstruction of the actual cup surface and the evaluation of both shape and dimensions of the worn polyethylene by comparison to the original spherical cup surface. The wear maps were compared to those generated by the mathematical model. The comparison shows similar patterns, with the maximum located near the cup superior borderline.

2. INTRODUCTION

The research dealing with the THA wear behaviour has acquired a renewed importance over the last few years. The materials commonly used to manufacture the prosthetic

Keywords: Hip arthroplasty, Acetabular cup, Polyethylene wear, Mathematical model

[1] Assistant Professor, Biomaterial Technology/ Bioeng. Dept., Politecnico di Milano, Milano, Italy
[2] Ph. D. candidate/ Bioengineering Dept., Politecnico di Milano, Milano, Italy
[3] Ph. D. candidate/ Structural Engineering Dept., Politecnico di Milano, Milano, Italy
[4] Professor, Structural Mechanics/ Structural Eng. Dept., Politecnico di Milano, Milano, Italy

wear coupling are currently metal or ceramic for the femoral head and UHMWPE for the acetabular cup. The wear of the UHMWPE articulating surface with consequent adverse tissue response to polymeric particulate debris has been identified as a significant cause of long term aseptic loosening [1], leading to failure of the implant. Increased durability of hip replacements is therefore related to a better understanding of the factors affecting the wear process and consequently to efforts to reduce the artificial joint wear.

Clinical wear studies are conducted on follow up radiographs, through the assessment of the femoral head penetration rate into the acetabular cup. Due to the low resolution of this method and to the presence of several uncontrolled variables affecting the phenomenon (e.g. third body damage, femoral head surface finish) clinical wear rate data are difficult to use in the evaluation of the factors affecting the wear process. The necessity of separating those variables has given the rationale to the development of hip simulators, in which laboratory wear tests are performed on a controlled system, mechanically similar to the prosthetic joint [2].

A mathematical approach may also be given to the problem, to evaluate the dependence of UHMWPE acetabular wear rate as a function of characteristic parameters of the patient and the hip prosthesis. The parametric model developed by the authors [3] couples contact stresses and sliding distances in whole-gait-cycles wear estimations. The parameters that are left free to vary independently and in given ranges are: patient activity and body weight, femoral head diameter and roughness, cup inclination angle. The results provide information about the influence of each parameter on the volume of wear debris produced over whole-gait-cycles. Moreover, iso-wear maps are generated at selected parameter combinations.

In this paper we describe an experimental work that has been carried out to measure, on retrieved cups, the wear volume. Experimental results will help in setting the mathematical model. We have, therefore, related iso-wear maps and wear volume estimations generated by the mathematical model to equivalent iso-wear maps and wear volume derived from 3D reconstruction after measurements on retrieved cups.

3. MATERIALS AND METHODS

3.1 The measurements

The experimental apparatus includes:
- a supporting frame on which the carriage that bears the transducers can move;
- a self-centring vise that can rotate around its axis for 360° and on which the acetabular cup is mounted;
- a carriage on which the slide bearing the "depth transducer" can move along the direction perpendicular to the axis of the vise: a "position transducer" measures the transversal position of the slide;
- two inductive displacement transducers (LVDT type): a "depth transducer" HBM WTK 10 (±10 mm stroke) for measuring the depth of the internal cavity of the acetabular cup, and a "positioning transducer" HBM WTK 20 (±20 mm stroke) for measuring the transversal position, as to the axis of the cup, of the first transducer; a conditioning station HBM DMCPLUS12 is used jointly with both the transducers.

The measurements refer to an origin O (0,0,0), a pair of axis Y and X which lie on the axis of the cup and on the diameter of the frontal plane, and an angular coordinate φ corresponding to meridian planes of the cup:

- the frontal plane of the component is identified by the upper surface of the flat rim that surrounds the internal cavity of the cup;
- the (0,0) origin is set as the centre of the cup, identified by the centre of a prosthetic head when it is aligned with the cup (through an appropriate aligning fixture) and inserted into the cup itself: by subtracting the distance between the (known) centre of the head and a marker on the head fixture from the distance between the frontal surface of the rim and the marker itself, we obtain the depth of the centre of the cup as to the surface of the rim. The datum plane Y=0 from which the cup depth is measured is set as the plane parallel to the frontal plane of the component through the centre of the cup;
- the axis of the self-centring vise, i.e. the axis of the cup, is referred to as the origin of the X coordinate (transversal position, along the diameter of the cup, of the depth transducer), measured by the position transducer;
- the origin of the angular coordinate φ is set in correspondence to the upper edge of the cup: from $\varphi = 0$ the positive versus for rotation of the cup is towards its posterior edge in the sagittal plane.

The acetabular cup is fixed in the vise so that its frontal plane is perpendicular to the axis of the vise itself, thus resulting parallel to the direction which the slide bearing the depth transducer moves along. The perpendicularity of the upper frontal plane as to the axis of the vise is controlled by measuring on the external rim, using the depth transducer, the maximum axial displacement in correspondence to a 360° rotation of the cup. The error on perpendicularity is acceptable when the maximum noticed displacement is less than 0,05 mm.

Acquisitions are taken by moving the depth transducer from the external rim up to the centre of the cup, i.e. in the radial direction of the component, and measuring both the depth of the inner surface of the cup (Y coordinate) and the radial position, i.e. the distance from the axis of the component (X coordinate): the (X,Y) couples of points depict the track of half of the contour of the internal cavity on a meridian plane. The cup is then rotated of +10° and subsequent measurements are taken. The entire procedure is repeated 36 times, furnishing 18 tracks of the cavity contour in meridian planes at 10° step.

The depth transducer has a spherical probe head (4,8 mm diameter) which is pressed against the acetabular inner surface by a spring of 0,1 N/mm elastic constant. To avoid the probe head detachment when a discontinuity due to wear or a high slope in the contour is encountered, a damper was mounted on the stylus of the transducer.

The bench and its parts were machined with the purpose of minimising errors in measurements. Maximum allowed tolerances were:

- error in parallelism between the axis of the vise and the axis of the bench in both horizontal and vertical planes: 0,02 mm along 300 mm
- error in orthogonality between the transversal slide and the axis of the vise: 0,02 mm along 300 mm

The maximum stated error of the displacement transducers is 0,04% full scale. Calibrations in the ranges of use were made and maximum noticed errors were:

- for the depth transducer: 0,056 mm
- for the position transducer: 0,080 mm.

3.2 The data processing

The measurements enabled 3D reconstruction of the actual cup surface and the evaluation of both shape and dimensions of the worn polyethylene by comparison to the original spherical cup surface. To this purpose the coordinate data were imported as a cloud of points (Fig. 1a) in a commercial solid modelling code. The 3D point interpolation was performed (Fig. 1b).

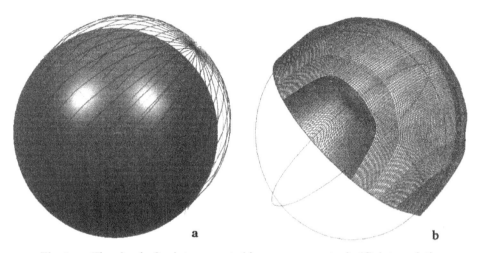

a   b

Fig. 1   a. The cloud of points generated by measurements; b. 3D interpolation

A solid model of the worn volume was obtained (Fig. 2). The surface-cloud difference outputs allowed to obtain the penetration profile point by point (Fig. 3). Volume changes and penetration profiles were obtained for each cup in the assumption that creep is negligible in the total penetration process.

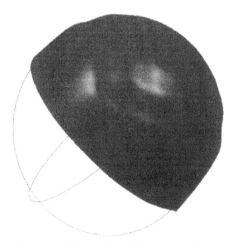

Fig. 2   Solid model of the worn volume

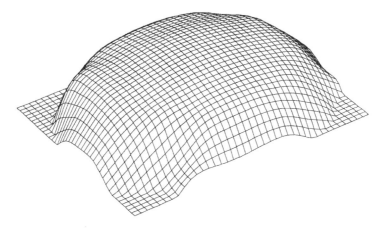

Fig. 3 Penetration profile of the retrieved cup

The parametric model was run, for each cup, selecting the appropriate parameter combination, in order to simulate the *in situ* mechanics of the prosthetic cup. The parameters used are:
- patient activity and body weight
- femoral head diameter and roughness
- cup inclination angle

4. RESULTS

The detailed protocol used to collect the cups and heads has been accepted and adopted by two orthopaedic centres which are involved in the research. Up to now five cups have been accepted for the measurements after inclusion in the protocol. The sampling of those is completed and the surface roughness has been measured on each relevant femoral head.

The wear volumes assessed after experimental measurements are shown for each cup in Table 1, together with the simulated comparison data.

Tab. 1 Wear rates assessed on retrieved cups and calculated through model simulations

sample	material	Ra [μm]	diameter [mm]	time *in situ* [y]	patient weight [kg]	measured wear volume [mm³/y]	[mm³]	calculated wear volume [mm³/y]	[mm³]
1	Co-Cr	0.051	32	10.00	60	202	2020	15.49	154.90
2	alumina	0.022	32	7.92	54	15.02	119	0.84	6.62
3	Co-Cr	0.107	28	1.33	65	84.21	112	5.65	7.53
4	alumina	0.026	32	8.25	63	25.09	207	0.97	8.04
5	alumina	0.022	32	1.67	63	77.84	130	0.96	1.62

Fig. 4 shows the iso-wear map assessed on a cup retrieved after 10 years *in situ*. Fig. 5 qualitatively shows the a whole-gait cycle wear pattern simulated for a comparable patient.

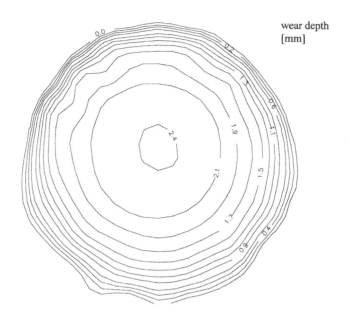

Fig. 4 Iso-wear map of a cup retrieved after 10 years *in situ*.

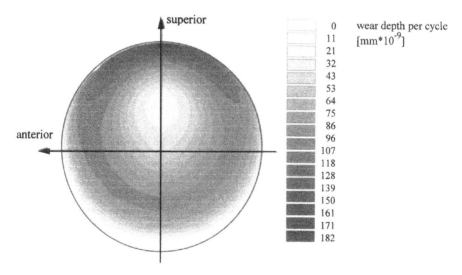

Fig. 5 Simulated whole-gait cycle wear pattern.

5. DISCUSSION

Our preliminary results suggest the following main observations.

The measured and calculated wear volume values reported in Table 1 appear in a very low degree of relationship though assessed for comparable *in situ* conditions. At present our first cycle wear outputs are multiplied into predictors of wear per year. This procedure is manifestly not realistic at the time being.

The wear map of Fig. 5 shows a marginal wear pattern with the maximum located near the cup superior borderline whereas the iso-wear map of Fig. 4, obtained after experimental measurements, shows a paracentral wear pattern with the maximum located superior to the cup polar point. As a matter of fact the simulated iso-wear map relates only to first gait cycles. As wear occurs the pressure distribution will be altered for subsequent cycles and eventually the situation will lead to a "worn in" pressure distribution that reflects the deviation from the author's assumed spherical geometry. The comparison between the measured and simulated wear patterns suggests that wear of UHMWPE acetabular cups is located on the bearing surfaces marginal portion, which initially corresponds to the actual marginal cup surface but continuously modifies as the femoral head advances into the polymer.

The mathematical model is therefore giving a qualitatively good description of the initial wear process but a substantial modification is needed to computationally include in the model the deviations from the assumed spherical geometry that affect the wear process when observed in the long term.

With respect to the experimental apparatus set up for the geometrical measurements on retrieved cups, our preliminary results indicate the feasibility of the proposed experimental method. One of the main difficulties of the procedure is the geometrical referencing of the acetabular cup with the measuring coordinate system. Some tests of repeatability show that the results are very sensitive to mispositioning. The authors are aware that the assessment of acetabular cup wear is generally very challenging in case of slight wear and the results may require comparison to other values obtained with different methods (e.g. Talyrond technique). If this is the case, measurement results should be referred to as an estimate of the worn volume order of magnitude [4]. Furthermore, when assessing the wear of explanted cups, the exact original geometry of the cup at time of manufacturing and implantation is unknown. The presented non thickness-based measuring method should therefore be favoured since it is less sensitive to manufacturer tolerances than the thickness-based methods (such as ultrasound) [5].

The research is intended to reach a significant number of cup samples considering the different parameters which characterise the wear process and have been included in the mathematical model.

6. AKNOWLEDGEMENTS

The authors are grateful to Mr. Gianni Bresciani (Integrazione per l'Industria S.r.l., Agrate Brianza, Milano) for his help in the experimental data processing with the solid modelling code CATIA (Dassault Systèmes, Paris).

7. REFERENCES

1. Harris W.H., The problem is osteolysis, Clin. Orthop. 1995, Vol. 311, 46-53.
2. Wroblewski B.M., Siney P.D., Dowson D. and Collins S.N., Prospective clinical and joint simulator studies of a new total hip arthroplasty using alumina ceramic heads and cross-linked polyethylene cups. J. Bone Joint Surg. [Br] 1996, Vol. 78-B N.2, 280-5.
3. Pietrabissa R., Di Martino E., Brugnolotti G., Chichi M. and Grappiolo G., Wear of acetabular component in total hip arthroplasty: a parametric mathematical model. In: Proc. ASME Int. Mech. Eng. Congr. 1996: Advances in bioengineering. Edited by S. Rastegar, Vol. 33, 361-2.
4. Derbyshire B., Hardaker C.S., Fisher J., Brummit K., Assessment of the change in volume of acetabular cups using a coordinate measuring machine, Proc. Instn. Mech. Engrs. Congr. 1994 Part H: Journal of Engineering in Medicine Vol. 208, 151-8.
5. Feikes J., Wimmer M.A., Berzins A., Schneider E., Galante J.O., Analysis of wear of retrieved acetabular components with a non-thickness-based method. Proc. 5th World Biomaterials Congr. 1996. Univ. of Toronto Press: Vol. II, 549.

FINITE ELEMENT ANALYSIS AND FATIGUE TEST PREDICTION APPLIED TO THE STANDARD FATIGUE TESTING OF HIP STEMS

H.L. Ploeg[1], H.W. Wevers[2], U.P. Wyss[3], M. Bürgi[4]

1. ABSTRACT

Fatigue test prediction tools can improve the development process of orthopaedic components. The objective of this study was to evaluate fatigue test prediction methods based on finite element analysis for the standard fatigue testing of hip stems. In order to simplify this initial study the standard hip stem fatigue test without torsion and a titanium alloy hip stem with a plain geometry were chosen for the evaluation. Two fatigue test prediction methods were investigated: total life and crack initiation life. Finite element analysis predicted a maximum local stress under static loading to within 3% of strain gauge results. However, the location of crack initiation during fatigue testing was further below the cement surface than predicted by the finite element analysis. The results of the fatigue test predictions demonstrated a high sensitivity to the material fatigue data and mean stress correction methods. Component fatigue strength was predicted to within two standard deviations of the mean fatigue test result. Life predictions ranged from a non-conservative factor of two to a conservative factor of 10. Successful component fatigue test prediction requires material fatigue data that is accurate to the component's conditions, including: environment, crystalline structure, alloy purity, and surface finish.

2. INTRODUCTION

The current success of total hip replacements is presenting new challenges to

Keywords: hip prosthesis, fatigue test prediction, finite element analysis

[1]PhD Student, Department of Mechanical Engineering, Queen's University, Kingston, ON, Canada and Project Manager, Biomechanics, Sulzer Orthopedics Ltd., Winterthur, Switzerland.
[2]Professor, Department of Mechanical Engineering, Queen's University, Kingston, ON, Canada.
[3]Professor, Department of Mechanical Engineering, Queen's University, Kingston, ON, Canada and Head of Research, Sulzer Orthopedics Ltd., Winterthur, Switzerland.
[4]Head of Biomechanics, Sulzer Orthopedics Ltd., Winterthur, Switzerland.

manufacturers due to a changing patient population and market. Younger active recipients demand a longer life from these components, while subjecting them to more rigorous loading. At the same time market competition is increasing with an exponential growth of new products. The manufacturer must deliver high quality products that meet endurance requirements and incorporate up-to-date technology and market trends at competitive prices.

2.1   The Standard Fatigue Testing of Total Joint Replacement Components

The development of total joint replacements (TJR's) combines three areas of investigation: clinical, experimental and theoretical. One aspect of this development process is the standard fatigue testing of TJR components. The first component fatigue test to be standardised was in response to the unacceptably high rate of *in vivo* hip stem fractures of the first generation. The standard, ISO 7206/3, specified vertical load application to the head of the hip stem, and fixation of the distal stem in bone cement. The stem was to be embedded to a level 80 mm below the centre of the head at an adduction angle of 10° causing bending in the coronal plane. This standard test was replaced with ISO 7206/4 to include out-of-plane bending and torsion, obtained by leaning the stem 9° anteriorly out of the vertical plane of load application. The tests are performed in a simulated physiological environment consisting of aerated, circulating Ringer's solution, 0.9% mg/L NaCl at 37°C. The stems are loaded with a sinusoidal waveform at a frequency of 6 Hz and a minimum to maximum load ratio, R, of 0.1. Hip stems are required to survive 5 million cycles of peak-to-peak loading of 3 kN, without torsion, and 2 kN, with torsion, according to ISO 7206/7 and /8.

Fatigue testing of TJR components is performed as part of design approval before their release into the clinic. Designs are tested to ensure that the minimum endurance requirements are fulfilled. The cost and time demands of the low frequency load application, and large sample sizes required to accommodate the scatter in component fatigue test results often prohibit determination of the actual fatigue strength of a component. However, component fatigue strength testing enables the development of a higher quality product through design optimisation including geometry, materials, surface finishes, coatings, and manufacturing methods. The process of component fatigue strength testing must be efficient. This study concerns optimising the fatigue testing of TJR components through its integration with finite element analysis (FEA) and fatigue test prediction methods.

2.2   Finite Element Analysis and Fatigue Test Prediction Methods

Fatigue is a process that starts with local yielding, eventually causing crack initiation, crack propagation and finally component fracture. It has been suggested that the accuracy of fatigue life prediction can be improved by modelling the sub-processes [1]. The fatigue life prediction methods examined in the present study were total life and crack initiation life. FEA can provide the relationship between global loading and local stress. The relationship between local stress and life (cycles to failure) is determined by the material's stress, or strain, versus life curve or damage curve. The complex process of fatigue is sensitive to many factors including: load spectrum, environment, alloy purity, crystalline structure, and surface finish. The material's damage curve must be adjusted to more accurately model the actual component and its environment.

The total life method is based on the material's stress life curve (Wöhler curve), which summarises the results of force controlled material fatigue tests. Crack formation in a specimen reduces its load-carrying cross-section, and under constant force, the crack tip stress intensity increases. There is therefore an acceleration of crack growth that results in specimen fracture. Life, on the stress life curve, is defined as total life: crack initiation, propagation and fracture.

The crack initiation life method is based on the material's strain life curve. Since strain can be measured, no inaccuracies are introduced through the assumption of linear elastic material properties as is required for the calculation of the stress life curve. The strain life curve is determined through displacement controlled material fatigue tests. Crack initiation reduces the specimen stress and crack growth does not accelerate making it possible to differentiate between crack initiation and propagation. Life, on the strain life curve, may be defined as crack initiation life or total life.

The classical approach to high cycle fatigue design is to assume a stress life curve asymptote at the material's endurance limit. A conservative mean stress correction calculation, for example the Goodman relation, and other correction factors, for example for surface finish, compensate for uncertainties in the factors affecting fatigue. A too conservative design approach, however, can hinder the development process. It is preferable to design for fatigue with more accuracy. This may be achieved through accurate stress analysis, that is FEA, and improved fatigue life prediction methods.

3. PURPOSE

The objective of this study was to investigate fatigue test prediction methods based on FEA for the standard fatigue testing of TJR components. The term fatigue test is used to include life prediction at a given stress level and fatigue strength prediction at a given life. Two approaches to fatigue test prediction were compared: total life and crack initiation life. The accuracy of the FEA and of the fatigue test prediction methods were evaluated through testing of strain-gauged components and fatigue testing.

4. MATERIALS AND METHODS

Care was taken to minimise the number of parameters for this initial study; therefore, a hip stem with a plain geometry and the standard hip stem fatigue test without torsion, ISO 7206/3 (see 2.1) were chosen for the evaluation of the test prediction methods. In favour of time, fatigue tests were performed to only 2 million cycles.

The Alloclassic® SLA Zweymüller®, from AlloPro Ltd., femoral hip stem, in the three sizes 1, 2 and 4, was the trial component. Sixteen stems per size were available for testing: one for the static test, three for pilot fatigue tests and twelve for the main fatigue tests. The SLA is a titanium alloy, ISO 5833/3: Ti 6Al 4V ELI, hip stem with a rough grit blasted surface and no geometric stress concentrations. The following material properties were assumed: an endurance limit of 500 MPa, stress amplitude and zero mean stress; elastic modulus of 110 GPa; and an ultimate tensile strength of 900 MPa [2,3,4]. The endurance limit was corrected for mean stress effects to 298 MPa, stress amplitude and a load ratio of 0.1, using the Goodman relation [5].

4.1   Experimental

Tests on strain gauged stems were performed in order to validate the FEA. Ten small strain gauges (HBM 2/120 KY 11) were applied to the hip stems at locations straddling the region susceptible to fracture, i.e. five above and five below the cement surface at 2 mm intervals. The hip stems were then embedded in bone cement according to ISO 7206/3 (see 2.1). Initial testing was performed in air at room temperature under quasi-static loading, that is, roughly 0.1 Hz. All hip stems were slowly loaded from 0 kN to 3 kN for 100 cycles in the first test, then in the second test from 0 kN to loads approaching the estimated mean component fatigue loads of 7 kN, 8 kN and 9 kN for sizes 1,2 and 4 respectively, also for 100 cycles. Further tests were performed in Ringer's solution under quasi-static loading and at 6 Hz to investigate effects due to environment and load rate.

Component fatigue tests were performed to evaluate the fatigue test predictions. The distributions of life at a given stress level and strength at a given life were determine. The fatigue tests were performed at two stress levels, referred to as high and low, at an estimated one standard deviation above and below the estimated mean fatigue strength. A log-Normal probability density function was fitted to the cycles to failure at the high stress level. The probability of failure at 2 million cycles for each stress level was determined therefore defining the mean and standard deviation for the Normal cumulative distribution function of fatigue strength. Fatigue tests were performed according to ISO 7206/3 (see 2.1) at 6 Hz until crack initiation or 2 million cycles. Crack initiation was detected by a decrease, 10-15%, in the vertical stiffness of the hip stem. Penetrant dye was used to investigate stem-cement debonding.

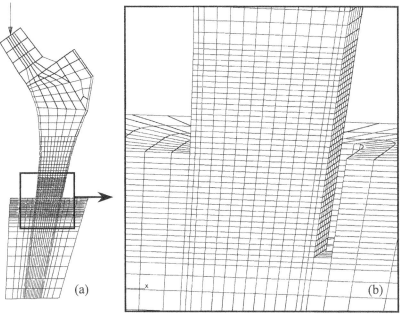

*Fig 1.   Finite Element Model. (a) hip stem size one according to ISO 7206/3,
            (b) stem-cement debonding was modelled by removing the failed elements.*

4.2 Finite Element Analysis

The FEA of stem size 1 (see Fig.1) was performed in two main steps. First a linear elastic analysis was performed to determine the area of stem-cement debonding. The cement elements next to the stem with tensile stresses above 10 MPa (the approximate tensile strength of the stem-cement interface [6]) were removed and the analysis was repeated using a large deflection analysis. A third analysis was performed to include cement plasticity. The area around the cement surface was modelled with a fine mesh of 8 noded brick elements. A vertical load of 7 kN (fatigue load at 50% probability of failure) was applied to the node located at the centre of the head. The stem was modelled as planar symmetric to take advantage of the symmetry in the geometry and load conditions. FEA was performed with the pre- and post-processor MSC/PATRAN® version 6.0 and the solver P3/ADVANCED FEA® version 1.5 from MacNeal Schwendler GmbH.

Table 1   Finite Element Model Summary

Material	Number Of Elements	Elasticity, E (GPa)	Poisson's Ratio
Stem	7,152	110	0.3
Cement	3,696	2.6	0.3
Total	10,848		

4.3 Fatigue Test Prediction

The strain life curve was defined according to the Manson-Coffin equation with mean stress correction according to Morrow (Eq. 1, Fig. 3a) and damage curve, which is independent of stress ratio, according to Smith, Watson, and Topper (Eq. 2, Fig. 3b) [1,2,5,7]. For this initial study, the stress life curve was calculated from the strain life curve (with zero mean stress) assuming linear elasticity (Fig. 4a) and corrected for mean stress using the Goodman relation (Fig. 4b). The Goodman and Smith Watson Topper relations are compared in Fig. 4b. The point of intersection with the line defining a load ratio of 0.1 is the endurance limit corrected for mean stress.

$$\varepsilon_a = \frac{\sigma_f - \sigma_m}{E}(2N)^b + \varepsilon_f (2N)^c \qquad (1)$$

$$P_{SWT} = \sqrt{(\sigma_a + \sigma_m) \cdot \varepsilon_a \cdot E} \qquad^1 \qquad (2)$$

5. RESULTS

The non-linear FEA including large deflection analysis, cement plasticity and debonding of the cement-stem interface found the maximum principal stress of 619 MPa in the size 1 hip stem 2.5 mm below the cement surface. The FEA slightly, 3%, overestimated the

---

[1]   b: fatigue strength exponent, c: fatigue ductility exponent, E: Young's Modulus, ε_a: strain amplitude, ε_f: fatigue ductility coefficient, N: cycles, P $_{SWT}$: Smith Watson Topper damage parameter, σ_a: stress amplitude, σ_f: fatigue strength coefficient, σ_m: mean stress

maximum stem stress, relative to strain gauge results of a stem under static loading in air. There was no difference between the stem strains measured in air and Ringer's solution; and, no significant difference between the stem strains measured under static and fatigue, at 6 Hz, loading. The location of the maximum principal stress agreed well with the strain gauge results at 2 to 3 mm below the cement surface. During fatigue testing, however, the stems failed a mean distance of 5.9 mm with a 2.3 mm standard deviation below the cement surface. FEA found a stem-cement debonding depth of 9 mm in comparison to a mean fatigue test result of 21 mm with a 11 mm standard deviation.

The component mean fatigue loads at 2 million cycles were: 7.0 kN for size 1, 8.4 kN for size 2 and 11.4 kN for size 4. Combining the results for all stem sizes, the probability of failure at the high stress level was 92% and 22% at the low stress level. The component fatigue tests found a wide distribution in fatigue life, with a dispersion (standard deviation/mean) of 97% at the high stress level, for all sizes combined (Fig. 2a). In contrast, the dispersion in component fatigue strength at 2 million cycles was small, 7% for size 1 (Fig. 2b). Fatigue life for hip stem size 1 at its mean fatigue strength of 7 kN, was overestimated by a factor of 2 with the Coffin-Manson equation (Fig. 3a), underestimated by a factor of 3 with the Smith Watson Topper damage curve (Fig. 3b), and underestimated by a factor of 10 with the stress life curve corrected for

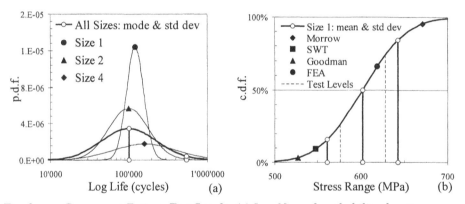

Fig. 2   *Component Fatigue Test Results (a) Log-Normal probability density functions (p.d.f.) of life at the high stress level for all sizes; (b) Normal cumulative distribution function (c.d.f.) of fatigue strength at 2 million cycles for hip stem size 1.*

Table 2   *Fatigue Test Predictions for Hip Stem, Size 1: Coffin-Manson equation with Morrow mean stress correction (ε-N), Smith Watson Topper Damage Curve (SWT) and stress life with Goodman mean stress correction (σ-N)*

Method	Life at Mean Fatigue Load, 7 kN (FEA: 619 MPa)		Mean Fatigue Strength, $\Delta\sigma$ at 2 Million cycles	
	Load	Million cycles	Load	$\Delta\sigma$ (MPa)
ε-N	$\Delta\varepsilon = 0.53\%$	4.2	$\Delta\varepsilon = 0.57\%$	671
SWT	$P_{SWT} = 461$ MPa	0.66	$P_{SWT} = 409$ MPa	548
σ-N	$\Delta\sigma = 619$ MPa	0.20	$\Delta\sigma = 527$ MPa	527
Testing	2 million cycles		602 MPa (± 41 MPa)	

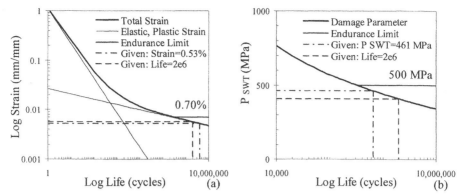

Fig. 3   *Fatigue Test Prediction Results: (a) Manson-Coffin equation with Morrow mean stress correction, Eq. 1: σ_f = 2069 MPa, ε_f = 0.8862, b = -0.1069, c = -.6982 [2]. Fatigue strength was overestimated by 11%. (b) Smith Watson Topper Damage Curve, Eq. 2. Fatigue strength was underestimated by 9%.*

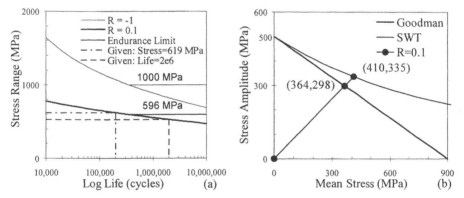

Fig. 4   *(a) Stress Life curve for no mean stress (R=-1) and with a Goodman mean stress correction for, R=0.1 (R: stress ratio). Fatigue strength was underestimated by 12%. (b) Goodman and Smith Watson Topper (SWT) relations for mean stress correction. Goodman is more conservative.*

mean stress with the Goodman relation (Fig. 4a, Table 2).   The fatigue strength predictions at 2 million cycles of life were all within two standard deviations of the mean fatigue test result (Fig. 2b). Scaling the FEA result to the endurance limit found a mean fatigue load for hip stem size 1 of 6.7 kN, 4% less than the mean of the experimental result, within one standard deviation of the mean.

6.   DISCUSSION AND CONCLUSIONS

A non-linear FEA including large deflection analysis, cement plasticity and simple debonding of the cement-stem interface accurately predicted the magnitude and location of maximum stem stress, relative to strain gauge results of a stem under static loading in air. Fatigue test conditions, however, were not modelled accurately. Stem strains were not significantly affected by testing medium, temperature or loading rate. However, the actual length of stem-cement debonding and location of crack initiation in the stems were twice as far below the cement surface than the FEA results. The duration of the fatigue tests allowed for cement creep and growth of the stem-cement debonding region. These processes were not modelled in the FEA. Further investigations to improve the FEA will be performed.

Component fatigue tests found a relatively narrow distribution in fatigue strength at 2 million cycles as opposed to the large dispersion found in fatigue life, typical for high cycle fatigue. Stress levels were below the material's yield strength, 800 MPa, justifying assumptions in linear elasticity. Since titanium alloys are notch sensitive, total fatigue life is dominated by crack initiation and the lack of crack propagation life in the model was justified.

The FEA accurately predicted local stress from global loading. The fatigue test predictions based on the FEA and published material fatigue data, however, could predict neither fatigue strength nor life to within one standard deviation of the mean test result. In high cycle fatigue, the double logarithmic damage curves are sensitive to small changes in stress level and mean stress correction. Life predictions to within a factor of 2-3 are considered acceptable. The strain life method was non-conservative with Morrow mean stress correction and conservative with Smith Watson Topper damage parameter. The stress life method with Goodman mean stress correction was the most conservative method. The endurance limit corrected for mean stress was a good predictor of component fatigue strength. Fatigue test predictions may be improved by using damage curves accurate to the component situation, including: load spectrum, alloy purity and crystalline structure, environment, and surface finish.

7. REFERENCES

1. Bannantine, J., Comer, J.J. and Handrock, J.L., **Fundamentals of Metal Fatigue Analysis**, Prentice Hall Inc., Englewood Cliffs, U.S.A., 1990.
2. Bäumel, A. and Seeger, T., **Materials Data for Cyclic Loading, Supplement 1**, Elsevier Science Publishers, Amsterdam, The Netherlands, 1990.
3. Semlitsch, M. and Weber, H., "Titanlegierung für zementlose Hüftendoprothesen", International Symposium die zementlose Hüftprothese, München, Germany, 1991.
4. Semlitsch, M. "Klassische und neue Titanlegierungen zur herstellung Künstlicher Hüftgelenke", International Conference on Titanium Products and Applications, Dayton, U.S.A., 1986.
5. Collins, J.A., **Failure of Materials in Mechanical Design: Analysis, Prediction and Prevention**, Second Edition, John Wiley & Sons, Toronto, 1993.
6. Verdonschot, N., **Biomechanical failure scenarios for cemented total hip replacement**, Ponsen & Looijen bv, Wageningen, 1995.
7. Ellyin, F., **Fatigue Damage, Crack Growth and Life Prediction**, Chapman & Hall, London, 1997.
8. Bury, K.V., **Statistical Models in Applied Science**, John Wiley & Sons, Toronto, 1975.

ACKNOWLEDGEMENTS: The author wishes to thank Robert Steger and Daniel Schmuki for their excellent technical support.

VALIDATION OF A THREE DIMENSIONAL FINITE ELEMENT MODEL OF A FEMUR WITH A CUSTOMIZED HIP IMPLANT.

B.Couteau[1], L.Labey[2], M.C.Hobatho[3], J.Vander Sloten[4], J.Y.Arlaud[5], J.C.Brignola[6]

1. ABSTRACT

The purpose of this study was to validate the three dimensional finite element model of a human femur with a customized hip implant. The experimental test concerned a strain gauge analysis and the numerical models were CT data based. Two numerical models were considered. The first one (theoretical model) was reconstructed by mean of the preoperative CT scan exam. The second one (real model) was issued from the postoperative CT scan exam.

The numerical results were compared with that of the experimental ones and the differences did not exceed 13% for the axial strains, and reached 50% for the transverse ones. This phenomenon was probably due to the anisotropic degree approximation and to the small range of magnitude measurements. Finally it has been demonstrated that this model was reliable to represent the femur with customized hip implant by a three dimensional finite element method.

2. INTRODUCTION

The observation time to criticize an orthopedic technique is generally ten years. But, the numerical tool have already demonstrated that it could bring some technical informations about the implant design parameters (1, 2, 3, 4, 5). In this way, it has been proved that the stem length had to respect a compromise to insure the stability of the implant and to limit the cantilever phenomenon. The ideal prosthesis material could be a composite material with some mechanical properties close to that of the cortical tissue.

Keywords: hip, implant, finite element modelling.

---

[1] MSc, Inserm U305, Centre Hospitalier Hôtel Dieu, 31052 Toulouse Cedex, France.
[2] MSc, Katholieke Universiteit Leuven, Celestijnenlaan 200A, B3001 Heverlee, Belgium.
[3] PhD, Inserm U305, Centre Hospitalier Hôtel Dieu, 31052 Toulouse Cedex, France.
[4] PhD, Katholieke Universiteit Leuven, Celestijnenlaan 200A, B3001 Heverlee, Belgium.
[5] MD, Centre Borely-Mermoz, 114 rue Jean Memoz, 13008 Marseille Cedex 08, France.
[6] MSc, Société Euros, ZI Athélia III, 13600 La Ciotat, France.

Also the collar seemed to privilege the local calcar load transfer.To sum up, the numerical tool is interesting in the orthopaedical development. However, most of the studies used it in a relative context. Few authors (6) studied the absolute aspect of the results and so the quantitative validation of the models.

The objective of our work was to develop a reliable modelling technique in order to predict the mechanical behavior of implanted human femur. In our study, a 3D Finite Element Method of a femur with a customized hip implant (Euros ®) has been developed and validated using mechanical testing of the same configuration

3. MATERIAL AND METHODS

3.1 Experimental test

An experimental compressive test of a human femur from a male subject with a customized hip implant was performed. Eight strain gauges were distributed on the lateral and medial cortical shaft along the prosthesis stem. The two superior gauges per face were 45° rosettes while the two inferior ones were axial gauges. By mean of a canal switch, the strain gauges were connected to the thermal compensatory gauge according to the Wheastone bridge schema.

The mechanical test simulated the static monopodal position by applying a compressive force on the femoral head and a tensile force on the lateral face of the great trochanter. The applied force on the body gravity center represented the body weight decreased of the ground limb weight (15.6% of the total body weight). For a subject weighted 70Kg, the gravity center force was 580N. However for technical reasons we had to limit the amplitude to 400N. This gravity force was obtained by a compressive force (Fc) on the femoral head of 1036N and a tensile force (Tf) of the adductor muscles of 652N (Fig.1).

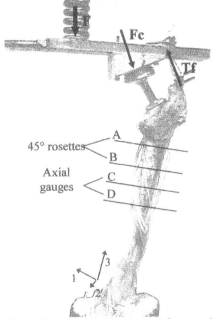

Fig. 1: Setup of the mechanical tests of the implanted femur with the gravity force (F) traduced by the compressive femoral head force (Fc) and the tensile force of the adductor muscles (Tf). The A, B, C, D levels indicated the gauges location along with the medial and lateral faces.

The gauge analysis permitted to obtain the axial strains (ε_3) along with the A, B, C, D levels, and the proximal principal strains (ε_m, ε_M) along with the A and B levels.

3.1 Numerical analysis

The models were CT data based and two models were developed. The theoretical model (Th.Mod.) was defined pre-operatively from the CT-scan exam of the bone and the IGES file format of the prosthesis. The real model (Re.Mod.) was constructed post-operatively from the CT scan exam of the bone with its implant.

Transverse CT images of the femur were performed on the cadaveric specimen with or without the customized hip implant. The CT images data were transferred to a workstation Indy (Silicon Graphics, Mountain View, CA, USA) by an interface software SIP305 developed in our laboratory. Each image was subjected to an edge detection to separate bone contour lines. The output file of the image treatment was a neutral file format containing the internal and external contours of the cortical bone. Then the 3D surfacic model was read via Patran3 Software (MSC Nastran, Los Angeles, CA, USA).

The mesh generation was identical for the two models. It was initiated by the determination of the prosthesis sections normal to the mean line of the implant. Then the implant was subdivided in geometrical volumes which were automatically meshed with 3D hexaedric elements with 8 nodes. The meshing of the prosthesis was followed by the sweeping of the elements in the bone structure. In accordance with the bone/prosthesis interface choice, the contact elements were created or not. Effectively, in order to quantify the influence of its definition two sorts of interface modelling were tested. The first choice planned a fully bonded interface where the exterior implant nodes were common to the adjacent bone ones. The advantages of a such representation lay in the low calculation times. The second type of the interface authorized the separation, the contact or the slipping of the two structures by the mean of specific 3D contact elements. The solver we used provided Inter3D elements which were hexaedric elements with four 2D Gap elements. The disavantages of the contact elements were the time calculation involved by the non linear analysis. This non linearity was imposed by the numerical management of the contact elements.

The mesh optimization was reached by the use of different size elements depending on the model. The average element volume was 70 mm^3 for the theoretical model and it was 150mm^3 for the real one. The theoretical model had twice more elements than the real one (Tab 1).

	3D hexaedric elements	3D contact elements
Theoretical model	7174	528
Real model	3453	224

Tab. 1: Description of the mesh compositions

The mechanical properties of the cortical bone were issued from the literature (7) and they represented an homogeneous orthotropic material. Initially, they represented the mid range of the published datas (Tab.2), but in order to fit the numerical results to the experimental ones, they had to be decreased (Tab.2). However, the anisotropic degree was respected.

	E_1 (GPa)	E_2 (GPa)	E_3 (GPa)	v_{12}	v_{13}	v_{23}	G_{12} (GPa)	G_{13} (GPa)	G_{23} (GPa)
Initially	9.8	10.3	16.7	0.350	0.194	0.188	3.3	4.2	4.5
After	6.0	8.1	11.4	0.172	0.107	0.110	2.3	3.0	3.1

Tab. 2: Mechanical parameters of the cortical bone before and after the model fitting.

The mechanical properties of the spongious bone were CT data based. A previous study elaborated the correlation (Eq.1) between the CT number in Hounsfield units and the cancellous Young modulus in MPa.

$$E = 3.85\ CT + 103 \text{ with } R^2 = 0.6 \qquad\qquad Eq.1$$

So the CT scan images of the bone permitted to obtain a three dimensional database of the spongious bone inhomogeneity. The assignment of the adequate properties to the spongious elements was automatically manage by a software developed in our lab (PM_QCT). To limit the number of spongious properties we define 21 various cancellous materials.

The contact elements of the non-linear interface integrated the static friction coefficient (μ). According to published results, the static friction coefficient between the bone and the titanium was found equal to 0.4. Moreover in the classical mechanic, 0.4 corresponded to a poor lubricated friction between a metallic material and a non-metallic material.

The loads and boundary conditions represented the experimental ones. To maintain the femur in the neutral position the condyle displacements were locked. Then, a compressive force was applied on the femoral head with an amplitude of 1036N and with an angle of 78° with respect to the transverse plane. The traction on the lateral face of the great trochanter simulated the adductor muscles. Its amplitude was 652N and the angle with respect to the diaphyseal axis was 33°.

The analyses were solved by the mean of Abaqus solver. The fully bonded interface implicated the linear formulation of the finite element method. But the contact elements involved constitutive non-linearities where the stiffness matrix became depending on the displacements. The resolution of such a problem used the incremental Newton-Raphson method, and the calculation time increased dramatically.

4. RESULTS

4.1 Axial strains

The experimental axial measurements showed medial compressive strains and lateral tensile strains. The medial strains increased between the A and B level then it decreased to the D gauge. The lateral tension decreased from the A gauge to the D one (Fig. 2).

By superimposing the numerical axial strains with the experimental ones, we noticed that the models had different precisions (Fig. 2).

This was emphasized by plotting the relative error for all gauge levels (Fig.3). Except for the medial level and the B lateral level of the theoretical model, the contact elements notably improved the models precision. In fact, without contact interface, errors often reached 15-20%. On the opposite, with the contact elements, the relative errors were never higher than 10%.

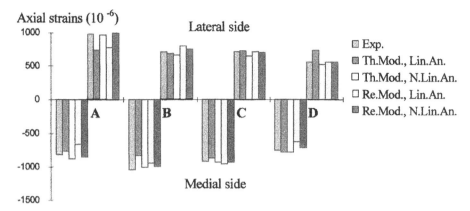

Fig. 2: Experimental axial strains (Exp.) along with the four levels of strain gauges (A, B, C, D) and numerical axial strains (Th.Mod., Re.Mod.) without (Lin.An.) and with the contact elements (N.Lin.An.).

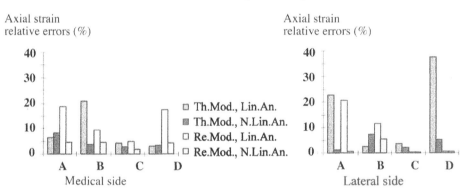

Fig. 3: Numerical axial strain relative errors (Th.Mod., Re.Mod.) along with the four levels of strain gauges (A, B, C, D) without (Lin.An.) and with the contact elements (N.Lin.An.).

4.2 Principal strains

The experimental principal strains were only measured at the A and B levels. However, by the compressive medial strains, they confirmed that the flexion acted in the medial sense (Fig.4).

In fact, the medial compressive strains were preponderant compared to that of the medial tensile ones. On the opposite, the lateral tensile strains plaid a major role.

About the numerical results, the non-linear analyses clearly improved the reliability of the preponderant pincipal strains i.e. the medial minimal and the lateral maximal ones. The transverse principal components were more difficult to obtain precisely (Fig.5).

In fact, the transverse relative error reached 80% for the linear analysis. By using contact elements, this error was decreased to around 30%. As previously, the axial principal strains provided an excellent accuracy with contact elements.

Finally, from a numerical point of view, we could observed that the theoretical model was in good agreement with the real one.

Principal strains (10⁻⁶)

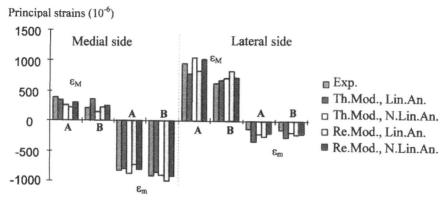

Fig.4: Experimental principal strains (Exp.) along with the two superior levels of gauges (A, B) and numerical axial strains (Th.Mod., Re.Mod.) without (Lin.An.) and with the contact elements (N.Lin.An.).

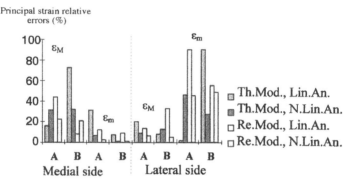

Fig.5: Numerical principal strain errors (Th.Mod., Re.Mod.) along with the two superior levels of strain gauges (A, B) without (Lin.An.) and with the contact elements (N.Lin.An.).

5. DISCUSSION

Our results showed an excellent accuracy in the axial direction . Globally, the axial strain error did not exceed 15% and it decreased to 3% with the contact elements. In the transverse direction, the error was notably increased. The best accuracy with the non-linear analysis was around 30% (Fig.6).

The transverse errors could be classified according to their origin. The first class could be explained from an experimental source. Firstly, it was difficult to obtain a perfectly plane surface intended to the rosettes bonding. Subsequently, the transverse strain measurements could be less accurate. Moreover, the transverse values were particularly low and so they certainly were more sensible to the error rate. Also, the constant anisotropic degree constituted an approximation for the numerical model. The interface modelisation was also another error cause. Finally, the pre-stress state was not simulated in our analysis because of its unknown mechanical effect.

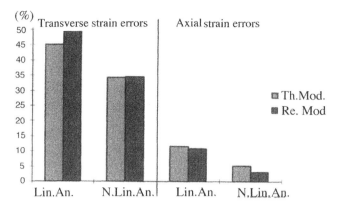

Fig.6: Transverse and axial strain error averaged on the eight gauge locations for the two numerical models with the two sorts of interface.

Initially, the cortical parameters corresponded to the mean range of the published datas. To fit the numerical results with the experimental ones, the mechanical properties had to be decreased of around 30%. According to a study of the literature (8), a difference of 23% was found between the metaphyseal and the diaphyseal region of cortical bone. So we supposed that our decreasing rate was closely linked to the variation of the cortical tissue along with the femoral axis.

The theoretical model appeared in agreement with the real one. We could explain this results by the fact that the theoretical positionning was respected. The prosthesis was only two millimeters deeper than provided by the preoperative model.
Only few studies dealed with the validation of numerical models of the femur with an uncemented (6) or cemented hip implant (9). However, experimental analysis concerning the influence of the uncemented implant (10, 11, 12) were available in the literature. The surgical fit of the prosthesis within the femoral canal seemed influence the bone strains. Like that a precise fit of the prosthesis at the isthmus appeared to produce the more realistic strain distribution.

Considering the load differences, the bone strains with an uncemented prosthesis appeared in coherent ranges of value. As shown in the table 3, more important loads implied higher compressive strains. On the opposite, a model with a cemented implant demonstrated lower compressive strains instead of widely higher applied loads.

Authors	Description	Loads	Axial medial strains (10^{-6})	Axial lateral strains (10^{-6})
(11)	Exp. Customized Impl.	Fc=2000N	$-1800 < \varepsilon_m < -600$	—
(9)	Exp. Cemented Impl.	Fc=2670N Tf=1973N	$-750 < \varepsilon_m < -250$	$50 < \varepsilon_M < 500$
Our study	Num. and Exp.	Fc=1036N Tf=652N	$-920 < \varepsilon_m < -820$	$620 < \varepsilon_M < 960$

Tab.3: Published datas concerning the strain ranges of value. The works were either experimental (Exp.) or numerical (Num.).

Finally, our modelling technique implied a reasonable accuracy of the strain distribution. The highest strain rate errors concerned exclusively low strain values.

Once this technique was validated, clinical cases could be performed in order to predict the mechanical behavior of implanted femur in vivo.

REFERENCES

1. Vichnin N.H., Batterman S.C., Stress analysis and failure prediction in the proximal femur before and after total hip replacement., ASME J. of Biomech. Eng., 1986, Vol.108, 33-41.
2. Rohlmann A., Mossner U., Bergmann G., Hees G., Kolbel R., Effects of stem design and material properties on stresses in hip endoprosthesis,J.Biomed.Eng., 1987, Vol.9, 77-83.
3. Prendergast P.J., Taylor D., Stress analysis of the proximo-medial femur after total hip replacement, J.Biomed.Eng., 1990, Vol.12/9, 379-382.
4. Ville J.A. ST., Ecker J.A., Winget J. M., Berghauer M.H., Tha anatomy of midthigh pain after total hip arthroplasty Johns Hopkins APL Technical Digest 1991, Vol.12/2, 198-214.
5. Cheal E.J., Spector M., Hayes W.C., Role of loads and prosthesis material properties on the mechanics of the proximal femur after total hip arthroplasty, J.Orthop.Res. 10/3, 405-422, 1992.
6. Verdonschot N.J.J., Huiskes R., Freeman M.A.R., Pre-clinical testing of hip prosthetic designs: a comparison of finite element calculations and laboratory tests. Proc.Instn. Mech. Engrs., 1993, Vol.207, 149-154.
7. Hobatho M.C., Rho J.Y., Ashman R.B., Atlas of mechanical properties of human cortical and cancellous bone, Proc. COMAC-BME II.2.6 meeting. Durham, UK, April 11-13, 1991, eds: G.Van Der Perre, G.Lowet, A.Borgwardt-Christensen.
8. Lotz J.C., Gerhart T.N., Hayes W.C., Mechanical properties of the metaphyseal bone in the proximal femur, J.Biomechanics, 1991, Vol.24/5, 317-329.
9. Rohlmann A., Mossner U., Bergmann G., Kolber R., Finite-element analysis and experimental investigation in a femur with hip endoprosthesis, J.Biomechanics, 1983, Vol.16/9, 727-742.
10. Walker P.S., Schneeweis D., Murphy S., Nelson P., Strains and micromotions of press-fit femoral stem prostheses, J.Biomechanics, 1987, Vol.20/7, 693-702.
11. Walker P.S., Robertson D.D., Design and fabrication of cementless hip stems, Clin.Orthop. and Rel. Res., 1988, Vol. 235, 25-34.
12. Jasty M., O'Connor D.O., Henshaw R.M., Harrigan T.P., Harris W.H., Fit of the uncemented femoral component and the use of cement influence the strain transfer to the femoral cortex, J.Orthop.Res.Soc., 1994, Vol.12, 648-656.

FINITE ELEMENT MODELLING OF BONE CEMENT FLOW DURING ACETABULAR COMPONENT INSERTION IN HIP REPLACEMENT

A.M.R. New[1], M.D. Northmore-Ball[2], K.E. Tanner[3]

1. ABSTRACT

Aseptic loosening is the most common cause of late failure of total hip replacement. Meticulous preparation of the bone bed and cement pressurisation are considered to be important to improve initial fixation of implants by facilitating effective interdigitation of cement and bone. In this study the finite element method has been used to predict the pressurisation and consequent penetration of bone cement into cancellous bone during component insertion in joint replacement operations. Parametric analyses showed cement penetration to be greater with flanged cups, reduced viscosity cement, higher insertion force and more permeable cancellous bone. Cement penetration varied from negligible (less than 0.5 mm) to more than 5 mm. The values for ultimate cement penetration were in good agreement with the range of values reported in the literature and appeared reasonable in comparison to typical post operative radiographic appearances.

Keywords: Bone cement, Fluid flow, Finite element modelling, Hip replacement

---

[1] Ph.D. Student, Unit for Joint Reconstruction, Robert Jones and Agnes Hunt Orthopaedic Hospital, Oswestry, Shropshire SY10 7AG, UK & IRC in Biomedical Materials, Queen Mary and Westfield College, Mile End Road, London E1 4NS, UK
[2] Consultant Orthopaedic Surgeon and Director, Unit for Joint Reconstruction, Robert Jones and Agnes Hunt Orthopaedic Hospital, Oswestry, Shropshire SY10 7AG, UK
[3] Reader in Biomaterials and Biomechanics, IRC in Biomedical Materials, Queen Mary and Westfield College, Mile End Road, London E1 4NS, UK

2. INTRODUCTION

Total hip replacement is a highly successful treatment for disabling disorders of the hip. Long term multicentre studies show the failure rate for well established implant designs to be around 10% after 10 years (1). However, current estimates put the annual number of hip replacements at 38,000 in the UK, 18% of which are revision operations (2). Similar trends are seen throughout the world and the number of revisions are forecast to rise for the foreseeable future. Revisions are complicated, costly and in general less successful than the primary operation (3). Further improvements to the long term performance of the primary hip are therefore highly desirable.

Aseptic loosening is the most common cause of late failure of total hip replacement. Analysis of the Swedish Arthroplasty Register has demonstrated that modern surgical techniques have lead to a reduction in the rate of aseptic loosening (4). The enhanced initial fixation brought about by careful preparation of the bone bed and cement pressurisation are considered to be responsible for the improvement. Various studies suggest that cement pressurisation and subsequent penetration of cement into cancellous bone produces a stronger interface (5,6). Prosthesis design is important for cement pressurisation (7,8) and the effects are clinically significant (9).

A model of cement flow during component insertion in joint replacement has been developed and then applied to quantify the effects of prosthesis design on cement penetration into cancellous bone in the acetabulum. The finite element method has been used to predict the pressurisation of bone cement and subsequent penetration into cancellous bone during component insertion in hip replacement operations.

3. THEORY

The incompressible isothermal flow of a Newtonian (linear) fluid can be completely described by the continuity (conservation of mass) equation

$$\frac{\partial \rho}{\partial t} + (\nabla \cdot \rho \bar{v}) = 0 \tag{1}$$

where ρ is the fluid density, t is time, $\bar{v}$ is the fluid velocity and by the Navier-Stokes (momentum) equation

$$\rho \frac{D\bar{v}}{Dt} = -\nabla P + \mu \nabla^2 \bar{v} \tag{2}$$

where P is pressure and μ the fluid viscosity.

The resistance to fluid flow produced by the porous cancellous bone was modelled by Darcy's law, which assumes a linear relationship between volumetric flow rate through and pressure gradient across the solid material

$$\bar{v} = -\frac{k}{\mu} \nabla P \tag{3}$$

the constant of proportionality being k, the permeability of the solid medium.

4. METHOD

The modelling of cement flow during insertion of a prosthesis into a bone bed is complicated by the changing shape of the fluid domain. To overcome this problem we used the method summarised in the flow chart (Figure 1), with the finite element method used to descretise the fluid domain and solve the fluid flow equations. In the finite element formulation the resistance to flow presented by the cancellous bone was implemented by incorporating extra resistance terms into the momentum equations in the elements representing cancellous bone, this formulation is equivalent to Darcy's law. The analysis was carried out using the ANSYS/FLOTRAN commercial finite element code (release 5.3) running under Windows 95 on an IBM compatible PC.

Prosthesis insertion was simulated by applying a constant velocity at all the boundary nodes associated with the exterior prosthesis surface. By calculating the steady state velocity and pressure solution to the Navier-Stokes and continuity equations and integrating the pressure over the same set of nodes, a linear relationship was established for insertion force vs. insertion velocity to allow the development of cement penetration based on a constant applied force, a more realistic simulation of a surgeon during an operation. A new solution was then calculated based on the cup insertion velocity extrapolated for the required insertion force, either 100 or 200 N, and the bone cement domain modified based on the velocity of the prosthesis and the exit velocity of the bone cement through the porous cancellous bone. Bone cement was assumed to behave as a linear viscous (Newtonian) liquid, with a viscosity of either 500 or 1000 Pa s. These values were representative of those measured by oscillating plate on plate rheometry for a number of commercial acrylic bone cements in their typical usage windows (10). Since cement penetration typically occurs soon after prosthesis insertion starts, no attempt was made to model the increase in viscosity with time for the curing cement. Cancellous bone was modelled as an isotropic porous material. It was further assumed that the flow was incompressible and isothermal, that gravitational forces could be neglected and that quasi steady state conditions were maintained (11). Laminar flow was a natural consequence of the fluid properties, fluid domain shape and insertion velocities/forces considered.

Axisymmetric finite element meshes of the acetabulum with an unflanged and flanged cup are shown in Figure 2. The deep region of the acetabulum was assumed to be covered by solid subchondral bone plate while the peripheral region was cancellous bone. At the cup surface velocity was specified as described above and in the flow chart. At the axisymmetry axis the velocity component across the axis was set to zero. On the exit boundaries, at the outer edge of the cancellous bone and between cup body and acetabulum (unflanged cup) and flange and acetabulum (flanged cup), pressure was specified as zero. Wall boundary conditions were assumed elsewhere (all velocity components zero).

Parametric analyses were carried out, varying cup insertion force and cement viscosity as described above. In addition the permeability of the cancellous bone was varied from 10^{-9} to 10^{-7} m^2 . The permeability of normal cancellous bone ranges from 2×10^{-9} to 10^{-8} m^2, corresponding to approximate wet densities of 1100 to 800 kg m^{-3} respectively. However pathological changes to cancellous bone as a result of arthritic conditions range from extensive cyst formation to appreciable densification (12). Due to the uncertainty of these changes the larger range of permeability was adopted.

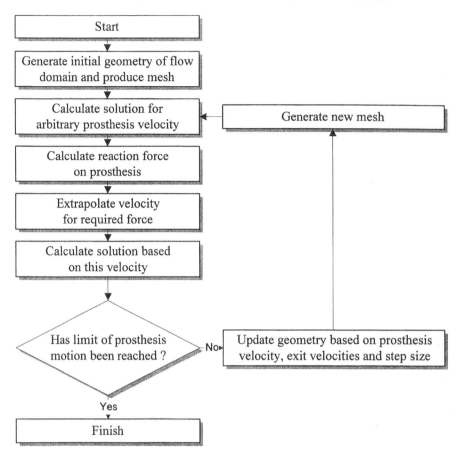

Figure 1: Flow chart showing modelling technique.

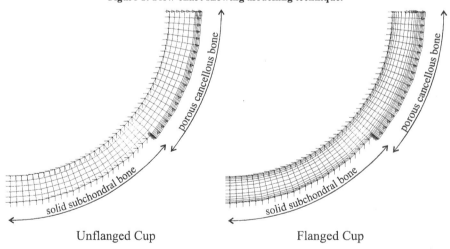

Figure 2: Finite element meshes of acetabula containing unflanged and flanged cups.

5. RESULTS

Figure 3 shows cement pressure distributions for two prosthesis positions and two prosthesis designs. As expected the pressure was always highest at the pole of the acetabulum. Closure of the gap between prosthesis and bone bed resulted in steeper pressure gradients.

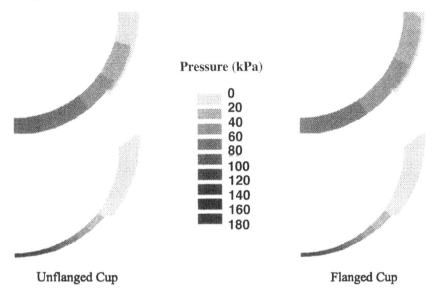

Pressure (kPa)

0
20
40
60
80
100
120
140
160
180

Unflanged Cup   Flanged Cup

Figure 3: Cement pressure distributions at two cup positions for an insertion force of 100N.

Further results are presented as cement penetration in the radial direction for the cancellous bone region. Figure 4-Figure 6 show the results of the parametric analyses. Cement penetration varied from negligible (Figure 6) to more than 5 mm (Figure 5).

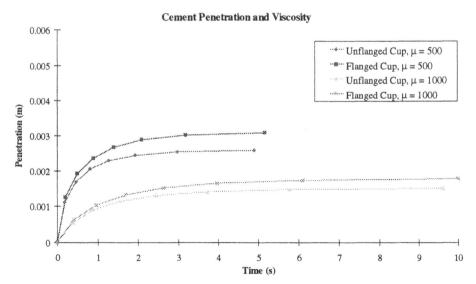

Figure 4: Predicted relationship between cement penetration and cement viscosity.
$(k = 10^{-7} \text{ m}^2, F = 100 \text{ N})$

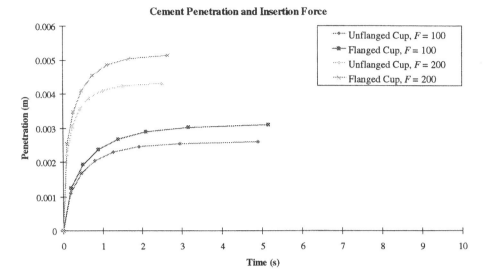

Figure 5: Predicted relationship between cement penetration and insertion force.
$(k = 10^{-7} \text{ m}^2, \mu = 500 \text{ Pa s})$

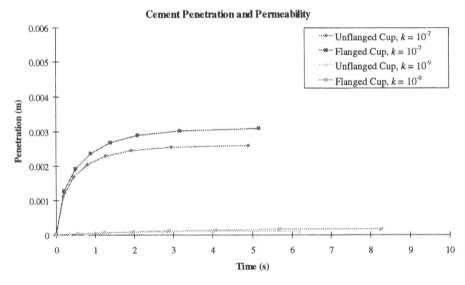

Figure 6: Predicted relationship between cement penetration and cancellous bone permeability.
$(F = 100 \text{ N}, \mu = 500 \text{ Pa s})$

6. DISCUSSION

Bone cement has an extremely complicated constitutive behaviour. In the curing period, up to 6 to 10 minutes from mixing, it changes from a labile fluid to an elastic solid. In this study the cement was modelled as a linear fluid. Previous studies have shown that bone cement can be adequately modelled as a Newtonian fluid during the time window in which it is used clinically (10). The dominant non-linear effect is increasing elasticity as the cement cures. Inclusion of the shear rate dependent viscosity noted by other authors (13,14) in the model was not found to be necessary to successfully predict experimentally determined cement pressures without penetration of cancellous bone. This is believed to be due to the limited range of shear rates experienced by the fluid in the modelled flow domains and the similar shear rates used in the determination of the linear viscosity of the bone cement. The parametric analyses showed cement penetration to be greater with flanged cups, reduced viscosity cement, higher insertion force and more permeable bone. The values for ultimate cement penetration are in good agreement with the range of values reported in the literature and appear reasonable in comparison to typical post operative radiographic appearances. All the parameters were varied between extremes of the ranges experienced clinically. Permeability appears to have a very strong effect simply because it was varied by a factor of 100 compared to factors of 2 for the insertion force and cement viscosity. It has been postulated that the final strength of the cement bone interface is principally determined by the strength of the cancellous bone (15), which is in turn related to the density (16) and hence permeability (13). It is not clear, therefore, how permeability would affect the interface strength, since less penetration into denser bone may produce a stronger interface.

7. ACKNOWLEDGEMENTS

The authors would like to acknowledge the Edward Smith Estate Charity for funding this work, which forms part of the Ph.D. project of AMRN. The IRC in Biomedical Materials is funded by the United Kingdom Engineering and Physical Sciences Research Council.

8. REFERENCES

1　Malchau, H., Herberts, P., Ahnfelt, L. and Johnell, O., Prognosis of total hip replacement, 61st Annual Meeting of the AAOS, San Francisco, 1993.

2　Report of the Biomaterials and Implants Research Advisory Group, United Kingdom Department of Health, 1996, 7.

3　Kershaw, C.J., Atkins, R.M., Dodd, C.A.F. and Bulstrode, C.J.K., Revision total hip arthroplasty for aseptic failure: a review of 276 cases, J. Bone Jt. Surg. 1991, Vol. 73B, 564-568.

4　Malchau, H. and Herberts, P., Prognosis of total hip replacement surgical and cementing technique in THR: a revision risk study of 134,056 primary operations. 63rd Annual Meeting of the AAOS, February 1996.

5　Krause, W.R., Krug, W. and Miller, J., Strength of the cement-bone interface, Clin. Orthop. Rel. Res., 1982, Vol. 163, 290-299.

6　Askew, M.J., Steege, J.W., Lewis, J.L., Ranieri, J.R. and Wixson, R.L., Effect of cement pressure and bone strength on polymethylmethacrylate fixation, J. Orthop. Res., 1984, Vol. 1, 412-420.

7　Beverland, D.E., Kernohan, W.G., Nixon, J.R., Orr, J.F. and Watson, P., Pressurisation of bone cement under standard, flanged and custom acetabular components for total hip replacement, Proc. I. Mech. E. Eng. Med., 1993, Vol. 207H, 19-23.

8 Shelley, P. and Wroblewski, B.M., Socket design and cement pressurisation in the charnley low-friction arthroplasty, J. Bone Jt. Surg. 1986, Vol. 70B, 358-363.

9 Hodgkinson, J.P., Maskell, A.P., Paul, A. and Wroblewski, B.M., Flanged acetabular components in cemented charnley hip arthroplasty - ten year follow up of 350 patients, J. Bone Jt. Surg. 1993, Vol. 75B, 464-467.

10 New, A.M.R., The relationship between cup fixation strength and cementation technique in the acetabulum in hip replacement, Ph.D. Thesis, University of London, 1997.

11 Tadmor, Z. and Gogos, C.G., Principles of polymer processing, John Wiley and Sons, 1979, 382-386.

12 Grennan, D.M., Rheumatology, Baillière Tindall, 1984, 89-103.

13 Beaudoin, A.J., Mihalko, W.M. and Krause, W.R., Finite element modelling of polymethylmethacrylate flow through cancellous bone, J. Biomech. 1991, Vol. 24, 127-136.

14 Krause, W.R., Miller, J. and Ng, P., The viscosity of acrylic bone cements, J. Biomed. Mat. Res. 1982, Vol. 16, 219-243.

15 Majkowski, R.S., Bannister, G.C., Miles, A.W., The effect of bleeding on the cement-bone interface - an experimental study, Clin. Orthop. Rel. Res., 1994, Vol. 299, 293-297.

16 Carter, D.R., Hayes, W.C., The compressive behaviour of bone as a two-phase porous structure, J. Bone Jt. Surg. 1977, Vol. 59A, 954-962.

NUMERICAL MODELLING OF CEMENT POLYMERISATION AND THERMAL BONE NECROSIS

G. R. Starke[1], C. Birnie[2] and P. A. van den Blink[3]

1. ABSTRACT

During the cement curing phase of a total joint replacement, the material undergoes a process of polymerisation, which causes the temperature within the cement to rise to over 100°C. This temperature increase can destroy the bone tissue adjacent to the cement. The extent to which bone cells are destroyed will be influenced by the temperature within the bone and the exposure time. A numerical model for the prediction of cement polymerisation and thermal bone necrosis has been developed and verified in order to examine the influence of cement thickness on the extent of thermal necrosis. The prediction of thermal necrosis was formulated as a damage model in which the normalised damage parameter was written as a function of the temperature history and the exposure time. The numerical model was implemented in two plain strain finite element models of the reconstructed proximal femur with minimum cement thickness values of 1.5 and 2.5mm.

2. INTRODUCTION

One of the problems associated with the use of bone cement is the dramatic increase in the cement temperature which is experienced during the polymerisation process. The temperature within the cement can increase to over 100°. This temperature increase can result in necrosis of the surrounding bone tissue. The death of the surrounding tissue can have an adverse effect on the long term success of the joint arthroplasty. During reconstruction of the hip the steel alloy femoral stem will act as a heat sink and thereby keep down the bone temperature to a certain extent. However, the ability of the heat sink to be successful will depend of the amount of cement present. The thicker the cement mantle, the greater the volume of material and hence the more heat that is required to be removed. However, the answer does not lie in simple reducing the thickness of the mantle, as the mechanics of the joint are affected by this. It is

Polymerisation, Thermal necrosis, Numerical simulation, Finite elements

[1] Researcher, Centre for Research in Computational and Applied Mechanics, University of Cape Town, Rondebosch, 7700, South Africa

[2] Former undergraduate student, Dept. of Mech. Eng., University of Cape Town, South Africa

[3] Former undergraduate student, Dept. of Mech. Eng., University of Cape Town, South Africa

therefore important to know what thickness of mantle can be tolerated without causing damage to the surrounding bone.

Determining if thermal necrosis has occurred is extremely difficult. This is because the surrounding bone experiences chemical damage due to monomer leakage, and also mechanical damage during the insertion process. The occurrence of thermal necrosis is, however, an accepted fact, and several studies have shown that thermal necrosis can indeed occur. Huiskes and Slooff (1981) found that thermal necrosis occurred in portions of the bone into which the cement penetrates. Mjoberg et al. (1984), in a study involving six humans, established conclusively that thermal injury to bone did take place. It has been suggested that thermal bone necrosis must be evaluated in two ways: by collagen protein denaturation and potential cellular death (Swenson et al., 1981). Generally protein coagulation occurs at between 56 and 70°C. Investigations on the time dependent temperature threshold behaviour of epithelial cells were performed by Moritz and Henriques (1947). They found that temperatures of 70°C killed the cells immediately. At temperatures of 55°C, cells were destroyed after 30s, and at 45°C cells had to be exposed for more than 5 hours to be affected. Jefferiss et al. (1973) reported the denaturation of young uncross linked collagen molecules at 45°C. However, permanent molecular rearrangement only occurred if such temperatures were maintained for more than an hour. At temperatures of 60°C they reported irreversible contraction of individual collagen fibres.

In order to establish the possibility of necrosis, transient temperatures are needed at the cement-bone interface. Several researchers have used experimental techniques for the measurement of bone temperature in vivo. However, the difficulty in taking these measurements has lead to the development of numerical models for the prediction of cement and bone temperature. Some early studies showed an inter-dependency of the temperature on the polymerisation process of the cement (Jefferiss et al., 1973). Baliga et al. (1992) proposed constitutive relations for the prediction of cement temperature during polymerisation. They proposed that the extent of polymerisation, α, could be expressed as the ratio of the amount of heat generated at the present time to the total heat generated at the end of polymerisation. Following the proposals of Piloyn (1966) and Kamal and Sourer (1973), Baliga et al. (1992) assumed that the rate of polymerisation could be represented by the kinetic relation which was written as a function of temperature and the extent of polymerisation. The resulting model was used to predict the temperature during polymerisation of Surgical Simplex bone cement. Although they found excellent agreement with the experimentally measured temperatures, the applicability of this model to other cements is not known.

In the current work a numerical model is developed for the purpose of predicting the heat generated in the cement due to polymerisation. The numerical model is validated by comparing the predicted temperatures to the experimentally measured temperatures in a commercially available bone cement. In addition a damage model is developed for the prediction of bone cell death as a function of temperature and exposure time. Subsequently the algorithms have been used to examine the influence of cement thickness on the temperature within the surrounding bone using two plane strain finite element models.

3. MATERIALS AND METHODS

A numerical model for the polymerisation of bone cement is developed in conjunction with the algorithm for the prediction of thermal necrosis which are set out in the following sections. An experimental model was also generated for the purpose of comparing the numerically predicted and experimentally measured values. In addition two plane strain finite element models have been generated for the purpose of examining the effect of cement thickness on the extent of thermal necrosis.

3.1 Cement Polymerisation

Based on the work of Baliga *et al.* (1992), a numerical model is developed which describes the heat produced within the bone cement, during the polymerisation phase. The model is developed in terms of the amount of polymerisation α, which is defined as the fraction of the amount of heat generated. The total amount of heat generated during polymerisation Q_{total} is a function of temperature.

$$\alpha = \frac{1}{Q_{total}} \int_0^t S \, dt \tag{1}$$

where S and Q_{total} are the instantaneous and the total amount of heat generated during the polymerisation process, respectively. However, for isothermal polymerisation (where the temperature is forcibly kept constant), at temperatures of between $30°C$ and $90°C$, its value was found to vary by less than 15% (Forbes, 1977, cited in Baliga *et al.*, 1992). Thus, in this analysis, it is assumed to be constant.

By differentiating (1) with respect to time it is seen that

$$S = Q_{total}\left(\frac{\partial \alpha}{\partial t}\right), \tag{2}$$

which is combined with (2) to give

$$S = \Re(T) \alpha^m (1-\alpha)^n, \tag{3}$$

where $\Re$ is a function of temperature.

$$\frac{\partial \alpha}{\partial t} = R(T)\alpha^m(1-\alpha)^n \tag{4}$$

where, m and n are temperature independent constants, while R is a function of the temperature T. The Baliga *et al.*, using the work of Forbes (cited in Baliga *et al.*, 1992), determined the parameters m, n and $\Re$ for Surgical Simplex PMMA. Forbes used a differential scanning calorimeter to experimentally determine the variation of S with time. The relation of S with time was then converted to one of S with α by using the trapezoidal method of numerical integration. Using the least squares method of curve fitting, the following parabola was fitted to the experimentally generated curves:

$$S = 4.4 \times 10^6 f(T)(\alpha - \alpha^2), \tag{5}$$

with the function $f(T)$ given by

$$f(T) = a_0 + a_1\left(\frac{T}{100}\right) + a_2\left(\frac{T}{100}\right)^2 + a_3\left(\frac{T}{100}\right)^3 + a_4\left(\frac{T}{100}\right)^4 + a_5\left(\frac{T}{100}\right)^5 \tag{6}$$

and

$$\begin{aligned}
a_0 &= -23.89 & a_1 &= 296.74 \\
a_2 &= -1352.97 & a_3 &= 2894.76 \cdot \\
a_4 &= -2806.62 & a_5 &= 1009.84
\end{aligned} \tag{7}$$

By comparing (4) and (5), it can be seen that the parameters m, n and $\Re$, for Surgical Simplex PMMA, are given by

$$m = n = 1 \text{ and } \Re = 4.4 \times 10^6 f(T). \tag{8}$$

The heat generation equations are incorporated into the finite element program ABAQUS 5.5, by the use of a FORTRAN subroutine. The finite element program uses an incremental formulation with a Newton-Raphson iterative solution algorithm. The routine must therefore be compatible with this two step solution process. The polymerisation and heat generation equations are therefore cast into an incremental form, by dividing the time domain into a sequence of intervals. The times at the start and the end of an arbitrary interval are given by t_n and t_{n+1}, respectively. Therefore the problem posed is as follows: at time t_n, given the temperature T_n, and the increment in temperature ΔT_{n+1} a solution is required for the amount of heat generated S_{n+1} and the rate of change of the heat generated with respect to temperature $\dfrac{\partial S_{n+1}}{\partial T_{n+1}}$.

Therefore, making use of a backward-difference numerical integration scheme, Equation (1) is rewritten in an incremental form as

$$\alpha_n = \sum_0^n \frac{S_n . \Delta t}{Q_{total}} \tag{9}$$

Therefore for the nth increment, α is rewritten as

$$\alpha_{n+1} = \alpha_n + \frac{S_{n+1} . \Delta t}{Q_{total}} \tag{10}$$

The value of α_{n+1} is stored as a state variable, which is used at each increment to update the amount of heat generated.

Equation (5) can be written in an incremental form as:

$$S_{n+1} = \Re(\alpha_{n+1} - \alpha_{n+1}^2). \tag{11}$$

Combining (10) and (11) gives

$$S_{n+1} = \Re\left[\left(\alpha_n + \frac{S_{n+1} . \Delta t}{Q_{total}}\right) - \left(\alpha_n + \frac{S_{n+1} . \Delta t}{Q_{total}}\right)^2\right]. \tag{12}$$

which can be solved using a variety of approaches. The rate of change of heat generation with respect to temperature is found by the partial differentiation of (5). Using the chain rule,

$$\frac{\partial S}{\partial T} = \frac{\partial S}{\partial \Re} \frac{\partial \Re}{\partial T} + \frac{\partial S}{\partial \alpha} \frac{\partial \alpha}{\partial T} \tag{13}$$

$$\frac{\partial S}{\partial \Re} = (\alpha - \alpha^2). \tag{14}$$

$\Re$ is given by (8) and therefore,

$$\frac{\partial \Re}{\partial T} = 4.4 \times 10^6 \left(\frac{a_1}{100} + 2a_2\left(\frac{T}{100}\right) + 3a_3\left(\frac{T}{100}\right)^2 + 4a_4\left(\frac{T}{100}\right)^3 + 5a_5\left(\frac{T}{100}\right)^4\right) \tag{15}$$

$$\frac{\partial S}{\partial \alpha} = \Re(1 - 2\alpha) \text{ and } \frac{\partial \alpha}{\partial T} = 0. \tag{16}$$

And finally,

$$\frac{\partial S}{\partial T} = 4.4 \times 10^6 \left(\alpha - \alpha^2\right) \left(\frac{a_1}{100} + 2a_2 \left(\frac{T}{100}\right) + 3a_3 \left(\frac{T}{100}\right)^2 + 4a_4 \left(\frac{T}{100}\right)^3 + 5a_5 \left(\frac{T}{100}\right)^4 \right). \tag{17}$$

The numerical model is implemented in the HETVAL subroutine (ABAQUS 5.5, Hibbit, Karlsson and Sorensen, 1996), which is called to provide the amount of heat generated, as well as the rate of change of heat generated with respect to time, at each material point in the cement elements.

3.2 Thermal Necrosis

From the experimental data of Moritz *et al.* (1947) which set out the amount of time required to destroy bone cells at a given temperature, Mazzullo *et al.* (1991) determined a function for the exposure time required for thermal necrosis, which they expressed as

$$t^*(T) = M \exp\left[\frac{\mu}{R(T - 310)} \right] \tag{18}$$

where t^* is the exposure time necessary to reach thermal necrosis at a given temperature T. R is the universal gas constant. The constants M and μ have been evaluated from a linear regression of the data from Moritz *et al.* (1947).

The possible death of bone cells can be written in terms of a normalised damage parameter ω, which is defined such that when $\omega = 0$ there has been no cell damage and when $\omega = 1$ thermal necrosis has taken place. Under non-isothermal conditions, the damage parameter can be estimated as the integral of the fractions of exposure time at assigned temperatures, and is therefore written as

$$\omega = \int_0^t \frac{dt}{t^*(T)} \tag{19}$$

where $0 \le \omega \le 1$.

The numerical implementation of the thermal necrosis equations follows the same form as the algorithm for the polymerisation where a backward-difference numerical integration scheme is employed. The algorithm has also been implemented in the HETVAL subroutine which provides the updated damage parameter at each time increment, at each material point in the cancellous and cortical bone elements.

3.3 Experimental Measurement of Cement Temperature

An experimental apparatus was constructed to measure the temperature in the bone cement during the polymerisation process (Figure 1). The experimental apparatus consisted of a 20mm diameter stainless steel tube, with a wall thickness of 2mm, into which the cement was placed. The 100mm length of tube was surrounded by a 150mm diameter sleeve of polystyrene foam. Seven thermo-couples were inserted into the cement via holes in the tube. Once the cement had cured the tube was sectioned and polished to determine the exact position of each of the thermo-couples. The experimental arrangement was simulated using an axi-symmetric finite element model. The numerically predicted temperature values, during cement curing, were compared to the experimentally measured values.

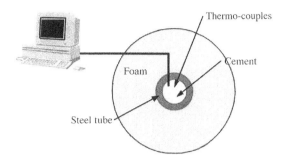

Figure 1 Experimental measurement of cement temperatures during polymerisation.

3.4 Finite Element Modelling

Two plane strain finite element models of the proximal femur reconstructed with the *Elite Plus* (Depuy International) femoral component have been used for the prediction of cement temperature and bone necrosis. The models have cement thickness values of 1.5 and 2.5mm and are shown in Figure 2.

(a)   (b)

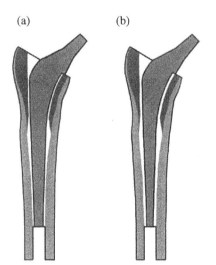

	Cond. (W/m °C)	Spec. heat (kJ/kg °C)	Density (kg/m^3)
Cancellous	0.293	2292.27	1333.00
Cortical	0.40	400.00	2000.00
Cement	0.20	$1250 + 6.50T$	1100.00
Implant	16.00	460.00	7800.00

Figure 2 Finite element models showing the implant, cement as well as the cancellous and cortical bone with (a) 1.5mm cement mantle, and (b) 2.5mm cement mantle. The inserted table gives the thermal properties used in the model.

The thermal properties of cancellous and cortical bone are not well known, and are extremely difficult to measure. The thermal properties used for the implant, the cement as well as the cancellous and cortical bone were taken from the literature (Swenson *et al.* 1981; Lundskog, 1972) and are set out in Figure 1. Although a thermal interface exists between the implant and the cement as well between the cement and the bone, the thermal nature of these interfaces are not clear and therefore they were assumed to be perfectly bonded. The positions (1) and (2) on the medial and lateral sides of the implants indicate points at which the temperature history within the cement has been recorded during the polymerisation process. The development of the experimental model and the finite element meshes of the proximal femur are set out in Birnie (1995).

4. RESULTS

A comparison of the predicted cement temperature and the experimentally measured values are shown in Figure 3. The experimentally measured temperatures have shown a very similar trend to the numerically predicted values. The experimental model was insulated using a thick layer of polystyrene foam, and therefore the final temperature values did not return to room temperature after the 2000 seconds for which the values were recorded. The peak temperature value occurs after approximately 600 seconds in both the experimental and numerical model. The experimentally measured peak temperature is approximately 105°C compared to the predicted value of 100°C. Both the numerical and the experimental model have shown that the polymerisation process is completed after approximately 1500 seconds.

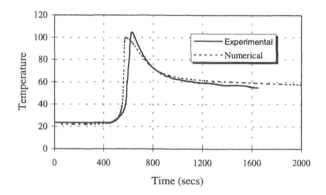

Figure 3 Comparison between the numerical predicted temperature history and the mean of the experimentally measured values.

The predicted temperature histories at the two positions in the lateral cement mantle, for each model, are shown in Figure 4 (1) and (2). Position (L1) is near the distal end of the mantle, while position (L2) is midway along the mantle where the cement is thicker. The solid lines show the temperature history for the 1.5mm mantle, while the dashed lines are for the thicker mantle. At position (L1), the peak temperature for the thicker cement mantle is approximately 50°C compared to 38°C for the thinner mantle. At position two, where the cement is thicker, both mantles experience higher peak temperatures, with the thicker cement reaching a peak of approximately 85°C compared to 73°C in the thinner mantle. At position (L1) the peak temperature is reached after approximately 220 seconds compared to approximately 240 seconds at position (L2).

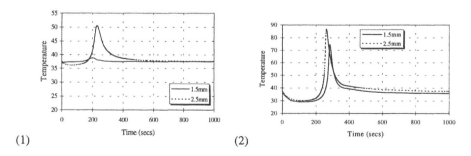

Figure 4 Temperature history at positions (L1) and (L2) on the lateral side for the two implant models.

The predicted temperature histories at the two positions in the medial cement mantle (adjacent to the lateral positions) are shown in Figure 5 (1) and (2). At positions (M1) the thicker mantle experiences a peak temperature of over 70°C compared to 45°C in the thinner mantle. The peak temperatures occur at the same time as at the corresponding position on the lateral side. On the medial side the mantles are very similar in thickness at position (M2) resulting in a very similar temperature history between the two models. At this point the peak temperatures peak at approximately 50°C.

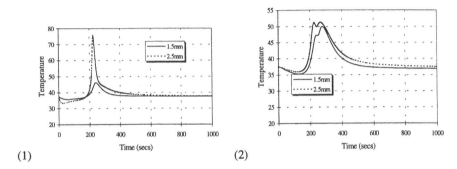

(1)   (2)

Figure 5 Temperature history at positions (M1) and (M2) on the medial side for the two implant models.

Contours of the temperature distribution in the two implant models after 266 and 316 seconds is shown in Figure 6 (a) and (b), respectively. After 266 seconds a temperature peak is seen on the medial side of the mid-mantle region in the thinner mantle, while in the thicker mantle there is a temperature peak in this region on both the lateral and medial sides. In addition, temperature peaks of close to 80°C are visible in the distal end of the mantle on both the lateral and medial sides. By 316 seconds, the peak temperature in the two models has dropped to 61°C, in both

(a)   (b)

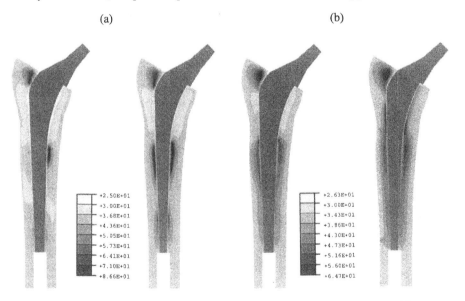

Figure 6 Cement and bone temperatures in the two models after (a) 266 seconds, and (b) 316 seconds. In each pair the left plot shows the 1.5mm mantle, while the 2.5mm mantle is shown on the right.

models. Also by this time the heat has been somewhat dissipated with less discernible peaks. At the proximal end of the mantle extremely high temperatures are predicted as no boundary condition is implemented in this region. In reality the heat from the cement would be transferred to atmosphere in this region.

The only regions in which thermal necrosis has been predicted is on the medial side in the mid-mantle region. Contour plots of thermal damage in this region are shown for the thinner and thicker mantles in Figure 7 (a) and (b), respectively. In this thinner mantle the amount of damage predicted is small, with a peak ω value of approximately 0.3. However, adjacent to the thicker mantle damage values of up to 0.8 have been predicted.

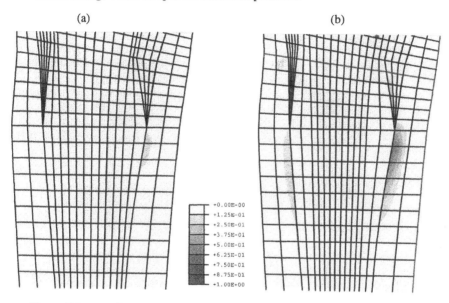

Figure 7 Extent of thermal necrosis at the end of the polymerisation process.

5. DISCUSSION

The current study sets out a numerical formulation for the prediction of cement and bone temperatures as a result of cement polymerisation. In addition, the temperature history is used to predict thermal necrosis in the cancellous and cortical bone using a damage formulation.

The comparison of the numerically predicted and the experimentally measured temperature values, show a deviation in the peak temperature of less than 5%. Furthermore, the predicted rates of temperature increase and decrease are very similar to those shown experimentally. The close correlation of the numerical and experimental results show that the numerical model approximates the temperature profile of a commercially available bone cement.

The temperature contours have shown that in the thicker cement mantle local *hot spots* of temperature develop distally, and in the mid-mantle region, on both the medial and lateral sides during the polymerisation process. On the lateral side a peak temperature of over 80°C was predicted. However these peaks are only noted in the mid-cement region of the thinner mantle model. The temperature contours show that the metal alloy implant acts as an effective heat sink, drawing heat out of the cement. The result of this is that the polymerisation of the cement occurs for the outer surface towards the femoral stem. This non-uniformity may have implications in terms of the properties of the resultant polymer.

Thermal bone necrosis has not been widely reported in joint replacement. This may be due to the fact that it is difficult to distinguish between damage as a result of the cement temperature, and that resulting from reaming and broaching during the surgical preparation. However, the amount of thermal necrosis prediction is minimal, with limited bone damage only being predicted in the thicker mid-mantle region. If the cement where kept at the required thickness of 1.5mm and 2.5mm it is likely that no damage would have been predicted in either mantle. The thermal necrosis model is, however, extremely simplistic and is based on one set of very old data which may not be valid for different bone types and densities. In addition there are many other factors, in addition to temperature, which may interact to determine the extent of bone necrosis.

In spite of the encouraging validation of the polymerisation simulation, the overall finite element model is subject to many limitations. Firstly, the thermal properties of cancellous and cortical bone are not well known and may vary considerably across a population sample. Secondly, the thermal boundary on the outer edges of the bone surface will determine the rate of heat loss. However, the amount of heat lost at this boundary is difficult to determine as this region is surrounded by a variety of different tissues. Thirdly, the position of the points at which the temperature histories were recorded are not necessarily the same for the two models, as the two meshes are different. Notwithstanding the simplifying assumptions, the present study represents a first step towards the successful prediction of thermal bone necrosis due to cement polymerisation.

6. ACKNOWLEDGEMENTS

The financial assistance of DePuy International (Leeds) is gratefully acknowledged.

7. REFERENCES

1. ABAQUS Version 5.5, Hibbit, Karlsson and Sorensen, Pawtucket, Rhode Island, USA

2. Baliga B R, Rose P L, "Thermal modelling of polymerising polymethylmethacralate, considering temperature-dependent heat generation", *J. Biomech. Eng.*, **114**, 251-258, 1992

3. Birnie C P, "Numerical modelling of the heat transfer due to the exothermic reaction in the curing bone cement", *B.Sc. (Mech) Eng. Final Year Project*, Dept. Mech. Eng., 1995

4. Huiskes R, Sloof T J, "Thermal injury of cancellous bone, following pressurised Penetration of Acrylic cement", *Trans. of the 17th annual meeting of Orthopaedic Research Society*, Nevada, pg.134, 1981

5. Jefferis C D, Lee A J C, "Thermal aspects of self-curing polymethylmethacralate", *J. Bone Joint Surg.*, **57-B**, 511-518, 1975

6. Kamal M R, Sourour S, "Kinetics and thermal characteristics of thermoset cure", *Polymer Eng. and Sci.*, **13**, 59-64, 1973

7. Lundskog J, "Heat and bone tissue", *Scand. J. Plast. Reconstr. Surg.*, **9**, 1972

8. Mjoberg B, Pettersson H, Rosenquist R, "Bone cement, thermal injury and the radiolucent zone", *Acta. Orthop. Scand.*, **55**, 597-600, 1984

9. Moritz A R, Henriques F C, "Studies of thermal injury II", *Am. J. Path.*, **23**, 695-700, 1947

10. Mazzullo S, Paolini M, Verdi C, "Numerical simulation of thermal bone necrosis during cementation of femoral prostheses", *J. Math. Biol.*, **29**, 475-494, 1991

11. Piloyn G O, Ryabchikov I D, Novikova O S, "Determination of activation energies of chemical reactions by differential thermal analysis", *Nature*, **212**, pg.1229, 1966

12. Swenson L W, "Finite Element Temperature Analysis of a Total Hip Replacement and Measurement of PMMA curing temperatures", *J. Biomed. Mater. Res.*, **15**, 83-95, 1981

BIOMECHANICAL REACTION OF DOUBLE COATING ON METAL IMPLANTS

I. Knets[1], J. Laizans[2], R. Cimdins[3], V. Vitins[3], M. Dobelis[3]

1. ABSTRACT

A plasma-sprayed hydroxyapatite (HAp) coating for titanium implant materials with intermediate layer of glass coating was used to eliminate the corrosion during prolonged implantation. Experimental implantations of the material proposed was carried out on rabbits. The finite element method (FEM) was used to simulate numerically both the axial loading of idealized bone with double coated implant and push-out test of the latter.

2. INTRODUCTION

Titanium and its alloys are finding increasing applications in the fabrication of surgical implants. There has been considerable interest, especially in the use of Ti-6Al-4V alloy for orthopaedic implants because of its biocompatibility and fatigue strength. A significant problem, however, still is to obtain a stable bonding strength between metallic implant and bone tissue at their interface. Different coatings on titanium implants are widely used to achieve good chemical bonding. A hydroxyapatite (HAp) coating ensures the bone ingrowth into it which mechanically attaches the bone to the metallic implant through this intermediate zone [Wolke et all., 1990]. The first results looked very promising because of very good bonding between bone tissue and the HAp

KEYWORDS

Titanium implant, Hydroxyapatite coating, Finite element method

---

[1] Professor, Biomaterials and Biomechanics Specialized Research Institute, Riga Technical University, Kalku iela 1, Riga, LV-1658, Latvia

[2] Research Assistant, Biomaterials and Biomechanics Specialized Research Institute, Riga Technical University, Kalku iela 1, Riga, LV-1658, Latvia

[3] Senior Researcher, Biomaterials and Biomechanics Specialized Research Institute, Riga Technical University, Kalku iela 1, Riga, LV-1658, Latvia

layer. Unfortunately, with time passing by it turned out that the weakest part in this system was the interface between metal implant and HAp coating causing the loosening of implant.

HAp coating did not protect enough the metallic implant against corrosion due to its porous structure. The corrosion resistant bioglass with specific composition was suggested to build an intermediate layer between implant and HAp coating [Maruno et all., 1992]. This composition provides good bonding with both the titanium implant and the HAp coating and improves the corrosion resistance at the interface with metal.

3. MATERIALS AND METHODS

The glass of chemical composition $35BaO-35B_2O_3-16Al_2O_3-9P_2O_5-5Nb_2O_5$ was synthesized in an electric furnace in the temperature range 1350-1400°C using Al_2O_3 crucibles and high purity staring materials. The glass melt was fretted by quenching in water with following milling of glass powder in a planetary ball mill in isopropyl-alcohol media. Fine milled glass powder (0.8-1.2 m^2/g) in the form of isopropanol suspension ($1.3-1.4\times10^3$ kg/m^3) was deposited on the surface of titanium alloy preheated to 700°C in a vacuum of 1×10^{-3} Pa with the following firing at temperature 870-900°C. To achieve as dense and pore-free as possible coating three consecutive layers of glass were used with a total thickness of 25-30 µm. The synthesized HAp powder was deposited on the previously created glass coating by plasma spraying process with total thickness of 60-80 µm, and consecutively heated at 500-710°C.

To evaluate the biocompatibility and bioactivity of different coatings 12 cylindrical specimens (5 mm diameter and 5 mm length) were made from Ti-6Al-4V alloy. Six of them were coated with plasma-sprayed HAp, the other six with both, the glass and the HAp. Experimental evaluation of these materials were performed on animal tests because it is impossible to create experiments on living human bone. Specimens were implanted into femur of 12 rabbits for 4 months. The bonding strength was determined by push-out test with constant speed 0.4 mm/min and calculated as push-out force to area of interface bone-implant. It was stated that the bonding strength between natural bone and composite coating was 12.3-15.9 MPa, but between bone and HAp coating - only 2.6-4.6 MPa.

The system of bone tissue-callus-composite coating and implant was simulated numerically. The finite element analysis was carried out using a program developed by numerically integrated elements of system analysis (NISA). The characteristics of the interface between bone and implant material in the millimeter range reflect the different interactions behaviour between tissue and material after implantation (Fig. 1).

Three-dimensional finite element model of 16 mm long idealized bone was designed. Elliptical cross-section of bone was assumed for model with dimensions of major and minor axis 12 and 8 mm, respectively. The thickness of cortical layer was 1.3 mm. The glass and HAp coating layers, and bone callus layer were modeled according to the layout in Fig. 1. The interfaces between different materials were assumed to have perfect bonding. 3D linear isoparametric hexahedral elements were used to generate 3D mesh of the model studied.

All the material properties were assumed to be isotropic, linearly elastic and homogeneous. Elastic properties of the materials used in calculations are presented in Table 1.

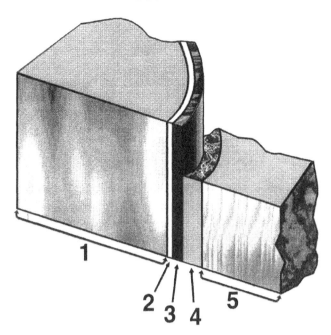

Fig. 1. Bone-implant interface layout used for computer simulation:

1 - titanium implant;
2 - glass coating (30 μm);
3 - HAp coating (80 μm);
4 - bone callus (90 μm);
5 - compact bone.

Table 1. Elastic properties of materials used for NISA calculations

Material	Young's modulus, GPa	Poisson's ratio
Bone	22	0.33
Bone callus	1	0.43
Titanium	106	0.3
Hydroxyapatite	4.5	0.3
Bioglass	7.5	0.3

4. BOUNDARY AND LOADING CONDITIONS

Two different loading simulations were used: axial loading of the bone diaphysis with implant and pushing out the implant from the bone. The final 3D model and axial loading scheme is presented in the Fig. 2.

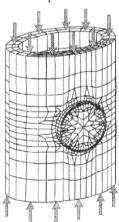

Fig. 2. 3D model of bone with cylindrical implant and axial loading scheme

4.1 Axial loading of bone

One end of the bone model was constrained in the superior-inferior direction. Only three additional constraints were added to provide nonsingularity of the global matrix. These boundary conditions were used to prevent rigid body translation and rotation of the whole model. The axial loading of the bone was simulated applying uniform displacement to the proximal cross-section of the bone. The value of displacement corresponded to 1% relative deformation of the bone in axial direction.

4.2 Push-out loading

To simulate the push-out test, the model was cut into two parts longitudinally. Only the part with implant and bone was used in the calculations. Because of symmetric 3D model push-out test simulation was performed on ¼ of the part considered. The neglected parts of the model were simulated by means of special boundary conditions. Push-out load was simulated applying uniform displacement to the cross-section of the implant. The value of displacement was 1 mm.

5. RESULTS

5.1 Axial loading of bone

Fig. 3 shows the distribution of the first principal stress in the system considered during natural loading conditions. Critical zones were revealed in the bone at the vicinity of implant border symmetrically to the loading direction and rotated by 45°. The values of stress in these areas exceed approximately by 17-25% the average value of stress. This could be a result of the peculiarities of the model used - the cylindrical implant with the coaxial bone callus layer. The stress concentration could be lowered by designing a different shape of implants.

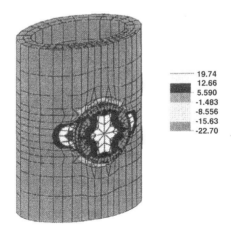

Fig. 3. First principal stress distribution in axial compression

5.2 Push-out test

Fig. 4. shows the maximal shear stress distribution during the push-out test in the simulated calculation model. It is interesting to note that maximum shear stresses occur at the very outer surface of compact bone. The maximum value of shear stress at these places exceed the average stress almost ten times.

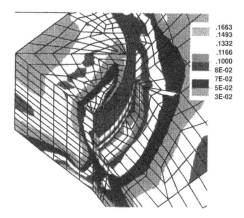

Fig. 4. Maximal shear stress distribution in
push-out test

6. DISCUSSION

The analysis of the results allowed to state that important factor is a shape of the implant - round, square or triangular. The implant should be fixed into the hole stiff enough to avoid micromotion with respect to the bone during the bone callus formation. Optimal form evens the stress distribution and reduces the stress values.

The analysis allows to state that the composite coating using the glass and the HAp coatings on the titanium alloy considerably improves not only the biocompatibility of prosthesis material but also provides strong bonding between bone and implant.

7. REFERENCES

1. Wolke J.G.C., Klein C.P.T., de Groot V. Thermal spray research and applications. Proc. of the Third National Thermal Spray Conference, Long Beach, CA, USA, 1990, p. 413-417.
2. Maruno S., Ban S., Wand Y., Iwata H., Itoh H. Properties of functionally gradient composite coating of HAp containing glass coated titanium and characters for bioactive implant. J. of the Ceram. Soc. of Japan, vol. 100, 1992, p. 363-367.

3. BONE ADAPTATION, STRUCTURAL MODELS AND ARCHITECTURE

BONE ADAPTATION: AN IDEA OF MULTISCALE MODELLING

J. M. Crolet

1. ABSTRACT

This paper presents the different components of a possible computational modelling which can be used for a better understanding of cortical bone adaptation: homogenization of the solid part, taking into account of fluid inside the channels, computation of flow and solute transport (calcium, potassium, ...) in such a medium. A result concerning the intrusion of calcium in the opposite sens of the flow is given and analysed.

2. INTRODUCTION

Bone is a vivant tissue which presents the particularity to change its properties or its architecture in function of the imposed stresses. So, depending on circumstances, it can change its mass (for example during the grouth), its mechanical properties (by remodelling) and its form (it is the morphogenis).

Several questions are of importance. What is the optimal structure? Is it possible to determine the bony mechanical behavior? Is it possible to understand such a functional adaptation? Is it possible to explain the bony pathologies? Different computational modellings can be used. Topological optimization (Bendsoe et *al.*) is certainly the most adapted theory to get, for given stresses, the optimal structure and perhaps to simulate the functional adaptation. But this adaptation can also be modelled with more fundamental explanations. The homogenization theory is, in my opinion, the only one allowing to get a map of the mechanical properties of bone (this theory is used for elasticity, but can also be used for permeability, ...). Nevertheless, there is, actually, no tool for the understanding of bony behavior in the pathological cases.

Professor, Laboratoire de Calcul Scientifique, Université de Franche-Comté
16 Route de Gray, 25 030 Besançon, France

The existence of several scales in bone is the major handicap for all these questions: indeed, each bone is a structure having its own behavior in the human body at the macroscopic scale; but the different transformations are induced by the cellular activity at a microscopic scale. Then the main difficulty is to understand how the informations are transmitted between these two scales.

3. THE HOMOGENIZATION OF BONE

3.1 Homogenization of the solid part

A first possibility to describe such a transmission is the homogenization theory: with given data for elementary components of bone at a microscopic scale (but not at the cellular scale), two kinds of results can be computed: firstly the macroscopic mechanical properties at each point of bone and secondly informations on microstresses or microstrains can be obtained. These informations at a microscopic scale are only available at the osteonal scale, and it is not quite simple to have similar results on the lowest levels.

The principal points of this theory in the framework of the elasticity can be summarized as follows. Let be an heterogeneous medium having a periodic structure, the basic cell (or the period) being made on several homogeneous media with well known mechanical properties. This theory gives the elasticity tensor of an fictive homogeneous medium, this medium having the same macroscopic mechanical behavior as the previous heterogeneous medium. This theory can be extended in the case of several levels of heterogenity and in the case of pseudo-periodic media. The first paper of such a theory applied to bone was published in 1989 (Crolet et al.) and several other papers have been published for cortical bone (Aoubiza et al., 1991, 1992, Crolet et al. 1993, Aoubiza et al., 1996 and Crolet et al. 1996) or for cancellous bone by Fyhrie et al. (1989, 1990) Hollister et al. (1989) and other authors.

The modelling for homogenization of solid part of cortical bone (Crolet et al., 1992) needs the knowing of Young's moduli and Poisson's ratii for collagen fibers and hydroxyapatite, the orientation of the collagen fibers in each lamella of the osteon (Fig 1.1), the architecture of the osteon itself according several types described by Ascenzi et al. (1968) or Katz (1980) (Fig 1.2) and the architecture of interstitial system made on "old" osteons (ie having an higher mineralization, ...). Then a value of the elasticity tensor can be computed in each point of bone.

Fig 1.1 Lamella   Fig 1.2 Osteon

Fig 1 Architecture modelized for compact bone

Many tests have been made on various configurations of periods with random simulations for the interstitial system and identical results were obtained for the homogenized coefficients in accordance with experimental data.

The most interesting result which was obtained deals with the micro stresses at the osteonal level. For a macrocopic stress given (for instance a tensor where all coefficients are equal to zero except the term Σ_{33} which has a value of 15 Mpa) the computed stress field in and around osteon is quite different (for example the component σ_{11} which presents a spectrum between -3 and 3 Mpa). But the main disadvantage of this modelling is the fact that the obtained behavior's law is only linear.

3.2 The taking into account of fluid

A first study taking into account of fluid was achieved in 92. This modelling was very simple because only a hydrostatic pressure was added in the previous computation. Of course, homogenized elastic coefficients were different, but the obtained behavior's law was always linear. The most significative result was an important increasing of the magnitude of the microstresses at the osteonal level. For example, for the same data as previously, the component σ_{11} has a larger non symetric spectrum between -11 and 19 Mpa).

A second investigation for taking into account the fluid has been pursued. The bone was considered as a biphasic medium: the solid part had the elastic properties previously computed and the fluid part, characterized by its velocity and its pressure, had the properties of a viscous incompressible and Newtonian fluid.

Concerning the homogenization of a set fluid - structure in the framework of this study, two important assumptions have been made: the fluid domain is connex and a period can be defined. The size of the channels can be chosen different in each direction. The following equations are considered inside the period:

-   the elasticity equation for the solid

$$\rho^S \frac{\partial^2 u^S}{\partial t^2} \ - \ \text{div} \ \sigma^S \ = \ 0 \qquad [1]$$

-   the Navier Stokes equation and the incompressibility condition for the fluid

$$\rho^F \frac{\partial v^F}{\partial t} \ - \ \mu \Delta v^F \ + \ \nabla p \ = \ 0 \qquad [2]$$

$$\text{div} \ v^F \ = \ 0 \qquad [3]$$

-   the continuity of velocities and normal component of the stress at the interface

$$v^F \ = \ \frac{\partial u^S}{\partial t} \qquad [4]$$

$$\sigma^S n^S \ + \sigma^F n^F \ = \ 0 \qquad [5]$$

It can be noted that the Navier Stokes equation is not available for all the period because it is not true in caniculae. This point is actually in development.

After various mathematical developments, an homogenized problem has been written: the displacement U^0 is solution to the equation :

$$\frac{\partial^2 U^0}{\partial t^2} - \operatorname{div} \Sigma^0 = 0 \qquad [6]$$

where Σ^0 is the behavior's law given by :

$$\Sigma_{ij}^0 = A_{ijkl}\, \varepsilon_{kl}(U^0) + B_{ijkl}\, \varepsilon_{kl}\left(\frac{\partial U^0}{\partial t}\right) + C_{ijkl} * \varepsilon_{kl}(U^0) \qquad [7]$$

with :

$$A_{ijkl} = Q_{ijkl}^{bone} \qquad [8]$$

$$B_{ijkl} = 2\,\mu\, \frac{Y^F}{Y}\, \delta_{ij}\, \delta_{kl} \qquad [9]$$

$$C_{ijkl} = \alpha \int_{Y^F} \varepsilon_{ij}\left[\frac{\partial}{\partial t}(\eta^{kl})\right] + \beta \int_{Y^S} Q_{ijmn}\, \varepsilon_{mn}(\eta^{kl}) + \gamma \int_{Y^F} \pi^{kl}\, \delta_{ij} \qquad [10]$$

η^{kl} and π^{kl} being auxiliary functions.

This relationships is a combination of 3 terms: the two first are the respective contributions of the solid part and of the fluid part; the last term is more complex and is effectively the consequence of the interaction fluid - structure. This is a viscoelasticity behavior's law with long memory. This result is perfectly in accordance with all experimental observations. Now it is possible to compute with enough precision the evolution in time of mechanical properties of cortical bone.

4. FLUID FLOW INSIDE THE BONE

The idea that the fluid has an important role in the process of bony evolution is conforted by recent advances in this domain (Owan et *al.*, 1996 and Brown et *al.*, 1996). So if the bony domain is considered as a porous medium, is it possible to compute the flow inside a bone ? This computation is based on the following remark: in a point of bone it is possible to get (by direct measurement or by computation) the time depending values for displacement, strains and stresses. The neightborhood of this point can be considered as a specific domain (which is different of the previously studied periods) and the deformation of this domain can be computed. This deformation is in fact the summ of two terms: the compressibility of the solid part which can be estimated via the Poisson's ratio and the motion of the incompressible fluid. So a flow (incoming or outgoing) can be computed on the boundaries of this domain. It is important to note that the fluid motion inside this domain cannot be determined directly.
If the bony area corresponding to this specific domain is considered as a saturated porous medium, the fluid flow is locally determined by computational methods. In fact a better numerical simulation can be made by computing simultaneously the flow and the transport of solute (calcium, potassium, ...) inside such a domain. This porous medium is

assumed to be saturated and the equations of conservation for the mass of fluid and for the mass of solute are considered. With conventional notations summarized in Annex, these equations are written :

$$S_S \frac{\partial p}{\partial t} + v \frac{\partial \rho^F}{\partial t} \frac{\partial C}{\partial t} + \nabla [-\rho^F \frac{\kappa}{\mu} (\nabla P - \rho^F g)] = 0 \qquad [11]$$

$$v \rho^F \frac{\partial C}{\partial t} - \nabla [v \rho^F (d_m I + D) . \nabla C] = f(C) \qquad [12]$$

It can be noted that the density of fluid ρ^F depends on the calcium concentration C. In this simulation, the Darcy's law has been considered, but there is no difficulty to use an other law (Biot's law for instance).

The written equations are non linear and strongly related but it is possible to prove the existence and the unicity of a solution which is computed in an iterative process by a finite element method based on hexaedrons with a P2 interpolation for the space discretization and an implicit scheme for the time discretization. Then, with adequate initial and boundary conditions, flow and calcium concentration are computed inside a given bony domain.

Consider the example of a parallepided with a given flow from upstream to downstream. An initial calcium concentration is given equal to zero inside the domain. If a strictly positive value is given on the upstream side, it is natural to get a non nul concentration inside, but if this positive value is imposed on the downstream side, it seems that the calcium concentration is always nul inside. In fact, because of the gravitational term in the Darcy's law and the variable density ρ^F, there is an intrusion of calcium in the domain with an other flow in the opposite sens of the imposed flow. This particular phenomenon is in accordance with the physical aspects of solute transport in porous media and could perhaps be used to explain some intrusion of calcium (or of other mineral solute) in porosities of microstructure (Crolet et al., 1995).

5. KEYWORDS

Computational method, Homogenization, Fluid - structure interaction, Porous media, Transport of mineral solute

6. REFERENCES

M.P. Bendsoe, J.M. Guedes, R.B. Haber, P. Pedersen and J.E. Taylor, An analytical model to predict optimal material properties in the context of optimal structural design, J. Appl. Mech., 1994, 61, 930-937.

J.M. Crolet and B. Aoubiza, Homogenization method, a tool of investigation in Biomechanics, Proceedings of the 5th International Symposium on Numerical Methods in Engineering. Eds R. Gruber, J. Periaux, R. P. Shaw. Computational Mechanics Publications. Springer-Verlag, 1989, Vol 2, pp 383-388.

B. Aoubiza and J.M. Crolet, OSTEON: Simulation of mechanical behavior of multiscale composite, New Advances in Computational Structural Mechanics. IACM and GAMNI Ed, 1991, pp 731-738.

B. Aoubiza, J.M. Crolet, A. Meunier and L. Sedel, A new process of mechanical characterization of compact bone, Recent advances in 'Computer Methods in Biomechanics and Biomedical Engineering'. J. Middleton, G.N. Pande and K.R. Wiliams Ed. Books and Journals International, 1992, pp 288-297.

J.M. Crolet, B. Aoubiza and A. Meunier, Compact bone: Numerical simulation of mechanical characteristics, J. of Biomech., 1993, Vol. 26, N°6, pp 677-687.

J.M. Crolet and B. Aoubiza, Présentation d'une simulation numérique du comportement mécanique de l'os compact, Archives Internationales de Physiologie, de Biochimie et de Biophysique, 1993, Vol. 101, Fasc. 5, pp 131-132.

B. Aoubiza, J.M. Crolet and A. Meunier, On the mechanical characterization of compact bone structure using the homogenization theory, J. Biomechanics, 1996, Vol 29, N°12, pp 1539-1547.

J.M. Crolet, B. Aoubiza and A. Meunier, Osteon's anisotropy, J. Biomechanics, 1996, Vol 29, N°12, pp 1675-1678.

D.P. Fyhrie, K.J. Jepsen, S.J. Hollister and S.A. Goldstein, Predicting trabecular bone strength and micro-strains using homogenization theory: application to a vertebra, Proceedings of the XIIth International Congress of Biomechanics, 1989, 173-174.

D.P. Fyhrie and S.J. Hollister, A tissue strain remodeling theory for trabecular bone using homogenization theory, Trans. 36th Ann. Meeting Orthop. Res. Soc., p 76, ORS Las Vegas 1990.

S.J. Hollister, D.P. Fyhrie, K.J. Jepsen, and S.A. Goldstein, An analysis of trabecular bone micro-mechanics using homogenization theory with comparison to experimental results, Proceedings of XIIth International Congress of Biomechanics, 1989, 110-111.

S.J. Hollister, D.P. Fyhrie, K.J. Jepsen, and S.A. Goldstein, Application of homogenization theory to the study of trabecular bone mechanics, J. Biomechanics, 1991, Vol. 4, 455-473.

A. Ascenzi and E. Bonucci, The compressive properties of single osteon, Anat. rec., 1968, 16, 377-392.

J.L. Katz, The structure and biomechanics of bone, In "Mechanical Properties of Biological Materials, 1980.

I. Owan, D.B. Burr, Ch. Turner, J. Qiu and R.L. Duncan, Osteoblasts do not respond to physiological levels of mechanical strain, but are responsive to fluid effects, Trans. 43th Ann. Meeting Orthop. Res. Soc., p 176, ORS San Francisco, 1996.

T.D. Brown, D.R. Pedersen, M. Bottlang and A.J. Banes, Fluid structure interaction as a determinant of nutrient medium reactive stress in mechanically stimulated cultures, Trans. 43th Ann. Meeting Orthop. Res. Soc., p 175, ORS San Francisco, 1996.

J.M. Crolet, F. Jacob, P. Lesaint and J. Mania, Tridimensional simulation of intrusion of salted water in an aquifer, in Computer modelling of seas and coastal regions, Vol. 2, Computational Mechanics Publications Ashurst Lodge, Southampton, 1995, pp 125-132.

7. NOTATIONS

ρ^S density of the solid part

ρ^F density of the fluid

u^S displacement of a solid point

σ^S stress tensor in the solid part

σ^F stress tensor in the fluid part

v^F velocity in the fluid

p pressure of the fluid

U^0 displacement in the homogenized problem

Σ^0 stress tensor in the homogenized problem

μ viscosity of the fluid

Y period of bone

Y^F fluid part of the bony period

Y^S solid part of the bony period

S^S storage coefficient

C concentration of a mineral solute

κ permeability tensor

g gravitation vector

ν porosity

d_m molecular diffusion

D dispersivity

THREE DIMENSIONAL MODEL OF BONE EXTERNAL ADAPTATION.

Fridez P[1,2], Rakotomanana L[1,2], Terrier A[1,2], Leyvraz PF[2]

1. ABSTRACT

Alteration of physiological stress within bone induces unbalance of bone cell activity. Two types of functional adaptation may then occur simultaneously: internal adaptation (relative density and/or anisotropy variation with a fixed geometry) and external adaptation (periosteal and endosteal apposition or resorption changing the geometry). In this study, a 3D model of bone external adaptation was developed. The mechanical stimulus controlling the change of bone shape was based on an invariant scalar function, based on the stress tensor, the orientation unit vector (anisotropy) and the external normal unit vector. The evolution law was based on a sigmoid-type function including an equilibrium zone. The time integration was performed by a forward Euler scheme. The model was first applied on 2D bone samples. A trabecular bone unit was simplified by a 2D square with a central hole. This unit was used to build a piece a trabecular bone and applied to study the adaptation of the internal surface of trabecular bone. From the pores geometry, two internal variables were defined: the relative density and the anisotropy direction. After several loading alterations, it was observed that the adaptation of relative density occurred more quickly than the adaptation of anisotropy. This adaptation model was also applied to evaluate the evolution of the external shape of a turkey ulna subjected to specific external load. The results were compared with experimental data with good correlation.

2. INTRODUCTION

After load alteration, response of bone tissue takes two forms: internal and external. The first one describes trabecular structure adaptation and the second one describes external

Keywords: bone, external adaptation, turkey ulna

---

1. Biomedical Engineering Laboratory, Federal Institute of Technology, 1015 Lausanne, Switzerland.
2. Orthopaedic Hospital, Av. Pierre-Decker 4, 1005 Lausanne, Switzerland.

shape adaptation. For the last twenty years, many aspects of bone adaptation have been studied; however, most attention was payed to internal bone adaptation. Indeed, several models were developed to simulate density (*e.g.* [1], [2], [3]) adaptation, anisotropy adaptation [4] or both of them [5] to mechanical stresses, while only a few models were developed for external adaptation (*e.g.* [6], [7]). This difference of interest may be explained by two main reasons: first, external adaptation occurs more slowly and is usually not so important than internal adaptation. In particular, after total hip arthroplasty, which is one of the major clinical applications of bone remodelling models, internal adaptation is predominant [8]. However, even in this case, X-ray studies showed a important cortical hypertrophy of the diaphysis changing the external shape of the femur [8] and by this way the mechanical properties of the femur. The second and certainly major reason is that external adaptation is very difficult to implement in a FEM algorithm. Indeed, as the external shape changes, external elements of the mesh do also change and may become incompatible with the FEM analysis. In the present work, a model of external adaptation was developed and a mechanical stimulus, based on plasticity criterion, was originally introduced. This model and its implementation were performed on the same basis than a previous model of internal adaptation [21], allowing to combine both internal and internal adaptation in the future.

3. METHODS

3.1. Model of bone external adaptation

By neglecting inertial effects and volume forces, the equilibrium equation completed by boundary conditions was used to calculate the stress within bone:

$$\begin{cases} \nabla S(x, t) = 0 , & \forall x \in \Omega \\ S(x, t) = \bar{S}(x) , & \forall x \in \partial\Omega_S \\ u(x, t) = \bar{u}(x) , & \forall x \in \partial\Omega_u \end{cases} \tag{1}$$

S being the 2nd *Piola-Kirchhoff* stress, u the displacement, Ω the bone volume, $\partial\Omega_S$ and $\partial\Omega_u$ the bone boundaries where respectively stress and displacement were applied. Bone was assumed to be homogeneous and transverse isotropic. Anisotropy might be characterized by the orientation unit vector $v(x)$ [9]. The stress-strain law wrote:

$$S = D(v){:}E \tag{2}$$

where D was the elastic stiffness and E the *Green-Lagrange* strain (*e.g.* [10]). Bone external adaptation was characterized by the evolution of the boundary projected on the normal unit vector $U_n \equiv u \cdot n$. The evolution law of external adaptation was then defined by a sigmoid function, $\forall x \in \partial\Omega_S$:

$$\frac{\partial}{\partial t} U_n(x, t) = \widehat{V}_n(\psi, \psi_e, s_1, s_2, s_3, U_r, U_a) \tag{3}$$

where ψ was the variable and $\psi_e, s_1, s_2, s_3, U_r, U_a$ the parameters characterizing the function (Fig. 1) [11]. The unit vector v was held constant with respect to time. The stimulus function ψ admitted as arguments the stress S, the orientation vector v and the normal unit vector n:

$$\psi(x, t) = \hat{\psi}[S(x, t), v(x, t), n(x, t)] \tag{4}$$

By applying the *Cauchy*'s theorem on representation of tensorial functions, we could

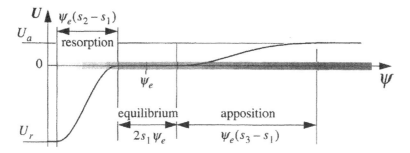

Fig. 1. The evolution law included a equilibrium zone, bounded by the lower limit $\psi_e(1-s_1)$ and upper limit $\psi_e(1+s_1)$. A maximal resorption rate U_r and apposition rate U_a were also introduced respectively below $\psi = \psi_e(1-s_2)$ and above $\psi = \psi_e(1+s_3)$. The grayscale levels correspond to stimulus intensity.

write the general form of stimulus function:

$$\psi = \widehat{\psi}(trS, trS^2, trS^3, v \cdot Sv, v \cdot S^2 v, n \cdot Sn, n \cdot S^2 n, v \cdot Sn, v \cdot S^2 n)$$

At this step, we preferred to relate the stimulus with physical invariants of stress tensor:

$$p = \frac{1}{3}trS \qquad \tau_{oct} = \sqrt{trS^2 - 3p} \qquad \sigma_{vv} = v \cdot Sv \qquad \tau_{vv} = \sqrt{v \cdot S^2 v - \sigma_{vv}^2}$$

$$p_{nn} = n \cdot Sn \qquad p_{nt} = \sqrt{n \cdot S^2 n - p_{nn}^2} \qquad p_{nv} = v \cdot Sn \qquad p_n p_v = v \cdot S^2 n$$

For the sake of simplicity, in the present work, we decomposed the stimulus into three additive functions (volume stimulus, surface stimulus, cross terms stimulus):

$$\psi = \widehat{\psi}_v(p, \tau_{oct}, \sigma_{vv}, \tau_{vv}) + \widehat{\psi}_s(p_{nn}, p_{nt}) + \widehat{\psi}_c(p_{nv}, p_n p_v)$$

Further, the cross terms stimulus was neglected: $\psi_c \equiv 0$. Moreover, the volume stimulus was set to the anisotropic plastic *Hill* criterion: $\psi_v = \psi_p$ [12]. For the surface stimulus, we adopted the simplest quadratic form as follows:

$$\psi_s = r_{p_{nn}p_{nn}} p_{nn}^2 + r_{p_{nt}p_{nt}} p_{nt}^2 \qquad (5)$$

We normalized ψ_s by using a similar approach as for elaborating the plasticity criterion. The surface plasticity criterion was chosen in such a way that the stimulus value was equal to 1 when stress reached the elastic limit at the surface. We then obtained:

$$r_{p_{nn}p_{nn}} = \frac{9}{2}p_1 + \frac{3}{2}p_2 \qquad r_{p_{nt}p_{nt}} = 2p_2 \qquad (6)$$

where p_i are the plastic coefficients developed in [12]. Finally, we proposed the stimulus function:

$$\psi = \alpha \widehat{\psi}_v(p, \tau_{oct}, \sigma_{vv}, \tau_{vv}) + (1-\alpha)\widehat{\psi}_s(p_{nn}, p_{nt}) \qquad (7)$$

In spite of previous assumptions, such a form of stimulus remained sufficiently general to include most of bone adaptation models (internal and external) [12]. The evolution equations (3) was completed by initial conditions $u(x, 0) = u_0(x)$.

3.2. Numerical strategies and parameters identification

For complex systems, a FEM-code [13] was used to calculate the stress field. The evolution law (3) was solved at each nodal point of the FEM-mesh by using an explicit Euler scheme. The mathematical model contained six unknown parameters. The identification of these parameters was performed with the experimental results published in the literature (e.g. [14]):

$$\psi_e = 0.02 \qquad s_1 = 0.35 \qquad s_2 = 0.8 \qquad s_3 = 2.7$$
$$U_r = -0.012\,\text{mm/day} \qquad U_a = 0.004\,\text{mm/day}$$

4. RESULTS

Three applications were retained: a 2D three points bending beam, a 2D hollow square cell and a 3D turkey ulna.

4.1. Three points bending beam

The model was applied to simulate the shape adaptation of a three point bending beam. After adaptation, the stimulus distribution reached the equilibrium zone and the shape of the beam was similar to that of iso-stressed beams [15].

Fig. 2. Three points bending of 2D beam. Grayscale level correspond to stimulus intensity: white for low value and dark for high value. See Fig. 1.

4.2. Hollow square cells

The model was also used to describe internal adaptation by simulating the adaptation of internal trabecular bone surfaces. First, a single 2D trabecular bone unit was modeled by a circular hole inside a square, hereafter named hollow square-cell (Fig. 3). Each trabec-

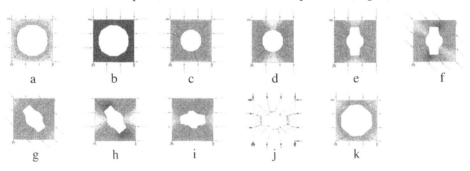

Fig. 3. Trabecular bone unit and successive loading conditions. Grayscale levels correspond to stimulus intensity: white for low value and dark for high value. See Fig. 1.

ulae (plain part of the hollow square cell) was considered homogeneous and isotropic [16]. α was set to 0.8 in (7). With the trabecular unit sample, successive loading conditions were applied. Initially, the circular pore in equilibrium under an iso-static stress (Fig. 3a). Next, the stress was doubled (Fig. 3b), inducing shrinkage of the pore and a new equilibrium geometry (Fig. 3c). A second imbalance was induced by applying a vertical compression (Fig. 3d). After bone adaptation, an orientation of the pore in the direction of the principal stress (Fig. 3e) was observed. The reaction was similar when the orientation of the compressive load was changed (Fig. 3f to Fig. 3i). At the end, the initial conditions (Fig. 3j) were again applied and the initial configuration was recovered (Fig. 3k). To compare the rate of density adaptation to the rate of anisotropy re-orientation, each pore was characterized by three variables: 1) the porosity, 2) the ratio of inertial moments, 3) the orientation of principal axis with respect to the horizontal. As displayed in Fig. 4, the density adaptation occurred in about 50 days in the first case (from b to c) and in less then 35 days in the final case (from j to k). Anisotropy adaptation was sensibly slower (about 250 days).

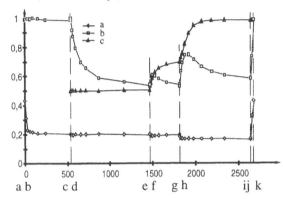

Fig. 4. Evolution of: a) porosity, b) inertial axis ratio, c) principal axes direction; during successive loading alterations. The letters under the figure correspond to Fig. 3.

Second, this hollow square-cell was periodically repeated to get a trabecular bone sample containing several pores (Fig. 5 and Fig. 6). This allowed us to study, in relatively simple configurations, not only the variation of density but also the trabecular re-orientation. Indeed, anisotropy and density were calculated from the shape and the area of the different cavities. The model was used to compare the adaptation of tissue at this trabecular level to the results obtained from a bone internal adaptation model with the continuum assumption. Simple loading conditions were applied on a rectangular bone sample: vertical compressive stress increased linearly (from right to left) on the upper side (Fig. 5a).

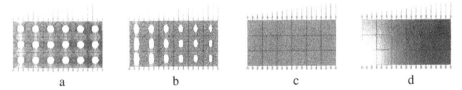

Fig. 5. Loading conditions and initial (a) and final (b) configurations of the rectangular sample. Grayscale levels correspond to stimulus intensity: white for low value and dark for high value (see Fig. 1). The contour plots (c and d) are the equivalent continuum model of fig. (a and b). Grayscale levels correspond to porosity: white for low value and dark for high value.

After adaptation (Fig. 5b), we noticed a decrease of the pore area in the region of higher stress (Fig. 5b). The porosity was then calculated and a contour plot was realized (Fig. 5c and Fig. 5d). The initial plot corresponded to an homogeneous density and the final plot was very similar to the results obtained with a model of internal adaptation [3] applied to the same problem. Third, to evaluate the re-orientation of the anisotropy with a bone sample composed of several trabecular units, more complex loads were applied to a square sample (Fig. 6). By this way, the principal stress direction was not uniform within the bone sample. In the initial situation (Fig. 6a), as the pores were circular, no orientation might be defined. After adaptation (Fig. 6b), the principal direction was calculated for each pore. We noticed a non-homogeneous orientation within the sample. The orientation of the anisotropy was approximately directed by the direction of principal stress tensor.

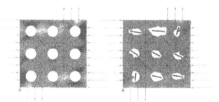

Fig. 6. Re-orientation of the anisotropy. Grayscale levels correspond to stimulus intensity: white for low value and dark for high value. See Fig. 1.

4.3. Turkey ulna

To compare the model with experimental 3D data, the results of Brown *et al.* [17] were used. In their study, the ulna of an adult turkey was surgically removed and kept alive, isolated from the organism. The ulna was then installed on a test machine (Instron) and cyclic stresses were applied. After eight weeks of loading, the ulna adaptation was evaluated from transversal slices. As the settled load was not physiologic, the results showed a geometry change. To simulate this experiment with the present model, a 3D-mesh of the ulna was reproduced. Boundary conditions similar to those of the experimental study were then applied. The results for the turkey ulna are presented in Fig. 7. Adaptation of the transversal slices was not uniform along the longitudinal axis. In the initial situation (Fig. 7a), an over-stimulated zone was present in the diaphysis. After adaptation, this zone (Fig. 7b) became thicker. At the opposite, the initially under-stimulated parts epiphysis became thinner. As displayed in the transversal sections, adaptation occurred mainly in the diaphysis accordingly.

Fig. 7. Turkey ulna. Before and after adaptation in longitudinal and transversal slices. Left/right correspond to cranial/caudal and up/down to proximal/distal. Grayscale levels correspond to stimulus intensity: white for low value and dark for high value. See Fig. 1.

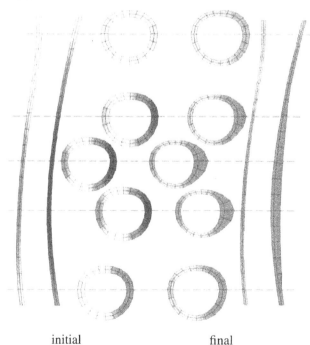

initial   final

5. DISCUSSION

5.1. Hollow square cell

Density evolutions calculated with the trabecular bone unit were quite realistic: densification was important at the beginning and then vanished at equilibrium. Moreover, the alignment of anisotropy to principal direction of stress was about 5 times slower than the adaptation of density. This results was also consistent with experimental observations [18]. During the anisotropy re-orientation, the density remained constant. The difference between the initial and final geometry was due to the equilibrium zone width, which permitted different similar configurations to be in equilibrium.

5.2. Trabecular bone

The results obtained with the rectangular trabecular sample were similar to the results obtained with a model of internal adaptation ([2], [3]). Indeed, the contour plot of the relative density, calculated from the relative hole area, corresponded to the relative density distribution obtained with the bone internal adaptation model. The complex load applied to the square sample showed the ability of the model to predict the re-orientation of anisotropy after modification of the external forces. Thus, the present model showed its ability to describe the microscopic nature of internal adaptation by relating the anisotropy direction ν and the relative density ϕ with the geometry.

5.3. Turkey ulna

The boundary condition used in the ulna simulation was a uni-axial stress. Because of the bone curvature, these external forces induced a bending moment, which was maximal in the diaphyseal part, were bone apposition was maximal. This apposition appeared only in the concave zones (caudal) of the periosteum and resulted in a diminution of the bone curvature and also of the internal torque. The resorption produced might be explained by several reasons. First, the static applied forces in this study were the mean value of the cyclic forces used by Brown and dynamic stress is less osteogenic than static one [19]. The corresponding static force should than be greater to produce a similar bone reaction. This effect was certainly important because of the apposition of disorganized fibrous bone on the extremities of the ulna. Second, the geometry of our mesh did not correspond exactly to the geometry of the real ulna. Therefore, the simple flexion that was applied might induce errors. Moreover, the position of the neutral axis on each transversal sections was different the one obtained by Brown. Despite the different assumptions underlying the present model, the results were qualitatively close to the experimental measurements.

5.4. Final remarks

The goal of this study was to build a model of external bone adaptation. The mechanical stimulus was defined by 2 contributions: volume and surface. The first one was based on the 4 (volume) mechanical invariants and second one the 2 (surface) mechanical invariants. This formulation was general enough to describe most of the stimuli met in the literature. The volume stimulus was set to the plasticity criterion of Hill. The surface stimulus was set to a normalized form which was built in a similar manner than a plastic criterion. Several outlooks are opened. First, this external adaptation model will be combined to an internal adaptation model. The problem of elements inter-penetration should be avoided by a re-meshing of the bone volume after each iteration. The whole stress history might also be used to obtain a time-delay response of bone reported in many studies ([20], [21]).

6. REFERENCES

1. Beaupré G. S., Orr T. E., and Carter D. R., An approach for time-dependent bone modeling and remodeling: Theoretical development, J Orthop Res, 1990, Vol. 8, 662-670.
2. Weinans H., Huiskes R., and Grootenboer H. J., The behavior of adaptive bone-remodeling simulation models, J Biomech, 1992, Vol. 25, 1425-1441.
3. Terrier A., Rakotomanana R., Ramaniraka N, and Leyvraz P., Adaptation models of aniso-tropic bone, Computer Methods in Biomechanical and Biomedical Engineering, 1997, Vol. 1, 47-59.
4. Cowin S. C., Sadegh A. M., and Luo G. M., An evolutionary Wolff's Law for trabecular architecture, J Biomech Eng, 1992, Vol. 14, 129-136.
5. Jacobs C. R., Simo J. C., Beaupré G. B., and Carter D. R., Anisotropic adaptive bone remo-deling simulation based on principal stress magnitude, *In* 41st annual meeting of Orthopae-dic Research Society, Orlando, 1995, 178.
6. Cowin S., Hart R., Balser J., and Kohn D., Functional adaptation in long bones: establishing in vivo values for surface remodeling rate coefficients, J Biomech, 1985, Vol. 18, 665-684.
7. Hart R. T., A theoretical study of the influence of bone maturation rate on surface remode-ling predictions: idealized models., J Biomech, 1990, Vol. 23, 241-257.
8. Engh C. A., McGovern T. F., Bobyn J. D., and Harris W. H., A quantitative evaluation of periprosthetic bone-remodeling after cementless total hip arthroplasty, J Bone Joint Surg, 1992, Vol. 74A, 1009-1020.
9. Boehler J. P., Applications of tensor functions in solid mecanics, Springer-Verlag, Udine, 1987.
10. Gurtin M. E., An Introduction to Continuum Mechanics, Academic Press, New York, 1981.
11. Fridez P., Modélisation de l'adaptation osseuse externe, *In* Physics Department, EPFL, Lau-sanne, 1996, .
12. Rakotomanana R. L., Leyvraz P. F., Curnier A., Heegaard J. H., and Rubin P. J., A finite ele-ment model for evaluation of tibial prosthesis-bone interface in total knee replacement, J Biomech, 1992, Vol. 25, 1413-1424.
13. Curnier A., Computational methods in solids mechanics, Kluwer Academic Publishers, Dor-drecht-Boston-London, 1994.
14. Carter D. R., Nauenberg T., Bouxsein M. L., and Mikic B., Using clinical data to improve computational bone remodeling theory, *In* Trans 39th ORS, San-Fransisco, 1993, 123.
15. Mirolioubov, and et al., Résistance des matériaux: manuel de résolution de problèmes, Mir, Moscou, 1971.
16. Mullender M. G., Huiskes R., and Weinnans H., A physiological approach to the stimulation of bone remodeling as a self organisational control problem, J Biomech., 1994, Vol. 27, 1389-1394.
17. Brown T. D., Petdersen D. R., Gray M. L., Brand R. A., and Rubin C. T., Toward an identifi-cation of mechanical parameters initiating periosteal remodeling: a combined experimental and analytic approach, J Biomech, 1990, Vol. 23, 893-905.
18. Fritton S. P., and Hart R. T., Simulation of in vivo trabecular bone adaptation, *In* Trans 41st ORS, Orlando, 1995, 176.
19. Odgaard A., and Weinans H., Bone Structure and Remodeling, *In* Recent Advances in Human Biology, World Scientific, 1994, .
20. Levenston M., Beaupre G., Jacobs C., and Carter D., The role of loading memory in bone adaptation simulations, Bone, 1994, Vol. 15, 177-186.
21. Terrier A., Rakotomanana R. L., and Leyvraz P. F., Effect of loading history on short term bone adaptation after total hip arthroplasty, *In* Computer Methods in Biomecanics and Bio-medical Engineering, Barcelona, 1997.

QUANTIFICATION OF THE VALIDITY AND ACCURACY OF AN INDIRECT LARGE SCALE FINITE ELEMENT APPROACH TO DETERMINING TRABECULAR BONE TISSUE MODULUS

C. R. Jacobs[1], B. R. Davis[2], E. M. Paul[3], A. M. Saad[4], and D. P. Fyhrie[5]

1. ABSTRACT

In the field of biomechanics, mathematical models are often used to indirectly compute material properties from test results of intact structures. Traditionally these models have been limited to simple structures such as columns, beams, and tubes. In the last few years the advent of large-scale computer methods in computational mechanics combined with automated voxel-based meshing techniques based on microtomographic datasets have made it possible to expand this approach to the determination of tissue or constituent material properties for large-pore foams such as cancellous bone. The goal of this project is to quantify the accuracy of this approach using a porous mechanical analogue to cancellous bone. The analogues are made of a thermoplastic material which has been treated with an additive to make it x-ray absorptive (commercially available bone cement). Voids were produced by including paraffin beads with the material as it is formed, which are later melted and removed. High resolution microCT scans are made of the analogues, and voxel-based finite element (FE) models prepared from the resulting data. When combined with direct mechanical testing of each sample, the FE models allow the constituent stiffness to be computed indirectly. The predicted stiffness can then be compared with the known stiffness of the void-free plastic as a measure of the accuracy of the method. Furthermore we evaluate the potential impact of testing boundary conditions, element size, and assumed constituent Poisson's ratio on

---

Keywords: Trabecular bone, Finite element modeling, Voxel-based meshing, MicroCT, Young's modulus

[1]Assistant Professor, Musculoskeletal Research Laboratory/Department of Orthopaedics and Rehabilitation, Pennsylvania State University, Hershey PA, 17033, USA
[2]Research Assistant, Musculoskeletal Research Laboratory/Department of Orthopaedics and Rehabilitation, Pennsylvania State University, Hershey PA, 17033, USA
[3]Senior Research Support Specialist, Musculoskeletal Research Laboratory/Department of Orthopaedics and Rehabilitation, Pennsylvania State University, Hershey PA, 17033, USA
[4]Research Assistant, Breech Research Laboratory, Henry Ford Hospital, Detroit MI, 48202, USA
[5]Head/Biomechanics Section, Breech Research Laboratory, Henry Ford Hospital, Detroit MI, 48202, USA

the accuracy of the method. We also compare the bone cement analogues to actual trabecular bone samples in terms of morphology. Although this novel combination of cutting edge technology from two distinct fields (large-scale FE analysis and microCT) holds great promise, this approach is untested and no quantification of its accuracy has been attempted using a material with known constituent properties. When validated, the technique will have an enormous impact in the field of orthopaedics as well as the study of large pore size foams in general. In our laboratory, the ability to indirectly determine cancellous bone tissue stiffness will greatly enhance our efforts to discover more about the mechanisms involved in bone adaptation and age-related osteoporosis.

2. INTRODUCTION

Cancellous bone can be considered as a two phase material made up of tissue and void space. The bone tissue is organized into a large pore open cell structure similar to a foam. In order to understand the function of cancellous bone tissue and the pathologies it is subject to, it is important to chracterize its mechanical behavior. This is distinguished from determining the apparent mechanical properties of cancellous bone which are a function of both the tissue behavior and its microstructural organization within the surrounding void space. Significant progress has been made in determining cancellous bone's apparent properties, however its tissue behavior has not been determined to a similar degree.

Applying traditional testing methods to determining tissue properties of cancellous bone involves machining and mechanically testing a small sample of bone [1]. Micro-specimens of trabecular bone have recently been testing in isolated experiments [2]. However, this approach is exceedingly labor intensive and problematic both due to possible damage caused by machining and a possible specimen size effect.

Constructing a mathematical model of the deformation of a given bone structure under load allows an investigator to predict the structure's elastic behavior based on assumptions regarding tissue material behavior and geometry. This effectively distinguishes the effect of tissue stiffness from the effect of geometric shape on structural stiffness. If this is combined with known experimentally measured structural behavior and the measured geometric configuration, it is possible to adjust the assumed tissue properties until the predictions of the mathematical model match the measured observations. This approach has been applied with great success in diaphyseal cortical bone due to its small pore size and relatively fine ultrastructure. Cancellous bone is not amenable to such an approach since a simple or analytical (closed form) solution for the complex geometry is not available.

Recent advances in computer power and in the field of computational mechanics have made it feasible to solve very large-scale finite element models, which can include highly detailed representations of complex shapes or geometry's [3, 4]. In order to construct a tissue level micromechanical model of cancellous bone (one that models the microstructure directly rather than using a continuum approximation), the three dimensional architecture must be determined. Several techniques are available, however micro-computed tomography has the advantage of being nondestructive, allowing for subsequent mechanical testing [5, 6]. MicroCT is essentially a miniaturized version of the common clinical CT scanner and uses mathematical reconstruction techniques to determine the ability of small regions of tissue (voxels) to absorb x-rays by repeatedly passing a small x-ray beam through the sample. These data can be directly and quickly converted to finite element meshes by treating each voxel judged to be part of the bone of interest as a finite element [7, 8]. The resulting mesh resembles a stack of small blocks or sugar cubes. Although significant errors are found at surfaces where the jagged boundary is a poor approximation, voxel-based meshes have been shown to be accurate in predicting overall structural behavior and interior deformations [9, 10].

Quantitative verification of the accuracy of voxel-based FE meshes was attempted by Hollister et al. [11] on 53 samples of human trabecular bone modeled with a 50μm element size. The models tended to underpredict the experimentally measured apparent stiffnesses with errors of 31% to 38%. Chavez and Keaveny [12] conducted a 2D analysis of three samples of bovine trabecular bone and found that the predicted apparent structural stiffness changed by as much as 134% when the mesh size was increased from 20μm to 100μm. Van Rietbergen et al. [13] found that the apparent stiffness changed by 19% in a single sample of whale vertebral bone when the mesh size was changed from 20μm to 100μm. This paper was the first to suggest that if the FE predicted apparent stiffness was combined with an experimental measurement of apparent stiffness, the tissue Young's modulus could be determined indirectly.

To date, there is no consensus in the literature as to the level of accuracy of large-scale voxel-based FE models of cancellous bone and how it is influenced by trabecular architecture and modeling parameters. Furthermore, no direct quantification of the validity of indirect tissue modulus determination by combining microstructural FE modeling with experimental measurements at the apparent level has been attempted. The goal of this project was to demonstrate the feasibility of this approach by quantifying its best-case accuracy in determining the tissue modulus of trabecular analogues fabricated of a material with known properties. Furthermore, the potential impact of the assumed tissue Poisson's ratio and boundary conditions was assessed.

3. METHODS

Our approach to verifying the practical accuracy of voxel-based FE optimization for determining the tissue stiffness of cancellous bone is to use a plastic bone analogue with a known tissue modulus. Voids or pores are introduced into a thermoplastic with homogeneous and isotropic material behavior, to produce a mechanical analogue to cancellous bone. Polymethyl methacrylate (PMMA) is commonly used as "bone cement" in orthopaedic surgery. When the two portions are mixed, the cement behaves as a thick viscous fluid for roughly two minutes during which time it can be forced or molded into and around small features. The final properties of bone cement vary from batch to batch and are dependant on the temperature and humidity during mixing.

Paraffin beads (approximately 2mm in diameter) were used to create a porous foam-like microstructure resembling that of cancellous bone. Beads were packed into cylindrical molds (5ml syringes) which have a narrow outlet to allow the release of air during injection without allowing the beads to escape. A 40g premeasured package of commercially available cement (Osteobond, Zimmer, Warsaw IN, USA) was thoroughly mixed under vacuum according to the manufacturers instructions to minimize any air bubbles. While still in a liquid state the cement was injected into the bead filled mold (figure). Also, a portion of the cement was injected into an empty mold to produce a non-porous control to account for variations in cement modulus from package to package and mixing conditions. After the cement had hardened in the mold, the mold was cut away freeing the sample. A low-speed irrigated diamond saw was used to prepare 5mm cubic specimens with precisely parallel faces. The samples (porous and non-porous) were baked at 120°C for six hours (1 hour with each face down) to remove the paraffin beads.

Three porous and one non-porous samples were prepared using this procedure and sent to the microCT facility of the Breech Research Center at Henry Ford Hospital. The samples were scanned with a 25μm voxel size. The microCT data was converted to a finite element mesh by selecting a CT value, below which voxels are considered to be void space. Histograms of CT value for these samples exhibit two sharp peaks corresponding to the two phases of the sample (cement and void) separated by a broad valley. A threshold was selected midway between the two phases that was effective at distinguishing cement voxels from void voxels. One specimen had to be rejected due to malalignment in the scanner.

Once the cement voxels were identified, the FE mesh was created using 50μm cubic hexahedral elements (Figure 1), each made up of eight microCT voxels. The resulting models were solved using an element-by-element solution scheme with an iterative pre-conditioned conjugant gradient solver implemented on a DEC 300 model 900 workstation. The boundary conditions of the model were specified to reproduce a standard parallel plate compression test, namely a downward vertical displacement applied to the top surface and no vertical displacement allowed at the bottom surface. The top and bottom boundary conditions were applied to a layer of nodes 400μm thick to avoid problems with small discontinuities and errors in alignment. Lateral displacement was allowed at

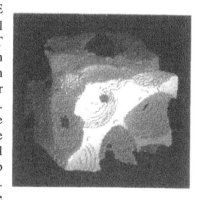

Figure 1: An FE mesh prepared for one of the PMMA analogues with a 50 μm element size.

both surfaces to produce frictionless or lubricated loading platen conditions. The resultant forces at the restrained nodes were summed and divided by the applied displacement, resulting in the predicted structural stiffness of the specimen.

Each specimen was then mechanically tested using a servohydraulic materials testing machine (Interlaken, Eden Prairie, MN, USA). The cubes were oriented identically in the microCT scanner, the FE model, and materials testing machine. The slope of the linear portion of the load-displacement curve was taken as the structural stiffness.

A finite element optimization procedure was applied to each specimen independently. The assumed Young's Modulus of the FE model was adjusted so that the computer predicted structural stiffness of the specimen matched the experimentally measured structural stiffness. Since the deformations are linear and elastic in this case the structural stiffness depends linearly on Young's Modulus, and only one modulus adjustment was required to achieve agreement. The modulus which resulted in agreement between the computer model and the experimentally determined structural stiffness was taken as an indirect measure of the tissue constituent modulus. The experimentally measured modulus for the solid specimen served as a gold standard for the indirectly measured stiffness of the porous specimens.

It is expected that the accuracy of this approach for modulus determination would depend on the morphology of the microstructure being analyzed. Therefore it is desirable to quantify and compare the microstructure of the bone cement analogues to that of actual trabecular bone. Specifically, accuracy is likely to depend on the ratio of element size to the microstructural feature size. To this end, mean trabecular thickness and spacing was measured from the microCT data for the bone cement specimens. For comparison, the measurement was repeated on ten rat vertebral bodies from another study. The L6 vertebra of male Fischer-344 rats were isolated and the end plates removed leaving a 4.2mm tall specimen with flat parallel top and bottom surfaces. The animals ranged in age from 4mo to 24mo.

In order to determine the sensitivity of the method to element size the microCT data can be artificially "coarsened" by replacing each group of eight adjacent 25μm voxels with a single 50μm voxel that has a CT value equal to the average of the eight it replaces. This approach was applied to the vertebral bone data to determine influence of element size when the method is applied to actual trabecular bone specimens.

Another possible source of error is the assumption of frictionless boundary conditions. It has recently been shown that the assumption of frictionless boundary conditions is suspect in compression testing of cancellous bone [14]. Despite efforts to lubricate the bone/platen interface, a frictionless boundary condition is likely to be difficult or impossible to achieve practically. Therefore, to assess the potential magnitude of error introduced by this assumption the 50 μm rat vertebral models were reanalyzed using a fully bonded boundary condition.

In this study cancellous bone tissue is modeled as an isotropic linear elastic material. As such two material properties must be specified, the Young's Modulus and Poisson's ratio. Since only one experimental value is available for matching with the FE results, it is only possible to determine one material property from the optimization procedure, namely the Young's Modulus. The Poisson's ratio must be assumed a priori. In the experiments above the Poisson's ratio of cortical bone was assumed to be 0.3. However, no data is available for the Poisson's ratio of cancellous bone tissue. Therefore we assessed the sensitivity of the method on the assumed Poisson's ratio by repeating the analysis with values of 0.0 and 0.45.

4. RESULTS

The Young's Modulus of the solid specimen was measured to be 1.98GPa. This falls within the accepted range of literature reported values [15]. The Young's Modulus which resulted in agreement between the FE predicted structural stiffness and the measured structural stiffness for the two porous specimens was 2.19GPa and 2.04GPa respectively. Thus, the indirect modulus determination method resulted in errors of 10.6% and 3.0%.

The results of the morphological measurements are presented in Table 1 for volume fraction, trabecular thickness, and trabecular spacing for both the bone cement analogues and the rat vertebrae. Note that the analogues are roughly two and one-half times denser than the vertebral bone with trabeculae that are twice as thick, however their spacing is comparable.

	Analogues (n=3)	Rat Vertebrae (n=10)
Volume Fraction	0.190 (0.025)	0.075 (0.011)
Trabecular Thickness (mm)	0.330 (0.055)	0.142 (0.018)
Trabecular Spacing (mm)	1.403 (0.046)	1.784 (0.396)

Table 1: Morphological measures for the bone cement analogues and rat vertebral bone; mean (S.D.).

The sensitivity results for the 10 rat vertebral samples are summarized in Table 2. A baseline configuration was taken as frictionless boundary conditions, a 50μm element size, and a Poisson's ratio of 0.3. For each of the other configurations the percent change from baseline is presented for the FE predicted structural stiffness (SS) and the resulting predicted Young's modulus obtained using the indirect method (YM).

Configuration	SS	YM
25μm	3.1%(2.4%)	-2.9%(2.2%)
Bonded BC's	9.9%(5.0%)	-10.5%(6.4%)
v=0.0	0.2%(0.4%)	-2.2%(3.7%)
v=0.45	3.9%(0.5%)	-5.7%(3.4%)

Table 2: Parametric results for the ten rat vertebra (mean (SD)). Results are presented as percent change from the baseline configuration for FE predicted structural stiffness (SS) and Young's modulus (YM).

5. DISCUSSION

The two bone cement analogues tested in this study are the first direct quantitative measure of the accuracy of the indirect tissue modulus determination method based on microCT and large-scale voxel-based finite element modeling. In both cases the predicted tissue modulus was acceptably close to the experimentally measured tissue modulus. This is convincing evidence that this method can feasibly yield accuracies that are comparable to other commonly employed experimental procedures. Of course, since actual cancellous bone tissue may be non-homogeneous and anisotropic, this represents a best case situation for the method and should be considered as an upper bound on the accuracy of practical measurements.

The quantification of bone volume fraction, trabecular thickness, and trabecular spacing allows a comparison of the morphology of the analogues to that of actual cancellous bone. Although the trabecular spacing was similar, the bone analogues were found to have a higher volume fraction and thicker trabecular structure. This is a function of the packing behavior of the spherical paraffin beads. Also, due to their spherical geometry the analogue's exhibit an isotropic microstructure which is inconsistent with the oriented microstructure of cancellous bone which can lead to anisotropic bulk properties [16]. Alternative bead geometries may be employed to yield both more realistic trabecular thicknesses and anisotropy. However, we expect the accuracy of FE models to scale with the average number of elements per trabecular thickness. Thus, it may be possible to extrapolate the results from the bone cement analogues to cancellous bone specimens if this ratio is maintained.

Several insights can be gained from the parametric studies of the vertebral models. First, the change in the FE results caused by decreasing the mesh size from 50μm to 25μm was relatively modest. When combined with the average trabecular thickness of the models of 142μm, this indicates that four elements per average trabecular thickness may be considered a safe guideline to obtain accurate results. This is a crucial finding since the mesh refinement from 50μm to 25μm increased the average number of nodes per model from 150 thousand to 1.3 million.

Changing the assumed boundary conditions from frictionless to fully bonded had a relatively large impact on the predicted modulus. This indicates that the assumption of a frictionless boundary condition has the potential to introduce significant errors into the procedure. The extent to which this occurs is determined by the level of friction in the parallel plate compression testing. Since practically achieving frictionless conditions may be difficult or impossible, bonding specimens to rigid end-caps for testing [14] may be a more appropriate testing configuration. Currently, we are evaluating the use of rigid yet x-ray transparent end-caps that can be bonded in place prior to microCT scanning. This approach has the added advantage of greatly simplifying positioning of the specimen in the scanner since the flat parallel top and bottom surface of the end-caps can be reproducibly mated to both the scanner and experimental testing machine in precisely the same configuration.

Finally, the results for changes in Poisson's ratio indicate that large-scale models of cancellous bone are relatively insensitive to changes in the assumed Poisson's ratio. This is understandable since the apparent Poisson effect of bulk specimens of cancellous bone is a consequence of trabecular microstructure rather than the Poisson's ratio of the bone tissue. This is an important and encouraging result since virtually no experimental data exist for the Poisson's ratio of cancellous bone tissue. Conversely, this implies that indirect methods based on FE modeling and matching with experimentally measured apparent properties would not be effective in determining tissue Poisson's ratio.

7. CONCLUSION

In summary, this study represents the first direct verification and quantitative measure of the accuracy of using large-scale microCT derived voxel-based finite element modeling in combination with experimental determination of apparent structural stiffness to determine the constituent or tissue Young's Modulus of porous materials. Additionally, we have quantified the relative contributions of mesh density, assumed boundary conditions, and assumed Poisson's ratio on the accuracy of this approach.

These are crucial first steps towards the widespread use of this method in the study of the mechanical behavior of cancellous bone tissue and its pathologies.

8. REFERENCES

1. Danielsen C. C., Mosekilde L., Andreassen T. T., Long-term effect of orchidectomy on cortical bone from rat femur: bone mass and mechanical properties. Calcified Tissue International , 1992, Vol. 50, 169-174.

2. Choi K., Kuhn J. L., Ciarelli M. J., Goldstein S. A., The elastic moduli of human subchondral, trabecular, and cortical bone tissue and the size-dependency of cortical bone modulus. [Review]. J Biomechanics, 1990, Vol. 23, 1103-1113.

3. Hughes T. J. R., Levit I., Winget J., An element-by-element solution algorithm for problems of structural and solid mechanics. Comp Meth Appl Mech Eng , 1983, Vol. 36, 241-254.

4. Papadrakakis M., Solving large-scale problems in mechanics. 1993, John Wiley & Sons, New York.

5. Feldkamp L. A., Goldstein S. A., Parfitt A. M., Jesion G., Kleerekoper M., The direct examination of three-dimensional bone architecture in vitro by computed tomography. J Bone & Mineral Res , 1989, Vol. 4, 3-11.

6. Fyhrie D. P., Fazzalari N. L., Goulet R., Goldstein S. A., Direct calculation of the surface-to-volume ratio for human cancellous bone. J Biomechanics, 1993, Vol. 26, 955-967.

7. Keyak J. H., Meagher J. M., Sinner H. B., Mote C. D., Jr., Automated three-dimensional finite element modelling of bone: a new method. J Biomed Eng, 1990, Vol. 12, 389-397.

8. van Rietbergen B., Weinans H., Polman B. J. W., Huiskes R., A fast solving method for large-scale FE-models generated from computer images, based on a row-by-row matrix-vector multiplication scheme. ASME CED, 1994, Vol. 6, 47-52.

9. Keyak J. H., Skinner H. B., Three-dimensional finite element modelling of bone: effects of element size. Journal of Biomedical Engineering, 1992, Vol. 14, 483-9.

10. Jacobs C. R., Mandell J. A., Beaupré G. S., A comparative study of automatic finite element mesh generation techniques in orthopaedic biomechanics. AMSE BED, 1993, Vol. 24, 512-514.

11. Hollister S. J., Brennan J. M., Kikuchi N., A homogenization sampling procedure for calculating trabecular bone effective stiffness and tissue level stress. Journal of Biomechanics, 1994, Vol. 27, 433-44.

12. Chavez O. T., Keaveny T. M. (1995) High-resolution finite element models of trabecular bone: the dependence of tissue strains and apparent modulus on imaging resolution. In: American Society of Biomechanics, pp 23-24, Stanford, CA.

13. van Rietbergen B., Weinans H., Huiskes R., Odgaard A., A new method to determine trabecular bone elastic properties and loading using micromechanical finite-element models. Journal of Biomechanics, 1995, Vol. 28, 69-81.

14. Keaveny T. M., Borchers R. E., Gibson L. J., Hayes W. C., Trabecular bone modulus and strength can depend on specimen geometry. J Biomechanics, 1993, Vol. 26, 991-1000.

15. Lautenschlager E. P., Stupp S. I., Keller J. C., Structure and properties of acrylic bone cement. In: Functional behavior of orthopedic biomaterials (eds. P Ducheyne, GW Hastings), 1984, pp 87-119. CRC, Boca Raton.

16. Goldstein S. A., The mechanical properties of trabecular bone: dependence on anatomic location and function. Journal of Biomechanics, 1987, Vol. 20, 1055-61.

DESIGN OF ANATOMICAL COMPUTER MODELS OF SKELETAL PARTS WITH SPECIAL EMPHASIS ON MECHANICALLY EQUIVALENT MODELLING OF SPONGY BONE

H. Druyts[1], J. Vander Sloten[1], G. Van der Perre[1], R. Gobin[1]

1. ABSTRACT

In most finite element analysis in orthopaedic biomechanics, the trabecular bone is modelled as a continuum. Inhomogenity of material properties and/or anisotropy may be introduced to account for the specific properties of the trabecular bone. However, if one wants to investigate the stability of the trabecular structure (e.g. after insertion of an orthopaedic implant), the propagation of cracks through a trabecular structure or the mechanics of pathologic or traumatic processes at a level where trabecular architecture is important (e.g. the development of osteoporosis), an other type of model of the trabecular structure is required.

A method is proposed by which the trabecular bone is modelled by means of discrete beam elements forming a space truss that is mechanically equivalent witch the spongy bone. Mapping the three dimensional architecture yields the necessary data to construct a detailed finite element model of this structure. To investigate the feasibility of this type of modelling, some idealised models of a vertebral body were constructed.

2. INTRODUCTION

Investigating processes at a trabecular level requires a more accurate and discrete model of the cancellous bone. The limitations of the continuum assumptions are described by Harrigan et al. (1988). Studying for the influence of bending and buckling, Pugh et al. (1973) were the first to introduce a crude model of the trabecular bone adjacent to the cartilage in the distal end of the femur. Raux et al. (1975) used a plate and rod model to calculate the stiffness of the cancellous bone in the human patella. A simple model of unidirectional cylindrically symmetrical structure was used by Williams and Lewis

Keywords: Spongy bone, Modelling, Computer model

[1] K.U.Leuven, Biomechanics and Engineering Design, Celestijnenlaan 200A, B-3001 Heverlee

(1982) to determine the values of the anisotropic elastic constants of approximately transversely isotropic cancellous bone. Klever (1984) constructed a structural model based on the plate/rod-like appearance of trabecular bone. The properties predicted by the model are in the range of actually measured material constants of trabecular bone. Gibson (1985) studied the mechanical behaviour of four types of cancellous bone (rod-like versus plate-like, asymmetric versus cylindrical symmetry) with idealised structures. A model of vertebral bone architecture was built by Jensen et al (1990). The results show the important influence of architecture on the global stiffness.

Recent techniques in the field of digital imaging and FEM have made it possible to construct full-scale Finite Element models of a small trabecular bone specimen with a huge number of elements (e.g. a 10x10x10 mm bone cube, comprising 511109 three dimensional brick elements, Van Rietbergen et al., 1995).

In this paper, a modelling method for cancellous bone is proposed that is more complex than the above mentioned discrete models, but contains far fewer elements than the full-scale models of small samples. The trabecular architecture is modelled using beam elements forming a space truss. Since the beams are, at each point, built parallel to the mean directions of the trabecular structure on that spot, and since the density is, at each point, equal in both the real bone and the equivalent structure the lacing is mechanically equivalent with the spongy bone. The mean directions are determined using the MIL technique (Whitehouse and Dyson, 1974, Cowin, 1986). Because this method creates complex structures which are difficult to evaluate in terms of displacements and stresses, idealised models of the vertebral body, with a simplified geometry and structure were created and tested.

3. MATERIALS AND METHODS

3.1 Acquisition of data

The bone specimen is divided into cubes of 10x10x10 mm. The central points of these cubes are situated on a regular three dimensional grid. The cube faces are stained with blue paragon, the marrow spaces are filled with alumina powder to enhance contrast. With a computer assisted image analysis system the sides of the cubes are analysed in order to determine the Mean Intercept Length as a function of the orientation. Also the areal density is measured. The results of the opposite faces are averaged and thus the data from three perpendicular planes are obtained for each cube.

With the data of the perpendicular faces, the properties of the three dimensional architecture (mean directions, their relative importance and the volume percentage) can be computed and represented using an ellipsoid (Cowin, 1986). The computations are made on PC using Matlab. The results are attributed to the central point of the sample, and hence, a three dimensional structure of points where the properties are known (=datum points) is obtained.

3.2 Data processing : classification

The mean directions in the different points need to be classified before the computation of the equivalent structure starts. Classification refers to the process of establishing groups of directions in different neighbouring datum points that will be used to build a continuous series of trabeculae (figure 1)

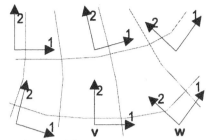

Figure 1 : Classified directions with equivalent structure

A mean direction 1 in point w is classified together with a mean direction 1 in the neighbouring point v if :

angle $(1_w, 1_v)$ < angle $(1_w, 2_v)$

angle $(1_w, 1_v)$ < angle $(1_w, 3_v)$

in which 2_v and 3_v are the other directions in point v.

After classification it is possible to compute in every arbitrary point in space, which is surrounded by 8 datum points, the local architecture as a weighted average of the properties in the surrounding points.

Since all datum points are situated inside the bone and since the building procedure (described below) always needs 8 surrounding datum points, the three dimensional structure with datum points has to be extrapolated in all directions beyond the endosteal surface of the cortical bone. Extrapolated values are calculated as weighted averages of the properties in the nearest datum points.

3.3 Building the equivalent structure

A three dimensional structure is then built. The equivalence with the trabecular bone structure can be obtained by building each beam element in accordance with the local architectural properties (local main directions and local volumetric density). Starting from a certain point, beams are built in a direction parallel to one of the mean principal directions of the spongy bone. The length of the beam is proportional to the MIL in that direction :

building step in direction x = K * MIL in direction x

in which x is mean direction 1, 2 or 3.

If K equals 1, the most refined model containing the highest number of elements, is obtained. If K increases, the number of elements in the model decreases significantly. The number of elements is proportional to $1/K^3$. K is denominated "proportionality factor".

The beam elements have a circular cross section with an area calculated to fit the measured local volumetric density.

Beam elements should arrive in the same node of the three dimensional structure, starting from different nodes and parallel to different directions. So the endpoint should be merged to one position. Depending on the way this is accomplished, the structure deforms. Two different algorithms were developed (figure 2).

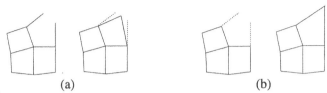

(a)   (b)

Figure 2 : Principles of merging algorithms

By averaging the endpoints (figure 2 a), the distortions are distributed over the elements. By choosing an intersection point (figure 2 b), there is less distortion of directions, but there is no control on deformation of length.
In order to correct the length deformations, elements that are too short can be joined and elements that are too long can be split.

The algorithm for the processing of data are implemented in the CAD program ANVIL-5000, using the GRAPL-IV language (Graphical Programming Language). The procedures are highly automated and require only manual interaction from the user at some crucial points in the model building process.

3.4 Trimming the equivalent structure

The consequence of the extrapolation of the datum points outside the endosteal envelope is that the equivalent structure pierces through the endosteal surface. So, in a last step of the CAD modelling, the structure has to be trimmed. This is done by calculating the intersections between the beams and the endosteal envelope and limiting the beam elements to the part inside the endosteal envelope. A result of an equivalent CAD-model of a vertebral body is shown in figure 3.

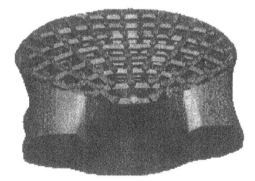

Figure 3 : 3D-view of an equivalent CAD-model of a vertebral body

3.5 Finite elements

The building procedure within the CAD program, constructs beam elements (forming the cancellous bone) containing all information necessary to create valid beam elements in the finite element analysis program ANSYS (coordinates of start and end node, cross sectional area, x-moment of inertia, y-moment of inertia, product of inertia, torsional constant, shear area factor K1 and K2). The cortical bone is modelled using solid elements (linear elements). Since both meshes have to be connected, the meshing forming the cortex should contain the nodes of the beam elements ending on the cortex. The division of a three dimensional surface into a mesh of triangular elements in which the minimal angle should not be less than about 20 degrees, based on predefined nodes, could not be realised by a commercially available mesh generator. A new mesh generator was developed.

3.6 Generation of an idealised model of a vertebra

Because the method described in this paper generates complex structures, which are difficult to interpret in terms of deformations and stresses, first some less complex models with a more regular structure were built, based on the same philosophy : a combination of a network of linked bars representing the trabecular bone surrounded by a cortical shell.

A continuous model and a discrete model with the same geometry, a cooling tower, representing the idealised geometry of a vertebral body (figure 4) were constructed. Only a quarter of the total geometry is modelled for symmetry reasons.

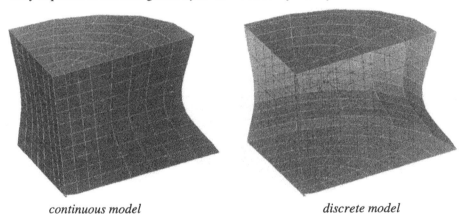

continuous model   *discrete model*
Figure 4 : idealised models of vertebral body (cooling tower)

Both models have a cortical shell, but in the discrete model the continuum within, representing the trabecular bone, is replaced by a space truss consisting of only vertical and horizontal beams. The beams were constructed using the nodes from the brick elements in the continuous model. This resulted in a relative higher number of beams in the middle of the model, but in the first instance all beams were given equal properties (cross section and Young's modulus). The overall volumetric density of this equivalent structure corresponds to the volumetric density of the continuous model, consisting of about 30 % bone material (Table 1).

	Discrete Model	Continuous Model
E_{cortex}	14,7 GPa	14,7 GPa
ρ_{cortex}	1300 kg/m^3	1300 kg/m^3
E_{trab}	0,49 GPa	0,16 GPa
ρ_{trab}	1300 kg/m^3	430 kg/m^3

Table 1 : material properties

4. RESULTS

From the results it appears that the discrete model forms a stable structure (i.e. it does not collapse). A good qualitative comparison can be made between the displacement in the z-direction of the continuous and the discrete model (figure 5).

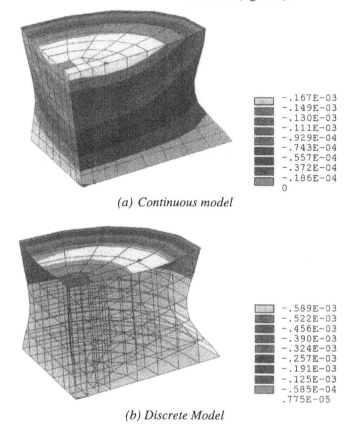

(a) Continuous model

(b) Discrete Model

Figure 5 : Displacement in z-direction

The interpretation of the stresses is more difficult because stress concentrations can occur on places where beams end on the trabecular surface. This effect manifests itself more clearly in the case of a relatively small number of beams and will disappear as more beams are used.

5. DISCUSSION

This paper introduces a new method to model larger parts of cancellous bone. The trabecular bone is modelled using beam elements forming a space truss. The beams are, at each point, built parallel to the local mean directions of the trabecular structure and the volumetric density is equal in the equivalent and the real trabecular structure, hence the lacing can be considered to be mechanically equivalent with the spongy bone. The technique developed here is rather applicable to rod-type trabecular bone than to plate-type trabecular bone.

The mechanical models thus obtained are not meant to replace continuum finite element models or microstructural models. Their strength is that they are lattice-like models of trabecular structures within larger skeletal parts, which can be used to simulate adaptive processes involving bone architecture, partly due to bone metabolic diseases. The algorithms developed have been used so far to build models of the proximal femur and vertebral body.

Preliminary results show that the obtained structure is stable, with a good qualitative agreement of the displacement with the continuous model. It appears that the equivalent structure is too stiff in the central region, but this can be explained by the fact that the volumetric density in the continuous and the discrete model, is not the same at each point. As mentioned in the generation of the idealised models, there is a relatively higher density of beams in the middle. This means that an overall equivalence of the structural parameters is not enough, and the cross section should be adapted to correspond to the local volumetric density, as it will be implemented in the full procedure (cfr par. 3.3).

This kind of model allows us to investigate the influence of different structural parameters, for instance thickness of the cortex, number of trabeculae, different Young's moduli etc. Thus it can be used to simulate some hypothetical mechanisms in the development of e.g. osteoporosis. It can also be used to investigate crack propagation in this kind of lattice-like structures.

In the future the method will be further validated with a real compression test on stereolithography models, similar to the models used above, which will also be simulated with finite elements.

6. REFERENCES

1. Harrigan, T.P., Jasty, M., Mann, R.W., Harris, W.H., Limitations of the continuum assumption in cancellous bone, 1988, J. of Biomechanics, Vol. 21, No. 4, 269-275
2. Pugh, J.W., Rose, R.M., Radin, E.L., A structural model for the mechanical behaviour of trabecular bone, J. of Biomechanics, 1973, Vol. 6, 657-670
3. Raux, P., Townsend, P.R., Miegel, R., Rose, R.M., Radin, E.L., Trabecular architecture of the human patella, J. of Biomechanics, 1975, Vol. 8, 1-7
4. Williams, J.L., Lewis, J.L., Properties and an anisotropic model of cancellous bone from the proximal tibial epiphysis, 1982, J. of Biomech. Eng., Vol. 104, 50-56
5. Klever, F.J., On the mechnics of failure of artificial knee joints, PhD. Thesis, 1984, Technische Hogeschool Twente
6. Gibson, L.J., The mechanical behaviour of cancellous bone, J. of Biomechanics, 1985, Vol. 18, No. 5, 317-328
7. Jensen, K.S., Mosekilde, Lis, Mosekilde, Leif, A model of vertebral trabecular bone architecture and its mechanical properties, 1990, Bone, Vol. 11, 417-423
8. Van Rietbergen, B., Weinans, H., Huiskes, R., Odgaard, A., A new method to determine trabecular bone elastic properties and loading using micromechanical finite element models, J. of Biomechanics, 1995, Vol. 28, No. 1, 69-81
9. Whitehouse, W.J., Dyson, E.D., Scanning electron microscope studies of trabecular bone in the proximal end of the human femur, 1974, J. of Anatomy, Vol. 118, 417-444
10. Cowin, S.C., Wolff's law of trabecular architecture at remodelling equilibrium, J. Of Biomechanical Engineering, 1986, Vol. 108, No 6/7, 83-88

DETERMINATION OF INDIVIDUALISED AND HOMOGENISED CHARACTERISTICS OF NORMAL HUMAN TIBIAE IN VIVO

É. Estivalèzes[1], M.C. Hobatho[2], G. Limbert[3], B. Couteau[4], R. Darmana[5]

1. ABSTRACT

The use of mathematical tools in biomechanical engineering, such as finite element modelling, requires the knowledge of the geometric properties as well as the mechanical properties. Eight adult human tibiae were analysed using computer techniques and finite element calculation to determine the individualised homogenised characteristics along the length of each bone. These homogenised characteristics represent tensile, bending and torsional stiffnesses of each cross section of the tibia. Heterogeneity of cancellous bone in the epiphyses has been considered and some rigidity relationships between cortical and cancellous bone have been derived.

2. INTRODUCTION

The aim of the study was to determine the individualised and homogenised characteristics of the human tibiae in vivo. These results should allow to get a better comprehension of the phenomenon of rotational abnormalities in children particularly the tibial torsion. Studies had been performed describing the natural tibial torsion and its mechanical behaviour with only geometrical parameters and approximate stiffnesses parameters (Piziali[1,2], Miller[3]). In order to assess parameters such as mechanical properties combined to the geometry, a method has been developed using a 2D finite element calculation of the individualised geometrical and mechanical properties of sections obtained from CT scans.

Keywords: Tibia, Cortical bone, Cancellous bone, Mechanical properties, F.E.M.

[1] Ph.D.,INSERM U305, Hôtel Dieu, 31052 Toulouse Cédex, FRANCE
[2] Ph.D.,INSERM U305, Hôtel Dieu, 31052 Toulouse Cédex, FRANCE
[3] MSc., INSERM U305, Hôtel Dieu, 31052 Toulouse Cédex, FRANCE
[4] MSc., INSERM U305, Hôtel Dieu, 31052 Toulouse Cédex, FRANCE
[5] MSc., Hôtel Dieu, 31052 Toulouse Cédex, FRANCE

3. MATERIALS AND METHODS

CT scans (Siemens, Somatom) were performed on left tibiae of 8 volunteer adults (mean age of 33 ± 5 years, 4M and 4F) every 2 cm from the proximal to the distal epiphysis with a slice thickness of 4 mm. The acquisition matrix was 256x256 for a mean field of view of 271x271 mm, then CT images were transferred to a workstation Indy (Silicon Graphics, Mountain View, CA, USA) using an interface software SIP 305 (© INSERM) developed in our laboratory. A total of nineteen slices was obtained. The different steps of the calculation of the homogenised characteristics were the following:

1) the reconstruction of the geometry for each section :
The software SIP 305 allowed the transfer and reconstruction of the geometrical contours of the bone based on an edge detection algorithm. An ASCII file recording data such as points, contour lines and surfaces was created. The different steps of the geometric reconstruction through SIP were:
- decoding and visualisation of scan or RMI images
- contour detection using a thresholded image
- construction of contour lines (parametric cubic curves)
- construction of surfaces

2) the visualisation and mesh for each section :
The reading of this ASCII file containing all the geometrical characteristics was performed via a pre and post processing software Patran5 (MSC Corp.). The meshing of a tibial cross section has been carried out with three node triangular elements in order to be compatible with the finite element software enabling the calculation of homogenised characteristics. Size elements has been optimised and number of triangular elements were about 1431 and 256 for the epiphyses and the diaphyses respectively.

3) the assignment of mechanical properties for each section :
Cancellous bone has been considered as an heterogeneous elastic isotropic material, and its mechanical properties has been assigned to each element of the meshed cross section by the use of a program PM_QCT (© Inserm) designed in our laboratory (Couteau[4]). This software enabled the study and the measurement of CT number (Unity Hounsfield) using ROI mapping (Region Of Interest) and the determination of the corresponding axial Young's modulus E_3 (MPa) (Rho[5]). The relation found between E_3 measured by an ultrasound technique and CT number (UH) was:

$$E_3 = 4.56 \times CT + 320 \tag{1}$$

To assign the different mechanical properties to the meshing, their spatial location have been used.
Mechanical properties of the cortical bone material were chosen as an isotropic transverse material and in fact, CT measurements did not show significant difference. These data have been assigned to the periphery of a slice and to all the nineteen slices of the studied tibia. The numerical values were obtained from the literature (Hobatho[6]) as 20700 MPa for the Young's axial modulus (E_3) and 12200 MPa for transversal plane moduli (E_1 and E_2). The Poisson's ratio were $\nu_{13} = \nu_{23} = 0.237$ and $\nu_{12} = \nu_{21} = 0.423$ and the shear modulus (G_{13}) was 5200 MPa.
Figure 1 shows a CT scan image of an epiphysis of a tibia. Its corresponding finite element model, displaying the distribution of cortical and cancellous bone, can be seen in figure 2. The material index indicates the different range of material properties contained in the cancellous bone.

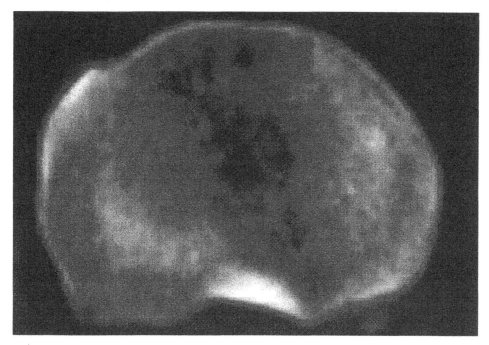

Fig. 1 CT scan image of a tibial epiphysis.

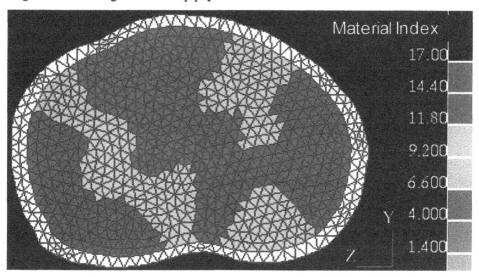

Fig. 2 Assignment of individualised mechanical properties on the individualised meshed geometry of a tibial epiphysis.

4) Calculation of the homogenised characteristics <ES>, <EIy>, <EIz> and <GJ> :
A finite element program CHAR_HOM (© Inserm), written in FORTRAN 77, has been developed in order to calculate the tensile stiffness <ES>, bending stiffness in the maximum <EIy> and minimum <EIz> plane of deformation, and the torsional stiffness <GJ> for each section. This 2D finite element program has been developed because at the moment, it does not exist any calculation code enabling the determination of the

torsional rigidity <GJ> of arbitrary shaped cross section. Input data to this program are taken from files, exported from Patran 5, containing informations like node coordinates, element connectivities, distribution of material properties. This program has been validated for simple geometries as crown shaped or rectangular cross sections (Estivalèzes[7]). The finite element program was based on the theory of a composite beam model with non uniform cross section and any orthotropic axis (Estivalèzes[8]). In the mechanical formulation, the composite beam was supposed to be constraint on one end, the other end undergoing loading such as tensile, bending and torsion. According to the beam theory, stresses in the plane of the cross section were negligible with respect to the stresses related to the longitudinal axis of the beam. It was also supposed that beam characteristics (geometry, material) were constant in any cross section. When subjected to the previously defined loading, the cross section was supposed to undergo a translation, a rotation and a warping. The solution was obtained by an energetic method based on the principle of virtual work combined with the Lagrange multipliers technique. Actually the calculation is done on each tibial cross section which is considered to be the cross section of a uniform composite beam of 200 mm lengthwise as shown in figure 3. A total of 152 finite element models was obtained.

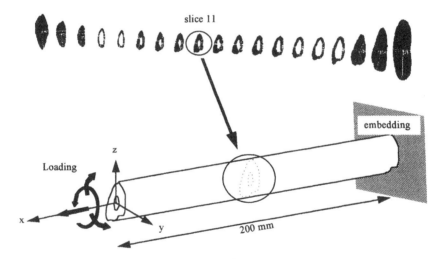

Fig. 3. Composite beam model of the eleventh slice

4. RESULTS

The results of homogenised characteristics : tensile stiffness <ES>, bending stiffness in the maximum <EIy> and minimum <EIz> plane of deformation and torsional stiffness <GJ> for each tibial section are shown graphically in figures 4-7.

Furthermore, some other calculation have been done on one single human tibia to obtain approximate relationships of homogenised characteristics. By determining the equivalent axial Young's modulus of the cancellous area as :

$$< E >_{cancellous} = \frac{\sum\limits_{i=1}^{21} E_i \cdot S_i}{\sum\limits_{i=1}^{21} S_i} \tag{2}$$

and by using the principal moments of inertia Iy and Iz for each of the six epiphysis sections, it has been possible to approximate the bending stiffness in a cross section defined as :

$$< EI_Z >_{app} = < E >_{cancellous} \cdot Iz_{cancellous} + E3_{cortical} \cdot Iz_{cortical} \qquad (3)$$

and

$$< EI_Y >_{app} = < E >_{cancellous} \cdot Iy_{cancellous} + E3_{cortical} \cdot Iy_{cortical} \qquad (4)$$

By comparison with the theoretical values of homogenised rigidities on each section, the use of approximated formulae may be done with an error percentage varying from 2.5 to 13.5% for <EIz> and from 4.3 to 13.2% for <EIy>. When neglecting the cancellous bone, that is to say when only considering cortical bone, the error percentage is then varying from 8.1 to 15.1% for <EIz> and from 8.2 to 15.5% for <EIy>.

Concerning an approximation formulae of the torsional rigidity, we have made a comparison with the calculation of torsional stiffness using polar moment of inertia, as it is often found in the literature (Piziali[1,2], Hsu[9], Harrington[10]). In this case, it was supposed that the cancellous bone was isotropic with a Poisson's ratio of 0.3 and by setting that :

$$< GJ >_{app} = < G >_{cancellous} \cdot Ip_{cancellous} + G13_{cortical} \cdot Ip_{cortical} \qquad (5)$$

where

$$< G >_{cancellous} = \frac{< E >_{cancellous}}{2(1+\upsilon)} \qquad (6)$$

The results has shown that there was a great lack of accuracy when using this approximation as the error percentage was varying from 15 to 45%. This was because in such an approximation, the warping function, usually introduced in calculation of beam with non circular cross section, was not taken into account.

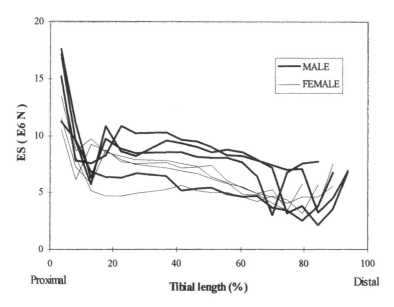

Fig. 4. Tensile stiffness <ES> function of the tibial length

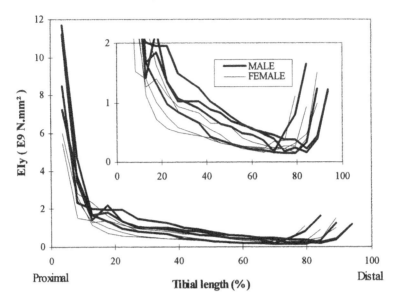

Fig. 5. Bending stiffness <EIy> in the maximum plane of deformation function of the tibial length

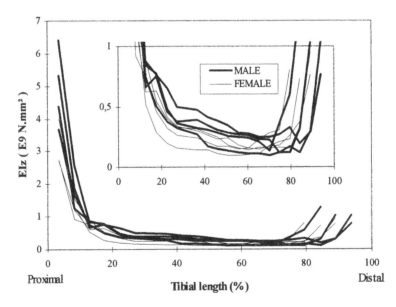

Fig. 6. Bending stiffness <EIz> in the minimum plane of deformation function of the tibial length

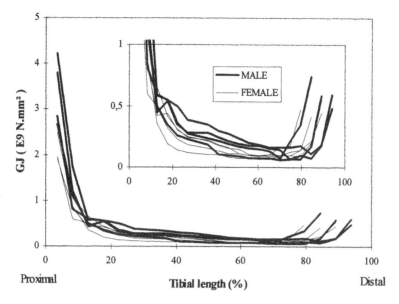

Fig. 7. Torsional stiffness <GJ> function of the tibial length

5. DISCUSSION

The homogenised characteristics were found to be higher at the proximal (0 to 20%) and distal (>80%) epiphyses and decreasing from 20 to 70% of the tibial length. This fact is explained as the epiphyseal surfaces, composed of cortical and cancellous bone, were larger than that of the diaphysis. But it must be noted that, at the epiphyses, on a geometrical point of view, though the cancellous bone was predominant because its surface was larger than the cortical, on a mechanical point of view, this phenomenon was reversing because the mechanical properties of the cortical bone were much higher than the cancellous. The evolution of these characteristics were similar to that of the geometrical parameters defined in the literature (Piziali[1,2]). To our knowledge no numerical data are available for comparison. In fact, our calculation of the torsion constant J uses the out of plane warping function often neglected, and usually this constant J is approximated with the polar moment of inertia, which is accurate only for circular cross sections. Influence of different material properties at the epiphyses leaded to the following relationships :

$$<EI>cortical+cancellous \approx <EI>cortical + <EI>cancellous \qquad (7)$$

Similar relations were found for <GJ>. It must be noted that cortical bone stiffness represented 80% of the global stiffness at the epiphyses but also that the effects of the cancellous bone were non negligible. Values of <EIy> were typically twice and three times higher than <EIz> and <GJ> respectively. The clinical application of such studies performed on children with torsional abnormalities showed that their deformities do not alter their bone stiffness (Limbert[11]).

6. REFERENCES

1. Piziali R. L., Hight T. K., An analysis of long bones : application to the human tibia, J. Biomech., 1976, Vol. 9, 695-701.

2. Piziali R. L., Hight T. K., and Nagel D. A., Geometric properties of human leg bones, J. Biomech., 1980, Vol. 13, 881-885.

3. Miller G. J. and Purkey W. W., The geometric properties of paired human tibiae, J. Biomech., 1980, Vol. 13, 1-8.

4. Couteau B., Hobatho M. C., Lowet G., Van Der Sloten J., Development of a three finite element model of the human proximal femur with individualised geometrical and mechanical properties. Proceedings of the symposium Toulouse May 5-6th, 1994. "Bone mechanical properties and finite element modelization". *Ed. Hobatho MC. et Darmana R*, 1996; 115-127.

5. Rho J. Y., Hobatho M. C., Ashman R. B., Relations of mechanical properties to density and CT numbers in human bone, Med.Eng.Phys., 1995, Vol. 17, 347-355.

6. Hobatho M. C., Rho J. Y. and Ashman R. B., Atlas of mechanical properties of human cortical and cancellous bone, J. Biomech., 1992, Vol. 25, 668.

7. Estivalèzes É., Étude et développement d'un élément de poutre composite pour un revêtement déséquilibré, *PhD. Thesis*, I.N.S.A. Toulouse, 1995.

8. Estivalèzes É. and Barrau J. J., Finite element calculation for composite beams, Proceedings of ESDA, 1996, 96-ESDA-2, 1-7.

9. Hsu E. S., Patwardhan A. G., Meade K. P., Light T. R., Martin R. M., Cross-sectional geometrical properties and bone mineral contents of the human radius and ulna, J. Biomech., 1993, Vol. 26, 1307-1318.

10. Harrington M. A., Keller T. S., Seiler J. G., Weikert D. R., Moeljanto E., Schwartz H. S., Geometric properties and the predicted mechanical behavior of adult human clavicles, J. Biomech., 1993, Vol. 26, 417-426.

11. Limbert G., Détermination des caractéristiques homogénéisées personnalisées du tibia humain in vivo - Application à l'étude du comportement mécanique du tibia, 1996, Master's thesis.

BONE DENSITY AND MICROSTRUCTURE - NEW METHODS TO DETERMINE BONE QUALITY AND FRACTURE RISK

Dieter Ulrich[1], Bert van Rietbergen[2], Andres Laib[1], and Peter Rüegsegger[3]

1. ABSTRACT

The diagnosis of bone fracture risk is usually based on measurements of bone density alone, since this is the only parameter that can be routinely measured *in vivo*. However, the fracture risk is also determined by the bone microarchitecture and loading conditions.

Recently, new imaging techniques and new large-scale finite element (FE) techniques have been developed. The imaging techniques allow the assessment of the bone microarchitecture, whereas the FE-techniques allow the assessment of mechanical properties and tissue stresses of bones. In this paper, it is aimed to evaluate the feasibility of these new techniques for the assessment of bone quality and fracture risk. Two applications of these new techniques are demonstrated. In a first study, a micro-CT scanner was used to generate computer reconstructions of 58 bone specimens at a resolution of 28 µm. Using these reconstructions as the geometry input for large-scale FE-analyses, it was possible to obtain all elastic properties for each specimen. In a second study, a 3-D pQCT scanner was used to generate a computer reconstruction of a radius end section at a resolution of 165 µm. Using a FE-model generated from this reconstruction, it was possible to calculate the tissue stresses during a fall. It is concluded that, compared to standard methods, the new techniques can provide unique and more complete information for the prediction of bone fracture risk.

Keywords: Bone Quality, Fracture Risk, FE Analysis, pQCT, micro-CT

---

[1]PhD student, Institute for Biomedical Engineering, University and ETH Zürich, Moussonstrasse 18, 8044 Zürich, Switzerland
[2]Post Doc, Institute for Biomedical Engineering, University and ETH Zürich, Moussonstrasse 18, 8044 Zürich, Switzerland
[3]Professor, Institute for Biomedical Engineering, University and ETH Zürich, Moussonstrasse 18, 8044 Zürich, Switzerland

2. INTRODUCTION

Bone fractures due to osteoporosis or other bone diseases constitute a major socioeconomic problem of our aging population. Such fractures are catastrophic events, the one year mortality rate from hip fractures is 25% and the probability of an older patient to regain the previous level of function after a hip fracture is less than 30% [1-3]. Hence the knowledge of the bone fracture risk in individual patients would be of considerable interest for diagnostic purposes as well as for the evaluation of the effectiveness of treatments. Therefore it is of great importance to establish an accurate evaluation of bone fracture risk.

The fracture risk is related to the bone 'quality' on the one hand and the external bone forces on the other. In this respect, we define bone 'quality' as the bone's capacity to carry load which, in turn, is determined by the bone internal and external morphology and the properties of the bone tissue material. Presently, standard methods such as DXA, US, MRI or QCT used for patient examinations provide merely bone density values in a direct or indirect way. Bone quality is predicted from these density measurements by the use of empirical relationships. It has been found, however, that relationships based on density alone can be inaccurate, such that it is possible that some of the patients at risk are not recognized whereas other receive unnecessary treatment.

For a more accurate prediction of bone quality, a more accurate description of the bone mechanical properties, in particular its anisotropy, is needed [4]. The anisotropic properties of bone are determined by its microarchitecture. It has been demonstrated that bone samples with the same bone volume fraction can have a variation in their mechanical properties of up to 53% [5]. Statistical analyses of bone specimens covering a large range of density values show that on average 15 - 20% of the variation of the mechanical properties are not explained by bone volume fraction [5-7]. On the other hand, a more accurate prediction of bone fracture risk requires more accurate information about the bone loading conditions, not only those during normal daily loading, but also those due to peak loadings such as a fall.

Recently, new techniques have been developed that could provide the information needed for a more accurate prediction of the bone fracture risk [8-11]. These techniques consist of new imaging techniques, that allow the assessment of the trabecular architecture, and new FE-techniques that allow the assessment of bone mechanical properties as well as bone tissue stresses. In the present chapter, it will be evaluated how these techniques can be used to reach a more accurate quantification of bone quality and tissue stresses.

3. NEW AND EVOLVING METHODS FOR THE EVALUATION OF THE MECHANICAL BONE QUALITY

3.1. Assessment of bone micro-architecture

Over the last few years, several techniques have been developed to digitize the three-dimensional micro-architecture of trabecular bone in full detail. All these techniques use a set of sequential high-resolution images to reconstruct the bone internal morphology in a three-dimensional voxel grid that can be stored in a computer. These reconstructions can be used to calculate morphological parameters describing the average trabecular orientation [12,13], connectivity [14] and other structural parameters e. g. trabecular thickness [15], trabecular spacing, and SMI [16]. The reconstructions can also be used as direct input to generate FE-models (see paragraph 3.2). Methods that have been used for the generation of high-resolution sequential images are the serial sectioning technique [17], micro-CT scanning [18,19], pQCT scanning [20] and, recently, MRI scanning [21]. Among these, the first two methods provide the highest resolutions, whereas the last two methods have the potential to be used in-vivo. In the following a micro-CT scanning and a pQCT scanning technique developed in our lab will be described in more detail.

3. 1. 1. Computer reconstructions based on micro-CT scanning images

The micro-CT system developed in our lab is based on a compact fan-beam type tomograph. A microfocus X-ray tube with a focal spot of 10 μm is used as a source. The filtered 40 kVp X-ray spectrum is peaked at 25 keV, allowing excellent bone versus marrow contrast due to the pronounced photoelectric contrast. Furthermore the system includes a detector to measure the intensity of the transmitted X-rays as well as acquisition electronics which digitize the detector signals and send them to a computer for reconstruction. The detector consists of a linear CCD-array with 1024 elements with a pitch of 25 μm. The bone biopsies are mounted on a turntable that can be shifted automatically in the axial direction.

The system provides a nominal isotropic resolution of 28 μm [18]. After measurement, a constrained 3-D Gaussian filter is applied to partly suppress the noise in the original volume data. Then, bone tissue is extracted and binarized with simple thresholding. For a realistic representation of the three-dimensional bone architecture (Fig. 1) an interpolating smoothing algorithm is used.

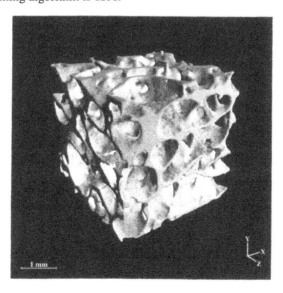

Fig. 1. 3-D computer reconstruction of a human trabecular bone specimen generated from sequential micro-CT images with a resolution of 28 μm.

3. 1. 2.   Computer reconstructions based on 3-D pQCT scanning images

The pQCT scanner developed in our lab is a special purpose low-dose three-dimensional peripheral quantitative computed tomography (3D-pQCT) system which is a miniaturized version of a translate-rotate type scanner. The system has a two-dimensional detector array and a 0.2 x 10 mm line-focus x-ray tube with an effective energy of 40 keV. This setup enables the simultaneous acquisition of a stack of parallel CT slices with a nominal resolution of 165 μm in plane. Measuring sites for standard pQCT are the forearm or the lower leg. A stack of up to 200 high resolution tomograms can be measured with a spacing in the axial direction of also 165 μm, thus delivering a 3-D representation of a bone end section with a voxel size of 165 μm in all three spatial directions. A sophisticated filter algorithm which detects the zero crossing of the second derivative in the fourier domain in combination with thresholding was used to binarize the data and to obtain a 3-D representation of the bone architecture (Fig. 2). For standard patient measurement stacks of 60 tomograms are taken. The skin radiation dose for such an examination is 0.8 mGy and the measuring time approximately 10 min. The arm or leg of the patient is immobilized during the examination in an anatomically formed carbon fiber shell.

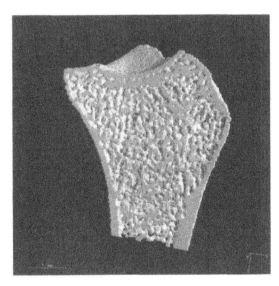

Fig. 2. 3-D computer reconstruction of a radius measured with the high resolution 3-D pQCT at a resolution of 165 μm. To see the trabecular network only half of the model is shown here.

3.2. Finite Element (FE) analyses of bone micro-architecture for the calculation of bone mechanical properties

Recently, two new automated techniques have been developed that use computer reconstructions as direct input to generate detailed FE-models of the trabecular architecture. With the first technique, also called the voxel conversion technique, bone voxels in the reconstruction are directly converted to brick elements in a FE-model [9]. With the second technique, the bone voxels are subdivided into tetrahedron elements using a marching cubes algorithm [24]. With both techniques, the preprocessor must check if all element are connected to the main structure; those that are not must be removed or supported by extra elements.

When using high-resolution images as generated by the micro-CT scanner, the FE-models will consist of a large number of elements. Typically on the order of 10^5-10^6 elements per cm^3 are needed to represent the architecture of trabecular bone at a resolution of 50 μm. At present, models of this size can only be solved when they are created using the voxel conversion technique, since FE-models created in this way allow the use of fast special purpose solvers. These use an iterative PCG solver, in combination with a very efficient matrix-vector multiplication scheme based on an Element-By-Element or Row-By-Row scheme [10]. By considering the fact that all elements in the FE-model have the same size, orientation and material properties, it is possible to drastically reduce the memory requirements needed for solving, in particular the global stiffness matrix is never actually computed or stored.

When using images with a resolution as provided by the pQCT scanner, the number of elements in FE-models generated with the voxel conversion technique is on the order of 10^4-10^5 elements per cm^3. In this case, commercial FE-software packages that implement iterative solvers can be used as well. The disk space, memory and CPU time requirements will be much larger though than for the special-purpose solvers. The obvious advantage of using commercial FE-software is that boundary conditions, material properties and elements can be chosen from the libraries provided with the code. For this reason, it is also possible to solve FE-models built from non-brick elements e. g. tetrahedral elements which are generated from the marching cube algorithm.

4. TWO EXAMPLES

4.1. Assessment of mechanical properties of femoral bone specimens

Using a large set of specimens from the BIOMED I [22] study it was investigated if the micro-CT-based FEA leads to a more accurate evaluation of bone quality than predictions based on apparent density alone.

58 human trabecular bone samples from the femoral head were measured with the micro-CT system providing a voxel representation of the bone microarchitecture with a resolution of 28 μm (Fig. 1). 5 mm cubes based on such representations were analyzed with FEA. Six FE-analyses were needed to obtain the orthotropic stiffness matrix and the principal directions of one specimen [11] which accounted for approx. 7 hours of CPU time and 60 Mbyte memory on a DECAlphaServer 8420.

It was found that all samples had near-orthotropic symmetries. After transformation of the stiffness matrix into the symmetry coordinate system, the three Young's moduli, shear moduli, and Poisson's ratios in the three orthotropic symmetry directions were calculated and sorted such that E_1 and G_{12} represent the largest and E_3 and G_{23} the lowest values, the Poisson's ratios divided by the moduli were sorted in descending order such that v_{12}/E_1 was the lowest value and v_{31}/E_3 the highest value. For the chosen tissue modulus of 5 GPa, the primary Young's moduli ranged from 170 to 1000 MPa, the tertiary moduli from 50 to 420 MPa (Fig. 3). Primary and tertiary shear moduli ranged from 50 to 350 MPa and 25 to 225 MPa respectively. The primary and tertiary

Poisson's ratio divided by the Young's moduli ranged from 0.0002 to 0.0015 and 0.0004 to 0.0034 respectively. Finally the degree of anisotropy E_1/E_3 ranged from 1.4 to 3.9, E_1/E_2 from 1.0 to 2.5, and E_2/E_3 from 1.0 to 2.6.

This study demonstrated that all elastic properties of bone specimens could be obtained using the detailed microstructural FE-techniques. Compared to predictions from bone density alone, new and subsequent information about the mechanical parameters can be obtained.

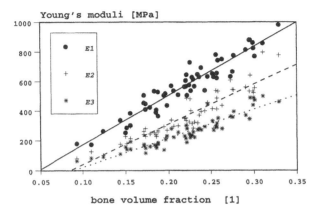

Fig. 3. Variation of Young's moduli for 58 femoral bone samples. The moduli were calculated with micro-CT-based FE-analysis.

4.2. Assessment of tissue stresses in a radius end section

The aim was to determine the fracture risk from tissue stresses that occur during a fall on the outstretched hand by using a finite element model that can represent the microstructure of the distal radius.

An excised human radius which showed normal bone density values was measured with the special high resolution 3-D pQCT scanner described before. The 3-D FE-model of the radius was generated by conversion of the voxels into finite elements. 350'000 hexaedral elements were required for the detailed 3-D representation of the radius architecture (Fig. 2). To simulate a realistic loading situation of the radius, physiological boundary conditions were applied to the joint surface [23].

The two contact areas were modeled as ellipses. The fall was simulated by applying a force of 500 N on the lunate and scaphoid each. The resulting boundary stresses were 13.5 and 9 MPa on lunate and scaphoid, respectively. All boundary stresses were chosen to be perpendicular to the joint surface. The bone was fully constraint on the proximal end. For solving 750 Mbyte memory and approx. 8 hours of CPU time were required, using the commercial FE-software Ansys 5.2 on a DECAlphaServer 8420.

Stresses and strains were calculated in all elements and a Von Mises equivalent stress contour plot of a cross-section through the radius was made (Fig. 4). The effect of the contact pressures from lunate and scaphoid can be observed. It is possible to see locations with thin connections which are highly loaded and likely to fail (Fig. 4, dark elements). It is concluded that it is now possible to calculate tissue stresses and strains for a piece of bone of the size of a radius end section.

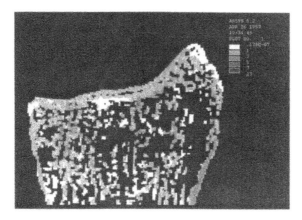

Fig. 4. Tissue stresses (Von Mises) showed in a cross section through the radius. The FE-model is based on a 3-D pQCT measurement at a resolution of 165 μm.

5. DISCUSSION

In this chapter it was aimed to evaluate the feasibility of new imaging and finite element techniques for the assessment of bone quality and fracture risk. We have focused on two new CT-based imaging techniques to quantify the three-dimensional trabecular architecture. The first, a micro-CT scanning technique, offers a high spatial resolution, but can only be used for small bone specimens. The second, a 3-D pQCT scanning technique combines a low radiation dose with a short scanning time and can be used *in vivo*. The resolution presently provided by this type of scanner is too low to image individual trabeculae in detail, but high enough to image the trabecular network and orientation. As demonstrated in the two examples, both CT-scanners can be used to generate the geometry needed for detailed finite element models. In the first example it was demonstrated how finite element techniques based on micro-CT images can be used to derive all elastic constants of bone specimens, thus to quantify the mechanical 'quality' of the specimens. In the second example it was demonstrated that whole-bone finite element models based on 3-D pQCT images can be used to quantify the bone tissue stresses during risk situations. It is thus concluded that the new techniques demonstrated here can provide the information needed for a more accurate prediction of the bone fracture risk.

Some of the limitations of the methods have to be discussed. First, both the calculated elastic properties of bone specimens and the tissue stresses depend on the tissue material properties. At present, however, there is not much consensus about the elastic properties for trabecular tissue material, and for this reason a unique value for the tissue modulus was assumed for all elements in the FE-models presented here. Nevertheless, the finite element models can account for variations in tissue properties as soon as this information is available. Second, since the resolution of the pQCT scanner used in this study was too low to model individual trabeculae, errors in the stress and strain calculation can be introduced. These errors can be due to errors in the geometry (in particular the trabecular thickness can be overestimated) and due to numerical errors related to the 'jagged' modeling of the trabecular surface. Earlier studies, however, have demonstrated that these errors are acceptable for most purposes [11,25]. Third, the methods rely on state-of-the-art imaging and computer equipment. Presently, only laboratory scanners can provide the required resolution and fast workstations or super-computers are needed to solve the large FE-models in a reasonable amount of time. Although these requirements inhibit routine patient examination at this stage, the methods can be used now in a research environment to improve our understanding of bone failure mechanisms and its determinants.

6. REFERENCES

1. Magaziner, J., Simonsick, E. M., Kasher, T. M., Hebel, J. R. and Kenzora, J. E., Survival experience of aged hip fracture patients, Am. J. Public Health, 1989, Vol. 79, 274-278.

2. Chrischilles, E. A., Butler, C. D., Davis, C. S. and Wallace, R. B., A model of lifetime osteoporosis impact, Arch. Intern. Med., 1991, Vol. 151, 2026-2032.

3. Cooper, C., Atkinson, E. J., Jacobsen, S. J., O'fallon, W. M. and Melton, L. J., Population-based study of survival after osteoporotic fractures, Am. J. Epidemiol., 1993, Vol. 137, 1001-1005.

4. Parfitt, A. M., Implications of architecture for the pathogenesis and prevention of vertebral fracture, Bone, 1992, Vol. 13, 41-47.

5. Ulrich, D., Hildebrand, T., Van Rietbergen, B., Müller, R. and Rüegsegger, P., The quality of trabecular bone evaluated with micro-computed tomography, FE-A and mechanical testing. In: Lowet, G., Rüegsegger, P., Weinans, H. and Meunier, A. (eds.), Bone Research in Biomechanics, pp. 97-112. ISBN: 90 5199 327 7. IOS Press, Amsterdam, 1997.

6. Rice, J. C., On the dependence of the elasticity and strength of cancellous bone on apparent density, J. Biomechanics, 1988, Vol. 21, 155-168.

7. Hodgskinson, R. and Currey, J. D., Young's modulus, density and material properties in cancellous bone over a large density range, J. Materials Science: Materials in Medicine, 1992, Vol. 3, 377-381.

8. Hollister, S. J., Brennan, J. M. and Kikuchi, N., A homogenization procedure for calculating trabecular bone effective stiffness and tissue level stress, J. Biomechanics, 1994, Vol. 27, 433-444.

9. Van Rietbergen, B., Weinans, H., Huiskes, R. and Odgaard, A., A new method to determine trabecular bone elastic properties and loading using micromechanical finite-element models, J. Biomechanics, 1995, Vol. 28, 69-81.

10. Van Rietbergen, B., Weinans, H., Polman, B. J. W. and Huiskes, R., Computational strategies for iterative solutions of large FEM applications employing voxel data, Int. J. Num. Meth. Eng., 1996, Vol. 39, 2743-2767.

11. Van Rietbergen, B. Odgaard, A., Kabel, J. and Huiskes, R., Direct mechanics assessment of elastic symmetries and properties of trabecular bone architecture, J. Biomechanics, 1996, Vol. 29, 1653-1657.

12. Turner, C., Cowin, S. C., Rho, J. Y., Ashman, R. B. and Rice, J. C., The fabric dependence of the orthotropic elastic constants of cancellous bone, J. Biomechanics, 1990, Vol. 23, 549-561.

13. Goulet, R. W., Goldstein, S. A., Ciarelli, M. J., Kuhn, J. L., Brown, M. B. and Feldkamp, L. A., The relationship between the structural and orthogonal compressive properties of trabecular bone, J. Biomechanics, 1994, Vol. 27, 375-389.

14. Odgaard, A. and Gundersen, H. J. G., Quantification of connectivity in cancellous bone, with special emphasis on 3-D reconstructions, Bone, 1993, Vol. 14, 173-182.

15. Hildebrand, T. and Rüegsegger, P., A new method for the model independent assessment of thickness in three-dimensional images, J. Microsc., 1997, Vol. 185, 67-75.

16. Hildebrand, T. and Rüegsegger, P., Quantification of bone microarchitecture with the structure model index, Comp. Meth. in Biomech. and Biomed. Eng., 1997, Vol. 1, 15-23.

17. Odgaard, A. and Linde, F., A direct method for fast three-dimensional serial reconstruction, J. Microscopy, 1989, Vol. 159, 335-342.

18. Rüegsegger, P., Koller, B. and Müller, R., A microtomographic system for the nondestructive evaluation of bone architecture, Calcif. Tiss. Int., 1996, Vol. 58, 24-29.

19. Feldkamp, L. A., Goldstein, S. A., Parfitt, A. M., Jesion, G. and Kleerekoper, M., The direct examination of three-dimensional bone architecture in vitro by computed tomography, J. Bone Min. Res., 1989, Vol. 4, 3-11.

20. Rüegsegger, P., The use of peripheral QCT in the evaluation of bone remodelling, The Endocrinologist, 1994, Vol. 4, 167-176.

21. Majumdar, S. and Genant, H. K., Assessment of trabecular structure using high resolution magnetic resonance imaging. In: Lowet, G., Rüegsegger, P., Weinans, H. and Meunier, A. (eds.), Bone Research in Biomechanics, pp. 81-96. ISBN: 90 5199 327 7. IOS Press, Amsterdam, 1997.

22. Dequeker, J., Assessment of quality of bone in osteoporosis-BIOMED I: Fundamental study of relevant bone, Clinical Rheumatology, 1994, Vol. 13 (suppl1), 7-12.

23. Viegas, S. F. and Patterson, R. M., Contact pressures within wrist joints, In: Schuind, F. et al. (eds.), Advances in the biomechanics of the hand, pp. 137-151. Plenum Press, New York, 1994.

24. Müller, R. and Rüegsegger, P., Three-dimensional finite element modeling of non-invasively assessed trabecular bone structures, J. Med. Eng. Phys., 1995, Vol. 17, 126-133.

25. Guldberg, R.E. and Hollister, S.J., Finite element solution errors associated with digital image-based mesh generation, ASME/BED, 1994, 147-148.

CONSTRUCTION OF FINITE ELEMENT MODELS ON THE BASIS OF COMPUTED TOMOGRAPHY DATA

G. Kullmer, [1] J. Weiser[2] and H. A. Richard[3]

1 ABSTRACT

In recent years the finite element method (FEM) has become the most appropriate tool for the stress analysis of the human musculoskeletal system in biomechanics. For the construction of finite element models from CT-scans we have developed a geometry orientated method ([FAM]GoFEG) and a voxel orientated method ([FAM]VoFEG). Both methods are presented and compared in this paper. At first the way of extracting the geometry of parts of the musculoskeletal system from CT-scans with the program [FAM]FiltraCT is described, which both methods have in common. With [FAM]GoFEG a CAD-model is built with the CAD-FE-program I-DEAS. This CAD-model can automatically be meshed with finite elements. [FAM]VoFEG utilises the special structure of CT-data, since every picture element of CT-scans represents a rectangular hexahedron. The finite elements are directly and automatically generated from CT-data. The direct transformation of every voxel of the CT-scan to a corresponding finite element results in a mesh with many elements and coarse surfaces. Therefore appropriate software tools have been developed to reduce the number of elements and to smooth the surfaces of the mesh. [FAM]GoFEG and [FAM]VoFEG are verified for a knee joint and a vertebral body, respectively.

Keywords: Finite Element Method, Mesh Generator, Computed Tomography Data

---

[1] Research Assistant, Institute of Applied Mechanics, University of Paderborn, Pohlweg 47-49, 33098 Paderborn, Germany
[2] Research Assistant, Institute of Applied Mechanics, University of Paderborn, Pohlweg 47-49, 33098 Paderborn, Germany
[3] Professor, Institute of Applied Mechanics, University of Paderborn, Pohlweg 47-49, 33098 Paderborn, Germany

2 INTRODUCTION

Biomechanics is an important field of Applied Mechanics. An essential question in this field is the stress analysis of the human musculoskeletal system. In recent years the finite element method (FEM) has become the most appropriate tool for this task since there are no restrictions concerning the geometry of the structure. For example the development of individually adapted implants and prostheses requires rapid and individual FE-mesh generation and stress analysis for the interesting part of the musculoskeletal system. Therefore we have developed two methods, a geometry oriented method ([FAM]GoFEG) and a voxel oriented method ([FAM]VoFEG), for computer aided mesh generation on the basis of computed tomography data. Both methods have in common that the geometry of the body parts is extracted from CT-scans. Young's modulus for every finite element is calculated from the respective mean Hounsfield density values using transformation formulas like that of Carter and Hayes / 1/ 2/.

3 SEPARATION OF BODY PARTS FROM CT-DATA WITH [FAM]FiltraCT

CT-scans are reproductions of the density distribution in a plane body layer with constant thickness. The smallest unit of CT-data is the single picture element. Keeping in mind the layer thickness one picture element represents a rectangular solid called voxel with constant dimensions. The voxel size is determined from the layer thickness, the matrix size and the diameter of the scan area / 3/. For every voxel the position co-ordinates can be calculated since the absolute co-ordinates of the first voxel, the dimensions and the arrangement of all the voxel are known. Every voxel is represented in the CT-data file through the mean density value in Hounsfield units of all the tissues contained in the voxel. For the construction of FE-models for specific parts of the body those voxel that belong to that part must be separated from the others. For this task the self made program [FAM]FiltraCT has been developed. At first a window technique is applied. All voxel with Hounsfield values not within the region of interest are hidden. The contrast of the contours is increased by assigning the mean density value to a voxel calculated from the density of the neighbouring voxel (Fig. 1a).

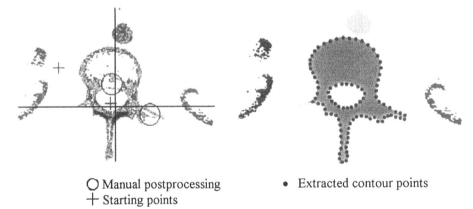

○ Manual postprocessing   • Extracted contour points
+ Starting points

a) CT-scan after applying the window- and b) Extracted voxel and contour points
the mean density technique

Fig. 1: Geometry extraction from CT-scans for a lumbar vertebra with [FAM]FiltraCT

After an occasional necessary manual closing of open contours or separation of merged bone contours in joint regions all the voxel not within the region of interest are eliminated just by setting starting points in the regions to be removed (Fig. 1a). The remaining voxel may be assigned to specific bones. These voxel groups are the input data for the voxel based mesh generation (^{FAM}VoFEG). In a further step all the co-ordinates of the contour points needed for the geometry based mesh generation (^{FAM}GoFEG) /4/ /5/ are determined using special self-made algorithms. For the improvement of the further steps and the smoothing of the contours the number of contour points is automatically reduced in a way that no appreciable loss in information occurs (Fig. 1b).

4 ^{FAM}GOFEG

The co-ordinates of the contour points extracted from CT-data for a single slice are imported in the CAD-software package I-DEAS^{TM}, where they are connected to closed B-spline curves. Afterwards these curves are transformed to cross section planes. Doing this for every slice of the bone structure results in an assembly of cross section planes as shown in Fig. 2. For the generation of the CAD-model from the cross section planes I-DEAS^{TM} offers the CAD-tools *surface by boundary*, *loft* and *mesh of curves* /6/.

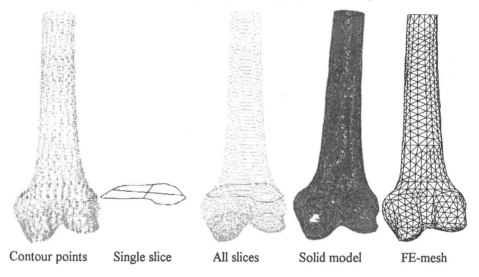

Contour points   Single slice   All slices   Solid model   FE-mesh

Fig. 2: Generating the CAD-model of the distal femur from CT-data with ^{FAM}GoFEG

Using the tool *surface by boundary* the free form surface is described only with the boundary curves. Parts created with *surface by boundary* can easily be meshed but topographic changes in the centre of the surfaces can not be recognised. This tool is valuable for the subdivision of parts into smaller partitions, which is sometimes necessary to improve mesh generation.

With *loft* a CAD-model may be generated through extrusion over several cross sections (Fig. 3). The resulting shape of the free form surfaces may be improved with connecting lines between the cross section planes fitted into the model at any place. But it is disadvantageous that every connecting line separates the over-all-surface of the CAD-model into single surfaces, because an increasing number of surfaces complicates the automatic mesh generation. An advantage of *loft* is, that solid models may directly be generated.

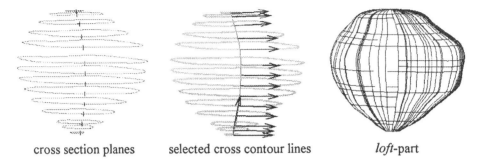

cross section planes   selected cross contour lines   *loft*-part

Fig. 3 Generation of the *loft*-part for the patella

With the third tool *mesh of curves* quadrilateral free form surfaces are created from base lines and cross section curves (Fig. 4). The further steps are similar to *loft*, but with this tool the surfaces are not divided into separate surfaces through additional base lines or cross section lines.

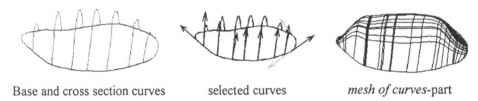

Base and cross section curves   selected curves   *mesh of curves*-part

Fig. 4: Generation of the proximal patella face with *mesh of curves*

The whole surface of the structure should be constructed with quadrilateral free form surfaces. Taking into account the advantages and the disadvantages of the three tools and depending on the topography of the single surface the most convenient method should be chosen. For an adaptive mesh generation it is sometimes necessary to divide the whole part into simple partitions.

For the meshing of three dimensional parts I-DEAS™ offers the options free mesh and mapped mesh generation / 6/. Free mesh generation is only possible for tetrahedral elements. Fig. 2 shows a tetrahedral mesh of the distal femur generated with the free mesh option. In principle there are no restrictions concerning the geometry of the part to be meshed. Nevertheless, there should not be to many surfaces enclosing the parts, since then the automatic mesh generator is not able to generate a mesh. This problem can be solved by dividing the whole part into partitions with fewer surfaces. Moreover the surfaces should not be too small, since they cause unreasonable small elements. Parts can be meshed with brick elements with the mapped mesh generation option. For the use of this option the whole part has to be divided into hexahedral or wedge-shaped partitions. Neighbouring surfaces of the partitions should neither enclose too sharp nor to blunt angles, because along such edges highly distorted elements are generated. Another advantage of mapped mesh generation, except of the fact that brick elements are used instead of tetrahedral elements, is the exact control over the size and the number of elements that are generated.

5 ^{FAM}VOFEG

For ^{FAM}VoFEG all the voxel that belong to the same part are extracted from CT-scans. The simplest method to generate a FE-mesh is to transform every voxel to a hexahedral finite element with eight nodes, taking advantage of the fact that every voxel represents a rectangular solid with constant dimensions. The height of the elements is set equal to the distance between neighbouring CT-scans. The in plane dimensions are set equal to the voxel grid size. With high resolution CT-scans, that are necessary for a satisfactory geometry reproduction, this results in a FE-mesh with an unreasonable large number of elements and coarse surfaces (Fig. 5). To improve the meshes we have developed software tools to reduce the number of elements and to smooth the surfaces.

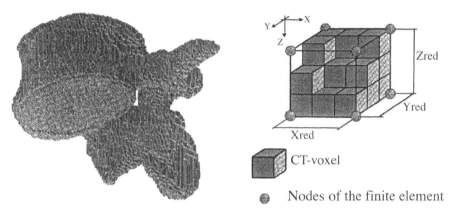

Fig. 5: FE-mesh of a vertebra with every voxel transformed to a hexahedral element

Fig. 6: Data reduction method

The arrangement of the voxel vertices represents a rectangular grid. For the reduction of the number of elements hexahedrons are generated with edge lengths measuring integer multiples of the grid size. The data reduction depends directly on the multiples of the grid size defining the element edge lengths. An element is generated if the volume fraction filled with voxel or if the number of voxel corner points in two element faces sharing the same edge exceed pre selected threshold values (Fig. 6). This results in meshes with a bad contour adaptation and very coarse surfaces as shown in Fig. 7.

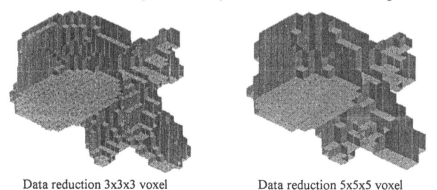

Data reduction 3x3x3 voxel   Data reduction 5x5x5 voxel

Fig. 7: FE-mesh of a vertebra after application of the data reduction method

For a better adaptation of the bone shape the surface nodes of the elements are projected step by step onto the vertices of the underlying voxel structure. Beginning with a starting element the surface nodes not yet lying on a voxel vertex are pushed one grid size towards the nearest voxel vertex in the inner of the element. Then this procedure is applied on the free nodes of all surface elements one after the other. This iteration loop is repeated until all surface nodes lie on voxel vertices. For a 2D-mesh the node projection order is shown in Fig. 8. The resulting 2D-mesh for one slice of a vertebra is shown in Fig. 9. The node projection method results in a good shape adaptation and in smoothed contours for 2D-meshes.

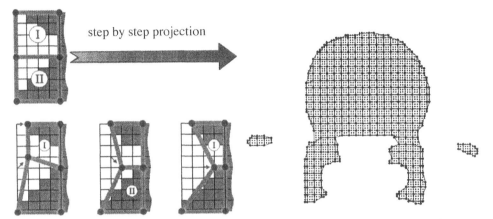

Fig. 8: Step by step projection of the surface nodes onto the vertices of the voxel in one plane

Fig. 9: Result of the contour adaptation for a vertebra in one plane

For 3D-mesh generation the following special features are additionally integrated into the node projection algorithm:
- During one iteration loop nodes are projected at most one grid size in each of all three directions towards the inner of the respective element they are related to.
- If the node movement will result in an inadmissible element distortion for one of the elements sharing this node the node remains unchanged.
- To avoid undesired element bridging at structural cavities nodes shared by only two elements are doubled and are free to move independently in either element.

Data reduction: 3x3x3 voxel   Data reduction: 5x5x5 voxel   Data reduction: 7x7x7 voxel
Threshold = 0.01   Threshold = 0.008   Threshold = 0.001
4359 elements; 6077 nodes   1220 elements; 1915 nodes   501 elements; 866 nodes

Fig. 10: FE-meshes for a lumbar vertebra automatically generated with ^{FAM}VoFEG

Fig. 10 shows three examples of FE-meshes for a lumbar vertebra automatically generated with ^{FAM}VoFEG. The examples clearly show the influences of the parameters for data reduction and the threshold value for element generation on the surface smoothing and the shape adaptation.

6 COMPARISON BETWEEN ^{FAM}GOFEG AND ^{FAM}VOFEG

The following two tables give brief lists of advantages and disadvantages of the two mesh generation methods for parts of the musculoskeletal system presented in this paper.

Table 1: Advantages and disadvantages of ^{FAM}GoFEG

^{FAM}GoFEG	
Advantages	Disadvantages
• Good shape approximation • Smooth surfaces • Simple mesh adaptation • Geometry based definition of boundary conditions possible • Boolean operations applicable for an easy modification of the model • Powerful graphical representation tools available • Direct access to manufacturing tools	• Low automation level • Time consuming generation of the CAD-model • Only tetrahedral elements available for free meshing of arbitrary volumes • Mapped meshing with hexahedral elements requires time consuming partitioning of the CAD-model • Missing direct correlation between voxel and finite element requires transformation software tool for the estimation of Young's modulus for each element

Table 2: Advantages and disadvantages of ^{FAM}VoFEG

^{FAM}VoFEG	
Advantages	Disadvantages
• Automatic mesh generation • Short runtime for mesh generation • Good shape approximation • Smooth surfaces • Different element sizes possible • Simple estimation of Young's modulus for each element due to direct correlation between voxel and element	• Only one element size per model implemented • Partial insufficient surface smoothing • Only element based definition of boundary conditions possible • Model modifications difficult

7 DISCUSSION

For the FE-mesh generation of parts of the human musculoskeletal two methods have been proposed. A geometry based method (^{FAM}GoFEG) and a voxel based method (^{FAM}VoFEG). For the application of ^{FAM}GoFEG the contours of the parts are extracted from CT-scans. ^{FAM}GoFEG utilizes the commercial software package I-DEAS Master Series for the construction of the CAD-model from the part contours and the meshing of the parts. For the application of ^{FAM}VoFEG self made software tools have been generated. With this method finite elements are created directly from voxel extracted from CT-scans. To improve the shape adaptation the surface nodes are pushed onto the underlying voxel structure. The comparison between the two methods shows that at present it cannot be stated which method overrides the other. Both methods have specific advantages and disadvantages. So it depends on the respective application which method should be preferred. Furthermore one of our aims is the improvement of the applicability of both methods to develop an almost automatic mesh generator that involves the advantages of both proposed methods. With such a tool one of our main goals the rapid production of individually adapted and optimized healing aids may be reached.

ACKNOWLEDGEMENT

The authors would like to thank the Heinz Nixdorf Stiftung for generous financial support of this research and Dr. Nöcker from the Brüderkrankenhaus St. Josef, Paderborn for providing us with CT-data and giving us technical support.

8 REFERENCES

/ 1/   Cowin, S. C.: The mechanical properties of cancellous bone. In: Bone Mechanics, Editor: Cowin, S. C., CRC Press, Second Printing 1991, S. 129-157

/ 2/   Carter, D. R. and Hayes, W. C., The compressive behaviour of bone as a two-phased porous structure, J. Bone Jt. Surg., 1977, Vol. 59(A7), 954-962.

/ 3/   Wegener, O. H.: Ganzkörpercomputertomographie. CD-ROM-Version 1.0, Blackwell Wissenschaftsverlag, Berlin 1995

/ 4/   Kullmer, G., Richard, H. A. and Weiser, J., Finite-Element-Analysen auf der Gundlage von Computertomographiedaten. In: FE-Workshop 1996, Die Methode der Finiten Elemente in der Biomedizin und angrenzenden Gebieten. Universität Ulm, Ulm, 1996, 2.1-2.10

/ 5/   Kullmer, G., Richard, H. A. and Weiser, J., Finite-element-analysis on the basis of computed tomography data. In: Advances in computational Engineering Science, Editors: Atluri, S. N., Yagawa, G., Tech Science Press, Forsyth, Georgia, 407-412.

/ 6/   I-DEAS^{TM} -SMART-VIEW, Structural Dynamics Research corporation (SDRC), Ohio, 1994.

DAMAGE AND REPAIR IN COMPACT BONE: SOME THOUGHTS ON THE SHAPE OF BASIC MINERALISATION UNITS

D.Taylor*

1. ABSTRACT

Compact bone is replaced and repaired through remodelling processes carried out by so-called Basic Mineralisation Units (BMUs). The method of operation of BMUs has been often described but is poorly understood. This paper takes as a starting point the physical shape of the BMU, which includes a resorption cavity and an apposition region. A simple hypothesis is proposed to explain the shape, based on signals emitted from a single growth point. It is shown that a BMU with the appropriate shape and properties can arise if cellular activity is controlled from this point. The point is then identified with a fatigue crack, the signal being diffusable substances emitted during its growth. A computer simulation is used to show that the solution obtained analytically is stable and capable of demonstrating BMU initiation, growth and movement. A number of other features can be explained, such as bifurcation. This hypothesis is attractive because it provides a natural link between the processes of damage and repair in bone.

2. INTRODUCTION

D'Arcy Thompson (1), in 1917, noted that the shapes of living things, though they may appear complex at first sight, can often be explained by relatively simple mathematical and physical laws. This paper aims to apply Thompson's principle to an entity which is of great importance to the integrity of compact bone: the BMU. The term Basic Mineralisation Unit was coined by Frost (2) to describe an assembly of cells whose function is to replace bone. It is believed that BMUs are essential to the mechanical integrity of bone because they remove damage in the form of small fatigue cracks, before these cracks are able to grow to cause failure. When BMU function is impaired (e.g. in osteopetrosis) bones become large but fracture easily.

Keywords: Bone, Repair, Damage, BMU
*Bioengineering Research Group, Mechanical Engineering Department, Trinity College, Dublin 2, Ireland.

BMUs operate by a combination of resorption of old bone (using specialised cells called osteoclasts) followed by the creation of new bone (known as apposition) by osteoblast cells. The result is a cavity which moves through bone at about 40μm/day, leaving in its wake a new osteon. The shape of the BMU (fig.1) consists of a relatively small, hemi-elliptical resorption cavity followed by a much longer region in which apposition is occurring. The movement direction is downwards (negative y-direction) in this diagram. The normal way of drawing the shape (on the left in fig.1) is a distortion, for clarity's sake, of what is in fact a very long, narrow tube. It is not clear why the shape takes this particular form; stress concentration is one reason, since BMUs tend to lie parallel to directions of maximum principal stress, but this highly elongated shape loses more in the creation of porosity than it gains in the reduction of stress. Other unresolved questions about the BMU are: (a) how does it know which way to go, and (b) how does it keep its shape whilst moving? Martin and Burr (3) have proposed a mechanism whereby BMUs follow lines of maximum principal stress by adjusting cellular activity according to local strain sensors, but their

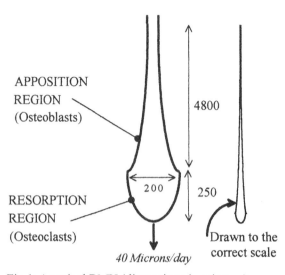

APPOSITION
REGION
(Osteoblasts)

4800

200   250

RESORPTION
REGION
(Osteoclasts)

40 Microns/day

Drawn to the
correct scale

Fig.1: A typical BMU (dimensions in microns)

mechanism only works in compression and the relevant sensors have not been identified. The problem of shape maintenance arises if we assume (as seems reasonable) that individual osteoblasts do not know which way to move, but simply remove bone in a direction normal to the surface at the place where they are attached. This implies that the activity (or number) of osteoblasts must decrease as we move away from the apex of the BMU, in order to avoid the situation where the cavity simply grows in size equally in all directions. The same applies to osteoblasts, except that, since we know that they are producing a lamellar structure, we assume that they lay down bone in the width (x) direction, though this assumption can be relaxed without greatly altering our predictions.

3. THEORY

3.1 A Hypothesis

The above problems can be resolved very simply if we assume that cellular activity (of both osteoclasts and osteoblasts) is controlled from a single growth-pont, G, located some distance ahead of the apex (fig.2). We imagine that some signal (which may be a diffusing chemical, an electrical field, etc.) is emitted from this point, giving rise to a signal strength, S, which thus varies as $1/g^2$, g being the distance from G to any point on

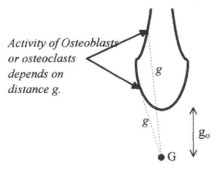

Fig.2: It is assumed that cellular activity depends on
the strength of a signal emitted from the point G.

the BMU. Assume that the activity level of osteoclasts (and therefore the amount of
resorption, r, achieved per day) is directly proportional to S, then:

$$r = R/g^2 \qquad (1)$$

...where R is a rate constant whose value is known since we know how fast the BMU
moves; when $g=g_o$ (the distance from G to the apex) then r=40μm/day. These simple
assumptions lead us to a unique shape for the BMU, which can be expressed as a
differential equation in x and y, because the requirement to maintain shape imposes
conditions on the value of the local slope, dy/dx. The result is:

$$\frac{dy}{dx} = \sqrt{\left[\frac{x^2 + (g_o + y)^2}{g_o^2}\right]^2 - 1} \qquad (2)$$

This equation can be solved numerically to obtain the description of shape, i.e. the
width x as a function of vertical distance, y. We assume that osteoblast activity is
controlled in a similar way except that now the amount of apposition, a, depends on
$(S)^{1/2}$, thus:

$$a = A/g \qquad (3)$$

...which leads to a simple result for the shape of the apposition region:

$$y = Ce^{-Dx} - E \qquad (4)$$

Here C, D and E are constants.

3.2 A Computer Simulation

Equations (2) and (4) give us an analytical result for the BMU shape, however it is very
useful to generate this shape using a computer simulation based on the activity

equations, (1) and (3). Apart from testing our mathematics, this allows us to investigate the stability of the solutions. Eqns (2) and (4) may be correct but could be positions of unstable equilibrium which might degerate as the BMU moves. More generally, it should be possible to show that the correct shape will evolve if we start from any other initial shape. Figs (3) and (4) demonstrate this evolution starting from a small spherical cavity. Here g_o=94μm and the growth point G is located vertically below the cavity as in earlier figures. Points on the surface of the cavity move in directons normal to the local surface according to eqns (1) and (3); apposition or resorption is chosen depending on whether a>r. Only one other condition is imposed; apposition is not allowed if x<20μm, in order to form a central space for the blood vessel in the finished osteon. Fig.3 shows the initial development and fig.4 (on which the y-scale has been compressed for clarity) shows full development of a stable shape over a period of 150 days. The shape of the resorption cavity is soon established and grows unchanged thereafter, gradually creating behind it the long tail of the apposition region, which then also moves forwards in a stable manner.

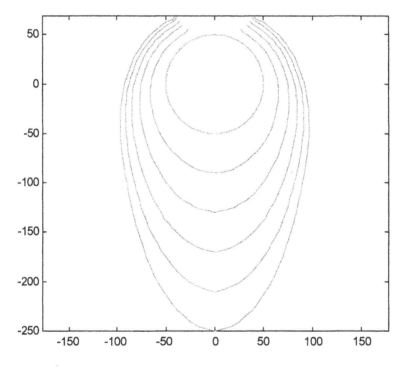

Fig.3: Initial results from the computer simulation, starting with a spherical cavity of radius 50μm. The lines show the outline of the cavity after periods of one day. The growth point is situated vertically below the cavity (6 o'clock position). Dimensions in microns.

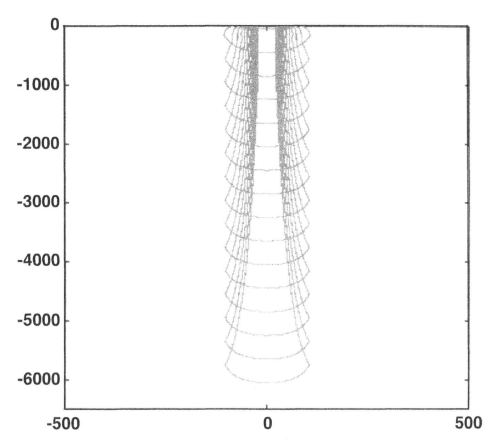

Fig.4: Results of the computer simulation beginning with a spherical cavity, as in fig.3. This figure shows the result after 150 days, plotting the cavity shape every ten days. Dimensions in microns; note the compression of the vertical scale.

4 COMPARISON WITH EXPERIMENTAL DATA

Fig.5 compares the shape derived from the simulation with experimental results from the literature (3,4,5). Average values were chosen from the measurements, which of course vary from one BMU to another and also by species, age, etc. The aim here is to show that our simple hypothesis is effective in predicting the general shape. Errors of the order of $10\mu m$ occur in the apposition region; the resorption space is very well predicted, though experimental data for this region is less precise. The theory allows only one adjustable parameter, g_0 , for which a value of $94\mu m$ gives the best fit. The only other constants - R and A - are rate constants whose value is fixed by the known rate of movement of the BMU.

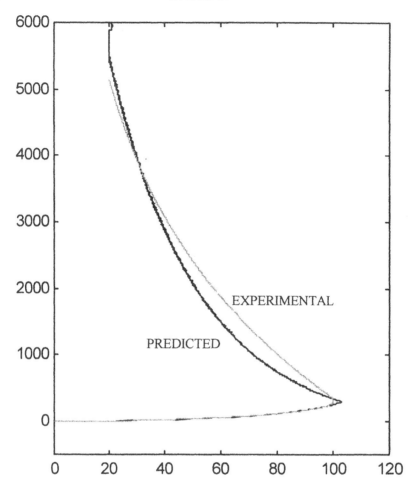

Fig.5: Comparison of BMU shape predicted using the simulation and measured experimentally. Dimensions in microns; note compression of the vertical scale.

5 DISCUSSION

This very simple hypothesis is capable of explaining the shape of the BMU, using a growth point, G, but what is the physical interpretation of this growth point? It is proposed that the actual source of the controlling signal is a fatigue crack, growing outwards from the apex. The crack will break through cells and consequently release many substances which, as they diffuse away from the crack, can act as the necessary signal. Many workers have already identified substances which can control activity levels of osteoclasts and osteoblasts, many of which are to be found within cells (6,7). Other theories have been developed to create a link between these chemical substances and mechanical loading (e.g.(8)). Since Frost (9), these theories assume the existence of a "mechanostat" - a device which is supposed to translate mechanical strain into a chemical signal. However, this mechanostat has not been found. The present approach avoids the need for a mechanostat; here the link between mechanical loading and cellular activity is provided by damage. This also provides us with a natural explanation for the BMUs assumed role as a remover of damage: the BMU will be initiated by the presence of a crack and will continue to grow until it removes the crack. It is possible that a dynamic equilibrium will be set up in this way, with the crack growing at the same rate as the BMU moves, thus maintaining the same crack length and therefore the same signal strength. The existance of this type of dynamic equilibrium between damage and repair has been proposed previously (10) without considering its physical nature.

The controlling equations which link cellular activity to signal strength - eqns (1) and (3) - have been arbitrarily chosen here, since we don't have any information about the actual chemical reactions involved. Eqn (1) was chosen in this form because it is the simplest possible relationship imaginable; S is assumed to be the concentration of a chemical which acts as a precursor to osteoclasts in a simple chemical reaction of the form:

$$S \rightarrow O'$$

...where O' represents an osteoclast (or its precursor). If, for example, the reaction was of the form:

$$2S \rightarrow O'$$

...then r would be proportional to S^2, eqn (1) would become $r=R/g^4$ and the shape of the resorption cavity would be different. As regards osteoblast activity, this must depend on S^n where $n<1$ so that osteoblasts are suppressed (relative to osteoclasts) at short distances from the apex, and dominate at long distances. The square root relationship was an arbitrary choice, and it may be that other relationships will give a more exact fit to the data.

This link between the fatigue crack and the BMU is attractive for two other reasons. Firstly, the crack will tend to grow in the direction of the weakest material, so the crack will act as a 'probe', directing the BMU towards material which should be removed. This makes the most efficient use of the BMU. Secondly, fatigue cracks of this kind will often bifurcate, creating a second branch crack. This would lead to a similar bifurcation of the BMU, which is a phenomenon often noted in the literature (3). Finally the theory is attractive because of its simplicity, and especially because it assumes no 'intelligence' on the part of individual cells; a given osteoclast, for example, does not have to know where it lies within the BMU or in which direction the BMU is travelling. In the present paper the mathematics was simplified by assuming that the signal S came entirely from a single point, assumed to lie at the centre of the crack. In future work a more realistic model will be developed, using the full three-dimensional shape of the crack and incorporating information about its rate of growth as a function of applied stress.

REFERENCES

1) Thompson, D.W., On growth and form, Publ.C.U.P.(UK) 1961

2) Frost, H.M., Bone remodelling dynamics, Publ. Charles C. Thomas (Springfield, IL, USA) 1963.

3) Martin, B.M. and Burr, D.B., Structure, function and adaptation of compact bone, Publ.Raven Press (New York) 1989.

4) Manson, J.D. and Walters, A.E., Observations on the rate of maturation of the cat osteon, J.Anatomy, London, 1965, Vol.99, 539-549.

5) Lee, W.R. , Apposition bone formation in canine bone: a quantitative microscopic study using tetracycline markers, J.Anatomy, London, 1964, Vol.98, 665-677.

6) Athanasou, N.A. Cellular biology of bone-resorbing cells, J.Bone Joint Surg. 1996, Vol.78-A 1096-1112.

7) Konttinen, Y.T., Imai, S. and Suda, A., Neuropeptides and the puzzle of bone remodeling, Acta Orthop.Scand. 1996 Vol.67 632-639.

8) Brighton, C.T., Fisher, J.R., Levine, S.E., Corsetti, J.R., Reilly, T., Landsman, A.S., WIlliams, J.L. and Thibault, L.E. The biochemical pathway mediating the proliferation response of bone cells to a mechanical stimulus J.Bone Joint Surg.1996, Vol.78-A, 1337-1347.

9) Frost, H.M. Intermediary organisation of the skeleton, Publ.CRC Press (Boca Raton, USA) 1986.

10) Taylor, D., Bone maintenance and remodelling: a control system based on fatigue damage, J.Orthop.Res., 1997, In press.

MATHEMATICAL MODELLING OF STRESS AND STRAIN IN BONE FRACTURE REPAIR TISSUE

T.N. Gardner, T. Stoll, L. Marks, M. Knothe-Tate

1. ABSTRACT

A method is described for examining the mechanical environment of a bone fracture, which arises from routine weight-bearing activity. A 2-D geometric model of a fracture site is developed from orthogonal radiographs which are scanned and digitised. Bridging tissue (callus) forming periostealy and endostealy across the fracture is separated into regions of common tissue histology using the digitised images. Region boundaries are identified from the disparity in pixel illumination intensity between regions. Initially, material properties are found for each callus region from tests reported in the literature on new bone tissue at the same temporal points in healing. A linear elastic FE analysis is performed using Cosmos/M, in which the initial properties are adjusted iteratively until measured relative displacements of the bone fragment ends, during two-legged stance, match forces and moments at the fracture calculated using a lower limb biomechanics model. Approximate tissue histology may be identified from its location, the temporal point in healing and from its material properties. Peak dynamic stresses and strains within the tissue are then calculated for maximum displacements that are measured during walking activity. Solutions of the model may be used to examine the influence of the mechanical environment on the observed pattern of healing in real fractures. The procedure is illustrated for a single subject with a mid-diaphyseal, tibial fracture.

2. INTRODUCTION

Previous research indicates that the mechanical environment of a bone fracture repair site profoundly influences the process and speed of healing [1,2]. This environment is

Keywords: Bone fracture, healing, model, callus, histology.

T. Gardner, Research Fellow, University of Oxford, NOC, Headington, Oxford, UK.
L. Marks, FE Consultant, Desktop Engineering, Long Hanborough, UK.
T. Stoll, M. Knothe-Tate, Research student, Supervisor, Swiss Federal Inst. of Technology, Zurich, Switzerland.

largely a consequence of the structure and stability of the fracture site and the degree of physical activity routinely practised by the patient [3,4]. Consequently, it is patient specific, with little commonality existing between subjects. Previous studies [5-8] have examined idealised models of generalised structure and stiffness, to which arbitrary forces or displacements are applied. Patterns of stress, strain and empirical healing indices are then examined in relation to common patterns of healing. A more direct approach is used in the present study which develops a custom model of an individual fracture. Temporal variations in the material properties of the different regions of callus are found. Additionally, stress and strain patterns occurring in the callus tissue are correlated with the observed pattern and timing of tissue differentiation and calcification throughout healing.

3. METHOD

3-D inter fragmentary displacement is monitored during two-legged stance and walking activity by attaching an instrumented spatial linkage [4] to the inner pair of bone screws across the fracture (Figure 1). The instrument uses Magnetic field Hall Effect sensors to measure displacement in six degrees of freedom. Displacements at the linkage are translated trigonometrically to monitor the 3-D motion of the distal bone fragment in relation to the proximal at the centre of the fracture. A test simulation using a vernier measurement jig established that displacements measured by the instrument are accurate to within ±0.025 mm and ±0.025°.

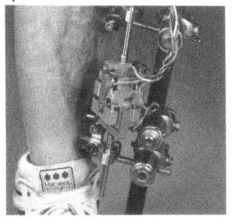

Figure 1 - The instrumented spatial linkage measures 3D inter fragmentary displacement at the fracture centre

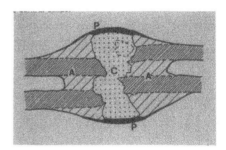

Figure 2 - Fracture repair callus is separated into 3 regions of common tissue histology (Sarmiento and Latta, 1995)

Forces and moments in the tibia at the level of the fracture are calculated from a biomechanics model of the lower limb musculo-skeletal system. A 7-camera VICON 370 motion analysis system is used with two AMTI floor mounted force plates and a 10-channel EMG system to collect data during activities performed by the subject. Simple isometric and two legged stance tests are performed to customise the mathematical model. EMG are used to ensure that agonistic and not antagonistic muscle activity is evoked in isometric tests and that abductors are not active in standing tests [9,10]. Validation tests of the biomechanics model have been carried out on two tumour patients with instrumented proximal femoral prostheses. Calculation of the axial forces in the femurs agreed well with the measured forces during standing using 2-D [9,10] and 3-D [11,12] analysis.

Radiographs are taken in the lateral and antero-posterior planes for scanning and digitising to provide a computerised image. The callus is divided into three regions of common tissue histology: adjacent, central and peripheral callus [13] (Figure 2). Adjacent callus is the tissue which is attached to the intact cortical bone periostealy and endostealy, where a loose trabecular structure of relatively mature tissue forms that rapidly changes in a few weeks from soft to hard tissue. Central callus is the tissue which bridges the proximal and distal fragments. It contains only avascular primitive tissue due to it's remoteness from a blood supply, until at some later stage endochondral ossification is able to replace the soft fibrous and cartilage tissue with hard calcified tissue. Peripheral callus tissue is a soft dense well orientated fibrous layer which is bounded externally by a new periosteum and progresses to new bone relatively quickly. Identification of the boundaries of these regions is accomplished by an analysis of the different ranges of grey scale intensity of the displayed image. It is assumed that the variations in pixel intensity are due to planar variations in x ray absorption, that are influenced by tissue density and atomic number. Intervals of grey scale range are displayed sequentially to establish the boundaries of regions with common image intensity and therefore common tissue histology.

2D geometric models are developed for FE analysis by mapping the internal and external boundaries of the bone and the three regions of callus in the coronal plane. Triangular elements are used with non-uniform mesh density, specified for each region, dependant on the complexity of their shape. Element thickness is determined from an 'equivalent width' of an equivalent rectangular section of bone (a transverse section of the bone with a full-width medullary space). This is found by estimating the second moment of area of the transverse section of the subject's tibia from orthogonal radiographs. The moment area of the tibial section is then matched with the moment area of the rectangular model section by adjusting model width to the 'equivalent width'. Figure 3 shows the model boundary conditions at the positions of the upper and lower bone screw insertion points. Displacements are applied at the fracture centre at the end of a rigid beam element connected to the upper insertion point, and reaction forces and moments are calculated at the fixed lower point.

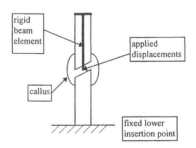

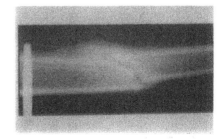

Figure 4 - The oblique, mid-diaphyseal, tibial fracture at 16 weeks post operation.

Figure 3 - Boundary conditions showing the fixed lower boundary and the displacement of the upper bone fragment applied at the fracture centre.

Initial values of material properties of the callus are obtained from the literature for each tissue region at each stage in healing. These are found from a power curve of elastic modulus versus density which has been generated from discrete measurements of bone modulus at different densities. Properties are iteratively adjusted until the calculated forces match the measured displacements for two legged-stance, using a linear elastic, plane stress analysis for isotropic material properties. Measured peak displacements arising from walking activity are applied to the FEM to calculate tissue stress and strain from the calculated tissue properties. Tissue histology is identified in each region from the regional location, the temporal point in healing and the material properties. Values are used to establish quantitatively the growth and maturation of repair tissue in response to the stress and strain environment imposed by the standing and walking activities.

3. RESULTS

The method is illustrated using data previously obtained from a subject with an oblique, mid-diaphyseal, tibial fracture (Figure 4) stabilised by a unilateral external fixator and measured at 4, 8, 12 and 16 weeks post operation. Each of the processes were carried out in the manner described above, except the calculation of tibial forces and moments at the fracture during two-legged stance. These were estimated from the measured ground reaction forces of the fracture subject, using a correlation between the two from the instrumented hip prosthesis data.

Figure 5 shows the irregular geometry of the callus and the different regions of tissue at the four stages in healing. Whereas the central and adjacent callus regions are apparent as early as week 4, the soft fibrous layer is not yet positioned at the periphery of the adjacent callus. It's development at the periosteal surface of the fragments appears to be inhibited, but by week 8 it is fully formed at the periphery of the callus.

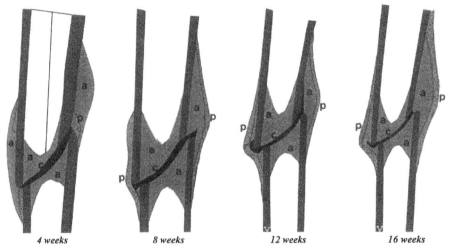

| 4 weeks | 8 weeks | 12 weeks | 16 weeks |

Figure 5 - Shows the regions of cortical bone and callus in the model (not all at the same scale) which have been identified from grey scale analysis of the computerised radiographic images. (a - adjacent, c - central, p - peripheral callus.)

Figure 6 shows equivalent peak strain calculations. Strain was uniformly low (dark) in the peripheral and adjacent callus regions and in the cortical bone. As expected it was high (light) in the central callus regions, reaching peaks at both lateral extremities. As a result of tissue calcifying and stiffening, strain reduced quickly from 69% (central), 14% (adjacent) and 22% (peripheral) at week 4, to 6% (central), 1% (adjacent and peripheral) at week 8. By week 16, strain was reduced to less than

Figure 6 - Equivalent strain patterns in the fracture repair callus

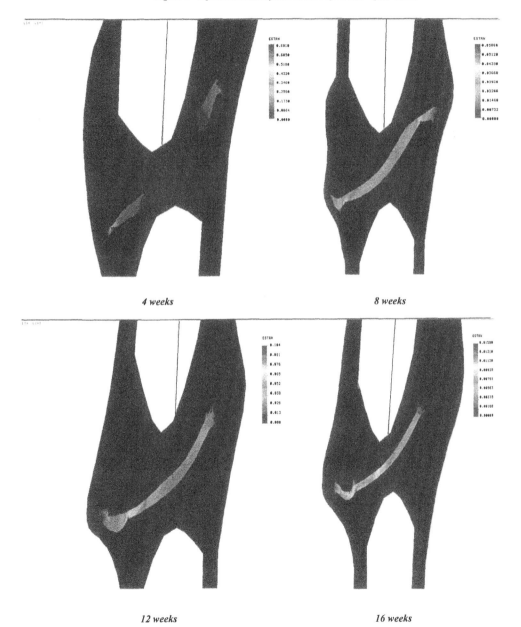

4 weeks

8 weeks

12 weeks

16 weeks

2%; tolerable for woven bone without incurring damage [14]. Elastic moduli in the central callus at week 4 indicated haematoma and granulation tissue [15], and at week 8 [15] through to 16 [16] it indicated fibrocartilage. The adjacent callus appeared to change from soft connective tissue at week 4 [17] to properties commensurate with dense fibrous tissue and trabecular bone by week 8 [18,19], and thence to cancellous bone [19]. The peripheral callus properties suggested early calcification by week 8 [19], which was well advanced by week 16 (a week before the fixator was removed).

Error analysis:
Adaptive mesh refinement by the H method concentrated refinement to local regions of stress discontinuity. These were shown to be confined locally to the junction between regional boundaries across which there was a large disparity in elastic properties [20]. When mesh density was determined regionally by complexity of model shape, this reduced stress errors in areas of stress concentrations to less than 10% in two iterations. Errors were calculated from the maximum difference between nodal stresses and the average stress at the node. The model confirmed that solutions are very sensitive to small perturbations in the geometry of the callus and in the value of material properties.

4. DISCUSSION

Since this modelling approach incorporates a detailing of the fracture geometry, and material properties of the repair callus are calculated rather than selected from an independent source, sensitivity to these parameters need not be considered. This approach enables us to examine the real stress state behind the subsequent expansion of the callus and the phase development of tissue in a single subject. However, further refinement of the model may need to be undertaken before this approach can be adopted with confidence for a large number of subjects. The following areas will be addressed.

As with other 2D models, the equivalent rectangular section is an attempt to incorporate some of the three dimensional features. A comparison between 2D and 3D models on the same subject is underway to examine the success of this. Nevertheless, the authors believe that their 2D approach is an improvement on the 2D methods used by others [5,6,21,22]. Although other approaches adopt a rectangular section of variable stiffness to represent the variability in model thickness, they are unable to include the endosteal callus which prevents a full examination of the adjacent callus region.

Influences on x ray absorption arising from variability in transmission path length through tissue leads to difficulty in identifying the different regions of tissue from planar radiographs. Even so, since depth changes are not as abrupt as the variations in material properties at boundaries between regions of dissimilar tissue, boundaries may be located with reasonable accuracy. To further improve accuracy, MRI images may be used to provide 3D information for modelling, or for correcting planar radiographic images by removing the spatial variation in pixel intensity which arises from variability in path length.

Material properties selected for the model during walking activity are consistent with a linear elastic, isotropic, monophasic, material. This is because gait cycle frequency is approximately 1 Hz, and physiological loading rates are probably too rapid for significant relative motion to occur between solid and fluid phases to produce a poroviscoelastic response[23]. A bulk, elastic, monophasic behaviour is probably more relevant at this rate of loading. Also, from a purely practical point of view, data for the model may not be available from the literature to define the variables required to simulate poroviscoelastic behaviour in each of the tissue regions at different temporal points in healing. Nevertheless, an attempt will be made at some time to redefine the material response to model poroviscoelastic behaviour in a transversely isotropic material. As well as examining the material response in the elastic range, account will be taken of yield since strains arising from weight-bearing activity are likely to cause damage to callus tissue.[3,4]

The contribution to loading support provided by the unilateral external fixator frame and also by the fibula as it heals and stiffens have not been accounted for within the model. This omission is acceptable within limits, since at 4 weeks post operation less than 10% of the predominantly axial tibial load is supported by the frame; the balance of load being transferred across the fracture because of inter fragmentary gap closure. The fixator contribution further reduces as the fracture heals and the callus stiffens.[24] Conversely, the load carried by the fractured fibula increases throughout healing, and it may carry up to 8% of tibial load by week 16.[25] Therefore, the combined contribution to tibial load bearing from both sources is expected to be less than 10% at any time during healing. To confirm this, an experimental simulation in the laboratory is proposed and a general stiffness matrix will be found which will be included in future models to simulate loading support by the frame and the fibula.

The model has been shown to provide useful information, but it may need further validation before it is adopted for the proposed study to correlate strain with healing.

5. REFERENCES

1. Kenwright J., and Goodship A. E., Controlled mechanical stimulation in the treatment of tibial fractures, Clin. Orth. and Rel. Res., 1989, Vol. 241, 36-47.
2. Goodship A. E. and Kenwright J., The influence of induced micromovement upon the healing of experimental fractures, J. Bone. Joint Surg., 1985, Vol. 67-B/4: 650-655.
3. Gardner T. N., Evans M. and Kenwright J., The influence of external fixators on fracture motion during simulated walking, J. Med. Eng. Phys., 1996, Vol.18:4, 305-313.
4. Gardner T. N., Evans M, Hardy J. R. W., Richardson J and Kenwright J., Dynamic inter fragmentary motion in fractures during routine patient activity, Clin. Orthop. Rel. Res., 1997, Vol. 336, 216-225.
5. Carter D. R., Blenman P. R., and Beaupre G. S., Correlations between mechanical stress history and tissue differentiation in initial fracture healing. J. Orthop. Res., 1988, Vol. 6, 736-748.
6. Blenham P. R., Carter D. R. and Beaupre G. S., Role of mechanical loading in the progressive ossification of a fracture callus, J. Orthop. Res., 1989, Vol. 7, No. 3, 398-407.
7. Cheal E. J., Mansmann K. A., DiGioia III A. M., Hayes W. C. and Perren S. M., Role of interfragmentary strain in fracture healing: ovine model of a healing osteotomy, J. Orthop. Res., 1991, Vol 9, No. 1, 131-142.

8. Biegler F. B. and Hart R. T. Finite element modelling of long bone fracture healing, 'Recent Advances in Computer Methods in Biomechanics and Biomedical Engineering' - Ed. J. Middleton, GN Pande and KR Williams, Books and Journals, International Ltd. 1992.

9. Lu T. W., Taylor S., Gill H. S., Walker P. S., Ling R. S. M., and O'Connor J. J., Gait analysis of patients with instrumented massive proximal femoral prostheses. Proc. Brit. Orth. Res. Soc., Spring 1995. J. Bone Jt. Surg. Vol. 77B: Suppl. III, p321.

10. Lu T. W., O'Connor J. J., Taylor S. and Walker P. S., Interpretation of isometric tests on patients with instrumented massive prosthesis, Proc. Europe. Soc. of Biomech., August 1996, p269.

11. Lu T. W., Taylor S.J.G., O'Connor J. J. and Walker P. S., Influence of muscle activity on the forces in the femur: an in vivo study. J. Biomech. (in press).

12. Lu T. W., O'Connor J. J. Taylor S.J.G. and Walker P. S., Influence of muscle activity on the forces in the femur: validation of a lower limb model. J. Biomech. (in press).

13. Sarmiento A. and Latta L. L., Functional fracture bracing. Tibia, humerus, ulna. Springer Verlag, Berlin, 1995.

14. Perren S. M. and Cordey J., Die Gewebsdifferenzierung in der Fracturheilung. Monatsschrift f. Unfallheilkunde, 1977, Vol. 80, 161-164.

15. Perren S. M. and Cordey J., 'Current concepts of internal fixation of fractures', 1980, Springer Verlag.

16. Abteilung Mess-, Regel- und Mickrotechnik, Universitat Ulm, Simulation einer fraktuirheilung, abstract anlasslich des Biomedizin Workshop Uniklinik Ulm, 13-14 April 1994.

17. Duck F. A., Physical properties of tissue - A comprehensive reference book, 1990 Academic Press Ltd., London, ISNB 0-12-222800-6.

18. DiGioia A. M., Three-dimensional strain fields in a uniform osteotomy gap, 1986, J. Biomech. Eng., Vol. 108, 273-280.

19. Davy D. T. and Connolly J. F., The biomechanical behaviour of healing canine radii and ribs, 1982, J. Biomech., 1982, Vol. 15, 235-247.

20. Pande G. N. and Lee J. S., A finite element homogenisation technique for bone and soft tissue interface analysis, 'Recent Advances in Computer Methods in Biomechanics and Biomedical Engineering' - Ed. J. Middleton, GN Pande and KR Williams, Books and Journals, International Ltd. 1992.

21. Rybicki E. F., Simonen F. A. and Weis Jr. E. B., On the mathematical analysis of stress in the human femur. J. Biomech., 1972, Vol. 5, 203-215.

22. Piziali R. L., Hight T. K. and Nagel D. A., An extended structural analysis of long bones - application to the human tibia, J. Biomech., 1976, Vol. 9, 695-701.

23. Carter D. R. and Hayes W. C., The compressive behaviour of bone as a two-phase porous structure, J. Bone Jt. Surg., 1977, Vol. 59-A, No. 7, 954-962.

24. Gardner T. N. and Evans M., Relative stiffness, transverse displacement and dynamisation in comparable external fixators, Clin. Biomech., 1992, Vol. 7, 231-239.

25. Lambert K. L., The weight-bearing function of the fibula, J. Bone Jt. Surg., 1971, Vol. 53(A), No. 3, 507-513.

Acknowledgements:

The authors wish to acknowledge valuable contributions to the project made by M. Evans and S Mishra (OOEC) for assisting with displacement measurements and illustrations, H. Simpson (Nuffield Dept. of Orth. Surgery) for clinical support, J. Hardy (Glenfield Hospital Leicester) for providing the test subject and HS Gill (OOEC) for assisting with the method of analysing radiographic images.

DEVELOPMENT OF A 2D FINITE ELEMENT MODEL FOR LONG BONES WITH ANY CROSS SECTION AND ORTHOTROPIC AXIS

É. Estivalèzes[1] and M-C. Hobatho[2]

1. ABSTRACT

The aim of this study was to develop a 2D finite element model for composite beams of any cross section, formed of orthotropic materials whose orthotropic axes were not necessarily orthogonal to the cross section. Actually, no code is able to determine the torsional rigidity of Saint Venant (J) of 2D arbitrary shaped cross sections. The determination of J requires a numerical calculation of the torsional warping function. Our finite element model will be used in particular to study cross sections of human long bones and more precisely to determine the homogenized characteristics of arbitrary shaped cross section (epiphyses or diaphyses).

2. INTRODUCTION

The aim of this paper was to set up a calculation method based on the finite element technique to study composite beams of arbitrary shaped cross section, formed of orthotropic materials whose principle axes were not necessarily aligned with the normal of the cross section. Strains and stresses of such a beam were analysed for different loadings such as : tensile force, bending moment and torsion. The problem was analysed using the principle of virtual work combined with the Lagrange multiplier technique. The resolution was carried out by a finite element method using an "economical" meshing of the cross section. The formulated hypotheses as well as the employed calculation methods were validated by comparison with a three-dimensional finite element program. A great advantage of the proposed method was that, compared

Keywords: Bone, Finite element Method, Composite beam, Orthotropic axis, Torsional warping function

---

[1] Ph.D., INSERM U305, Hôtel Dieu, 31052 Toulouse Cédex, France
[2] Ph.D., INSERM U305, Hôtel Dieu, 31052 Toulouse Cédex, France

to a three-dimensional formulation, the number of elements required to discretise the cross section was significantly lower (90%) and then the calculation was faster.

A great number of structures can be treated as beams if they are sufficiently slender, like long bones for example. Usually, such structures can be studied using three-dimensional finite element software, but this technique is often quite expensive and the form of the results is not necessarily easy to interpret.

Hodges[1] pointed out that a structural theory sufficiently general to treat composite beams, of arbitrary shaped cross section, whose mechanical behaviour is subject to coupling terms between strains due to various loads and to cross section warping effects, does not yet exist. By analogy with the classical theory of homogeneous isotropic beams, some works have been done on composite beams by reducing the study of a three-dimensional model to a one dimensional model by using integral parameters or the homogenisation method, Barrau[2], Crolet[3], Hollister[4]. Of course composite beams have already been treated in various works like for example, in Nouri[5], Puspita[6], Whitney[7], Wörndle[8]..., but with the hypothesis that an orthotropic axis of the beam constituents was aligned with the longitudinal axis of the beam. However, optimisation of beam behaviour leads to use composite materials whose orthotropic axes are no longer orthogonal to the cross section. Consequently, previous theories are no longer valid due to coupling terms. In two previous papers (Estivalèzes[9,10]), we have developed a finite element calculation method based on variational principles in order to study composite beams having a symmetric cross section.

Finally, this paper presents a calculation model for composite beams of any cross section with arbitrary orientation of orthotropic axes with respect to the beam longitudinal axis. This calculation model was particularly interesting for the study of long bones. It allowed the determination of stresses and strains in tibial cross sections geometrically and mechanically reconstructed from CT scan images.

In a previous paper (Estivalèzes[11]), simple sections have been considered in order to verify the developed theories. In this article, we present the study of the cross section of the diaphysis of human tibia and the comparison of our model with a three-dimensional finite element calculation software : MSC/Patran.

In a general way, the developed model was able to treat beams made of orthotropic materials, where the fibres were oriented by an angle α, for example, with respect to the beam longitudinal axis. The different materials considered were assumed to be perfectly stuck together with similar Poisson's ratio. Therefore, it was possible to suppose that the stresses σ_{yy}, σ_{zz}, τ_{yz} were negligible with respect to the stresses σ_{xx}, τ_{xy}, τ_{xz}, Wörndle[8].

3. MATERIAL AND METHODS

A typical beam made of N orthotropic phases whose orthotropic axes were not parallel to the beam's longitudinal axis is presented in Figure 1. The beam of span L was clamped at one end S_0 (x=0) and was free to warp. The beam was subjected to different loadings like tensile force N_x, bending moment M_f or torsion M_t at S_1 (x=L).

For each loading, a calculation model was determined by using the principle of virtual work and according to the cases, the Lagrange multiplication technique.

In any cross section, beam characteristics (geometry, materials) were constants.

It was presumed that a cross section S between S_0 and S_1 underwent a translation, a rotation and a warping (Rehfield[12]).

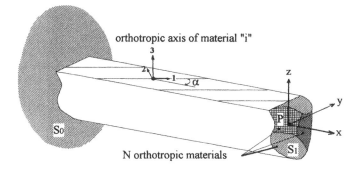

orthotropic axis of material "i"

N orthotropic materials

Fig. 1 Beam Characteristics

For each loading case N_x, M_f and M_t, since these force or moments were not depending on x, and according to previous hypothesis, the chosen displacement field was:

$$\begin{cases} u = u_0(x) - y.\theta_z(x) + z.\theta_y(x) + g(y,z) \\ v = v_0(x) - z.\theta_x(x) + f(y,z) \\ w = w_0(x) + y.\theta_x(x) + h(y,z) \end{cases} \quad (1)$$

where $\begin{cases} u_0(x): \text{longitudinal displacement on x} \\ v_0(x): \text{transversal displacement on y} \\ w_0(x): \text{transversal displacement on z} \end{cases}$ and $\begin{cases} \theta_x(x): \text{rotation of S / x} \\ \theta_y(x): \text{rotation of S / y} \\ \theta_z(x): \text{rotation of S / z} \end{cases}$

$g(y,z)$ was the seeking out of plane warping function, $f(y,z)$ and $h(y,z)$ were two in plane warping functions.
The strain field became :

$$\begin{cases} \varepsilon_{xx} = \dfrac{du_0(x)}{dx} - y.\dfrac{d\theta_z(x)}{dx} + z.\dfrac{d\theta_y(x)}{dx} \\ \gamma_{xz} = \dfrac{dv_0(x)}{dx} + \theta_y(x) + y.\dfrac{d\theta_x(x)}{dx} + \dfrac{\partial g(y,z)}{\partial z} \\ \gamma_{xy} = \dfrac{dw_0(x)}{dx} - \theta_z(x) - z.\dfrac{d\theta_x(x)}{dx} + \dfrac{\partial g(y,z)}{\partial y} \end{cases} \quad (2)$$

Beam characteristics (geometry and material) were independent of x, N_x, M_f and M_t were also independent of x, so we supposed that strains were independent of x.

$$\begin{cases} \dfrac{du_0(x)}{dx} = a & \dfrac{d\theta_z(x)}{dx} = d \\ \dfrac{dv_0(x)}{dx} + \theta_y(x) = b & \dfrac{d\theta_y(x)}{dx} = e \\ \dfrac{dw_0(x)}{dx} - \theta_z(x) = c & \dfrac{d\theta_x(x)}{dx} = f \end{cases}$$

where a, b, c, d, e and f were constants.
After integration and by taking into account the boundary conditions at the clamped section, finally, the strain field could be written as:

$$\begin{cases} \varepsilon_{xx} = a - y.d + z.e \\[2mm] \gamma_{xz} = b + y.f + \dfrac{\partial g(y,z)}{\partial z} \\[2mm] \gamma_{xy} = c - z.f + \dfrac{\partial g(y,z)}{\partial y} \end{cases} \qquad (3)$$

Supplementary hypothesis for M_t :
Warping did not work on the initial cross section (S_0). This means that the work done due to tensile force and bending moment was zero for a displacement resulting from torsion. We then had the following relations :

$$\iint_S \sigma_{xx}(N_x).u_{(x\,=\,0)}dS = 0$$

$$\iint_S \sigma_{xx}(M_y).u_{(x\,=\,0)}dS = 0 \qquad (4)$$

$$\iint_S \sigma_{xx}(M_z).u_{(x\,=\,0)}dS = 0$$

where $\left.\begin{array}{c} \sigma_{xx}(N_x) \\ \sigma_{xx}(M_y) \\ \sigma_{xx}(M_z) \end{array}\right\}$ were normal stresses found for previous loadings N_x, M_y and M_z.

In the principal coordinate system (P,x,y,z), the relation between stresses and strains was :

$$[\sigma]_{(x,y,z)} = [P][C][P]^T[\varepsilon]_{(x,y,z)}$$

where $[P]$ was the transformation (6x6) matrix linking the orthotropic coordinate system to the principal one, and $[C]$ the stiffness (6x6) matrix in the orthotropic coordinate system. According to previous hypothesis : σ_{yy}, σ_{zz}, and τ_{yz} were considered negligible with respect to σ_{xx}, τ_{xz} and τ_{xy}, so the stress field in a material "i" could be written as (Estivalèzes[13]) :

$$\begin{cases} \sigma_{xxi} = Q_{11i}\varepsilon_{xx} + Q_{15i}\gamma_{xz} + Q_{16i}\gamma_{xy} \\[1mm] \tau_{xzi} = Q_{15i}\varepsilon_{xx} + Q_{55i}\gamma_{xz} + Q_{56i}\gamma_{xy} \\[1mm] \tau_{xyi} = Q_{16i}\varepsilon_{xx} + Q_{56i}\gamma_{xz} + Q_{66i}\gamma_{xy} \end{cases} \qquad (5)$$

4. RESOLUTION

For each loading case, the solution was obtained by the writing of an equivalent functional allowing the calculation of the warping function g(y,z) on the meshed cross section using the finite element method.
For N_x, M_y and M_z, the chosen functional was the total potential energy defined as:

$$U = \frac{1}{2} \iiint_V (\sigma_{xx} \cdot \varepsilon_{xx} + \tau_{xz} \cdot \gamma_{xz} + \tau_{xy} \cdot \gamma_{xy}) dV - V_{Fext} \qquad (6)$$

where V_{Fext} was the potential energy due to the applied forces.

For M_t, we chose the following functional :

$$U = \frac{1}{2} \iiint_V (\sigma_{xx} \cdot \varepsilon_{xx} + \tau_{xz} \cdot \gamma_{xz} + \tau_{xy} \cdot \gamma_{xy}) dV + \lambda \iint_S \sigma_{xx}(N_x) \cdot u_{(x=0)} dS \cdots$$

$$\cdots + \beta \iint_S \sigma_{xx}(M_y) \cdot u_{(x=0)} dS + \delta \iint_S \sigma_{xx}(M_z) \cdot u_{(x=0)} dS - V_{Fext} \qquad (7)$$

The minimisation of U with respect to the unknowns and the Lagrange multipliers (λ, β, δ) gave solution.

5. RESULTS AND DISCUSSION

Validation was done on the cross section of the diaphysis of human tibia. This was a uniform beam of length L = 200 mm (Fig. 2). It was formed of only one material : cortical bone.

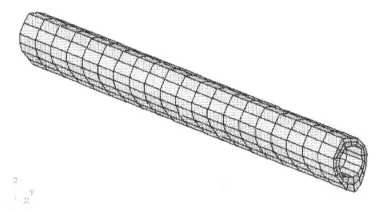

Fig. 2 Studied Beam

The figure 3 shows a meshing of the cross section with three node triangular elements.

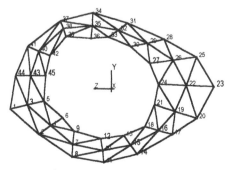

Fig. 3 Meshing of the cross section

The characteristics of a Cortical bone Material (Hobatho[14]), were :

$$E_1 = 8800\text{MPa} \quad G_{12} = 2960\text{MPa} \quad \upsilon_{21} = 0.464$$
$$E_2 = 9600\text{MPa} \quad G_{13} = 4400\text{MPa} \quad \upsilon_{31} = 0.461$$
$$E_3 = 17600\text{MPa} \quad G_{23} = 5120\text{MPa} \quad \upsilon_{32} = 0.4$$

For each loading, results were compared with those of a three-dimensional finite element program MSC/Patran.
For each loading, we calculated stresses and strains.
The following figures 4, 5, 6 and 7 shows the stresses and strains obtained, on one hand with our proposed model, and on the other hand those given by a three-dimensional software MSC/Patran for the loading case $M_t = M_x$, as this loading case requires at first the calculation of the three others : N_x, M_y and M_z.

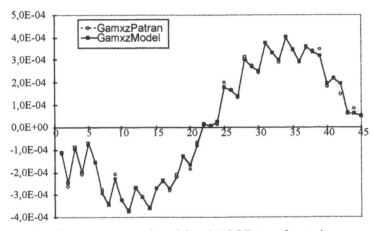

Fig. 4 Comparison between proposed model and MSC/Patran for strain γ_{xz}

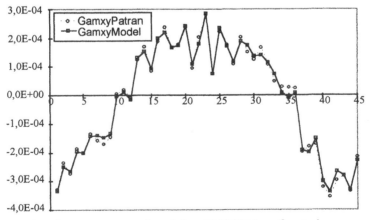

Fig. 5 Comparison between proposed model and MSC/Patran for strain γ_{xy}

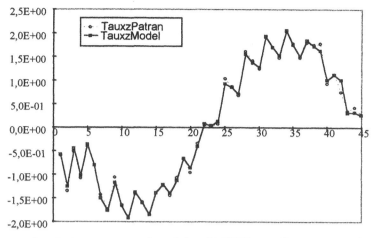

Fig. 6 Comparison between proposed model and MSC/Patran for stress τ_{xz} in MPa

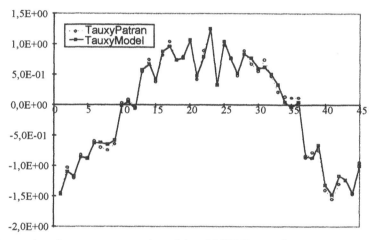

Fig. 7 Comparison between proposed model and MSC/Patran for stress τ_{xy} in MPa

As the position and number of node were the same in the proposed model and in MSC/Patran, the horizontal axis represents the number of each node of the meshing of the cross section, and the vertical axis represents strains or stresses with respect to the number of node.

All other strains and stresses have not been represented as they were close to zero.

For a small number of elements, the average global error committed by the proposed model was inferior to 7% compared with MSC/Patran software.

6. CONCLUSION

This calculation model has enabled to treat composite beams and particularly long bones of any cross section subjected to different loadings, and as the orientation of orthotropic axis of material constituent was arbitrary with respect to the beam longitudinal axis, coupling effects were considered.

The use of the principle of virtual work combined to the Lagrange multiplier technique was well adapted to solve this kind of problem and to determine the torsional warping function. The comparison between the obtained results and that of a three-dimensional finite element calculation software has enabled to validate the modelization of a human diaphysal tibia. The formulated hypotheses as well as the method used have been demonstrated. Furthermore, this type of calculation could be achieved with a rather small number of element, as the meshing was only done on the cross section, and the solution was obtained quickly.

7. REFERENCES

1. Hodges, D. H., Review of composite rotor blade modelling, 29th Structures, Structural Dynamics and Materials Conference, 1988.

2. Barrau, J-J., Chambard, O., Gay, D., and Nuc, M., Homogénéisation en torsion d'une poutre composite, Actes du Troisième Colloque Tendances Actuelles en Calcul de Structures, 1985, 283-296.

3. Crolet, J. M., Homogenization : mathematical method applied to haversian cortical bone structure, 1st World Congress of Biomechanics, 1990, 156.

4. Hollister, S. J., Brennan, J. M., Kikuchi, N., A homogenization sampling procedure for calculating trabecular bone effective stiffness and tissu level stress, J. Biomechanics, 1994, Vol. 27, 433-444.

5. Nouri, T., Homogénéisation et calcul des contraintes de cisaillement dans les poutres composites à sections quelconques et à constituants orthotropes, PhD. Thesis, 1993.

6. Puspita, G., Barrau, J-J., Gay, D., Computation of flexural and torsional homogeneous properties and stresses in composite beams with orthotropic phases, Comp. Structure, 1993, Vol. 24, 43-49.

7. Whitney, J. M., Kurtz, R. D., Analysis of orthotropic laminated plates subjected to torsional loading, Comp. Engineering, 1993, Vol. 3, 83-97.

8. Wörndle, R., Calculation of the cross section properties and the shear stresses of composite rotor blades, Vertica, 1982, Vol. 6, 111-129.

9. Estivalèzes, E., Barrau, J-J., Ramahefarison, E., Détermination des contraintes normales et de cisaillement dans la section droite d'une poutre composite, 11° Congrès Français de Mécanique, 1993a, 201-204.

10. Estivalèzes, E., Barrau, J-J., Ramahefarison, E., Determination of homogenised characteristics for composite beams, 6th European Conference on Composite Material, 1993b, 211-216.

11. Estivalèzes, E., Hobatho, M. C., Barrau, J-J., Gay, D., Development of a finite element composite beam model with a non uniform cross section and any orthotropic axis, Application to human long bones modelization, Bone Mechanical Properties And Finite Element Modelization (Ed. Hobatho-Darmana), 1996, 109-114.

12. Rehfield, L., Hodges, D. H., Atilgan, A. R., Some consideration on the non-classical behaviour of thin-walled composite beams. American helicopter society national form proceedings, 1988, Vol. 88.

13. Estivalèzes, E., Etude et développement d'un élément de poutre composite pour un revêtement déséquilibré. PhD. Thesis, 1995.

14. Hobatho, M. C., Ashman, R. B., Darmana, R., Morucci, J. P., Assessment of the elastic properties of human tibial cortical bone by ultrasonic measurements. In vivo assessment of bone quality by vibration and wave propagation technique, (Ed. Van Der Perre-Lowet-Borgwardt Christensen), 1991, 45-56.

SWELLING AS AN APPROACH TO THE SIMULATION OF CORTICAL BONE REMODELLING

W. R. Taylor & S. E. Clift

ABSTRACT

Bone remodelling is the process whereby bone can grow and resorb in order to adapt to the mechanical demands of the surrounding environment. Mathematical descriptions of the relationship between the mechanical stimulus and the net loss or gain of bone have been widely published in the literature. These mathematical remodelling theories have generally required specific computer codes, usually based on finite-elements, to be developed in order to produce predictions of changes in bone density and distribution. This study has, however, explored an alternative method for representing three dimensional bone remodelling changes using a standard finite-element swelling routine. In its simplest form the model generates physical expansion of a finite-element in a single direction, defined as normal to its free surface. When used to represent the displacement of the periosteal or endosteal surface, the code examines the local strain environment and calculates the magnitude of the remodelling stimulus. The code then uses the calculated stimuli to swell the free surface elements proportionately, thus simulating long-term remodelling growth or resorption. Preliminary results suggest a good correlation of predictions with experimental turkey ulna data from the literature.

INTRODUCTION

The maintenance of bone mass is dependent upon a balance between the level of activity of osteoblasts and osteoclasts. Under so-called "normal" physiological conditions, external forces acting on the body produce stresses and strains which may vary locally but which do not vary to such an extent that the equilibrium between osteoblastic and osteoclastic activity is disturbed. However, if there are significant changes in the local bone stress patterns, for example those produced as the result of the implantation of a prosthesis, then a mechanical stimulus is be transduced into a biochemical stimulus which may preferentially activate osteoclasts or osteoblasts,

Keywords: Bone Remodelling, Swelling, Growth

Department of Mechanical Engineering, University of Bath, Bath, BA2 7AY, England, UK.

leading to a change in bone mass. If the magnitude of the local remodelling potential is higher than that required to maintain balance in cellular activity, then bone mass would be increased locally. Conversely if the potential is too low then the local bone mass would be reduced.

Studies of mathematical modelling of the bone adaptation process have appeared in the literature. Early studies by Hart et al[1] and Huiskes et al[2] examined both cortical and cancellous bone remodelling in two dimensions. Beaupré et al[3] followed this work by modelling the density changes that occur in the proximal femur. Their work included time dependency and hence a rate of bone deposition. Perhaps the most notable contribution, however, was by van Rietbergen et al[4] whose work compared the bone growth found in canine femurs to computer predictions. Their model projected growth patterns around the implants by translating nodal positions although the requirement for mesh refinement made this process lengthy.

Swelling is a phenomenon associated with the change in volume of a solid through absorption of moisture or surrounding fluid. Mathematically, swelling is simply a measured physical change in volume in response to any physical quantity or parameter. This paper demonstrates that the growth of cortical bone can be represented using the Finite Element Method (FEM) by imposing swelling strains. In addition, the modelling of cancellous bone as a simple homogenous structure can easily be achieved using the same code.

METHODOLOGY

The basic model for the development of the remodelling algorithm was created using the ABAQUS commercial finite element package[5]. It consisted of four, twenty noded elements, constructed in a quarter circle, with symmetrical boundary conditions to model a full circle (Fig 1). The geometry was chosen in order to enable the possibility of simultaneous endosteal and periosteal bone growth or resorption.

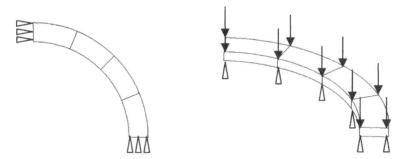

Figure 1 - Basic Analysis Model

The model was homogenous and isotropic, and the material properties were taken from Cowin[6] using an elastic modulus of 18GPa and a Poisson's Ratio of 0.28. The loading was applied as a distributed load over the upper surface.

A method of allowing progressive loading cases was devised, in which each step consisted of a number of loading increments (Fig 2). Each of these incremental loads could represent a set of conditions seen in the 'normal' physiological environment; running, walking or climbing stairs, for example. For the purposes of simple loading for the model, a pressure of $200N/mm^2$ was applied. The rate of strain seen in each element in response to the applied loading was then calculated in each increment to represent the mechanical stimulus[7-9].

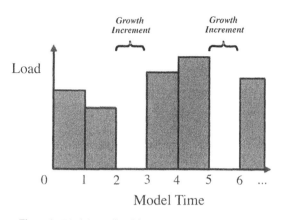

Figure 2 : Model Loading History

The remodelling stimulus, equivalent to the biological growth potential, was calculated using the standard tri-linear remodelling curve, Figure 3 [2-4, 10]. The dead zone represents the period in which no preferential remodelling occurs since the bone is under normal physiological loading conditions. Large values of strain rate (towards point D) generate positive or growth stimuli. Likewise resorbing or negative stimuli are calculated for low levels of strain rate, where a maximum magnitudes of 1/N may be achieved in each increment (over N loading increments). In this manner, a maximum growth of ±1 was obtained.

The final increment in each step was a 'growth increment' in which no load is applied but the growth requirements for the whole step were summed to create a total remodelling stimulus. This value for the total remodelling potential was transferred into a growth capacity when multiplied by the growth constant that defines the maximum possible growth. The

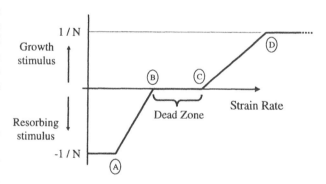

Figure 3 : Tri-linear Remodelling Curve

elements are then swelled proportionately in a single direction, defined by the node orientation, using the standard ABAQUS swelling subroutine. Obviously, 'growing' the elements by imposing swelling strains creates a self-perpetuating growth environment since the remodelling stimulus is a function of the elemental strain. It is therefore necessary to remove the swelling strains from the calculation of functional loading strains, which is achieved by remodelling in response to the elastic strains only and excluding their inelastic counterparts.

In addition to the technique described here for growth of cortical bone, the code allows development to include modelling the changes that occur in bone density associated with internal cancellous bone remodelling. The process of determining the growth potential thus remains identical, incorporating both the loading history and the tri-linear remodelling curve. In place of the elemental swelling that occurs in the cortical bone modelling, however, the elements are simply subjected to a change in Young's Modulus or density, proportional to the growth potential. This becomes similar to the method of density predictions used by Orr, Beaupré and co workers [3,10].

RESULTS AND DISCUSSION

Growth of the elements was achieved in response to the applied loading. The growth occurred in the third increment of each step and continued in progressively decreasing amounts until the Remodelling Stimulus in the elements had entered the Dead Zone region of the curve (Figure 3), and no further expansion occurred. The initial growth patterns displayed a rippling effect on the surface of the elements, which was found to occur as a result of the inadequate definition of the expansion direction.

As a direct result, the ORIENTATION subroutine was used to determine the expansion direction for each material point. Vectors defining the four edges of the element in the direction of growth were calculated using the co-ordinates of the corner nodes. This allows the code to be valid for both eight and twenty noded brick elements (Figure 4).

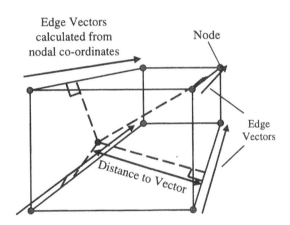

Figure 4 : Material Point Orientation

The material point direction vector was defined using the perpendicular distance from the material point to each edge vector, and was calculated as follows :

$$\text{Material Point Direction Vector} = \sum \frac{\text{EdgeVectors}}{\text{Distance to Vectors}}$$

In this manner, every material calculation point had its own growth direction. The results are shown in Figure 5. These growth patterns demonstrate the potential of the methodology to expand elements normal to the free surface.

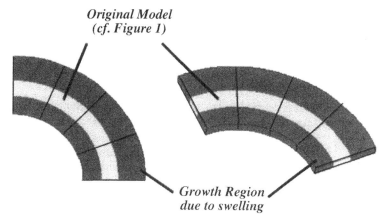

Original Model
(cf. Figure 1)

Growth Region
due to swelling

Figure 5 : Model Growth after Swelling

The progressive growth of the elements is displayed in Figure 6, in which the peaks appear during the growth increment of the loading cycle. The exponential decay in the model growth was due to progressively reducing stress and strain patterns and hence remodelling potential, caused by the increasing size of the elements.

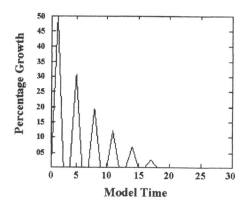

Figure 6 : Graph of Progressive Growth

Additional testing of the process was performed to demonstrate the ability of the elements to vary their growth patterns at all the material calculation points. For this to occur, a linearly increasing distributed load was applied to the upper surface of the elements. The magnitudes of the remodelling values were set such that the inner surface would resorb, the centre section would be in the dead zone and have no remodelling associated with it, and the outer surface would grow. The result was a slight outward migration of the elements, although the motion ceased due to the increasing load causing the inner surface to grow.

This study has displayed that modelling of highly complex bone growth patterns is possible using the ABAQUS software. It demonstrates the capability to expand elements normal to the free surface.

As in any model, however, simplifications have been made. The growth orientation is a matter of further investigation, especially for modelling geometries more complex than the one presented here. The continual shape changes that could occur might alter the direction of the surfaces normal. As a result, there may be the requirement to recalculate the growth orientation.

In addition, there is continued discussion in the literature concerning the nature of the mechanical remodelling stimulus that relates the biological bone growth to the functional loading conditions. It may be that the remodelling stimulus is a function of a great number of mechanical signals that all interact in an intricate manner. The code has been constructed in such a way that the calculation of the stimulus may be easily altered, and hence various combinations may be examined against animal experimental data.

It must be noted that the magnitudes of the results gained from this model are not significant. The process is being elucidated here, since all the final remodelling parameters will vary according to the location and the actions being modelled. These values need to be produced by modelling of natural bones under normal physiological conditions, since it can be assumed that these bones are in remodelling equilibrium. The position of the Dead Zone could thus be considered defined from the limiting values of the natural model. Bone remodelling would therefore occur when the growth stimuli exceed these bounding values.

CONCLUSIONS

Expansion of elements has been achieved in a manner consistent with the long-term growth patterns seen in bone remodelling. The standard ABAQUS swelling subroutine has been successfully used to enlarge the elements in response to a mechanical stimulus, calculated, in this instance, from the rate of strain. The method allows the growth patterns to respond to both endosteal and periosteal remodelling since the growth potential is calculated internally to the elements. Due to the loading regime, the growth of the elements occurs progressively as an iterative process, with no requirement for mesh refinement. In addition, modelling of the cancellous bone adaptation process can be easily included in the code.

ACKNOWLEDGEMENTS

The authors would like to thank HKS for the provision of the ABAQUS software package. Valuable discussions with Dr. Greg Starke (University of Capetown, SA) and Matthew Warner (University of Bath, UK) are gratefully acknowledged.

REFERENCES

1.Hart R. T., Davy D. T. and Heiple K. G., Mathematical modelling and numerical solutions for functionally dependent bone remodelling, 1984, Calcif Tissue Int, Vol. 36, S104 - S109

2. Huiskes R., Weinans H., Grootenboer H. J., Dalstra M., Fudala B. and Slooff T. J., Adaptive bone-remodelling theory applied to prosthetic-design analysis, J. Biomechanics, 1987, Vol. 20, No. 11/12, 1135 - 1150

3. Beaupré G. S., Orr T. E. and Carter D. R., An approach for time-dependent modelling and remodelling – Application : A preliminary remodelling simulation, J. Orthop Res, 1990, Vol. 8, No. 5, 662 - 670

4. van Rietbergen B., Huiskes R., Weinans H., Sumner D. R., Turner T. M. and Galante J. O., The mechanism of bone remodelling and resorption around press-fitted THA stems, J. Biomechanics, 1993, Vol. 26, No. 4/5, 369 - 382

5. Hibbitt, Karlsson and Sorensen, ABAQUS (standard) version 5.5, User Manuals, 1995 Pawtucket, RI, USA

6. Cowin S. C., The mechanical properties of cortical bone tissue in *'Bone Mechanics'*, CRC Press Inc, 1989, Chap. 6, 97 - 127

7. O'Connor J. A. and Lanyon L. E., The influence of strain rate on adaptive bone remodelling, J. Biomechanics, 1982, Vol. 15, No. 10, 767 - 781

8. Rubin C. T., Lanyon L. E., Osteoregulatory nature of mechanical stimuli: Function as a determinant for adaptive remodelling in bone, J. Orthop Res, 1987, Vol. 5, 300 - 310

9. Carter D. R., Mechanical loading histories and cortical bone remodelling, Calcif Tissue Int, 1984, Vol. 36, S19 - S24.

10. Orr T.E., Beaupré G. S., Carter D. R. and Schurman D. J., Computer predictions of bone remodelling around porous-coated implants, J. Arthroplasty, 1990, Vol. 5, No. 3, 191 - 200

A NON-LINEAR FINITE ELEMENT FORMULATION FOR MODELING VOLUMETRIC GROWTH

V. Srinivasan, R. Perucchio, R. Srinivasan and L. Taber

1 Abstract

We present a finite element formulation for growth in the myocardial tissue during early embryonic morphogenesis. To model growth, we focus on its two important physical criteria—increase in tissue-mass, and kinematic equilibrium of the grown configuration. We have incorporated these criteria in a three-dimensional nonlinear finite element code for incompressible, anisotropic hyperelastic materials previously developed for numerical modeling of myocardial muscle activation. We have successfully tested our formulation for growth on several problems, ranging from one to several elements subjected to various growth-strains, boundary conditions, and combinations of growing and non-growing regions.

2 Introduction

Growth is a phenomenon pivotal to the generation of biological form, and is defined as a process of mass-addition. The current work focuses on modeling the kinematics of growth during early embryonic morphogenesis. The myocardial tissue is modeled as an incompressible, anisotropic, hyperelastic material, which shows a growth-induced nonlinear stress and displacement behavior.

Keywords: Biomechanics, Growth, Finite Element Analysis, Kinematics

Department of Mechanical Engineering, University of Rochester, Rochester, NY 14627

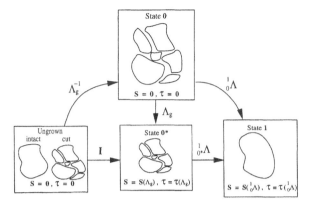

Figure 1: Global states of the solid – initial ungrown state (cut and intact), grown cut state **0**, compressed intermediate state **0***, final equilibrium state **1**. **S** is the Second Piola-Kirchoff stress tensor and τ is the Cauchy stress tensor, and Λ represents the deformation tensor.

By the use of shift-ratios (stretch-ratios) and methods of continuum mechanics, we build the mathematical framework for the growth phenomenon. We then cast our equations into a nonlinear finite element formulation, and compute the spatial-temporal motions of a growing tissue. Finally, we test our formulation on several models subjected to different growth parameters and boundary conditions.

3 Theory

The fundamental idea is that the strain measures in the constitutive (stress-strain) equations of each material element, rather than being referred to an unloaded configuration, must be referred to a stress-free configuration; where there is no stress in the absence of strain—a state with zero strain energy [1, 2, 3]. We will define four *global states* for the tissue: the initial *ungrown* tissue, a hypothetical no-stress state **0**, a stressed intermediary **0***, which has the same geometry as the input configuration, and the final, grown, equilibrium state **1**. All the four aforementioned states are depicted in Fig. 1.

3.1 Kinematics of growth

The intact, ungrown myocardial tissue-mass is dissected into isolated tissue-elements by making appropriate imaginary cuts. These elements undergo a volume expansion when they grow and the growth may be isotropic or anisotropic. Note that, in this macroscopic description, each element may consist of one or several cells, with cell division and cell enlargement being mechanically equivalent. The imaginary state **0** represents the collection of these grown elements which are yet isolated from each other. The elements therein are related to their cut, ungrown configuration through a Cauchy-Green deformation tensor, denoted by Λ_g^{-1}, where Λ_g is given by

$$\Lambda_g = \left[\boldsymbol{\lambda}^T \boldsymbol{\lambda}\right] . \tag{1}$$

In the above equation, λ is a deformation gradient which characterizes growth and is called a shift-tensor [1]. We characterize the shift-tensor by specifying shift-ratios λ_i in three *principal* material directions. Because the elements in state **0** grow without any boundary influences—"free expansion," none of them is under any stress.

The separated, grown tissue-elements (state **0**) cannot in general be assembled back into an equilibrium configuration without any additional deformation. As a first step, each grown element of state **0** is "forced" back to its ungrown, cut geometry. State **0*** represents this enforced state, where the tissue-elements are under a compressive stress and have the same geometry as the input. The elements are now allowed to seek out their equilibrium configuration. We represent the deformation tensor that connects state **0*** to the equilibrium state **1** as $_{0^*}^1\Lambda$, and the one linking the zero strain-energy state **0** to the equilibrium state **1** as $_0^1\Lambda$. The *global* deformation tensor, which relates the ungrown, intact state to the final, equilibrium state, is the product of the identity tensor **I** and $_{0^*}^1\Lambda$, and hence in effect is $_{0^*}^1\Lambda$ only.

State **0*** is just an intermediate step, albeit a necessary one from the implementation standpoint, because the grown, cut configuration being a hypothetical state, we cannot determine the geometry of an element in state **0**. Hence, to find the final equilibrium state, we have to use the ungrown, input configuration as our reference because of its known geometry. The imaginary *cut* configurations of grown and ungrown states are contrived only to pick a convenient thread of reason to explain the rationale behind applying a start-up compressive stress on the input geometry. Fig. 1 delineates the relationship between the different states and also indicates the stress-state in each. The three Cauchy-Green deformation tensors Λ_g, $_0^1\Lambda$ and $_{0^*}^1\Lambda$ are related as follows:

$$_{0^*}^1\Lambda = {}_0^1\Lambda\,\Lambda_g^{-1}. \tag{2}$$

Incompressibility. The mass-increase associated with the "free" growth of dissected tissue-elements is modeled as a compressible phenomenon. But the deformations necessary to assemble the freely grown elements to form a continuum in equilibrium have to preserve the element volumes, hence the process is incompressible. Thus, the third invariant of $_0^1\Lambda$ has to be necessarily equal to unity; no such restrictions will apply for Λ_g, which represents the volume expansion.

Solution approach. We use an *Updated Lagrangian* approach to seek out the equilibrium configuration. The most recent equilibrium state is used as the reference configuration. It is important to note that the stress terms in the constitutive relations are first computed using the deformation tensor $_0^1\Lambda$ and then updated to the current reference state.

4 Basic Formulation

We assume the existence of a potential function W which represents the strain energy per unit volume in the reference state **0** [4, 5, 6] for hyperelastic materials.

For incompressible hyperelastic materials the potential function is augmented by an additional incompressibility constraint term, which is expressed by setting the value of the third invariant, I_3, of Cauchy-Green deformation tensor, $_0^1\Lambda$, equal to unity. Thus W is given by [4, 7]

$$W = W_e + P(I_3 - 1) \tag{3}$$

where W_e and $P(I_3 - 1)$ are, respectively, the elastic and hydrostatic parts of the strain energy potential, and P is a Lagrange multiplier related to the hydrostatic pressure in the element, hereafter referred to as hydrostatic pressure for convenience.

The total potential energy of the element is given by

$$\Pi(\epsilon_{ij}, P) = \int_{t_v} [W_e + P(I_3 - 1)] \, dv + V \tag{4}$$

where V is the work performed by the external loads. The state of strain is described by the Green-Lagrange strain tensor ϵ_{ij}, and is measured with respect to state 0. The stationary condition of the total potential energy requires that

$$\delta\Pi = \frac{\partial\Pi}{\partial\epsilon_{ij}}\delta\epsilon_{ij} + \frac{\partial\Pi}{\partial P}\delta P = 0 \, . \tag{5}$$

Since this equality has to hold for any variation of ϵ_{ij} and P, which are independent from each other, the two variations have to vanish separately yielding the conditions

$$\frac{\partial\Pi}{\partial\epsilon_{ij}} = 0 \, , \tag{6}$$

$$\frac{\partial\Pi}{\partial P} = 0 \, . \tag{7}$$

The former equation leads to the nonlinear *equation of motion* of the element, while the latter leads to the *incompressibility equation*.

4.1 Derivation of Finite Element Matrices

We convert the nonlinear equation of motion into a linear one by incremental decomposition [8, 7]. Then the governing linearized equation is dicretized at the nodes of the finite element model through shape functions [8, 7]. The present isoparametric formulation is based on a trilinear Lagrangian interpolation for displacements and constant interpolation for pressure. The nodal variables in the element are: three Cartesian components of the incremental displacements, and one incremental hydrostatic pressure. The element-displacement discretization is represented as

$$\Delta\mathbf{u}(r, s, t) = \mathbf{H} \, \Delta\hat{\mathbf{u}} \tag{8}$$

where $\mathbf{H}$ is the displacement interpolation function matrix and $\Delta\hat{\mathbf{u}}$ is the vector of incremental nodal displacements. In the present case the hydrostatic pressure is assumed *constant* in each element and, therefore, no interpolation functions for the pressure are required.

The linear and nonlinear parts of the incremental strains are expressed in terms of incremental displacements by employing the appropriate strain-displacement transformation matrices ${}^t\mathbf{B}_L$ and ${}^t\mathbf{B}_{NL}$ respectively [8]. Within each element, the equation of motion is expressed in matrix form as

$$
\left\{ \int_{{}^t_v} {}^t\mathbf{B}_L^T \left[{}_t\mathbf{C}^{(e)} + {}^t\mathbf{P}\,{}_t\mathbf{C}^{(h)} \right] {}^t B_L \, dv + \int_{{}^t_v} {}^t B_{NL}^T \left[{}^t\tau^{(e)} + {}^t P\,{}^t\tau^{(h)} \right] {}^t B_{NL} \, dv \right\} \Delta\hat{\mathbf{u}}
$$
$$
+ \left\{ \int_{{}^t_v} {}^t\mathbf{B}_L^T\,{}^t\tau^{(h)} \, dv \right\} \Delta\hat{\mathbf{P}} \tag{9}
$$
$$
= {}^{t+\Delta t}\mathbf{R} - \int_{{}^t_v} {}^t\mathbf{B}_L^T \left[{}^t\tau^{(e)} + {}^t\mathbf{P}\,{}^t\tau^{(h)} \right] \, dv
$$

where $\Delta\hat{\mathbf{u}}$ and $\Delta\hat{\mathbf{P}}$ are the unknown incremental nodal displacements and element-hydrostatic pressure respectively, and ${}^t v$ is the element-volume and ${}^t P$ is the element-hydrostatic pressure at time t. ${}_t\mathbf{C}^{(e)}$ and ${}_t\mathbf{C}^{(h)}$ are the material property tensors. Note that any anisotropy present in the problem enters the finite element formulation through the two tensors ${}_t\mathbf{C}^{(e)}$ and elastic part of the Cauchy stress tensor ${}^t\tau^{(e)}$. ${}^t\tau^{(h)}$ represents the contribution from the incompressibility constraint to the Cauchy stress.

The incompressibility condition arising from Eq. 7 is cast in incremental form. The final matrix form of the incremental incompressibility condition is given by

$$
\left[\int_{{}^t_v} \left({}^t\tau^{(h)} \right)^T {}^t\mathbf{B}_L \, dv \right] \Delta\hat{\mathbf{u}} = \int_{{}^t_v} \left(1 - {}^t I_3 \right) \, dv . \tag{10}
$$

Eqs. (9) and (10) form the mixed elemental system of equations, with the incremental displacements and incremental hydrostatic pressure as unknowns. In matrix form, these are expressed as

$$
\begin{bmatrix} \mathbf{K}_u & \mathbf{K}_p \\ \mathbf{K}_p^T & 0 \end{bmatrix} \begin{bmatrix} \Delta\hat{\mathbf{u}} \\ \Delta\hat{\mathbf{P}} \end{bmatrix} = \begin{bmatrix} \mathbf{F}_u \\ \mathbf{F}_p \end{bmatrix} \tag{11}
$$

where $\mathbf{K}_u$ is the $N \times N$ coefficient matrix of the incremental displacements $\Delta\hat{\mathbf{u}}$ in Eq. (9), $\mathbf{K}_p$ is the $N \times M$ coefficient matrix of the incremental hydrostatic pressure $\Delta\hat{\mathbf{P}}$ in Eq. (9), $\mathbf{K}_p^T$ is the $M \times N$ coefficient matrix of the incremental displacements $\Delta\hat{\mathbf{u}}$ in Eq. (10) (also the transpose of $\mathbf{K}_p$), $\mathbf{F}_u$ is the vector of nodal forces and the right hand side expression in Eq. (9), and $\mathbf{F}_p$ is the right hand side coefficient in Eq. (10).

Computer Implementation. We use a penalty procedure—perturbed Lagrangian form—to avoid the zero diagonal term. The pressure variable is condensed out and the element is now treated as a standard displacement element. This approach is known as *mixed* or *consistent penalty method*.

5   Numerical Examples

In all the examples, the input geometry is the ungrown, intact configuration. The value of the growth shift-ratio (stretch-ratio) is 0.89—induces about 12% mass-addition. There is not much biological basis for such a choice, rather it's the

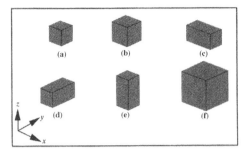

Figure 2: Growth of one-element tissue model: (a) original ungrown state. The rest are the final grown states, state **1**: (b) symmetric boundary conditions on three mutually adjacent faces, (c) free only in the x-direction, (d) free only in the y-direction, (e) free only in the z-direction, (f) same conditions as in (b) but applied stretch-ratio in each coordinate direction is 0.5.

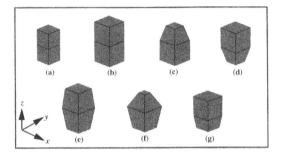

Figure 3: Growth of two-element model: (a) original ungrown state. The rest are the final grown states, state **1**, under different boundary conditions: base on rollers–(b) both elements grow, (c) only bottom element grows, (d) only top element grows; base clamped–(e) both elements grow, (f) only bottom element grows, (g) only top element grows.

minimum value we could impose on our model for invagination. The shift-ratios in the three principal directions are the same and are applied uniformly in the elements which are allowed to grow.

One-element model. All the different cases are analyzed using one element. The initial shape of the element is shown in Fig. 2a. In all the tests the nodal displacements, final volumes and stresses are as we expect. The results for all cases are shown in Fig. 2.

Two-element model. The purpose of various tests made with the two-element model is to put to further test the validity of the formulation with respect to inter-element kinematics, study the effect of elastic boundary conditions and the interactions at the interface of the growing and non-growing areas. The results of the different tests are shown in Fig. 3.

Some cases tested above, however, have discrepancies in the stress patterns and final element-volumes. We believe this is a fallout of using a linear element and

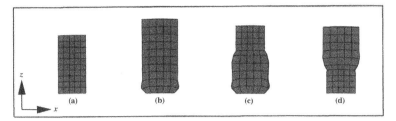

Figure 4: Kinematics of the two-element domain under a finer mesh: The base is clamped for boundary conditions; (a) finer mesh of the two-element domain, (b) whole domain grows, (c) only bottom half grows and (d) only top half grows.

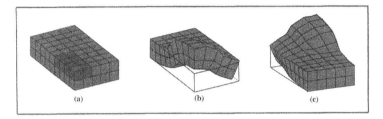

Figure 5: Invagination model of a tissue slice: (a) the ungrown shape and only the elements in darker shade will grow, (b) final grown shape of the tissue where the marked elements have grown by 12%, (c) final shape from another viewpoint. The outlines in (b) and (c) mark the initial shape of the element.

approximating the element-hydrostatic pressure with one variable. To overcome these effects we mesh the same two-element domain with 250 cube elements, and apply the same growth conditions. For boundary conditions we clamp the base of the model. This time, the displacements and stress patterns do match our predictions. The simulations are illustrated in Fig. 4.

Invagination. Invagination is essential to many kinds of morphogenetic processes. It is the formation of a small pocket in an initially flat piece of tissue. Here we present a three-dimensional analysis of invagination. In most cases the shape changes associated with this phenomenon are axisymmetric with respect to the center of the initial depression. Because of the symmetry, we model here only one-quarter of the cell-sheet [9] (Fig. 5): the sheet is clamped at one end and symmetry conditions are applied at the opposite and adjacent sides. The dimensions of the model is 5 by 2 by 10 units and is meshed with 100 unit-cube elements. Fig. 5 illustrates the results of this test.

6 Conclusion

We have introduced a technique to model the mechanical characteristics of growth, wherein we specify the growth parameters. The current facility can be used to model non-uniform growth in anisotropic tissues. We have presented here the results of preliminary tests on simple models. The results prove that if the growth-strains of an element of an originally stress-free body are geometrically compatible, then the body remains stress-free after growth occurs. Invagination process has been simulated on a tissue-slab successfully. The next step will be to model the interaction of stress and growth—growth parameters will be computed according to the growth laws proposed in the literature [2, 10, 11, 3].

References

[1] L. A. Taber. On nonlinear theory for muscle shells: Part II. Application to the active left ventricle. *J. Biomech. Eng.*, 113:63–71, 1991.

[2] E. K. Rodriguez, A. Hoger, and A. D. McCulloch. Stress dependent finite growth in soft elastic tissues. *J. Biomech. Eng.*, 27:455–467, 1994.

[3] L. A. Taber. Biomechanics of growth, remodeling and morphogenesis. *Appl. Mech. Rev.*, 48(8):487–545, 1995.

[4] J. T. Oden. *Finite Elements of Nonlinear Continua.* McGraw-Hill, New York, 1972.

[5] A. E. Green and J. E. Adkins. *Large Elastic Deformations and Nonlinear Continuum Mechanics.* Oxford University Press, London, 1960.

[6] A. E. Green and W. Zerna. *Theoretical Elasticity.* Oxford University Press, London, 2nd edition, 1968.

[7] R. Srinivasan. *Geometric modeling, meshing and nonlinear finite element analysis of the embryonic heart during early cardiac morphogenesis.* PhD thesis, University of Rochester, 1996.

[8] K. J. Bathe. *Finite Element Procedures.* Prentice-Hall, 1996.

[9] G. W. Brodland and D. A. Clausi. Embryonic tissue morphogenesis modeled by FEM. *Transactions of the ASME*, 116:146–155, 1994.

[10] L. A. Taber and D. W. Eggers. Theoretical study of stress-modulated growth in aorta. *J. Theor. Biol.*, 180:343–357, 1996.

[11] I. Lin and L. A. Taber. A model for stress-induced growth in the developing heart. *J. Biomech. Eng.*, 117:343–349, 1995.

CT-SCAN DATA ACQUISITION TO GENERATE BIOMECHANICAL MODELS OF BONE STRUCTURES

Marco Viceconti[1], Cinzia Zannoni[2], Fabio Baruffaldi[3], Luisa Pierotti[4],
Aldo Toni[5] and Angelo Cappello[6]

1. ABSTRACT

The conversion of a CT scan dataset in an accurate and manageable three dimensional model is a basic procedure in almost every branch of computational biomechanics. However, little attention is usually given to the CT scan acquisition protocol to be used in these cases. In this work a set of procedures is described which were developed to improve the quality and the usefulness of the CT scan datasets to be used for modelling purposes. The proposed protocol addresses three specific aspects of the CT data collection: calibration, scan planning

Keywords: Computed Tomography, Bone and Bones, Finite element analysis

[1] Researcher, Laboratorio di Tecnologia dei Materiali, Istituti Ortopedici Rizzoli, via di Barbiano 1/10, 40136 Bologna, Italy.

[2] PhD Student, Dipartimento Elettronica Informatica e Sistemistica, Università di Bologna, via Risorgimento, 40136 Bologna, Italy.

[3] Researcher, Laboratorio di Tecnologia dei Materiali, Istituti Ortopedici Rizzoli, via di Barbiano 1/10, 40136 Bologna, Italy.

[4] Researcher, Servizio di Fisica Sanitaria, Policlinico S. Orsola-Malpighi, 40139 Bologna, Italy.

[5] Confirmed Researcher, Orthopaedic Clinic of University of Bologna and Head, Laboratorio di Tecnologia dei Materiali, Istituti Ortopedici Rizzoli, via di Barbiano 1/10, 40136 Bologna, Italy.

[6] Associate Professor, Dipartimento Elettronica Informatica e Sistemistica, Università di Bologna, via Risorgimento 2, 40136 Bologna, Italy.

and reconstruction of metallic parts. All these techniques, when properly used, allows an accurate data collection, which is the preliminary requirement for the creation of every good biomechanical model.

2. INTRODUCTION

In many published FEM studies on skeletal biomechanics CT scan datasets are used to define the geometry of bone segments with or without implanted devices; see for example [1-3]. The conversion of a CT dataset into a three dimensional geometric model is a complex, multi-step procedure. Each of these step is a potential source of errors. When these errors propagate, minor procedural inaccuracies may results in significant global errors.

In a preliminary study, a three dimensional (3D) solid model of human femur replica, with a prosthetic stem implanted into the medullary canal, was generated starting from CT scan data. Reference geometries were available both for the intact femur and for the stem; Cross-measurements showed local errors up to 6 mm. Error analysis demonstrated that this resulting error was the propagation effect of many minor inaccuracies.

Scope of the present work is to evaluate geometry and density errors induced by the CT scan acquisition and to identify a set of procedures and methods to minimize such errors. The topic will be discussed in particular from three perspectives: the acquisition protocol, the presence of metallic inserts and the segmentation procedure.

3. CT SCAN PLANNING

3.1 Problem description

The accuracy of a 3D reconstruction is related to the scanning plan adopted during CT data acquisition. In the hospital daily routine, the position of the CT scans is decided by the radiologist according to the observations made on a pre-scanning scout image, an X-ray projection of the bone segment under examination. In particular situations, i.e. when the patient needs a custom-made prosthesis, an accurate reconstruction of the geometry and bone density of the proximal femur is needed. In this case radiologists simply apply standard scanning protocols for long bones, as proposed by the producers. These protocols typically suggest a 5 mm scanning step in the smaller trochanter region, and 10 mm all over the femur for a total number of 50-60 slices (Fig. 1a). They are standardized for all patients ignoring inter-femoral geometric and densitometric differences. To improve 3D reconstruction accuracy a new method which automatically locates a limited number, N, of CT scans according to local changes in bone structure geometry and density has been developed.

3.2   Materials and methods

The 3D data sets and scout images of three femurs, a composite femur (Pacific Research Labs, Vashon Island, Washington, USA) and two human femurs *ex-vivo,* have been collected. The femurs have been scanned with a 1 mm scanning step on a GE Sytec 3000 scanner.

The scanning plan optimization method (DIEOM) is based on an Discard-Insert-Exchange technique and it is applied to an anteroposterior X-ray projection of the bone segment [4]. Fig. 1 shows an example of the position of the slices in the standard radiological protocol and in the optimized scanning sequence. As it could be expected, an higher number of slices is set in the proximal region where geometric and densitometric gradients are higher.

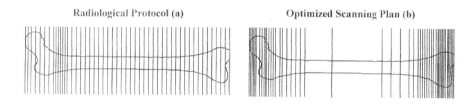

Fig. 1 - Position of the CT scans in the Radiological Protocol for custom made prosthesis (a) and in the Optimized Scanning Plan (b).

Applying the optimization method on a 2D image yields a scanning plan which optimizes the 2D reconstruction error. The procedure has, thus, been validated by comparing the errors obtained by reconstructing the 3D data set from the 3D-optimized and the 2D-optimized scanning plan. The 3D-optimized plan has been evaluated *in-vitro*, applying the optimization procedure directly on the 3D data set.

3.3   Results

As shown in Fig. 2, direct 3D optimization yields root mean square reconstruction errors which are only 5% lower than the 2D-optimized plan. This proves that 2D-optimization, performed on the AP scout image, provides a good sub-optimal scanning plan for 3D reconstruction of the bone structure. Further on, 3D reconstruction errors given by the optimized scanning plan and a standard radiological protocol for long bones have been compared. Results show that the optimized plan yields from 20% to 50% lower 3D reconstruction errors, showing the importance of the definition of the scanning plan during CT data acquisition.

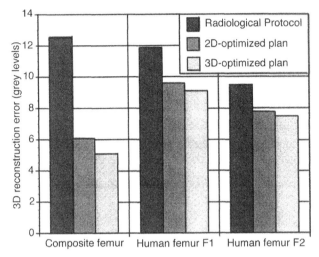

Fig. 2 - 3D reconstruction errors obtained from the Radiological protocol, 2D-optimized and 3D-optimized scanning plans.

4. EFFECT OF METALLIC IMPLANTS

4.1 Problem description

In many cases, the bone segment to be scanned is near to or in contact with a metallic device. Various classes of errors and artefacts can degrade image accuracy when metal objects are scanned. Among these, X-ray scattering, beam hardening, and partial volume averaging, are of greatest concern in determining the geometry of the metal implant and the density of the host bone [5]. Evidences can be found in literature of an apparent increase in bone mineral density produced by intramedullary titanium implants [6]. More over, CT images of a titanium implant acquired with an old generation GE Sytec 3000 scanner have shown distortions in geometric features reconstruction up to 5 mm. In this study CT accuracy in titanium implant geometric features and surrounding bone tissue density reconstruction has been investigated. A HiSpeed Advantage GE CT scanner which periodically undergoes calibration procedures, and allows direct access to digital images has been used. Direct access to digital data is critical for the correct evaluation of stem geometric features. The digitization of the CT film, in fact, introduces diffraction artefacts which further degrades image information.

4.2   Materials and methods

The analysis of CT scanner accuracy have been investigated on two titanium prosthetic stems AnCAFit (Cremascoli, Milano, Italy) of calliper 17 and 9.5 mm, which represent the typical range in stem dimensions. The cortical bone surrounding the implant was simulated with the aid of plasticine. This cortical bone substitute has an attenuation coefficient equal to 1338 HU, comparable to that of human cortical bone 1000÷1800 HU [7]. Six images for each stem have been acquired with a 10 mm scanning step covering the stem geometric variations. The stem contour was extracted from each image using a Conjugate Gradient (CG) Algorithm and was compared with the contour of the corresponding coronal section of the stem solid model provided by the producer. The difference between the corresponding curves have been evaluated measuring the maximum contour distance, $e(z)$. To evaluate the repeatability of the measuring process four data acquisition sessions with the same scanning plan was performed.

The effects of the metal stem on the bone density evaluation was investigate over six images (three for each stem), with the metal area ranging from 200 to 440 mm^2. The bone density was evaluated in each crown of pixels surrounding the implant.

4.3   Results

The error $e(z)$ has shown no significant dependency from the z variable. The standard deviation of the mean value of e has been used as the index of repeatability of the measuring procedure.

Stem	Average error	Maximum error	ε90
AnCAFit 17	0.41±0.16	0.75	0.62
AnCAFit 9,5	0.56±0.14	0.93	0.67

Table 1 - Titanium stem reconstruction accuracy measurements. ε90 is defined as the smallest error equal or higher than the 90% of all the measured errors.

The results reported in Table 1 show that global accuracy in the reconstruction of an AnCAFit with calliper ranging from 9.5 to 17 mm is equal to 0.4÷0.6 mm, corresponding to 1.5 and 2 pixels (pixel size 0.3 mm). This value is comparable with the measurement sensitivity (±1pixel), showing that modern CT scanners allows to reconstruct titanium stems with an error lower than 1 mm.

The bone density measured in the region surrounding the stem shows an apparent increase of 2÷3% of the cortical bone substitute (1338 HU) comparable with the results found in literature [6]. An high correlation (r^2=0.96) between the eccentricity of the stem section and the density perturbation has been found. Fig. 3 shows the density perturbation as a function of the distance from the stem. As expected in the first crown of pixels the average error is quite high. Anyway at a distance of 3 and 9 pixels for the AnCAFit 17 and the AnCAFit 9.5 stem respectively the tissue density is included within normal values.

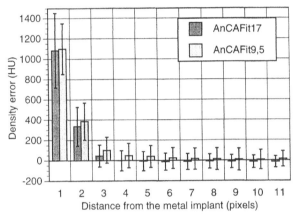

Fig. 3 - Average density error in each pixel crown surrounding the stem.

Table 2 shows the effect of the bone tissue density perturbation due to the metal implant over the assessment of the material properties of the elements of a FE mesh with central symmetry.

Element dimension	Δρ% corona 1	Δρ% corona 2	Δρ% corona 3
20 pixels (6 mm)	(8±4)%		
10 pixels (3 mm)	(11±9)%	(0±5)%	
7 pixels (2 mm)	(17±11)%	(0±6)%	(0±4)%

Table 2 - Evaluation of the error in the assessment of the material density associated to the FE mesh elements

Mesh with a smaller element dimension allows to confine the error in the first crown of elements. The simplest way to correct the error is to reduce of a known value the density of the 2 pixels crown closer to the implant. For example, in the 7 pixels case, the reduction of the density of the first and second crown of 1000 HU and 500 HU respectively decreases the mean element error down to 0.88%.

5. SURFACE EXTRACTION

5.1 Problem description

The creation of biomechanical models usually requires the extraction of the bone geometry from a CT dataset (surface extraction). There are many ways to extract bone geometry from a CT dataset. Probably the simplest is 2D segmentation. Each slice is processed independently and inner and outer bone contours a detected. The contours are stacked in 3D and used as reference to create a solid model usually through skinning operations. Although very simple this approach is not adequate to handle bifurcations such as femoral condyles and it usually requires a significant human effort to create a reasonable model. 3D segmentation can be obtained using the Marching Cube (MC) algorithm or derived methods [8]. This approach allows the automatic extraction of iso-density surfaces from a 3D dataset; the surface is represented by a connected set of triangles (hereinafter called *tiled surface*). To investigate the use of such approach, a femur was reconstructed using the 2D segmentation and the 3D segmentation.

5.2 Materials and methods

An intact human femur analogue (Mod. 3103, Pacific Research Labs, USA) was scanned with a 1 mm slice thickness using an HiSpeed Advantage CT scanner (General Electric, USA). The final dataset was made of 512 x 512 x 451 voxels. The voxel size was 0.3 x 0.3 x 1.0 mm. The dataset was then sub-sampled at 2 mm; the remaining slices were used as a control. Tissues classification was achieved using a simple thresholding procedure.

Each image of the dataset was segmented using a Conjugate Gradient (CG) algorithm. Preliminary measurements demonstrated that for well contrasted objects, CG contours may be considered as reference curves, having an accuracy lower than a pixel. The accuracy of solid models created from stacked 2D contours in general depends on the modelling strategy and on the object complexity. For a cylinder with an axial conic cavity it is possible to reach an out-of-tomographic plane accuracy of 1.4% of the largest diameter [9]. In a femur model, however lower accuracies would be expected.

The femur dataset was also segmented using the Discretized Marching Cube (DMC). The DMC method is derived from the MC original formulation, but it resolves the topology problems and produce a much smaller number of triangles [10].

While some applications may use these tiled solids, most modelling activities require a polynomial solid model. Thus, the CG contours and the DMC tiled surfaces were imported in a solid modeller (EMS, Intergraph ISS, USA) and here used as reference to create a Non Uniform Rational B-Splines (NURBS) solid model of the femur.

To estimate to geometric accuracy of the solids the slices not used in the dataset were segmented and their contours were used as reference. The resulting solids were intersected with a set of twenty sampling coronal planes evenly spaced along the femur. The intersection curves were compared to the reference contours. The two curves were compared measuring the maximum curves distance, $e(z)$. To express a global accuracy was used the $\varepsilon 90$ parameter, defined as the smallest error equal or higher than the 90% of all measured errors.

5.3 Results

The two solid models and their error distributions are showed in Fig. 4.

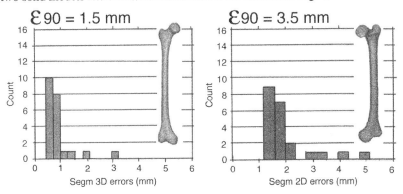

Fig. 4 - Error distribution of the two solid models of the femur, obtained from 3D (left) and 2D (right) segmentation.

Although computationally intensive, the creation of the solid model from the tiled surface was much easier than that from the stacked contours. The main difference was the modelling of the bifurcations, which was much easier with the 3D segmentation. The model based on the 3D segmentation was more accurate: the solid model based on the DMC tiles surface had an $\varepsilon 90 = 1.5$ mm where the solid model created from the stacked contours had a $\varepsilon 90 = 3.5$ mm. These values are the upper boundary of the reconstruction error; in most control sections the error was less than a millimeter.

6. ACKNOWLEDGEMENTS

This work was carried out in the frame of the "Prometeo Project", a joint research program between Istituti Ortopedici Rizzoli, C.I.N.E.C.A., University of Bologna and Cray Research.

7. REFERENCES

1.   McNamara B.P., Viceconti M., Cristofolini L., Toni A. and Taylor D., Experimental and numerical pre-clinical evaluation relating to total hip arthroplasty. In: Computer Methods in Biomechanics and Biomedical Engineering, edited by Middleton, J., Jones, M.J. and Pande, G.N. Amsterdam: gordon and Breach, 1996, pp. 1-10.

2.   Mertz B., Lengsfeld M., Müller R., Kaminsky J., Rüegsegger P. and Niederer P. Automated generation of 3D FE-models of the human femur; comparison of methods and results. In: Computer Methods in Biomechanics and Biomedical Engineering, edited by Middleton, J., Jones, M.J. and Pande, G.N. Amsterdam: gordon and Breach, 1996, pp. 125-134.

3.   Mehta B. V., Mulabagula R. and Patel J. V. Finite element analysis of the human skull considering the brain and bone material porperties. In: Computer Methods in Biomechanics and Biomedical Engineering, edited by Middleton, J., Jones, M.J. and Pande, G.N. Amsterdam: gordon and Breach, 1996, pp. 217-227.

4.   Viceconti M., Cappello A., Massari B., Bertozzi A. and Zannoni C., Optimal slice positioning for long bone CT reconstruction, proc. IEEE-EMBS, Amsterdam, Nov. 1996, no. 3.1.5-2.

5.   Moss A. A., Gordon G., Genant H. K., eds. Computed tomography of the body with magnetic resonance imaging, 2nd ed. Philadelphia: Sounders, 1992, pp. 1355-83.

6.   Sutherland C. J. and Gayou D. E., Artifacts and thresholding in X-Ray CT of a cortical bone and titanium composite, J. Comput. Assist. Tomogr., 1996, 20, pp. 496-503.

7.   Saulgozis J. and Ozolanta I., Complex biomechanical investigations of the proximal part of human femur, proc. XVth Congress International Society of Biomechanics, Jyväskylä (Finland), July 1995, pp. 806.

8.   Lorensen W. and Cline H. Marching Cubes: A high resolution 3D surface construction algoritm, ACM Computer Graphics, 1987, Vol. 21, pp.163-170.

9.   Baruffaldi F., Lucenti M., Viceconti M., Toni A. and Giunti A. Reliability of 3D reconstruction by use of CT images on film, Proceedings of the XIV Congress of the International Society of Biomechanics, 1993, ISB Press, Vol. 1, pp. 210-212.

10. Montani C., Scateni R., Scopigno R. Discretizied Marching Cubes. In: Bergeron, R. D. and Kaufman, A. E., editors, Proceedings of the "Visualization '94" Congress, 1994, IEEE Computer Society Press, pp. 281-287.

MODELLING OF BONE-IMPLANT INTERACTION

S. Pietruszczak[1], D. Inglis[2] and G.N. Pande[3]

1. ABSTRACT

This paper outlines a methodology for analysing problems involving bone-implant interaction. An elastic formulation for cortical and trabecular bone is presented wherein bone is considered to be an inhomogeneous, anisotropic material. The formulation incorporates the concept of 'directional porosity' as a measure of the material microstructure. Subsequently, a simplified procedure for the identification of material parameters is proposed. The procedure employs Eshelby's solution to estimate the elastic constants while the principal directions of anisotropy are derived from Wolff's hypothesis of trabecular architecture. The formulation for trabecular bone is later extended to the elastoplastic regime and the functional form of the failure criterion is proposed. The mathematical formulation is illustrated by a numerical example. In particular, some preliminary results of a finite element analysis of a hip implant system are discussed.

2. INTRODUCTION

Finite element simulations of bone-implant interaction have a significant potential of contributing to the bone prosthesis design. The reliability of a numerical solution depends mainly on the accurate modelling of the geometry, boundary and loading conditions, as well as material behaviour. Earlier simulations of bone-implant systems were restricted to isotropic [1,2] or transversely isotropic [3,4] material models. However, the literature indicates that bone is a strongly inhomogeneous material which displays orthotropic symmetry [5,6].

In this paper, a constitutive framework is developed in which bone is described as an inhomogeneous, anisotropic material. Both elastic and elastoplastic formulations are provided incorporating a tensorial measure of bone architecture, known as a fabric tensor.

keywords: anisotropy; fabric tensor; failure criterion; implant; femur.

[1] Professor, Dept. of Civil Engineering, McMaster University, Hamilton, Ont., Canada L8S 4L7
[2] Grad. Student, Dept. of Civil Engineering, McMaster University, Hamilton, Ont., Canada L8S 4L7
[3] Professor, Dept. of Civil Engineering, University of Wales Swansea, Swansea SA2 8PP, U.K.

The coefficients of the fabric tensor at a specific bone site may be evaluated by morphometric measurements [6,7]. In general, determination of the anisotropy characteristics throughout a bone by invasive measures is not feasible. An alternative approach is based on the use of non-invasive scanning procedures such as computed tomography (CT). Such an approach is both efficient and possibly of clinical value for optimal prosthesis design. The methodology adopted here, for examining bone-implant interaction, is based on estimates of porosity distribution derived from apparent density data generated by CT scan, a similar technique used by other researchers [2,4,8].

3. ELASTICITY FORMULATION FOR CORTICAL/TRABECULAR BONE

Both cortical and cancellous bone represent an inhomogeneous anisotropic material. The anisotropy effects are due primarily to the geometric arrangement of the porous microstructure, while the matrix material itself may be considered as isotropic. In order to describe the bone architecture, the formulation presented here incorporates the concept of 'directional porosity' (after refs. [9,10]), which is considered as an orientation-dependent measure of 'porosity', i.e. a scalar quantity defining the void space fraction.

In order to define the 'directional porosity', consider a unit sphere enclosing a representative volume of the material. Let the sphere be intercepted by a set of test lines of length $\bar{L}$ and orientation v_i relative to the fixed Cartesian coordinate system. The fraction of $\bar{L}$ occupied by voids can be defined as

$$L(v_i) = l(v_i)/\bar{L}; \qquad l(v_i) = \sum_k l_k(v_i) \tag{1}$$

where $l(v_i)$ represents the total length of interceptions of this line with the void space. The mean value of the quantity $L(v_i)$, averaged over the domain S, is

$$L_{av} = \frac{1}{4\pi} \int_S L(v_i) f(v_i)\, dS; \qquad \frac{1}{4\pi} \int_S f(v_i)\, dS = 1 \tag{2}$$

where $f(v_i)$ is a scalar valued function describing the spatial distribution of test lines (for uniformly distributed test lines $f(v_i) = 1$). It can be shown that, the first integral in eq.(2) is a measure of the average porosity of the material, n, whereas the lineal fraction occupied by pores is an unbiased estimator of the volume fraction of voids in the direction v_i, i.e.

$$n = L_{av}; \qquad \bar{n}(v_i) \equiv L(v_i) \tag{3}$$

The function $\bar{n}(v_i)$ can be represented by the generalized double Fourier series. The desired best fit approximation can be established by the 'least square' method leading to a representation in terms of symmetric traceless tensors Ω_{ij}, Ω_{ijkl}.... (cf. [11])

$$\bar{n}(v_i) = n(1 + \Omega_{ij} v_i v_j + \Omega_{ijkl} v_i v_j v_k v_l + ...) \tag{4}$$

The higher rank tensors Ω_{ijkl},... relate to the higher order fluctuations in void space distribution. Thus, in order to describe a smooth orthogonal anisotropy it is sufficient to employ an approximation based on the first two terms of the expansion (4). In such a case, the function $\bar{n}(v_i)$ may be defined as

$$\bar{n}(v_i) \approx 3\,n\,A_{ij}\,v_i\,v_j\;; \qquad A_{ij} = \frac{1}{3}\,(\delta_{ij} + \Omega_{ij}) \;\Rightarrow\; A_{ii} = 1 \qquad (5)$$

where A_{ij}, referred to as 'fabric tensor', is a non-singular measure of the spatial distribution of voids. If the eigenvalues of A_{ij} are distinct, then $\bar{n}(v_i)$ reflects a smooth orthogonal anisotropy. On the other hand, if $\Omega_{ij} = 0$ then the material is said to be isotropic.

In the elastic range, the constitutive relation may be written in the following form

$$\sigma_{ij} = D_{ijkl}\,\varepsilon_{kl}\;; \qquad D_{ijkl} = D_{ijkl}\,(A_{pq}, n) \qquad (6)$$

in which the components of D_{ijkl} are a function of the fabric tensor and the average porosity. The simplest representation of the elasticity tensor is that proposed in ref. [12]

$$d_\alpha = B_{\alpha\beta}\,k_\beta\,(n) \qquad (7)$$

where d_α denotes the individual non-zero components of D_{ijkl}, k's are functions of n only and $B_{\alpha\beta}$ depends on the basic invariants of the fabric tensor. The representation (7), although attractive, has a serious limitation. Namely, the operator $B_{\alpha\beta}$ is singular implying that there is no unique set of k's describing given material properties [13]. In recent years other approximations have been developed (e.g. [14], [15]). However, most of them are restrictive, as they neglect certain terms in the general representation which may be of significance.

4. ON THE IDENTIFICATION OF MATERIAL PROPERTIES

In general, the directional distribution of porosity in a bone can be derived from consecutive high resolution CT images. Unfortunately however, such results are not readily available. Thus, given the limitations of representation (7) on one hand, and the lack of sufficient experimental data on the other, a simplified procedure is suggested here for identification of the elastic constants. The procedure is based on estimates derived from Eshelby's solution to the inclusion problem.

Consider a comparison material with a set of randomly dispersed elliptical voids whose geometry is consistent with the porosity distribution $\bar{n}(v_i)$. For simplicity, refer the geometry of the problem to the principal directions of anisotropy e_i. Invoking Eshelby's solution, the average elastic properties of the composite medium can de described as (c.f. [16])

$$\varepsilon_{ij} = C_{ijkl}\,\sigma_{kl}\;; \qquad C_{ijkl} = C_{ijmn}^{o}\,[\delta_{mnkl} + n(\delta_{mnkl} + S_{mnkl})] \qquad (8)$$

Here C_{ijkl}^{o} defines the isotropic properties of the matrix material, δ_{ijkl} is a forth order unit tensor and S_{ijkl} is Eshelby's tensor, whose components are a function of the geometry of the inhomogeneity. The general procedure for estimating the average anisotropic properties of bone material can be summarized as follows:

- given the basic material properties and the spatial distribution of porosity , i.e. C_{ijkl} and $\bar{n}(e_1)$, $\bar{n}(e_2)$, $\bar{n}(e_3)$, at one specific location, define S_{ijkl} and determine the isotropic properties of the matrix C_{ijkl}^{o}
- given C_{ijkl}^{o} and S_{ijkl} estimate C_{ijkl} over the entire domain.

The principle directions of anisotropy can be estimated by invoking Wolff's hypothesis of trabecular architecture. The latter states that the principal stress axes coincide with the principal trabecular directions in cancellous bone at remodelling equilibrium. In mathematical terms this implies that the matrix multiplication of σ_{ij} and A_{ij} is commutative, i.e.

$$\sigma_{ip} A_{jp} = \sigma_{jp} A_{ip} \tag{9}$$

or, since A_{ij} and Ω_{ij} are coaxial,

$$A_{ip} \Omega_{jp} = A_{jp} \Omega_{ip}; \quad \Rightarrow \quad \sigma_{ip} \Omega_{jp} = \sigma_{jp} \Omega_{ip} \tag{10}$$

The procedure involves a numerical analysis of an intact bone (i.e. a femur) under an average physiological load P_i. The latter may be defined as a weighted average corresponding to typical physical activities

$$P_i = \sum_{\alpha} \frac{t^{\alpha}}{t_o} P_i^{\alpha} \tag{11}$$

where t^{α} is the time interval, within t_o, associated with an activity resulting in P_i^{α}. The problem can be solved using a Newton-Raphson iterative scheme under the constraints specified in Eq. (9) or Eq. (10). It should be emphasized that the simplified procedure, as outlined above, needs to be implemented only in the case when sufficient experimental data is unavailable for the specification of the function $\bar{n}(v_i)$. Otherwise, the principal directions of anisotropy coincide with those of Ω_{ij}.

5. ELASTOPLASTIC FORMULATION FOR TRABECULAR BONE

In the formulation presented above, both cortical and cancellous bone are considered as the same tissue with different porosity characteristics. Given the fact that the porosity of the cancellous bone may be very high, it is reasonable to expect that this material may exhibit some irreversible (plastic) deformations. It is recognized that in an intact bone subjected to typical physiological loads, the response of the trabecular network will still be predominantly elastic. However, for a certain class of problems, in particular, the analysis of bone-implant interactions, irreversible deformations are likely to occur and should be properly taken into account. In this section, a simple procedure is outlined for describing the anisotropic properties of trabecular bone in the elastoplastic regime.

For an isotropic material, the yield function, f, is typically assumed as an isotropic scalar-valued function of the state of stress, plastic deformation history and a set of material constants. Choosing the uniaxial compressive strength, f_c, as a representative material parameter, the yield criterion can be expressed as

$$f = f(\sigma_{ij}, \kappa, f_c) = 0; \quad \kappa = \kappa(\varepsilon_{ij}^p) \tag{12}$$

In order to extend this formulation to the anisotropic case, one can assume that f_c is a scalar-valued function of the orientation of the sample relative to the 'loading direction', l_i. In particular, the spatial distribution of f_c may be related to the density/porosity of the material by employing the representation analogous to that of Eq. (4), i.e.

$$f_c = f_{co} \left(\frac{\rho}{\rho_o} \right)^\alpha = f_{co} \left(\frac{1 - \bar{n}(l_i)}{1 - n_o} \right)^\alpha ; \quad 1 \leq \alpha \leq 2 \tag{13}$$

where

$$\bar{n}(l_i) = n (1 + \Omega_{ij} l_i l_j) \tag{14}$$

In order to define the 'loading direction', consider a differential element of the material in the form of a parallelepiped and invoke the classical definition of the stress vectors acting on each face of this element. Since the dimensions are said to be infinitesimal, one can consider the conjugate stress vectors as concurrent and specify the resultant stress vector as

$$T_i = \sigma_{ij} (e_i^{(1)} + e_i^{(2)} + e_i^{(3)}) = \sqrt{3} \sigma_{ij} m_j \tag{15}$$

where $m_i = \{1,1,1\}/\sqrt{3}$ and e_i's are the base vectors associated with an arbitrarily chosen frame of reference. The loading direction l_i is now defined as being collinear with the resultant stress vector, i.e.

$$l_i = \frac{T_i}{\|T_i\|} = \frac{\sigma_{ij} m_j}{(\sigma_{kl} \sigma_{kp} m_l m_p)^{1/2}} \tag{16}$$

The specification of the functional form of the failure criterion for bone is rather speculative at this stage, as there is little experimental information available. At the same time however, the architecture of bone structure bears some analogy to that of a class of cemented granular mixtures. Recognizing this fact, it seems rational to assume that the conditions at failure are affected by all three basic stress invariants. A reasonable compromise between the complexity and accuracy of the formulation may be achieved by adopting the functional form typically used for brittle-plastic materials (after ref. 17])

$$F = \bar{\sigma} - g(\theta) \bar{\sigma}_c = 0 \tag{17}$$

where

$$\bar{\sigma}_c = \frac{-a_1 + \sqrt{a_1^2 + 4a_2(a_3 + I/f_c)}}{2a_2} f_c \tag{18}$$

In the above equations $I = -I_1$, $\bar{\sigma} = (J_2)^{1/2}$, $\theta = 1/3 \sin^{-1} (3\sqrt{3} J_3 / 2\bar{\sigma}^3)$, where I_1 and (J_2, J_3) are the basic invariants of the stress tensor and the stress deviator, respectively. Moreover, the parameters a_1, a_2 and a_3 represent dimensionless material constants and $f_c(l_i)$ denotes the uniaxial compressive strength of bone, which is defined by Eq. (13). In the principal stress space, Eqs. (17) and (18) describe an irregular cone with smoothly curved meridians and a non-circular convex cross-section defined by $g(\theta)$.

The simplest functional form of the yield criterion may be obtained by postulating a geometric similarity to representation (17), i.e.

$$f = \bar{\sigma} - \beta(\xi) g(\theta) \bar{\sigma}_c = 0 \tag{19}$$

In the above equation $\beta(\xi)$ is designated as the hardening function and, for a stable material, may be chosen in a simple hyperbolic form

$$\beta(\xi) = \frac{\xi}{A + \xi} ; \qquad \dot{\xi} = (\dot{e}_{ij}^{p} \dot{e}_{ij}^{p})^{1/2} \tag{20}$$

where $\dot{e}_{ij}^{p}$ is the deviatoric part of the plastic strain rate. It should be noted that $0 \leq \beta \leq 1$ and $\xi \to \infty \Rightarrow \beta \to 1$. The latter signifies the local failure of the material associated with the formation of microcracks, i.e. fracture of individual trabeculae. Thus, in the context of a boundary-value problem, the distribution of β will provide a clear indication of the extent of structural damage in the system.

Finally, the constitutive relation can be formally derived by adopting the standard plasticity procedure. Satisfying the consistency condition and invoking the additivity of elastic and plastic strain rates, the following relation is obtained

$$\dot{\sigma}_{ij} = D_{ijkl} \dot{\varepsilon}_{kl} ; \qquad D_{ijkl} = D_{ijkl}^{e} - \frac{1}{H} D_{ijpq}^{e} \frac{\partial Q}{\partial \sigma_{pq}} \left(\frac{\partial f}{\partial \sigma_{rs}} + \frac{\partial f}{\partial f_{c}} \frac{\partial f_{c}}{\partial \sigma_{rs}} \right) D_{rskl}^{e} \tag{21}$$

where D_{ijkl}^{e} is the elastic tensor, Q=const. is the plastic potential and H is the hardening modulus.

6. NUMERICAL EXAMPLE

In this section the preliminary results of a finite element analysis pertaining to a femur-endoprosthesis system are reported. The geometry of the problem is shown in Fig. 1. CT imaging and post-processing software were used to extract the surface geometry of an adult human left femur. The neck and head of the femur had been resected according to surgical procedure for a total hip arthroplasty. Bone contours were connected into four-sided surface patches to form a closed volume using the COSMOS/M finite element analysis system[4]. Three quarters of the proximal femur were modelled using 30,308 four-noded tetrahedral elements. It is important to emphasize that according to the proposed formulation, both cortical and trabecular bone were considered as the same material with different porosity characteristics. This eliminated the need to explicitly define the endosteal surface geometry, which is particularly ambiguous and difficult to delineate near the articulating joint structures.

An anatomic prosthesis (Howmedica implant # 6280-1-010 NDXUB) with porous coated collar was modelled using 2,439 tetrahedral elements. The implant was made of a forged cobalt chromium alloy (Vitallium, E = 220GPa, ν = 0.32). At this stage, complete bonding was assumed at the bone-implant interface. Future analyses will incorporate frictional gap elements to study interface micro motions and to properly model the effect of cementless fixation.

A uniformly distributed load was applied over the end surface of the prosthesis to simulate the one-legged stance phase of gait. The forces of the hip abductor muscles were omitted for this preliminary study. The force resultant was directed along the long axis of the femur and had a magnitude of three times body weight (BW = 750N). Nodes at the distal end of the femur were completely restrained.

---

[4] Structural Research & Analysis Corp.

Fig.2 presents the distribution of average porosity over a coronal section. The values were obtained from the CT data used to define the model geometry. Each element was assigned a unique average porosity value based on a point inclusion and volume averaging procedure. Apparently, low values of porosity (<10%) are observed in the region adjacent to the periosteal contour and signify the presence of cortical bone. Fig.3 shows the porosity distribution of an axial cross section, 75mm from the apex of the greater trochanter. For comparison, the corresponding CT image is also shown. In general, the porosity gradients are quite high, with very low porosity values near the periosteal surface and progressively increasing values in the direction of the medullary canal.

Given the space limitations, only the results corresponding to isotropic material with non-homogeneous porosity distribution are reported here. The main objective is to demonstrate that the interaction between the bone and the implant will, in general, involve generation of plastic deformations. A continuum distribution of the material properties of the femur was obtained from the porosity estimates using the procedure outlined in section 4. The components of C_{ijkl} were determined directly from representation (8), with Eshelby tensor corresponding to the case of randomly dispersed spherical inhomogeneities. Values for Young's modulus and Poisson's ratio of the matrix material were E = 17 MPa, ν = 0.3. The failure criterion (17) was evaluated using f_{co} = 100 MPa which was then scaled using a quadratic form of Eq. (13).

The main results of the numerical analysis are presented in Figs. 4 an 5, which show the distribution of the damage factor β as defined in Eq. (19). As mentioned before, $\beta \to 1$ signifies a local failure within the trabecular architecture, i.e. formation of micro-cracks. Fig.4 shows the distribution of β over a coronal section, while Fig.5 presents the same distribution in an axial cross section, 75mm from the apex of the greater trochanter (compare with Fig.3). It is evident that in the region along the lateral side of the bone-implant interface, the values of β are as high as 0.9, indicating high stress concentrations and the generation of significant irreversible (plastic) deformations. In general, given the fact that the porosity of the material in the neighbourhood of the implant is very high, as seen in Fig.2, the threshold value of β at which plastic deformation is likely to commence may be very low.

7. CONCLUDING REMARKS

Based on the hypothesis that both cortical and cancellous bone represent the same material with continuously varying porosity characteristics, an elastic description of their mechanical response has been presented. A procedure for identification of material parameters has been proposed, based on Eshelby's solution of randomly dispersed elliptical cavities and Wolff's hypothesis of trabecular architecture of bone. A tentative elastoplastic formulation for trabecular bone, in which the uniaxial compressive/tensile strength varies with orientation, has been developed. A numerical example involving an anatomic prosthesis with proximal porous coating has been given. The distribution of average porosity was obtained from CT scan data. The results indicate that irreversible plastic deformations take place in the region adjacent to the implant and their importance is significant in the evaluation of the implant performance.

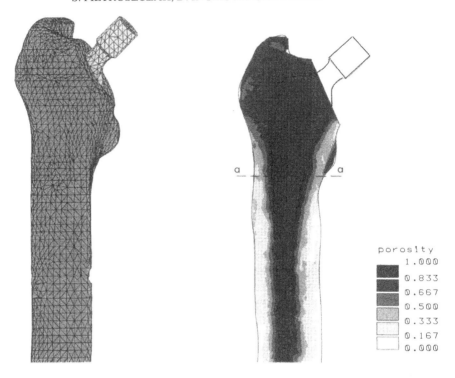

FIGURE 1 Problem geometry and finite element discretization (100mm shown).

FIGURE 2 Distribution of average porosity over a coronal section (implant outlined).

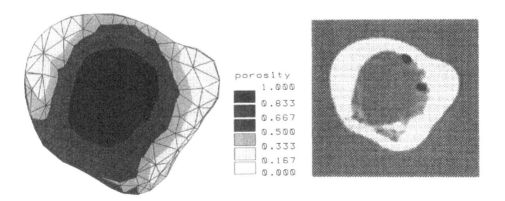

FIGURE 3 Porosity distribution in an axial cross section (a-a) with corresponding CT image.

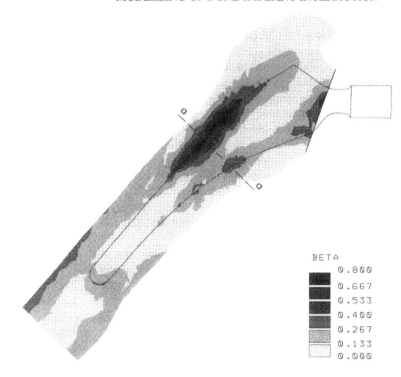

FIGURE 4 Distribution of damage factor β, Eq. (13), over a coronal section.

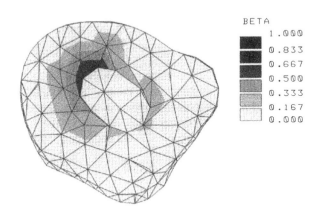

FIGURE 5 Distribution of damage factor β, Eq. (13), in an axial cross section (a-a).

8. ACKNOWLEDGEMENTS

The authors wish to acknowledge the contribution of Dr. P. Cheng (formerly a Postdoctoral Fellow at McMaster University) to the procedure outlined in Section 4 and its numerical implementation. Special thanks to Dr. Chris Gordon, Dr. Colin Webber, and Dr. Claude Nahmias, of the Department of Nuclear Medicine, McMaster University Medical Centre, for their assistance with CT imaging, data post-processing, and lending of the femur-implant system.

9. REFERENCES

1. Van Rietbergen B., Huiskes R., Weinans H., Sumner D. R., Turner T. M. and Galante J. O., The mechanism of bone remodeling and resorption around press-fitted THA stems, J. Biomechanics 1993, Vol. 26, No. 4/5, 369-382.

2. Mann K. A., Bartel D. L., Wright T. M. and Burstein A. H., Coulomb frictional interfaces in modeling cemented total hip replacements: a more realistic model, J. Biomechanics 1995, Vol. 28, No. 9, 1067-1078.

3. Vichnin H. H. and Batterman S. C., Stress analysis and failure prediction in the proximal femur before and after total hip replacement, Jour. Biomech. Eng. 1986, Vol. 108, 33-41.

4. Rubin P. J., Rakotomanana R. L., Leyvraz P. F., Zysset P. K., Curnier A. and Heegaard J. H., Frictional interface micromotions and anisotropic stress distribution in a femoral total hip component, J. Biomechanics 1993, Vol. 26, No. 6, 725-739.

5. Ashman R. B., Cowin S. C., Van Buskirk W. C. and Rice J. C., A continuous wave technique for the measurement of the elastic properties of cortical bone, J. Biomechanics 1984, Vol. 17, No. 5, 349-361.

6. Turner C. H., Cowin S. C., Young Rho J., Ashman R. B. and Rice J. C., The fabric dependence of the orthotropic elastic constants of cancellous bone, J. Biomechanics 1990, Vol. 23, No. 6, 549-561.

7. Odgaard A., Kabel J., Van Rietbergen B., Dalstra M. and Huiskes R., Fabric and elastic principal directions of cancellous bone are closely related, J. Biomechanics 1990, Vol. 30, No. 5, 487-495.

8. Merz B., Niederer P., Müller R. and Rüegsegger P., Automated finite element analysis of excised human femora based on precision-QCT, Jour. Biomech. Eng. 1996, Vol. 118, 387-390.

9. Pietruszczak S. and Krucinski S., Description of anisotropic response of clays using a tensorial measure of structural disorder, Mech. Mater. 1989, Vol.8, 327-249.

10. Pietruszczak S. and Krucinski S., Considerations on soil response to the rotation of principal stress directions, Comp. Geotech. 1989, Vol. 8, 89-110.

11. Kanatani Ken-Ichi , Distribution of directional data and fabric tensor, Int. J. Eng. Sci. 1984, Vol. 22, 149-161.

12. Cowin S.C., The relationship between the elasticity tensor and the fabric tensor, Mech. Mater. 1985, Vol. 4, 137-147.

13. Pietruszczak S., Bone modelling; description of aging and functional adaptation, Comp. Meth. Biomech. Biomed. Eng. 1996 (submitted).

14. Rubin P.J., Rakotomanana R., Leyvraz P.F., Zysset P.K., Curnier A. and Heegaard J.H., Frictional interface micromotions and anisotropic stress distribution in a femoral total hip component, J. Biomechanics 1993, Vol. 26, 725-739.

15. Cowin S.C., Sadegh A.M. and Luo G.M., An evolutionary Wolff's law for trabecular architecture, Jour. Biomech. Eng. 1991, Vol. 114, 129-136.

16. Mura T., Micromechanics of defects in solids, Martinus-Nijhoff, Dordrecht, 1987.

17. Pietruszczak S., Jiang J and Mirza F.A., An elastoplastic constitutive model for concrete. Int. J. of Solids & Structs. Vol 24. 1988.

NUMERICAL MODELING OF THE BIOMECHANICAL REACTION OF HUMAN TIBIA UNDER COMPLEX LOADING STATE

I. Knets[1], J. Laizans[2], M. Dobelis[3]

1. ABSTRACT

The finite element method (FEM) is used to calculate analytically the extreme values of stresses appearing in different zones of cross-section of human tibia diaphysis during the gait cycle. The results are analyzed with respect to experimental data on heterogeneity of material properties and functional adaptation of bone to external loads.

2. INTRODUCTION

The different bones of the human skeleton system have been subjected to varying loads during the lifetime. The internal structure and external form or architecture of bones reflect their biomechanical response to these loads. The main functions of human lower leg bones are to support the standing posture and the walking movement. The origin of normal and shearing forces acting on the tibia are body weight and tensile forces of different muscle groups.
It was stated earlier (Ashmuth R., 1981) that in the normal tibia, the area moments of inertia decrease from proximal to distal end. In the normal tibia the moments of resistance of the cross-sections are the highest dorsally and the smallest ventrally. As revealed by photon densitometry and microradiography, bone density is higher in the dorsal cross-section areas of tibia than in the anterior one. In tibia osteons with a

KEYWORDS

Finite element method, Tibia, Load simulation

[1] Professor, Biomaterials and Biomechanics Specialized Research Institute, Riga Technical University, Kalku iela 1, Riga, LV-1658, Latvia
[2] Research Assistant, Biomaterials and Biomechanics Specialized Research Institute, Riga Technical University, Kalku iela 1, Riga, LV-1658, Latvia
[3] Senior Researcher, Biomaterials and Biomechanics Specialized Research Institute, Riga Technical University, Kalku iela 1, Riga, LV-1658, Latvia

prevalence of longitudinal arranged collagen fibers seem to predominate in the anterior cortex, whereas osteons with a prevalence of transversally arranged fibers seem to predominate in posterior cortex. This tends to the assumption that the tibia is adapted by triangular cross-sectional shape, distribution of bone density and mineralisation, and collagen fiber arrangement in osteons to preponderant dorso-concave bending in sagittal plane.

3. MATERIALS AND METHODS

Human tibia has rather specific form of its cross-section. In diaphysis it has almost triangular cross-section and can be divided into six zones: three corner zones and three intermediate zones (Fig. 1). The biomechanical reaction of each zone to the mechanical loading is different and reflects the biomechanical adaptation of tibia to physiological conditions of loading. It was found that during the lifetime there were changes not only in the absolute values of characteristics of the mechanical properties of bone tissue in these zones, but there were also changes in the redistribution of them over the zones (Knets et all., 1980).

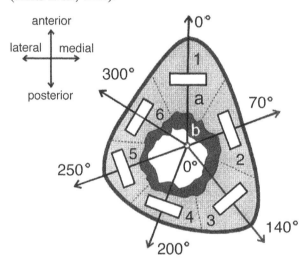

Fig. 1. Triangular form of human tibia and location of test samples in six zones (1-6) of the diaphysis of its cross-section with compact (a) and cancellous (b) bone layers

Three-dimensional finite element model of 100 mm long human tibia diaphysis was created using dimensions from the five consecutive cross-sections along the length of 34 year old male left tibia. Finite element meshes were defined by the boundaries of the different materials determined on the micrographs of the sections mentioned above. Six zones of compact bone and one homogeneous inner layer of cancellous bone were used in this model. The compact bone-cancellous bone interface was assumed to have perfect bonding. 3-D linear isoparametric eight-node hexahedral elements were used to generate the 3-D mesh of the tibia diaphysis model studied. The model consisted of 1800 solid elements with 13416 nodes.

The compact bone material was assumed to be orthotropic, linearly elastic and heterogeneous. Values of nine independent constants were used in this study and derived from the experimentally determined results of compact bone tissue for the age group 19-44 years (Knets et all., 1980). The distribution of these constants (Young's moduli E_i, shear moduli G_{ij} and Poisson's ration μ_{ij}) along the six zones of tibia cross-section is presented in Table 1. The experimental values of the modulus of elasticity E_1 in the table correspond to the E_Z in 3-D model. The orthotropic material axis were aligned with the 3-D finite element model global axes.

Table 1. Independent constants for compact bone tissue and cancellous bone

Material constants	Compact bone in zones of cross-section						Cancellous bone
	1	2	3	4	5	6	
E_1 (GPa)	19.55	18.50	17.52	15.98	18.40	20.18	0.10
E_2 (GPa)	8.23	8.13	9.18	8.21	9.47	7.80	0.10
E_3 (GPa)	6.66	6.16	7.83	6.15	7.60	6.13	0.10
G_{12} (GPa)	4.99	5.16	5.22	4.40	5.12	4.60	0.04
G_{13} (GPa)	3.38	3.43	3.93	3.00	4.53	3.06	0.04
G_{23} (GPa)	2.22	2.53	2.40	2.20	2.69	2.51	0.04
μ_{12}	0.295	0.301	0.313	0.308	0.323	0.300	0.30
μ_{13}	0.308	0.306	0.322	0.313	0.331	0.308	0.30
μ_{23}	0.600	0.591	0.578	0.562	0.767	0.631	0.30

4. BOUNDARY AND LOADING CONDITIONS

The distal outer cortical layer of this model was constrained in the superior-inferior direction. Only three additional constraints in other directions were added to provide nonsingularity of the global matrix. These boundary conditions were used to prevent rigid body translation and rotation of the whole model.

The bone loading conditions used in this study simulated different complex loading states which appear from fluctuating bending stresses during different phases of normal walking. These loading conditions were idealized and included only three extreme phases from the gait cycle. These loading states in model were reached applying the compression load eccentrically with respect to the central axis of symmetry of the bone (Fig. 2). Changing the value and direction of eccentricity it was possible to analyze the effect of nonuniformity of the mechanical properties over the zones of cross-section on the distribution of stresses and strains in tibia.

The loading in the 3D-model was applied uniformly along the whole cross-section of the bone in the way that upper plane only rotates along appropriate axis. Loading case I corresponds to the condition when the whole cross-section is deformed parallel to the initial state. The maximum displacement chosen in the calculations corresponded to 1% of strain. Loading case II accounts for bending in the medial-lateral direction. The maximum displacement value in the outer layer of lateral compact bone tissue simulated 1% of strain while it remained zero in the medial layer. Loading case III simulates maximum bending in the anterior-posterior direction. The maximum displacement value in the outer layer of anterior compact bone tissue simulated 1% of strain and remained zero in the posterior layer. The 1% strain value was considered to be the extreme deformation value which may encounter during the natural loading conditions.

5. RESULTS

Using the finite element model the numerical calculations were performed with the NISA software. The normal stress distribution in bone volume was determined and analyzed in

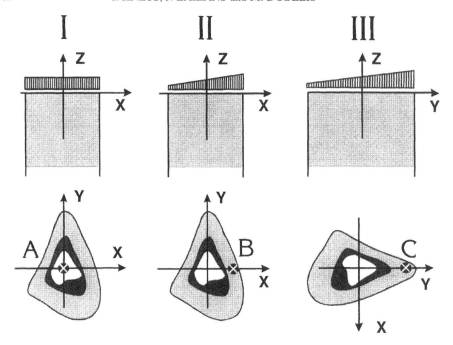

Fig. 2. Three different displacement forms (1-3) of tibia diaphysis and location of corresponding conditional loading points (A-C).

the different zones of the diaphysis of human tibia model. The calculated ranges of stresses in the middle section of tibia were transformed to a colour spectrum to display the results for easier comparison. It should be pointed out that when comparing the results in different loading conditions, a particular colour code represents slightly different stress level in the component area. The results presented in the study should be treated as being of a qualitative rather than of a quantitative nature. This is because of a numerous assumptions used and limitations of the FEM.

5.1 Loading case I

The highest values of normal compression stresses in the diaphysis of the cross-section of tibia (Fig. 3) were found in the anterior-lateral zone 6. The highest value is reached at the very outer layer -- 45 MPa (label **a** in Fig. 3). The compressive stresses within zones 1 through 5 are distributed relatively uniform and their values range from 0.19 to 20 MPa (label **b** in Fig. 3). No tensile stresses were revealed for this state of loading.

5.2 Loading case II

The medial-lateral bending causes the highest values of compressive stresses in the medial-posterior zones 2 and 3 (Fig. 4a). The values of compression stress range from 7.8 to 21 MPa in the posterior-medial zone 3, and up to 34 MPa in the medial zone 2. Practically almost zero stress values or small tensile stresses were revealed throughout the whole cancellous bone layer which is not a load carrying structure at all.

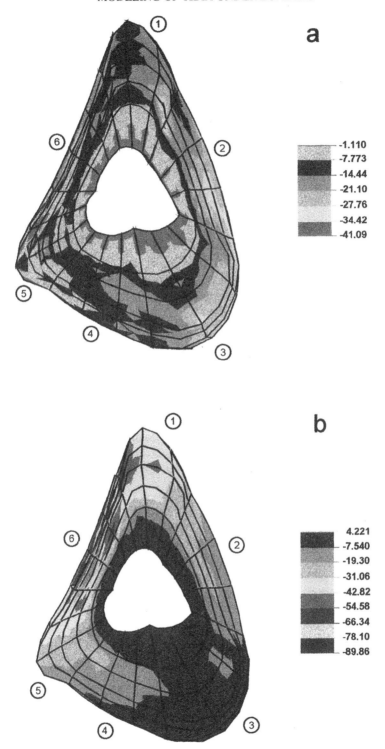

Fig. 4. Bone normal stresses σ_z in the cross-section of the diaphysis of the tibia in loading case II (a) and case III (b). Numbers 1-6 indicate zones of cross-section.

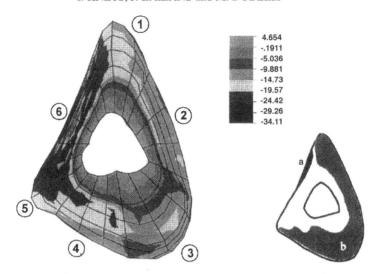

Fig. 3. Bone normal stresses σ_z (MPa) in the cross-section of the diaphysis of the tibia in loading case I. Numbers 1-6 indicate zones of cross-section. See explanation for labels **a** and **b** in text.

5.3 Loading case III

The anterior-posterior bending causes the highest values of compression stresses in the outer layer of anterior-lateral zone 6 (Fig. 4b). In a very narrow layer of this zone the stresses reach 90 MPa. The posterior-medial zone 3 along with the inner cancellous bone layer is relatively unloaded - the stresses are close to zero.

It should be pointed that the lowest theoretically calculated normal stresses were stated in the same zones where the lowest experimentally determined values of ultimate tensile strength were noticed. This proves the bone material functional adaptation capability to the external loads.

6. DISCUSSION

There have been widely used isotropic and homogeneous mechanical properties of human tibia in a numerous of FEM studies. The present analysis considers orthotropic material properties and heterogeneous model structure instead. This should be regarded as much closer approximation with respect to other 3-D tibia models used so far. Very important is a factor of actual load simulation in the tibia model. In further investigations this should be taken into account when modeling the whole tibia.

7. REFERENCES

1. Knets I., Pfafrod G., Saulgozis J. (1980) Deformation and fracture of hard biological tissue. Zinātne, Rīga, 320 p. (In Russian).
2. Ashmuth R., Amtmann E. On the functional morphology of normally formed and rachitically deformed human tibiae and fibulae. Z. Morph. Anthrop., 1981, Bd 72, 23-46.

COMPUTATIONAL MODELLING OF INFLUENCE OF CLAMP WITH SHAPE MEMORY ON STRESS-STATE OF BONE

Jitka Jírová [1] , Josef Jíra [2] , Michal Micka [3]

1. ABSTRACT

The paper deals with the research of the stress state of the bone caused by application of a clamp with shape memory. Such clamps are used in orthopaedy for healing of fractures. The paper presents the philosophy of the generation of a computation model for the application of the FEM in the ANSYS programme. The computation model is based on experimental results. Using the model we can analyse the stress state generated by the clamp not only on the surface of the bone but also in the region of the fracture.

2. INTRODUCTION

The aim of our research was to describe the stress state at the bone surface in the area where the osteosynthetic clamp was fitted. Such clamps are used in orthopaedy for healing of fractures. Recently a number of new materials have appeared in orthopaedic surgery [1]. Among them alloys with shape memory, which have the ability to remember the shape they had before mechanical deformation. The deformation brought about at low temperature vanishes and the original form is restored under the influence of heating if temperature reaches or overpasses the transformation temperature. During this recovery process some of these alloys are able to exhibit considerable forces and recover considerable deformations (up to 10%). This is the shape memory effect.

Keywords: osteosynthetic clamp, shape memory effect, method of photoelastic coating, strain gauge measuring, computational model (FEM)

[1]Chief Research Worker and Head of Lab./ Institute of Theoretical and Applied Mechanics, Prosecká 76, 190 00 Prague 9, Czech Republic
[2]Assoc.Professor/ Faculty of Transportation Sciences of the Czech Technical University, Konviktská 20, 110 00 Prague 1, Czech Republic
[3]Pricipal Research Worker and Deputy Head of Dept./ Institute of Theoretical and Applied Mechanics, Prosecká 76, 190 00 Prague 9, Czech Republic

While if in martensitic state material can easily be deformed it is very resistant in the austenitic state. If it is deformed in the martensitic state and then heated the deformation disappears.

The developed axial pressure and the resulting fixation at the contact surfaces of bones have a favourable effect on fracture healing [2], [3]. The material with shape memory exerts the desirable pressure on bones due to increase of temperature only. The maximum force which the clamp is able to develop during the restoration of the original form can be influenced not only by the composition of the material but also by the clamp design.

Fig. 1: Two different shapes of clamps

3. MATERIAL AND METHODS

3.1 Numerical analysis

Using a computation model we can analyse the stress state generated by the clamp not only on the surface of the bone but also in the region of the fracture. The mathematical model links up with the results of experimental research. On the base of experimental results we can verify the correctness of the mathematical model and further use it for determination of the magnitude of the force generated by the clamp after warming and for analysing of the stress state in the fracture section.

The calculation was made using the Finite Element Method by applying the ANSYS programme version 5.0. The geometrical shape of the tibia is described by 17 cross sections in distances 20 mm and the basic spatial model is created using direct generation of nodes and elements (Fig.2). Both the external and internal circumferences of each cross section were divided into the same number of parts and using APDL programme language the basic network of the bone was created. The mathematical model of a bone was generated by direct generation of three-dimensional elements in the right-hand cylindrical co-ordinate system. The axis x proceeds in radial direction, the axis y in peripherial direction, the axis z in the direction of cylindrical axis. In the selected cylindrical co-ordinate system the nodal points are generated in the

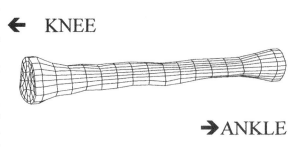

Fig. 2. Mathematical model

individual tibia cross sections by the selected method. These points are the nodes of the individual 20-node three-dimensional isoparametric elements SOLID95. It can tolerate irregular shapes without as much loss of accuracy. SOLID95 elements have compatible displacement shapes and are well suited to model curved boundaries.

There are 18 of these elements around the circumference. The mathematical model comprises 2794 nodes and 470 spatial isoparametric elements SOLID95. The termination at the knee and the ankle is of complicated shape; therefore a certain simplification has been selected. At the end of the bone level bottoms of the constant thickness have been generated the edges of which are identical with the first and the last cross sections.

To enable the modelling of the effect of the osteosynthetic stem on the stress / strain diagram in the cortical bone, it is necessary to provide a model of the hole in the bone shell in which the clamp is fixed. For this purpose a set of 24 elements was generated in two rings instead of the 8 elements of the basic model, 12 of them modelling a small cylindrical inclusion of a different material and 12 modelling the surroundings of the cortical bone. The cylindrical inclusion is actually a hole drilled in the bone shell. Its axis is perpendicular to the tibia axis. The thickness of the inclusion cylinder is identical with the cortical bone thickness in the given part of the ring.

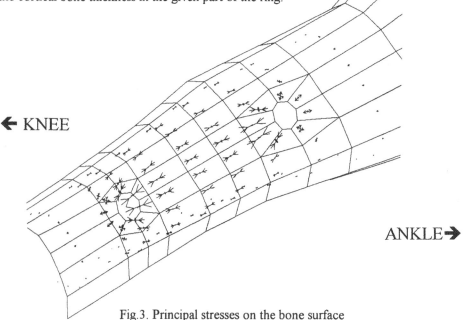

← KNEE

ANKLE→

Fig.3. Principal stresses on the bone surface

The force generating after warming of the clamp, i.e. the contracting effect is simulated by two equal forces applied in the centre of the inclusion bases. The direction of these forces is parallel with the longitudinal axis of the tibia. Between the inclusion and the cortical bone continuous stress and strain flow (and not a contact of two separate volumes) has been considered. The effect of the whole clamp is modelled by two rings with an inclusion in which the forces act in opposite directions. This expresses the contraction produced by the ends of the clamp in the bone. Between these two rings a ring of the basic model is inserted with a denser network which contributes to a more accurate stress and strain computation in the area between clamp ends.

In the initial phase of the numerical solution we designed a coincident model with experimental one, i.e. we analysed the stress state of the intact bone. In the cortical bone the stress and strain between two stem ends were computed using a part of the basic model of the whole tibia, i.e. the part in which two rings were replaced with the rings with an inclusion and the intermediate ring with a denser network. The boundary conditions have been so generated as to model the bearing in three points about the circumference of the first and the last cross sections in cylindrical co-ordinates. In the first section the radial displacement is prevented in two points and peripheral and radial displacements in third point. In the last section the radial displacement is prevented in two points and the displacements in all three directions is prevented in the third point. These conditions actually correspond with the bearing of the model as a simply supported beam with the prevention of torsion. The magnitude of the contracting force is taken as 1 N in the first step. After the computation this value was corrected linearly in accordance with the results of experimental measurements. The actual value of this force was ascertained as 40 N and the problem was recomputed.

The stress (Fig.6) and strain (Fig.7) in the cortical bone between two clamp ends with a transverse bone fracture in between was computed, once again, using a part of the basic model of the whole tibia as in the case of the intact bone. The boundary conditions have been so generated as to model the bearing in the three points of the first and last cross section in cylindrical co-ordinates. In the first section the displacement in peripheral and radial directions is prevented in all three points. In the last section the displacement in all three directions is prevented in all three points. In the cross section with the fracture the peripheral displacement is prevented in all three points. These conditions actually correspond with the bearing of the model as a simply supported bear with the prevention of torsion and the prevention of buckling in the fracture. The contraction force of the clamp of the material with shape memory is introduced in the same way as in the preceding case. In the fracture where two separate parts of the cross section are in contact the contact elements CONTAC 49 are defined. These elements are used to define the contact area of both parts in the course of the computation and to compute the contact stress in the fracture cross section.

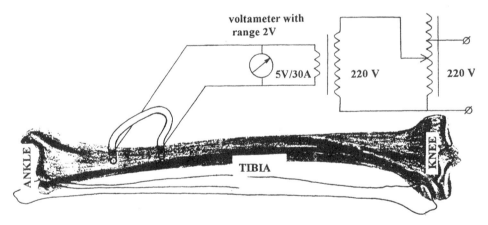

Fig.4. Scheme of clamp fitting and heating

3.2 Experimental research

We investigated the stress state at the bone surface [4], [5] caused by the fitting of a clamp made of an alloy prepared at the Technical University in Bratislava. This experimental prototype of a clamp was designed and produced at Poldi Company in Kladno [6]. Using a special fixture the clamp was prolonged about 2 mm in a water bath mixed up with ice. The predeformed clamp was fixed into two holes drilled in the bone at the distance of 35.5 mm (Fig.5). After fixing, the clamp was slowly warmed up to the temperature of 80 °C in four steps. The first measuring was done under the room temperature (+20°C) and the other ones at 40°C, 60°C and 80°C.

To ascertain the stress state we have used the reflection photoelastic method. Using this method we can find directions and differences of principal stresses at the investigated bone surface including stress concentrations. A reflex layer 2 mm thick with optical sensitivity constant K = 13,176 N.mm^{-1} and a modulus of elasticity E = 3200 MPa was applied to the surface of the bone. The measurements were carried out by means of the type 030 reflex polariscope of the firm Vishay.

To ascertain the stress values in two principal directions at the central point we have used the strain gauge measurement. One strain gauge XY 11 3/120 HBM with 2 measuring grids in perpendicular directions was applied to the central point between two holes on the bone. The strain gauge used has the temperature characteristic close to the bone tissue and was protected against weather water at room temperature. The results of the strain gauge measurement were used for the evaluation of the photoelastic measurement (Fig.5).

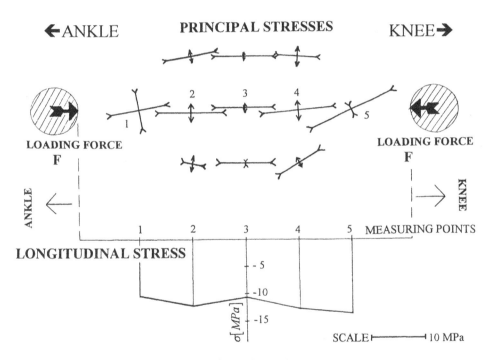

Fig. 5. Results of experimental measurements

4. RESULTS AND CONCLUSIONS

In the Fig.4 the directions and values of the principal stresses on the bone surface between the inclusions of the clamp ends are given to be compared with experimental results (Fig.5). In the Fig.6 longitudinal stress SIG Z - distance diagram is given with emphasized part corresponding to the diagram of experimental results (Fig.5). Both diagrams have the same character, i.e. the lowest compression value SIG Z is the middle part of investigated field and in the directions of clamp fittings the values of this stress increase. In the knee direction agreement of experimental and numerical results is more favourable than in the ankle direction. In the case of experimental results we have measured in the ankle direction first the gentle increased compression value and further to the ankle almost the same value as in the middle point. In this place it is necessary to take the shapes of cross sections very precisely and also the results depend on the part in spongy and cortical bone tissues.

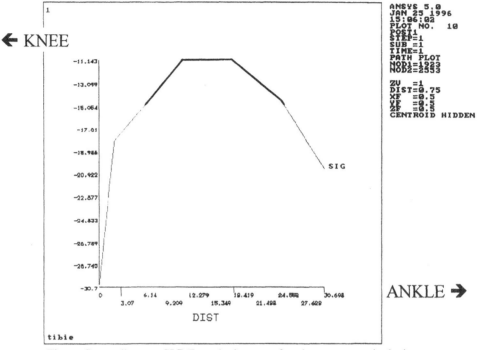

Fig.6. Stress diagram SIG Z on the bone surface between two inclusions

We can state good agreement of experimental (Fig.5) and numerical results (Fig.6). Therefore we can use the computation model for stress state analysis both in any cross section of the intact bone and in contact planes of the fracture. In the initial phase of solution we have analysed the middle cross section of intact bone. On the side of the bone cross section where no clamp was fitted we have found tension stress which could be eliminate by application another clamp. In the case of fractured bone we calculated in the contact planes only compression stresses (Fig.7). On the basis of these results the location and number of osteosynthetic clamps can be planned for the given case of fracture healing.

Acknowledgements: The authors wish to acknowledge the support of this work by project No. 103/97/0729 of the Grant Agency of the Czech Republic.

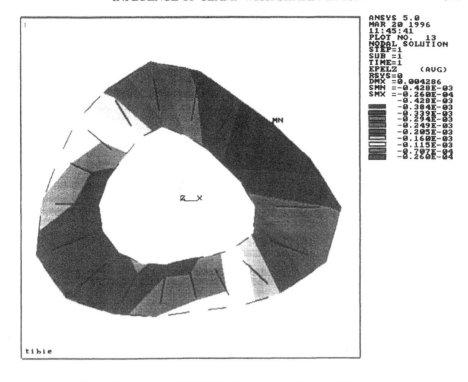

Fig.7. Strain state EPEL Z in the contact planes of fracture

REFERENCES

1. Kafka V., Jírová J., Smetana V., Materials with shape memory: theory, application and clinical experience, Inženýrská mechanika, 1995, No.2, Vol.2, pp. 99-106 (in Czech)

2. Huiskes, R., Dalstra, M., Venne, R.v.d., Grootenboer, H., Sloof, T.J., A hypothesis concerning the effect of implant rigidity on adaptive cortical bone remodelling in the femur. Proc.5th Meeting ESB, 1986, Germany

3. Huiskes, R., Bone remodelling around implants can be explained as an effect of mechanical adaptation, Total Hip Revision Surgery, ed.Jorge O. Galante, Aaron G. Rosenberg and Johl J. Callaghan, Raven Press, Ltd., New York, 1995, pp.159-171

4. Jirová J., Kafka V., Štěpánková H., Stress field on tibia surface fitted with shape memory osteosynthetic clamp, Proc.5th international Conference Biomechanics of Man '94, 1994, pp. 52-55, Benešov, Czech Republic

5. Jírová J., Kafka V.: Stress-state at bone surface caused by a clamp with shape memory, Österreichische Ingenieur- und Architekten-Zeitschrift, 1995, Vol. 140, pp.303-304

6. Květ, I., Development of osteosynthetic clamp with shape memory and analysis of its properties, Degree theses, 1991 (in Czech)

FINITE ELEMENT METHOD FORMULATION FOR THE INTERACTIONS BETWEEN VARIOUS ELASTIC-VISCOELASTIC STRUCTURES IN BIOMECHANICAL MODEL

S. Piszczatowski[1], K. Skalski[2], G. Slugocki[3], A. Wakulicz[4],

1.ABSTRACT

The mechanical approach to the structural bone-implant analysis by Finite Element Method (FEM) of the elastic - viscoelastic structures will be given in this paper. On the basis of the physical model of the bone-implant system taking into account non-homogeneous and viscoelastic bone material properties, quasi-static boundary value problem (QSBVP) describing the interactions between elastic and viscoelastic material coupled with the proper conditions on the area of contact will be formulated. The variational formulation, necessary to solve the problem will be also presented as an example of the applied approach to QSBVP. Some numerical results of strain, stress and energy density distributions will be shown in different points of the bone-implant system.

2.INTRODUCTION

Nonhomogeneous structures have as components not only typical construction materials but also viscoelastic properties (e.g. plastics, fibre glasses, etc.) and these properties are of a growing interest in mechanics. For the purposes of strength analysis one replaces, in some cases, the composite materials by an homogeneous material whose properties and parameters are properly designed.

Keywords: Viscoelasticity, Finite Element Method, Variational formulation,

---

[1] Assistant, Mechanical Department, Bialystok University of Technology, Wiejska 45A, 15-351 Bialystok, POLAND

[2] Professor, Department of Manufacturing Technology, Warsaw University of Technology, Warsaw, POLAND

[3] Researcher, Institute of Aeronautics and Applied Mechanics, Warsaw University of Technology, Warsaw, POLAND

[4] Researcher, Polish Academy of Sciences, Institute of Mathematics, Warsaw, POLAND

However, such a method cannot be applied when the effects resulting from interactions between the elastic and viscoelastic media must be considered. Such situations occur, among others, in biomechanical problems like the strength analysis of a bone-implant system (e.g. the hip joint implant). In such a system one has in general 2 types of materials: elastic (endoprosthesis) and viscoelastic (bone) [1], [2]. The one of most troublesome problems is the assertion of the optimal interaction conditions in the bone-implant system. Despite continuing progress in implant design, the results obtained are not satisfactory, and still a considerable percentage of hip joint alloplastics lead to premature failure. Damages occurring in the implanted joint are effects of the disadvantageous time-dependant changes in its structure. Therefore taking into account the viscoelastic properties of bone tissue in strength analysis allows us to observe the changes occurring in the bone-implant system and can lead to improved knowledge of the causes of damage.

3. BIOMECHANICAL MODEL OF THE BONE-IMPLANT SYSTEM

3.1. Notation

In such a model one uses the viscoelastic modelling of the bone material properties with its anisotropy. The implant is treated as a pure elastic and isotropic material. The modelled bone implant system is presented in Fig.1. Let us denote:

1. V_I, V_{II} -domains of bone I and implant II respectively; I -viscoelastic, II -elastic,
2. S_I, S_{II} -boundaries of the domains V_I and V_{II},
3. S_C -curve (surface) dividing the domains V_I and V_{II},
4. S_{DI}, S_{FI}, S_{DII}, S_{FII} -boundaries subsets of the domains V_I and V_{II} on which the boundary displacement (D) and stress (F) conditions are defined.

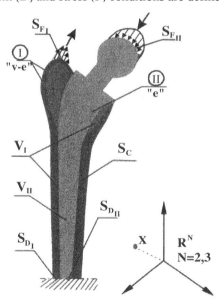

Fig. 1. Biomechanical model of the bone-implant system

3.2. Formulation of the problem

The problem consists in determining following functions: displacements $u^I(t,x)$ and $u^{II}(t,x)$, strains $e^I(t,x)$ and $e^{II}(t,x)$, stresses $s^I(t,x)$ and $s^{II}(t,x)$ satisfying the given boundary conditions on $S_{D_I}, S_{D_{II}}$ and $S_{F_I}, S_{F_{II}}$ and belonging to the following spaces:

$$u \in \mathcal{U} \ (R \times V) := \left\{ v: (v_i) \in W_2^1 \left[V, W_1^1(R_+) \right]^N \right\}, \tag{1}$$

$$\varepsilon \in \mathcal{E} \ (R \times V) := \left\{ e: (e_{ij}) \in L^2 \left[V, W_1^1(R_+) \right]^{N \times N} \right\}, \tag{2}$$

$$\sigma \in \mathcal{S} \ (R \times V) := \left\{ s: (s_{ij}) \in W_2^1 \left[V, L^1(R_+) \right]^{N \times N} \right\}, \quad i, j = 1 \dots N, \quad N = 2 \text{ or } 3, \tag{3}$$

where: L^p -the Banach space of functions integrable in Lebesgue sense with p-power, W_m^p -the Sobolev space (m=1, p=1,2).

3.3. Quasi-static boundary value problem (QSBVP)

The constitutive equations of viscoelastic material has the following form [3]:

$$\sigma_{ij}(t,x) = \int_{-\infty}^{t} G_{ijkl}(t - \tau, x) \frac{\partial}{\partial \tau} \varepsilon_{kl}(\tau, x) d\tau, \quad x \in V, \tag{4}$$

or taking into account the Stieltjes convolution definition:

$$\sigma_{ij}(t,x) = \left(G_{ijkl} * d\varepsilon_{kl} \right)(t,x), \quad \text{where: } i, j = 1, \dots, N \tag{4a}$$

and the generalized tensor of material both elastic and viscoelastic properties is defined below:

$$\widetilde{G}_{ijkl}(x,t) = \begin{cases} 0, \ t \leq 0 \\ D_{ijkl}(x), \ t > 0, \ x \in V_{II} \\ G_{ijkl}(t,x), \ t > 0, \ x \in V_I \end{cases} \tag{4b}$$

The equilibrium equations for the quasi-static state can be expressed by:

$$\sigma_{ij,j}(t,x) + f_i(t,x) = 0, \quad \text{where: } f_i = \begin{cases} f_i^I, \ x \in V_I \\ f_i^{II}, \ x \in V_{II} \end{cases} \tag{5}$$

Compatibility equations for the small deformations model are as follows:

$$\varepsilon_{ij}(t,x) = 0.5 \left[u_{i,j}(t,x) + u_{j,i}(t,x) \right], \quad x \in V = V_I \cup V_{II} \tag{6}$$

and boundary conditions with the given displacement function φ_i (displacement boundary conditions):

$$u_i(t,x) = \varphi_i(t,x), \quad \text{where: } \varphi_i = \begin{cases} \varphi_i^I, \ x \in S_{D_I} \\ \varphi_i^{II}, \ x \in S_{D_{II}} \end{cases} \tag{7}$$

and with the given load distribution function ψ_i (stress boundary conditions):

$$\sigma_{ij} n_j(t,x) = \psi_i(t,x), \quad \text{where: } \psi_i = \begin{cases} \psi_i^I, \ x \in S_{F_I} \\ \psi_i^{II}, \ x \in S_{F_{II}} \end{cases} \tag{8}$$

n_j -being j-th component of the external normal unit vector

Finally, reaction conditions occurring between the domains V_I and V_{II} on the surface S_C have the form:

$$u_i^I(t,x) = u_i^{II}(t,x), \text{ for } x \in S_C \tag{9}$$

$$\sigma_{ij}^I n_j^I(t,x) = \sigma_{ij}^{II} n_j^{II}(t,x), \text{ for } x \in S_C \tag{10}$$

3.4. Variational formulation for QSBVP

The defined quasi-static boundary value problems can be effectively solved [4] by using the variational formulations and next the FEM approximation. The formulated variational functional is treated jointly for viscoelastic and elastic bodies. For the sake of simplicity let us introduce the following additional notation:

$$S_D = S_{D_I} \cup S_{D_{II}}, \quad S_F = S_{F_I} \cup S_{F_{II}}, \quad V = V_I \cup V_{II}$$

$$\mathbf{V} = V \times R_+, \quad \mathbf{S}_D = S_D \times R_+, \quad \mathbf{S}_F = S_F \times R_+, \quad \mathbf{S} = S \times R_+,$$

The Washizu-Reissner hybrid, potential energy functional J_V with the Lagrange's multipliers has the form [5], [6]:

$$J_V(u, \varepsilon, \sigma) = \int_{\mathbf{V}} \left[0.5\widetilde{G}_{ijkl} * d\varepsilon_{ij} * d\varepsilon_{kl} - \sigma_{ij} * d\varepsilon_{ij} - (\sigma_{ij,j} + f_i) * du_i \right] d\mathbf{V} +$$
$$+ \int_{\mathbf{S}_D} (\sigma_{ij} n_j * d\varphi_i) d\mathbf{S} + \int_{\mathbf{S}_F} (\sigma_{ij} n_j - \psi_i) * du_i d\mathbf{S} \tag{11}$$

and due to it, one gets the solution of the QSBVP since the following THEOREM takes place:

THEOREM:

The QSBVP equations are satisfied while the u,e,s searched functions are a critical (stationary) point (u, ε, σ) of the J_V functional, i.e.

$$\left\langle DJ_V(u,\varepsilon,\sigma), [v,e,s] \right\rangle = 0; \quad \forall [v,e,s] \in \mathcal{U} \times \mathcal{E} \times \mathcal{S} \tag{12}$$

or

$$\frac{d}{d\tau} \left[J_V(u + \tau v, \varepsilon + \tau e, \sigma + \tau s) \right]_{\tau=0+} = 0 \tag{12a}$$

Introducing the following bilinear form:

$$a([u,\varepsilon,\sigma],[v,e,s]) = \int_{\mathbf{V}} \left[0.5\widetilde{G}_{ijkl} * d\varepsilon_{kl} - \sigma_{ij} \right] * de_{ij} d\mathbf{V} + \int_{\mathbf{V}} \sigma_{ij} dv_{i,j} d\mathbf{V} +$$
$$- \int_{\mathbf{V}} \left[d\varepsilon_{ij} - 0.5(du_{i,j} + du_{j,i}) \right] * ds_{ij} d\mathbf{V} - \int_{\mathbf{S}_D} u_i * d(s_{ij} n_j) d\mathbf{S} \tag{13}$$

and the following linear functional:

$$\left\langle f, [v,e,s] \right\rangle := \int_{\mathbf{V}} f_i * dv_i d\mathbf{V} - \int_{\mathbf{S}_D} \varphi_i * d(s_{ij} n_j) d\mathbf{S} + \int_{\mathbf{S}_F} \psi_i * dv_i d\mathbf{S}, \tag{14}$$

then (12) may be written as follows:

$$\left\langle DJ_V(u,\varepsilon,\sigma), [v,e,s] \right\rangle = a([u,\varepsilon,\sigma],[v,e,s]) - \left\langle f, [v,e,s] \right\rangle = 0 \tag{15}$$

4. APPROXIMATION OF VARIATONAL EQUATION BY USING FINITE ELEMENTS METHOD

The problem consists in approximating the joined space of displacement, stress and strain fields $\mathcal{U} \times \mathcal{E} \times \mathcal{S}$ and thereafter to build the functional using the appropriate shape functions for each mentioned field. We can make the following assumptions:

let $\mathcal{T}_h(V)$ be a finite element division of V,

let us consider W_h as an approximation of the space $W_2^1(V, W_1^1(R_+))$; displacement $(1 \times N)$, strain $(N \times N)$, stress $(N \times N)$ fields are approximated by:

$$\mathbf{W}_h \equiv \left\{ [W_h]^N \times [W_h]^{N \times N} \times [W_h]^{N \times N} \right\}. \tag{16}$$

Then the approximating space W_h may be defined as following:

$$W_h := \left\{ \Theta_h : \Theta_h(t, x) = \sum_{q=1}^{T_h} \sum_{p \in \Sigma_h} \Theta_h^{(q,p)} \alpha_q(t) \gamma_p(x) \; ; \; \Theta_h^{(q,p)} \in R \right\} \tag{17}$$

where the basis functions related to time are:

$$\alpha_q(t) := \begin{cases} 0, \; t \notin (t_{q-1}, t_{q+1}) \\ \dfrac{t - t_{q-1}}{t_q - t_{q-1}}, t \in (t_{q-1}, t_q) \\ \dfrac{t_{q+1} - t}{t_{q+1} - t_q}, t \in (t_q, t_{q+1}) \end{cases} \tag{18}$$

and the basis space shape functions are:

$$\gamma_p(x) \in \Gamma_h := \{ w_h \in W_2^1(V) : w_h \big|_E \in P(E) \} \; p \in \Sigma_h, \tag{18a}$$

where: $\Sigma_h := \bigcup_{E \in T_h(V)} \Sigma_E$, $\{E, P(E), \Sigma_E\}$ Ciarlet notion of finite element.

$P(E)$ is a space of the approximating polynomials on $E, E \in T_h(V)$ being simplexes in R^N.

Determining of the basis of the W_h space is the next step in the given approach.

Since the basis of W_h is given by: $\{\gamma_p \alpha_q\}_{(p,q) \in Q_h}$, $Q_h := \sum_h x \{1, .., T_h\}$.

Therefore, basis of $[W_h]^N$ and $[W_h]^{N \times N}$ are defined by

$$\{\gamma_p \alpha_q e_i\}_{(p,q) \in Q_h}, \; i \in \{1, ..., N\} \quad , \quad \{\gamma_p \alpha_q e_i e_j^T\}_{(p,q) \in Q_h}, \; i,j \in \{1, ..., N\} \tag{19}$$

respectively, where : $e_i = (\delta_{i1}, ..., \delta_{iN})^T$, δ_{ik}-Kronecker symbol.

Finite dimensional approximation of the variational equation (15) has the form:

$$[u_h, \varepsilon_h, \sigma_h) \in \mathbf{W}_h: \quad a([u_h, \varepsilon_h, \sigma_h], [v_h, e_h, s_h]) = <f, [v_h, e_h, s_h]>, \tag{20}$$

$$\text{for all} \quad [v_h, e_h, s_h] \in \mathbf{W}_h.$$

The variational equqtion is (20) is satisfied by the functions :

$$u_h = \sum_{i=1}^{N} \sum_{(p,q) \in Q_h} \underline{u}_{ipq} \gamma_p \alpha_q e_i$$

$$\varepsilon_h = \sum_{i,j=1}^{N} \sum_{(p,q) \in Q_h} \underline{\varepsilon}_{ijpq} \gamma_p \alpha_q e_i e_j^T \tag{21}$$

$$\sigma_h = \sum_{i,j=1}^{N} \sum_{(p,q) \in Q_h} \underline{\sigma}_{ijpq} \gamma_p \alpha_q e_i e_j^T$$

if only coefficients $\underline{u} = \{\underline{u}_{ipq}\}_{(p,q) \in Q_h}$, $i=\{1,...,N\}$, $\underline{\varepsilon} = \{\underline{\varepsilon}_{ijpq}\}_{(p,q) \in Q_h}$, $i,j=\{1,...,N\}$,

$\underline{\sigma} = \{\underline{\sigma}_{ijpq}\}_{(p,q) \in Q_h}$, $i,j=\{1,...,N\}$ are solutions of the following system of linear equation

$$A^{(1,1)}\underline{u} + A^{(1,2)}\underline{\varepsilon} + A^{(1,3)}\underline{\sigma} = f^{(1)}$$

$$A^{(2,1)}\underline{u} + A^{(2,2)}\underline{\varepsilon} + A^{(2,3)}\underline{\sigma} = f^{(2)} \tag{22}$$

$$A^{(3,1)}\underline{u} + A^{(3,2)}\underline{\varepsilon} + A^{(3,3)}\underline{\sigma} = f^{(3)}$$

Block matrices $A^{(i,j)}$ are defined as follows:

$$A^{(i,j)} = \left(\left[a^{(i,j)} \left(\left[g, ge_1^T, ge_1^T \right], \left[\bar{g}, \bar{g}\bar{e}_{\bar{l}}^T, \bar{g}\bar{e}_{\bar{l}}^T \right] \right) \right]_{g=\gamma_p \alpha_q e_k} , \bar{g}=\gamma_{\bar{p}} \alpha_{\bar{q}} e_{\bar{k}} \right) \tag{23}$$

$$(p,q,k,l), \quad (\bar{p},\bar{q},\bar{k},\bar{l}) \in Q_h x\{1,...,N\}^2, \text{ where:}$$

$$a^{(i,j)}\left([u,\varepsilon,\sigma], [\bar{u},\bar{\varepsilon},\bar{\sigma}] \right) = a\left([\delta_{i1}u, \delta_{i2}\varepsilon, \delta_{i3}\sigma], [\delta_{j1}\bar{u}, \delta_{j2}\bar{\varepsilon}, \delta_{j3}\bar{\sigma}] \right)$$

Right hand sides one can obtain from (14) in the form:

$$f^{(1)} = \left\{ \int_V f_i * d(\gamma_p \alpha_q) dv + \int_{S_F} \psi_i * d(\gamma_p \alpha_q) dS \right\}_{(p,q) \in Q_h, \quad i \in \{1,...,N\}}$$

$$f^{(2)} = 0 \tag{24}$$

$$f^{(3)} = \left\{ -\int_{S_D} \varphi_i * d(\gamma_p \alpha_q) n_j \right\}_{(p,q) \in Q_h, \quad i,j \in \{1,...,N\}}$$

5. RESULTS

The 3D implant-bone system model has been submitted to numerical analysis. The viscoelastic bone material properties and the bone tissue nonhomogeneities (cortical and stem bones) were taken into account [1], [2]. A constant load was applied to the system simulating the patient standing on one foot (Fig.2). The rigid joint between implant and bone was modelled as well. The Fig. 2 shows the stresses, strains and energy density values changes versus time in the chosen points of the bone. These results are expressed as quotients (in percentages) of the stress, strain, energy density at the time $t_1 = 100s$ and $t_2 = 10000s$ to their initial values at the time $t_0 = 0$ (adequate to the elastic behavior). A 30% strain increase rate is observed at some points of the system.

These rate changes are not identical at the total volume of bone and does not reach the strain characteristics for material creep under the applied constant loading. This fact results from the decreasing stresses in the bone tissue as the effect of the stress field redistribution occurring in the elastic-viscoelastic structure under load.

6. CONCLUSION

1) Application of the hybrid displacement-strain-stress variational formulation to the finite element analysis improves the accuracy of the numerical results.
2) The proposed constitutive equations (4) have the form which allows the QSBVP simplification by removing the reaction boundary constraints (9),(10).
3) The viscoelastic properties of the bone tissue cause not only the strain-creep but also stress field redistribution.
4) The rheological weakening of the bone mechanical properties causes the increase of stresses in implant. This observation may give some explanations of the stress fielding effect

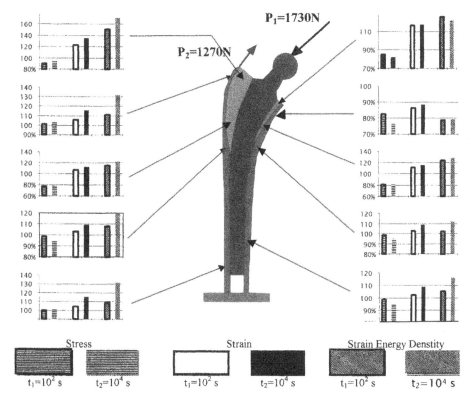

Fig. 2. Changes of effective strains, stresses and energy density (per %) for time 100 s and 10000 s related to their values at the initial time (elastic state)

Acknowledgements:

Grant of KBN (State Committee of Scientific Research) No.7T07A 008 10 is gratefully acknowledged.

REFERENCES:

1. Lakes,R.S., Katz,J.L., Sternstein,S.S. Viscoelastic properties of wet cortical bone - I - torsional and biaxial studies. J.Biomech., 1979, Vol. 12, 657-678.
2. Deligianni,D.D., Maris, A., Missirlis, Y. F. Stress relaxation behaviour of trabecular bone specimens, J. Biomech., 1994, Vol. 27, 1469-1476.
3. Christensen, R. M., Theory of viscoelasticity, An introduction, Academic Press, NY, 1971.
4. Fabrizio, M., An existence and uniqueness in quasi-static viscoelasticity, Quart. Appl. Math., 1987, Vol. 47, No. 1.
5. Washizu, K., Variational Methods in Elasticity and Plasticity, Pergamon Press, New York , 1968.
6. Giorgi,C., Marzocchi,A. New variational principles in quasi-static viscoelasticity, J. Elasticity, 1992, Vol. 29, .86-96.

4. SPINE AND VERTEBRA MECHANICS

THE INFLUENCE OF AGEING ON MECHANICAL BEHAVIOUR OF INTERVERTEBRAL SEGMENT

E.A. Meroi[1], A.N. Natali[1], H. Trebacz[1]

1. ABSTRACT

The loss of load carrying capacity of intervertebral segment with age depends both on disc degeneration process and on mechanics of cancellous and cortical bone of vertebral body. The aim of the present work is to compare the intervertebral segment response under axial loading in the healthy and degenerated conditions of hard and soft tissue in time. The problem is approached by using a hyperelastic formulation for the constitutive material of the intervertebral disc. Collagen fibres which reinforce the annulus fibrosus are represented by truss elements working only in traction. Disc degeneration, related to the loss in liquid content in disc tissue, is simulated by varying disc compressibility. A transversely isotropic law is assumed for cortical and cancellous bone in the vertebral body. Ageing and the correlated reduction in bone amount has a double effect on cortical shell, leading both to a decrease in thickness and material characteristics. Mechanical properties change with age even more in cancellous than in cortical bone and the load distribution pattern is consequently affected, especially if ageing in vertebral body is coupled with disc degeneration effects. A three-dimensional finite element model of the intervertebral segment is developed adopting a non-linear formulation. The normal and reduced load-carrying capacities of the intervertebral segment are analysed.

2. INTRODUCTION

One of the most remarkable clinical features of spinal osteoporosis is the changes in shape and size of vertebrae in elderly patients. Vertebral deformities are often observed even if the patients have never experienced a considerable trauma and do not have any symptoms except for the back pain in the acute stage. A deeper insight into the influence of age and disease related to bone loss on the strength of vertebra is allowed by the comparison of biomechanical investigations of intervertebral segment in elderly and young people.

The vertebral body consists of a cancellous bone core surrounded by bony shell as cortical ring and bony end-plates. The cancellous bone inside vertebra consists of a

Keywords: biomechanics, intervertebral segment, hyperelasticity.

[1] Center of Biomechanics, Università di Padova, Italy

trabecular network, assumed as isotropic in the horizontal plane. The trabecular network is inhomogeneous, being more dense and regular towards the end plates and more porous and irregular in the central part of the body [1]. The cortical shell of vertebral body does not resemble cortical bone histologically. Bony end-plate is thin and porous and could be characterised as condensed trabecular bone rather than true cortical bone [1-3]. After the peak bone mass is attained at the age of about 25 years, the human skeleton loses bone mass with increasing age. Ageing leads to a decrease in bone amount and a change in its structure. In vertebra, the degree of bone anisotropy increases markedly with age because of disproportionate loss of trabeculae in vertical and horizontal directions [4, 5]. Shell thickness decreases slightly with age, the rim, approximately 400 - 500 μm thick in young person, in an elderly individual has a thickness of only 200 - 300 μm [1-3]. There is a general age-related reduction in the material strength and stiffness of both cortical and trabecular bone. In addition, bone becomes increasingly brittle and fractures under lower strength [6]. The maximum compressive load-bearing capacity of the lumbar vertebral body decreases with age, being in elderly individual 0.8-1.50 kN [5, 7]. The results of experimental studies suggest that the cortical thickness is a main determinant feature for the in vitro compressive strength of the whole vertebral body [3] and that relative importance of the bony shell around vertebra increases with age [2, 8]. In elderly individuals, the shell, despite of its smaller thickness, contributes about 70% of the load bearing capacity of the whole vertebral body, since properties in trabecular bone decline much faster than in cortical bone [8]. The ageing pattern is different for men and for women but the sex differences with respect to age are small compared with the pronounced age-related changes [5]. The results of experiments suggest that an interdependency of trabecular bone properties and intervertebral disc properties may exist [9-12]. Disc degeneration influences the stress state in vertebra and the mechanism of the vertebral fracture [9, 12]. This behaviour may also contribute to vertebral fragility by microdamage accumulation, especially in osteopenic vertebrae with low bone mass [10].

The objective for the present analysis is to compare the load distribution within the normal, healthy vertebra of a young person with the osteopenic vertebra of an elderly person. Specifically, we investigate the distribution of load between trabecular and cortical bone within the vertebra and the differences in load distribution in young and old individuals and the role of disk degeneration in load distribution. A three-dimensional finite element model of the intervertebral segment is analysed under an axially imposed displacement in a geometric and material non-linear formulation. A further non linearity arises since a hybrid formulation is adopted for treating the incompressible or almost incompressible response of disc material: the deviatoric and volumetric components of strain energy are distinguished within the formulation, by adopting a Lagrange multiplier technique in the definition of internal work over the single element. The normal and reduced load-carrying capacity of the intervertebral segment is analysed.

3. COMPUTATIONAL ANALYSIS

3.1. Numerical model

The biomechanical model consists of an intervertebral segment composed of two vertebral bodies L4-L5 with an interposed intervertebral disk. Because of symmetry considerations, the finite element model is assumed as in Figure 1 [15, 16].

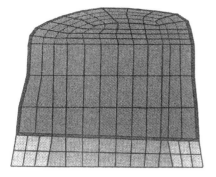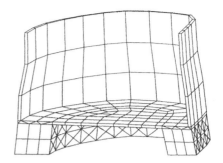

Figure 1: Finite element model

A transversely isotropic linear elastic constitutive law is assumed for cortical and cancellous bone in the vertebral body. Isotropy is assumed for bony end-plate. Ageing and the correlated decrease in bone amount have a double effect on cortical shell, leading both to a variation in shape, in terms of a reduction in thickness, and in structure, with also a variation of anisotropy ratio. Mechanical properties change with age even more in cancellous than in cortical bone and the load distribution patterns itself is consequently affected. The material constants assumed for lumbar vertebrae of young and old persons are given in Table 1.

		trabecular bone		cortical bone	
		young	old	young	old
E_{33}	(Mpa)	300	100	12000	10000
$E_{11}=E_{22}$	(Mpa)	150	33	8000	6700
G_{12}	(Mpa)	58	13	2900	2400
G_{13}	(Mpa)	79	44	3600	2900
v_{12}		0.3	0.3	0.4	0.4
v_{13}		0.2	0.2	0.3	0.3
v_{23}		0.2	0.2	0.3	0.3

Table 1

The values of material constants for bony shell around the vertebral bodies are deduced from the lowest values given for cortical bone reported from experimental studies [6, 17-19]. The changes of bony ring material during ageing are chosen from the data for ageing cortical bone [6, 17, 20-21]. Taking into account that bony endplate should be characterised as condensed trabecular bone [1-3], intermediate properties between trabecular and cortical bone are assumed. Since bony shell thickness influences the strength of the vertebra [3, 22] and thinning of the shell during ageing is experimentally confirmed [2, 3], a different cortical ring thickness is assumed for young and old vertebra.

The material constants for trabecular bone in lumbar vertebra and the changes of bone properties during ageing are deduced from [3, 4, 5, 8, 11, 13, 23, 24]. We did not consider changes in overall shape and size of the vertebral body during ageing.

A hyperelastic formulation is adopted for the disc ground material and for the nucleus, that undergoes large deformations behaving like a rubber material, showing incompressible or almost incompressible behaviour. The collagen fibers are modelled as elastic truss elements not resistant to compression. The use of a hyperelastic approach avoids preliminary assumptions and allows for a quite general approach to the investigation of the decreasing mechanical resistance of disc ground material. Disc degeneration, related to the loss in liquid content, is simulated by varying the compressibility in annulus fibrosus and nucleus pulposus.

3.2. Hyperelastic formulation

For a hyperelastic material, a Mooney-Rivlin constitutive law can be defined as a polynomial function of the three strain invariants, that is given in terms of right Cauchy-Green strain tensor $\mathbf{C} = \mathbf{F}^T \mathbf{F}$:

$$E = \sum_{i,j,k=0}^{N} c_{ijk} \left(I_1 - 3\right)^i \left(I_2 - 3\right)^j \left(I_3 - 1\right)^k \tag{1}$$

with

$$I_1 = \operatorname{tr} \mathbf{C}, \qquad I_2 = \left(\frac{1}{2} I_1^2 - \operatorname{tr} \mathbf{C}^2\right), \qquad I_3 = \det \mathbf{C} \tag{2}$$

When facing incompressible behaviours ($I_3=1$), since incompressibility is a kinematic constraint which gives rise to a reaction stress which does not work in any motion compatible with the constraint, the hydrostatic pressure, $-p\mathbf{I}$, is not related to constitutive equations, but only evaluable from equilibrium and boundary conditions. To take the derivatives of E with respect to all kinematic variables, an augmented formulation is used introducing the Lagrangian multiplier p to impose the constraint $I_3=1$:

$$E = E\left(I_1, I_2\right) - \frac{1}{2} p\left(I_3 - 1\right) \tag{3}$$

A mixed approach is to be preferred also in the case of an almost incompressible material because its effective bulk modulus can be so large compared with its shear modulus that the stiffness matrix can nearly show singularities.

The volume change $\bar{J}$ is then constrained in its variation, by means of the Lagrangian multiplier p, to be equal, over each element, to the Jacobian determinant J:

$$E = E\left(\hat{I}_1, \hat{I}_2, \bar{J}\right) - p\left(J - \bar{J}\right) \tag{4}$$

where $\hat{I}_i$ are the strain invariants evaluated from the deformation gradient with the volume change eliminated:

$$\hat{\mathbf{F}} = J^{-\frac{1}{3}} \mathbf{F} \tag{5}$$

The general form then becomes:

$$E = \sum_{i+j=1}^{N} c_{ij} \left(\hat{I}_1 - 3\right)^i \left(\hat{I}_2 - 3\right)^j + \sum_{k=1}^{N} \frac{1}{d_k} \left(\bar{J} - 1\right)^{2k} \tag{6}$$

The influence over results of c_{ij} and of compressibility coefficients d_k is presented, in order to simulate different disc configurations in dependence on age and disc degeneration.

4. RESULTS

The analysis pertains to normal and aged intervertebral segment. Osteopenia is physiological result of ageing and pathologically increased osteopenia can result in osteoporosis. Some general indications can be given as result of the analysis. The average value of load supported by cortical ring is approximately 60% and a slight increase of load transmitted by cortical shell is observed, about 10% more in thinner shell for old person. This is in agreement with experimental testing: cortical thickness has a primary role for the in vitro compressive strength of the whole vertebral body [3]. Even if shell thickness decreases slightly with age, considerable evidence indicates that decrease in trabecular bone mass and strength are the primary causes of age-related reductions in vertebral strength, and that the relative contribution of the shell to vertebral strength actually increases with age [2]. Mosekilde also observed that the relative load bearing effect of the cortical ring increases with age as the trabecular bone strength decreases [8]. Compact bone fails in compression with strains in the range 1.4% to 2.1%, but begins to yield at strains between 0.6% and 0.8%. Given that yielding involves rapid accumulation of microdamage within the bone, it seems adequate to base skeletal safety factors on the yield strain, rather than the ultimate failure strain of bone tissue [8] and 0.006 could be assumed as a limit value for strain in old bone. The maximum value for ultimate strain in vertebral core is about 2% [13]. In our model, at the same equivalent vertical load level, correspondent to different vertical imposed displacement, maximum compressive strain varies from 0.26% in young to 0.5% in old cortical bone and from 0.93% in young to 2% in old trabecular bone. It seems that the critic values are reached almost at the same time for cortical and trabecular bone, but when trabecular bone reaches the limit value, the cortical part is still under its limit strain. This is in agreement with the fact that in symptomatic osteoporosis the fractures often occur in the central trabecular bone without affecting the surrounding cortical ring. Different conditions of disc degeneration are considered, in terms of compressibility and hyperelastic coefficients for disc ground material. The importance of a proper description of disc behaviour is evidenced by its influence on stress distribution patterns and on deformed configuration, and this is in accordance with the observed relation between the degree of disk degeneration and fracture type [9]. If the disc response in time is under investigation, a poroelastic model can be adopted [26, 27].

In our model there are three sets of properties affecting the results. The first set regards vertebra properties: the influence of three different levels of bone characteristics are reported in these results. The first level refers to the vertebral body of a young man with a shell thickness of 0.5 mm; the third to the vertebral body of an old man with a shell thickness of 0.3 mm, the second is an intermediate condition with the same bone characteristics of the old man, but with the shell thickness of the young one. The second set regards disc ground properties: coefficients c_{ij} are chosen to represent young, intermediate and old conditions. The third set regards disc compressibility: compressibility coefficients are assumed as 0.0, 0.05 and 0.1 to simulate three different levels of disc dehydration.

A constant displacement is imposed at the top level of the vertebral body, reaching a correspondent maximum compressive load of 3 kN, chosen in accordance with data

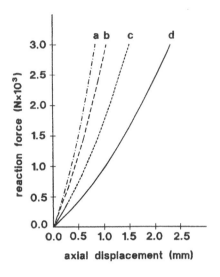

Figure 2: Axial displacement versus equivalent applied load.

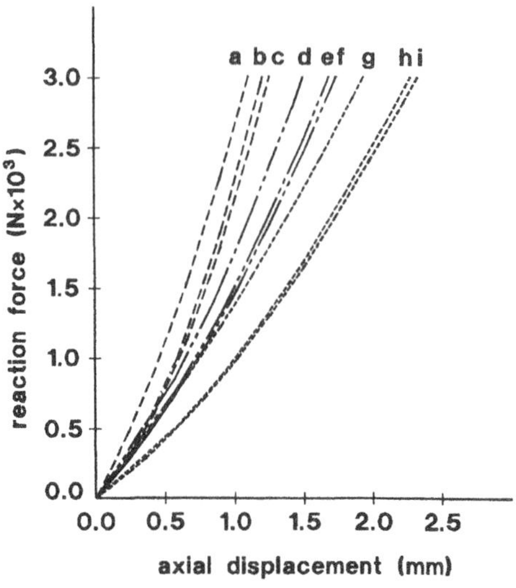

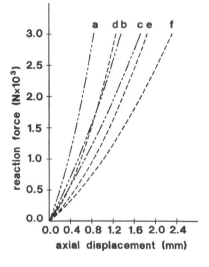

Figure 3: Influence of vertebral body conditions on overall vertical stiffness.

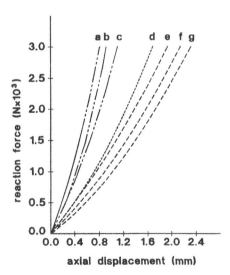

Figure 4: Influence of disc-ground material properties on the overall vertical stiffness.

Figure 5: Influence of disc compressibility on the overall vertical stiffness.

reported in literature [28, 29].

In Figure 2 axial displacement versus equivalent applied load is given. The extreme curves correspond to the best (a) and worst (d) conditions, to show the range of the overall vertical stiffness of the whole intervertebral joint. The others curves are for the cases in which intermediate properties are assumed for vertebral body with a young disc (b) and intermediate ground properties and compressibility are adopted for disc, with a young vertebral body (c). In Figure 3, it can be seen how vertebral body conditions affect the overall vertical stiffness of the model with a young (a, b, c), intermediate (d, e, f), and old disc (g, h, i). Figure 4 evidences the influence of disc-ground material properties on the overall vertical stiffness of the whole model. Curves a, b, c are the three different responses for a configuration of incompressible disc and young vertebra, curves e, f and g are for a compressibility equal to 0.1 for disc with old vertebra, and curve d is obtained giving intermediate characteristics to all three sets of properties. Figure 5 evidences the relevant influence of disc compressibility for the intervertebral joint of a young (a, b, c) and old (d, e, f) person.

6. CONCLUSION

The objective of the present analysis is to compare the load distribution under axial loading in the healthy intervertebral segment of a young person with the correspondent one in a degenerated phase depending on age. The overall response is related to the variations in the mechanical properties of the constitutive materials. The present approach seems particularly effective in representing intervertebral joint response with age, by using a flexible formulation that allows for an evaluation of the influence of material parameters which affect the results.

7. REFERENCES

1. Mosekilde Li. Vertebral structure and strength in vivo and in vitro. Calcif Tissue Int 1993, 53 S 1: 121-126.
2. Silva MJ, Wang C, Keaveny TM, Hayes WC. Direct and computed tomography thickness measurements of the human, lumbar vertebral shell and endplate. Bone 1994, 15: 409-414
3. Vesterby A, Mosekilde Li, Gundersen HJG, Melsen F, Mosekilde Le, Holme K, Sorensen S. Biologically meaningful determinants of the in vitro strength of lumbar vertebrae. Bone 1991, 12: 219-224
4. Galante J, Rostoker W, Ray RD. Physical properties of trabecular bone. Calcif Tiss Res 1970, 5: 236-246
5. Mosekilde Li, Mosekilde Le. Sex differences in age-related changes in vertebral size, density and biomechanical competence in normal individuals. Bone 1990, 11: 67-73.
6. Martin RB. Aging and strength of bone as a structural material. Calcif Tissue Int 1993, 53 S 1: 34-39.
7. Granhed H., Jonson R., Hansson T. Mineral content and strength of lumbar vertebrae. Acta Orthop Scand 1989, 60: 105-109.
8. Mosekilde L, Mosekilde L. Normal vertebral body size and compressive strength: relations to age and to vertebral and iliac trabecular bone compressive strength. Bone 1986, 7: 207-212.

9. Hansson T, Keller T, Jonson R. Fatigue fracture morphology in human lumbar motion segments. J Spinal Disord 1988, 1: 33-38.

10. Hasegawa K, Turner CH, Chen J, Burr DB. Effect of disc lesion on microdamage accumulation in lumbar vertebrae under cyclic compression loading. Clin Orthop 1995, 311: 190-198.

11. Keller TS. Predicting the compressive mechanical behavior of bone. J Biomech 1994, 27: 1159-1168.

12. Shirado O, Kaneda K, Tadano S, Ishikawa H, McAfee PC, Warden KE. Influence of disc degeneration on mechanism of thoracolumbar burst fractures. Spine 1992, 17: 286-292.

13. Hasegawa K, Takahashi HE, Koga Y, Kawashima T, Hara T, Tanabe Y, Tanaka S. Mechanical properties of osteopenic vertebral bodies monitored by acoustic emission. Bone 1993, 14: 737-743

14. McNally DS, Adams MA, Goodship AE. Can intervertebral disc prolapse be predicted by disc mechanics? Spine 1993, 18: 1525-1530

15. Meroi EA, Natali AN. The mechanical behaviour of bony endplate and annulus in prolapsed disc configuration, J Biomed Eng 1993, 15: 235-239.

16. Natali A, Meroi E. Nonlinear analysis of intervertebral disk under dynamic load, ASME J Biomech Eng 1990, 112: 358-363.

17. Carter DR, Spengler. Mechanical properties and composition of cortical bone. Clin Orthop Rel Res. 1978, 135:192-217.

18. Lotz JC, Gerhart TN, Hayes WC. Mechanical properites of metaphyseal bone in the proximal femur. J Biomech 1991, 24: 317-329.

19. Natali AN, Meroi EA. A review of the biomechanical properties of bone as a material. J Biomed Eng 1989, 11: 266-276.

20. Courtney AC, Hayes WC, Gibson LJ. Age-related differences in post-yield damage in human cortical bone. Experiment and model. J Biomech 1996, 29: 1463-1471.

21. McCalden RW, McGeough JA, Barker MB, Court-Brown CM. Age-related changes in the tensile properties of cortical bone. The relative importance of changes in porosity, mineralization, and microstructure. J Bone Jt Surg-Am 1993, 75A: 1193-205.

22. Bryce R, Aspden RM, Wytch R. Stiffening effects of cortical bone on vertebral cancellous bone in situ. Spine 1995, 20: 999-1003.

23. Britton JM, Davie MW. Mechanical properties of bone from iliac crest and relationship to L5 vertebral bone. Bone 1990, 11: 21-28.

24. Fyhrie DP, Schaffler-MB. Failure mechanisms in human vertebral cancellous bone. Bone 1994, 15: 105-109.

25. Biewener AA. Safety factors in bone strength. Calcif Tissue Int 1993, 53: 68-74.

26. Meroi EA, Natali AN, Schrefler BA. A poroelastic approach to the analysis of spinal motion segment, Computer Methods in Biomechanics & Biomedical Engineering, Gordon and Breach Publishers, Amsterdam, 1996, 325-338.

27. Meroi EA, Natali AN. Schrefler BA. A partially saturated model of intervertebral segment undergoing large deformations, Joint Conf. Italian Group of Comput. Mech. and Ibero-Latin Am. Assoc of Comp. Meth. in Eng., Padova, Sep 1997.

28. Nachemson AL. Disc pressure measurements. Spine 1981, 6: 93-97.

29. Magnusson M, Granqvist M, Jonson R, Lindell V, Lundberg U, Wallin L, Hansson T. The loads on the lumbar spine during work at an assembly line. Spine 1990, 15: 774-9.

THREE-DIMENSIONAL FINITE ELEMENT MODELLING OF THE MECHANICAL BEHAVIOUR OF HUMAN VERTEBRAL CANCELLOUS BONE

H. Walter[1], F. Lbath[2], D. Mitton[1], E. Cendre[3] and C. Rumelhart[4]

1. ABSTRACT

An idealised, structural model of vertebral cancellous bone has been proposed in order to study its elastoplastic behaviour. A new automated mesh generator was developed to generate a three-dimensional finite element model based on architectural data. Finite element analysis was performed using the code ABAQUS®, assuming the trabecular tissue behaviour perfectly plastic, in order to simulate a physiological static load (compression test in the vertebral column axis). The stress-strain curve of the model was similar to those obtained experimentally on aged human vertebral cancellous bone specimens (a first linear elastic region followed by a second phase of collapse).

2. INTRODUCTION

Vertebral body is one of the trabecular bone sites where age-related fractures most frequently occur. So the knowledge of the mechanical behaviour of the cancellous bone may enhance our understanding of this problem. Cancellous bone, however, is a complex and heterogeneous material and its mechanical properties are influenced by its architecture. Moreover its architectural features vary widely with the anatomical site, the direction of loading and the age.

Different studies of the central part of the vertebral body have shown that the trabecular architecture typically resembles a lattice of thick vertical columns and thinner horizontal struts [1],[2]. Furthermore, in a description of the structure of cancellous bone, Gibson reported that in bones where the loading is largely uniaxial, such as vertebrae, the trabeculae develop a "columnar" structure with cylindrical symmetry [3]. So human vertebral cancellous bone is transversally isotropic, the longitudinal axis corresponding to

Keywords: vertebral cancellous bone, 3D finite element model, elastoplastic behaviour

[1] Ph.D. student, Laboratoire de Mécanique des Solides, INSA, 69621 Villeurbanne, FRANCE
[2] Ph.D., Laboratoire de Mécanique des Solides, INSA, 69621 Villeurbanne, FRANCE
[3] Ph.D. student, Laboratoire de Contrôle Non Destructif, INSA, 69621 Villeurbanne, FRANCE
[4] Professor, ARPTAL, Laboratoire de Mécanique des Solides, INSA, 69621 Villeurbanne, FRANCE

the anatomical inferior-superior axis [1], [4].

Jensen & al. [2] proposed a model of human vertebral cancellous bone in order to investigate the influence of age and architecture on mechanical behaviour. The architecture of this model is based on a network of horizontal and vertical cylinders, which can be generated with irregularities in order to be more realistic. This model accounts reasonably well for age-related changes in mechanical properties but only in the elastic range.

Another modelling strategy is to construct a large-scale three-dimensional finite element model based on the real structure of a sample of cancellous bone using three-dimensional serial reconstruction techniques or HRCT (High Resolution Computed Tomography) [5],[6]. The real architecture of cancellous bone is taken into account but important means are required and the results of the analysis are only valid for the specimen reconstructed. Moreover all models in the literature are analysed in the elastic range.

In this paper a three-dimensional idealised structural model of human vertebral cancellous bone is presented. With variable architectural parameters, this model takes into account the trabecular bone volume fraction and the architectural data of a real structure. Finite element simulations of compression tests in the longitudinal direction were carried out and the results in the elastic and plastic ranges were compared with experimental data obtained on aged human vertebral bone specimens. An analysis in the elastic range was carried out in order to study the elastic anisotropy (longitudinal/transversal apparent Young's modulus ratio) by simulating compression tests in longitudinal and transversal directions.

3. MATERIALS AND METHODS

3.1. Experimental tests

The samples studied were obtained from 20 subjects aged between 47 and 95 years (mean 79 years).

Machining

The anatomical vertebral axes were marked on specimens. Cylinders were taken from the left central part of the vertebral body with a 14 mm core drill. 9mm cubes (Figure 1b) were obtained with a low speed diamond blade saw (Isomet). Once the machining was done, samples were kept in a 50% saline-ethanol solution [7] at 4°C for 3-4 days before testing. Each specimen was placed two hours at room temperature prior to mechanical tests.

Testing

A universal screw-driven machine (Schenck RSA 250) fitted with a 5000N load cell (TME, F 501 TC) was used for the test. The displacement was measured with a new immersed extensometer which was placed directly at the end of the specimen, in order to avoid correction with machine stiffness. Tests were run at $0.5mm.min^{-1}$ to minimise viscous effects. The tests were performed in the physiological saline bath (0.9% NaCl) non buffered maintained at 37°C (±0.5°C) by a regulator (MINISTAT) equipped with a platinum probe (PT 100) [8]. In order to improve the compression test, 10 preloading cycles were applied to the samples to reach a steady state [9] and the Young's modulus was determined on the last cycle. Specimens were tested in uniaxial compression in each of the three-orthogonal directions (anterior-posterior (A-P), medial-lateral (M-L) and inferior-superior (I-S)). Then the maximal compressive strength (σ_{max}) and the strain at this strength ($\varepsilon_{\sigma max}$) were defined in a test to failure in the I-S direction.

3.2. The three-dimensional model

The three-dimensional geometry of the model is obtained by stacking unit open cells to form a connected network of vertical columns and horizontal struts. The length and the thickness of the columns and the struts can be different in the three directions and the thickness in the middle of trabeculae can be thinner than at a connection (Figure 1a). Histomorphometric and stereologic studies have quantified the architecture of cancellous bone in terms of specific parameters such as bone volume /tissue volume (BV/TV (%)), mean trabecular thickness (Tb.Th. (µm)) and mean trabecular separation (Tb.Sp. (µm)) [10]. These parameters, obtained by HRCT, are used to establish the dimensions (length and thickness) of the model. But it is clear that the individual trabeculae are not quite parallel to the vertical or horizontal axes. In order to create a more realistic model, the cells can be inclined in the vertical plane (2,3) and the position of some lattice joints can be randomly modified (Figure 2).

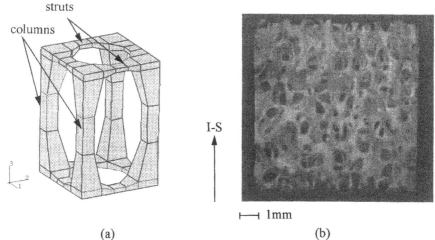

(a)   (b)

Figure 1 : Geometry of a unit open cell (a) - Sample of vertebral cancellous bone (b)

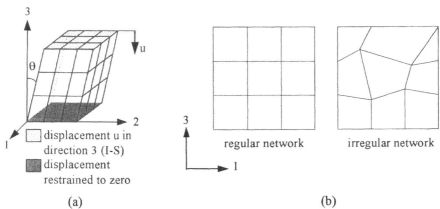

(a)   (b)

Figure 2 : Inclination of the model with boundary conditions for a compression test in direction 3 (I-S) (a) - Modification of the lattice joint positions (b)

A transversally isotropic model can be obtained by making the geometrical parameters (thickness and length) equal in both horizontal directions.

3.3. The finite element analysis

The finite element analysis was performed with the code ABAQUS® (HIBBIT KARLSSON & SORENSEN). Eight-node linear brick elements with identical material properties were used. The trabecular bone material was assumed isotropic, this does not play an important role in the overall mechanical properties [11] and its behaviour perfectly plastic with a Young's modulus, E, equal to 1.5GPa [12], a yield stress, σ_e, of 20MPa and a Poisson ratio of 0.3.

Uniaxial compression tests in the direction 1 or 3 (3 being the I-S direction) were simulated by restraining the nodes located in the bottom plane to zero whereas the nodes located in the top plane were displaced until a global strain of 0.6% was reached in the case of an elastic analysis or until a strain of 2.4% in the case of an elastoplastic analysis (Figure 2a). For an elastic analysis, the apparent Young's modulus of the specimen in the α-direction, $E^a_{\alpha\alpha}$ (MPa), was calculated from the formula:

$$E^a_{\alpha\alpha} = \frac{\sigma^a_{\alpha\alpha}}{\varepsilon^a_{\alpha\alpha}} \qquad (1)$$

where $\varepsilon^a_{\alpha\alpha}$ is the apparent strain in the α-direction (which is imposed to 0.6%) and $\sigma^a_{\alpha\alpha}$ (MPa) the apparent stress in the α-direction calculated from:

$$\sigma^a_{\alpha\alpha} = \frac{F_{\alpha\alpha}}{S} \qquad (2)$$

where $F_{\alpha\alpha}$ (N) is the total reaction force at the top face and S (mm^2) the compressed surface.

An efficient pre-processor was developed in order to automatically generate a finite element mesh (Figure 1a) based on architectural data such as trabecular length, trabecular thickness, inclination of the longitudinal trabeculae, degree of lattice disorder and the number of cells to stack.

For the models with a stack of 3x3x3 cells and more, the results are the same. So all the models were tested with 3x3x3 cells in order to reduce the CPU time. This model has 1188 elements defined by 2800 nodes and requires 8 hours of CPU time on a HP 712/60 computer for an elastoplastic analysis.

4. RESULTS

A set of trabecular dimensions corresponding to the mean values of the specimens experimentally tested was used in order to generate a model which is representative of the set of these samples. The results obtained for this model in the elastic and plastic ranges are compared to the experimental data (Table 1 and Figure 3).

	BV/TV (%)	Tb.Th (µm)	Tb.Sp (µm)	E^a_{33} (MPa)	E^a_{33}/E^a_{11}	σ_{max} (MPa)	$\varepsilon_{\sigma max}$ (%)
experiment	17.78	259.81	1354.26	135	2.3	1.6	1.7
sample 1	18.8	267	1150	133	1.9	1.34	1.41
model	18.78	173	1053	170	2.66	1.78	1.4

Table 1 : Comparison between the experimental results (mean values; sample 1: data and results close to mean values) and numerical ones (model based on data of sample1)

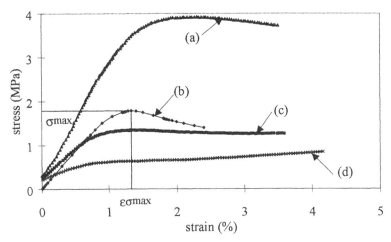

Figure 3 : Stress-strain curves for a compression test in the I-S direction :
(a) : upper experimental curve - (d) lower experimental curve
(c) : experimental curve of sample 1 - (b) curve of the model

The distribution of the equivalent plastic strains of the model compressed in the longitudinal direction for a strain of 2.4% (Figure 4) shows that yielding occurs typically in transversal trabeculae. The equivalent plastic strain, ε^p_{eq}, is defined by the following formula, where $d\varepsilon^{pl}$ is the tensor of small plastic strains :

$$\varepsilon^p_{eq} = \int \sqrt{\frac{2}{3} d\varepsilon^{pl} : d\varepsilon^{pl}} \tag{3}$$

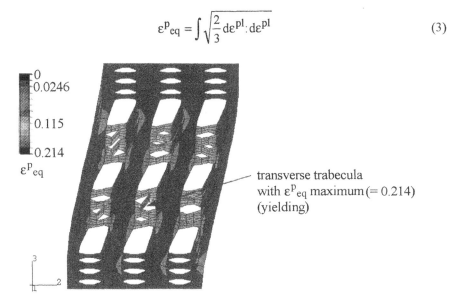

Figure 4 : Distribution of equivalent plastic strains, ε^p_{eq}, of the model for a deformation of 2.4% in the I-S direction (direction 3)
With: ε^p_{eq} maximum in the columns = 0.0246 and ε^p_{eq} maximum in the struts = 0.115 except for the one pointed on the figure

5. DISCUSSION

5.1. The results

The numerical results agree well with the experimental ones in the elastic and plastic ranges (Table 1). The stress-strain curve of the model (Figure 3) in compression in the I-S direction is similar to the mean experimental one and shows a first linear elastic region followed by a second phase of collapse as described by Fyhrie & Schaffler [13]. In fact they studied the behaviour of human vertebral cancellous bone in compression in the I-S direction until a strain of 15%. Their typical stress-strain curve shows an initial linear portion followed by a region of material softening (caused by damage) and by a plateau region, as does our curve. Although not shown, sufficient compression would close the pores in the material causing a rapid material stiffening called consolidation region [3], [13]. Moreover they observed that the broken elements were horizontal (transverse to the applied load) and the vertical trabeculae remained structurally intact but showed regions of microdamage within the bone matrix, as was noticed by ourselves (Figure 4).

5.2. The architecture of the model

The accuracy of methods that use a cancellous bone model depends on their capability to describe realistic trabecular architecture. Although the model presented is not a copy of the exact trabecular geometry of a real cancellous bone specimen, it describes very well the elastoplastic behaviour of one, based on architectural data. This is due to high flexibility and to the different types of irregularities which can be introduced.
The inclination of the studied models was equal to $5°$. This choice seemed to be the best compromise between an inclination of $0°$ where the structure was stiffer than the experimental one (E^a_{33}=203MPa) and the elastic anisotropy was overestimated (E^a_{33}/E^a_{11}=3.21), and between an inclination of $10°$ where both parameters were low (E^a_{33}=117MPa, E^a_{33}/E^a_{11}=1.85). But for these two cases, the stress-strain curves did not present a region of material softening observed experimentally (Figure 5). Moreover in the case of an inclination of $0°$, the equivalent plastic strains were equal to zero in the horizontal trabeculae and were maximum in the vertical ones which does not agree with the experimental observations [13].

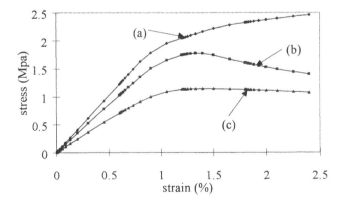

Figure 5 : Influence of the inclination θ of the model : (a) θ=0°; (b) θ=5°; (c) θ=10°

In order to investigate the influence of trabeculae with variable thickness, a simulation of a model generated from the same architectural data (BV/TV, Tb.Th, Tb.Sp, inclination...) but with a constant thickness of trabeculae was carried out. If the results were the same in the elastic range, the phase of collapse did not appear on the stress-strain curve.

It is clear that the architecture of the model is idealised and its main fault is a lack of termini (horizontally oriented) in the structure. In fact the age-related changes of the cancellous bone are a loss of bone mass and of continuity of the trabecular lattice. For vertebral bodies, primarily horizontal trabeculae are lost, leaving the vertical ones which might even get thicker [1], [4], [14]. Because the structure of the model is highly connected, the apparent Young's modulus E^a_{33} is slightly overestimated with regard to the experimental one.

5.3. Trabecular tissue behaviour

The model chosen for the behaviour of trabecular bone material is idealised because the knowledge of the trabecular tissue properties is very limited due to the small size of the trabeculae and the difficulties of performing experimental tests on them. Young's modulus E, equal to 1.5GPa and similar to the one found by Williams & Lewis [12], is low with regard to other values in the literature (between 1.3GPa [12] and 14.8GPa [15] for the human trabecular bone material) but, as we explained in the previous part, the structure of the model is idealised and highly connected. So in order to approach the experimental apparent stiffness of the global structure, a low value of Young's modulus E in the range of the data available was chosen.

The Poisson ratio and the yield stress, σ_e, were arbitrarily chosen because no data on these two parameters could be easily found. Choi & Goldstein determined fatigue properties of human single trabeculae from tibia using four-bending cyclic tests [16]. Their median S-N curve (S: outer fiber tensile stress (MPa); N: number of cycles to failure) can give an indication of the ultimate stress of trabecular material by taking the value of S for the lowest value of N (S=160MPa) and of the yield stress by tacking the value of S for a high value of N, which corresponds to the fatigue strength to N cycles. The fatigue strength to 10^6 cycles is equal to 90MPa, which can seem to be high in comparison with the value chosen for the yield stress, σ_e, in our model but Choi & Goldstein found a mean Young's modulus of trabecular bone tissue (5.72GPa) four times higher than the one we used. Moreover their trabecular specimens were obtained from a pair of human tibia (59 years old, healthy male) when our cancellous bone specimens were obtained from human vertebrae (mean age: 79 years).

Furthermore, with the finite element code ABAQUS®, it is not possible to introduce an ultimate stress in this behaviour law and the failure of the trabeculae can only be predicted based on the observation of the distribution of the equivalent plastic strains.

6. CONCLUSION

A three-dimensional structural model of human vertebral cancellous bone was presented and its mechanical behaviour was representative of the real one, even if its architecture was idealised. This model was generated by an efficient pre-processor from stereologic parameters of a real structure, such as bone volume /tissue volume, mean trabecular thickness and mean trabecular separation. The various parameters studies have shown the effectiveness of this model for reproducing the stress-strain curve of a real specimen, based on architectural data in the elastic and plastic ranges, making this model original. In fact the models are usually tested in the elastic range due to the fact that no data on trabecular bone tissue properties are available, except for Young's modulus.

This model could be adapted to other anatomical sites which have a trabecular structure with the same type of symmetry such as tibial cancellous bone.

7. REFERENCES

1. Mosekilde Li., Sex differences in age-related loss of vertebral trabecular bone mass and structure-biomechanical consequences, Bone, 1989, Vol. 10, 425-432.

2. Jensen K.S., Mosekilde Li., Mosekilde Le., A model of vertebral bone architecture and its mechanical properties, Bone, 1990, Vol. 11, 417-423.

3. Gibson L.J., The mechanical behaviour of cancellous bone, J. Biomechanics, 1985, Vol. 18, 317-328.

4. Mosekilde Li., Mosekilde Le., Danielsen C.C., Biomechanical competence of vertebral trabecular bone in relation to ash density and age in normal individuals, Bone, 1987, Vol. 8, 79-85.

5. Muller R., Ruegsegger P., Three-dimensional finite element modelling of non-invasively assessed trabecular bone structures, Med. Eng. Phys., 1995, Vol. 17, 126-133

6. Van Rietbergen B., Weinans H., Huiskes R., Odgaard A., A new method to determine trabecular bone elastic properties and loading using micromechanical finite element models, J. Biomechanics, 1995, Vol. 28, 69-81.

7. Ashman R.B., Cowin S.C., Van Buskirk W.C., Rice J.C., A continuous wave technique for the measurement of the elastic properties of cortical bone, J. Biomechanics, 1984, Vol. 17, 349-361.

8. Mitton D., Hans D., Rumelhart C., Meunier P.J., Density and test conditions effects on compression and shear strength of cancellous bone from the lumbar vertebrae of ewe, Med. Eng. Phys., 1997, in press.

9. Linde F., Pongsoipetch B., Frich L.H., Hvid I., Three-axial strain controlled testing applied to bone specimens from the proximal tibial epiphysis, J. Biomechanics, 1990, Vol. 23, 1167-1172.

10. Parfitt A.M., Mathews C.H.E., Villanueva A.R., Kleerekoper M., Frame B., Rao D.S., Relationships between surface, volume and thickness of iliac trabecular bone in aging and in osteoporosis, J. Clin. Invest., 1983, Vol. 72, 1396-1409.

11 Kabel J., Van Rietbergen B., Odgaard A., Dalstra M., Huiskes R., Tissue anisotropy is not important for cancellous bone elastic properties, Bone, 1996, Vol. 19, 1455

12. Williams J.L., Lewis J.L., Properties and an anisotropic model of cancellous bone from the proximal tibia epiphysis, J. Biomech. Eng., 1982, Vol. 104, 50-56.

13. Fyhrie D.P., Schaffler M.B., Failure mechanisms in human vertebral cancellous bone, Bone, 1994, Vol. 15, 105-109.

14. Vesterby A., Mosekilde Li., Gundersen H.J.G., Melsen F., Mosekilde Le., Holme K., Sorensen S., Biologically meaningful determinants of the in vitro strength of lumbar vertebrae, Bone, 1991, Vol. 12, 219-224.

15. Rho J.Y., Ashman R.B., Turner C.H., Young's modulus of trabecular and cortical bone material: ultrasonic and measurements, J. Biomechanics, 1993, Vol. 26, 111-119.

16. Choi K., Goldstein S.A., A comparison of the fatigue behaviour of human trabecular and cortical bone tissue, J. Biomechanics, 1992, Vol. 25, 1371-1381.

The authors wish to thank Professor BEJUI (Laboratoire d'Anatomie de la faculté LAENNEC) for providing bone samples.

COMPUTATIONAL MODELLING OF THE INTERVERTEBRAL DISC USING ABAQUS

J. S. Tan[1], S. H. Teoh[2], G. W. Hastings[3], C. T. Tan[4] and Y. S. Choo[5]

1. ABSTRACT

In this paper the intervertebral disc in the lumbar segment (L_{2-3}) of the spine was modelled and analysed using the finite element software, ABAQUS®. Two modes of loading conditions, axial compression and axial torsion, were used. The graph of axial compressive load against axial displacement and the graph of applied axial torque against angular displacement were compared against similar plots by other researchers and were found to be consistent with published results. Analysis was carried out using this model to determine the significance of the presence of the annulus fibrosus and the nucleus pulposus in the intervertebral disc. Results revealed that the nucleus pulposus and the annulus fibrosus play important roles in load bearing, under axial compression and axial torsion. The computational modelling serves as a useful tool in the preliminary design of an intervertebral disc prosthesis (IVDP).

Keywords: Intervertebral, IVDP, Prosthesis, Finite element, ABAQUS

---

[1] Research Assistant, Centre for Biomedical Materials Applications and Technology (BIOMAT), Department of Mechanical and Production Engineering, National University of Singapore, 10 Kent Ridge Crescent, Singapore 119260

[2] Deputy Director, Institute of Materials Research and Engineering (IMRE), & Assiociate Professor and Director, Centre for Biomedical Materials Applications and Technology (BIOMAT)

[3] Programme Director, Institute of Materials Research and Engineering (IMRE) & Visiting Professor, Centre for Biomedical Materials Applications and Technology (BIOMAT)

[4] Senior Consultant, Department of Orthopaedic Surgery, Singapore General Hospital, Outram Road Singapore

[5] Director, CAD/CAM/CAE Centre, Faculty of Engineering, National University of Singapore

2. INTRODUCTION

The normal human spine consists of numerous elements. It provides structural support for the musculskeletal torso, protects the sensitive spinal cord and gives flexibility to many complicated motions [1]. Disruption of the normal spine can be discomforting and sometimes fatal [2]. This can arise from injuries, tumors, muscle dysfunction and spine degeneration which lead to common complaints such as low back pain, slip discs and spondylolisthesis. With the advent of high speed computational tools, it is now possible to design implant prosthesis and test out its performance in the computer before actual fabrication. In this paper the intervertebral disc in the lumbar segment (L_{2-3}) of the spine was modelled and analysed using the finite element software, ABAQUS®. Two modes of loading conditions namely, axial compression and axial torsion, were used.

3. CONSTITUENTS OF A NORMAL INTERVETEBRAL DISC

The intervertebral disc is made up of two constituents, the nucleus pulposus and the annulus fibrosus which are sandwiched between two endplates. It is the disc that allows for the movement of a vertebral body, and at the same time restricts the movements allowed. It also transmits forces from one vertebral body to another. Each of these two constituents of the intervertebral disc performs a different role under different loading conditions. From the mechanical point of view, the nucleus pulposus, which can be considered as a ball of fluid, transmits forces exerted on it hydraulically and evenly in all directions. The annulus fibrosus, with its uniquely orientated fibers, opposes the axial rotation of one endplate with respect to the other. It surrounds the nucleus like a rubber band to restrict its bulge. As a whole, the intervertebral disc is able to sustain different kinds of loading conditions, thus allowing for the motion of the adjoining vertebral bodies. Under axial compressive load, the force is absorbed by the nucleus, which is in turn transmitted hydraulically to the annulus fibrosus. Under torsion, the fibers resist the force exerted in the circumferential direction and hence restricting the relative motion of the adjacent vertebrals.

4. FINITE ELEMENT MODEL

The geometry of the finite element model used in this paper was taken from a publication by Shirazi et al [3]. Their model was based on a L_{2-3} specimen of a 29-year-old woman. Dimensions of the model used in this report follow closely as reported in that paper [3] (Fig. 1). The height of the disc was assumed to be uniform over its cross-sectional area at 11mm (Fig. 1d). The thickness of the anterior and lateral regions of the annulus were measured to be 1.2 times that of the posterior region with the posterior surface of the disc considered flat (Fig. 1b). The nucleus pulposus was assumed as a ball of incompressible fluid occupying 46% of the cross-sectional area of the disc and the annulus fibrosus was modelled as a series of bands surrounding the nucleus (Fig. 1c). Each band of the annulus fibrosus was modelled to consist of strong collagenous fibers embedded in a ground substance. Four concentric bands of ground substance and eight layers of collagenous fibers were modelled. The fibers were arranged in a crisscross pattern, making an average angle of 60 degrees with the spinal axis.

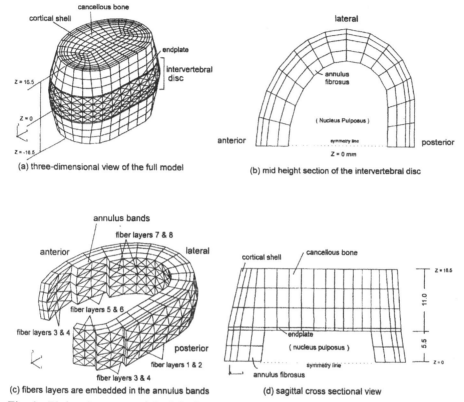

(a) three-dimensional view of the full model

(b) mid height section of the intervertebral disc

(c) fibers layers are embedded in the annulus bands

(d) sagittal cross sectional view

Fig. 1: Finite element model of the intervertebral disc

The entire finite element model of the motion segment consisted of the intervertebral disc described above, sandwiched between two endplates and two half vertebral bodies (Fig. 1a). The motion segment could be assumed to be symmetrical about two horizontal planes cutting through the mid section of the vetebral bodies. Since the vertebral bodies were not expected to deform significantly under loading as compared to the intervertebral disc, only half of the top and bottom vertebral bodies were modelled. The posterior facets were not modelled in this analysis as we were only interested in the intervertebral disc.

The finite element model (Fig. 1a) was made up of a total of 2911 nodes and 3572 elements, consisting of 2320 "8-noded linear brick elements", 896 "2-noded linear truss elements" and 356 "4-noded linear hydrostatic fluid elements". The brick elements were used to model the cortical bone, the cancellous bone, the endplate and the ground substance. The linear truss elements were used to model the fibers of the annulus (Fig. 1c) and the 4-noded elements were used to define the fluid.

4.1. Loading Conditions

To simulate an intervertebral disc under compressive loads, all the nodes located on the top face of the model were specified with a fixed displacement in the negative Z direction. The total resultant axial force acting on these nodes was the load applied. As there is symmetry in the loading conditions and the geometry of the intervertebral disc about 2 planes, only a quarter of the disc needed to be modelled for the compression test (Fig. 2). The analysis was carried out to a maximum displacement of 4 mm, with incremental displacements of 1 mm.

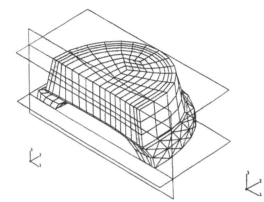

Fig. 2: Symmetry planes were used for compression analysis. The nodes on the top plane were uniformly displaced in the negative Z direction

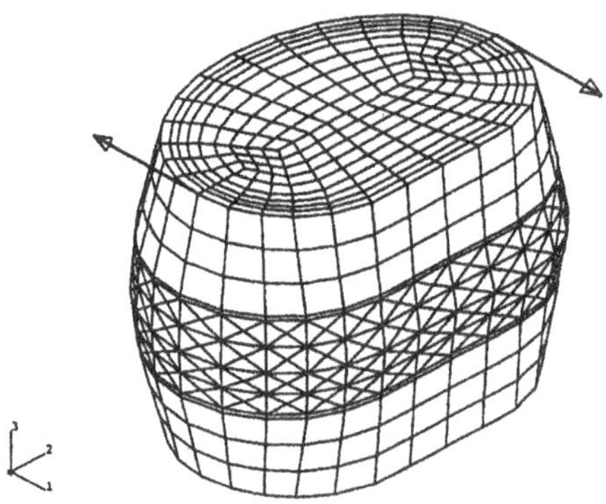

Fig. 3: Two point forces were used to simulate an intervertebral disc under torsion.

To simulate an intervertebral disc under torsional loads, two concentrated forces of equal magnitude but opposite directions were applied on the top edge of the motion segment (Fig. 3). The perpendicular distance between the points of action of these two forces multiplied by the magnitude of the force gave the torque. The angular displacement was calculated from the geometry of the displaced disc with respect to the undeformed disc. The analysis was carried out to a maximum applied force of 450 N (13.14 Nm) using step increment of 50 N (1.46 Nm).

4.2. Boundary Conditions

For the compression analysis, the quarter model of the motion segment was used (Fig. 2). Nodes located in the bottom horizontal Z plane were symmetrical about the Z plane and were restricted in U_z, ϕ_x and ϕ_y, whereas nodes located in the vertical Y plane were symmetrical about the Y plane and were restricted in U_y, ϕ_x and ϕ_z. A single node in the vertical Y plane on the top surface was constrained in U_x to prevent rigid body motion.

For the torsion analysis, the full model of the motion segment shown in Fig 1a was used. As there was symmetry in the loading conditions and the geometry of the intervertebral disc about the bottom plane, nodes located in the bottom horizontal Z plane (Fig. 3) were considered symmetrical about the Z plane and were restricted in U_z, ϕ_X and ϕ_Y. A single node on the centre of the bottom surface was constrained in U_X and U_Y to prevent rigid body motion.

5. RESULTS AND DISCUSSIONS

5.1. Comparison of Present Results with Published Work

The finite element model created was compared using the load-displacement plots against similar plots by researchers using the finite element method [3], [4], [5] & [6] and against similar plots by researchers using *in vitro* testing [7], [8], [9], [10] & [11] (Fig. 4 & Fig. 5). In both comparisons, the results predicted using the present model fell within the range of curves by other researchers. The predicted results using ABAQUS laid within the range reported by Brown [7] (Fig. 4) and Lee [6] (Fig. 5). The results also compared well with other experimental results by Markolf and Morris [8], Markolf [9], Virgin [10] and Hirsch and Nachemson [11]. In comparison with other finite element models by Goel [1] and Shirazi [3] & [4], the present results laid very close to theirs. The present results could therefore be considered to compare reasonably well with results by other researchers, taking into account the differences in formulation, geometry, and material properties used by the various researchers.

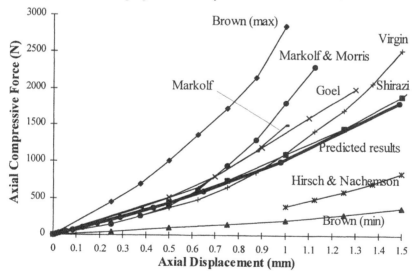

Fig. 4: Comparison of present results with similar plots using the finite element method [3], [4], [5] & [6] and in vitro testing [7], [8], [9], [10] & [11] by other researchers, for a disc under compression.

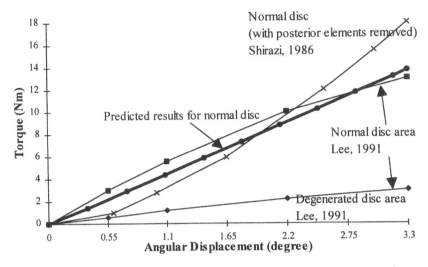

Fig. 5: Comparison of present results with similar plots by other researchers using the finite element method, for a disc under torsion

5.2. Load Bearing Role of the Fibers and the Nucleus

The load bearing roles of the fibers and/or the nucleus of the intervertebral disc under compression or torsion were determined by carrying out a finite element analysis of an intervertebral disc without any fibers and/or nucleus under the action of an axial compressive force or an axial torque and comparing the resultant force-displacement curve with the curve for a normal disc. The plots of axial compressive force against axial displacement (Fig 6) and plots of torque against angular displacement (Fig. 7) were obtained for four types of discs: normal disc, disc without nucleus, disc without fibers, and disc without nucleus and fibers.

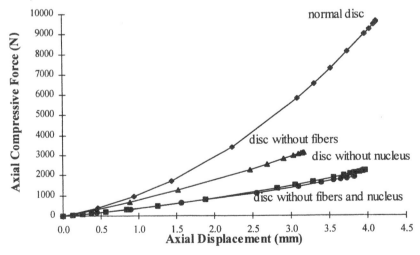

Fig. 6: Plots of axial compressive force against axial displacement for normal disc, disc without nucleus, disc without fibers, and disc without nucleus and fibers

From Fig. 6 and Fig. 7, the significance of the nucleus and the fibers under the loading of only an axial compressive force and under the loading of only an axial torque were determined. It was observed from these graphs that the nucleus played a significant role in bearing axial compressive loads and the fibers plays a big role in bearing torsional loads.

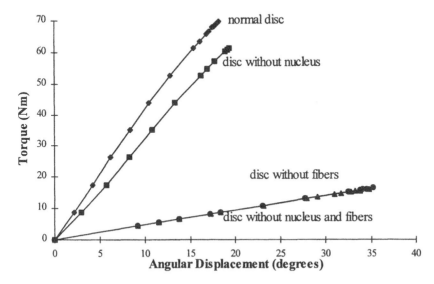

Fig. 7:   Plots of torque against angular displacement for normal disc, disc without nucleus, disc without fibers, and disc without nucleus and fibers

6. CONCLUSION

In this project, a finite element model of a L_{2-3} disc motion segment of the human spine was developed using ABAQUS®. The model accounted for the geometrical variations in the posterior, anterior and lateral regions of the disc. The relative volumes of the nucleus pulposus and the annulus as well as the volume fraction within the annulus fibrosus and the composite structure of the annulus were taken into account in modelling the intervertebral disc. The stiffness of the model was compared against results by other researchers under compression and torsion and was found to match reasonably well with these works. The results also revealed that the nucleus pulposus and the annulus fibrosus played important roles in load bearing, under axial compression and axial torsion.

7. REFERENCES

1.   Goel V.K. and Weinstein J.N., Biomechanics of the spine clinical and surgical perspective, Boca Raton, FL, CRC Press, 1990.
2.   White A.A. and Panjabi M.M., Clinical biomechanics of the spine, 2nd ed., Philadelphia, J.B. Lippincott, 1990.
3.   Shirazi-Adl A., Shrivastava S.C. and Ahmed A.M., Stress analysis of the lumbar disc-body unit in compression, Spine, 1984, Vol. 9, 120-134.
4.   Shirazi-Adl A., Ahmed A.M. and Shrivastava S.C., Mechanical response of a lumbar motion segment in axial torque alone and combined with compression, Spine, 1986, Vol. 11, 914-927.
5.   Goel V.K., Kong W.Z., Han J.S., Weinstein J.N. and Gilbertson L.G., A combined finite element and optimization investigation of lumbar spine mechanics with and without muscles, Spine, 1993, Vol. 18, 1531-1541.
6.   Lee C.K., Langrana N.A., Parsons J.R. and Zimmerman M.C., Development of a prosthetic intervertebral disc, Spine, 1991, Vol 16 (Supplement), S253-S255.
7.   Brown T., Hansen R.J. and Yorra A.J., Some mechanical tests on the lumbosacral spine with particular reference to the intervertebral discs, J. Bone Joint Surg., 1957, Vol. 39A, 1135-1164.
8.   Markolf K.L. and Morris J.M., The structural components of the intervertebral disc, A study of their contributions to the ability of the disc to withstand compressive forces, J. Bone Joint Surg., 1985, Vol. 56, 675-687.
9.   Markolf K.L., Deformation of the thoracolumbar intervertebral joints in response to external loads, J. Bone Joint Surg., 1972, Vol. 54A, 511-533.
10. Virgin W.J., Experimental investigations into the physical properties of the intervertebral disc, J. Bone Joint Surg., 1951, Vol. 33B, 607-611.
11. Hirsch C. and Nachemson A., New observations on the mechanical behaviour of lumbar discs, Acta. Orthop. Scand., 1954, Vol. 23, 254-283.

A MEDICAL IMAGE BASED TEMPLATE FOR PEDICLE SCREW INSERTION.

K. Van Brussel[1], J. Vander Sloten[1], R. Van Audekercke[1], B. Swaelens[2], L. Vanden Berghe[3], G. Fabry[3]

1. ABSTRACT

The placement of pedicle screws in spinal fixation still implicates a high complication risk and the success depends strongly upon the skills and experience of the surgeon. In order to enhance the precision and the safety of pedicle screw insertion, a mechanical drill guide is proposed to indicate the direction of the optimal path for the insertion of the screw. The aim of the project is to develop a patient dependent template that exactly fits some selected areas on the posterior part of a spinal segment incorporating drill holes indicating the correct position and orientation for the pedicle screws. The template can only be placed on the posterior part of one particular vertebra in one unique way. Using CT-images of the patient's spine, pre-operative planning software enables the surgeon to determine the optimal position and orientation of the pedicle screw relative to the vertebrae. A CAD/CAM program uses the morphologic information of the vertebrae and the 3D information of the planned drill paths to design the template. The template is produced by stereolithography as rapid prototyping technique. The use of a Zeneca experimental resin which is sterilizable and USP Class VI guarantees evaluation during clinical trials. Four concepts of parametrized models have been completed based on CT-images of a patient's L3-L5 segment and currently being evaluated.

2. INTRODUCTION

Human vertebrae can be damaged due to trauma or as a result of repetitive overloading. These injuries are treated by bridging the defective vertebrae using osteosynthesis material together with bone grafts to repair the damaged vertebral body or even to allow a fusion of

keywords: drill guide, pedicle screws, template, computer aided design, rapid prototyping

[1] K.U.Leuven, Division of Biomechanics and Engineering Design, Celestijnenlaan 200A, B-3001 Heverlee
[2] Materialise N.V., Kapeldreef 60, B-3001 Heverlee, Belgium
[3] K.U.Leuven, Division of Orthopaedics, Weligerveld 1, B-3212 Pellenberg, Belgium

this body with the neighbouring vertebrae. Also spinal deformations (scoliosis) require correction and stabilization. Classical instrumentation consists of rods (Cotrel-Dubousset, Harrington, Luque-Galveston, ..) attached to the bone by means of hooks or wires. More recently, transpedicular screws are introduced as an alternative bone/implant interface. The insertion of such pedicle screws implicates a relatively high complication risk and the success depends strongly on the experience of the surgeon [1]. Incorrectly drilled holes or malplacement of the screws can result in nerve root injuries and fracture of the pedicle. The precision and hence the safety of pedicle screw insertion can be enhanced by a certain degree of automation of the critical actions. The approach developed at KULeuven, in the framework of the European Brite-Euram project on Personalized Implants and Surgical Aids (PISA), proposes a mechanical guide to indicate the direction of the optimal path for the insertion of the pedicle screw. In this way an advanced surgical tool is put at the disposal of the surgeon, leaving the medical responsibility in his hands.

3. METHODS

The development of the drill guide implicates the construction of a complete design system or environment. An overall design environment must be developed which enables the engineers as well as the surgeons and radiologists to have a complete view on each aspect of the drill guide concept. Therefore the template design principles are considered as well as the overall data handling during development.

3.1. Template Design Principles

The template is a rigid body and provides two mechanical interfaces: the first interface realizes a connection between the drill and the drill guide. The second interface connects the template to the vertebra, such that a unique, correct and stable position is achieved. Implementing both interfaces in a drill guide design assumes a fully personalized approach. Therefore, a 3D spine reconstruction by means of CT-scans is used as a basis for the template design.

3.1.1. Design options

Three prototypes have been developed considering following design options:

1. A basic option in the design is the number of vertebrae one template must fit at once. Generally the acquisition of CT-data is performed on patients in supine position whilst the patient lies in prone position during operation. Therefore, pre-operative CT-data of the spinal segment of interest is not reliable for template designs covering several vertebrae at once. Consequently, for all following designs one template per vertebra is chosen as a basic design option.
2. At the bone/implant interface the number of contact points must be chosen. Assuming the area of one contact region is small compared to the overall size of the contacting bodies, at least three contact regions are needed to create a stable connection between two solid bodies. Two prototypes are developed using three contact regions, one using four contact regions, depending on clinical requirements described in the next paragraph.

3. The type of contact between the template and the vertebra must be considered: the use of large contact areas as interface introduces unique vertebral surface information into the template design. In order to deal with the occurrence of soft tissue remainders, knife-edge or laminar contact points may be preferred as an alternative to large contact areas. Two concepts use large contact areas. The third concept has knife-edge or laminar contact points.

3.1.2. Parametrization

The prototypes exist as a fully parametric design in the a computer aided design environment. Mainly, three groups of parameters may be distinguished:

1. One set of parameters consist of the position and orientation of the optimal drill path. These parameters reflect the information of the pre-operative planning of the surgeon.
2. A second parameter set consists of the surface information of the contact areas on the vertebra by means of contour information or nurbs surfaces.
3. Thirdly, a limited set of parameters characterize the dimensions of the drill guide.

If a drill guide concept has been approved to be suitable for clinical application, each patient can be treated by personalizing these parameter sets.

3.2. Data Handling

In general four consecutive steps (three pre-operative, and one per-operative) can be distinguished in the development of a personalized drill guide (fig. 1).

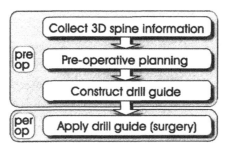

Fig. 1: General flowchart for the production of a personalized mechanical drill guide.

1. In order to enable a 3D reconstruction, morphologic information must be collected about the spinal segment of interest.
2. Based on a visual representation of this information the surgeon performs a pre-operative planning and determines the optimal drill position and orientation through the pedicle.
3. The last pre-operative step is the manufacturing of the drill guide according to the pre-operative planning.
4. Finally, the surgeon applies the drill guide during the intervention.

This general procedure is applicable to all types of personalized surgical tools and implants and will be explained in more detail for the case of the drill guide for pedicle screw insertion.

Figure 2 illustrates the procedure in the form of a complete data flowchart.

- The morphologic information of the patient's spine is acquired by a CT-scanner.
- The raw CT-data need to be segmented in order to extract the 3D information of the posterior periosteal cortical surface of the spine. The segmentation is performed by Mimics 4.1 software of Materialise N.V. Belgium.
- To enable transfer to a design environment, the information is converted into an IGES description of contours (CAD/CAM/CAE neutral format). This step is performed by CT-Modeler software of Materialise N.V.

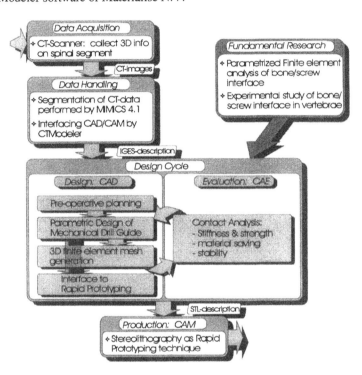

Figure 2: Complete data flowchart for the development of a mechanical drill guide for pedicle screw insertion.

- The 3D contour information of the vertebral segment is presented to the design environment that consist of a CAD/CAM package (EDS Unigraphics v11) at one hand and a finite element software (MARC k62) at the other hand.
- Based on the contour data the surgeon determines the optimal position and orientation of the pedicle screw relative to the vertebrae. For this pre-operative planning the CAD/CAM software is used. In a later stage of the project a user-friendly interface will be developed for this purpose.
- The Solid Modeler of Unigraphics uses the morphologic information of the vertebrae and the 3D information of the planned drill paths to design a template that exactly fits selected areas of vertebral processes and that incorporates holes which indicate the correct drill paths.

- An early evaluation of the drill guide may be of interest in order to optimize the concept in terms of stiffness, strength and stability. For this purpose the CAD modeler is directly linked to a non-linear finite element software package MARC k62/MENTAT 3.2 of MARC Analysis Research Corporation. The combination of both Unigraphics and MARC software results in a complete computer assisted design environment with pre-surgical simulation capabilities.

- The result of the design cycle Unigraphics/ MARC is a virtual parametric design of a mechanical drill guide for pedicle screw insertion. This virtual product is converted into a real product by the use of stereolithography as rapid prototyping technique [2]. According to the latest developments in stereolithography manufacturing times of 3 hours are possible. The use of a Zeneca experimental resin which is sterilizable and USP Class VI (temporary use inside the body allowed) guarantees evaluation during clinical trials.

- Additionally, the system as a whole is supported by fundamental research on the quality of the bone/screw interface. This research consists of finite element analyses of screws placed in the verteberal pedicles. The screws as well as the vertebra exist as fully parametrized bodies in the CAD/CAM package Unigraphics. The influence of bone density and screw shape on quality of the interface is investigated. The results of these analyses will be validated by in vitro experiments that will be carried out on cadaveric vertebrae in the near future.

4. RESULTS

4.1. Prototypes

Three prototypes have been developed, according to the template design principles described in previous paragraph. Each of the prototypes deals with specific clinical requirements.

A first drill guide, named prototype A (fig. 4), has been completed based on CT-images of a patient's L3-L5 segment (fig. 3 left). The guide is able to assist the insertion of two pedicle screws in L4.

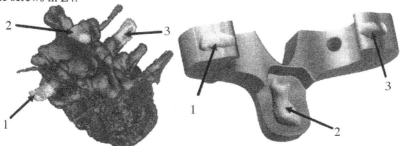

Figure 3. Left: three support points on the vertebra for prototype A. Right: three support points incorporated in prototype A as an interface to the spinal cortex.

As support surfaces three areas of one vertebra (L4) are used, specifically the two transverse processes and the top of the spinal process. For each support point a large contact area guarantees the uniqueness of fit.

In order to evaluate the stability of this three point contact fit and to permit design optimizations, finite element contact analyses have been performed. To make a first non virtual evaluation of the quality of the fit between the guide and the vertebra, also the spinal segment L3-L5 is reconstructed using stereolithography. A rendered image of the drill guide is shown in figure 3 on the right side.

The main disadvantage of this prototype is the use of the upper part of the spinal process as support surface which urges the removal of the posterior spinal ligament. In most of the

scoliotic cases this ligament is removed and no problem occurs, but in some traumatic cases it is preferable not to perform any posterior ligament removal. In this case the top of the spinal process is not suitable as contact area.

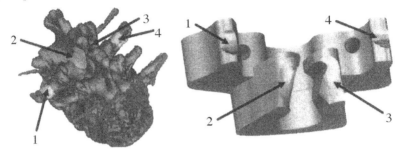

Figure 4: Left: four support points on the vertebra for prototype C. right: three support points incorporated in prototype C as an interface to the spinal cortex.

Therefore, a second prototype has been developed using the two side faces of the spinal process rather than the top of it (figure 4 left). Due to the limited attainability of the transverse processes during operation, these support surfaces are reduced. Consequently, four instead of three contact points are used. For each support point a large contact area guarantees the uniqueness of fit. Following the same procedure as shown in figure 2 this concept converged to prototype C (figure 4 right).

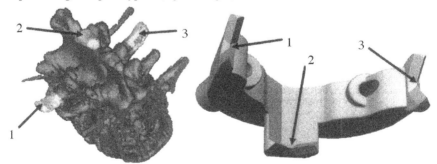

Figure 5: Prototype D uses three knife-edge support points as interface to the cortex.

A third concept deals with the occurrence of soft tissue on the support areas. Large contact areas, as used in previous prototypes, have the disadvantage of being sensitive to errors in the contact surface description. Since soft tissue may cause such errors, it may be desirable to use knife-edge or laminar contact. Laminar support enhances the pressure in the contact region and reduces the resistance caused by soft tissue. Prototype D uses knife-edge contact on the same sites as applied in prototype A (three region contact) (fig. 5).

4.2. Concept Evaluation

To evaluate the concepts before production, finite element analyses are performed using the MARC k6.2 software. Stiffness, strength and stability are tested. A typical output of a finite element analysis of prototype D is shown in figure 6 on the left, where the equivalent Von Mises stress is calculated under simulated finger pressure by the surgeon. As load a three finger pressure is assumed, each finger pressing 50 N (strong surgeon, overestimation) normal to the contact face. A tetrahedral element mesh is generated by Unigraphics GFEMFEA module. Each concept turned out to be stiff (max. deformations < 0.1 mm) and strong enough (max. Von Mises stress << yield stress = 62 MPa).

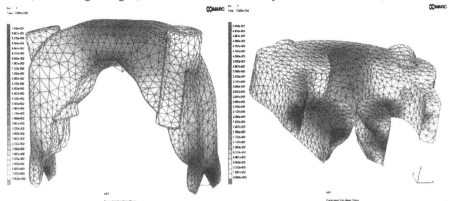

Figure 6: Left: result of a finite element analysis of prototype D on a vertebra (the vertebra is blanked). Right: result of a finite element analysis of prototype C on a vertebra (the vertebra is blanked). Mainly three of the four contacts are used due to the hyperstatic nature of a four point contact structure.

The simulation of the stability of a drill guide when placed on the vertebra under simulated finger pressure by the surgeon is more complex. In these analyses a finite element mesh (tetrahedral elements) of the drill guide is placed on a finite element mesh (tetrahedral elements) of a vertebra.

The meshes are generated using the Unigraphics GFEMFEA module. Figure 6 on the right illustrates the result of prototype C placed on a simplified model of a vertebra (contact areas are accurate). In this case the figure shows that mainly three of the four contacts are used due to the inherently hyperstatic structure of the concept. Subjective evaluation on the stereolithographic model of the spinal segment confirms the instability.

5. DISCUSSION

Two concepts (A, D) use three point support whilst prototype C uses four point support. In terms of stability three points of support are preferred. Moreover, the collaborating orthopedic surgeons prefer prototypes A and D to prototype C even if this implicates an additional per-operative action (removal of the posterior ligament).

Two types of contact are developed:
1. a fit between drill guide and vertebra using large areas of contact,
2. laminar or knife-edge contact, developed in order to cope with remainders of soft tissue.

Yet, based on the simulations, no final decision has been made which type of contact is most suitable for the drill guide application. In this case a per-operative evaluation by the surgeons is needed to decide which concept will be elected for further development. Further development may take advantage of both contact types: knife edges will be used on the transverse processes whilst the spinal process will guarantee a unique position of the guide using a large contact area. In order to perform an objective evaluation of the stability of the drill guides, cadaver studies are planned and a protocol has been established for limited clinical trial.

Compared to fully computer assisted spine surgery [3][4][5] the mechanical drill guide has advantages.
- It only requires a low investment since no expensive computer equipment and 3D optical tracking hardware is needed in the operation room.
- The drill guide does not increase the complexity of the surgery.
- Since the position of the guiding system upon the posterior part of the vertebrae is unique, no pre-operative referencing is needed, which is one of the main difficulties of fully computer assisted navigation techniques.

6. ACKNOWLEDGEMENTS

The authors wish to thank R. Assaker and J.-F. Kulik of Centre Hospitalier Regional et Universitaire de Lille for their interest and support.

7. REFERENCES

1. K. Van Brussel, J. Vander Sloten, R. Van Audekercke, G. Fabry, *Internal fixation of the Spine in Traumatic and Scoliotic Cases. The Potential of Pedicle Screws.* Technology and Health Care, Vol. **4**, pp 365-384 (1997)
2. B. Swaelens, J.-P. Kruth, *Medical Applications of Rapid Prototyping Techniques,* Proceedings of the Fourth International Conference on Rapid Prototyping, Dayton, June, pp 107-120 (1993)
3. L.-P. Amiot, H. Labelle, J. A. Deguise, M. Sati, P. Brodeur, C.-H. Rivard, *Computer-Assisted Pedicle Screw Fixation, A Feasibility Study,* Spine, Vol. **20**, Number **11,** pp 1208-1212 (1995)
4. L.-P. Nolte , L. J. Zamorano, Z. Jiang, Q.Wang, F. Langlotz, E. Arm, H. Visarius, *A Novel Approach to Computer Assisted Spine Surgery,* Proc. 1st Int. Symp. on Medical Robotics and Computer Assisted Surgery, 323-328 (1994)
5. S. Lavallée, P. Sautot, J. Troccaz, P. Cinquin, P. Merloz, *Computer-assisted spine surgery: a technique for accurate transpedicular screw fixation using CT data and a 3-D optical localizer.* J Image Guid Surg, 1:1, 65-73, (1995)

MECHANICAL PROPERTIES OF LUMBAR DURA MATER AND BIOMECHANICS OF SPINAL ANESTHESIA PROCEDURE

R.Pietrabissa[1], S.Mantero[1], V.Quaglini[2], R.Contro[2], M.Runza[3]

1. ABSTRACT

Spinal anaesthesia has experienced a great increase in use in the last decades. Many needles are available for this procedure but the most commonly used are the so called atraumatic needles. They have a rounded tip with a lateral hole to deliver the anaesthetic. This particular shape guarantee the simple divarication of the dura mater fibers hence a lower degree of trauma.

Samples of both human and bovine lumbar dura mater have been used in this study. The samples have been cut to obtain specimens for mechanical testing. The stress-strain tests of the dura in longitudinal an circumferential direction have been performed using a MTS 858 Mini Bionix test machine. No significant difference between human and bovine samples has been noted as regard the maximum stress.

To simulate the dura behavior during puncture a purposely designed equipment has been mounted on the MTS machine. The equipment allows to use the dura as the upper wall of a small tank contained a buffered saline physiological solution at 40 cmH_2O of pressure which is the mean pressure of the cerebrospinal fluid for a seated patient. The needle is moved and when the dura is penetrated the force drop down thus stopping the needle movement. Both the maximum force and displacement are recorded and allow to describe the mechanism of dura penetration and to correlate it with dura structure and mechanical properties.

Keywords: Dura mater, Mechanical properties, Scanning electron microscopy

[1] Dipartimento di Bioingegneria, Politecnico di Milano, Piazza Leonardo da Vinci, 32. 20133 Milano, Italy

[2] Dipartimento di Ingegneria Strutturale, Politecnico di Milano, Piazza Leonardo da Vinci, 32. 20133 Milano, Italy

[3] Servizio di Anestesia e Rianimazione, IRCCS Ospedale Maggiore di Milano, Milano, Italy

2. INTRODUCTION

Spinal anesthesia is one of the simplest technique available in "regional anesthesia" and it's use has dramatically increased in the last decade for elective surgery.

The procedure is performed by puncturing the dura mater and injecting a "local anesthetic" into the dural sac, which affords a direct access to the exposed spinal nerves ensuring a rapid, dense and predictable anesthesia. The procedure is performed using special spinal needles which penetrated through the epidural space, beyond the dura mater and the arachnoid membrane, reaching the lumbar cerebrospinal fluids. The spinal needles have a lateral hole from which the anaestetic can reach the cerebrospinal fluid. After the dura elastic recovery, the lateral exit hole of the needle may lay completely under the dura surface or astride of the dura. The position of the lateral hole of the needle could be an important characteristic in order to decide whether or not is necessary to proceed with the needle after the dura penetration.

Equipments and techniques relating to dural puncture differ according to needle shape, bevel orientation, tip design, angle of penetration and patient position. These variables together with anatomic and biomechanical properties of lumbar dura influence the incidence of post-dural puncture headache (PDPH) which is related to the cerebrospinal fluid leakage due to the size and/or shape of perforation hole. It is supposed that needles of smaller gauge (i.e smaller diameter) and rounded/ogival needle tip (atraumatic needles) will reduce the leakage of cerebrospinal fluid by reducing the size of the hole produced on the punctured dura.

Despite the decrease in PDPH, which has occurred with the introduction of atraumatic needles, this complication continue to occur and uncertainties regarding the cause of PDPH continue as the exact anatomic composition and mechanical properties of human dura remain controversial.

Although fiber content has been clearly identified (collagen and elastin fibers in an intercellular ground substance) it remains disagreements regarding their orientation (1,2,3,4,5).

Despite the publication of numerous studies, some authors describe the dura as a longitudinally organized fibro-elastic membrane (2,3,4,5) while other describe the dura as a loose connective tissue or a tissue having collagen fibers running in different directions (1)

Last but not least, mechanical properties gives us an idea of the physiologic function of this membrane in vivo, underlying that dura mater is subject to high longitudinal tensile stress rather than to transverse, "hoop", stress too. Anatomical features, including orientation of structural components has to be considered.

Aim of this study was to evaluate the structural anatomy of the lumbar dura mater by using the scanning electron microscopy (SEM) and its biomechanical properties by experimental mechanical characterization.

The mechanical characterization of the dura (human and bovine), both in longitudinal and circumferential direction, allows to evaluate its isotropicity. A quantitative evaluation of the mechanical behavior of the dura during needle penetration has been obtained using a simulator in order to compare different needles functionality.

3. MATERIAL AND METHODS

3.1 Microscopic observations

Scanning Electronic Microscopy (SEM) was performed on human dura mater samples obtained at autopsy.

The stabilization of the dura mater membrane was performed fixing the samples with 2.5% glutaraldehyde solution in 0.1 cacodylate buffer (pH 7.2), washed with the same buffer, dehydrated through a graded series of ethanol (up to 100%) and critical point-dried using CO_2. The purpose of this treatment was to promptly block all cellular, molecular and macromolecular components of the biological tissue. After fixation and dehydratation, the samples were mounted on an aluminium stub, sputter-coated with gold and examined with a Leica-Cambridge Stereoscan 440 Microscope at 5-10 keV acceleration voltage.

3.2 Mechanical tensile test

Samples (10 cm length) of lumbar dura mater were obtained at autopsy for 5 human subjects. Samples of 5 bovine dura mater, obtained at slaughtering, were also tested for comparison.

In order to evaluate the biomechanical properties of the dura mater, both longitudinal and circumferential specimens have been obtained. Figure 1 shows the shape, dimensions and orientation of the dura mater specimens.

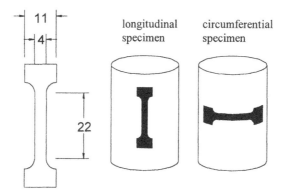

Fig.1 Shape, dimensions and orientation of dura mater specimens.

The human dura mater samples, placed in normal saline solution, have been studied within one hour from explantation or frozen at -4°C for 24h before testing. The bovine dura mater samples were immediately tested or frozen for 96h at -4 °C or stored at 4°C for 96h before testing. The thickness of the specimens was measured using a digital Palmer micrometer (Mitutoyo). The mechanical tests were executed using a MTS Mini Bionix 858 (25N load cell). The cross bar speed was 10 mm/min, the temperature 20±1°C and the relative humidity 60%.

During each test the load-displacement curve was recorded.

3.3 Spinal anaesthesia simulation

Figure 2 shows an illustration of the experimental set of the spinal anaesthesia simulator which has been used to evaluate the dura mater retraction (displacement) after needle puncturing and the force necessary to perforate the dura itself.

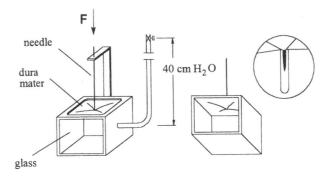

Fig.2 Experimental setup of the spinal anaesthesia simulator.

The simulator is composed of a chamber filled with ringer solution and sealed at the upper wall with a flat sample of dura mater. The pressure of the ringer solution has been set in order to simulate the cerebrospinal fluid (20 cmH$_2$O to simulate laying patient and 40 cmH$_2$O to simulate sitting patient). The chamber and the needle have been mounted on the MTS Mini Bionix. The needle constrained to the cross bar in a vertical position, has been moved against the dura until it completely perforated the dura.

4. RESULTS

Figure 3 shows a photograph of a transversal section of the dura mater together with a magnification: the dura mater appears organized in successive parallel layers made up of collagen fibers joined in bands and running mostly in a perpendicular direction with respect to the point of view of the photograph. Figure 4 shows a photograph of the dura mater in which the longitudinal orientation of the collagen fiber is well documented.

Table I resumes the results of the mechanical characterization performed on the human dura mater in term of ultimate stress and strain and incremental modulus referred to collagen response.

Figure 5 shows in a stress/strain diagram the morphology of the mechanical response of the human dura mater to both longitudinal and circumferential stretch.

Figure 6 shows the result of the puncture simulation using an atraumatic needle, in term of needle displacement versus force.

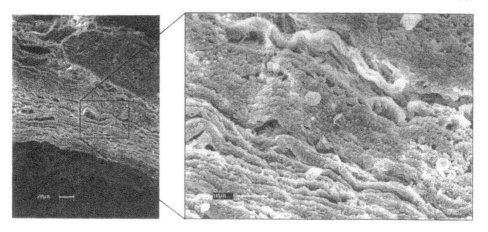

Fig.3 Microscopic images of a transversal section of human dura mater.

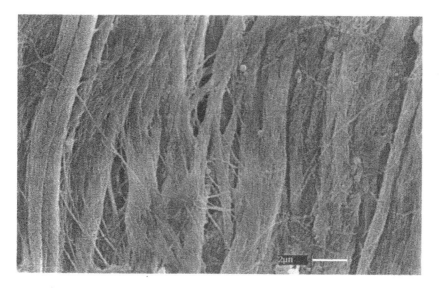

Fig.4 Microscopic image of longitudinal oriented collagen fibers.

Table I Results of the mechanical characterization performed on the human dura mater

	σ_{max} (MPa)	ε_{max} (%)	E (MPa)
Fresh dura	Longitudinal direction		
1 FL	17.9	17	87,6
2 FL	8,4	6,4	56,3
3 FL	9,9	9,4	59,4
4 FL	15,7	13,2	102,7
5 FL	12,8	20,0	53,3
6 FL	20,2	17,9	82,2
7 FL	11,3	10,8	73,4
8 FL	8,1	5,3	66,0
Frozen dura (-4° C)	Longitudinal direction		
1 CL	13,2	13	76
2 CL	14,1	12,9	67,7
3 CL	19,7	11,2	134,5
4 CL	8,5	8,7	45,9
Fresh dura	Circumferential direction		
1 FC	3,1	33,9	6
2 FC	4,2	51,7	6
3 FC	3,2	41,8	6

Stress (MPa)

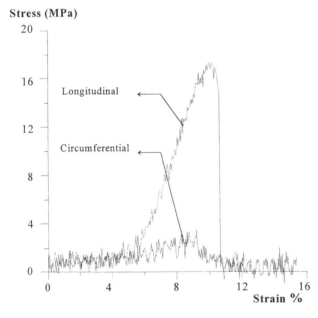

Fig.5 Stress/strain response of the human dura mater in longitudinal and circumferential direction of strain.

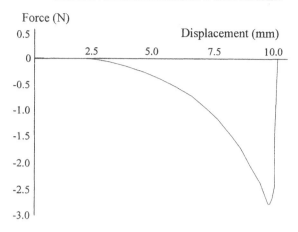

Fig.6 Needle force versus displacements during a test of puncture simulation.

5. DISCUSSION

The dura mater anatomic configuration, and its related biomechanical properties can be evaluated by mechanical tests linked with a microscopic analysis. The results of the biomechanical tests, which showed clearly that the tissue exhibits greater tensile strength and stiffness if stretched in the longitudinal direction, is consistent with its micro-architecture, which leads to a ortotropic structure in which the primarily direction of the collagene fibers, which routinely sustains longitudinal stress and deformation, is longitudinal. From a physiological point of view, the dura appears configured to allow changing volume by developing of transverse stress associated with this physiologic function, but it must also withstand greater longitudinal stress during physical activity. The anatomy and the microscopic structure of dura mater is under debate: some authors (1) indicates the high isotropicity of the dura mater as organized fibroelastic membrane in the longitudinal direction, others (2,3,4) concludes that the tissue does not exhibits any isotropicity as a loose connective tissue having a concentric orientation fibers contradicting the description of the dura mater having parallel, longitudinally placed collagen fibers. This biomechanical study confirms that human dura mater has a predominant longitudinally oriented structure as originally described by Green (6). Due to the fact that most of the bundles of fibers are arranged longitudinally, the atraumatic needles would shift but not break the fibers leaving a hole which would disappear when the needle is extracted.

The experimental model used in this study to evaluate the force carried by the needle to puncture the dura mater tries to mimic the in vivo operation. The limit of this experimentation is related to the fact that a segment of dura similar in size to those that we punctured would have a slight curvature in vivo, would have external supporting structures (epidural vessels, fat, laminae and ligaments) and might not have uniform wall tension. Nevertheless, the puncture simulator can be a useful experimental set-up to evaluate the different performance, in term of dimension of the holes, degree of recovery of the dura mater after puncturing and characterization of different needles functionality.

6. REFERENCES

1. Patin, D.J.,Eckstein, E.C., Harum, K., Pallares, V.S., Anatomic and biomechanical properties of human lumbar dura mater. Anesth. Analg., 1993, Vol. 76, 535-40.
2. Fink, B.R., Walker, S., Orientation of fibers in human dorsal dura mater in relation to lumbar puncture, Anesth. Analg., 1989, Vol. 69(6), 768-72
3. Reina, M.A., Lopez-Garcia, A., Dittmann, M., De Andres, J.A., Analisis estructural del espesor de la dura madre humana mediante microscopia electronica de barrido, Rev. Esp. Anestsiol. Reanim, 1996, Vol.43, 135-137.
4. Reina, M.A., Lopez-Garcia, A., Dittmann, M., De Andres, J.A., Analisis de la superficie externa e interna de la duramadre humana mediante microscopia electronica de barrido, Rev. Esp. Anestesiol. Reanim., 1996, Vol. 43,130-134.
5. Reina, M.A., Dittmann, M., Lopez-Garcia, A., van Zundert, A., New perspectives in the microscopic structure of human dura mater in the dorsolumbar region, Regional. Anesthesia, 1997, Vol.22(2), 161-166.
6. Green, H.,M., Lumbar puncture and the prevention of postpuncture headache, JAMA, 1926, Vol.86, 391-392.

A THREE DIMENSIONAL FINITE ELEMENT MODEL OF A CADAVERIC SECOND CERVICAL VERTEBRA (THE AXIS)
I:- Modeling

E C Teo[1] and J P Paul[2]

1. ABSTRACT

The purpose of this first part of the study was to develop a three-dimensional finite element model of the human second cervical vertebra. The dimensions of the model was obtained by digitizing the outer surface profile of an embalmed, adult second cervical vertebra. The overall geometric shape and size of the vertebra was a representative of an adult cadaveric vertebra. The unique bony structure and its distinguished tooth-like process of the vertebra was modeled consisting of 2772 nodes and 2044 isoparametric 8-noded and 6-noded solid elements, comprising an layer of element of cortical bone encasing the inner cancellous bone. The model should offer potential for the later analytical study of the injury mechanisms of either the hangman or odontoid fracture of the second cervical vertebra. These patterns of fracture were commonly reported for victims involved in head injury under traumatic accidents.

2. INTRODUCTION

In human, the second cervical vertebra (the axis) is a unique bony structure and easily distinguished by its blunt tooth-like process (the dens or the odontoid). Its articulation with the superior vertebra and the inferior vertebra below permit the wide range of the

Keywords: Cervical vertebra, Modeling, Odontoid fracture, Hangman's fracture

[1] Lecturer, Engineering Mechanics/ School of Mechanical & Production Engineering, Nanyang Technological University, Nanyang Avenue, Singapore 639798
[2] Professor, Biomechanics/ Bioengineering Unit, University of Strathclyde, Glasgow G4 0NW, Scotland, UK

motion of head in all directions [1-3]. The dens together with its attachments of ligaments prevents excess translation of the superior vertebra while permits the head and the superior vertebra to rotate around the dens. The vulnerability of the axis to fracture of either hangman's type [4] or odontoid type [5] is common for victims involve in traumatic situations such as vehicular collision, diving into shallow water, fall, etc.

The small oval articular facet at the front of the dens forms a joint with the back of the anterior arch of the superior vertebra, direct impact load to the head could transmit a shear force at the base of the dens and also induce bending moment at the pars interarticularis results in fracture specified pattern [6,7].

In our previous study [8], our goal was to produce a finite element model of the axis, which will allow the investigation of the stress distribution in the axis in a qualitative manner. To our knowledge, this is the first analytical model representation of the axis. demonstrating the stress were localized either at the dens or the par-interarticularis under one unit load, depending on the direction of the loading.

The purpose of this study was to re-model the three-dimensional model of the axis including the representation of the outer layer and inner layer of cancellous bony structures. It was also aimed that the model would allow future investigation of stress distributions in the axis under different boundary conditions and load application for the study of the fracture mechanism.

3. MATERIALS AND METHODS

A three-dimensional model of the human second cervical vertebra including the presentation of outer layer of cortical bone encasing the cancellous bony structure was developed. The geometric dimensions is based on an embalmed, adult second cervical vertebra as shown in figure 1.

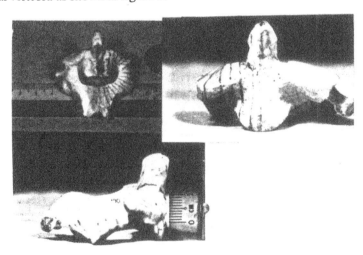

Fig 1. Posterior, anterior and lateral views of the 2nd cervical vertebra used for finite element modeling

The axis is of unique geometry, small in size and of complex shape, the use of ordinary measuring devices such as vernier calipers to obtain the coordinates of the outer profile would involve great difficulties, lengthy time and result in data of doubtful accuracy. Hence a computer controlled Coordinate Measuring machine (CMM) (UMC550, Carl Zeiss, Germany) was used instead; it allows the digitizing of points on the surface of the vertebra with ease and accuracy. Half of the vertebra was marked with lines using pencil over the surface, dividing it into eighteen sections of approximately equal interval apart; ensuring that the geometric shape of the model to the anatomic structure was maintained.

The vertebra, with a allen screw tighten on it, was then clamped in its upright position to a rig resting on the platform of the CMM machine. The Zeiss CMM machine connected with five pointer probe was then used to digitize around the lines marked in an orderly manner (Figure 2).

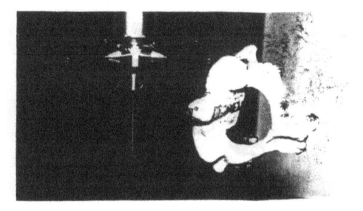

Fig 2. The set-up of the measurement configuration showing the probe and the vertebra

The coordinates of each point digitized were automatically registered once the probe's sensing pressure was reached. These coordinates were then processed and read in phase one of a commercial available finite element software, PAFEC FE. During the interactive session, all the points were connected using beam elements to create the outer profile of each section as shown in figure 3. Within these sectional views, additional nodes were generated within each section to allow the creation of finite element mesh between sections. The complete meshing of all the finite elements generated between the first section to the last section led to the formation of the three-dimensional model of the vertebra. Figure 4 shows the three-dimensional finite element model of the left half of the vertebra. It consists of 1022 solid elements, mixture of quadrilateral 8-noded (brick) and triangular 6-noded (wedge) finite elements. The vertebra was assumed to be bilateral, and the full finite element model was generated by mirroring about the mid-sagittal plane. The consistent discretization into finite number of elements from section 1 to 18 allows the mechanical properties of cortical and cancellous bones to be defined based on previous studies [9,10] for the outer layer of finite elements encasing the inner layers of elements, respectively. The articular cartilage layer and the intervertebral disc between the vertebra and its inferior vertebra can also be simulated using spring elements of stiffness defined in the normal direction and axial and anterior shear directions, respectively.

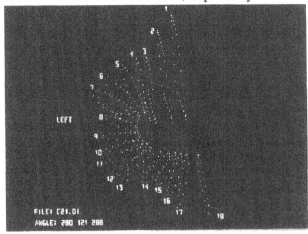

Fig 3. Generated sectional view of the 2nd cervical vertebra

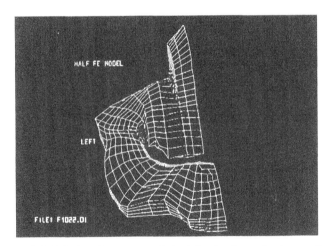

Fig 4. 3-d finite element model of the 2nd cervical vertebra

4. DISCUSSION

The finite element model developed in the current work is a three dimensional representation of the human second cervical vertebra. Seven specimens obtained from elderly subjects died of various diseases, with minor variations in the overall dimensions, were used to quantify the geometrical shape and size of the second cervical vertebra. The embalmed eighth specimen was used for the digitizing and modeling of the mesh. Table 1 showed the relevant data regarding the cadaveric specimens used.

Table 1: Relevant data on cadaveric specimen

Specimen No	Sex	Age	Weight (N)	Height (m)	Cause of death
1	F	76	528	---	Ischaemic heart disease
2	M	73	396	1.52	Ischaemic heart disease
3	M	73	378	1.68	Bronchopneumonia
4	F	80	305	1.50	Acute myocardial infraction
5	M	64	695	1.62	Liver cirrohosis
6	M	54	340	1.57	Ischaemic heart disease
7	M	51	404	1.70	Bronchopneumonia
8	--	--	--	--	-----

The second cervical vertebra of these cadavers were obtained to determine the overall width and depth of the vertebra, the odontoid process, spinal canal and end-plate. The results were tabulated in Table 2 below.

Table 2: Second cervical vertebrae dimensional data (in mm)

S.No	H_o	D_o	W_o	H_d	D_d	W_d	D_c	W_c	D_b	W_b
1	31	46	45.5	15	7.1	7.5	19	23	16.5	19.5
2	32	46.5	45	15	8.5	8.5	20	27	15.5	19.5
3	33	52.5	44.5	14.5	11	10.5	21	24	19.0	20.5
4	32	42.5	44.5	13	10.5	9.5	17	22	15	17.5
5	32	48.5	44.0	16.5	7.2	8.5	20.0	23.5	15	17.5
6	31.5	47.5	42.2	13.5	10.5	9.2	19.5	23	16	18.5
7	36	47.5	44.0	15	11.5	9.5	20.5	25.5	17	19
8	28	44.0	43.0	12.5	10.5	8.5	18.0	23	15.5	18.2
Mean	31.94	46.88	44.09	14.38	9.6	8.96	19.4	23.88	16.2	18.78
Ratio	H_o/D_o		W_o/D_o		H_d/H_o		W_d/D_d		W_c/D_c	W_b/D_b
**	0.68		0.94		0.45		0.93		1.23	1.16

H: Height (S-I) D: Depth (A-P) W: Width (Lateral)

o: Overall dimension d: Odontoid process (dens) c: Spinal canal b: Base of body

** Mean ratio

The linear dimensions of the vertebral parts measured compared favorably with those of the study by Panjabi, et al (11). In their studies of twelve fresh autopsy spine specimens of subjects had an average age of 46.3 years (range, 19-59 years), weight of 678 N (range, 540-850 N), height of 1.678 m (range, 1.57-1.78 m), and male-to-female ratio of 8 to 4, the spinal canal measured 21 mm and 24.5 mm, while the end-plate measured 17.5 mm and 15.6 mm in width and depth, respectively.

The embalmed second cervical vertebra (specimen no. 8) used for the finite element modeling was a idealized size and shape of human adult specimen. The final form of the vertebral model ensured that the three dimensional geometry was maintained with avoidance of sharp transitions in the element shape or size. Mesh refinement of the critical regions can be incorporated independently if required, to investigate a more detailed description of the pattern of the response in a particular analysis.

The possibility to enhance the geometric and mechanical equivalence of the finite element model to a real object, and to hence the mechanical behavior of the model approximate to those for a real object using the Finite Element Analysis has been applied successfully in the modeling of the second cervical vertebra.

The various stages involved in the modeling of the vertebra are described in sufficient detail to give an appreciation of the finite element modeling. The geometric representation of the vertebra model in this work was developed precisely to ensure the reliability of the analyses carried out in the future.

5. REFERENCES

1. Panjabi MM, Dvorak J, Duranceau J, et al (1986) Three-dimensional movements of the upper cervical spine. Spine 13(7):726-730

2. Panjabi MM, White AA & Johnson RM (1975) Cervical spine mechanics as a function of transection of components. J Biomech 8:327-336

3. Panjabi MM, Summers DJ, Pelker RR, et al (1986) Three-dimensional load-displacement curves due to forces on the cervical spine. Orthop Res 4:152-161

4. Schneider RC, Livingston KE , Cave AJE, et al (1965) Hangman's fracture of cervical spine. J Neurosurg 22:141-154

5. Anderson LD & D'Alonzo RT (1974) Fractures of the odontoid process of the axis. J Bone & Joint Surg 56(A):1663-1674

6. Cornish BL (1968) Traumatic spondylolisthesis of the axis. J Bone & Joint Surg 50(B):31-43

7. Mouradian WH, Fietti VG, Cochran GV, et al (1978) Fractures of the odontoid -A laboratory and clinical study of mechanisms. Orthop Clin North Am 9:985-1001

8. Teo EC, Paul JP & Evans JH (1994) Finite element stress analysis of a cadaveric 2nd cervical vertebra. J Med & Bio Engrg & Computing 32(2):236-238

9. Evans FG (1973) Mechanical properties of bone. Thomas, Springfield

10. Yamada H (1970) Strength of biological materials. Williams & Wilkin, Baltimore

11. Panjabi MM, Duranceau J, Goel V, et al (1991) Cervical human vertebra - Quantitative three-dimensional anatomy of the middle and lower region. Spine 16(8): 861-869

MODELLING THE TRUNK RESPONSES TO LUMBAR MANIPULATIVE FORCES

Michael Lee[1] and Grant P Steven[2]

1. ABSTRACT

To optimise the use of spinal manipulation for maximum effectiveness and safety, it is necessary to investigate the kinematics and dynamics of current procedures. The objective of our investigations is to model the responses to the slow application of manipulative force to the bony prominences in the midline of a prone subject. A finite element (FE) model was developed to allow the prediction of forces and deformations in the various anatomical elements of the spine, ribcage and pelvis. The validity of the modelling procedure was examined by comparing the model's predictions of behaviour with the responses demonstrated by living subjects. Studies of the average responses to low lumbar forces showed that the model was a good predictor of the observed behaviour of living subjects. To more broadly examine the model validity, an investigation was carried out to evaluate the degree to which the model could predict variations in mechanical behaviour between 20 healthy individuals. Because of the clinical relevance, the prediction of responses to low lumbar forces (at L4) was the main focus of this experiment although the behaviour in response to forces at 4 other spinal locations (L1, T10, T7, T4) was also studied. In the case of the low lumbar (L4) forces, good correlation between model and living subjects was found (r=0.67), but at other locations the model showed substantially lower correlations with living subject data.

2. INTRODUCTION

For many centuries low back pain has been treated with manipulative therapy[1]. From around the time of Hippocrates a treatment method that involved the application of force

Keywords: spine, manipulation, model, Finite Element Analysis

[1] Lecturer, School of Exercise & Sport Science, University of Sydney, East St, Lidcombe, NSW, 2141 Australia

[2] Professor, Department of Aeronautical Engineering and Finite Element Analysis Research Centre, University of Sydney, 2006, Australia

to the patient's spine has been used. Modern methods involve the manual application of force, usually directed towards the bony prominences of the spine; the spinous and transverse processes[2]. Despite the long history of the use of manipulative therapy, there is little information available about the mechanical effects of this type of treatment.

Part of the process of advancing our understanding of manipulative procedures and optimising their use for maximum effectiveness and safety, is to investigate the kinematics and dynamics of current procedures. To establish safe practice of manipulative therapy it is necessary to know the tissue loads that are produced by the application of a given therapeutic force and to compare these loads with the maximum load carrying capacity of the tissue. For maximum clinical effectiveness, a strategy for spinal loading is needed that focuses the stress in certain tissues while avoiding other tissues. Further, knowledge of the relative movements between the vertebrae of the spinal column is required. To achieve the objectives of safety and effectiveness requires information about movements and stresses that cannot be readily obtained by experiments with living subjects.

The objective of our investigations is to understand the mechanical responses to one type of manipulative procedure. There are many different manipulative procedures but the procedure we have chosen to study involves the slow application of force to the bony prominences in the midline (the spinous processes) while the subject is lying prone on a treatment table. The manual force is usually applied to the lower lumbar vertebrae (L4 and L5) as these are the areas of greatest clinical problems. Modelling was performed under the assumption that the spinal posture in the prone position is the same as that in the standing position, and this is supported to some extent by previous research[3].

3. METHODOLOGICAL ISSUES

The patient's force-displacement response consists of an initial non-linear toe region, followed by an approximately linear region[4]. The response can be adequately described by a linear function provided the applied force is above 30 N and the range of forces considered is moderate, less than 100 N[5]. Currently, clinicians place great importance in the movement stiffness at the point of loading, as an indicator of the mechanical behaviour of the spine[2]. Hence, the gradient of the linear portion of the force-displacement relation, the stiffness coefficient K, is the main parameter of clinical interest. However, the actual sensitivity of this parameter to various factors, in particular the spine mechanical properties, has not yet been determined. Clinicians hypothesise that K will be quite sensitive to changes in the stiffness of intervertebral joints adjacent to the point of loading. It has been demonstrated that K changes in association with changes in the level of low back pain[6].

Another aspect of the spinal response that is of interest to clinicians is the size and nature of intervertebral movement at the intervertebral joints near the point of loading. Many patients come for treatment with symptoms thought to be due to pressure on a nerve root at its point of entry into the spine from the periphery, the intervertebral foramen. Certain intervertebral movements are associated with changes in the intervertebral foramen size and these changes may have direct effects on the nerve root. Hence knowledge of the intervertebral movements associated with particular manual loading patterns will help the clinician to predict the effect of treatment on the size of the intervertebral foramen and hence on nerve root-related symptoms. Further, knowledge of intervertebral movements, combined with existing information about the anatomical connections of tissues of interest, will allow the clinician to make qualitative predictions about the possible effects of a manipulative force on symptoms arising from particular tissues. For example the prediction of an increase in the amount of intervertebral flexion is likely to be associated with an increase in the amount of tension in the supraspinous ligament and may cause an increase in symptoms arising from this structure. Lastly, the question of safety can be addressed initially by considering the therapeutic tissue loads in relation to the failure loads, or more detailed examination of the stress distributions within tissues can be conducted.

There are a number of possible modelling approaches to providing the information needed by clinicians. Use of the FE method is valuable because of the ultimate need to examine stress distributions within irregular elements. However, the traditional FE approach of detailed modelling of all the members[7] involved is difficult to apply because of the large number of separate components. Since a large number of finite elements would be used for each anatomical structure then the total structure, comprising many anatomical components, would require excessive solution times and storage needs. One alternative approach is to study only a small region of the spine with appropriate boundary conditions to represent the restraint provided by the connected tissues. The difficulty with this approach is that the boundary conditions are hard to accurately simulate. In experiments with human subjects we have shown that movement of the spine occurs some distance away from the point of loading[8] and so the model must be relatively large to extend to a boundary that can be modelled as showing no displacement. On the other hand, if movement at the boundary is to be allowed, there must be sufficient data available to accurately describe the movement.

Because of the lack of knowledge about behaviour within the human body, we have adopted the approach of building an FE model that is extensive in its coverage of the human body, yet with each anatomical element represented as simply as possible.

One of the major problem areas in simplifying the FE representation of the body is the representation of each intervertebral joint. These joints are complex structures comprising discs, paired zygapophyseal joints and a number of ligaments. Although cadaver studies have found that the intervertebral joints behave in a highly non-linear fashion[9], the extent to which this behaviour exists in the living is not known. We have chosen to represent the behaviour as linear on the basis that for the joints of greatest interest, the loads vary within the moderate range and so possible non-linearities at the upper and lower loads are less important.

We have developed an approach[10], also used by Gardner-Morse, Stokes and colleagues [11], of representing the intervertebral joint complex by a single beam. Previously published results of experiments with cadaver spinal units have been used to establish the approximate linear stiffness of each intervertebral joint. A disadvantage of this method is that the observed linear stiffness matrix does not follow the pattern of a standard beam behaving in accordance with the "Engineers Theory of Bending". We have dealt with this problem by working in conjunction with the software supplier (*STRAND6* supplied by *G+D Computing*) to alter the FEA software. The modified software reads the experimentally determined element stiffness matrix directly for each intervertebral beam, rather than using the conventional method of assembling the element stiffness matrix from the basic geometric and material properties.

4. SIMULATION OF RESPONSES OF AVERAGE MALE AND FEMALE

4.1 Scope

For our initial analyses the vertebrae, sternum and pelvic bones are represented as three-dimensional rigid bodies. The three-dimensional representation of these bones facilitates the application of loads to particular parts of each bone [manipulative forces or muscle forces], as well as the calculation of displacements of key points on the vertebra for qualitative consideration of loading of certain anatomical elements. Each of the ribs is initially modelled as a series of 6 standard beams, again providing an economical method of representation. The skin and subcutaneous fat is represented with linearly elastic springs interposed between the manipulative force and the vertebra, as well as between the pelvis and the treatment table.

4.2 Geometry

The geometry (Figure 1) was selected to produce an average male and female model. Previously published data for the spine were obtained to describe the overall spinal curvature as well as the dimensions of the individual vertebrae[12, 13, 14]. The ribcage geometry was based on published data[15] but the data were then scaled to match the spinal column dimensions. The dimensional data for the pelvis was obtained by direct measurement of a cadaver specimen and the data were also scaled to match the spinal dimensions. Average values of superficial soft tissue thickness were obtained by referring to published data[16, 17].

4.3 Material properties

As previously mentioned, the data for the material properties of the intervertebral joints were based on published data[11, 18, 19, 20], assuming the responses to be linear. These data come from experiments that have involved relatively small numbers of cadavers and so their relation to responses of living subjects is not clear. Rib and costal cartilage property data were based on those values developed by Roberts & Chen[15] but were assumed to be constant along the length of any one rib. The sacro-iliac joints were each modelled with 2 beams and their properties adjusted to match the behaviour of the joints that has been observed[21]. The properties of the superficial soft tissues were obtained by fitting a linear approximation to the published compressive behaviour[16, 17] corresponding to the approximate range of pressures expected in this context.

4.4 Boundary conditions

The subject was initially assumed to be lying on a rigid bed. Therefore the ribcage was assumed to be supported so that no anterior movement of the most anterior points was allowed due to the relatively thin layer of soft tissue overlying the sternum and nearby ribs (in females the breasts were assumed to be able to move to allow good contact between sternal area & bed surface).

The bony pelvis was assumed to be supported at the pubic symphysis, but separated from the rigid bed surface by a compressible layer of skin and subcutaneous tissue. The pelvis was also assumed to be constrained from free rotation in the sagittal plane by a torsional spring, whose behaviour was matched to the resistance observed in a study of living subjects performed in our laboratory[22].

4.5 Validation study

To indicate the validity of the model as a representation of human behaviour, the model behaviour was compared with living human responses that have been observed in our laboratory. A number of different experimental situations were examined to allow different aspects of the validity to be studied. The behaviour of the model was found to be close to the average human responses in the following contexts: (i) with the subject lying prone, loads were applied to L3, L4 or L5 vertebrae, with the movement stiffness at the point of loading used as the outcome measure; (ii) subject lying prone with load applied at the mid-lumbar spine, L3 - the skin surface displacement at L3 and at several points along the spine was used as the outcome measure; (iii) ribcage compression with subject lying supine - movement stiffness at the point of loading was used as the variable for comparison between the model and human responses.

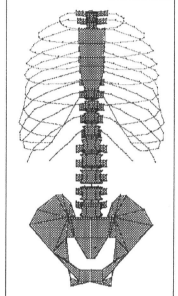

Figure 1. The STRAND6 model of average male subject (front view)

4.6 Results

Studies of the responses to middle-to-low lumbar forces have shown that the model was a good predictor of the average behaviour of living subjects[23]. Not only did the overall structure stiffness at the point of loading match the average human subject stiffness, but the movement of remote points, the pelvis and the thoracic vertebrae, were also well predicted. In addition, the compressibility of the model ribcage was well matched our experimental data from living subjects[24]. In almost all cases the model's predicted responses were within one standard deviation of the mean values observed in living subjects, suggesting that the model is a good representation of average behaviour. These experiments focussed on situations in which the load was applied in the mid-to-lower lumbar spine because this is where clinical manipulative therapy forces are most commonly applied.

5. INDIVIDUAL VARIATIONS

5.1 Background

One of the aims of modelling the responses to manipulative forces is to give a greater understanding of the variables that determine the variations between individuals in their responses to a manipulative force. Therefore, although the model is a good predictor of average responses, if it is to be useful for clinical purposes it must also be capable of accurately predicting the diversity of responses that have been observed among a group of individuals. Hence the next phase of the modelling was to produce a series of models that were matched. to 20 individuals. Each individual's measured responses to manipulative forces were then compared with the model's predictions.

5.2 Method

The model was matched to each individual subject on the basis of gender-based bony geometry (ribcage and pelvis), gross dimensions, skinfold thicknesses, and abdominal compressibility. Because of the clinical relevance, the prediction of responses to low lumbar forces (at L4) was the main focus of this experiment. However, to examine reasons for possible failure of the model to predict an individual's behaviour, the behaviour in response to forces at other spinal locations (L1, T10, T7, T4) was also studied.

5.3 Results

The correlations between the model predictions and the observed responses are shown in Figure 2. In the case of the low lumbar (L4) forces, good correlation between model and living subjects was found (r=0.67). At other locations the model showed substantially lower correlations with living subject data (r=0.07 to r=0.34) suggesting a significant role for subject variables not modelled. This result is consistent with recent findings from another study conducted in our laboratory[25]. An examination of the anthropometric predictors of responses to manipulative forces applied to each lumbar vertebral level found that

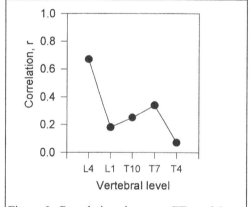

Figure 2. Correlations between FE model predictions and observed responses.

skinfold thickness was a very strong predictor of low lumbar responses but that the strength of prediction decreased for upper lumbar levels. Because the subject's skinfold thickness is included as a variable in our FE model, it could be expected that the variability in response attributable to skinfold thickness would be manifest in the predictions of the model. The weaker relation between skinfold thickness and the response to manipulative force at upper lumbar levels could therefore be one major factor explaining the lower correlation between model predictions and living subject behaviour at upper lumbar levels. Other possible reasons for the poor predictions of responses at certain locations include: inadequate representation of the abdominal contents, failure to account for variations between individuals in spinal material properties, and deficient representation of the individual variations in boundary conditions.

6. CONCLUSIONS

We conclude that, for simulation of slow loading over the L4 vertebra, our FE model is a valid predictor of both average and individual responses. However the failure of our model to predict the variation of responses at other vertebral levels suggests the presence of variables not accounted for by the model.

Our research in the immediate future will be aimed at improving the model's ability to predict variations in responses to loads in the upper lumbar and lower thoracic spine. Possible variables that may be incorporated in the model will be: individual spinal material properties and inclusion of more details of the nature of the support of the body ribcage and pelvis at the model boundary.

7. ACKNOWLEDGMENT

We would like to acknowledge the contribution of Assoc. Prof DW Kelly to the development of the FE model.

8. REFERENCES

1. Schiotz EH, Cyriax J: Manipulation Past and Present. William Heinemann Medical Books, London, 1975
2. Maitland GD: Vertebral Manipulation, 5th Ed. Butterworths, London, 1986
3. Knight C, Pitman E: Comparison of lumbar posture in standing and in prone lying in normals. Unpublished project report for BAppSc (Physio), Lidcombe, Cumberland College of Health Sciences, 1989.
4. Lee M., Svensson N.L., Measurement of stiffness during simulated spinal physiotherapy. Clin. Phys. Physiol. Meas., 1990, Vol. 11, 201-207.
5. Lee M, Latimer J, Maher C: Normal response to large postero-anterior lumbar loads - a case study approach. J Manip Physiol Ther. In press, 1997
6. Latimer J., Lee M., Adams R. et al., An investigation of the relationship between low back pain and lumbar posteroanterior stiffness, J. Manip. Physiol. Ther. 1996, Vol. 19, 587-591.
7. Shirazi-Adl A., Ahmed A.M., Shrivastava S.C., A finite element study of a lumbar motion segment subjected to pure sagittal plane moments, J. Biomech., 1986, Vol. 19, 331-350.
8. Lee M., Svensson N.L., Effect of frequency on response of the spine to lumbar posteroanterior forces, J. Manip. Physiol. Ther., 1993 Vol. 16, 439-446.
9. Schultz A.B., Warwick D.N., Berkson M.H., Nachemson A.L., Mechanical properties of human lumbar spine motion segments - Part 1: Responses in flexion, extension, lateral bending and torsion. J. Biomech. Eng., 1979, Vol. 101, 46-52
10. Lee M. Response of the spine to one manipulative physiotherapy procedure, MBiomedE thesis, 1990, University of New South Wales.
11. Gardner-Morse M.G., Laible J.P., Stokes I.A.F., Incorporation of spinal flexibility measurements into finite element analysis. J. Biomech. Eng., 1990, Vol. 112: 481-483.

12. Stagnara, P., de Mauroy, J.C., Dran, G., Gonon, G.P., Costanzo, G., Dimnet, J. and Pasquet, A. Reciprocal angulation of vertebral bodies in a sagittal plane: Approach to references for the evaluation of kyphosis and lordosis. Spine, 1982, Vol. 7, 335-342.

13. Nissan, M. and Gilad I., Dimensions of human lumbar vertebrae in the sagittal plane. J. Biomechanics, 1986, Vol. 19, 753-758.

14. Schultz, A.B., Belytschko, T.B., Andriacchi, T.P. and Galante, J.O., Analog studies of forces in the human spine: mechanical properties and motion segment behaviour J. Biomechanics, 1973, Vol. 6, 373-383.

15. Roberts, S.B. and Chen, P.H., Elastostatic analysis of the human thoracic skeleton. J. Biomechanics, 1970, Vol. 3, 527-545.

16. Himes, J.H., Roche A.F., Siervogel, R.M., Compressibility of skinfolds and the measurement of subcutaneous fatness. Am. J. Clin. Nutr., 1979, Vol. 32, 1734-1740.

17. Brožek, J., Kinzey, W., Age changes in skinfold compressibility. J. of Gerontology, 1960, Vol. 15, 45-51.

18. Panjabi, M.M., Brand, R.A. and White, A.A., Three-dimensional flexibility and stiffness properties of the human thoracic spine. J. Biomechanics, 1976, Vol. 9, 185-192.

19. Berkson, M.H., Nachemson, A. and Schultz, A.B., Mechanical properties of human lumbar spine motion segments - Part II: Responses in compression and shear; Influence of gross morphology. J. Biomech. Engng., 1979, Vol. 101, 53-57.

20. McGlashen, K.M., Miller, J.A.A., Schultz, A.B. and Andersson G.B.J., Load displacement behaviour of the lumbo-sacral joint. J. Orthop. Res., 1987, Vol. 5, 488-496.

21. Miller, J.A.A., Schultz, A.B. and Andersson, G.J.B., Load-displacement behaviour of sacroiliac joints. J. Orthop. Res., 1987, Vol. 5, 92-101.

22. Lee, M., Lau, T. and Lau, H., Sagittal plane rotation of the pelvis during lumbar posteroanterior loading. J. Manip. Physiol. Ther., 1994, Vol. 17, 149-155.

23. Lee M, Kelly D.W. and Steven G.P., A model of spine, ribcage and pelvic responses to a specific lumbar manipulative force in relaxed subjects. J. Biomechanics, 1995, Vol. 28, 1403-1408.

24. Lee, M., Hill, S. and Scullin, J., Ribcage compressibility in living subjects. Clin. Biomech., 1994, Vol. 9, 379-380.

25. Viner A., Lee M., Adams R., Postero-anterior stiffness in the lumbosacral spine: The relationship of stiffness between adjacent vertebral levels, Spine, 1997, In press.

DYNAMIC THREE-DIMENSIONAL FINITE ELEMENT MODEL OF A SITTING MAN WITH A DETAILED REPRESENTATION OF THE LUMBAR SPINE AND MUSCLES

B. Buck[1], H. P. Woelfel[2]

1. ABSTRACT

A three-dimensional dynamic Finite-Element-Model (FEM) of sitting 50-percentile man is presented. The model is based on a close representation of human anatomy with special focus on the lumbar spine and muscles. The lumbar spine model consists of a detailed representation of the discs and vertebras with six different materials (including orthotropic and viscoelastic constitutive laws), non-linear ligament models and a non-linear contact model in the articular facets. The muscle model contains the dynamic properties of passive as well as active muscle tissue. Forces in muscles are calculated using an optimization approach. The model is completed with rigid-body models for the upper torso with neck, head and arms, the pelvis and the legs and the viscera. The model is validated starting with the disc model and ending with the whole body model by numerous static and dynamic experimental results taken from literature. It is used to evaluate the influence of muscles on whole body dynamics. Forces in the lumbar spine are calculated which are necessary to assess the potential risk of whole body vibrations (WBV) for the lumbar spine.

2. INTRODUCTION

Low-back pain due to long-time exposure to WBV is a widespread problem with professions like truck or bulldozer drivers or helicopter pilots. In Germany it is accepted as an occupational disease for which compensation can be claimed. The aetiology of the desease is not yet fully explained. Especially mechanical aspects are difficult to assess

Keywords: WBV, Lumbar Spine Model, Muscle Model, Internal Forces

---

[1] Research assistant   Fachgebiet Maschinendynamik, Technical University Darmstadt,
[2] Professor   Petersenstrasse 30, 64287 Darmstadt, Germany

due to the limitations in measuring loads and displacements of the lumbar spine in vivo. Nevertheless an assessment of already existing degenerations and industrial working places is urgently needed. Additionally, the influence of muscles on whole body dynamics is interesting, since muscle fatigue could change the human dynamic behaviour under WBV. Existing dynamic models with detailed lumbar spine [1, 2, 3, 4, 5, 6, 7, 8, 9] are insufficient in modelling the viscera, muscles and energy dissipation. Therefore, a new dynamic, anatomy based FEM including the three-dimensional whole body dynamics of sitting man as well as a detailed representation of the lumbar spine and muscles is presented.

3. METHOD

The Finite Element approach is chosen, enabling the detailed modelling of structures with a wide variety of materials under large 3D-movements. Due to the dynamic character of the problem not only geometrical and stiffness properties have to be modelled, but also the mass distribution and most importantly the damping or energy dissipating characteristics. The modelling concept has to be based on anatomy since phenomenological models do not allow to extrapolate measured forces or accelerations at the seat or head to internal forces in the lumbar spine. The model basically consists of five components: a detailed model of the lumbar spine, a rigid-body model of the upper torso with neck, head and arms, a detailed model of the lumbar muscles, a rigid-body model of the pelvis, legs and gluteal tissue and a visceral model.

3.1 Lumbar spine model

First, the geometry of the lumbar spine has to be described with a FE mesh. Since non-linear dynamic calculations have to be made, the number of nodes must be kept as small as possible. Therefore the widespread approach to digitize a CT-image and mesh the volume by an automeshing algorithm (e.g. [10]) cannot be used. Instead the geometry is generated starting from a basic model of the functional unit (FU) which contains information about material distribution as well as meshing. This basic model is geometrically transformed into any of the lumbar FUs by parametric scaling, rotating and translating, see Figure 1. 18 parameters are used to determine the geometry of each FU. The parameter values were taken from literature to represent a 50-percentile man (for details see [11,12]).

The materials of the lumbar spine were modelled by six material descriptions: a linear isotropic constitutive law with four different sets of parameters for compact bone, spongy bone, the bone of the vertebral arch and the cartilage endplate, a linear orthotropic material for the annulus and a linear viscoelastic almost incompressible representation of the nucleus of the disc. Energy dissipation is modelled by material-dependent Raleigh-damping except in the nucleus where it is inherent in the viscoelastic material description. Material parameters are determined from material testing experiments taken from literature wherever possible. When reliable information is not available, material parameters are determined by fitting the models behaviour to experimental investigations of the whole disc or FU. Especially the parameters for energy dissipation have to be determined in this way

Figure 1: Basic model of FU and transformed model of vertebra L3. Different materials are shown.

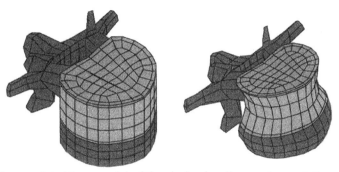

The lumbar spine model is completed by a model of the six lumbar ligaments consisting of non-linear springs with individual force-deformation-characteristics taken from literature [13]. Additionally the articular facets are modelled by three contact elements on each side which orientation in space varies for each vertebral level. Energy dissipation of these components is represented by linear dashpot elements.

3.2 Muscle model

The anatomy of the lumbar muscles is modelled by 18 different muscles. Each muscle is divided into several muscle cords to describe the lines of action of fan-like muscles. A total of 102 muscle cord is used. The muscle cords connect the attachment points on the skeleton by a straight line. The geometry of the pelvis, thoracic spine and rib cage is modelled for this task. For an example for muscle representation see Figure 2.

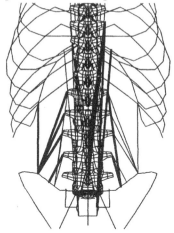

Figure 2: Example of muscular anatomy representation: M. quadratus lumborum (bold lines on the left) and M. longissimus dorsi (bold lines on the right)

Each muscle cord has to be assigned mechanical properties. In the case of muscle tissue the properties consist of the properties of the passive muscle tissue, which are known from autopsy experiments, and the properties of the activated muscle, which must be determined in living organism. Existing whole body models neglect the activated part of tissue properties. The force in an active muscle depends on length of the muscle, its lengthening velocity and the activation level of the muscle. It is therefore necessary to determine the activation level in each muscle prior to determination of its mechanical properties. The mechanism by which the neuro-muscular system is responding to a vibration input is not yet known. Hence, the following assumption is used: the activation of the muscles is constant over time and is determined by the static load, gravity and external loads put on the upright sitting man. This assumption is likely to be fulfilled for random WBV and harmonic WBV with frequencies above about 7 Hz. In these cases muscles do not show any activation response which is correlated to the mechanical WBV input [14].

Since the equilibrium of the upper torso under gravity is a statically overdetermined problem no unique solution exists: six equations of equilibrium can be used to determine the 18 muscle forces and three reaction forces in the lumbar spine. The reaction moments in the lumbar spine can be assumed to be zero, since in upright position the spine cannot bear any noticeable amount of moment. This problem is solved using an optimization approach. The cost function used is the square of sum of the stress in each muscle. This cost function is used by several other authors [7,15,16,17] and can be justified by experimental results on muscle fatigue [18]. The optimization scheme lead to muscle forces, which are 1 to 8 % of maximal muscle forces for upright sitting with hands on a steering wheel without external loads.

Little is known about the mechanical properties of activated muscle subjected to vibration. Most experimental investigations are made on animal muscle which is fully activated and subjected to harmonic or fast extension movements. Since mechanical properties at low activation levels are needed, the mathematical muscle model of Hatze [19, 20, 21, 22, 23] is used to calculate the muscular response to harmonic length changes of amplitudes characteristic for WBV. The model was verified using the existing experimental results at high activation levels. The calculated force responses to the excitation were transformed in the frequency domain with the help of FFT. They are split into an in-phase part and into an out-of-phase part. The first can be described as a stiffness coefficient the latter as a damping coefficient. Both coefficient depend on the frequency of the excitation and the activation level of the muscle. The dependency on excitation amplitude can be neglected in the range of amplitudes used here. As an example, the characteristic diagram for the stiffness is shown in Figure 3. The resulting coefficients were used for calculations in the frequency domain, where small frequency bands were calculated using the mechanical muscle properties of the centre frequency.

Figure 3: Characteristic diagram of relative muscle stiffness (i.e. the stiffness relative to maximal force divided by optimal muscle length)

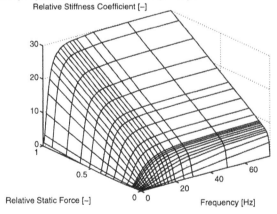

3.3 Rigid body models of remaining components

The upper torso, neck and head are modelled as rigid bodies connected by linear springs and dashpots. The parameters were taken from literature [24]. The arms are modelled as rigid beams connected by frictionless pin joints.

The visceral model is a chain of nine vertically aligned point masses interconnected by linear springs and dashpots. Additionally, the visceral masses are connected with the lumbar spine, the pelvis and the upper torso. Parameters were determined by evaluating cross sectional and volume data obtained from literature [25] and our own CT-images.

The pelvis is modelled as rigid body. The gluteal tissue is represented by linear translational and rotational springs and dashpots. Their parameters were determined by fitting the models results to experimental results for WBV. The modelling approach for the legs is similar to the arms. The complete model of sitting man is shown in Figure 4.

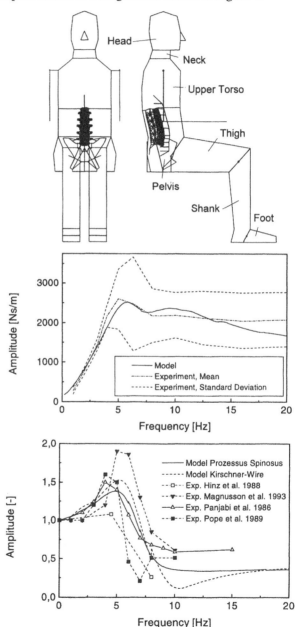

Figure 4: Complete model of sitting man

Figure 5: Verification:
top: Amplitude of the vertical driving-point impedance at the seat, experimental data from [26, 27],
bottom: Amplitude of the transfer-function from the seat to processus spinosus L3, experimental data from [28, 29, 30, 31]

4. VERIFICATION

The model and its components is verified using a great number of static and dynamic experimental results taken from literature (for details, see [11]). All components, for which experimental data is available (disc, functional unit, lumbar spine, muscle model), are first verified separately using autopsy results. Then, the whole body model is verified using data gained on experiments with living subjects. As an example, figure 5 shows model and experimental results for the vertical driving-point impedance at the seat and the transfer-function from the seat to the processus spinosus of vertebra L3.

5. RESULTS

Many calculations are made regarding the influence of non-linear model components, sensitivity of the model to single parameters, dynamic behaviour of the viscera, influence of the muscle model and internal forces in the lumbar spine. Figure 6 shows the influence of the muscle model exemplified as changes in driving-point impedance when the muscle model is removed. Significant changes can be observed in the higher frequency range above 6 Hz. Calculations are made with an algorithm, which linearises the equations of motion about a nonlinearly calculated prestressed state (gravity).

Figure 6: Influence of muscle model on the amplitude of vertical driving point impedance

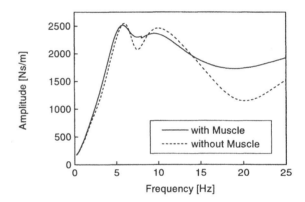

Real-life accelerations-time-functions measured in mining industry were used to calculate internal force-time-functions in the disc L3/L4, see Figure 7. The algorithm used accounts for all non-linear model components.

Figure 7: Internal forces in the L3/L4 disc under excitation measured in mining industry: dynamic part, static force = 411,6 N

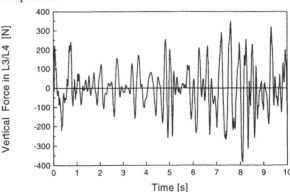

6. DISCUSSION

The model presented here is the first WBV model which combines a close anatomic representation of the lumbar spine with dynamic models of the upper torso, head and upper extremities, the pelvis and lower extremities, the viscera and a dynamic muscle model, which represents the mechanical properties of passive as well as active muscle tissue. Results show, that the influence of the muscle model is significant above about 6 Hz, which corresponds with the experimental results of Pope et al. [32], who found effects of different levels of muscle contraction in the same frequency range. The calculated internal forces in the lumbar spine can be used to assess the risk of specific working places. For example, the force-time-function shown in figure 7 is above the fatigue limit for elderly workers, when the compression strength of 2000 N reported by Jäger et al. [33] is used. The exact value of compression strength is still discussed, so the results obtained here have to be interpreted with care. On the other hand, the necessity of sound strength values is underlined by the relevance of lumbar spine fatigue shown by the results presented in this paper.

7. LITERATURE

1. Belytschko, T., Schwer, L. , Schultz, A.: A Model for Analytic Investigations of Three-Dimensional Head-Spine Dynamics, AMRL-TR-76-10, 1976
2. Belytschko, T., Privitzer, E.: Refinement and Validation of a Three-Dimensional Head-Spine Model, AMRL-TR-78-7, 1978
3. Belytschko, T., Privitzer, E.: Theory and Application of a Three-Dimensional Model of the Human Spine, Aviation, Space and Environmental Medicine, 158 - 165, 1978
4. Belytschko, T., Rencis, M. , Williams, J.: Head-Spine Structure Modelling: Enhancements to secondary loading Path Model and Validation of Head-Cervical Spine Model, AAMRL-TR-85-019, 1985
5. Privitzer, E., Belytschko, T.: Impedance of a Three-Dimensional Head-Spine Model, Mathematical Modelling, Volume 1, 189 - 209, 1980
6. Williams, J.L., Belytschko, T.B.: A Three-Dimensional Model of the Human Cervical Spine for Impact Simulation, J. of Biomechanical Engineering, 105, 321 - 331, 1983
7. Dietrich, M., Kedzior, K. , Zagrajek, T.: A biomechanical model of the human spinal system, IMechE 1991, Part H: J. Eng. Med., Proc Instn Mech Engrs, Vol. 205, 1991
8 Dietrich, M., Kedzior, K. , Zagrajek, T.: Biomechanical Modelling of Human Spine System, in: M. Dietrich, Lect. Notes ICB Sem. Biomech, Warschau, 38 -59, 1992
9. Luo, Z., Goldsmith, W.: Reaction of a Human Head/Neck/Torso System to Shock, J. Biomech., Vol. 24, 499 - 510, 1991
10. Lavaste, F., Skalli, W., Robin, S., Roy-Camille, R. , Mazel, C.: Three-Dimensional Geometrical and Mechanical Modelling of the Lumbar Spine, J. Biomech., Vol. 25, 1153 - 1164, 1992
11. Buck, B.: Ein Modell für das Schwingungsverhalten des sitzenden Menschen mit detaillierter Abbildung der Wirbelsäule und Muskulatur im Lendenbereich, Dissertation (eingereicht), TH Darmstadt, 1997
12. Buck, B., Wölfel, H.: A Dynamic Model for Human WBV with Detailed Representation of the Lumbar Spine, 10th Conf. of the European Society of Biomechanics, Leuven, Abstracts, 338, 1996
13. Nolte, L.-P., Panjabi, M.M. , Oxland, T.R.: Biomechnical Properties of Lumbar Spine Ligaments, in: G. Heimke, U. Soltesz, A. Lee: Clinical Implant Materials, Adv.

in Biomaterials, Vol. 9, Proc. 8th European Conf. on Biomaterials, Heidelberg, 1989

14. Seroussi, R., Wilder, D.G., Pope, M.H., Cunninghham, L.: The Added Torgue Imposed on the Lumbar Spine During Whole Body Vibration Estimated Using Trunk Muscle Electromyography, RESNA '87, Proc. of the 10th Ann. Conf. on Rehab. Techn., San Jose, 811 - 813, 1987

15. Dietrich, M., Kedzior, K., Zagrajek, T.: Modelling of Muscle Action and Stability of the Human Spine, in: J.M. Winters, S.L.-Y.Woo, Multiple Muscle Systems: Biomechanics and Movement Organization, Springer Verlag, 1990

16. Dietrich, M., Kedzior, K. Zagrajek, T.: Modelling of Spinal Muscles Co-Operation, IVth Int. Symp. on Comp. Simul. in Biomech., Paris, BMG1-6 - BMG 1-9, 1993

17. Pandy, M.G., Zajac, F.E., Sim, E., Levine, W.S.: An Optimal Control Model for Maximum-Height Human Jumping, J. of Biomech., Vol. 23, 1185 - 1198, 1990

18. Crowninshield, R.D., Brand, R.A.: A Physiological Based Criterion of Muscle Force Prediction in Locomotion, J. Biomech., Vol. 14, 793 - 801, 1981

19. Hatze, H.: A Myocybernetic Control Model of Skeletal Muscle, Biological Cybernetics, Vol. 25, 103 - 119, 1977

20. Hatze, H.: A complete set of control equations for the human musculo-skeletal system, J. Biomechanics, Vol. 10, 799 - 805, 1977

21. Hatze, H.: A General Myocybernetic Control Model of Skeletal Muscle, Biological Cybernetics, Vol. 28, 143 - 157, 1978

22. Hatze, H.: Neuromusculoskeletal Control Systems Modeling - A Critical Survey of Recent Developments, IEEE Transact. Autom. Contr., Vol. AC-25, 375-385, 1980

23. Hatze, H.: Myocybernetic control models of skeletal muscle, University of South Africa Press, Pretoria, South Africa, 1981

24. Amirouche, F.M.L., Ider, S.K.: Simulation and Analysis of a Biodynamic Human Model Subjected to Low Accelerations - a Correlation Study, J. Sound Vib., 123(2), 281 - 292, 1988

25. Eycleshymer, A.C., Schoemaker, D.M.: A cross section anatomy, Appleton-Century-Crofts, New York, 1970

26. Kinne, J., Melzig-Thiel, R.: Optimierung und Prüfung von Schwingschutzsitzen mit mechanischen Mensch-Modellen -Mittlere Frequenzgänge der Eingangsimpedanz von sitzenden Menschen (Teil 1), ErgoMed, 17. Jahrgang, 5, 154 - 157, 1993

27. see [26], (Teil 2), ErgoMed, 17. Jahrgang, 6, 191 - 193, 1993

28. Hinz, B., Seidel, H., Bräuer, D., Menzel, G., Blüthner, R.: Bidimensional accelera tions of lumbar vertebrae and estimation of internal spinal load during sinusoidal vertical whole-body vibration: a pilot study, Clin. Biomech., 3, 241 - 248, 1988

29. Magnusson, M., Pope, M. , Rostedt, M., Hansson, T.: Effect of backrest inclination on the transmission of bertical vibrations through the lumbar spine, Clin. Biomech., Vol. 8, 5 - 12, 1993

30. Panjabi, M.M., Andersson, G.B.J., Jorneus, L., Hult, E., Mattson, L.: In Vivo Meas urements of Spinal Column Vibrations, J. Bone Joint Surg., 68 A, 695 - 702, 1986

31. Pope, M.H., Broman, H., Hannson, T.: The dynamic response of a subject seated on various cushions, Ergonomics, Vol. 32, 1155 - 1166, 1989

32. Pope, M.H., Broman, H. ,Hansson, T.: Factors Affecting the Dynamic of the Seated Subject, Journal of Spinal Disorders, Vol. 3, 135 - 142, 1990

33. Jäger, M.: Assessing Spinal Load during Manual Material Handling via the Compressive Strength of Cadaeveric Lumbar-Spine Elements, 10th Conf. of the European Society of Biomechanics, Leuven, Abstracts, 174, 1996

A 3-Dimensional Large Deformation FEA of a Ligamentous C4-C7 Spine Unit

F. Heitplatz[I] , S. L. Hartle[II] and C. R. Gentle[III]

1. ABSTRACT

This study was conducted to develop a three-dimensional finite element model of the human cervical spine structure using data from the Visible Human Project. The developed model is the first step in an attempt to simulate the three-dimensional movement of the cervical spine during whiplash accidents in order to predict the strain inside the spinal ligaments, with a view to supporting the development of car restraint systems. The finite element model of the 4 vertebrae, C4-C7, consisted of solid elements, contact surfaces, non-linear springs and spar elements. Appropriate material properties from the literature were used. The progressive non-linear deformation behaviour of the intervertebral discs ,however, was derived from an analytical function and applied to non-linear spring elements. Since the development of the model is targeted on the prediction of spinal motion in transient dynamic simulations, compromises have been made on the topological accuracy to reduce the model complexity. The motion characteristics of the spine segment have been compared with experimental results where available. The results show that reduction of the topological structure to its mechanically significant features and programming of non-linear springs with analytically derived force / deformation characteristics to model the intervertebral discs, are effective measures for modelling the spine with a minimum of degrees of freedom.

Keywords : 3-D FEA, cervical-spine, large deformation, non-linear

I Research Student, Biomechnaics / Department of Mechanical and Manufacturing Engineering at Nottingham Trent University. Burton Street NG1 4BU Nottingham GB
E-Mail: f.heitplatz@domme.ntu.ac.uk
II Senior Lecturer, Department of Mechanical and Manufacturing Engineering at Nottingham Trent University
III Professor, Department of Mechanical and Manufacturing Engineering at Nottingham Trent University

2. INTRODUCTION

The problem of so-called " whiplash " injuries in car accidents due to either posterior or anterior impact has been discussed for many years but many questions remain unanswered concerning the mechanism of injury and the exact cause of sometimes chronic pain. It is widely accepted that the pain following whiplash injuries is mainly caused by damage to the soft tissue. However, the common belief that whiplash is a hyperextention/hyperflexion injury fails to explain many of the victim's complaints.

Recently a new theory has been developed which describes whiplash as a hypertranslation[1,2] injury where relative translation between head and trunk in the initial phase of the relative movement potentially leads to high strains of the alar and transverse ligament as well as the anterior ligament at the C0-C2 complex. It is the long term aim of this research project to develop a computer model capable of simulating whiplash situations in order to determine stress and strain in these ligaments during the initial stages of whiplash and hence test this theory.

The use of computer models to simulate spinal mechanics is widely accepted. Most models, particularly FE-Models, concentrate on the mechanics of the lumbar spine[3] with particular interest in the mechanics of the intervertebral disc[4,5]. However, several computer models of the cervical spine have been published. Early studies of the cervical spine where undertaken with the use of the finite difference method[6] but subsequently FE models of parts of the spine have been developed[7,8]. There is only one study of the full three dimensional cervical spine as an FE-Model which has been used to investigate the effects of whiplash[9], and this model does not include non-linear material properties for the ligaments and the intervertebral disc and it is not very detailed in its analysis of the C0-C2 complex. It is, therefore, the aim of this study to develop a non-linear FE-Model of the cervical spine, which is capable of dynamically simulating the cervical spine in whiplash accident situations. There are several requirements which have to be considered to achieve this goal; the model has to have a representative topology which includes all those features of the cervical spine which are important for the motion characteristics, as well as a representative model stiffness incorporating the material non-linearities. The number of degrees of freedom in the model has to be held to a minimum to allow full transient non-linear FEA with limited computing resources. A part solution for such a model is here presented.

3. METHODS

The presented model of the C4-C7 part of the cervical spine is based on the full-colour male dataset of the Visible Human Project[10]. The bony structure of the cervical spine was extracted using image processing software. Point arrays were created and transferred to the CAD-software Pro/ENGINEER®. A parametric model was built. During digitisation, simplifications were made to the vertebral bodies and spinosus processes in order to allow mapped meshing with hexagonal 8 noded FE-elements. The transverse processes were not included because they are not connected to any spinal ligaments and have therefore no relevance for the model in its current state. The overall dimensions of the vertebrae were preserved and a special focus was given to the topology of the facet joints to allow realistic motion characteristic of the joints under

3-dimensional large deformation sliding contact. A 3-dimensional tension only spar element was used to model the ligaments. The details of the final model, in terms of the type of elements, the number of elements used to replicate various regions of the C4-C7 model and the material properties assigned to various structures, are provided in Table 1. Material properties for the ligaments are taken from the lumbar spine[11] due to a lack of experimental data for the cervical spine. The bony structure is seen as homogeneous and nearly rigid.

Spinal Elements	FE Element Type	Material Properties	No. of Elements
4 Vertebrae	3-D 8 node solid	E=12000 $\nu = 0.3$	856
3 Intervertebral Discs	3-D 8 node solid	E= 3 $\nu = 4.2$	36
	3-D non-linear springs	$F(\delta)$ see equation(1)	60
3 Facet Joints	3-D symmetric surface to surface contact	k = 1200	1410
Ligaments:			
3 Ligamentum Nuchae (funicular portion)	3-D tension only cable	E = 8.0	9
3 Inter-Spinous	3-D tension only cable	E = 10.0	51
3 Ligamentum Flavum	3-D tension only cable	E = 15.0	9
3 Anterior Longitudinal	3-D tension only cable	E = 7.8	6
3 Posterior Longitudinal	3-D tension only cable	E = 10.0	6
Total number of elements			2443
Total number of nodes			1702
Total number of DOF			4605

E = Young's modulus (N/mm^2), ν = Poisson's ratio, k = Contact Stiffness (N/mm)
Table 1: Details of the FE-Model

Stiffness of the intervertebral disc is modelled by a combination of a linear elastic base material with embedded non-linear springs. Due to a lack of experimental data on the material properties for the intervertebral disc in the cervical spine, the stiffness of the discs had to be calculated using an analytical model which has been described fully previously[12]. Briefly the algorithm is based on the assumption of material incompressibility, and therefore constant disc volume, which has been shown to be justified in experiments. The function describes the non-linear force/deformation behaviour of human discs of known shape under uniform axial loading and is based on a model which pictures the disc as a bulging cylindrical volume with The shape of the bulge described by a hyperbolic function. While the disc is deflected parallel to the transverse plane, the 3 dimensional deformation is predicted. The disc is therefore modelled as a combination of water and soft tissue in a shell of fibre material which acts as a tensile `skin` surrounding the outside of the disc. The shell has a non-linear progressive stiffness[13] and therefore the stress/strain behaviour of the shell is modelled by a hyperbola function.

The axial force requred to produce a deformation of δ in the disc is given by:

$$F(\delta) = ((A_{disc} - A_{shell}) \cdot E_{disc} \cdot \varepsilon(\delta)_{disc}) + (A_{shell} \cdot E(\delta)_{shell} \cdot \varepsilon(\delta)_{shell}) \tag{1}$$

Symbol	Description	Unit
$F(\delta)$	Force normal to horizontal mid-plane of the disc	N
δ	Deformation of disc in axial direction	mm
A_{disc}	Cross-section area of disk in mid plane	mm^2
A_{shell}	Cross-section area of tensile shell in mid plane	mm^2
E_{disc}	Linear Young's-Modulus of base material (annulus fibrosus ground material)	N/mm^2
E_{shell}	Non-linear Young's-Modulus of tensile skin (Sharpey's fibers[14])	N/mm^2
ε_{disc}	Strain of the base material	
ε_{shell}	Non-linear strain of the tensile skin depending on the disc bulge	

Table 2: Description of symbols used in equation (1)

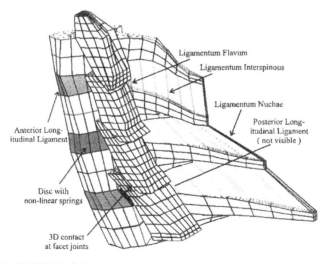

Figure 1: C4-C7 FE-Model

4. RESULTS

All calculations in the model were undertaken assuming symmetry at the mid-sagittal plane, therefore all node points located on the mid-sagittal plane of the half model were restricted in the lateral direction. An iterative solution was applied including the large deformation theory with an updated transformation matrix. The frontal solver of the ANSYS5.3® FE-package was used.

4.1 Stiffness

To evaluate the model stiffness in comparison with *in vitro* test data the model was reduced to the C4-C6 level, because of availability of experimental data[15]. The response was obtained by applying uniform axial compression at the top nodes of the model i.e. at the top of the vertebral body of C4, while the inferior nodes of C6 were constrained. In other words the superior nodes of the C4 vertebral body were displaced downwards in the vertical direction by applying a maximum load of 1000 Newton in 10 steps.

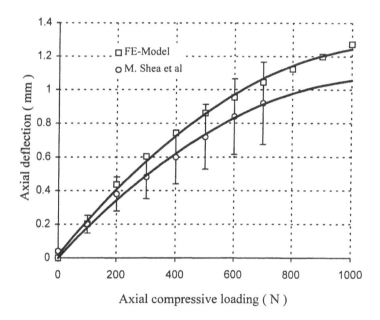

Figure 2: C4-C6 part of the model under axial compression, and compared with experimental data from Shea *et al.*

It can be clearly seen that the quantitative characteristic of the deformation curve compares well with the independent experimental data. It is impossible to make an exact quantitative comparison due to the large standard deviation of ±26.7% found in Shea's experiments. However, particularly under bending where compressive load on the disc is high, it is clear that the special mechanical nature of the disc cannot be neglected. Therefore incorporation of non-linear springs into the system has proven to be an effective measure to obtain a realistic model stiffness for FE-Models of the cervical spine.

4.2 Deformation of the C4-C7 spine unit under anterior loading

To show the model performance under large deformation the following boundary conditions were applied: the inferior aspect of the C7 vertebra body was restricted in all directions and a force of 400 N (i.e. 200 N for the half model) was applied in the anterior direction of the transverse plane at the superior aspect of the C4 vertebral body. Figure 3 shows the deformation of the model between the unloaded and the fully loaded states. Notice the 3D-sliding contact at the facet joints resulting in rotational movement of the vertebrae and therefore large strain at the inter-spinous ligament and the nuchal ligament. Last can only be fully modelled when the model is expanded to the occiput where it connects at its superior border. However this simple experiment shows the possibility of an involvement of the nuchal ligament in whiplash-type injuries.

There is no direct experimental data against which to compare this deformation but an indication of approximately 25° under a load of 400 N is not unrealistic. Furthermore it is clear that the discs do not undergo a large compression at the anterior aspect, as it would be the case in a linear modelling approach.

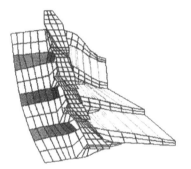

Figure 3a: FE-model unloaded   **Figure 3b:** FE-model with 400 N in anterior direction

The high level of simplification and the use of 8 node solid elements did resulted in only 4605 degrees of freedom.

5. CONCLUSION AND DISCUSSION

A non-linear, 3-dimensional large deformation finite element model of the cervical spine has been developed. The model, as far as was practical, was validated by comparing the predicted data against experimental values. The model performed well but further improvements are needed to fully realise its potential as a complement to *in vivo* and *in vitro* experimental studies. The model does not include muscles (active or passive effects) or overlaying soft tissue. The biggest obstacle in cervical spine modelling, however, is the lack of experimental data for this region of the spine. It should be emphasised that the principle aim of the present investigation was the development of a simplified large deformation model, suitable for transient loading, which describes the gross motion of this part of the spine with realistic model stiffness. The viscoelastic effects of the discs are not included because no significant dependency on loading speed for the stiffness of cervical spines has been reported, but hyperelastic modelling of discs and ligaments is seen as an option to further improve the model. The model is not suitable for determining stress inside the bony structure and the intervertebral discs, as this is not the target of this study. The feasibility of the modelling technique has been shown and the model can now be extended to include the Occipital-Atlanto-Axial complex (C0-C2) and passive muscular response, leading to a model which will be suitable for predicting ligamentous strain. This will help to answer a biomechanical question which is impossible to investigate with volunteer or *in vitro* experimental studies.

6. ACKNOWLEDGEMENT

We thank the National Library of Medicine (USA) for the Visible Human Data Set and Dipl.-Ing. Markus Beckman for his help in generating the CAD model.

7. REFERENCES

1. Penning L., Hypertranslation of the head backwards: part of the mechanism of cervical whiplash injury. Orthopaede, 1994, Vol. 23, 268-274.

2. Worth D. R., Cervical Spine Kinematics. PhD Thesis: Flinders University of South Australia 1985.

3. Stokes I. A .F. and Gardner-Morse M., Lumbar spine maximum efforts and muscle recruitment patterns predicted by a model with multijoint muscles and joints with stiffness. J. Biomechanics. 1995. Vol. 28, 173-186.

4. Wu J. S. S. and Chen J. H., Clarification of the mechanical behaviour of spinal motion segments through a three-dimensional poroelastic mixed finite element model. Med. Eng. Phys., 1996, Vol. 18, 215-224.

5. Sharizi-Adl A. and Drounin G., Nonlinear Gross Response Analysis of a Lumbar motion Segment in Combined Sagittal Loading. J. Biomechanical Engineering, 1988, Vol. 110, 216-222.

6. Merrill T., Goldsmith W. and Deng Y. C., Three-dimensional response of a lumped parameter head-neck model due to impact and impulse loading J. Biomechanics, 1984, Vol. 17, 81-95.

7. Saito T., Yamamuro T., Shikata J., Oka M. and Tsutsumi S., Analysis and prevention of spinal column deformity following cervical laminectomy, pathogenetic analysis of post laminectomy deformities. Spine, 1991, Vol. 16, 494-502.

8. Yoganandan N., Kumaresan S., Voo L. M., and Pintar F. A., Finite element modeling of the C4-C6 cervical spine unit. Med. Eng. Phys., 1996, Vol. 18, 569-574

9. Kleinberger M., Application of finite element techniques to the study of cervical spine mechanics. Proceedings of the 37th Stapp Car Crash Conference, San Antonio, Texas, November 7-8, 1993, 261-72.

10. http://www.nlm.nih.gov/research/visible/visible_human.html

11.Goel V. K., Park H. and Kong W., Investigation of the vibration characteristics of the ligamentous lumbar spine using the finite element approach. J. Biomechanical Engineering., 1994, Vol. 116, 377-383.

12.Heitplatz F. and Gentle R. C., Theoretical investigation into the axial strength of intervertebral discs based on a non-linear progressive shell model. Submitted for publication in J. Engineering in Medicine.

13.Green T. P., Adams M. A. and Dolan P., Tensile properties of the annulus fibrosus. Eur Spine J., 1993, Vol. 2, 209-214.

14.White A. A. III and Panjabi M. M., Clinical Biomechanics of the Spine. ed. 2, J.B. Lippincott, Philadelphia, 1990

15.Shea M., Edwards W. T., White A. A. III and Hayes W.C., Variations of stiffness and strength along the human cervical spine. J Biomechanics, 1991, Vol. 24(2), 95-107

INVESTIGATING THE ROLE OF THE DYNAMIC CURVATURE OF THE HUMAN SPINE USING A COMPUTER-BASED MODEL

S. L. Grilli[1] and B. S. Acar[2]

1. ABSTRACT

A high incidence of low back pain has been associated with activities such as manual materials handling and nursing which involve bending of the spine whilst lifting large loads. Numerous mathematical models have been developed to investigate this problem, but are based upon representing the spine as a 'lever'. This is concerned with balancing the flexion moments generated by anterior body weight and loads lifted, by extension moments generated predominantly by the spinal muscles. A high level of anatomic detail for these muscles has been included. However, an alternative representation of the spine as an arch structure has been proposed by Aspden (1). This involves both muscle forces and spinal curvature acting together to withstand the loads applied. The predictions from Aspden's model favour a reduction in the predicted compression at the lower lumbar intervertebral jonts, although the anatomic detail in these studies has been limited. The current paper therefore considers how the anatomic detail of this model might be improved in order to investigate the role of the precise curvature of the spine as a load bearing element. This includes consideration of the spine as a whole, with the application of distributed body weight, and a quantification of the individual muscle forces.

2. INTRODUCTION

Numerous mathematical models of the spine have been developed to investigate the mechanics of lifting activities without the risks and costs involved with experimental

Keywords: Spine, Mathematical Model, Computer, Curvature, Arch

[1] PhD Student/ Department of Computer Studies, Loughborough University, Loughborough, Leicestershire. LE11 3TU, UK
[2] Lecturer/ Department of Computer Studies, Loughborough University, Loughborough, Leicestershire. LE11 3TU, UK

techniques. These models have been aided considerably by computer techniques which allow the automatic calculation of results, thereby allowing the modeller to focus upon the interpretation of the results rather than the tedious processing.

The majority of these models have been based upon the concept of the lever model. This calculates the flexion moments produced by body weight and loads lifted acting anterior to the spine. The model then determines the extension moment required to be produced predominantly by posterior muscle forces, so that flexion and extension moments are in equilibrium. These moments are generally considered about the lower lumbar levels, (L5/S1 or L4/L5) at which moments and loads are predicted to be greatest. Equilibrium of the structure is ensured if both forces and moments are in equilibrium. For equilibrium of forces, a resulting reaction at the intervertebral joint, with compression (C) and shear (S) components occurs. Figure 1 illustrates this concept, with the force due to body weight represented by a single force (W), load lifted (L), and a collective muscle force (M).

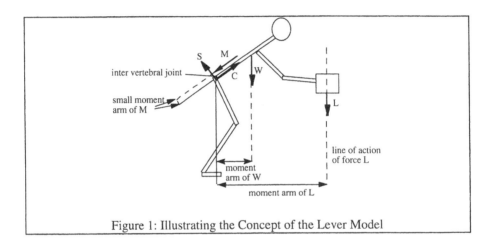

Figure 1: Illustrating the Concept of the Lever Model

These models have investigated the loads on the spine for various static postures (2), and lifting techniques (3, 4). The resulting compression at the lumbar intervertebral joint predicted by these models has been used to define standards for lifting activities. However the majority of lever models have been restricted to the analysis of loads at a single level of the spine, and the curvature of the spine has largely been ignored. It has been suggested the curvature of the spine increases the distance between the intervertebral joints and the lines of action of the muscles forces, thereby increasing the moment arm of the forces, and consequently the moments they exert (5). However, an alternative mechanism, representing the spine as an arch structure has been proposed by Aspden (1). For this mechanism, both the muscle forces and the curvature of the spine act to withstand the loads applied.

A comparison made between the arch and lever models using the same loading conditions as those used in a lever model of Morris et al. (6), showed the arch model favoured considerable reduction in loads at the intervertebral joint (1). However, investigations into lifting activities using the arch model, and the anatomic detail of the model have not been pursued to the depths of the lever model. In a previous paper we discussed the features of both models (7), and proposed that the development of the arch model may be critical towards gaining a greater understanding of the role of the curvature of the spine in load bearing, and the effect it may have upon the incidence of low back pain. This paper therefore considers how, aided by computer techniques, such a model might be developed so that further investigations can be performed.

3. THE ARCH MODEL

Aspden (1) has proposed that the curved configuration of the vertebrae of the spine resists loading in a similar way to a masonry arch. The vertebrae are compressed together allowing the spinal curve to retain its shape under loading in a similar way that the voussoirs (or brick elements) are compressed.

The supports at either end of a masonry arch generate reaction forces which have components parallel and perpendicular to the reference line (an imaginary line joining the two ends of the arch.) The parallel component provides the 'compressive thrust' (H) needed to compress the elements together, thereby providing strength and stiffness to the arch. Opposing directions of thrust at either end ensures equilibrium in this direction, while the perpendicular force component (R) equilibrates with those of the external loads (W1, W2, W3) acting on its convex surface (Figure 2). However, unlike a masonry arch, the spine is a structure fixed at one end with the sacrum and pelvis, but free at the other. Aspden therefore proposed that the sacrum acts as an end support, while a combination of body weight and muscle and ligament forces provide the necessary compressive thrust at the other end.

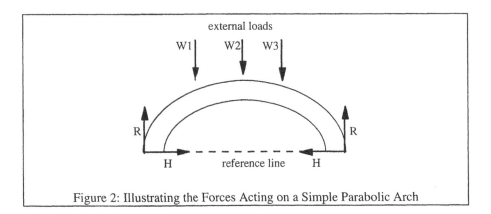

Figure 2: Illustrating the Forces Acting on a Simple Parabolic Arch

The resultant forces acting at each point of loading on the arch collectively follow a path known as the thrustline. The existence of a thrustline indicates the arch is in equilibrium at every section, while stability of the arch is achieved when these forces can be transmitted through its entire length. For this to occur, the thrustline must lie within the cross section of the arch, or for the spine, approximately within the region of the vertebral bodies. Thus whereas the lever model considers equilibrium of forces and moments at a single intervertebral joint, the arch model ensures equilibrium and stability at every level of the spine.

For loads acting on the convex surface of the arch, in a direction predominantly perpendicular to the reference line, the thrustline is inclined relatively steeply. In contrast loads acting mainly parallel to the arch reference line result in a flatter thrustline. A curvature which can accommodate a steep thrustline can therefore tolerate greater loads applied perpendicular to its convex surface. For the spine, large loads can be accommodated by a large curvature, or by increasing the muscle forces to increase the compressive thrust provided and to flatten the thrustline so that stability is ensured.

In the flexed posture, Aspden (8) has proposed body weight acts almost perpendicular to the convex surface of the curved thoraco-lumbar spine. The compressive thrust required by the muscle forces must therefore be increased, resulting in an increase in compression at the intervertebral joint. In contrast in the erect posture it was proposed body weight acts almost parallel to the lordotic arch of the lumbar spine. The force required by the muscles in this situation is therefore reduced, and a relatively flat thrustline path is obtained, with lower compression at the intervertebral joint.

However while arch theory suggests the curvature of the spine has a role in load bearing which results in a significant reduction in the muscle force required and the resulting compression at the intervertberal joint, previous investigations are limited by the anatomical detail included.

4. LIMITATIONS OF THE ARCH MODEL

i) Consideration has been made only for the lumbar and thoraco-lumbar regions of the spine. However, stability of one particular region of the spine, such as the lumbar region, does not imply stability in other regions or indeed the whole spine. Realistically therefore stability of the whole spine must be considered.

ii) Aspden's investigations (1) were based upon the same loading conditions as those for the lever model of Morris et al. (6), so that a direct comparison between the two models could be made. This involved representing the force due to the weight of the head and upper body applied at T2, the force due to the weight of the trunk applied at T12, and the force due to external loads carried applied at T6. However, the arch model considers the action of forces relative to the precise curvature of the spine. The thrustline determined by a few collective forces may be significantly different to the thrustline determined by a distributed mass along the entire curvature of the spine. Determining stability for the spine therefore requires the application of a distributed mass along the entire curvature.

iii) In addition to the simplified representation of body weight, Aspden's investigations (1, 8) involved a collective force applied at the upper end of the arch. This represented the compressive thrust provided by the muscle and ligament forces and components of body weight. However, realistically the compressive thrust provided by the muscle forces (and ligaments and body weight) is generated between the vertebrae along the entire length of the spine.

5. DEVELOPMENTS OF THE DYNAMIC ARCH MODEL

5.1 Whole Spine

The muscle elements of the spine are active and may change their contractile state to suit the level of compressive force required by the arch structure. Thus unlike the masonry arch whose shape has been calculated and fixed in order to ensure stability, the spine may alter configuration and adjust to the new level of compressive thrust required by changing muscle recruitment patterns. The spine is therefore a dynamic structure which may adjust its curvature and muscle patterns for a range of loads and postures to ensure equilibrium and stability. Muscle forces acting along the entire length of the spine are responsible for generating the necessary forces to ensure the vertebrae maintain their position and withstand the loads applied (Figure 3). In this way, the muscle forces may supplement the free end of the spine so that large reaction forces at the cervical end are not required. Stability of the spine as a whole can therefore be considered, in a variety of postures.

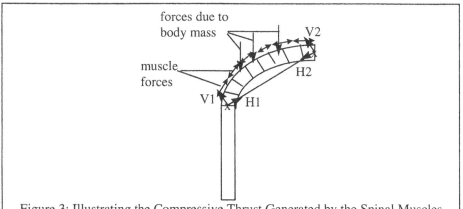

Figure 3: Illustrating the Compressive Thrust Generated by the Spinal Muscles between Vertebrae for a Curved Portion of the Spine.

5.2 Weight Distribution

The thrustline determined by a distributed mass along the spine will be necessary for a realistic investigation of stability. A more precise distribution of body weight can be obtained based on the models of Takashima et al. (9) for the thoraco-lumbar regions, and Merril et al. (10) for the head and cervical spine. Using these data will allow the loading pattern to be matched to the precise curvature of the spine, so that a more realistic indication of stability than reported in the literature can be determined.

5.3 Role of the Muscles

While both ligaments and muscles generate forces along the spine, the ligaments are passive structures which must be stretched before a tensile force can be generated. In contrast the muscles are active elements which may adjust their activation levels according to the compressive thrust required. Initial development of the model is therefore to consider the muscle forces.

5.3.1 Force Direction

Muscle forces acting mainly parallel to the spine have been described by Aspden (5) and are proposed to provide the necessary thrust for the spinal arch. However, in addition to the compressive component provided by the muscle forces, a component of force perpendicular to the arch reference line is also generated (Figure 4). Indeed McGill et al. (11) reported that representation of the muscle forces as a single force parallel to the spine does not account for the shear forces which they may provide. These perpendicular components may be necessary for the precise adjustment of the thrustline so that its path follows that of the vertebral bodies, ensuring stability. The direction of the muscle forces, and the precise quantification of these forces in terms of their individual components is therefore required.

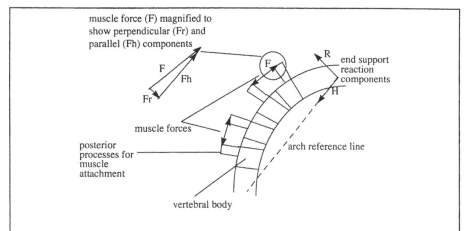

Figure 4: Illustrating Both Perpendicular and Parallel Components of Muscle Force Relative to the Arch Reference Line for a Portion of the Spinal Curve

5.3.2 Moments

The end support reactions for an arch are determined by consideration of the moments generated by each individual force about these ends, so that the entire forces and moments in the arch are in equilibrium. Muscles acting along the curved surface of the spine must be considered as individual forces which are eccentric to the end supports. The majority of muscle groups attach to the posterior processes of the vertebrae (spinous and transverse). However, some muscles such as longus colli in the cervical region and psoas in the lumbar region attach anteriorly to the vertebral bodies. The eccentric action of each muscle force thus varies, which in turn will affect the resulting arch end support reactions and the resulting thrustline path.

5.3.3 Dynamic Behaviour

Changes in spinal curvature are brought about by contraction of the muscle forces. However the muscles themselves undergo mechanical changes as a consequence. Muscles acting on the spine are attached to the vertebrae mostly by the transverse and spinous processes. A change in vertebral position may result in an alteration in the relative locations of the points of attachment, and a consequent change in the line of action of the muscle force. The components of force perpendicular and parallel to the arch reference line are consequently adjusted. In addition the moment arm of the muscle force, and consequently the moment generated about the intervertebral joints, can be altered. Further the magnitude of the components of the external forces acting (and indeed any force) may be altered as the joint axis direction changes. A consequent change in end support reactions and resulting thrustline may then occur. The dynamic properties of the spinal arch therefore need to be considered.

6. COMPUTER BASED MODEL

The developments of the arch model require the quantification of the effects of individual muscle forces and a distributed loading pattern for body weight. This includes a vast number of forces each acting with different magnitude, direction and point of application. The direction of the forces determines the proportion of the force components perpendicular and parallel to the arch reference line, which in turn determine the reactions at the supports, and the path of the resulting thrustline. A systematic way of defining the forces relative to the reference line of the arch, and determining the resultant thrustline is therefore required.

The computer based modelling application 'ADAMS' (Automatic Dynamic Analysis of Mechanical Systems) is a suitable tool for defining the magnitude, direction and point of application of each force. Representation of the vertebrae of the whole spine can be achieved by defining rigid bodies whose spatial orientation and location determine the overall configuration of the spine to be investigated. Defining the point of application, magnitude and direction of each force relative to the local co-ordinate system of the appropriate vertebra allows the changes in force properties to be automatically accounted for upon alteration of spinal configuration. The relationship between each force and the reference line of the arch can then be defined in order to calculate the resulting thrustline. In this way, for a given configuration of the spine, with a given system of forces acting, the thrustline can automatically be determined. Graphical display of this thrustline then provides an immediate indication of the regions of the spine which may be unstable. Automation of these calculations allows the numerous muscle forces of the entire spinal column to be considered. In this way the curvature of the spine and its relationship with the action of muscle groups can be investigated in a detailed manner.

7. SUMMARY

Development of the arch model may be critical towards gaining a greater understanding of the role of the curvature of the spine in load bearing, and the effect it may have upon the incidence of low back pain. This paper considers how this model might be developed, and suggests a whole spine model, with a distributed force representing body weight, and the individual quantification of muscle forces are essential factors when considering the effect the precise curvature of the spine may have on stability. In particular, the magnitude, direction, and point of application of each force must be considered. The computer based modelling application 'ADAMS' allows forces to be determined relative to vertebral configuration so that the changes in force properties due to alteration of spinal configuration can be automatically accounted for. Graphical representation of the resulting thrustline provides a direct indication of areas of instability within the spine. In this way the role of the dynamic curvature of the spine as a load bearing element can be investigated. This is the approach employed by the authors in order to research further into back pain.

8. REFERENCES

1) Aspden, R. M., Intra Abdominal Pressure and its Role in Spinal Mechanics, Clinical Biomechanics, 1987, Vol. 2, 168-174

2) Chaffin, D. B., A Computerised Biomechanical Model - Development of and Use in Studying Gross Body Actions, J. Biomech., 1969, Vol. 2, 429-441

3) Garg, A. and Herrin G. D., Stoop or Squat: A Biomechanical and Metabolic Evaluation, AIIE Transactions, 1979, December, 293-302

4) Leskinen, T. P. J., Stalhammar H. R., Kuorinka I. A. A and Troup J. D. G., A Dynamic Analysis of Spinal Compression With Different Lifting Techniques, Ergonomics, 1983, Vol. 26, No. 6, 595-604

5) Aspden, R. M., Review of the Functional Anatomy of the Spinal Ligaments and Lumbar Erector Spinae Muscles, Clinical Anatomy, 1992, Vol. 5, 372-387

6) Morris, J. M., Lucas D. B. and Bresler B., "Role of the Trunk in Stability of the Spine", J. Bone and Joint Surgery, 1961, Vol. 43 A, No. 3, 327-351

7) Grilli, S. L. and Acar B. S., Human Spine Modelling Using Engineering Design Tools, Conference Proceedings, UTMIK 7th International Machine Design and Production Conference, September, 1996, METU, Ankara, Turkey, 915-926

8) Aspden, R. M., The Spine as an Arch. A New Mathematical Model, Spine, 1989, Vol. 14, 266-274

9) Takashima, S. T., Singh S. P., Haderspeck K. A. and Schultz A. B., A Model for Semi-Quantitative Studies of Muscle Actions, J. Biomech., 1979, Vol. 12, 929-939

10) Merrill, T., Goldsmith W. and Deng Y. C., Three Dimensional Response of a Lumped Parameter Head - Neck Model Due to Impact and Impulsive Loading, J. Biomech., 1984, Vol. 17, No. 2, 81-95

11) McGill, S. M. and Norman, R. W., Effects of an Anatomically Detailed Erector Spinae Model on L4/L5 Disc Compression and Shear, J. Biomech., 1987, Vol. 20, No. 6, 591-600

A MATHEMATICAL MODEL OF IDIOPATHIC SCOLIOSIS

J.M. Brown[1], M.I.G. Bloor[2], R.A. Dickson[3], P.A. Millner[4] and M.J. Wilson[5]

1. ABSTRACT

Idiopathic Scoliosis, the most common form of spinal deformity in children, is defined as the lateral curvature of a spine, where the spine has no underlying neuromuscular or structural abnormalities (hence the name 'idiopathic'). In addition to producing an unsightly hunched-back appearance, often giving psychological distress, there can be harmful long term effects on the cardiorespiratory system. This paper describes the initial results of the development of a mathematical model for the complex three-dimensional shapes of the deformity using the novel 'PDE Method'. The PDE Method has been used for the design and description of surfaces for CAD (Computer-Aided Design), some of which have been analysed computationally and manufactured using Rapid Prototyping technologies. PDE surfaces are basically produced by defining their boundaries (i.e. edges) followed by the generation of surfaces between them which are solutions of elliptic partial differential equations (a boundary-value approach). Collaboration between applied mathematicians and orthopaedic surgeons has resulted in these concepts and techniques being applied to the production of a computational model for the complex geometries of scoliotic spines which embody the necessary detailed geometry and growth features of a real spine.

2. INTRODUCTION

The aim of this work is to capture, in a mathematical model, the essential changes that occur in the shape of the vertebral bodies during the progression of Idiopathic Scoliosis.

Keywords: Idiopathic Scoliosis, Geometric Model, Spine, Vertebra, Surface

[1] Research Fellow, Department of Applied Mathematical Studies, University of Leeds. LS2 9JT. U.K.
[2] Professor, Department of Applied Mathematical Studies, University of Leeds. LS2 9JT. U.K.
[3] Professor of Orthopaedic Surgery, St. James's University Hospital, Beckett Street, Leeds. LS9 7TF. U.K.
[4] Senior Lecturer in Spinal Surgery, St. James's University Hospital, Beckett Street, Leeds. LS9 7TF. U.K.
[5] Reader, Department of Applied Mathematical Studies, University of Leeds, Leeds. LS2 9JT. U.K.

The spinous processes were not included in the model at this stage as it is believed that they have a minor role to play in the development of the condition, but there is the potential for them to be included in future models. The model that has been produced will provide a firm basis from which to build a complete geometry model of the vertebral bodies in a spine, which will be extended to model how the deformity develops over time.

The novel feature of this work is the use of the 'PDE Method' to generate the geometry of the vertebral bodies. This method has been previously used in the area of computer-aided design with the benefit of being able to quickly, efficiently and intuitively generate complex shapes. It is possible to build into these geometry models parameters that can be varied to produce changes in the overall shape of the objects. Such parameters will be used to produce the shape of individual vertebral bodies from a generic template.

Idiopathic Scoliosis, the PDE Method and the application of the PDE method to Idiopathic Scoliosis will be described in more detail below.

3. IDIOPATHIC SCOLIOSIS

Idiopathic Scoliosis, defined as the lateral curvature of a spine, where the spine has no underlying neuromuscular or structural abnormalities, is a common condition that affects children from infancy to adolescence. It accounts for more than 95% of spinal deformities detected in the community, and has a prevalence of up to 1%. With time and growth the individual vertebrae within the curvature become wedge-shaped and rotate; thus the deformity becomes a complex three-dimensional entity.

From a clinical perspective, the deformity of idiopathic scoliosis has been extensively documented in the literature in terms of quantitative study [1,2,3,4], development [5,6], aetiology and pathogenesis [1,2,7], biomechanics [8] and treatment [9,10]. Owing to the complex three-dimensional nature of the deformity, mathematical descriptions or models have proved challenging, and have often been confined to deformations in two planes only, ignoring axial rotation. A number of techniques have been employed in this context, ranging from an analytical approach [11,12,13] to an approach making use of computer graphics [14].

The combination of three-dimensional computer graphics with the increasing sophistication of medical imaging processing has given rise to work which involves not only the quantification of the deformity from patient data, but also the ability to reconstruct geometric models of spinal shape for examination using computer graphics [15]. Furthermore, shape reconstruction using medical imaging has been developed to the point where it is possible to generate a mesh for the finite element stress analysis of a spine [16,17].

The present work seeks to go beyond that described above to produce a truly three-dimensional approach, which takes into account deformations in the sagittal, coronal and transverse planes, and has the potential to account for changes in size, shape and position in 3D-space with time. The paper describes how this aim has begun to be met by producing a PDE model of the vertebral bodies, the building blocks of the spine, and how they may be assembled to produce the spinal shape.

4. A GEOMETRIC MODELLER - THE PDE METHOD

The PDE Method is a developing computer-based method of defining objects with complex surface shapes. The method is somewhat unconventional in that a surface, although defined parametrically in terms of u and v, is generated in 3-space by using an elliptic partial differential equation as the mapping from parameter space to 3-space. In other words, we view u and v as co-ordinates of a point in the domain Ω of the u,v-plane, and any point $\underline{X}(u,v)$ on the surface is obtained as the solution of a boundary value problem in Ω, using data defined around the boundary Γ (see Fig. 1). The curves that form the 'edges' of the surface along with appropriate derivative conditions, determine the boundary conditions on Γ. The surface is produced as a solution to an elliptic partial differential in Ω - hence the name PDE Method. The boundary conditions are imposed by the designer, though not necessarily directly in mathematical terms, in order to produce a particular surface shape. Complex surfaces can be produced by defining a series of 'character-lines' in three-dimensional space, which form the boundaries between distinct subregions of the overall surface which are filled out using the PDE Method [18,19,20].

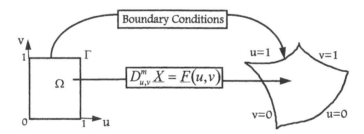

where D is an Elliptic Partial Differential Operator

Fig. 1 The PDE Method

One of the advantages of the method over conventional CAD techniques for surface definition [21,22], is that it can efficiently parameterize complex surface shapes. In this context, the parameters are shape parameters so that the method can produce complex surfaces using only a limited number of design variables. This facilitates the ease with which the shape of the object's surface may be changed or numerically optimised to improve its performance. The method has been used in the computer-aided design of a wide variety of objects, e.g. ship hulls [23], marine propellers [24], aircraft [19], and internal combustion engines [20] (see Fig. 2).

Fig. 2 Various CAD Objects Produced By the PDE Method

5. A GEOMETRIC MODEL OF IDIOPATHIC SCOLIOSIS

Before attempting to produce a simple geometric representation of a vertebral body, it is essential that an assessment is made of the way in which such geometries might be used to model physical processes which occur, e.g. growth, and the synthesis of these basic elements in the overall structure of the spine. As has been mentioned, the parametric nature of the PDE surface model is the distinctive feature that allows a basic generic model to capture the variety of shapes of the individual vertebrae present in normal and scoliotic spines. The shape parameterization chosen is therefore critical to this process and while it is not sensible to give the detailed algebra used in the formulation of the solution of the boundary value problem, it is important and instructive to give a general description of the method used.

The generic model of a vertebral body, from which all the variants arise, is generated from two PDE surface patches which join at approximately the mid-plane in the vertical direction of the vertebral body (see Fig. 3). The following fourth order elliptic partial differential equation (1) (where a is constant) is used to generate the surface between the boundaries.

(1)

$$\left(\frac{\partial^2}{\partial u^2} + a^2 \frac{\partial^2}{\partial v^2} \right)^2 \underline{X} = 0$$

The contour defining the cross-sectional shape on this plane forms a boundary condition for each of the patches and is a line u=1 say for the lower patch. The derivative conditions applied on this contour ensure that the complete surface of the body has tangent plane continuity across the joining line. The magnitude of the derivative influences strongly the shape of the resulting surface. The boundary u=0 for the lower patch is simply a point where the end-plate meets what the axis of the vertebral body. The derivative conditions here first of all ensure that a tangent plane to the surface is defined and the magnitude chosen to mimic the shape of the cross section at the waist of the vertebra. Any asymmetries in the shape would of course be reflected not only in the shape at the waist but also in this derivative distribution.

In order to incorporate into the model the provision for deformed vertebrae in a sufficiently general way, allowance must be made for the specification of the tangent plane at u=0 and the plane at u=1 to account for wedging and also any twist must be quantified. The natural way to do this is to introduce Euler angles [25] as parameters in the following way. For a particular vertebra, a Cartesian co-ordinate system is set up with u=0 for the lower patch at the origin. Then, regarding the tangent plane at the 'axis' of the upper surface patch as the end of a right cylinder, the Euler angles of the imaginary cylinder are used to specify the deformation of the vertebral body, the boundary conditions at the mid-plane being obtained through interpolation. This approach has the advantage of an unambiguous parameterization of the deformation whilst setting up a new frame of reference for the next vertebra in the spine.

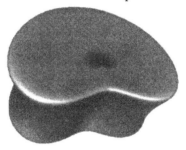

Fig. 3 The Generic Model of a Vertebral Body

Parameters were chosen that produce the shape changes that are found in the vertebrae of normal and scoliotic spines as illustrated below in Fig. 4.

Fig. 4 Examples of Wedging and Asymmetries Captured by the PDE Model

The models of the vertebrae can be viewed using three-dimensional computer-graphics. It is possible to rotate the vertebrae in real time in all three planes to enable them to be viewed from any direction. It can be seen from Fig. 2 to Fig. 4 that the surfaces generated by the PDE Method are inherently smooth. The fairness of the PDE surfaces has been investigated and their comparative smoothness confirmed in [26]. As it is straightforward to generate a closed surface mesh (Fig. 5), when performing many computational analyses such smooth meshes have significant advantage over the 'bumpy' meshes generated from scan data.

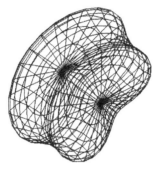

Fig. 5 Closed Surface Mesh of A Vertebra

Part of a spine has been built using the vertebral body building blocks outlined above. Fig. 6 shows the case in which all the vertebrae are normal, apart from the fourth from the top for which a wedging has been incorporated.

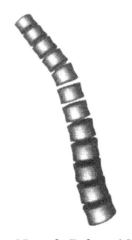

Fig. 6 Part of a Deformed Spine

6.CONCLUSIONS

Structural scoliosis is not a matter of right/left asymmetry; rather the underlying problem appears to be front/back asymmetry with relative anterior spinal overgrowth. To try and gain further understanding of the condition a full three-dimensional geometric model of the spine is being developed using the PDE Method. The first step described in this paper has been to produce a efficient parameterised model that is able to capture the shape of normal vertebral bodies and those found in scoliotic spines. Key advantages of the PDE method are that the surfaces generated are smooth and closed surface meshes can easily be generated that allow physical analysis of the mechanical properties.

Future work will seek to incorporate data from surveys on normal and scoliotic spines to produce a model of the geometry of a scoliotic spine that incorporates the change in shape of the vertebral bodies over time due to both growth and deformation. Using the full model the possible link between abnormal vertebral growth and the development of the mechanically unstable lordosis which is thought may trigger the deformity [27] will be investigated. From the understanding gained, it is hoped to identify spinal shapes that are vulnerable to idiopathic scoliosis; quantifying their potential for developing a serious deformity. Another aim is to the improve the user interface to allow the surgeons to 'play' with the model and give feedback on its accuracy. With such improvements the software could be used as a teaching tool. If desirable the spinal processes may also be incorporate in the model using the PDE method.

It is also hoped to apply the PDE Method to other medical applications. The method has the potential to be of use in many other areas including soft tissue modelling and customised prosthesis design using the potential of the PDE Method to automatically optimise the designs [23].

7. REFERENCES

1. Dickson, R.A., The Aetiology of Spinal Deformities, The Lancet, May 21, 1988, 1151-1156.
2. Dickson, R. A., The Etiology and Pathogenesis of Idiopathic Scoliosis, Acta Orthopaedica Belgica, 1992, Vol. 58, Suppl. 1, 21-25.
3. Kojima, T. and Kurokawa, T., Quantitation of Three-Dimensional Deformity of Idiopathic Scoliosis, Spine, 1992, Vol. 17, No. 3S, S22-S29.
4. Skalli, W., Lavaste, F. and Descrimes, J.-L., Quantitation of Three-Dimensional Vertebral Rotation in Scoliosis: What are the True Values, Spine, 1995, Vol. 20, No. 5, 546-553.
5. Perdriolle, R. and Vidal, J., Thoracic Idiopathic Scoliosis Curve Evolution and Prognosis, Spine, 1985, Vol. 10, No. 9, 785-791.
6. Perdriolle, R. and Vidal, J., Morphology of Scoliosis: Three-Dimensional Evolution, Orthopedics, 1987, Vol. 10, No. 6, 909-915.
7. Dickson, R. A., Lawton, J. O., Archer, I. A. and Butt, W. P., The Pathogenesis of Idiopathic Scoliosis, J. Bone J. Surg., 1984, Vol. 66-B, No. 1, 8-15.
8. Deacon, P., Archer, I. A. and Dickson, R. A., The Anatomy of Spinal Deformity: A Biomechanical Analysis, Orthopedics, 1987, Vol. 10, No. 6, 897-903.
9. Dickson, R. A., 'Idiopathic Scoliosis: foundation for physiological treatment, Annals R. Coll. Surg. Eng., 1987, Vol. 69, 89-96.

10. Piggot, H., Growth Modification in the Treatment of Scoliosis, Orthopedics, 1987, Vol. 10, No. 6, 945-952.

11. Hierholzer, E. and Luxmann, G., Three-Dimensional Shape Analysis of the Scoliotic Spine using Invariant Shape Parameters, J. Biomechanics, 1982, Vol.15, No.8, 583-598.

12. Stokes, I. A. F. (Chair), Three-Dimensional Terminology of Spinal Deformity, Spine, 1994, Vol. 19, No. 2, 236-248.

13. Stokes, I. A. F., Bigalow, L. C. and Moreland, M. S., Three-Dimensional Spinal Curvature in Idiopathic Scoliosis, J. Orthop. Res., 1987, Vol. 5, No. 1, 102-113.

14. Howell, F. R. and Dickson, R. A., The Deformity of Idiopathic Scoliosis made visible by Computer Graphics, J. Bone. J. Surg., 1989, Vol. 71-B, No. 3, 399-403.

15. Andre, B., Dansereau, J. and Labelle, H., Effect of Radiographic landmark identification errors on the accuracy of three-dimensional reconstruction of the human spine, Medical and Biological Engineering and Computing, 1992, Vol. 30, No. 6, 569-575.

16. Breau, C., Shirazai-Adl, A. and de Guise, J., Reconstruction of a human ligamentous lumbar spine using CT images - a three-dimensional finite element mesh generation, Annals of Biomedical Engineering, Vol. 19, No. 3, 291-302.

17. Goel, V. K. Monroe, T., Gilbertson, L. G. and Brinckman, P., Interlaminar Shear Stresses and Laminae Separation in a Disc, Spine, 1995, Vol. 20, No. 6, 689-698.

18. Bloor, M. I. G. and Wilson, M. J., Using Partial Differential Equations to Generate Free-Form Surfaces, Computer-Aided Design, 1990, Vol. 22, 202-212.

19. Bloor, M. I. G. and Wilson, M. J., The Design of a Generic Aircraft Geometry using the PDE Method, J. Aircraft, 1995, Vol. 32, No. 6, 1269-1275.

20. Bloor, M. I. G. and Wilson, M. J., Complex PDE Surface Generation for Analysis and Manufacture, Computing, 1995, Supplementum 10, 61-77.

21. Bezier, P., Example of an Existing System in the Motor Industry: The UNISURF System, Proc. Roy. Soc. Lond., 1971, A321, 207-218.

22. Woodward, C. D., Cross-Sectional Design of B-spline Surfaces, Comput. and Graph., 1987, Vol. 11, 193-201.

23. Lowe, T. W., Bloor, M. I. G. and Wilson, M. J., The Automatic Design of Hull Surface Geometries, J. Ship. Res., 1994, Vol.38, No. 4, 319-328.

24. Dekanski, C., Bloor, M. I. G., Nowacki, H. and Wilson, M. J., The Geometric Design of Marine Propeller Blades using the PDE Method, Fifth International Symposium on the Practical Design of Ships and Mobile Units, eds. Caldwell, J. B. and Ward, G., 1992, 596-609.

25. Rutherford, D. E., Classical Mechanics, Oliver and Boyd, 1957, 150-153.

26. Brown, J. M., Bloor, M. I. G., Bloor, M. S., Wilson, M. J. and Nowacki, H., Properties of the B-spline Represensations of PDE Surfaces that are Generated Using the Finite Element Method, Computer-aided Surface Geometry and Design, The Mathematics of Surfaces IV, 1994, 335-348.

27. Stokes, I. A. F. and Labile, J. P., Three-dimensional analysis of right thoracic idiopathic scoliosis by asymmetric growth, J. Biomechanics, 1990, Vol. 23, No. 6, 589-595.

REQUIRED STIFFNESS DISTRIBUTION IN A MATTRESS FOR AN OPTIMAL CURVATURE OF THE HUMAN SPINE DURING BEDREST

[1]B. Haex, [1]J. Vander Sloten, [1]R. Van Audekercke

1. ABSTRACT

Incorrect body support and positioning during sitting, working and sleeping often cause low back pain. The fact that people spend a large part of their lives in bed justifies the need for an objective and scientifically sound method to determine the right sleeping system for each individual.

A parametric finite element model of the combination individual-mattress was developed to predict the curvature of an individual's vertebral column when lying on a specific sleeping system. The resulting model represents the spine as a succession of quadrangular elements, each with its specific simplified geometric and material properties. The rest of the human body is modeled by rigid skeleton elements linked by joints and surrounded by soft tissue.

Simultaneously two experimental setups were used to validate the model by measuring vertebral positions during bed rest. The first setup is a camera system detecting reflecting bullets which are mounted a) on the spinous processes (lateral recumbency) and b) on pins which are pierced through the mattress (posterior recumbency). A second experimental setup was built to perform indirect but fast measurements by measuring not the position of the spinous processes, but the position of the mattress surface.

Both measurements and F.E. analyses were performed for 30 people on 5 different mattresses. It was possible to decide whether and where mattress stiffness had to be adjusted. Further one could decide which mattress (out of 5) offered the best supporting qualities based on F.E. predictions. A numerical shape prediction of the vertebral column was not yet possible, which justifies the need for more reliable (e.g. 3D) analyses.

Keywords : Spine, Mattress, Finite element analysis, Sleep

---

[1] Division of Biomechanics and Engineering Design, Katholieke Universiteit Leuven, Celestijnenlaan 200A, B-3001 Heverlee.

2. INTRODUCTION.

Incorrect body support and positioning during sitting, working and sleeping often cause low back pain. The fact that people spend a large part of their lives in bed justifies the need for an objective and scientifically sound method to determine the right sleeping system for each individual. Considering the fact that people sleep on the average 20 % of the night on their back and 75 % on their side, the best mattress has to be chosen by assigning different weight factors to a) its supporting qualities and b) its pressure relieving qualities for each body carriage. This study will focus rather on the ergonomic qualities than on the pressure relieving qualities.

2.1 Ergonomic qualities

Most specialists suggest that the optimal system has to support the human spine such that it adopts its natural position, which is assumed to be the same as it takes in the upright position.[1-2] Others suggest the lumbar lordosis has to be flattened.[3-4] If the first thesis is presupposed then an optimal body support gives rise to the spinal column being a straight line for lateral recumbency (fig 1). For posterior recumbency an optimal support gives the spinal column the same thoracal kyfosis and lumbar lordosis as in the upright position, yet slightly smoothened by the loss of body weight working in longitudinal direction on the spinal column. Consequently a small prolongation of the spine occurs, as happens in weightlessness.

Fig. 1 : Optimal body support and poor body support for lateral recumbency.

2.2 Pressure relieving qualities

The duration and the intensity of pressure has to be limited to improve the blood and oxygen support and thus to prevent decubitus ulcers.[5-7] An increasing number of authors mentions the importance of shear forces to be minimized to prevent stretching and angulation of the bloodvessels.[8-9]

3. METHODS

3.1 Modeling

A parametric 2D finite element model of the combination individual-mattress was developed to predict the curvature of an individual's vertebral column when lying on a specific sleeping system. The finite element analysis software used (MARC®/MENTAT®) is particularly suited for non-linear and contact problems. Although previous studies of the body-mattress combination were not found back in literature, studies in the early seventies observing and modeling spinal behavior in an ejector seat[10] were helpful. Little information was found about spinal stiffness parameters[11], weight distribution[12-14] and rib cage influences[15]. More recent finite element models[16] were useful but too detailed in modeling only the vertebral column and could thus not be implemented in a total body-on-mattress analysis.

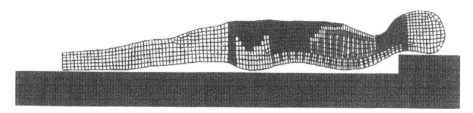

Fig. 2 : Two-dimensional finite element model before implementation of gravity.

The resulting model (fig 2.) represents the spine as a succession of quadrangular elements, each with its specific simplified geometric and material properties. The rest of the human body is modeled by rigid skeleton elements linked by joints and surrounded by soft tissue. Geometrical parameters consist of body width each 0.05 m, spinous process coordinates and vertebral size. Both the vertebra and intervertebral disc sections are assumed to be quadrangular; geometrical properties are obtained from litterature[10] and own anatomical data, their stiffness parameters are derived from Schultz[11]. The human skeleton is assumed to be rigid and is represented by a simplified shoulder and pelvic girdle, a rib cage, head and legs. Geometrical properties were based on RX data and Andriacchi.[16] The soft tissue stiffness parameters were assumed to remain constant over the body length and were approximated by linear properties. Weight distribution was based on geometrical properties and on litterature.[12-14]

The non-linear mattress stiffness parameters were obtained from measurements performed in the research laboratory of the mattress company RECTICEL. On a compression bench the force-displacement characteristics of mattresses with different densities were measured. This mattress compression was analyzed by an axisymmetric F.E. model. The force-displacement output of the model was brought in accordance with the actual measurements by adjusting the stress-strain input of the model. These values were used as an input for the body-mattress model.

3.2 Measurements

At the same time an experimental setup was built to measure the lumbar and thoracal spinous process positions during bedrest. For lateral recumbency 17 reflecting markers (12 thoracal and 7 lumbar) which are glued to the skin covering the spinous processes are detected by a MAC-Reflex camera system. Two reference markers permit to position these markers in the vertical plane, approximating the vertebral curvature (fig 3).

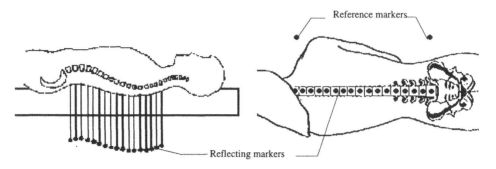

Fig. 3 : Experimental setup for posterior and lateral recumbency.

For posterior recumbency a system with pins which are pierced through the mattress measures the spinous process positions. A conducting strip glued to the spinous processes and an electrical circuit permit to control permanent contact between the pins and the spinous processes. Two reference markers allow to recalculate the positions of the markers mounted on the lower side of the pins to the actual spinous process positions. Thus an approximation of the vertebral curvature is set up for posterior recumbency.

A second experimental setup was built to perform indirect but fast measurements by measuring not the position of the spinous processes, but the position of the mattress surface. Comparison between the two measurement systems proved the second one to be reliable enough for posterior but not for lateral recumbency, which already implies a huge saving of time.

To implement an individual's entire geometry in the finite element model a third setup (fig 4.) was built. A laser system which can slide along a vertical axis scans the surface and gives a precise idea about a) the contours and b) the thoracal kyfosis and lumbar lordosis.

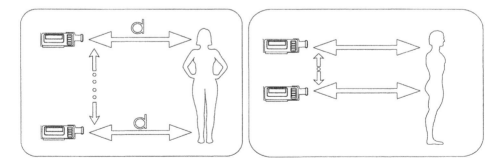

Fig 4. : Measurements of the entire geometry and detailed spine measurements.

4. RESULTS

Both measurements and F.E. analyses were performed for 30 people on 5 different mattresses for a) posterior and b) lateral recumbency. An F.E. output can be seen on the figure below.

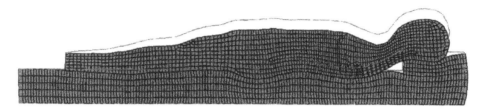

Fig 5. : Finite element output : deformation after application of gravity.

Starting from nodal displacements and reaction forces the resulting vertebral curvature and the resulting interface pressures were calculated. These values were compared to the measurements. One can focus on mattress deformations or on spinal behaviour as is shown below.

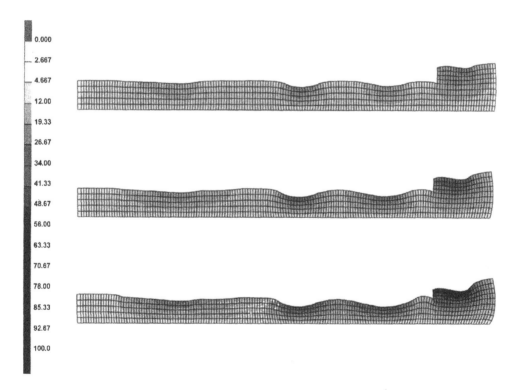

Fig 6. : Compression of three different mattresses supporting the same subject.

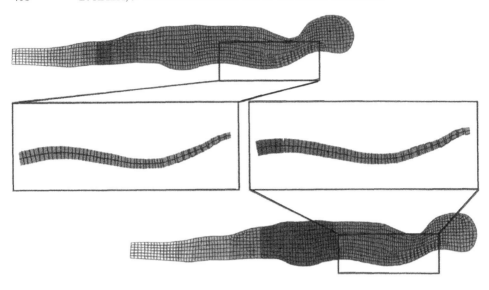

Fig 7. : Spinal deformation of an individual lying on two different mattresses.

In spite of inaccuracies in the measurement system and insufficient knowledge of body stiffness and weight distribution parameters it was possible to decide whether and where mattress stiffness had to be adjusted for a) posterior and b) lateral recumbence. For most people the hip zone of the mattress must be stiffened to prevent the pelvis from canting forward in posterior recumbency; the shoulder zone must be soft to prevent lateral recumbency scoliosis.

Further one could decide which mattress (out of 5) offered the best supporting qualities based on F.E. predictions. For each of the 30 persons the best of the five mattresses was determined taking the natural spine position - defined by the laser measurement system - as a standard and considering the fact that people sleep averagely 20 % of the night on their back and 75 % on their side. These findings correlated well with objective camera and laser measurements : in most cases the F.E. model assigned the same ranking to the mattresses as a specialist would do. This good correlation was partly caused by the large differences between the different mattresses.

Quantitative comparisons between the measured and the predicted values for both sleeping positions allowed us to improve and to refine the finite element model. Nevertheless the correlation between predicted and measured spinous process positions was in most cases not high enough to define the optimal sleeping system. The need for a future quantitative determination of an individual's optimal sleeping system by a numerically correct shape prediction of the vertebral curvature justifies the development of new more reliable (e.g. 3D) analyses.

5. DISCUSSION

The large differences between the analyzed sleeping systems allowed us to predict easily the best one for each individual. The rather low correlation between predicted and measured spinous process positions was mainly caused by the 2D approximation and inaccuracies in the body stiffness parameters, in spite of the feed back from measurements. To get a more precise idea of spinal behavior during bedrest new full 3D finite element analyses will be necessary. Nevertheless the model can already be put into practice both to facilitate new mattress development and to advise individuals about their optimal sleeping system.

6. ACKNOWLEDGEMENTS

The authors wish to thank the test persons for their time and the company RECTICEL® for their financial and practical support.

7. REFERENCES

1. Oliver J., Middleditch A., Functional anatomy of the spine, chapter 10 : posture, 1991, 294-305.
2. Pheasant S., Ergonomics, work and health, chapter 5 : posture, 1991, 98-107.
3. Adams M.A., Hutton W.C., the effect of posture on the lumbar spine, J. Bone surg., 1985, 625-629.
4. Cailliet R., Correction of faulty mechanics in therapeutic approach to low back pain, Low back pain syndrome, chapter 5, 1976, 107-142.
5. Allen V., Ryan D.W., Murray A., Potential for bed sores due to high pressures : influence of body sites, body position, and mattress design, Br. J. Clin. Pract., 47:4, 1993, 195-197.
6 Clark M., Cullum N., Matching patient need for pressure sore prevention with the supply of pressure redistributing mattresses, J. Adv. Nurs., 1992, 310-316.
7. Rondorf-Klym L.M., Langemo D., Relationship between body weight, support surface and tissue interface pressure at the sacrum, Decubitus, 1993, 22-30.
8. Bennett L., Kavner D., Lee B.Y., Trainor F.S., Shear versus pressure as causative factors in skin blood flow occlusion, Phys. Med. Rehabil., 1979, 309-314.
9. Goossens R.H.M., Snijders C.J., Design criteria for the reduction of shear forces in beds and seats, J. Biomech, 1995, 225-230.
10. Orne D., King Liu Y., A mathematic model of spinal response to impact, J. Biomech., 1971, 49-71.
11. Schulz A.B., Analog studies of forces in the human spine : mechanical properties and motion segment behaviour, J. Biomech., 1973, 373-383.
12. Y. King Liu, Inertial properties of a segmented cadaver trunk : their implications in acceleration injuries, Aerospace Medicine, 197, 650-657.
13. The visible human project, http://www.nlm.nih.gov/research/visible/visible_human.html
14. S.E. Erdmann, Gos T., Density of trunk tissues of young and medium age people, J. Biomech., Vol 23, No 9, 1990, 945-947.

15. T. Andriacchi, A model for studies of mechanical interactions between the human spine and rib cage, J. Biomech., 1974, 497-507.

16. V.K. Goel, L.G. Gilbertson, Applications of the F.E. method to thoracolumbar research - past, present and future, Spine, 1995, 1719-1727.

5. RECONSTRUCTIVE SURGERY, VIRTUAL REALITY
AND IMPLANT ANALYSIS

BIOMECHANICS OF THE HAND: USE OF THE *SIGMA* GLOVE

E. Barnes[1], J. M. T. Penrose[2], N. W. Williams[3] and E. A. Trowbridge[4]

1. ABSTRACT

Assessment of surgical outcomes after reconstructive surgery of the hand is a difficult problem. Static goniometric measures of joint angles and subjective evaluation of specified tasks form the core of this appraisal. A custom built glove has been constructed at the University of Sheffield – Sheffield Instrumented Glove for Manual Assessment – (SIGMA) from carbon ink bend sensors and interfaced with a computer. From the 28 sensors placed within the glove real time measures of joint angles that are associated with different manipulative assignments can be collected, displayed and stored for evaluation. To visualise this procedure a virtual hand has been constructed from digitised images. The virtual hand moves in synchrony with the fingers that are enclosed in the *SIGMA* glove.

The studies undertaken to calibrate, validate and assess the reliability of the new computer interfaced measuring device are reported. A metal hand with fingers that could be positioned at known angles was used for calibration. The *SIGMA* glove was validated on this hand against the 'gold standard' of static goniometry using the method of agreement advocated by Bland and Altman. Reliability was tested using a proforma produced from plaster of paris, and was assessed by intraclass correlation using ANOVA with repeated measures. An individual subject gripped the proforma to leave an imprint and then, wearing the *SIGMA* glove this device was gripped and released 20 times.

KEYWORDS: Goniometry, Hand, Biomechanics, Glove

---

1. Principal lecturer, Centre for Health & Medical Research, University of Teesside, Middlesbrough, Cleveland. TS1 3BA, England.
2. Research Assistant, Department of Medical Physics and Clinical Engineering, University Of Sheffield, Royal Hallamshire Hospital, Glossop Road, Sheffield. S10 2JF, UK
3. Specialist Registrar, Department of Plastic and Reconstructive Surgery, Northern General Hospital NHS Trust, Herries Road, Sheffield S5 7AU, UK
4. Reader, Department of Medical Physics and Clinical Engineering, University Of Sheffield, Royal Hallamshire Hospital, Glossop Road, Sheffield. S10 2JF, UK

2. INTRODUCTION

The measurement of range of motion (ROM) has been recognised as a valuable tool in therapeutic interventions for many years. ROM quantifies the limitation of motion and thus the extent of impairment. In addition it provides information regarding the aetiology and treatment of joint problems and a means of monitoring changes or evaluating the effectiveness of treatment modalities[1].

ROM is traditionally assessed using goniometry which has been referred to as a 'fundamental part of the "basic science" of physical therapy'[2]. Studies into the technique of goniometry and, subsequently, the validity and reliability have been conducted from as early as 1949 [3]. Despite the limitations of the technique of goniometry it is still the most widely used means of assessing ROM.

The reliability of goniometry is dependent on a number of factors. In assessing intra-tester reliability, Hamilton & Lachenbruch [4] reported variance ranging from 1.5° for one tester to .945° for another. However, the manner in which this was calculated was not reported and the percentage variance is lower than that reported by other researchers. Hellebrandt et al[3] reported intra-tester reliability for highly skilled observers to be within 3°, but this was only for 70% of trials. Inter-tester reliability has proved to be more problematic. Solgaard et al.[5] reported that intra-tester Coefficients of Variance (CV's) were 5% to 8% whereas inter-tester (CV's) ranged from 6% to 10%. Low[6] found that the mean error between observers was much larger than the error between different measurements taken by the same observer. (Range 3.3°- 6.3° and 0.6° to 3.3° respectively). In addition it was found that where the average of several measurements is used, reliability increases.

It has been suggested that the problems with reproducibility of results may be associated with the standardisation of the process. Ekstrand et al.[7] compared two methods of measuring ROM in the lower extremities, one including explicit details for standardisation of procedure. It was found that with these clear procedures the intra-tester CV was reduced from 7.5%±2.9% to 1.9%±0.7%. This is supported by Bell-Krotoski et al.[1] who state that the performance of goniometry is technique dependent.

The validity of ROM for assessing 'hand function' has been questioned with relation to the fact that ROM assesses a one-dimensional hinge movement without taking account of the three-dimensional structure of the finger joints[8]. However, within this limitation Gajdosik & Bohannon[2] suggest that ROM closely approximates movement around a central point and that the measurement is clinically valid. The assessment of the goniometer as a valid tool was reported by Boone et al.[9] where the accuracy was assessed against five known angles, but no results were reported. No other papers reported assessing the validity of the goniometer.

Whilst goniometry may be acceptable clinically, it is a static measure of dynamic function and can only be used to assess one joint at a time. In addition the position of the proximal joints can influence the range of motion achieved at the measured joint because a muscle can be a two-joint muscle where the tendon action over one joint will influence movement at the other joint[10]. Also, goniometry cannot be used to measure abduction and adduction movement with any real accuracy. Dynamic assessment of the hand has been conducted using a variety of techniques including biplanar radiographic recording[11] and multi-camera reconstruction[12]. However, these techniques are not viable for clinical use.

The use of goniometry has its limitations and does not meet the selection criteria for assessment programmes as described by Dent & Orr[13] and Nichols[14] which include: quick to administer; repeatable; relevant to the needs of the patient, and objective.

3. THE *SIGMA* GLOVE

A digital glove input device has been constructed at Sheffield for clinical usage. It is the *SIGMA* (Sheffield Integrated Glove for Manual Assessment, Figure 1). It has 28 carbon ink bend sensors sewn into a Lycra glove that connect, via a bundle-cable at the wrist, to an interface box. This houses a constant current source for each sensor and interfaces to the parallel port of a PC. The sensors (Patent #5.086.785, Abrams Gentile Entertainment, USA) consist of a flexible circuit board with a strip of specialised carbon ink printed on one side of it. Printed circuit tracks run from both ends of the ink strip to wire connectors at one end of the sensor. When the sensor is bent with the carbon side of the sensor on the outside of the bend, the carbon ink is stretched and a corresponding change in resistance can be measured. The resistance change varies almost linearly with the angle of bend, and independently of the radius of the bend. The base resistance of each sensor is of the order of 10 KΩ and the change in resistance over 90° bend is approximately 200%. The sensors can also be cut to size (as long as any resultant track breaks are repaired) resulting in only an overall change in base resistance. Each current source has a gain setting which can be adjusted to give a sensor voltage reading of between zero and 3-4 volts. This is then fed to a 32-channel Analogue to Digital converter which is then interrogated by the specialised software in the computer.

Figure 1. The SIGMA glove and interface box.

The sensors are placed over the dorsal aspects of each finger joint: the four distal interphalangeal joints (DIP's), proximal interphalangeal joints (PIP's) and metacarpophalangeal joints (MCP's) of the fingers, to measure flexion. They are also placed between the fingers to record abduction. Additional sensors are placed to detect ulnar drift, MCP joint hyperextension and wrist flexion/extension and abduction/adduction. The thumb has 3 sensors, one each to measure flexion and extension at the interphalangeal joint, and one to measure flexion at the MCP joint.

The glove has been developed in three different sizes (small, medium and large) and is designed with the palmar aspect of each finger, as well as the majority of the palm, left open. The glove is secured over the hand by Velcro® tape fasteners. This feature, combined with the elastic nature of the Lycra, ensures that the glove can be used with a range of mild deformities and fits snugly about most shapes of hand.

4. RELIABILITY

Reliability was assessed on five healthy subjects with no hand pathologies, four male and one female. A plastercast proforma was made for each subject wearing the glove to minimise the effect of variation in grip posture. Each subject gripped and released the proforma twenty times and the voltage was recorded for flexion and extension of the DIP's, PIP's and MCP's for the four fingers and the thumb.

Test-retest reliability was determined using intra-class correlation[15] (see Table 1). This provides estimates of systematic and error variance between trials. It was determined from a two-way ANOVA with one between-subjects factor and one within-subjects factor (trials). The intra-class correlation coefficient R was calculated by:

$$R = (MS_s - MS_e) / MS_s$$

where MS_s= the mean squares for subjects and MS_e= the mean squares for error.

	DIP joint	PIP joint	MCP joint
2nd finger (index)	0.996	0.978	0.996
3rd finger (long)	0.985	0.989	0.997
4th finger (ring)	0.982	0.976	0.979
5th finger (pinky)	0.998	0.992	0.957
Thumb		0.942	0.986

Table 1. Intraclass correlation coefficient R for repeated grasping of proforma

All values are above .90 and therefore consistency across trials is high[16]. However, in recognition of the limitations of intraclass correlation for assessment of reliability[17] further analysis was conducted to determine a value for variance.

The extent of random variance of each sensor for each grip (see Table 2) was determined using a within-subjects coefficient of variation[18].

$$100 * (\sqrt{MS_e}/Mean)$$

	Dip Joint	PIP Joint	MCP Joint
2nd Finger (index)	15.8	4.4	7.1
3rd Finger (long)	6.8	7.9	5.6
4th Finger (ring)	4.6	4.15	5.6
5th Finger (pinky)	4.1	9.3	7.5
Thumb		16.1	5.9

Table 2. Percentage variance at each joint

These results suggest that although there is a high correlation the variance is high. The extent to which the variance is due to differences in grip posture or to sensor variation cannot be determined. However, it has been noted that the glove does not fit as snugly in some areas as others and this can cause movement of the sensor. This was particularly evident for the thumb during flexion where the sensor was prone to lateral movement.

Reliability was also assessed between sessions, where the glove was removed between each trial (see Tables 3 and 4). Each subject repeated the protocol ten times.

	DIP Joint	PIP Joint	MCP Joint
2nd Finger (index)	0.977	0.984	0.992
3rd Finger (long)	0.936	0.992	0.988
4th Finger (ring)	0.944	0.978	0.976
5th Finger (pinky)	0.943	0.979	0.942
Thumb		0.975	0.974

Table 3. Intra-class correlation R for between sessions repeated grasping of proforma.

	DIP Joint	PIP Joint	MCP Joint
2nd Finger (index)	14.7	2.9	5.1
3rd Finger (long)	7.5	4.8	7.9
4th Finger (ring)	7.6	4.5	4.4
5th Finger (pinky	7.4	8.3	4.7
Thumb		16.8	7.4

Table 4. Percentage Variance for each grip between sessions

Similarly although there is a high correlation there is a high level of variance and indeed an increase in variation of the sensor readings associated with removing and readorning the glove, particularly for the DIP joints. It was noted that the DIP Velcro straps were affected most by the variation of hand size and glove fit and the glove may need modifying to compensate. However, there are clinical implications here where particular attention will be needed in the adorning of the glove to ensure measurement consistency.

5. *SIGMA* GLOVE CALIBRATION

For use in the clinical setting each sensor required calibrating in degrees. Measuring an angle on solid objects with goniometry is associated with a lower error (1°), than that recorded where the interface is with relatively soft finger tissue (5°) [8]. To ensure accuracy for calibration an 'ideal' aluminium hand was built (see Figure 2). This consists of rectangular cross-section 'bone' segments, which are adjustable in length, connected together via lockable hinges. The ball and socket type joints connect the 'metacarpal' segments to a fixed 'wrist' block, thus allowing some abduction/adduction and rotation of the individual fingers. Thus the metal hand can conform to almost any size and posture attainable by the human hand, and can be fixed in posture for testing purposes.

For calibration of the glove, a range of wooden blocks were machined with two surfaces at known angles: 0°; 10°;20°; 30°; 50°; 60°; 70°; 80°; 90°; 110°. The glove was placed on the metal hand and each block in turn placed on the palmar side of the joint. The joint was conformed with the apex of the angled block in the 'palmar crease' to achieve the set angle. The sensor over the dorsal surface was parallel to the palmar surface and therefore produced the same angle.

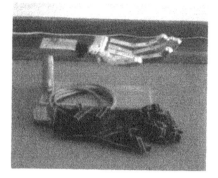

Figure 2. The SIGMA glove and *Ideal* aluminium hand.

It was assumed that there was a direct mapping between the voltage output of each sensor and the angle bend of that sensor. Three recordings were made of each joint on each block and a series of points relating the sensor voltage to joint angle plotted. A cubic polynomial regression was applied to produce an equation for each sensor directly relating voltage to angle of flexion. The coefficients of each equation were stored to allow conversion from voltage to angle automatically as the voltages were read from the glove.

6. VALIDITY OF THE *SIGMA* GLOVE

To validate the glove, an alternative 'Gold Standard' method had to be selected for comparing joint angles. Goniometry was selected as it is the most frequently used assessment tool. The glove was placed on the metal hand and the joints flexed over the wooden blocks as for calibration. The glove was removed and the process repeated with the angle of each joint assessed with the goniometer over the dorsal aspect.

Traditionally, Pearson's correlation[15] has been used to assess validity. This gave excellent results with R>0.9 in all cases. However in this case, the use of this method may be unsuitable. It does not yield a quantative difference between the devices, and the static goniometer is not a true 'Gold Standard' as it has a degree of variation associated with it. The results for DIP2, PIP2 and MCP2 were then analysed using the 'Agreement' technique. This method quantifies the differences between the two forms of measurement in terms of the mean difference and the limits within which the differences will lie[19].

	Mean Difference	Standard Deviation of Difference	95% Agreement Limit
DIP2	-6.2°	6.4°	-19.4° ↔ 6.6°
PIP2	-1.86°	4.47°	-10.8° ↔ 7.1°
MCP2	1.79°	4.51°	- 7.2° ↔ 10.8°

Table 5. Agreement limits for *SIGMA* glove and goniometry on the metal hand.

Again the major bias appears to be on the DIP where the glove is reading significantly higher than the goniometer (see Table 5). This has been associated with the open finger end design of the glove, as the glove does not fit as snugly on bending the DIPs. It may be appropriate to modify the design to encapsulate the finger end, although this may have consequences for the adaptability of the glove to different pathologies.

7. CONCLUSIONS

This study shows that the *SIGMA* glove is a viable proposition as a goniometric tool. Whilst there is a comparable level of error between the glove and the static goniometer, the extent to which these variations are acceptable will be a clinical as opposed to a statistical decision. However, with design modifications, the *SIGMA* glove will offer an improved alternative to goniometry. It has the advantages of the ability to: assess static and dynamic ROM; monitor all joints of the hand simultaneously and thus can be used to assess specific tasks; monitor joint velocity/angular movement with time and so assess joint 'stiffness'.

The data collected from the *SIGMA* glove is passed to a virtual hand which provides a visualisation tool. Previously, relatively crude means have been used to demonstrate results of therapy; for example, a wire bent around the finger and then transferred to the notes by drawing round the shape of the wire[20]. Photography has also been used, but poses problems such as variations in the position of the camera[2]. Therefore the glove has potential in terms of visual feedback both for the therapist and the patient. It also provides a useful research device for analysis of movement.

There is scope to improve the design and thus the accuracy of the glove. Also, validity has yet to be established on the human hand, which is not topologically or kinematically the same as the metal hand. It is also anticipated that because of its documented variation, an alternative 'gold standard' to goniometry is desirable. To this end, work will be conducted in exploring the viability of using cameras and a digitising system for 3D analysis.

8. Acknowledgements

Thanks must go to Jean Watson, Technical Instructor at the Northern General Hospital, for her help with the making of the gloves.

REFERENCES

1. Bell-Krotoski, J. A., Breger-Lee, D.E., Beach, R.B., Biomechanics and evaluation of the hand, in: Hunter, J. M., Mackin, E. J., Callahan, A. D. (eds.), Rehabilitation of the hand: Surgery and Therapy 1, 4th edition (1990), Mosby, St. Louis, Missouri, 153-184.
2. Gajdosik, R.L., and Bohannon, R. W., Clinical measurement of range of motion. Review of goniometry emphasising reliability and validity, *Physical Therapy*, 1987, Vol. 67 (12), 1867-1872.

3. Hellebrandt, F. A., Duvall, E.N., Moore, M.L., The measurement of joint motion: Part III. Reliability of goniometry, *Physical Therapy Reviews*, 1949, Vol. 29, 302-307.

4. Hamilton, G.F. and Lachenbruch, P. A., Reliability of goniometers in assessing finger joint angle, *Physical Therapy*, 1969, Vol. 49 (5), 465-469.

5. Solgaard, S., Carlsen, A., Kramhoft, M., et al, Reproducibility of goniometry of the wrist, *Scand. J. Rehab. Med.*, 1986, Vol. 18, 5-7.

6. Low, J. L., The reliability of joint measurement, 1976, *Physiotherapy*, Vol. 62 (7), 227-229.

7. Ekstrand, J., Wiktorsson, M., Oberg, B., et al, Lower extremity goniometric measurements: a study to determine their reliability, 1982, *Arch. Phys. Med. Rehab.*, Vol. 63, 171-175.

8. Cantrell, T., and Fisher, T., The small joints of the hands, 1982, *Clinics in Rheumatic Diseases*, Vol. 8 (3), 545-557.

9. Boone, D.C., Azen, S. P., Lin, C., et al, Reliability of goniometric measurements, 1978, *Physical Therapy*, Vol. 58 (11), 1355-1360.

10. Hurt, S. P., Considerations in muscle function and their application to disability evaluation and treatment, 1947, *Am. J. Occup. Therapy*, Vol. 1, 69-73.

11. Chao, E. Y. S., An, K., Cooney III, W. P., Linscheid, R. L., Biomechanics of the hand: a basic research study, 1989, World Scientific Publishing Co., Singapore, 73-96.

12. Lee, J. W., and Rim, K., Measurement of finger joint angles and maximum finger forces during cylinder grip activity, 1991, *J. Biomed. Eng.*, Vol. 13, 152-162.

13. Dent, J. A. and Orr M. M., 1993, Surgeon's workshop Which tests to choose when assessing hand function, *J. Roy. Coll. Surg. Edin.*, Vol. 38, 315-319.

14. Nichols, P. J. R., Are ADL indices of any value? 1976, *Occupational Therapy*, Vol. 39, 160-163.

15. Thomas, J. R., and Nelson, J. K., Research methods in physical activity, 1996, Human Kinetics Books, Illinois,

16. Vincent, W. J., Statistics in kinesiology, 1995, Human Kinetics, Illinois, 167-183.

17. Mullineaux, D. R., Scott, M. A., Batterham, A. M., The use of 'agreement' for assessing digitiser reliability/objectivity, 1994, proceedings of the biomechanics section of the British Association of Sport and Exercise Sciences (edited by J. Watkins), Leeds, 17-20.

18. Bland, J. M., An introduction to medical statistics,1995, Oxford Medical Publications, Oxford University Press, Oxford, 265-290.

19. Bland, J. M., and Altman, D. G., Statistical methods for assessing agreement between two methods of clinical measurement, 1986, *Lancet*, i, 307-310.

20. Glanville, H. J., Objective assessment of treatment of hand injuries-a new method of measurement, 1964, *Annals of Physical Medicine*, Vol. 7, 304-306.

COMPUTATIONAL MODELLING OF STRESS STATE DURING HAND TREATMENT

Josef Jíra[1] , Jitka Jírová[2]

1. ABSTRACT

The article concerns two problems of the hand treatment solved by the ANSYS programme. The contractures result from profound burns, scald, electric shock, and from secondary healed wounds. There are two ways of treatment flexion contractures of the proximal interphalangeal joint: (i) the conservative treatment by gradual splintage, (ii) traction using a reposition device. In order to find out the actual stress state of the articulate cartilage it was necessary to create the computation model of behaviour of the joint structure during treatment.

The further aim of the contribution is to submit experience with the computational modelling of the behaviour silastic Swanson's prosthesis of the wrist of the original and modified type. The computational models of these types were derived on the basis surgery's experience. Computational studies of two deformation states of the wrist were carried out for flexion 20 degrees and extension 30 degrees. Obtained mechanical quantities for two types of the wrist joint implant were compared and results were summarised.

2. INTRODUCTION

2.1 Treatment flexion contractures of PIP

Flexion contractures are one of major therapeutical problems in orthopaedics, plastic and reconstructive surgery of the hand. In most cases the contractures result from profound

Keywords: Flexion contractures, Swanson's prosthesis of the wrist, computational modelling

[1]Assoc.Professor/ Faculty of Transportation Sciences of the Czech Technical University, Konviktská 20, 110 00 Prague 1, Czech Republic
[2]Chief Research Worker and Head of Lab./ Institute of Theoretical and Applied Mechanics, Prosecká 76, 190 00 Prague 9, Czech Republic

burns, scald, electric shock, and from secondary healed wounds. The study compares two ways of treatment flexion contractures of the proximal interphalangral joint (PIP):

- the conservative treatment by gradual splintage, using compressive bandage;
- traction using a reposition device.

Both treatments mentioned above involve the gradual recovery of joint movement by stretching the shortened structures. The purpose of the study was to evaluate the stress state inside the joint structures in other words the articular cartilages, which suffer the most during reconstruction.

2.2 Silastic Swanson wrist joint implant

The key influence of the carpal joint on the biomechanics and physiology of the hand is indisputable. Pathological states resulting in the disturbance of carpal geometry, particularly in the reduction of the so-called carpal height and in the lateral carpus deviations deteriorate the function of distal anatomical structures, particularly finger muscles and tendons. Moreover, a painful, unstable and axially deformed carpus causes the patient considerable difficulties and restricts his mobile ability. For these reasons it is understandable that the problems of carpal joint replacement have become an interesting and significant branch of endoprosthetics not only abroad, but recently also in the Czech Republic. In the surgery and rheumato-surgery of the hand [1] the problem of the carpus endoprosthesis represents an important element which may restore the axial position of the carpus, eliminate the joint instability and pain and restore the mobility within the limits needed for everyday current activities of the person concerned.

The origin and development of the silastic replacement of the carpal joint can be monitored since 1973 [2, 3], when Prof. Alfred Swanson from the USA published his first experience with his implant. In 1974 the material of the implant was improved by the application of silicon elastomer complying better with the requirement of resistance against the damage of the implant surface by abrasion or fissure. Also the implant was modified to its present shape. At present the implant consists of a silicon rubber casting (Fig.1.) the flexural zone of which is of oval form. From this zone two stems of rectangular cross section tapering towards the ends project. The stems are introduced proximally into a canal drilled in the radium diaphysis and distally into a canal made in os capitum and in the basis of the third metacarpus (Fig.1.). The implant is introduced approximately in the axis of the so-called central carpal pillar. Its flexibility

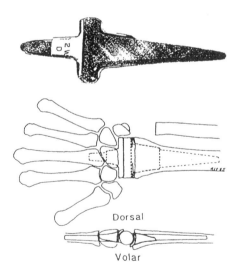

Fig.1. Silastic Swanson Wrist Joint Implant

is facilitated by the so-called piston effect which is a free motion of the stems in the canals in the direction of longitudinal axis taking place during implant flexion and extension. This movement reduces considerably the strain of the flexural zone and increases the flexibility of the joint replacement.

3. NUMERICAL COMPARISON OF BOTH WAYS OF TREATMENT FLEXION CONTRACTURES OF PIP

3.1 Frame Structures

The difference in load between the two methods of joint reconstruction is outlined by the Fig. 2a and b.

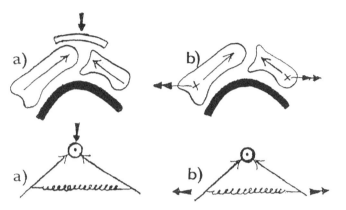

Fig.2. Scheme of two loading types

By calculation of the frame structure (Fig.3) we calculated for the same deflection of the joint in the direction of axis y, i.e. for the same value of stretch on the tendon (F=14.4 N) approximately a triple force on the joint in the case of the first "conservative" method (Fig.3a) in comparison with using a reposition device (Fig.3b).

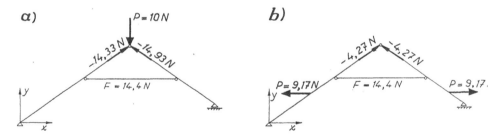

Fig.3. Frame structures

We represented the phalanxes as hollow tubes a modulus of elasticity comparable with cortical bone E=17 GPa and μ=0.25. Beam elements (hollow tubes) were mutually hinged connected with no regard to cartilage. The tendon was again substituted by a bar hinged connected to both beams with modulus of elasticity E=20 MPa which was experimentally measured [4]. We considered elastic behaviour of all structures, we did not calculate with viscoelastic behaviour of tendon which is in nature significant. It means that the results are correct only for start of loading. The frame structure was calculated as statically determinate structure with fixed bearing on one side and mobile bearing on the other side (Fig.2).

3.2 Plane FE Model

In order to find out the actual stress state of articulate cartilage we have constructed the Finite Element model. The plain model was constructed on the basis of the radiograph of fingers [5], when the co-ordinates of nodes were transferred into the computer by using ScanJet IIp. The stress state was determinated numerically using the ANSYS 5.0 programme. Both cortical and spongy bone tissues were described by two-dimensional plane elements. This element is defined by four nodes having two degrees of freedom at each node: translations in the nodal x and y directions. The element was used for plane strain analysis. The soft tissue (tendon and skin) was substituted by a link.

Articular joint was constructed using contact elements which might be used to represent contact and sliding between two surfaces in two dimensions. The element geometry is a triangle with the base being a line between two nodes on one of the surfaces and the opposing vertex being a node on the other surface. The coefficient of dry friction between the articulate cartilages of a joint is very low (can be as low as 0.0026, about the lowest of any known solid material); hence no friction was considered between the cartilage surfaces.

We have used for the solution of the problem a superelement which was a group of previously assembled ANSYS elements of different types and material properties. This superelement is treated as a single element. The reason for substructuring are to reduce computer time and to allow solution of very large problems with limited computer resources. In nonlinear analysis, which is the contact problem, we can substructure the linear portion of the model so that the element matrices for that portion need not be recalculated every equilibrium iteration. Since the contact elements require an interactive solution, substructuring the whole structure can result in a saving in computer time.

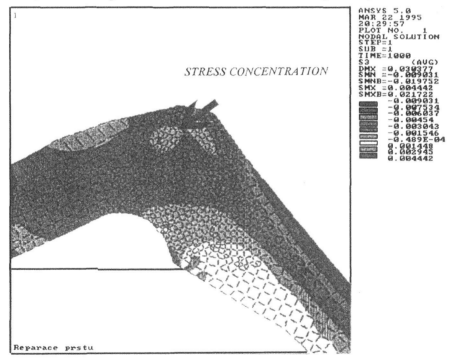

Fig.4. Detailed field of compressive stresses

As an example the solution of the simplify problem is given in the Fig.4 when we considered spongy bone tissue for the both bone structure. The problem was reduced into one superelement and 1624 contact elements. The single matrix of the superelement was calculated from two different element types: 1292 plane elements (bone tissue and cartilage) and one link (soft tissue) and from three different material properties (spongy bone, cartilage and tendon).

4. COMPUTATIONAL MODELLING OF BEHAVIOUR OF ORIGINAL AND MODIFIED TYPE OF WRIST JOINT IMPLANT

By means of a computational model we have ascertained the stress state particularly on the volar and the dorsal sides, where the greatest changes due to the modification can be expected. We have used a plane computation model with the assumption of plane strain state. The computation was made by the FEM and the ANSYS programme. As it concerned only active comparison of two models of analogous boundary and loading conditions and not actual stress values, a unit modulus of elasticity E=100 MPa was used.

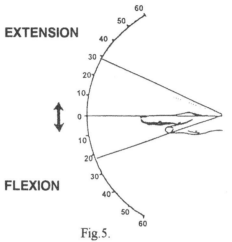

Fig.5.

The computations considered two cases of deformation by the prescription of displacement of the distal stem end fitted into the body of os capitulum and the base of IIIrd metacarpus: (i) flexion 20°, (ii) extension 30°. They are the average flexion and extension values of post-operation state. The 20° flexion produces the displacement of the distal end of the short stem of 10.18 mm in volar direction and of 1.80 mm in perpendicular direction. The 30° extension produces the displacement of 14.90 mm in dorsal direction and 3.99 mm in perpendicular direction. The fitting of the longer stem in the distal part of the radium was simulated by zero displacement of nodal points situated along the dorsal and the volar sides in the dorsal-volar direction with the possibility of any displacement in perpendicular direction. This simulation of boundary conditions replaces the possibility of free stem motion in the bone canal in the direction of its longitudinal axis.

Fig.6. FE Model of classic silastic Swanson carpal replacement with flexion

The mathematical model of the classic silastic Swanson carpal replacement consists of 4 297 nodal points (Fig.6) and of 4 199 isoparametric PLANE 42 elements. In the flexural zone of the endoprosthesis, i.e. in the area where the modification would produce the greatest change, the density of nodal points and elements was increased.

allows for improved movement of the wrist. From the mechanical view-point, the design modification results in a changed stress state with both extension and flexion even if no additional external load is considered. For this purpose, an analysis of the stress state obtained had to be undertaken and the points of peak stresses identified. In the area of the cut on the palmar side which is to improve carpus motility the nodes and elements were removed. The modified model of the wrist implant (Fig.7) consists of 4 183 nodal points and 4 061 elements.

Fig.7. FE Model of modified type of Swanson carpal replacement with flexion

5. RESULTS AND CONCLUSIONS

5.1 Conservative treatment by gradual splintage

We have calculated the stress state in the interphalangeal joint for the conservative method of treatment only (Fig.2a), which was determinated in the paragraph 3.2, as the considerably unfavourable one. In Fig.4 we can see the stress state of compression in the joint which is substantial for analysing of this problem. The whole loading of the articulate cartilage, which has been calculated for the frame structure (Fig.3a), is in fact concentrated on the relatively small area, i.e. it is distributed non-uniformly. The results have confirmed the experiences of orthopaedic surgeons, who often report in cases of conservative method of treatment pain and swelling of the joint. In this way the articulate cartilage is considerably even unphysiologically overstressed (Fig.4) during the conservative treatment.

5.2 Original and modified type of silastic Swanson wrist joint implant

For evaluation purposes the concentrations of tensile stresses, decisive for the ascertaiment of the bearing capacity of the given material, are most important. As we have been comparing the change of carpus motility due to replacement modification, another important computed value was the maximum force F_y which must be produced for the achievement of the carpus maximum flexion of 20° and maximum extension 30°. The third assessed quantity was the reaction R_y, i.e. the force produced on the contact surface of the resected bone and the endoprosthesis stem during flexion at the volar side and extension at the dorsal side. However, the reaction is not of substantial significance for the assessment of the bearing capacity of the endoprosthesis, as its silicon surface is protected with the titanium protection plate in the area of transition of the flexural zone into the stem.

Tables I and II give the maximum values of tensile stresses SIG 1 for flexion on the dorsal side and extension on the volar side, the maximum required force F_y which must be produced on the volar side, the maximum reaction R_y at the contact between the bone and the replacement, once again on the side of its origin .

Table I. Maximum Values of Structural Characteristics for Flexion 20°

Replacement type	max. SIG 1	F_y	R_y
Classic	10.827 MPa	1.6643 N	3.9022 N
Modified	10.678 MPa	1.0810 N	1.7539 N

Table II. Maximum Values of Structural Characteristics for Extension 30°

Replacement type	max. SIG 1	F_y	R_y
Classic	22.028 MPa	2.4336 N	3.4545 N
Modified	11.856 MPa	1.5816 N	1.7432 N

The value of max. SIG 1 for modified replacement in Table II occurs in the same place as in the classic replacement. The actual maximum SIG 1 stress occurs in the cut vertex of the modified replacement. It is of relative magnitude of 25.625 MPa and depends on the geometry of the notch.

The values given in the above tables yield the following conclusions:

1. The achievement of the flexion 20° requires in the case of the classic replacement a force F_y which is 1.5 times (i.e. by 54%) higher than the required by the modified replacement; the reaction R_y on the contact surface with the resected bone in the case of the classic replacement is practically twice as large (exactly by 122.5%) than in the case of the modified replacement.

2. The achievement of extension 30° requires, in the case of the classic replacement, again a force F_y 1.5 times (by 54%) higher than that required by the modified replacement. The reaction R_y originating on the contact surface with resected bone of the classic replacement is twice as high (by 98%) as that originating in the case of modified replacement. The maximum tensile stresses in the modified replacement occur in the vertex of the cut; for this reason this detail must be afforded particular attention during replacement manufacture lest the production inaccuracies should produce stress concentrators.

It follows that the J. Pech's modification of the Swanson carpal replacement enables better motility of operated carpus. The probable concentration of tensile stresses which may arise in the vertex of the notch during extension due to its geometry can be eliminated by the technological procedure of the modified implant manufacture.

Acknowledgements: The authors wish to acknowledge the support of this work by project No. 103/96/0268 of the Grant Agency of the Czech Republic.

REFERENCES

1. Pech, J., Sosna, A., Rybka, V., Pokorný, D., Wrist arthrodesis in rheumatoid arthritis, The J. of Bone and Joint Structure, 1996, No.5, Vol.78-B, pp.785-786
2. Swanson, A. B., Flexible Implant Arthoplasty for Arthritic Distabilities of the Radiocarpal Joint, Orthop.Clin.North Amer., 1973, Vol.4, pp.383-394

3. Swanson, A. B, Flexible Implant Arthoplasty in the Hand and Extremities, St.Louis, Mosby, 1973

4. Jírová, J., Kafka, V., Experimental Modelling of Rheological Behaviour of Tendon. Proc.of 31st Conference of Experimental Stress Analysis (EAN), 1993, Měřín, Czech Republic, pp.151-154 (in Czech)

5. Chao, E. Y. S., An, K.-N., Cooney III, W. P., Linscheid, R. L., Biomechanics of the Hand, World Scientific, Singapore•New Jersey•London•Hong Kong, 1989

6. Pech, J., Wrist Endoprosthetics, Schola Nova - Komenium, Prague, 1996

KINEMATIC AND STRESS ANALYSIS OF METACARPOPHALANGEAL JOINT IMPLANTS

J.M.T. Penrose[1], N.W. Williams[2], D.R. Hose[3], E.A. Trowbridge[4]

1. ABSTRACT

Rheumatoid arthritis and post-traumatic osteo-arthritis are very common conditions in the joints of the finger. They are often very painful and are also associated with deformity and a reduction in function. At present, flexible one-piece Silastic implants are the most popular solution to relieve the pain in the joint, correct deformity, restore function and improve the general look of the hand. These implants are designed primarily for flexion/extension but do allow some abduction/adduction as well. Long term studies show that these implants tend to fail after around 3-4 years, mostly due to fracture. This places further limits on the function, but the implants are not usually replaced, instead being left to act as simple spacers between the bones.

To improve the design of implants, kinematic and stress responses can be studied in computer models that closely mimic the functional environment. In this study, simple two and three dimensional finite element analyses of a prosthesis under load have been performed. The kinematic behaviour of the implants were simulated, and areas of undue stress and friction identified. Results were visualised using a novel interactive process that involved a Virtual Hand.

KEYWORDS

Metacarpophalangeal Joint, Prosthesis, Finite Element, Visualisation

1. Research Assistant, Department of Medical Physics and Clinical Engineering, University Of Sheffield, Royal Hallamshire Hospital, Glossop Road, Sheffield. S10 2JF, UK
2. Specialist Registrar, Department of Plastic and Reconstructive Surgery, Northern General Hospital NHS Trust, Herries Road, Sheffield S5 7AU, UK
3. Lecturer, Department of Medical Physics and Clinical Engineering, University Of Sheffield, Royal Hallamshire Hospital, Glossop Road, Sheffield. S10 2JF, UK
4. Reader, Department of Medical Physics and Clinical Engineering, University Of Sheffield, Royal Hallamshire Hospital, Glossop Road, Sheffield. S10 2JF, UK

2. INTRODUCTION

Rheumatoid arthritis and post-traumatic osteo-arthritis are very common conditions affecting the metacarpophalangeal (MCP) joints in the human hand. These conditions are often very painful and are also associated with deformity and a reduction in function. Whilst permanent fixation of the joint (arthrodesis) is often performed in the distal and proximal interphalangeal joints, it is contraindicated at the metacarpophalangeal joints due to the importance of an adequate range of motion for many manual tasks. Various prostheses and implants have been developed for the reconstruction of the metacarpophalangeal joint, all designed to restore a functional range of motion as well as correct existing, and prevent further, deformity. They also help reduce pain in the joint and give a cosmetic improvement.

Today the most common mode of rheumatoid MCP reconstruction is by implantation of a flexible one-piece endoprosthesis. These implants are fabricated from Silastic, medical grade poly-di-methyl-siloxane, and they allow flexion and extension, and some abduction and adduction. Two particular designs are now predominantly used in clinical practice; the Swanson design[1,2] which has changed little since its introduction in 1966, and the more recent Sutter design[3] (1987). The main difference between these two is the shape of their central hinge mechanisms (see figure 3).

The implantation process involves the removal of the metacarpal head with an oscillating saw and the reaming of the intramedullary canals of both the proximal phalanx and metacarpal with wedge-shaped files. The surrounding soft tissue structures are released, and the fingers realigned to correct joint deformities such as palmar subluxation and ulnar drift. In this process, the collateral ligaments are often sacrificed. A size of implant is then selected and the implant stems are then inserted into the reamed canals. The implants are not cemented into the cavities as the process is exothermic and can cause cellular necrosis. The implant is stabilised by the formation of a fibrous capsule around the joint, a process known as encapsulation[4]. After reconstruction, in the short term, a range of motion up to 60 degrees is observed[4]. However, in the longer term, fibrosis around the implant means a range of motion of around 30 degrees is usually retained[5,6].

Experience with these implants has made clinicians realise that they are not without their own problems with respect to failure and recurrence of deformity. Implanted MCP prostheses usually last around 3-4 years before failure and are also known to sublux[5]. Reported failure rates vary, but figures quoted for the Swanson design are 10-20% in the longer term studies[5,7,6], and 45% after 3 years[7] for the Sutter design. Occasionally the patient is unaware that the implant has fractured, and as replacement is inefficient the implant is often left to act as a simple spacer between the metacarpal and proximal phalanx, preventing osseosynthesis.

Little is known about the dynamic behaviour of these implants within the joint or whether failure is due to high stress loading or local abrasive wear on the stem, or both. More recently, titanium grommet collars have been supplied with the Swanson implant to prolong their lifespan. These slip over each stem and fit snugly up against the central hinge piece, but little evidence exists as to their effectiveness. However, any design that is inherently resistant to wear and failure would be at an advantage. Ideally, any new implant design would be thoroughly tested in a dynamical in-situ simulation before mass production and clinical testing.

In this study, the Finite Element Analysis method was employed to compute the kinematic and stress responses of two and three dimensional models of each implant design. These models were built into environments that would mimic the loads placed on an actual implant 'in-situ'.

3. MATERIALS AND METHODS

3.1. Software and Hardware

The construction and loading of the finite element meshes, as well as the solving and post-processing, was performed using the ANSYS 5.3 (Swanson Analysis Systems, Inc.) package. This was run on a 166Mhz Pentium-PC with 48Mb RAM. The solver uses an implicit procedure and can perform the non-linear large deformation analysis (via a modified iterative Newton-Raphson method) appropriate to this modelling situation.

3.2. 2-D Finite Element Models

Scaled models of both the Swanson and Sutter design implants were constructed in 2-D using 8-node quadrilateral elements. These elements feature midside nodes which allow for quadratic displacements across the element edges. They are more efficient for representing the complex stress fields anticipated here. The implants are modelled with their stems inserted into C-shaped bone ends. The cavities of these bones are modelled so that their shape conforms to the simple reaming process by which they are created. The outer geometry of each bone was not considered important and so was not shaped to any degree of accuracy. No stress calculations were to be performed on these bone areas, and so were meshed with more simple 4-node quadrilateral elements.

Fig. 1. The 2-D implanted Sutter (left) and Swanson (right) finite element models.

Two tensile spar elements were placed between the two bone ends, one on the dorsal aspect and one on the palmar aspect, across the implant. These were included to provide stability in the model, and some degree of compression on the implant, as exerted by the joint encapsulation. Contact elements were set up between the outer surfaces of each stem and the inner surfaces of the corresponding bone end. These elements were introduced to create frictionless surface-to-surface contact detection and interaction.

Fig. 2. The 2-D implanted Swanson finite element model, with proximal collar fitted.

An additional Swanson design model was created where the proximal stem was fitted with a titanium grommet collar. The lateral sides of the grommet were modelled by creating single elements which spanned the implant mesh connecting the palmar and dorsal grommet sections (see figure 2). These lateral sections were assigned a stiffness one order of magnitude less than the rest of the grommet ring to represent the bending stiffness under diametrically opposed forces.

3.3. 3-D Finite Element Models

Scaled 3-D finite element models of both the Swanson and Sutter design implants were constructed using 20-node hexahedral (brick) elements (see figure 3). These elements also feature midside nodes.

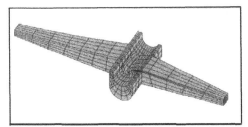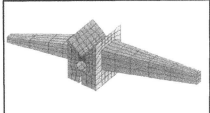

Fig. 3. The 3D Swanson (left) and Sutter (right) implant design finite element meshes.

The mesh density is similar for the two models, but is not considered adequate for absolute calculation of stress values. Thus these models are considered as 'stiffness models', used primarily for the analysis of kinematic response.

Each model was considered 'clamped' at the proximal stem. This was achieved by constraining edge nodes further than 3.5 mm from the centre of the hinge portion to have zero displacement in all three axial directions. The implants were made to flex by constraining the most distant distal stem corner nodes to have a finite displacement in the palmar axis plane.

3.4. Material Properties

The material properties of Silastic[8] were taken to be: a Young's Modulus of 2.41 MPa, a density of 1140 kg.m^{-3} and a Poisson's ratio of 0.45. The bone region properties[9] were taken to be: a Young's Modulus of 100 MPa and a Poisson's ratio of 0.2. These properties represent a linear elastic, isotropic material model, and do not consider any anisotropy, or viscoelasticity that bone or Silastic may posses. The spar elements were assigned a Young's Modulus of 0.05 MPa, a cross-sectional area of 5 mm^2, and an initial strain of 7.5% to provide the joint compression. These values were chosen to be large enough not to cause ill-conditioning of the finite element stiffness matrix, but not so large as to compress the implant unrealistically. Lastly, the material properties of titanium were taken to be: a Young's Modulus of 500 MPa, a density of 4500 kgm⁻³ and a Poisson's ratio of 0.34.

4. RESULTS

Figure 4 shows the stress intensity contours on the surface of each flexed 3-D implant model. Numerical contour legends are not given as it was felt that the mesh designs were only suitable for qualitative assessment, not accurate stress calculation.

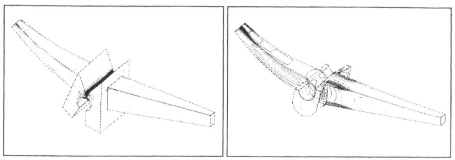

Figure 4. Surface stress contours for the two 3-D implant models flexed through 26°.

It should be noted that the highest stress concentrations for the Sutter design occurred on the upper and lower aspects of the central 'hinge' portion of the implant. Also, high stress concentrations for the Swanson design were found on the inner aspect of the central portion of the implant, and on the hinge-stem interface.

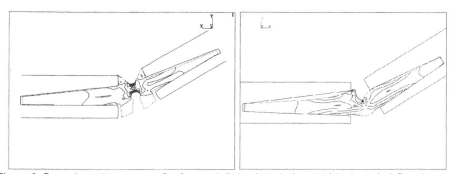

Figure 5. Stress intensity contours for the two 2-D implant designs within the 'joint' flexed palmarly through 30°.

Figure 5 shows the stress intensity contours throughout each implant within the flexed joint 2-D models. Similarly, figure 6 shows the stress intensity contours throughout the 2-D Swanson implant design fitted with a proximal grommet collar. The outline of the bone ends are also drawn on these figures to clarify the implant positions. Again, numerical contour legends are not given due to the low mesh densities used.

Figure 7 shows a novel interactive approach to the visualisation of the implant analysis data. It shows the 30-sensor SIGMA glove (Sheffield Instrumented Glove for Manual

Assessment) linked via Virtual Reality technology to an articulated skeletal hand model[10]. This model had a Sutter implant inserted into the second MCP joint.

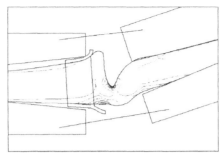

Figure 6. Stress intensity contours for the Swanson implant design fitted with a proximal grommet collar. The 'joint' is flexed palmarly through 15°.

Thus, as the gloved hand is manipulated, the skeletal model responds in real-time. When the reconstructed joint is flexed the resulting stress contours in the implant can be seen in a window at the bottom of the screen.

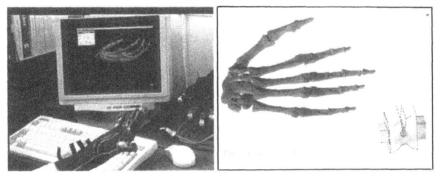

Figure 7. The SIGMA glove (left) and the *Virtual Hand* (right) with implanted Sutter prosthesis.

5. DISCUSSION

Prediction of the behaviour of an implanted joint prosthesis is of paramount importance to the designer. The movement of the implant within the joint and resultant stress during its use should be thoroughly evaluated as a function of the implant's material properties and geometry.

Figure 4 shows that the 3-D results for both implant designs indicate stress concentrations at the hinge/stem interface, and across the central hinge portions themselves. The central concentration was most apparent with the Sutter design and this coincides with reported fracture incidence with this design[6].

The Swanson design showed greater concentrations at the hinge/stem interface; this also coincides with reported regions of high fracture incidence[7]. The mechanism of this fracture may have something to do with the unusual motion of the central portion observed when successive sub-steps of the flexion are animated. These show the central portion 'rocking' as the distal stem is flexed palmarly, (also observed by Gillespie et al.

[11]). It indicates that the centre of rotation of the flexion may be more proximal than the central hinge, nearer the stem interface, thus increasing the stress in this area.

It was also noted that the Swanson design was more resistant to flexion[12]. That is, it was 'stiffer' as a replacement joint and this may have implications when considering the suitability of the implant for each patient.

Figure 5 shows the results from the 2-D analyses and, like the 3-D models, stress concentrations in the central hinge areas can be seen. These 2-D models, however, are only plane-stress representations and the mesh density is low, hence the concentrations at the hinge/stem interfaces are not so apparent. But since these implants have been modelled in a more realistic 'joint' environment, they give more information about the motion of the implants during flexion. They show that with the Sutter design, the proximal and distal faces of the central portion come in to and out of contact with the bone ends during flexion. This can cause cortical erosion and increased stress at the hinge/stem interfaces. Similarly, with the Swanson design, the point of contact between the inner bone corners and the implant stems moves during flexion, and proximally, the dorsal bone can come out of contact altogether. Again, this can lead to localised erosion both on the bone and the implant. These sites are indicators for possible implant fracture.

It can also be seen, from animations of both 2-D design flexion sub-steps, that the implant stems 'piston' in and out of the bone cavities. Whilst this movement was originally intended as a feature of the Swanson design in order to reduce stresses throughout the implant[11], the motion of the implant stems over the inner surface of the intramedullary canals could cause erosion to both the implant and the inside of the bone. This may cause further complications such as increased lateral movement of the implant allowing ulnar drift to re-occur.

Figure 6 shows how the implant could be protected from local wear by the use of the titanium grommet collar. However, whilst this may also reduce stress at the hinge/stem interface, it can be seen that the collar moves with the implant during flexion, rather than remaining stationary. This may mean the stresses are simply transferred to the bone ends and corners, causing damage. Coincidentally, surgeons generally do not fit grommets where rheumatoid patients are concerned due to their lack of adequate bone stock.

6. CONCLUSION

This study shows that finite element analysis can provide qualitative evidence of advantage and disadvantage in prosthetic design. It can also indicate possible mechanisms of implant wear and fracture whilst in situ, supplementary to the surgeon's experience and long term studies. These techniques can be used to advance the development of new implant designs in advance of the prototyping stage.

7. Acknowledgements

The authors would like to thank ANSYS (Europe) for the supply of the software license, and Avanta Orthopaedics for the supply of the implants. Thanks also go to Mr Paul McArthur of the Northern General Hospital Plastic and Reconstructive Surgery Unit, the

Trent Regional Health Authority, the Royal College of Surgeons of England and the British Association of Plastic Surgeons for their support.

REFERENCES

1. Dow-Corning-Wright. Data sheet: Metacarpal joint implant arthroplasty, July 5-9 1986.
2. Swanson, AB. Silicone rubber implants for replacement of arthritic or destroyed joints in the hand. *Surgical Clinics of North America*, 1968, 48, 1113-1127.
3. Bechenbaugh, RD and Linscheid ,RL. Arthroplasty in the hand and wrist. In: 'Operative Hand Surgery'. Green, DP (ed.). Churchill Livingstone, New York, 1993. 159-172.
4. Swanson, AB. Finger joint replacement by silicone rubber implant and the concept of implant fixation by encapsulation. *Annals of the Rheumatic Diseases*, 1969, 28(5), 47-55.
5. Beevers, DJ, and Seedhom, BB, Metacarpophalangeal joint prosthesis: A review of the clinical results of past and current designs. *Journal of Hand Surgery*, 1995, Vol. 20B(2): 125-136.
6. Wilson, YG Sykes, PJ and Niranjan, NS. Long-term follow-up of Swanson's silastic arthroplasty of the metacarpophalangeal joints in rheumatoid arthritis. *Journal of Hand Surgery*, 1993, 18B(1), 81-91.
7. Bass, RL, Stern, PJ and Nairus, JG, High implant fracture incidence with Sutter silicone metacarpophalangeal joint arthroplasty. *Journal of Hand Surgery*, 1996, Vol. 21A(5): 813-818.
8. Lewis, G and Alva, P. Stress analysis of a flexible one-piece type first metatarsophalangeal joint implant. *Journal of the American Podiatric Medical Association*, 1993, 83(1), 29-38.
9. Akagi, T Hashizume, H Inoue, H et al. Computer simulation analysis of fracture dislocation of the proximal interphalangeal joint using the finite element method. *Acta Medica Okayama*, 1994, 48(5), 263-270.
10. Williams, NW Trowbridge, EA La Hausse-Brown, TP and Caddy, CM. The Virtual Hand and Joint Implants. In: Association of Surgeons in Training Yearbook, 1996: 58-60.
11. Gillespie, TE Flatt, AE Youm, Y and Sprague, BL. Biomechanical evaluation of metacarpophalangeal joint prosthesis designs. *Journal of Hand Surgery*, 1979, 4(6), 508-521.
12. Penrose,JMT Williams,NW Hose,DR and Trowbridge,EA, An examination of one-piece metacarpophalangeal joint implants using finite element analysis. *Journal of Medical Engineering and Technology*, 1996, Vol. 20(4/5): 145-150.
13. Penrose,JMT Williams,NW Hose,DR and Trowbridge,EA, In-situ simulation of one-piece metacarpophalangeal joint implants using finite element analysis. *Medical Engineering and Physics*, 1997, (in press).
14. Vahvanen, V and Viljakka, T. Silicone rubber implant arthroplasty of the metacarpophalangeal joint in rheumatoid arthritis. *Journal of Hand Surgery*, 1986, 11a, 333-9.

3D FINITE ELEMENT MODELLING OF A CARPOMETACARPAL IMPLANT COATED WITH HYDROXYAPATITE

F. Lbath[1], C. Rumelhart[2]

1. ABSTRACT

A three-dimensional finite element model of a total cementless carpometacarpal prosthesis coated with hydroxyapatite was developed in order to study the coating and the bone/implant interface in two clinical situations for key pinch. A semi-automatic mesh generator was developed. Finite element analysis was carried out using the commercial ABAQUS® package. Hydroxyapatite mechanical characteristics were determined, in order to take into account this biomaterial behaviour in the finite element model. The hydroxyapatite/trabecular bone interface was either perfectly bonded representing a long term situation where the implant is perfectly osteointegrated, or considered in contact with friction at bone/implant interface in the early post-operative situation. The finite element model predicted, immediately after surgery (non linear interface) and in the long term after bony ingrowth (fully bonded interface), large stresses in the hydroxyapatite coating.

2. INTRODUCTION

Among the methods used to treat osteoarthritis of the carpometacarpal joint, located at the thumb base, is total joint prosthesis which is a ball-and-socket implant like the hip endoprosthesis. The implant under study is the ARPE® non-cemented prosthesis (Merck Biomaterial France and Pr J.J. COMTET). The stem, coated with plasma-sprayed hydroxyapatite, is press-fitted into the first metacarpal bone.

The hydroxyapatite coating is a bioactive material giving a lifelong biological bond with the surrounding bone [1], [2], [3]. Clinical investigations on retrieved implants (hip

Keywords: Finite element modelling, Carpometacarpal implant, Hydroxyapatite, Non linear interface, Medical imaging

[1] Ph. D., INSA, LMSo, Bât. 304, 20 av. A. Einstein, 69621 Villeurbanne Cedex, FRANCE
[2] Professor, ARPTAL, INSA, LMSo, Bât. 304, 20 av. A. Einstein, 69621 Villeurbanne Cedex, FRANCE

prosthesis) (between 8 weeks and 15 months after surgery) showed an early bone deposition on the coating as well as on the cancellous bone, leading to early bridging of bone between them [4], [5]. However reduction in the thickness of the hydroxyapatite coating has been recently observed [1], [2], [5],

The knowledge of these resorption phenomena is far from being complete and the evolution of this resorption in the long term is unknown as is also its clinical effects.

In order to obtain a better understanding of the stress transfer across this layer, which could affect the material behaviour, finite element analyses were undertaken. Most finite element models of stem fixation assumed full bonds across the interface between the different materials [6], [7].

The fixation of cementless prostheses via bony ingrowth is a long term process and perfect bonding does not represent the early postoperative situation. Non linear finite element analyses were performed in order to take into account the sliding and the friction existing between the stem and the bone [8], [9].

A finite element model was developed based on accurate descriptions of the bone geometry, typical surgical positioning of a well-fitted prosthesis within the bone, the nonlinear bone/hydroxyapatite interface conditions representative of the early post-operative situation and a realistic definition of anatomical loading. A three-dimensional finite element analysis was performed in order to study the Von Mises equivalent stresses and relative micromotions of the hydroxyapatite coating and the bone/implant interface in two clinical situations: the early post-operative case, where there is contact with friction and the long term one, where the implant is perfectly osteointegrated.

2. MATERIAL AND METHOD

2.1. Geometry acquisition

The geometry of the first metacarpal bone was obtained from 60 contiguous transversal 1mm thick Computed Tomography (CT) scan slices of a cadaver hand specimen. The numerical bone data were computed from 21 representative CT scans, obtained from an Elscint 2400 CT (General Hospital, Dijon, France). Our in-house interactive boundary detection software was used to extract the periostal and endosteal contours of each CT images [10]. The three-dimensional geometry was reconstructed by stacking the contours of adjacent images. Based on these surfaces, the appropriately sized ARPE prosthesis was added to fit the model of the metacarpal bone.

2.2. Mesh generation

A semi-automatic mesh generator was developed: the cortical bone and the implant were meshed as well as the 100 μm thick hydroxyapatite coating. The implant and the cortical bone were ideally connected by another element layer representing the trabecular bone. The finite element code is ABAQUS® (HIBBIT KARLSSON & SORENSEN). From the outer to the inner layer, the cortical, cancellous and hydroxyapatite layers were meshed into three-dimensional isoparametric eight-node brick elements and the prosthesis into six-node triangular prism elements and four-node tetraedric ones.

The x-axis represents the radial-ulnar direction, the y-axis the dorsal-palmar direction and the z-axis the longitudinal axis oriented distally. The origin is the center of the spherical head of the metacarpal implant (Figure 1).

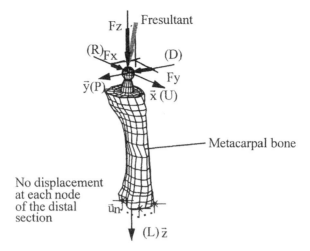

Figure 1: The finite element model and its loading.

(P: palmar, U: ulnar, D: dorsal, R: radial)

The hydroxyapatite/trabecular bone interface was either perfectly bonded representing a long term situation where the implant is perfectly osteointegrated, or in contact with friction in the post-operative situation. The non linear interface was modeled by contact elements (INTER8) based on the standard Coulomb friction law.

In order to approach the coefficient of friction μ, defined as the ratio of the tangential load and the normal load, friction tests between ewe vertebral spongy bone (2 samples kept frozen at -18°C until machining, then at -4°C for 12 hours and finally at room temperature 2 hours before testing) and hydroxyapatite coating of a titanium alloy bar (4 x 5 x 50 mm^3) were undertaken. The normal load F_n was applied to the bone by a 10 N mass attached to it. The titanium bar was fixed to the frame of a horizontal traction machine (Amsler DK f 10m) used to apply to the bone the tangential load F_t which was measured by a force transducer (ATEX FA 5000) (experiments were repeated 3 times). The friction coefficient was computed and the mean value was 0.21 (± 0.02).

2.3. Material properties

Materials were assumed to be homogeneous, linearly elastic and isotropic. The elastic modulus of the titanium alloy implant, cancellous bone, cortical bone were taken as 110 GPa, 1 GPa and 17 GPa respectively. The yield strengths were 900 MPa, 15 MPa and 150 MPa respectively. A Poisson ratio of 0.3 was assigned to each material.

The mechanical properties of the hydroxyapatite coating were investigated. Tensile tests were undertaken on composite thin bars (steel bar - section = 0.4 x 3.7 mm^2- coated with 100 µm hydroxyapatite) with strain gauges (CEA-06-032-UW-120) stuck on both faces to eliminate bending [11]. Total strain versus force was recorded. The hydroxyapatite Young's modulus was computed using *sandwich material theory*, global stiffness being equal to the sum of each material stiffness :

$$E_{hydroxyapatite} = \frac{E_{specimen} h_{specimen} - E_{steel} h_{steel}}{h_{specimen} - h_{steel}} \quad (1)$$

The experimental mean value was 40 Gpa (±15). The Poisson coefficient was assumed to be 0.3.

Moreover a four-point bending test on titanium bars (4 x 5 x 50 mm^3) coated with hydroxyapatite (thickness: 180 μm) was performed in order to determine the hydroxyapatite tensile strength. The applied load was recorded and the experimental mean value of the tensile strength was 105 MPa (±17) [11].

2.4. Load and boundaries

To determine the carpometacarpal constraining force and muscle forces (direction, point of load application and magnitude) during key pinch, a new method was introduced from Magnetic Resonance Imaging (MRI) images obtained from a living hand (with no history of injury or pathology), using a Siemens Manetom Impact-1 Tesla- unit (Desgenettes Hospital, Lyon, France) [12]. The thumb was studied as a mechanical system, where the carpometacarpal joint was modeled as a universal joint and muscles as unknown forces acting as inextensible cables [13], [14]. The constraining key pinch force was computed for a 100N key pinch load magnitude applied to the tip of the thumb (corresponding to about the maximal key pinch). The magnitude of the resultant force is 685 N with the following components: 39 N, 108 N, 675 N on x-, y-, z-axis respectively: (Figure 1).

The loading is simulated by restraining the nodes located in the distal plane to zero whereas the nodes located on the spherical head of the implant are displaced until the constraining key pinch load is reached.

Finally, the overall mesh consisted of 1104 elements and 2718 nodal points. All computations were performed on a HP700. The typical CPU time required per run was approximately 25 minutes.

3. RESULTS

The postoperative and long term models predict different distribution of the Von Mises equivalent stresses. The magnitude of these stresses in the proximal (5 mm below the basis the prosthesis neck) and the distal (32 mm below the basis the prosthesis neck) transversal cross sections (from the outer to the inner part: the cortical bone, cancellous bone, hydroxyapatite layer and the titanium stem) of finite element models are illustrated with a colour scale in Figure 2.

Globally, in the distal bone region, close to the end of the stem, the Von Mises equivalent stresses are higher than in the proximal one. The bone ulno-dorsal part in the proximal section and the radio-palmar part in the other cross-section are the areas where there is the most constraint. The maximum values however are different in bone for both models. Actually stresses are higher in the postoperative situation than in the long term one.

In the postoperative situation, the maximum stresses are developed in the compact bone (45 MPa in the ulno-dorsal proximal section, 125 MPa in the radio-palmar distal one). Contrary to the compact bone and for both sections, the cancellous bone is most constrained in the ulno-dorsal part (about 8 MPa).

In the long term situation, whereas the proximal bone section is not stressed, the distal metacarpal compact bone is the most constrained (maximum 115 MPa in the palmar part) and the Von Mises stresses are about 16 times higher than in the cancellous bone (7 MPa).

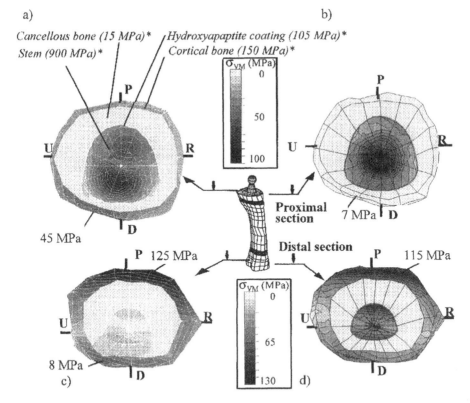

Figure 2: Study of the Von Mises stresses in proximal (a, b) and distal (c, d) cross sections in the early post-operative situation (a, c) and the long term one (b, d). (P: palmar, U: ulnar, D: dorsal, R: radial), ( )* yield strength

In Figure 3, the Von Mises equivalent stresses (σ_{VM}) versus the longitudinal position (Z) along the stem are reported in the hydroxyapatite coating for both clinical cases. The stresses are evaluated in the dorsal and palmar region and are slightly higher in the dorsal than in the palmar part of the hydroxyapatite layer. The curves showed that in the postoperative situation, the stresses are maximum at the end of the stem (50 MPa) whereas in the long term situation the highest stresses are reached at about the two thirds along the length of the stem (80 MPa). The maximum stress peak is about 1.6 times higher in the osteointegrated model than in the other one. Moreover stresses are low in the proximal part of the coating.

At bone/implant interface in the postoperative situation, it is observed that the maximum normal micromotions (microgap) are higher than the maximum tangential ones (microslip). The maximum peak value is reached in the distal palmar part (200 μm) and proximal radial part (150 μm) for normal and shear micromotions respectively.

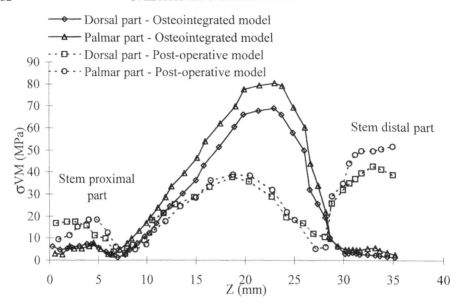

Figure 3: Von Mises stresses versus the longitudinal postion (Z) in the dorsal and palmar part of the hydroxyapatite layer in the post-operative and the long term (osteointegrated) situation.

4. DISCUSSION

4.1. The postoperative model

After surgery, with the hypothesis where the key pinch is achieved at its maximum magnitude, the Von Mises stresses are large in the cortical and the cancellous bones and the hydroxyapatite coating, which could lead to the yielding of the bone and damage to the coating. However maximum key pinch should not be reached following surgery because the patient's thumb is immobilized for several weeks. Therefore the postoperative model is more constraining than the real clinical situation.

As mentioned by Rubin et al. [8], a realistic description of bone/implant interface is probably situated between fully bonded interface and frictionless one. Our postoperative model described the bone/implant interface by taking into account the friction between the stem and the surrounding bone. Since the magnitude of the micromotions depends on its value [8], the considered friction coefficient was not taken arbitrarily. For pratical reasons, however, it was approached using ewe cancellous bone, instead of metacarpal cancellous bone.

For a key pinch of 100 N, the relative micromotions are relatively high (maximum equal to 200 μm) and their influence must not be neglected and may act on bone remodelling [2].

4.2. The osteointegrated model

Bone remodelling around the implant was considered in this study. The 'theoretical' osteointegration of the implant leads to a complete new stress distribution. Moreover the levels are smaller in the long term model than after surgery.

However the stresses in the cortical bone are high and higher in the distal part than in the proximal one. Overconstraining the bone is harmful and could lead to remodelling of the bone which has been observed in clinical situation [11].

The stresses in the hydroxyapatite coating are quite high and are relatively close to the yield point.

4.3. Material properties

The results presented for hydroxyapatite take into account a 40,000 MPa hydroxyapatite Young's modulus and the 105 MPa tensile strength. These values were determined using a small number of samples. The quantitative results should therefore be studied with care, knowing the influence of the choosen material properties on the results.

Though the material is brittle, only the tensile strength is considered and the Von Mises criterium is used. The latter should then only be considered as an indicator of stresses.

The bone properties found in literature are considered constant and isotropic along the bone which is only a rough approximation of the reality. It is known that compact and cancellous bones are anisotropic and heterogeneous and that mechanical properties (Young's modulus) of cancellous bone are dependent on bone morphology and on apparent density [15]. In the future models should take into account more realistic mechanical properties, for instance, by computing bone data from CT scans and by integrating a bone remodelling model [16].

During surgery the stem is press-fitted into the medullary canal by compacting the cancellous bone, leading to a modification of the mechanical properties of the surrounding bone. Bone data from CT scans should be studied after surgery. Until now, however, medical technology can not image the implant in situ because of metallic artefacts.

5. CONCLUSION

A three-dimensional finite element model of a total cementless carpometacarpal prosthesis coated with hydroxyapatite was presented. A specific pre-processor was developed using accurate anatomical data obtained from metacarpal CT slices. A current life gesture (key pinch) was studied, from MRI, to apply a typical load. Coated hydroxyapatite samples were tested in order not to consider arbitrary mechanical properties (Young modulus, tensile yield strength and coefficient of friction). The hydroxyapatite/trabecular bone interface was either perfectly bonded, representing a long term situation where the implant is perfectly osteointegrated, or considered in contact with friction in the early post-operative situation. The finite element model predicted, immediately after surgery (non linear bone-implant interface) and in the long term after bony ingrowth (fully bonded interface), large stresses in the hydroxyapatite coating and bone, which based on our knowledge of the mechanical properties of materials, implicate the reliability of arthroplasty.

6. REFERENCES

1. Dhert, W. Plasma-Sprayed Coatings and Hard-Tissue Compatibility: A Comparative

Study on Fluorapatite Magnesiumwhitlockite and Hydroxylapatite. Phd: University of Leiden Netherland, 1993. 145 p.

2. Soballe, K. Hydroxyapatite ceramic coating for bone implant fixation: Mechanical and histological studies in dogs. Acta Orthopaedica Scandinavica, 1993, Vol. 64, Suppl. N° 255, p. 1-58

3. Munting, E. Bone apposition and hydroxyapatite coating resorption in hip joint replacements in dogs after up to five years implantation. Proceedings of the 10th European Conference on Biomaterials, Davos, 8th to 11th september 1993, p. 98.

4. Nourrissat, C et al. Hydroxyapatie et ostéo-intégration, mythe ou réalité? Analyse des résultats à 4 ans de la prothèse de hanche A.B.G. Actes du Congrès E.F.O.R.T.(1er Congrès Européen de Chirurgie Orthopédique), Avril 1993, N°400

5. Geesink, R. Five year results of hydroxyapatite coated total hip replacement. Actes du Congrès E.F.O.R.T.(1er Congrès Européen de Chirurgie Orthopédique), Avril 1993, N°399.

6. Tanner, K.E. et al. Is stem length important in uncemented endoprostheses? Med. Eng. Phys., 1995,Vol. 17, N°4, p.291-296

7. Brown, T.D. et al. Global mechanical consequences of reduced cement/bone coupling rigidity in proximal femoral arthroplasty: a three-dimensinal finite element analysis. Journal of Biomechanics, 1988, Vol. 21, N° 2, p. 115-129.

8. Rubin, P.J., Rakotomanana, R.L., Leyvraz, P.F., Zysset, P.K., Curnier, A., Heegaard, J.H. Frictional interface micromotions and anisotropic stress distribution in a femoral total hip component. Journal of Biomechanics, 1993, Vol. 26, N° 6, p. 725-739.

9. Keaveny, T.M., Bartel, D.L. Fundamental load transfer patterns for press-fit, surface-treated intramedullary fixation stems. Journal of Biomechanics, 1994, Vol. 27, N° 9, p. 1147-1157.

10. Lbath-Lassays, F., Rumelhart, C., Comtet, J.J. A three-dimensional biomechanics study of the carpometacarpal joint and a prosthesis using medical imaging. World Scientific Publishing Co Pte Ltd (Singapour) sous la direction du Professeur F. Schuind, 1997.

11. Lassays-Lbath, F. Etude biomécanique 3D d'un implant trapézo-métacarpien à l'aide de techniques d'imagerie médicale, d'éléments finis et de caractérisation mécanique. Thèse de mécanique: Institut des Sciences Appliquées, 1996, 289p.

12. Lbath, F., Rumelhart, C., Comtet, J.J. Etude biomécanique in vivo de l'articulation trapézo-métacarpienne à l'aide de l'IRM. Détermination de l'effort résultant et modélisation articulaire. La main, 1996, 1: 13-21.

13. Cooney, W. P. et al. Biomechanical analysis of static forces in the thumb during hand function. Journal of Bone and Joint Surgery, 1977, Vol. 59A, No 1, p 27-36

14. Giurintano, D.J. et al. A virtual five-link model of the thumb. Med. Eng. Phys., 1995, vol. 17, N°3, p. 297-303

15. Hvid, I. et al. B. X-ray computed tomography: the relations to physical properties of proximal tibial trabecular bone specimens. J. Biom., 1989, n° 8/9, p. 837-844.

16. Van Rietbergen, B. et al. The mechanism of bone remodelling and resorption around press-fitted THA stems. J. Biom., 1993, n° 4/5, p. 369-382

Thanks to Drs Chabaud and Legros (H^{al} Desgenettes), Dr Grange-Gellé (H^{al} Dijon), S^{té} Merck Biomaterial.

COMPUTER SIMULATION OF A MOBILE BEARING KNEE PROSTHESIS

A. Imran[1] , J.J. O'Connor[2] and T.W. Lu[3]

1. ABSTRACT

A computer graphics based sagittal plane model of the knee is developed for a prosthesis with fully unconstrained and congruous surfaces, which require retaining the cruciate and collateral ligaments of the knee. The aim of this study is to validate the model results with experimental observations, and to analyse the effects on joint mechanics of surgical malplacement of the implanted components. A single radius femur, a flat tibia and a congruous unconstrained meniscal bearing formed the model prosthesis components. The ligaments were modelled as arrays of elastic fibres which buckled when slack. The model muscles were represented by straight lines except where they wrapped around the bones. A bi-articulating model of the patello-femoral joint was used. A graphic interface was developed. The results show close agreement between the model calculations and the available experimental observations. The model can be used for quantitative and visual analysis of the effects of surgical errors in component placement.

2. INTRODUCTION

Clinical experience suggests that unicompartmental meniscal bearing knee replacement requires careful patient selection and precise surgical technique (1). Surgical malplacement of the prosthesis components can result in stretched ligaments or in increased joint laxity. The purpose of the present study is to use a sagittal plane model of a prosthetic knee with a mobile bearing and congruous surfaces with ligaments intact, to

KEY WORDS: Knee Model, Meniscal Bearing Knee, Surgical Malplacement, Graphic Interface.

---

[1] Postgraduate student, Department of Engineering Science and Orthopaedic Engineering Centre (OOEC), University of Oxford, Oxford, England.
[2] Professor and Director of OOEC, University of Oxford, Oxford, England.
[3] Postgraduate student, OOEC, University of Oxford, Oxford, England.

quantify ligament forces and antero-posterior (A/P) laxity during flexion, resulting from the surgical malplacement of prosthesis components, or resulting from using incorrect bearing thickness. The model calculations are compared with the experimental observations available in the literature. A computer-graphics based user interface was developed for visual analysis of the effects of surgical malplacement in the sagittal plane of the prosthesis components.

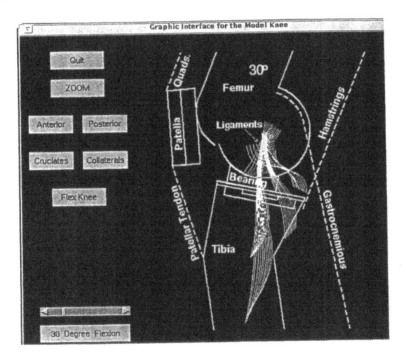

Fig 1. The model knee at 30° flexion. The control buttons provide various options on computer. Continuous passive flexion can be analysed be clicking the button marked 'Flex Knee'. Alternatively, a flexion angle can be chosen by using the slider button at the lower left corner, and 'Anterior' or 'Posterior' drawer test can be studied by clicking on the corresponding button. The buttons marked 'Cruciates' and 'Collaterals' provide the option to set the corresponding ligaments 'off' (hide) or 'on' (reveal).

3. METHODS

We used a computer-based mathematical model of the knee in the sagittal plane, after joint replacement (Figure 1). The femoral surface was represented by a part of a circle and tibial surface was flat. A fully congruous meniscal bearing, concave above and flat below, was used between the femoral and tibial components. With this arrangement, the mobility of the natural joint can be restored while the low contact stresses associated with congruous surfaces give very low wear rates (2). Anterior and posterior cruciate ligaments (ACL and PCL), and lateral and medial collateral ligaments (LCL and MCL) were represented by arrays of elastic fibres, which buckled when slack (3,4). The model

MCL was divided into long superficial fibres and short deep fibres to take account of the different fibre layers observed in the MCL (5).

3.1 Neutral Position of the Components:

A neutral position of the components with an appropriate thickness of the bearing was defined at $0°$ as that which maintained more or less a constant length of selected fibres of the model cruciate ligaments (referred to as *neutral fibres*) (3,6). For this position of the components, all other fibres of the ACL, MCL, LCL and the posterior fibres of the PCL remained just tight in extension and slack during passive flexion. The anterior fibres of the PCL were just tight at $120°$ flexion and slack during passive extension.

3.2 Model Muscles:

Straight line representation was used for the lines of action of the model quadriceps, hamstrings and gastrocnemius muscles except where they wrapped around the bones. A bi-articulating model of the patello-femoral joint was used (7). The quadriceps force (Q) was transmitted to the tibia through the patellar tendon force (P). P and Q remained in equilibrium with a patello-femoral contact force (CP) (8). The ratio of P to Q can be calculated if the angles which P and Q make with CP are known (8). During flexion, the relative positions of the bones and positions and orientations of the model muscles and ligaments were calculated from the model.

3.3 Antero-Posterior Laxity Test:

An antero-posterior (A/P) laxity test, such as the Lachman test, was simulated at each flexion angle by applying an A/P force and a moment to the tibia. In the absence of the muscle force, any external A/P force is resisted mainly by the cruciate ligaments of the knee (9-11). Anterior forces acting on the tibia are resisted mainly by the ACL and MCL and posterior forces by the PCL and LCL (9-11). When the ligaments are slack, tibial movement is unresisted. When the ligaments are stretched, the joint stiffness (external force required for small tibial displacements) is increased.

3.4 Effects of Malplacement:

The effects on joint laxity or resistance to A/P movement of component malplacement about the neutral position can be analysed for anterior/posterior or proximal/distal positions of the femoral component and proximal/distal position of the tibial component with full extension as reference. In this paper, the results are presented for the neutral placement; distal placement of the tibial component, by choosing too thin a meniscal bearing; and proximal placement of the tibial component, by choosing too thick a bearing.

3.5 Graphic Interface:

A computer based graphics interface was developed for visual analysis of passive flexion, and for visual analysis of the A/P laxity test. Some results are presented in this paper in the form of computer-generated diagrams in black and white (Figures 1, 2). In the computer graphics, different colours are used to distinguish the ligaments. The part of the ligament which is stretched is represented by thick straight lines, while the part of the ligament which is slack is shown by curved fine lines showing buckling of the fibres (4). Therefore, it is relatively easier by using graphical methods to appreciate the changing shapes of the ligaments and the portions of the ligaments which are tight. On the computer screen, control buttons give the options of flexing the knee, selecting a flexion angle and analysing the loading of the ligaments after relative movement of the bones. The purpose is to develop a user-friendly environment which can be used to analyse the behaviour of the knee after joint replacement.

4. RESULTS AND ANALYSIS

The neutral fibres of the cruciate ligaments remained almost isometric throughout flexion, becoming slightly slack during the mid-range. Maximum slackening was at 60°: 0.9% for the ACL neutral fibre and 1.1% for the PCL neutral fibre. This slackness accounted for about half a mm A/P laxity at 60°. Compared to the cruciate ligaments, the model collateral ligaments make little or no contribution to the A/P force.

During 0 to 90° flexion, the model bearing moved posteriorly by 10.2 mm, which is comparable with the *in vitro* average measurements in this flexion range of 12.8 mm in the medial and 15 mm in the lateral compartments of 4 knees implanted with a meniscal prosthesis (8). The effects of tibial rotation, not included in the model, can be judged from the fact that the lateral bearings moved more than the medial bearings during flexion.

Figure 2 gives a graphic representation of the model knee during flexion. The part of the ligament which is stretched is shown by thick lines, while the part of the ligament which is slack is shown by curved lines. The anterior and posterior laxities of the joint are represented by the thick lines extending from the sides of the bearing. The fibres of the MCL and LCL are not shown for clarity.

Figures 2(a) shows the components in *neutral position* at 0°, 60° and 120°. The bearing moves posteriorly during flexion. The shapes of the ligaments change, fibres cross and uncross, slacken and tighten due to relative rotation and translation of their attachment areas. Figures 2(b) show that a bearing 2 mm thinner than neutral slackened the ligaments, increasing joint laxity. Figures 2(c) show that a bearing 2 mm thicker than neutral stretched the ligaments. However, the stretch or laxity in the ligaments is flexion dependent. For the thicker than neutral case, the bearing and the femur, at every flexion angle, slide on the tibial component to the point where the net ligament force parallel to the tibial surface is zero. However, the tibio-femoral contact force no longer passes through the intersection of the ligament forces, except at 68°, giving rise to an unbalanced moment. To hold the joint in any flexion position, an extending moment is required for flexion angles less than 68° and a flexing moment for angles greater than 68°. The unbalanced moment is generally small, about 0.5 Nm. However, there is significant resistance near extension, possibly a contributory factor to the presence of a residual flexion deformity in some patients.

Figure 3 shows antero-posterior force-displacement curves calculated at 30° flexion. The use of a 2 mm thinner than neutral bearing increased joint laxity, while 2 mm thicker than

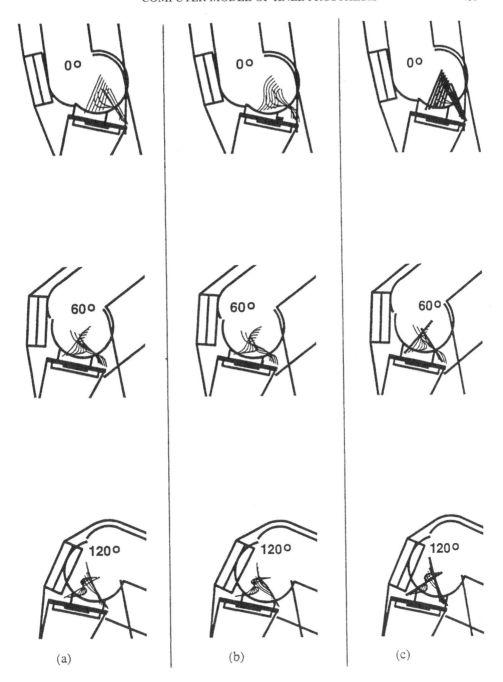

(a) (b) (c)

Fig 2. The model at 0°, 60° and 120°: (a) components in the neutral position; (b) the effect of using a bearing 2 mm thinner than neutral; (c) the effect of using a bearing 2 mm thicker than neutral.

neutral bearing increased resistance to A/P movement and displaced the bearing and femur posterior to their neutral position. These effects of bearing thickness varied in magnitude with flexion angle.

Figure 4 shows the orientation of the patellar tendon with the posterior direction, plotted over the flexion range. The values from model calculations are within the range reported previously (13-15).

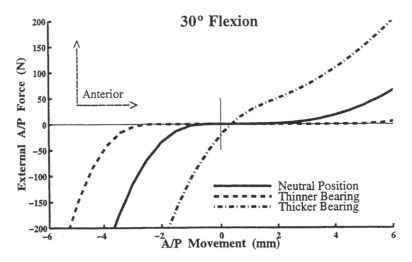

Fig 3. Antero-posterior force-displacement curves calculated at 30° for *neutral* placement of the components and for incorrect bearing thickness.

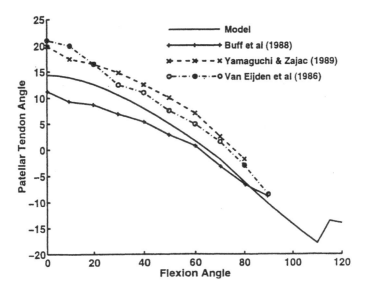

Fig 4. Orientation of the patellar tendon with the long axis of tibia plotted against flexion. The discontinuity in the model calculations at high flexion angles is due to the change of patello-femoral contact from patellar surface to tibio-femoral articular surface.

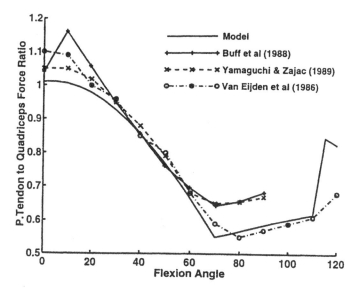

Fig5. The ratio of patellar tendon to quadriceps force plotted against flexion. The ratio drops from about 1 at 0° to 0.55 at 68° and then increases. This sudden change occurs with the quadriceps wrapping around the front of the femur at 68°.

In Figure 5, the ratio of patellar tendon to quadriceps force is plotted over flexion. Again, the model calculations show good agreement with other studies (13-15).

5. CONCLUSIONS

The results show that in an unconstrained knee prosthesis with retained ligaments, a single radius femoral and a flat tibial components, with a mobile bearing, if positioned correctly, can keep the selected fibres of the cruciate ligaments near isometric during flexion. These fibres are mainly responsible for controlling the kinematics of the unloaded knee in the sagittal plane (6,8,12). Figures (2a, 4 and 5) show that such a prosthesis design can reliably restore the mechanics of the knee in the sagittal plane.

6. CLINICAL SIGNIFICANCE

This analysis demonstrates that two millimetre differences in the thickness of the bearing can alter significantly joint laxity or joint stiffness after replacement. The analysis shows that appropriate mathematical models can be useful tools for developing prosthesis designs, analysing their effects after implantation, and understanding the role of surgical techniques.

Figures 4 and 5 provide theoretical support for the proposition that restoration of the *kinematics* of the tibio-femoral joint will restore the *mechanics* of the patello-femoral joint and avoid the patellar complications which are one of the major remaining problems of the modern knee replacement.

The model developed for this study can be used for quantitative and visual analysis of the interaction between component placement and the resulting joint laxity or ligament forces over flexion.

7. ACKNOWLEDGEMENTS

The authors would like to thank Mr John Goodfellow and Mr David Murray for their advice. This work was supported by a Felix Scholarship for research at Oxford University.

8. REFERENCES:

1. Carr A, Keyes G, Miller R, O'Connor J and Goodfellow J. Medial unicompartmental arthroplasty: a survival study of the Oxford Meniscal Knee. Clin Orth Rel Res, 1993; 295: 205-13.
2. Psychoyios V, Crawford R, O'Connor J and Murray D. Polyethylene wear in congruent meniscal bearing unicompartmental knee replacements. Trans Combined 7th Conference of European Orthopaedic Research Society (Barcelona), April 22-23, 1997; p-107.
3. Zavatsky AB and O'Connor JJ. A Model of Human Knee Ligaments in the Sagittal Plane, Part 1: Response to Passive Flexion. Proc Inst Mech Eng (Part H), J Engng Med, 1992; 206(H3): 125-34.
4. Lu TW and O'Connor JJ. Fibre recruitment and shape changes of knee ligaments during motion: as revealed by a computer graphics-based model. Proc Inst Mech Eng (Part H), J Engng Med, 1996; 210: 71-9.
5. Warren LF and Marshall JL. The Supporting Structures and Layers on the Medial Side of the Knee. J Bone Jt Surg [Am], 1979; 61(1): 56-62.
6. Fuss FK . Anatomy of the Cruciate Ligaments and their Function in Extension and Flexion of the Human Knee Joint. Am J Anatomy, 1989; 184: 165-76.
7. Gill HS and O'Connor JJ. A Bi-articulating Two Dimensional Computer Model of the Human Patellofemoral Joint. Clinical Biomechanics, 1996; 2(2): 81-9.
8. O'Connor JJ, Shercliff TL, Biden E and Goodfellow JW. The Geometry of the Knee in the Sagittal Plane. Proc Inst Mech Eng (H), J Engng Med, 1989; 203: 223-33.
9. Brantigan OC and Voshell AF. The Mechanics of the Ligaments and Meniscii of the Knee Joint. J Bone Jt Surg [Am], 1941; 23 (1): 44-66.
10. Kapandji IA. The Physiology of the Joints, volume 2. The Lower Limb. Churchill Livingstone, Edinburgh, 1970.
11. Markolf KL, Mensch JS and Amstutz HC. Stiffness and Laxity of the Knee : The Contributions of the Supporting Structures. J Bone Jt Surg, 1976; 58 (5-A): 583-94.
12. Goodfellow JW and O'Connor J. The Mechanics of the Knee and Prosthesis Design. J Bone Jt Surg [Br], 1978; 60 (B): 358-69.
13. Van Eijden TMG, Kouwenhove E, Verburg J and Wejis WA. A methematical model of the patello-femoral joint. J Biomech, 1986; 19: 219-29.
14. Buff H-U, Jones LC and Hungerford DS. Experimental determination of forces transmitted through the patello-femoral joint. J Biomech, 1988; 21: 17-23.
15. Yamaguchi GT and Zajac FE. A Planar Model of the Knee Joint to Characterise the Knee Extensor Mechanism. J Biomech, 1989; 22: 1-10.

A KNEE ARTHROSCOPY TRAINING TOOL USING VIRTUAL REALITY TECHNIQUES

Hollands, R.J.[1] and McCarthy A.D.[2]

1. ABSTRACT

Arthroscopy is the form of minimally invasive surgery concerned with joints. The majority of arthroscopic procedures are for inspection or repair of the knee joint. Unlike other forms of keyhole surgery, arthroscopic surgery uses rigid tools in extremely confined cavities. These cavities contain structures which can be damaged easily by mistakes made during training; these mistakes can result in permanent joint damage. Current training methods include a variety of techniques from simple 'black-box' trainers to complex physical models, however the majority of the training takes the form of supervised and sometimes unsupervised operations on patients. This paper describes the use of the computer-based graphical technology of virtual reality to provide a viable alternative to existing training methods.

2. INTRODUCTION

The minimally invasive techniques of endoscopy offer advantages over traditional open surgery. Patient trauma is minimised and the associated improvements in recovery times can lead to shorter stays in hospital, making it attractive to both patients and healthcare providers alike due to reduced patient intervention costs. However, endoscopy requires a different set of skills to those learned for open surgery and leads to additional requirements for surgical training. Arthroscopy is the form of endoscopy concerned particularly with joints.

Keywords: Arthroscopy, Virtual Reality, Simulation, Knee

[1] Research Associate / Dept. Automatic Control & Systems Engineering, University of Sheffield, Mappin Street, Sheffield, S1 3JD, U.K. email: r.hollands@sheffield.ac.uk
[2] Research Associate / Dept. Medical Physics & Clinical Engineering, University of Sheffield, Royal Hallamshire Hospital, Glossop Road, Sheffield, S10 2JF U.K. email: a.d.mccarthy@sheffield.ac.uk

Similar to other forms of endoscopic or "keyhole" surgery, it accesses the operation site through small incisions (or portals) in the skin. Unlike the freedom of movement available in open surgery, arthroscopic instruments are restricted to insertion through, and rotation around, the fulcrum of the portal. An arthroscope, comprising a rigid fibre-optic camera and light source, is used to view inside the joint and its display is shown on a nearby monitor. Distinct from other forms of keyhole surgery, arthroscopic surgery requires manipulation of the limb during surgery to enable access to and inspection of the confined joint cavities.

3. CURRENT TRAINING METHODS

Training for arthroscopic surgery involves a number of techniques. The skill of triangulation, where the instruments are manipulated to bring them into view of the camera, is often taught using simple 'black-box' models.[1] Cadavers can be used for training; however, their use is limited due to ethical restrictions, the associated costs of cadavers, and because the joints are altered by the preservation process. Therefore, simple physical models of the internal structures of the joint tend to be used. Although these models provide a greater sense of realism than the abstract 'black-box' models, they reproduce only some of the key internal structures and are less challenging of technique than a real joint. In addition, practising destructive surgical procedures will damage the physical model, requiring the insertion of replacement parts and making them expensive in both components and time. Hence, much of a surgeon's training takes the form of supervised surgery on patients.

It has been estimated that it can take as many as 500 operations for a trainee surgeon to achieve competence in diagnostic arthroscopy alone.[2] During this learning period, patients are at risk, because accidental damage cannot usually be repaired. The surfaces of the joints are particularly vulnerable, as accidental contact can scuff the articular cartilage, increasing the patient's risk of premature osteo-arthritis.[3] A requirement exists therefore, for a flexible, realistic and economical arthroscopy training tool that is capable of providing the majority of early training tasks. The trainer should enable a trainee surgeon to enter the operating theatre much higher on the learning curve than allowed by current training methods. The virtual reality arthroscopic training tool under development at Sheffield has been designed to meet as many of these objectives as possible, in a convenient computer-based system.

4. OVERVIEW OF THE VIRTUAL REALITY ARTHROSCOPIC TRAINER

Virtual reality (VR) is a technology that allows the creation of three-dimensional environments within a computer that can be experienced in real time. In VR's 'purest' form, the user's head position and orientation is monitored to generate an appropriate view of the virtual environment displayed on a head-mounted-display. However, a simpler version of the system, known as desktop VR, utilises the normal computer monitor as a 'window' onto the virtual world, and allows the user to locate this window at any position or orientation within the virtual environment.

It is important in a surgery training system for the equipment to feel as natural and realistic as possible, because this minimises problems in transferring the skills acquired from the VR trainer back to actual surgery. The desktop VR system is an ideal approach for the trainer since the computer monitor can be used in exactly the same way as the video monitor viewed in real surgery. The essential structures within the knee joint can be modelled geometrically within the computer to generate a view of anywhere within the knee joint. In real surgery, the surgeon chooses what to view within the joint by positioning the arthroscopic camera appropriately. If an artificial arthroscope is fitted with a tracking device capable of monitoring the arthroscope's position and orientation, it can be used to control the view within the computer-generated joint in the same way as a real arthroscope.

To allow the trainee surgeons to practise manipulating the instruments within the joint, a simple, physical leg model is used in conjunction with replica instruments. To aid surgical realism, it is essential that the physical leg model be jointed realistically, since to facilitate joint inspection the surgeon must open joint cavities by manipulating the lower limb. There are two important differences between the leg model used in this VR system and the traditional physical models. Firstly, the whole lower limb, not merely the knee joint, has been modelled. Secondly, the VR leg is completely hollow; the key structures within the knee are generated completely within the computer system, and are not physical replicas within the physical knee model.

The virtual reality approach gains its unique advantage over traditional training methods by modelling as much as possible purely within the computer. After developing a suitable database of pathologies, a software menu could be used to select from the database. Alternatively, it would be feasible for the computer system to present the trainee with a random pathology, for example, to test identification or diagnostic skills. The addition of appropriate kinematic simulations to the geometric model will allow the trainee to practise surgical procedures. Where these procedures are destructive, requiring the repair or replacement of a traditional physical model, a virtual model could be restored to its original form simply by resetting the software.

Economic viability is an important final consideration with any system. Whilst computer based training systems or simulations may promise more features than traditional techniques, these features become redundant if they can only be offered at a prohibitive price. State-of-the-art surgical simulations exist, but they require the use of multiple high-end graphics workstations costing hundreds of thousands of pounds.

The virtual reality knee arthroscopy trainer being developed by the Virtual Reality in Medicine and Biology Group (at the University of Sheffield, UK), attempts to provide all of the benefits outlined above, whilst using a relatively inexpensive PC as its operating platform. The rest of this paper describes the Sheffield system in more detail.

5. SYSTEM HARDWARE

The structure of the system hardware can be seen in Figure 1. The current system will run satisfactorily on a entry-level Pentium 133MHz, equipped with 16Mb RAM, sound card, and low-end 3D graphics card. However, in readiness for the next phase of the system development, the low-end platform has recently been upgraded to an Intergraph TDZ310 Pentium Pro machine, equipped with 64Mb RAM and a mid-range graphics accelerator. The upgrade was necessary since the next development phase will introduce a number of graphically intensive routines, such as tissue deformation. It is expected, given the current rate of computer development, that this computer system will become entry level within the next two years.

Although it is possible to use any normal computer monitor for training purposes, a 50 cm screen has been chosen for the prototype system since it most closely matches the size of the video monitor used in operating theatres. The keyboard and mouse are used as part of normal computer operation and are not required during a training session, except to operate any additional training features, such as those described in the section 5.3.

5.1 The Tracking System
The primary interface to the replica instruments and artificial leg is an electromagnetic position/orientation tracking system, the Polhemus FASTRAK. The transmitter generates a low-intensity magnetic field to allow movements of the receivers to be monitored by the precision-wound coils within the receivers. The system has no mechanical linkages to obstruct instrument manipulation, and does not require a clear line of sight between the transmitter and receiver.

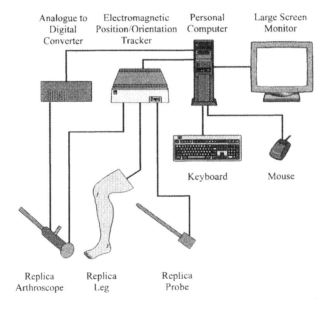

Figure 1 - System Hardware

The receivers are around 2.5 cm³, and can be easily attached to the replica devices. Although there is, by necessity, a cable running from each receiver to the control unit, the resulting system does not encumber the user; it allows the replica instruments and camera to be utilised in much the same way as their real counterparts.

Tracking performance is critical to creating the illusion of operating within a real knee. The maximum tracking sensitivity of the FASTRAK system is only up to 70 cm between the transmitter and receiver. Thus, to maximise tracking performance, the transmitter unit is situated close to the receivers in the upper thigh of the artificial leg.

5.2 The Artificial Leg

A realistically jointed physical knee is an obvious requirement for an arthroscopic training system, although, surprisingly, some knee arthroscopy simulators do not include one.[4,5] In the Sheffield system, the model leg is the key feature of the trainer. The replica instruments and camera are introduced via small holes in the artificial leg, as in real surgery, and the joint cavities can be opened to enable joint inspection by manipulating the leg appropriately. Unlike traditional physical models, the replica leg in this system is completely empty. To simplify it, movements have been restricted to the two primary degrees of freedom of flexion-extension and adduction-abduction. Although more complex kinematics can be modelled in the software simulation, flexion-dependant knee abduction is modelled physically, using a cam mechanism.

The degree of flexion and abduction of the knee is monitored using the same type of sensor as used to track the instruments. As described in section 5.1, the transmitter unit is mounted within the thigh of the artificial leg. In addition to maximising tracking performance, this ensures that measurements made are always with respect to the leg - regardless of how the leg is positioned in the real world. This provides great freedom in the design of the mechanism needed to anchor the leg to the bench. To allow movements similar to that found at the hip, the Sheffield system uses a ball joint that can have the degree of resistance to movement set by the user. This system provides a solid fixing for normal use, but enough flexibility to perform manoeuvres such as the 'figure of four', used by surgeons to open the lateral compartment of the joint.

Although a full force-feedback system was not in the original design specifications for the system, it is intended that the physical components should feel realistic. While the flexion-dependant abduction of the knee goes some way towards this in a dynamic sense, it is intended to weight the artificial leg to match that of an average adult's leg. The 'skin' medium through which the instruments are passed will also be improved. The current system uses a 6 mm neoprene cover which is flexible enough to flex with the knee, but solid enough to provide a robust entry fulcrum and additionally provides a degree of resistance to axial movement of the instruments.

5.3 Artificial Instruments

To afford a sense of realism, the arthroscope and instruments used in the Sheffield system are physical replicas of real equipment. Because ferrous metal could interfere with the electromagnetic tracking system, the replicas are made from plastic. In addition to the six degrees of freedom measured by the electromagnetic sensor, the arthroscope requires an additional sensor to measure the camera roll around the arthroscope axis. This is provided by a rotational potentiometer that is monitored using a separate analogue to digital converter.

One of the advantages of using a virtual reality system over any of the traditional training methods is apparent in the use of the physical replicas. A real arthroscope is extremely susceptible to damage by careless handling in the early stages of training, whereas the plastic replica is cheap enough to be considered disposable.

6. THE SYSTEM SOFTWARE

The software is written in Visual C++ running under Windows NT, using a set of virtual reality libraries called WorldToolkit (WTK) from Sense8. The use of WTK libraries accomplishes the mundane tasks of creating a virtual environment, leaving the programmer free to develop the specific requirements of the simulation. This particular software also allows the program to be recompiled to run on a variety of computing platforms including high-end Silicon Graphics workstations. However, with the increased power available from Windows NT workstations, it is unlikely that a change in platform will be required.

6.1 Graphical Objects

The geometrical structures within the virtual knee were created by digitising plastic replicas using the Polhemus 3Draw digitising system. In VR, each geometry (or object) is formed by a number of polygons. If fewer polygons have to be drawn per frame, the frame rate will appear faster and more realistic, however, a fast frame rate will be at the expense of geometric detail. In contrast, over-specification of detail, as can occur when using MRI or CT reconstructions, will result in slower simulations. By digitising the replicas by hand, it was possible to obtain detail only where it was required, such as around the articular surfaces, thereby producing simple objects that allowed the simulation to run at a realistic rate. Although the model used has only a few hundred polygons in each object (Figure 2a), the use of smooth shading makes the object look much more detailed (Figure 2b). The low resolution of the data set is only revealed when looking at the object profile. Surgeons who have used the system have not reported the level of detail to be a problem.

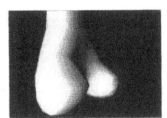

a) Wireframe   b) Smooth shaded

Figure 2 - Graphical object display

6.2 Kinematic Simulation

The initial model of knee movement developed was an extension of the four bar link model used by O'Connor and Zavatsky.[6] The four links in the model represented the two cruciate ligaments, assumed to always be under tension, and their two pairs of connection points, (see Figure 3). The two cruciate ligaments actually provide the limits to the flexion-extension movement of the knee, with the posterior cruciate ligament under maximum tension at full flexion, and the anterior cruciate ligament under maximum tension at full extension. The kinematic model used in the trainer provides a similar action to O'Connor and Zavatsky's model. However, it has been extended to enable the cruciate ligaments to undergo changes in tension, and includes the flexion-dependant, abduction-adduction limiting effects of the collateral ligaments.[8]

6.3 The Interface

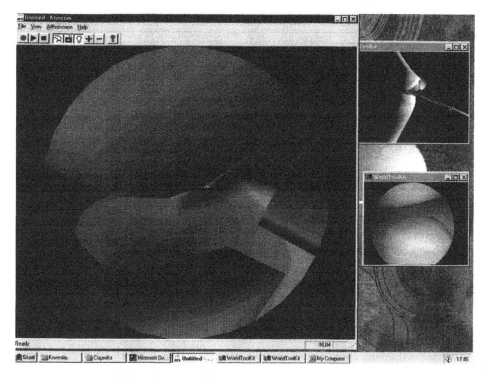

Figure3. Typical view of the simulator window.

When used purely for its primary purpose as an arthroscopic simulator, no direct interaction with the computer is required. However, additional features are available for which a standard Windows graphical interface is used (Figure3). During the early stages of training, the trainee surgeon may find it difficult to orient the position of the camera and tools within the knee. Therefore, it is possible to open a second 'overview' window, to provide a macroscopic view of the key components of the knee and the position of the tools and camera within them. To combat the idealised nature of the synthetic images, a third window may be opened which displays an image captured from real surgery. The image is chosen from a database of such pictures, and uses the current position of the camera and tools to provide the closest match to that which is being artificially generated.[8]

The software-based nature of this simulator also provides a great deal of flexibility in the hardware being simulated. The trainee can use the menu to choose a camera with or without a roll indicator, with different inclination angles, and different fields of view. It is also a simple matter to change tool specifications, or even to swap the virtual leg a right leg to a left one and vice-versa. The computer system can record a trainee's progress, or, it can record and replay an expert's approach to a procedure.

7. CONCLUSIONS

The ability of VR to recreate complex three-dimensional environments is ideal for reproducing the topology of human anatomy. Compared with other training techniques, a VR-based system can be more realistic, flexible, and convenient to use. Current VR training systems fail to compete with traditional training systems because of the need to purchase expensive high-powered graphics workstations. The surgical simulator described in this paper has been designed to run on a standard entry-level PC, thus making it financially competitive with non computer-based systems.

The prototype system described here has been tested by a number of surgeons from the Royal Hallamshire and Northern General Hospitals in Sheffield, and as part of orthopaedic workshops. All the surgeons have been impressed by the challenging nature of the trainer, afforded by its collision-detection algorithm, and also by realism afforded by the virtual environment, enhanced by the inclusion of the jointed leg model. Development of the virtual training system is fully supported by Smith and Nephew Healthcare Ltd who have expressed an interest in incorporating the final trainer in their arthroscopic courses and training centres throughout the country.

8. REFERENCES

1. Meyer R.D., Tamarapalli J.R. and Lemons J.E., Arthroscopy training using a "black box" technique, *Arthroscopy: The journal of arthroscopic and related surgery,*1993, Vol. **9**, 3, 338-340
2. Miller W.E., Learning Arthroscopy, Southern Medical Journal 7, 1985 Vol. **8**, 8, 935-940
3. Bamford, D. J., Paul S.A., Noble J. and Davies D.R., Avoidable complications of arthroscopic surgery, *Journal Royal College Surgeons of Edinburgh*, 1993, Vol. *37*, 92-95
4. Logan I., Virtual reality training simulator, http://www.enc.hull.ac.uk/CS/VEGA/medic/surgery.html, 1995
5. Zeigler R., Fischer G., Müller W. and Göbel M., Virtual reality arthroscopic training simulator, *Journal of Computers in Biology and Medicine,*1995, Vol. **25**, No.2, 193-203
6. O' Connor J.J. and Zavatsky A., Anterior cruciate ligament function in the normal knee, In *The Anterior Cruciate Ligament: Current and Future Concepts* (D W Jackson *et al.* Eds), Raven Press Ltd, New York, 1993, 39-52
7. Hollands R.J. and Trowbridge E.A., A PC-based virtual reality arthroscopic surgical trainer, *Proceedings of Simulation in Synthetic Environments*, New Orleans, 1996, 17-22
8. Hollands R.J., Trowbridge E.A., Bickerstaff D., Edwards J.B. and Mort N. "The particular problem of arthoscopic surgical simulation – a preliminary report", In *Computer Graphics: Developments In Virtual Environments* (Earnshaw R.A. & Vince J.A., Eds) Academic Press, London, 1995, 475-482

The virtual arthroscopic simulator project is supported by Virtual Presence Ltd., and Smith and Nephew Healthcare Ltd., and is funded by a grant from the Engineering and Physical Sciences Research Council.

3D FINITE ELEMENT STUDY OF GLENOID IMPLANTS IN TOTAL SHOULDER ARTHROPLASTY

C. Baréa[1], MC Hobatho[2], R. Darmana[3], M. Mansat[4]

1. ABSTRACT

The purpose of this study was to develop and validate a methodology based on finite element modelling techniques in order to analyse the mechanical behaviour of the gleno-humeral joint both with and without prosthetic implantation, so that the mechanism of glenoid implant loosening can be better understood and new prosthetic implant designs can be optimised.

The models revealed contact pressures, stress and strain data which can be compared to the experimental results. Numerical results showed good qualitative agreement with the experimental measurements. The scapula stress and strain following prosthetis implantation were lower than in the intact one.

1. INTRODUCTION

A review of the literature indicates that the humeral head mobility is responsible for implant loosening by the « rocking horse » phenomenon [1]. In order to elucidate this mechanism, two different prosthetic implant designs have been evaluated.

Finite element models were computed and validated with an experiment which simulated loads through the joint. Contact surfaces were measured with pressure sensitive Fuji film and deformations were recorded at 4 points on the scapula surface with strain gauges.

Experiments and models were performed on 3 shoulders: A normal one, one implanted with the standard Neer II glenoid prosthesis, and one with a new anatomically shaped implant.

Keywords: gleno-humeral joint, implant, validation.

---

[1] Msc, INSERM U305, Centre hospitalier hôtel-Dieu, 31052 Toulouse, France

[2] PhD, INSERM U305, Centre hospitalier hôtel-Dieu, 31052 Toulouse, France

[3] Msc, INSERM U305, Centre hospitalier hôtel-Dieu, 31052 Toulouse, France

[4] Professor, CHU Purpan, service d'orthopédie traumatologie, 31059 Toulouse, France

For the three-dimensional modelling, geometries of the different parts (humeral head, implant, cement, and scapula) were obtained from CT scan images processed by an in-house program. A previous study gave the anisotropic properties of the glenoid cancellous bone by using correlations between CT numbers and Young's moduli.

2. MATERIALS AND METHODS

2.1. Experimental testing

Experiments and models were performed on the same fresh cadaveric shoulder. The scapula and the humerus were cemented with the apparatus and the shoulder was frozen between each experiment.

The first study was performed on a normal shoulder which was then implanted with a standard Neer II glenoid prosthesis for the second study. The implant was then carefully removed and replaced with a new anatomic implant for the final study.

The experiment simulated abduction of the arm with a load through the joint (cf. fig. 1). Popen et al. [2] reported that the maximum load in the gleno-humeral joint was approximately one body weight, so a 700 N load was applied.

Contact surfaces were measured with pressure sensitive Fuji film and deformations were recorded at 4 points with strain gages (cf. fig. 2).

Experiments were performed 3 times at each level of abduction for the 3 shoulders. T-tests were performed in order to evaluate the influence of implants, of abduction, and strain gauge positions on the overall results.

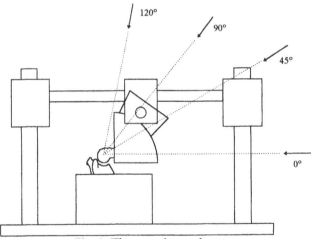

Fig. 1: The experimental setup.

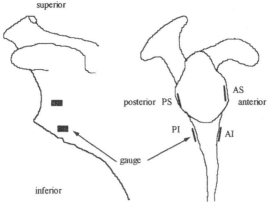

Fig. 2: Strain gauge positions on the scapula.

2.2. Finite element modelling

Prior to the experimental analysis, a CT scan was performed on the scapula (cf. fig. 3). Geometries of the different parts (humeral head, implant, cement, and scapula) were obtained from the CT scan images processed by an in-house program (SIP ©INSERM U305). These geometries were integrated in the pre and post treatment software PATRAN (MSC/PATRAN Corp.) and finite element analysis was performed using ABAQUS (cf. fig. 4). The load and boundary conditions replicated the experimental testing conditions.

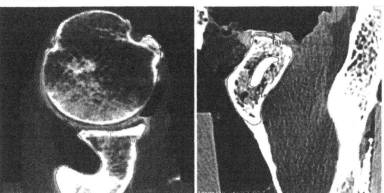

Fig. 3: CT scan images.

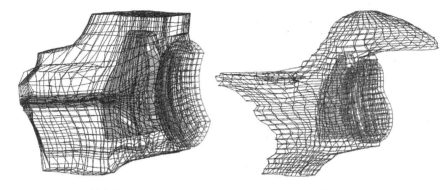

Neer II   Anatomic
Fig. 4: Finite element meshes of scapulae with cement and glenoid implants.

A previous study provided the anisotropic properties of the glenoid cancellous bone [3]
by using correlations between CT numbers and Young's moduli (cf. table 1). Direction
1 was medial-lateral, direction 2 was posterior-anterior and direction 3 was superior-
inferior.

$$E_1 > E_2, E_3 \ (p < 0.01) \text{ and } E_2 = E_3 \tag{1}$$

(n=117)	E_1 (MPa)	E_2 (MPa)	E_3 (MPa)	ρ (kg/m^3)	QCT
mean	342	213	194	269	205
±SD	±164	±77	±94	±66	±48
(min.-max.)	(71-876)	(93-573)	(27-745)	(112-483)	(92-354)

Table 1 : Summary of mechanical properties of glenoid cancellous bone [3].

For the cortical bone, Young's modulus reached the value of 16 GPa.

3. RESULTS

Statistical analysis showed that globally experimental strains on the 4 gauges were not
modified by the angle of abduction ($p > 0.05$). The experimental results showed
compression on the posterior side and tension on the anterior side of the scapula.
Significant differences ($p < 0.001$) were found between the three studies (cf. fig. 5).

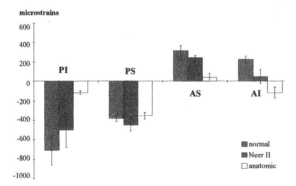

Fig. 5: Experimental results for the 3 shoulders.

The experimental contact study showed slight differences due to the movement of the
head on the glenoid for different positions of abduction. Within the three cases, contact
areas also moved: for the normal and Neer II implanted scapulae, contact areas were

mainly on the posterior side of the articular surface. With the anatomic implant, contact areas were more centred. The finite element modelling on the normal shoulder also provided information on contact pressures which can be compared to the experiment (cf. fig. 6).

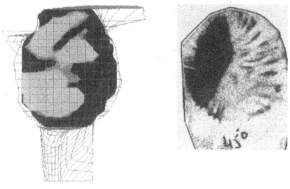

Fig. 6: Comparison between computed contact area and Fuji film experimental result on the normal shoulder.

For the normal and Neer II implanted scapula, compression and traction on the posterior and anterior faces of the scapula were found (cf. fig. 7 and 8). With the new anatomic implant, the more centred contact area decreased the observed stresses on the surface of the scapula in comparison to the two other models (cf. fig. 9).

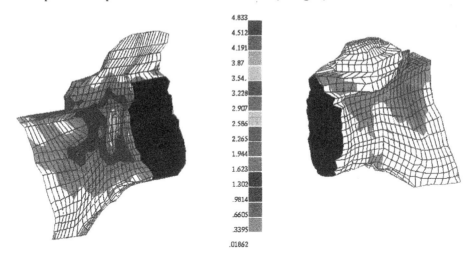

Fig. 7: Computed Von Mises stresses on normal shoulder.

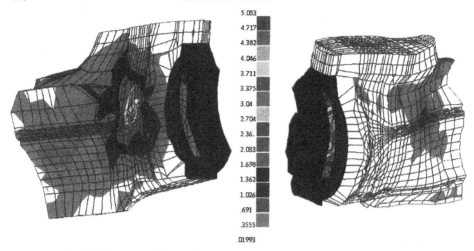

Fig. 8: Computed Von Mises stresses on Neer II implanted shoulder.

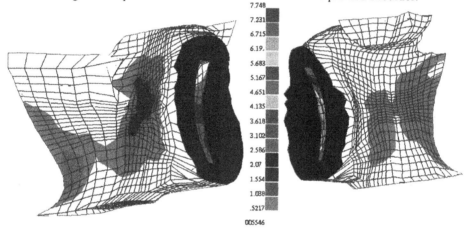

Fig. 9: Computed Von Mises stresses on anatomic implanted shoulder.

4. DISCUSSION

Globally, scapula strain and stress levels were lower after prosthetis implantation compared with the intact shoulder (especially for the anatomic implant).

With our scapula setup position, the Neer II glenoid prosthetis resulted in an eccentric applied load while the new anatomic prosthetis centred it. This was confirmed by the observations from the Fuji films.

Numerical results showed good *qualitative* agreement with the experimental ones. These results showed a strong influence due to the load application area on the measured strains. With implants the calculated strains were lower than the normal shoulder:

Quantitatively, there was good agreement for the normal shoulder (cf. fig. 10): Errors between experiment and numerical results varied from 20 to 68 µstrains (or 6 to 46 percent relative error).

For the Neer II and the new anatomic implant, errors on the four gauges were higher (cf. fig. 11): From 46 to 336 μstrains. The lack of precision of gauge positions on numerical measurement nodes explains these results. There is a need to increase the precision of the gauge positions in order to improve the validation of models.

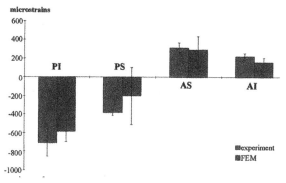

Fig. 10: Comparison between experimental and numerical results on the normal shoulder.

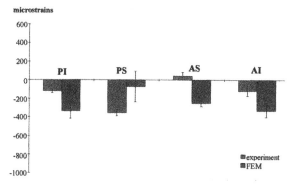

Fig. 11: Comparison between experimental and numerical results with the anatomic implant.

5. CONCLUSION

There was good agreement between experimental and numerical results. For the same experimental setup with the same cadaveric specimen, the glenoid implants in the scapula diminished the stresses in the bone. The volume of cement used for the fixation of the implants may be an important factor in explaining these lower values: The cement "stress shielded" the bone so that the surface strains after implantation were lower.

The Neer II glenoid implant produced stresses closer to the normal shoulder even if this implant seemed to be more sensitive to the « rocking horse » phenomenon.

The new anatomic implant seemed to be less sensitive to the off-centred load but the stresses found in the bone were lower than in the normal scapula.

For the four different cases, differences were shown due to experimental conditions: Implantation, gauge positions on the scapula.

It is important to note that our results *should not be extrapolated* to the physiological sistuations since our load and boundary conditions were experiment dependant. The principle objective of the experiment was to validate the finite element models. Also geometric properties (obtained with CT images) were individualised.

In the future, this method can be used for in vivo situations when a total shoulder arthroplasty is necessary for pathological cases, such as rheumatoid arthritis, osteonecrosis.

6. REFERENCES

1. WEIMANS H, HUISKES R Quantitative analysis of bone reactions to relative motions at implant-bone interfaces *J. Biomechanics,* 26 (11), p.1271-1281 (1993)

2. POPEN NK, WALKER PS Forces at the gleno-humeral joint *Clinical Orthopaedics and Related Research* , 135, p.165-170 (1978)

3. BAREA C, MANSAT P Mechanical properties of the glenoid cancellous bone *43st Ann. Meet., Orth. Res. Society,* February 9-13 (1997)

A WIRELESS TELEMETRY SYSTEM FOR MEASUREMENT OF FORCES IN MASSIVE ORTHOPAEDIC IMPLANTS *IN VIVO*

S.J.G. Taylor[1] and P.S. Walker[2]

1. ABSTRACT

A wireless telemetry system is described for measurement of forces acting upon massive orthopaedic implants, in man, to determine the load distribution in the fixation during routine activities and how this changes over long periods. The system comprises: the permanently implanted instrumented prosthesis, modified to enclose strain gauges and electronics, a single inductive link for powering the implant and telemetering the data, a microcontroller-based signal processor, a UHF radio link, and lap-top PC for real time data logging. The subject wears a small battery-powered inductive energiser, induction coil and microcontroller during the measurements. The strain gauges and implanted instrumentation are hermetically sealed inside cavities within the prosthesis, and are connected via a feedthrough to a small implanted induction coil. This coil is located outside the body of the prosthesis to maximise the power coupling efficiency, and is encapsulated using silicone rubber. Four prostheses have so far been implanted and data recorded over 1.5-3 years for each to date. This paper will focus on the electrical aspects of the telemetry system. Extracts of the axial force, bending moment and axial torque data will also be presented for the first Mk2 instrumented distal femoral replacement, during walking.

2. INTRODUCTION

Massive orthopaedic implants are those where a segment of the bone shaft (for example the femur or tibia) as well as the joint itself, is replaced with a metal prosthesis in cases of bone tumour. This is in contrast to a total joint implant (for example the hip or knee joint) where only the joint itself is replaced. An intramedullary (IM) stem, integral with the prosthesis shaft, is cemented into the medullary cavity of the remaining bone. This method of fixation has been used for the last 40 years at Stanmore, and has been largely successful. A serious problem remains however in the medium to long term, because of

Keywords: Telemetry, Forces, Strain Gauges, Prosthesis, Inductive Coupling.

[1]Senior Research Fellow, [2]Professor of Biomedical Engineering and Head of Department, Centre for Biomedical Engineering, University College London, Royal National Orthopaedic Hospital Trust, Brockley Hill, Stanmore HA7 4LP, UK.

aseptic loosening of the prosthesis in the bone and often eventual failure of the fixation, experienced as pain and loss of support. Bone remodelling at the transection site starts to occur after insertion of the prosthesis due to the altered loading conditions, and the accompanying bone loss and interface changes combine to weaken the fixation.

The current telemetry work was initiated to investigate the mechanism of load transfer in massive fixations, and to provide a quantifiable means of testing various methods of enhancing fixation. Direct measurement of forces acting upon and distributed throughout the prosthesis would provide reliable *in vivo* data over extended periods, enabling testing of improvements in methods and materials to prolong the life of the fixation. Two generations of instrumented prosthesis (Mk1 and Mk2) were developed for measurement of forces and moments in the prosthesis during activity. Each prosthesis has permanently implanted instrumentation for strain measurement, and power is supplied (and data telemetered) using magnetic coupling.

The Mk1 instrumented prosthesis has 2 strain channels, measuring axial force at the mid-shaft level and at the tip of the IM stem. The ratio of these forces is of particular interest in determining the loosening mechanism (Taylor et al 1997). The Mk2 instrumented prosthesis has 6 strain channels to telemeter bending moments and axial torque in the shaft as well as the forces. In addition, both devices measure temperature and humidity within the cavities occupied by the electronics. The Mk1 instrumented prosthesis has been implanted into two subjects (in July and December 1991), and the Mk2 device implanted into two further subjects (in February and November 1995).

3. DESIGN CONSTRAINTS AND CHOICES

The implanted portion of the instrumentation was the most critical part of the telemetry system design because of the restricted space available, and because of its inaccessibility and vulnerability. Key requirements were that it should be small, light, robust, consume low power and operate reliably over long periods of time. Key aspects of the implant instrumentation design were the method of power supply, the location of strain sensors and of the implanted electronics required for amplification, signal conditioning and telemetry. The choice of any one option for each of these aspects has a bearing upon the choice of the others. The key features of the type of implant power supply and the type and location of the strain gauges and instrumentation will first be addressed.

3.1 Implant power supply options

The implanted prosthesis should be able to remain functional for several years, although needing to be powered only periodically. Several options exist for its power supply:

1   A connection to an external power supply, by direct wires,
2   An implanted battery, continuously powering the implant,
3   An implanted battery, supplying power only when measurements are required,
4   An inductive power supply.

The first option requires a transcutaneous connection. For long term measurements to be possible, the subcutaneous part would need to be exposed during measurements and remain implanted at all other times without risk of repeated infection. Furthermore, the leads to the strain gauges would require excellent long term protection against mechanical damage and moisture ingress if the very small leakage currents (which would cause large and unpredictable measurement inaccuracies) were to be prevented. This method has only been used over short periods (Rydell 1966), and longer term measurements could not be carried out with confidence about maintained accuracy.

The second option is non-invasive but offers only a finite useful lifetime for the implanted instrumentation. Even if the implant current were as low as 100 microamps, a 300mAh 9V battery would only provide an implant life of 125 days. There is also the possibility of excess current being drawn due to leakage paths. A further disadvantage is that batteries contain highly toxic substances with the possibility of evolution of gas.

The third alternative is to connect the battery to the implant circuit via a magnetic device such as a reed switch, figure 1 (top). This method has been used by others, but problems have been experienced with failure of the reed switch in the 'on' state, causing the battery to be exhausted rapidly (Davy et al 1990). Possible solutions to this problem might include a timer to limit the maximum 'on' time or the use of two switches in series. However, this all adds additional hardware.

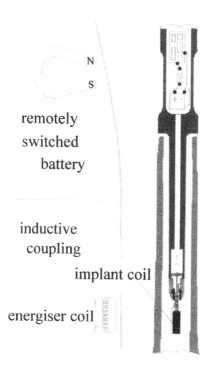

remotely
switched
battery

inductive
coupling

implant coil

energiser coil

Figure 1: Two alternative methods of powering the implanted prosthesis: a (remotely switched) implanted battery (top), and inductive coupling (bottom).

The fourth alternative, inductive coupling, is the most attractive option in terms of reliability, safety and required space (for a small implant coil). Furthermore, if operated at radiofrequency, sufficient bandwidth is available for implant data to be telemetered over the same link in the opposite direction. Inductive powering requires two coils, one connected to the implant and the other externally applied, to be sufficiently coupled electromagnetically so as to enable enough power to be induced for the implant to operate, figure 1 (bottom). This method was chosen for powering the implant. The inductive link used here consists of one implanted and one external tuned circuit having the same resonant frequency, 1.4MHz. The external circuit is tuned to the implant circuit to operate the link. Figure 1 also shows the location and size of the implant coil used in the Mk1 proximal femoral replacement.

3.2 Options for location of strain gauges and implant electronics

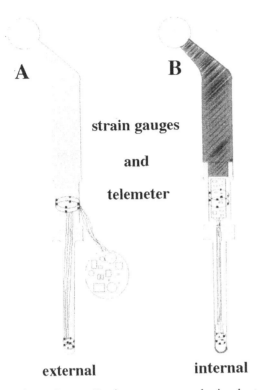

strain gauges

and

telemeter

external   **internal**

Figure 2:   Strain gauges and implant
electronics outside the prosthesis (A), and
hermetically sealed inside the prosthesis (B).
The latter option was chosen for the
instrumented prostheses.

An obvious and easy method of instrumenting the prosthesis is to attach the strain gauges to the external surface, figure 2(A). The instrumentation and telemeter could then be sited in soft tissue, separated from the prosthesis, and connected to the strain gauges with an umbilical cable. The advantages of such a method are that:

a)   no structural modifications to the implant are required;

b)   strain gauges could easily be bonded and inspected;

c)   the telemeter could be subcutaneous, and therefore close to the external coil, giving better coupling.

However, if the gauges were protected only by soft adhesive encapsulant several likely problems remain:

a)   The gauges and their connecting wires are exposed to forces applied by muscles;

b)   The umbilical cable is vulnerable to shearing movements of muscles over the surface of the prosthesis causing fatigue fracture;

c)   The gauges are at risk of corrosion, or shunting by ionic leakage currents, due to the possibility of liquid condensation at the encapsulant-gauge interface resulting from poor bonding to either the gauge alloy or its connecting wires. This would be observed as strain gauge drift, before eventual failure.

These disadvantages are made more severe because of the need to make long-term measurements, over several years. It would be impossible to distinguish between a gradual change in strain due to bone remodelling and a real or apparent change in electrical resistance of the gauge due to corrosion or leakage currents developing over a period of time.

These constraints on the accuracy of the measurements and the integrity of the implant were considered unacceptable, and an alternative scheme was adopted: the strain gauges and electronics would be housed in welded cavities within the body of the prosthesis, figure 2(B). This would greatly alleviate doubt about the integrity of the strain gauges, particularly if they could have a dry environment, and if the humidity was monitored to confirm this. The main penalty for this method is that the gauges must be bonded to the internal cavity walls, unless an insert is used. However, such 'down-hole' gauging techniques are well established. Creating a large cavity inside the prosthesis reduces the bending stiffness and strength, but the design can still be kept within the safe limits for the maximum anticipated loads. Both the gauges and electronics would thus be housed in a mechanically secure, low humidity environment. Wire connections between the gauges and electronics would be short and secure mechanically and electrically, and the implanted electronics kept dry and at constant temperature.

3.3 Options for the location of the implanted coil

If the implant coil was placed within a cavity inside the prosthesis (in order to hermetically seal it) the shielding effect of the titanium would result in very poor coil coupling, even if a high permeability material was used to concentrate the magnetic flux. This would preclude the use of a portable battery-powered energiser even if the implant current were as low as 1 mA. Furthermore, such high permeability materials are usually most efficient at audio frequencies, which severely restricts the signal bandwidth if the same link is used to telemeter the data. The data must in that case be telemetered from the implant by a separate radiofrequency channel, which requires an electromagnetic 'window' through which to pass. This type of design is used in one implanted telemetric hip which has a ceramic head, allowing radio propagation (Bergmann et al 1990). Since massive prostheses are usually made at Stanmore with a cobalt-chrome femoral head, a feedthrough would be required for the aerial in this case.

Despite the clear advantage of having the implant coil integral with the electronics inside an hermetic cavity, it was considered most desirable for the patient to be unencumbered by a heavy trailing power lead. If the coil could be encapsulated outside the titanium body, a vast improvement in coupling could be achieved, allowing battery powering, and a radio frequency passive telemetry system (Donaldson 1986) to be used for data transmission. The telemetered data could then be retransmitted by UHF radio to a nearby receiver. A biocompatible coil, wound on a ferrite former, was encapsulated in void-free silicone rubber, the arrangement designed to resist breakdown by moisture over extended periods. An hermetic feedthrough was used to connect the coil to the implanted electronics, and an adhesive rubber was used to encapsulate the feedthrough insulator between the pin (connected to one end of the coil) and the prosthesis body (connected to the other end of the coil).

The location of the implant coil on the Mk1 prosthesis was different to that of the Mk2. In the case of the Mk1, the coil was positioned 5mm from the stem tip, cast in silicone rubber and PMMA bone cement, figure 1. This enabled the coil to be located away from the metal prosthesis, and also enabled a smaller diameter external coil since the thigh girth reduces towards the knee. Both of these features maximised the magnetic power efficiency. In the case of the Mk2, the use of thin film strain gauges necessitated a different mechanical arrangement at the stem tip, and the more severe bending moment acting on the distal femoral stem tip required the tip cavity to be shorter than the Mk1. Both of these considerations meant that less space was available at the tip both for electronics and for the mechanical arrangement required for the interface to the coil. The coil was therefore resited in the main shaft region, closer to the main electronics cavity, figure 3. This had the additional advantage that the external coil was then closer to the knee, maximising the coil coupling. The implant and external coils were aligned to be coplanar, and despite the large disparity in areas a useful power transfer efficiency (power into the load divided by power input to the energiser) of 10% was obtained.

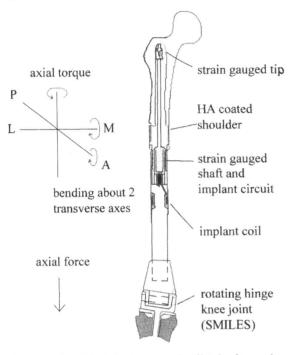

axial torque

P

L ——————— M

A

bending about 2
transverse axes

axial force

strain gauged tip

HA coated
shoulder

strain gauged
shaft and
implant circuit

implant coil

rotating hinge
knee joint
(SMILES)

Figure 3: Mk2 instrumented distal femoral fixation, with the shaft measurements and directions. As well as axial forces at the shaft and stem tip, the Mk2 measures the two transverse bending moments and axial torque.

3.4 Possible strain gauges types

For the Mk1 device, it was thought desirable to have the simplest combination of mechanical geometry and strain gauges, to avoid complicating the transducer. This was achieved with a single solid tubular shaft and foil gauges mounted on the inside walls, figure 2(B). Thin film strain gauges were used in the Mk2 design instead of foil gauges. This was for two main reasons: possible drift of foil gauges, and reduced implant power consumption. Foil gauges are bonded to the implant surface using a thin adhesive layer. Under certain environmental conditions laboratory tests have shown that gauge drift can occur over time (Taylor & Donaldson, 1992). Thin film gauges have no adhesive layer, both the insulator and conductor layers being formed on the substrate by sputtering. They are therefore believed to be more stable than foil gauges at high strain and/or high humidity. (However, the low strain, low humidity environment of the Mk1 gauges ensured that there was only a small likelihood of drift having occurred in the Mk1 devices.) No measurable drift was found after fatigue testing the Mk2 design (2.3kN peak load for 15M cycles), which gave confidence about the stability of the gauges over multiple load cycles at physiological strain levels. A further reason for choosing thin film gauges was the high values of resistance achievable due to the combined effects of the high resistivity alloy, the thin conductor film and the long and narrow conductor pattern produced by laser etching. In order to maximise the sensitivity to each stress type it was desirable to retain the existing basic Mk1 circuit topography, in which each strain channel had its own fully contributive bridge of 4 strain gauges. The use of thin film gauges allowed $20k\Omega$ gauges to be used for all 5 strain channels, thereby almost halving the total bridge current of the two $5k\Omega$ Mk1 channels. Since thin film gauges must be deposited onto an external surface, the Mk2 shaft was constructed as two overlapping cylinders, figure 3, and the gauges deposited onto flats on the outer surface of the inner part. These were then welded together after wiring the gauges to the electronics.

4. SYSTEM OPERATION AND SIGNAL PROCESSING

Figure 4 shows the electrical system schematic. The system is in three physically separate parts: the implant, a portable patient box and external coil, and a remote receiver / data logger. When the energiser is active, the voltage induced across the implant tuned circuit is rectified, smoothed and applied to a precision voltage reference, whose output supplies the strain gauge bridges and implant electronics. Measured strains, together with temperature and humidity signals, are amplified and converted to a serial pulse-position modulated signal and telemetered via impedance modulation of the same inductive link. This method of 'passive signalling' is possible because, even when the two tuned circuits are loosely coupled, the impedance of the implant circuit is reflected in the energiser circuit and forms a significant part of the total energiser impedance. Abruptly modulating the implant circuit impedance with the serial data causes a corresponding reflected impedance change in the external coil. This signal is detected and amplified, and periods of the signal corresponding to data are then digitised using a fast counter controlled by a microcontroller. The encoded data stream is serially transmitted as RS-232 at 9600 Baud from the subject over a 418 MHz UHF radio link to a receiver linked to a 486/100MHz PC. The data capture software, written in C++, decodes the serial data stream and displays the forces and moments dynamically in real time according to algorithms derived during calibration. In the Mk2 version of the system, an inverted calibration matrix is used to correct for small cross-sensitivities in the shaft channels found during calibration.

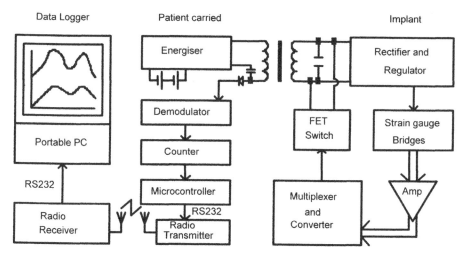

Figure 4:   The telemetry system electrical system schematic. The inductive link operates both as a power source for the implant and as the transcutaneous telemeter. Strain gauge bridge outputs are amplified and converted to a serial pulse stream, which modulates the impedance of the implant circuit. This signal is detected in the energiser circuit, the data encoded using RS-232 protocol, and radiotransmitted to a remote PC.

Each strain channel is sampled at 200Hz (Mk1) or 100Hz (Mk2), the timing controlled by a crystal in the implant. In the event of failure of any channel, clamp circuitry ensures that all other channels are unaffected. The implant voltage is telemetered to ensure that the induced voltage is sufficient to operate the voltage reference, and a warning is flagged if this voltage falls below a threshold at any time during a capture.

5. DATA FOR WALKING FROM MK2 SUBJECT

Forces and moments are captured over periods of 5 to 15 seconds during controlled activities. A typical data sample is shown in figure 5. This single gait cycle was captured during a 15 second walk at 1.2m/s, at 2 years post-op. The subject walked unaided with no obvious gait abnormality. The shaft force peaked 3 times during stance: at heelstrike (1050N), mid-stance (1500N) and late stance (2050N). The tip force showed 3 corresponding peaks (900N in late stance). The bodyweight was 67kg. The bending moment about the A-P axis peaked at 45Nm (varus) in late stance, the bending moment about the M-L axis peaked at 33Nm (hip flexion) in mid-stance, and the axial torque peaked at 10Nm (internal rotation) in late stance.

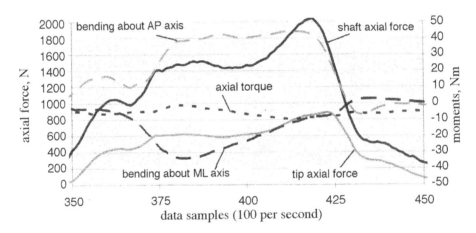

Figure 5:   Forces and moments recorded during walking at 1.2m/s

REFERENCES

1. Taylor S.J.G., Perry J.S., Meswania J.M., Donaldson N., Walker P.S., Cannon S.R. Telemetry of forces from proximal femoral replacements and relevance to fixation. J. Biomechanics 1997, **30**, 225-234.
2. Rydell N.W. Forces acting on the femoral head prosthesis. Acta Orthop Scand Suppl., 1966, **88**.
3. Davy D.T., Kotzar G.M., Berilla J., Brown R.H. Telemeterized orthopaedic implant work at Case Western reserve University. In 'Implantable Telemetry in Orthopaedics', April 26-28, 1990. Eds Bergmann, Graichen, Rohlmann. Publ. Freie Universität Berlin. ISBN 3-927433-45-4, pp205-219 (1990).
4. Bergmann G, Graichen F, Rohlmann A. Instrumentation of a hip joint prosthesis. Ibid. pp35-63.
5. Donaldson, N de N. Passive signalling via inductive coupling. Med. & Biol. Eng. & Comput., 1986, **24**, 223-224.
6. Taylor S.J., Donaldson N. Strain telemetry in orthopaedics (Developments in prospect). In 'Strain Measurement in Biomechanics', Eds. Miles, Tanner. Chapman & Hall, 1992., p101.

MEDICAL IMAGING AND RECONSTRUCTION TOOLKIT AS A MEAN FOR DESIGNING GEOMETRIC MODELS OF HUMAN JOINT ELEMENTS

W. Święszkowski [1], K. Skalski [2], K. Kędzior [3], S. Piszczatowski [4]

1. ABSTRACT

The medical imaging and reconstruction program (TOMCOMP) which allows CAD/CAM systems and CT (Computerized Tomography) system integration is elaborated in the present work. The main TOMPCOMP program tasks are: data reading from CT with reconstruction and visualisation of the separate tomograms, external and internal shapes generation of the bone elements by the use of image processing methods and algorithms, determination of the xyz characteristic bone shape points coordinates and their export to the CAD/CAM systems. By the surfaces in the NURBS space technique appliance, the reconstructed bone modelling was made in the CAD system. The results of the application of the above described system are presented as an example of radial bone head identification by the CT used to examine the object in the Unigraphics system.

2. INTRODUCTION

Graphic data visualisation and its interpretation has become one of the most important problems in surgical diagnostics of human organs (Woolson 1989) in contemporary medicine. A proper shapes and dimensions identification of the bones in human joints (elbow and hip joints, etc.) can be the most decisive factor in the design of a proper endoprosthesis.

Keywords: Computerized Tomography, reconstruction, filtering, geometrical modelling

[1] Postgraduate student, Faculty of Power and Aeronautical Engineering, Warsaw University of Technology, Warsaw, Poland
[2] Professor, Faculty of Production Engineering, Warsaw University of Technology, Warsaw, Poland
[3] Professor, Faculty of Power and Aeronautical Engineering, Warsaw University of Technology, Warsaw, Poland
[4] Assistant, Faculty of Production Engineering, Bialystok University of Technology, Bialystok, Poland

Computerized Tomography (CT) (Packer et all 1987, Hariss 1993, MacDonald 1989) is the most advanced technique used nowadays in non-invasive clinical diagnosis. One common characteristic of the CT is that it is similar to cutting the body in many slices, making a picture of each slice (Deniris et al. 1997). A different representation of the information contained in the bone cross-section images is not easy to analyse but it is necessary to obtain the correct visualisation of the bones. The use of a computer can lead to adequate presentation of the anatomical object by 3D - modelling. There are many computer systems for visualising medical images but many of them cannot define a geometric solid or surface model of the identified object. This possibility is necessary in designing an endoprosthesis appropriate to an individual patient's needs. Nowadays, the effective designing and modelling of the endoprosthesis geometry needs the use of CAD/CAM systems (MacDonald 1989). Nevertheless, these systems don't have the ability of direct data acquisition from diagnostically devices like CT. There is a need, therefore, for a scientific description of a computer system which allows the precise reconstruction of images obtained from the identification of bones with the help of computerized tomograph, and which further allows us to prepare data necessary for a fast and appropriate geometric modelling of bone in the CAD/CAM - Unigraphics system. This text tries to give tools for integrating CT and CAD/CAM systems, allowing the geometric modelling of bones as well as the design of a new generation of endoprostheses conforming to the individual needs of the patient.

3. METHODS

The acquisition and reconstruction of data from computerized tomography as well as its preparation of geometric modelling in CAD/CAM system are basic tasks of a designed and integrated computer system. These system tasks are linked with the execution of stages in the visualization process shown in the diagram (Fig.1). The methodology of computerized system design is described in the following diagram.

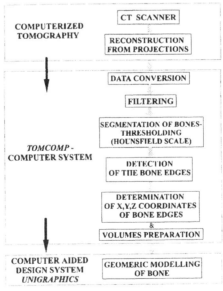

Fig. 1. Global sketch of the visualisation procedure in the computer integrated system

3.1. CT identification and reconstruction from projection

CT scans of the identified object were made using the Siemens DR2 tomograph controlled by a PDP11 computer with an RT11 operating system. In the CT scanner, information from the attenuation of X-rays as they pass through the object is used to reconstruct the attenuation coefficient map of the object, leading to characterisation of different tissues in the body. When the fan beam of intensity I_{in} passes through an object of attenuation coefficient $\mu(x,y)$ its intensity as it comes out of the object (Fig. 2) is given by

$$I_{out} = I_{in} \exp\left\{ - \int_{r_k(n)} \mu(x,y)dxdy \right\}$$

(1)

where a projection distance $r_k(n)$ is the ray from the source k to the detector element n for that source. From intensity measurements of I_{in} and I_{out}, it is possible to estimate the attenuation coefficient as the line integral of $\mu(x,y)$ along $r_k(n)$:

$$\int_{r_k(n)} \mu(x,y)dxdy = P_{k(n)} = \ln\left\{ \frac{I_{in}}{I_{out}} \right\}$$

(2)

where $P_k(n)$ is called the projection ray sum. The aim of reconstruction is to compute the $\mu(x,y)$ according to integral equation from the knowledge of a finite number of values of $P_k(n)$ for various values of k and n. One method for solving this problem is the filtered back projection method given by Radon (Fager et all. 1993). This method has found applications in CT scanners.

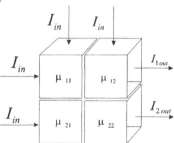

Fig. 2. Obtaining projection data

The results obtained from the reconstruction from projection are not directly applicable to the three - dimensional (3D) modelling realised by means of a CAD system. The information should be processed and written in a format compatible with the operating system running the designer's computer (DOS operating system). These tasks are performed in the developed computer system which is called TOMCOMP.

3.2. TOMCOMP computer system - medical imaging and reconstruction toolkit

The program tomcom.exe was written in the C++ language utilizing a Microsoft Visual C++ v. 4.0 compiler. The program runs on the MS Windows 95 system.
The program performs processing tasks that include data conversion, filtering, segmentation of bone, detection of the bone edges, determination of x,y,z coordinates of bone edges and preparing data for geometric modelling in CAD system.

Data Conversion

The data conversion is realised in TOMCOMP in two stages:

1. data transfer from the PDP tomographic station with RT11 operating system to the PC computer with DOS operating system,
2. the acquisition of the images.

The tomographics device used produces a stack of two-dimensional images. Usually stacks consist of 30 to 100 images, each image having a resolution of 256^2 to 512^2 pixels with up to 4096 possible greyvalues per pixel. For the subsequent processing in the TOMCOMP, these 4096 greyvalues are mapped to different greyvalue-ranges (e.g. 256) and written in BMP format.

Filtering

In many cases the data coming from the imaging device contains so called artefacts. Artefacts are various erroneous information contained in an image. The lost information can be reconstructed using some adequate algorithms.

In the TOMCOMP computer system these algorithms are called filtering-algorithms and they are contained in the "Calculate" module of the program.

The filtering can be thought as the convolution of the image with a filtering mask (3), which equals a multiplication of the images with the transform of the filtering mask in the frequency domain. For instance, when filter with following filtering mask:

$$\text{Filtr} = \begin{matrix} f_{-1,-1} & f_{0,-1} & f_{1,-1} \\ f_{-1,0} & f_{0,0} & f_{1,0} \\ f_{-1,1} & f_{0,1} & f_{1,1} \end{matrix} = \begin{matrix} 1 & 1 & 1 \\ 1 & 1 & 1 \\ 1 & 1 & 1 \end{matrix} \tag{3}$$

is applied to filtering image the new pixel value equals:

$$\begin{aligned} \text{new\_pixel\_value} = & \left(f_{-1,-1} \cdot s_{-1,-1}\right) + \left(f_{0,-1} \cdot s_{0,-1}\right) + \left(f_{1,-1} \cdot s_{1,-1}\right) + \\ & + \left(f_{-1,0} \cdot s_{-1,0}\right) + \left(f_{0,0} \cdot s_{0,0}\right) + \left(f_{1,0} \cdot s_{1,0}\right) + \\ & + \left(f_{-1,1} \cdot s_{-1,1}\right) + \left(f_{0,1} \cdot s_{0,1}\right) + \left(f_{1,1} \cdot s_{1,1}\right) \end{aligned} \tag{4}$$

where: f - element of the filtering mask, s - element of the filtered image.

Linear and non linear filters were applied to images filtering in the program. The linear gaussian filter removes noise and it smoothes edges as well. Another group of filters applied in program, called non linear filters, are minimum filter, maximum filter and median filter. They are more effective in removing single spots distributed all over the image and retaining the original edges of the bone. The pair of filters minimum and maximum as a method of closing gives good results in removing noise which has a detrimental influence in the later process of edge detection. The median filter also gives good results when used for noise removal. Theoretically, the harmful side effect of this filter, namely the removal of data similar in size to one pixel, is in this case to its advantage, as picture elements of such size are considered as noise and should be removed. After noise removal and the resulting edge enhancement, we employ binarization. Due to the varying quality of scanning it was decided to build in brightness and contrast controls. The proper adjustment of these two parameters helps to bring out those elements of the image which interest us, and also allows us to perform binarization without the necessity of additional algorithms. In the case of lower-quality tomographs, the program contains graphics tools for manual image correction.

Following is a fragment of the program function declaration responsible for the image filtration process.

```
void CCalc::RunFilter(UINT nIDResource, int br, int contr)
{
        //Filters declaration
// Gaussian Filter
        int        fd_Gauss[3][4]=    {{ 1, 2, 1},
                                        { 2, 4, 2},
                                        { 1, 2, 1}}; fd_Gauss[0][3] = 16;
// Minimum Filter
        ..........
        //end of filters declaration
        //CTomComDoc* pDoc = GetDocument();
        int i, j, flt1[3][4];
        unsigned int x, y;
        float temp;
        long int l1=0, l2=0, l3=0;
        int flt[3][4]={{ 0, 0, 0, 1},
                        { 0, 0, 0, 0},
                        { 0, 0, 0, 0}};
        switch(nIDResource)
        {
        case1:
                {for(i=0; i<3; i++)
                        for(j=0; j<4; j++)
                                flt[i][j] = fd_Gauss[i][j];            break;}
        case 2:
                {for(i=0; i<3; i++)
                        for(j=0; j<4; j++)
                                flt[i][j] = fd_minimum[i][j];          break;}
        ..........
        default: {}
        }
```

Segmentation

Segmentation aims at the location of a segment of the bone tissue in the CT image. The normalised acquisition of CT data according to the Hounsfield scale was applied as the bone segmentation criterion. The TOMCOMP program makes it possible to overlay the CT scans with the windows representing specified intervals of Hounsfield's coefficients (C_T). It was proved (Sumner et al. 1989), that best results in the segmentation can be obtained when setting up different intervals of Hounsfield's coefficient in measurements of the cortical bone external shape and its marrow cavity, respectively. A proper choice of the interval makes it possible to attain the accuracy levels of 2.6% of the external diameter and 7.4% of the marrow cavity diameter, respectively.

Detection of the bone edges

Bone edges detection aims at locating the bone edges in the image and defining them as closed contours (bone external shape and its marrow cavity). A Laplacian Filter (5) was applied for edge detection in the developed program.

$$Filter = \begin{matrix} 0 & -1 & 0 \\ -1 & 4 & -1 \\ 0 & -1 & 0 \end{matrix} \tag{5}$$

Determination of x, y ,z coordinates of bone edges

The final task which the program is to designate coordinate points lying on selected outer and inner contours of bone tissue. Fig.3 shows the methodology employed in the determination of bone contour coordinates. The coordinates (x_i, y_i) were determined on the basis of selected measurements of cortical bone borders in each section examined z_j (Fig. 3b). The measured points are distanced from each other by a constant value of angle $\alpha= 10°$. The number of points, their angular distance, and the direction of their designation should all be the same for each bone section examined. This is connected with the properties of the methods of describing the surfaces, which will be utilized in the three-dimensional reconstruction of the object in the CAD system. The third missing coordinate, z_j, is obtained on the basis of information on the distances between each segment, ΔZ_j (Fig. 3a).

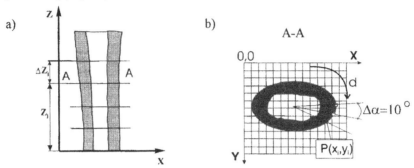

Fig. 3. Methodology of obtaining coordinates of measured points: a) determination of coordinate z, b) determination of coordinates x, y

Data Export to CAD System
Data comprising the coordinate points of bone contours come from the transformation of each tomogram (bone cross-section scan) put together into one set containing the coordinates of all points representing the scanned object. The file is written in the format compatible with the CAD/CAM system called Unigraphics.

3.3. Geometric modelling in CAD/CAM system

The geometric modelling of the bone was realised by using the CAD - Unigraphics system. The following stages were realised in the process of geometric modelling of the bone:
- Reading in and preparing the point coordinates to the 3D modelling of the object,
- Bone surface approximation using the NURBS (Non-Uniform Rational B-Splines) method of surface description,
- Generating the one patch surface spread over the net points,
- Visualisation and modification of the model shape by means of the Unigraphics system.

4. RESULTS

The elaborated, integrated TOMCOMP system is employed in the reconstruction and geometric modelling of the head of the radius in the elbow joint. The results of research using CT (Siemens DR-2 with matrix resolution of 512x512 pixels) were a series of 26 images, cross-sections of the head and neck of the patient's radius. Scans were made at intervals of 1mm for 26mm from the elbow joint. A sample image from this tomographic study appears as Fig. 4.

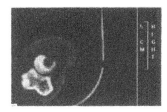

Fig. 4. A sample image, cross-section of the radial head

For the further transformation of data, the TOMCOMP program was used, as described in this text. Data were transferred from the format used in the RT11 system (the system serving the PDP tomograph computer) to the format readable by a PC operating system. Next the lost information in images were reconstructed with using some adequate algorithms. Fig.5. shows the effects obtained from the tomogram filtration process after using the filters: a)Gaussian, b)median, c) maximum and d)minimum.

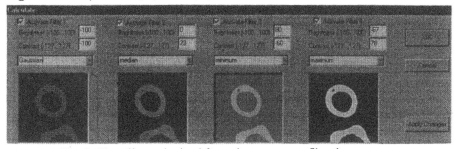

Fig. 5. The effects obtained from the tomogram filtration process

The next stage of the system's operation was the segmentation of CT images in order to clearly determine areas of occurrence of cortical bone. Then Laplacian Filter was applied for the bone edges detection (Fig. 6a).

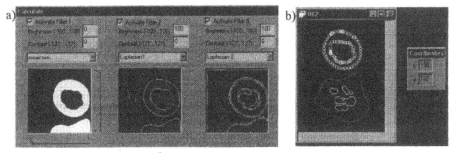

Fig. 6. a) Segmentation and bone edges detection, b) determination of x,y coordinates

The coordinates x, y, z from all measured points forming the outer contours of the head of the radius, as well as its marrow cavity, obtained automatically by the operation of the TOMCOMP system (Fig. 6b) were collected in a parametric database and transferred to the Computer Aided Design (CAD) system (Święszkowski et al 1996). The Unigraphics CAD system was used for modelling of the radial head geometry. The NURBS method (Non-Uniform Rational B-spline) for a 3D description of the surfaces was used. 3D geometrical models representing real shapes of the tested bones have been obtained (Fig. 7).

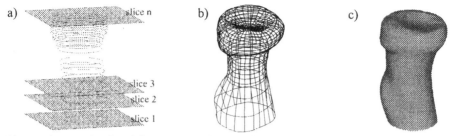

Fig. 7. a) Net of points lying on each slice, surface (b) and solid (c)representation of the radial head

5. CONCLUSIONS

The developed TOMCOMP program permits the integration of the technique of bone element identification - computerized x-ray tomography (CT) with computer-assisted design system (CAD). This allows the modelling and creation of data bases for biological objects as complex as human bones. A result of the graphic reconstruction of tomograms of the head of the radius with the help of the developed program was data obtained which allowed the correct modelling of the identified object in the Unigraphics system.

ACKNOWLEDGEMENTS

This research has been accomplished as part of grant no. 7 T07A 010 13 sponsored by the State Committee for Scientific Research.

REFERENCES

1. Woolson S.T., Dev P., Fellingham L., Vassiliadis A., Three-dimensional imaging of bone from computerized tomography. Clin. Orthop., 1989, 202, 239-248.
2. Packer D.L., Cayton P.D., Computed Tomography: The revolution in computer based medical imaging. In: C.J. McDonald, (Ed.), Images, signal and devices, Springer, Berlin 1987, 78-90.
3. Harris L.D., Computed Tomography and Three-Dimensional Imaging, Biodynamic Research Unit, Rochester 1993.
4. MacDonald W., Trum C.B, Hamilton S.G.L., Designing an implant by CT scanning and solid modelling, J. Bone Jt Surg., 1989, 68-B, 208-212.
5. Demiris A., Mayer A., Meinzer H., 3-D Visualization in medicine: an overview. Contemporary Perspective in Three-Dimensional Biomedical Imaging, C. Roux and J.-L.Coatrieux (Eds.) IOS Press, 1997, 79-105.
6. Fager R.S., Paddanarappagari K.V., Kumar G.N., Pixel-based reconstruction (PBR) promising simultaneous technique for CT reconstruction, IEEE Trans. Med. Imaging, 1993, vol. 12, no.1, 4-9.
7. Sumner D.R., Olson C.L., Freeman P.M., Andriacchi T.A., Computed tomographic measurement of cortical bone geometry, J. Biomechanics, 1989, no. 6/7, 649-653.
8. Święszkowski W., Skalski K., Kędzior K., Werner A., Pomianowski S. Geometric modelling of the radial head and its endoprosthesis, Proc. of the XV-th Polish Conference on Theory of Machines and Mechanism, Bialystok, 1996, 514-521.

SENSITIVITY ANALYSIS OF A FLEXIBLE FOOT PROSTHESIS USING FINITE ELEMENT MODELLING TECHNIQUES

P. Allard[1], J. Dansereau[2], M. Duhaime[3], R. Herrera[4], F. Trudeau[5]

ABSTRACT

Flexible foot prostheses were developed to restore gait propulsion during the push-off period of the stance phase of walking. A finite element model of an asymmetrical keel was developed and a sensitivity analysis carried out on eight different geometrical models to determine which configuration provided the most energy storage during stance. An symmetrical keel yielded the highest mechanical energy (17.4 J) compared to a symmetrical shaped keel (5.8 J). The covering material had little effect on the deformation of the keel. Fifteen below-knee amputees were fitted with the new design and their gait pattern was compared to the one developed with their usual foot prosthesis. A simultaneous bilateral 3-D gait analysis revealed a 7% increase in walking speed when using the new design and less knee and hip muscle powers at push-off. These results support the use of an asymmetrical keel design.

Keywords : Prosthesis, Foot, Finite element modelling, Gait analysis

[1] Professor/Department of Physical Education, University of Montreal, CP 6128, Succursale Centre-Ville, Montréal, PQ, H3C 1J7, Canada
[2] Professor/Department of Mechanical Engineering, École Polytechnique de Montréal, Montréal, PQ, Canada
[3] Professor/Orthopaedic Surgery, Surgeon-in-Chief, Shriner's Hospital, 1579 Cedar Avenue, Montréal, PQ, H3G 1A6, Canada
[4] Researcher/Research Department, Laboratoire d'orthèses et de prothèses Médicus, 5165 10th Avenue, Montréal, PQ, H1Y 2G7,Canada
[5] Research Assistant/Department of Physical Education, University of Montreal, CP 6128, Succursale Centre-Ville, Montréal, PQ, H3C 1J7, Canada

INTRODUCTION

Lower limb amputees lack the powerful plantar flexors to propel themselves forward during gait. They have to compensate their lack of ankle propulsion by a stronger pulling action of the hip flexors. Up to the mid 1970s, prosthetic feet were essentially a passive appendage enabling the amputee to fit a shoe to the artificial leg. A prosthetic foot generally consists of a keel which is the structural element of the foot, a covering or filler material which give the appearance of a foot and a thin skin which protects the covering material from wearing to rapidly. In some feet, such as the Solid Ankle Cushion Heel (SACH) and the Single-Axis (articulated at the ankle), the heel portion of the covering material is more flexible, simulating ankle plantar flexion at heel-strike. Though the cushioned heel favoured a better transition between the sound and the affected limbs it did not contribute to the amputee's propulsion. These feet can be considered as the first generation by Allard et al.[1].

New polymers and composite materials caused considerable changes in the design of prosthetic feet. The covering material now gives a life-like appearance to the artificial foot and the colour of the foot can be matched to that of the skin. The rigid wooden keel is replaced by a flexible structure capable to store deformation energy during the early part of stance and release it during push-off. The efficacy of these second generation prosthetic feet to absorb and release mechanical energy has been documented in laboratory test by Contoyannis[2]. However, Torburn et al.[3] reported no clinical advantages to any of the four energy storing feet tested on lower limb amputees (STEN, Carbon Copy, Seattle Foot and Flex Foot) as well as to the SACH foot. It appears that lower amputees do not fully take advantage of the mechanical properties of their flexible foot prostheses. This can be attributed in part to the lack of medio-lateral control and propulsion due to a symmetrical keel design.

An symmetrical keel design would deform in both bending and torsion allowing a three-dimensional propulsion to the foot prosthesis. The objectives of this work are to demonstrate that a) an asymmetrical keel can store more strain energy than a symmetrical keel, b) the mechanical properties of the covering material can influence the deformation of the keel and c) an asymmetrical keel foot prosthesis provides a clinical advantage to lower limb amputees.

DESIGN CONCEPT

The asymmetrical keel consists of three basic curvatures as shown in Fig. 1. The heel portion has an internally concave curvature which controls body weight transfer from the sound to the affected limb at heel-strike while the central portion of the keel has an arch which acts as a leaf-spring. Finally, the toe-end is curved internally to provide a forward and medial propulsion at the end of the stance phase. The function of these curvatures have been described in Prince et al.[4]. This prosthetic foot is known as the Space Foot because of its asymmetrical shaped keel.

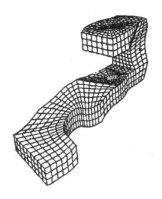

Fig. 1. Geometric model of the asymmetrical keel of the Space Foot.

METHODS

A finite element model of the keel was developed and a sensitivity analysis was carried out on the keel to determine if all three curvatures are necessary. This was followed by a finite element model of the covering materials to establish their effect on the deformation of the keel. Finally, the Space Foot was fabricated and fitted to 15 below-knee amputees and a 3-D gait analysis was performed to compare the performance of the Space Foot to that of the amputees' usual foot prosthesis.

FEM model of the keel and sensitivity analysis

The FEM model of the keel which corresponded to a US size 10 foot was developed using IBM CAEDS software on an IBM RISK 6000 workstation. The mesh configuration consisted of 490 brick elements with 1019 nodes. The keel was made of a monolithic block of Delrin 150 SA having a modulus of elasticity of 3100 MPa for a loading of speed of 5.1 mm mim^{-1} and a Poisson coefficient of 0.35.

The FEM keel model of the Space Foot was validate from a 3-D gait analysis. Six reflective markers were located on the lateral side of the bare keel at 30 mm intervals. Three 60Hz video cameras of a Motion Analysis Corporation Expert Vision system were position along an arc of 60° and about 3 m from an AMTI force plate. Cameras and force plate were synchronised by the system. The keel was rigidly fixed to the sole of an able-bodied subject. After camera calibration, the subject was asked to walk as normally as possible in front of the cameras and step on the force plate. Once the data were filtered according to Winter[5], the maximum deformation angle of the keel was 16.5° for a load of 700N. For the same force orientation and magnitude, the FEM model predicted a deformation of 17.8° which we found acceptable considering the measurement and reconstruction errors of the video system.

Once validated, the FEM model of the keel was modified to take into account any combinations of the three basic curvatures as well as a symmetrical keel. The maximum deformation, the frontal torsion angle and the strain energy were estimated for each of the eight keel combinations. The load consisting of a vertical force of 684N, a medial force of 41N and a anterior force of 100N was applied to the toe end of the keel. It was determine from a 3-D gait analysis of a below-knee amputee fitted with the Space Foot at the time when the vertical force was at a maximum during the push-off period.

Combined FEM model of the covering material and keel

The FEM model of the covering material consisted of a heel section, a filler to give a foot shape-like form to the keel and a protective skin. The combined keel-material FEM model had 5529 elements and 7669 nodes. Table 1. summarises the number of elements and the modulus of elasticity of each part of the Space Foot. The same boundary conditions were applied to the keel and covering material models.

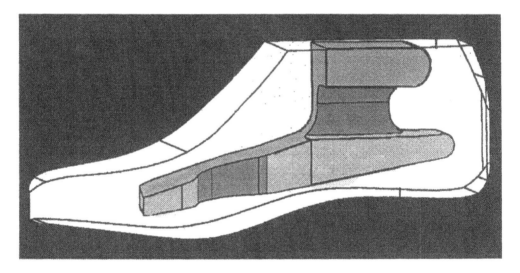

Fig. 2.   Asymmetrical keel of the Space Foot.

To determine the effect of the covering material on the maximum displacement of the keel, the modulus of elasticity of each part of the covering material was increase and decrease by an order of magnitude and the results compared to the standard values determined experimentally. Furthermore, simulations were also carried out for a soft and hard SACH foot as well as for a Seattle Foot.

Three-dimensional gait analysis

Simultaneous bilateral three-dimensional gait analyses were carried out to determine the performance of the Space Foot with respect to the amputees' usual foot prosthesis. Nineteen able-bodied men and 15 amputees participated in this part of the study. Four amputees were wearing a Flex Foot, five a SACH and six a Seattle Foot. All the amputees were consecutively fitted with the Space Foot.

Table 1 : Number of elements and the modulus of
elasticity of each part of the Space Foot
and for the SACH and Seattle Foot.

Space Foot	Number of elements	Standard values for the moduli of Elasticity (MPa)
Filler	2421	47
Heel	983	473
Skin	233	817
Keel	1992	3100
SACH (soft)	5000	239
SACH (hard)	5000	2 221
Seattle Foot	5000	708

Four cameras of a Motion Analysis Corporation system were place on each side of the subject at a about a 4 m distance from the two AMTI force plates. Three reflective markers per segment were used to define pelvis, thighs, legs and feet. After camera calibration, the subjects were asked to walked at their natural free walking speed and step consecutively on each plate plates. Video data was captured at 90 Hz while force plate data were collected at 360 Hz.

The video data were filtered before the 3-D co-ordinates were calculated and an inverse dynamic approach was used to determined the 3-D muscle powers as described in Allard[6].

RESULTS AND DISCUSSION

The strain energy, the frontal torsion angle and the maximum displacement of the toe-end of the keel estimated by the finite element model of each of the 8 keel configurations are reported in Figure 3. The symmetrical keel always displayed the lowest values and more importantly no frontal torsion. The asymmetrical keel of the Space Foot with the three basic curvatures had the highest values of the 8 keel configurations with three times more strain energy than that of the symmetrical keel. Figure 4 illustrates the three-dimensional deformation of the keel of the Space Foot measured at the toe-end. Though the heel section only had the highest strain energy (10.7 J) of the three curvatures, it had no frontal torsion angle. The highest frontal torsion angle (4.6°) of the three basic curvatures measured at the toe-end was brought about by the toe-end. The heel-arch and heel-toe-end results were similar in terms of toe displacement and strain energy. However, the arch model and the heel-arch combinations had no frontal torsion. The arch section was eliminated in the final design because of it contributed little in terms of strain energy storage and in frontal torsion while simplifying the design and fabrication of the keel.

Before addressing the results from the FEM model of the combined keel and covering materials, it is important to note that the sensitivity analysis was carried out by varying the modulus of elasticity of one material (heel, filler or skin) by one order of magnitude while the keel's and the other two materials' modulus of elasticity were kept constant. Generally, the covering material slightly affected the deformation of the keel. The filler had the greatest influence on the keel's displacement due to the large volume it occupies in giving a foot form shape. When it was the stiffest, the keel's displacement was reduced by 4.27%compared to 4,19% and 3.69% for the heel and skin respectively. When the filler was the least stiff, the displacement of the keel was the least reduced to 2.17%. The Seattle Foot' behaviour would be similar to that of the stiffest filler while the SACH foot would reduced the keel's deflection by 9%.

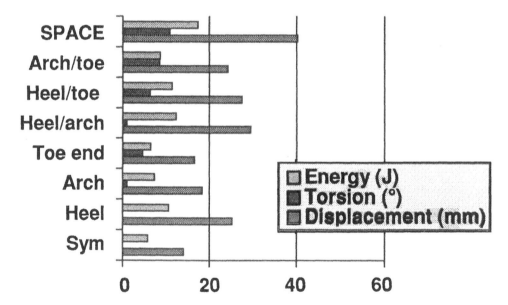

Fig. 3   Strain energy, frontal torsion angle and maximum displacement of the toe-end of the keel of each of the 8 keel configurations.

Though the heel section was modelled with three different moduli of elasticity, the softer one can not be practically used because it would bottom out as the amputee put his or her weight on it at heel-strike. Thus, the best combination of materials would be a soft filler, a soft skin and a standard heel. This arrangement would only reduce the keel's deflection by 1.95%.

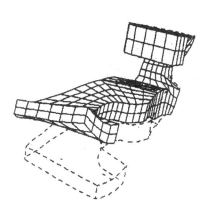

Fig. 4.   Geometric model of the deformed keel of the Space Foot. The dashed lines represent the unloaded keel.

To test the asymmetrical keel design, 15 below-knee amputees were fitted with the Space Foot and after a one month adaptation period their walking pattern evaluated by means of a 3-D gait analysis. Compared to their usual foot prosthesis, the amputees' walking speed increased by 7.7% when they were fitted with the Space Foot. This improvement was the more apparent with the Seattle Foot (10.6%) and the Flex Foot (7.5%). The amputees wearing a SACH foot improved their walking speed by only 0.9%. This can be explain in part that these amputees were slow walkers (1.12 m/s) compared to the others (1.22 m/s). Providing them with a better foot (Seattle, Flex or Space) would not necessarily improve their walking speed by a substantial amount.

Hip flexion and ankle plantar flexion are responsible to propel the lower limb forward. The amputees usually compensate their lack of plantar flexion by a stronger pull of the hip flexors. The use of a flexible keel foot prosthesis should provide them with some plantar flexion push-off resulting in a reduction in their hip activity. When comparing ankle propulsion defined by the muscle power[6] generated at push-off, the Flex Foot developed the highest value (1.5 W/kg) whereas the SACH and the Space feet the lowest at about 0.25 W/kg. However, the hip action developed by the Space Foot (1.6 W/kg) was similar to that of the Flex (1.7 W/kg) but both were smaller than the Seattle (1.9 W/kg). The long blade of the Flex Foot combines excellent ankle motion with propulsion whereas the Seattle and the Space feet are fixed at the ankle. The three-dimensional action of the Space Foot compensated in part the lack of ankle propulsion and motion but not to the point of matching the Flex Foot performance. A greater stability at heel-strike provided by the double curvature in the heel component of the Space Foot may be responsible for

---

[6] Power is the product of the net muscle moment by the joint angular velocity. Power is generated when the polarity of the muscle moment and the angular velocity are the same otherwise power is absorbed.

the amputees increase in walking speed. It allowed the amputee to make a better weight transfer from the sound to the affected limb without reducing the amputee's overall speed.

CONCLUSION

Simulation using FEM modelling techniques revealed that an asymmetrical keel deformed both in bending and in torsion and stored more strain energy than a symmetrical keel. Furthermore, it was shown that the covering material slightly reduced the deformation of the keel. The best arrangement was a soft filler and skin materials combined with a standard density heel. The three-dimensional actions of the asymmetrical keel design was manifested by a 7.7 % increase in the amputees' walking speed though it had the least ankle push-off and hip pulling actions. The asymmetrical keel provided both control and propulsion functions to the amputees. Asymmetrical keel design can be considered as the third generation of prosthetic feet.

ACKNOWLEDGEMENTS

This study was founded by the Natural Science and Engineering Research Council of Canada, the Ministère de l'Industrie, du Commerce, de la Science et de la Technologie du Québec et les Laboratoires d'orthèses et de prothèses Médicus. The authors express their gratitude to Mr. Jean Arrache for his technical assistance.

REFERENCES

1. Allard, P., Trudeau, F., Prince, F., Dansereau, J., Labelle, H. and Duhaime, M. Modeling and gait evaluation of an asymmetrical keel foot prosthesis. Medical and Biological Engineering and Computing, 1995, Vol. 33, 2-7.
2. Contoyannis, B. Energy storing prosthetic feet. Bull. Vet. Affairs, Central Development Unit, Melbourne, Australia, 1987.
3. Torburn, L., Perry, J., Ayyappa, E., Shanfield, S.L. Below-knee gait with dynamic elastic response prosthetic feet : a pilot study. J. Rehab. Res., 1990, Vol. 27, 369-384.
4. Prince, F., Allard, P., McFadyan, B.J. and Aïssaoui, R. Comparison of gait between young adults fitted with the Space Foot and nondisabled subjects. Archives of Physical Medicine and Rehabilitation, 1993, Vol. 74, 1369-1376.
5. Winter, D.A. The biomechanics and motor control of human movement. Wiley and Sons, 277p.
6. Allard, P., Lachance, R., Aissaoui, R. and Duhaime, M. Simultaneous bilateral 3-D gait analysis. Human Movement Science, 1996, Vol. 15, 327-346.

6. SOFT TISSUE STRUCTURES, CONTACT AND BIOFLUID MECHANICS

POROHYPERELASTIC-TRANSPORT-SWELLING FINITE ELEMENT MODELS: APPLICATIONS AND MATERIAL PROPERTIES FOR LARGE ARTERIES

B. R. Simon,[1] M. V. Kaufmann,[2] J. Liu,[3] and A. L. Baldwin[4]

ABSTRACT

This communication summarizes a presentation at the Third International Symposium on Computer Methods in Biomechanics and Biomedical Engineering,[1] and describes the development and applications of porohyperelastic-transport-swelling (PHETS) formulations and finite element models (FEMs) for the analysis of soft, hydrated rabbit aortic tissues. These PHETS models include both material and geometric nonlinearity and coupled transport of mobile tissue fluid and species dissolved in the fluid. Corresponding Eulerian and Lagrangian field theories allow identification of material property functions and form the basis for the FEMs. Data-reduction procedures provided intimal and medial elasticity, permeability, and diffusion-convection transport properties. PHETS FEM results compare well with data from our laboratory and data presented in the literature for pressure-radius response, free tissue fluid flux, and diffusive-convective flux of labeled mobile albumin in the walls of rabbit aortas.

1. INTRODUCTION

Porohyperelastic-transport-swelling (PHETS) models allow detailed analysis of biological structures where coupled finite deformation and fluid/species transport are of interest. There are numerous works in the literature that report the use of poroelastic and mixture theories for the study of soft tissue mechanics. Finite element models (FEMs) have been developed to implement such theories using porohyperelastic (PHE) or biphasic (BIP) formulations. FEMs using PHETS, triphasic (TRI), or multiphasic formulations are reported in the literature. Here, we illustrate the PHETS theory and a FEM to carry out a structural/transport analysis of the rabbit aorta.

Keywords: Finite element models, porohyperelastic, transport, arteries.

[1]Professor, Aerospace and Mechanical Engineering, The University of Arizona, Tucson, AZ, 85721, USA
[2]Research Engineer, The Hewlett-Packard Company, Palo Alto, CA, 94304, USA
[3]Research Assistant, Aerospace and Mechanical Engineering, The University of Arizona, Tucson, AZ, 85721, USA
[4]Associate Professor, Physiology, The University of Arizona, Tucson, AZ, 85721, USA

2. FIELD EQUATIONS

2.1. Fundamental fields

In this paper, the field theory is summarized in one-dimensional form for a single neutral species. However, these equations can be extended to a three-dimensional form that includes the effects of multiple charged species. The solid displacements and velocity are $u = x - X$ and $v^s = v = \dot{u} = du/dt$, and fluid/species relative velocities are $v^{fr} = n(v^f - v^s)$ and $v^{cr} = n(v^c - v^s)$, where the average fluid and species velocities are v^f and v^c at x and t. The porous solid material is incompressible and is saturated by an incompressible mobile fluid so that the porosity is $n = 1 - J^{-1}(1 - n_0)$ with $n_0 = dV_0^f/dV_0$. The species concentration is defined as $c = dm^c/dV^f$. Deformation measures are the volume strain, $J = dV/dV_0$, deformation gradient, $F = \partial x/\partial X$, and Green's strain, $E = 1/2(FF - I)$. When both solid and fluid materials are incompressible, the true densities, $\rho_T^S = dm^S/dV^S$ and $\rho_T^f = dm^f/dV^f$ are constant and $\partial v^{fr}/\partial x + \partial v/\partial x = 0$. Stress measures include the total Cauchy (true) stress, $\sigma = dF_{int}/dA$, and pore fluid stress, $\pi^f = -$ (pore fluid pressure).

2.2. Phenomenological equations—equivalence of PHETS and TRI models

The equivalence of the PHETS and TRI models is based on the phenomenological equations. In the TRI model, these equations relate mechanical and chemical potential gradients ($\partial \pi^f/\partial x$, $\partial \mu^f/\partial x$, $\partial \mu^c/\partial x$) to drag forces associated with differences in *absolute* constituent velocities and form the basis for the TRI model. The PHETS phenomenological equations are associated with *relative* velocities and have the form

$$\frac{\partial \sigma}{\partial x} = 0 \tag{1}$$

$$n^2\left(\frac{\partial \pi^f}{\partial x} - \rho_T^f \frac{\partial \mu^f}{\partial x}\right) = a^{ff}v^{fr} - a^{fc}v^{cr} \tag{2}$$

$$-n^2 c \frac{\partial \mu^c}{\partial x} = -a^{cf}v^{fr} + a^{cc}v^{cr} \tag{3}$$

where a^{ff}, $a^{fc} = a^{cf}$, and a^{cc} are phenomenological coefficients. In the mixed PHETS model, u, π^f, and c are the primary fields. v^{fr} and v^{cr} are eliminated from the theory by solving (3) for $v^{cr} = (a^{cc})^{-1}[a^{cf}v^{fr} - n^2 c(\partial \mu^c/\partial x)]$ and then substituting this expression for v^{cr} in (2), yielding the generalized Darcy law (equation 7 below), and in the relative species flux definition, $j^{cr} = cv^{cr}$, yielding equation (8) below.

2.3. Mixed PHETS field equations

The Eulerian mixed PHETS field equations are the *conservation equations* (momentum, solid-fluid mass, and species mass),

$$\frac{\partial \sigma}{\partial x} = 0 , \qquad \frac{\partial v^{fr}}{\partial x} + D = 0 , \qquad \frac{\partial j^{cr}}{\partial x} + \frac{\partial}{\partial t}(nc) + \frac{\partial}{\partial x}(ncv) = 0 \tag{4}$$

the *kinematic equations* (strain rate and gradients),

$$D = \frac{\partial v}{\partial x} , \qquad e^{\pi} = \frac{\partial \pi^f}{\partial x} , \qquad e^c = \frac{\partial c}{\partial x} \qquad (5)$$

and the *constitutive equations* (hyperelastic, including effective stress, σ^e; Darcy law; and relative species flux),

$$\sigma = \sigma^e + \pi^f I \qquad (6)$$

$$v^{fr} = k^{ff}\left(e^{\pi} + \frac{\partial \pi^c}{\partial x}\right) , \qquad \frac{\partial \pi^c}{\partial x} = g^c e^c \qquad (7)$$

$$j^{cr} = -d^{cc} e^c + b^{cf} c v^{fr} \qquad (8)$$

These equations are to be solved subject to initial and boundary conditions for u, π^f, and c. The principle of virtual velocities serves as a basis for FEMs, as well as providing correspondence rules between Eulerian and Lagrangian formulations. In the Lagrangian formulation, the fundamental fields are dependent on X and t, i.e., $u = u(X, t)$, $\pi^f = \pi^f(X, t)$, and $c = c(X, t)$. The Lagrangian PHETS model is similar to the equations above with corresponding material properties.

3. PHETS MATERIAL PROPERTIES

Consider a single, neutral species and assume no explicit strain dependence in μ^f and μ^c. Then we have corresponding sets of Eulerian and Lagrangian material properties for the PHETS model(s). The Eulerian material properties are as follows: (a) solid, generalized hyperelastic strain energy, $U^e = U^e(E, c)$ in $\sigma^e = J^{-1}F(\partial U^e/\partial E)F$; (b) fluid, permeability, k^{ff}, and osmotic coefficient, g^c; and (c) species, diffusion, d^{cc}, and convection, b^{cf}, coefficients.

4. FINITE ELEMENT MODEL (FEM)

The rabbit aorta was modeled using a "mixed" Lagrangian PHETS-FEM (described in detail by Kaufmann[2]). A modified Petrov-Galerkin method provided accurate solutions for convection flux associated with relatively large Peclet numbers. Interpolations in each finite element are $u = \mathbf{N}_u \bar{\mathbf{u}}$ (quadratic) and $\pi^f = \mathbf{N}_\pi \bar{\pi}$ and $c = \mathbf{N}_c \bar{\mathbf{c}}$ (both linear). Then, the quasi-static assembled global system of equations for the FEM are

$$\mathbf{C}\dot{\mathbf{r}} + \mathbf{R}^{int} = \mathbf{R}^{ext} \qquad (9)$$

Boundary and initial conditions are prescribed and a time integrator is applied with a Newton-Rapheson iteration to determine $\bar{\mathbf{u}}$, $\bar{\pi}$, and $\bar{\mathbf{c}}$. The FEM represented a segment of rabbit aorta as a layered cylinder (intima and media) in a state of axisymmetric, plane strain and was subjected to initial and boundary conditions representing internal pressure, axial stretch, and prescribed species concentrations at the internal and/or external surfaces. The adventitia was removed in the experiments and was not included in the FEM.

5. DATA REDUCTION AND EXPERIMENTS

The data-reduction methods were based on a generalized least-squares approach in which an error, $e = \sum_{M=1}^{N_{data}} (f_M^{model} - f_M^{exp})^2$ was minimized to determine material parameters. In this expression, $f^{model} = f^{model}$ (material parameters) and f^{exp} is the corresponding experimental value. Here, f^{model} is either analytical or obtained from the PHETS-FEM. The experimental data were obtained from rabbit aortas subjected to axial stretch, inflation, and immersion in bath(s) of labeled albumin. The apparatus was described by Simon et al.[3] Mobile water motion in the arterial wall was determined from the velocity, v_B, of an air bubble introduced in the inflation tube circuit. Values for v_B were measured at various pressure levels in order to determine arterial tissue permeability.

We quantified the isotropic generalized Fung form for $U^e = (1/2)C_0(e^{\phi} - 1)$ with $\phi = C_1'(\overline{I_1} - 3) + C_2'(\overline{I_2} - 3) + K'(J - 1)^2$ using undrained and drained inflation data to provide C_0, C_1', C_2', and K' values. Intimal and medial permeabilities were determined from steady-state tests on intact and de-endothelialized vessels to provide v_B versus P data to determine a constant value for k_{MED} and the nonlinear form for k_{INT}. The constant intimal and medial diffusion coefficient values, d_{INT} and d_{MED} were determined from measured labeled albumin profile(s) after 25 min of immersion of the aortas in internal and external baths with concentration c_B and $P = 0$ (no convection). Once d_{INT} and d_{MED} were determined, we used labeled albumin concentration profiles reported by Tedgui and Lever[4] in our data-reduction procedures to determine optimal values for b_{INT} and b_{MED} (both assumed *constant*). These experiments combined diffusion and convection by determining albumin concentration profiles for rabbit aortas at $P = 70$ and 180 mm Hg. Both damaged and intact vessels were tested, and the labeled albumin was only in the *external* bath at concentration c_B. Figure 1 compares labeled albumin concentration profiles obtained using these material parameters in PHETS-FEMs of Tedgui and Lever's[4] experiments.

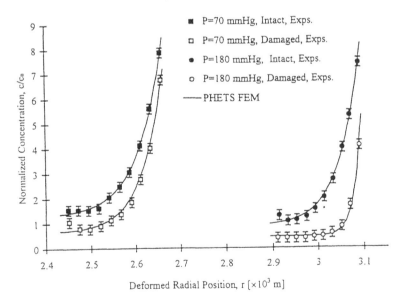

Fig. 1. Steady-state labeled albumin concentration profiles (c/c_B) in the deformed arterial wall: finite element results compared with Tedgui and Lever's[54] experimental data; c_B = bath concentration of albumin.

6. CONCLUSION

In this communication we have highlighted a simplified, one-dimensional formulation for the PHETS theory and a mixed PHETS-FEM that were used to analyze coupled transport processes of a (neutral) species in the arterial wall. The three-dimensional theory including charge effects is given by Kaufmann.[2] The basic phenomenological equations were manipulated into the PHETS form (either Eulerian or Lagrangian) so that FEMs could be developed. This PHETS formulation also allowed the identification of material properties that have physical significance as hyperelasticity, permeability, and diffusion/ convection coefficients. This paper also has indicated specific experiments and data-reduction methods that can be used to determine material parameters for the PHETS-FEMs of large arteries. We have pointed out the equivalence between the PHETS and the TRI models using the phenomenological equations and have noted direct mathematical relations between material properties. This means that either PHETS or TRI material properties can be introduced in the PHETS-FEMs of a hydrated soft tissue where water and species transport are of interest.

We intend to extend this research in a number of directions in the future. Our isotropic models will lead to anisotropic theories and FEMs once appropriate experimental data are available for large arteries. We will also introduce the effects of active smooth muscle and prestress conditions associated with tissue remodeling (opening angle) in the PHETS-FEMs. A detailed consideration of the complex behavior of the intima (barrier transport function, etc.) at the microscopic level and consideration of the adventitia are anticipated in our next models. We are currently investigating the role of charged species and $\partial \pi^c/\partial x$ in the theoretical and FEM development.

We anticipate specific application of the PHETS theory and FEMs in the study of atherogenesis in the walls of arteries at sites where this disease has been observed to initiate. PHETS-FEMs will be combined with computational fluid dynamic models of the arterial system in order to more fully understand the etiology of atherosclerosis. We are also developing PHETS-FEMs of local drug delivery systems in which the balloon catheter and transport processes in the arterial wall are being simulated. Tissue-engineered arterial grafts are currently being studied in our laboratory using experiments and PHETS-FEMs to begin quantifying the processes of tissue in-growth and entholelialization. Our models may also be useful in identifying mechanical/transport graft responses that may serve as initiators of hyperplasia in these arterial grafts.

ACKNOWLEDGMENTS

NSF Grant BES-9410571 and the Office of International Affairs, The University of Arizona, Tucson, Arizona.

REFERENCES

1. Simon, B. R., Kaufmann, M. V., Liu, J., and Baldwin, A. L., "Porohyperelastic-Transport-Swelling Finite Element Models—Applications and Material Property Determination for Large Arteries," 3rd International Symposium on Computer Methods in Biomechanics and Biomedical Engineering, Barcelona, Spain, May 7-10, 1997.

2. Kaufmann, M. V., 1996, "Porohyperelastic Analysis of Large Arteries Including Species Transport Swelling Effects," Ph.D. Thesis, Aerospace and Mechanical Engineering, The University of Arizona, Tucson.

3. Simon, B. R., Kaufmann, M. V., McAfee, M. A., Baldwin, A. L., and Wilson, L. M., 1997, "Identification and Determination of Material Properties for Porohyperelastic Analysis of Large Arteries," *J. Biomech. Eng.* (accepted).

4. Tedgui, A. and Lever, M. J., 1985, "The Interaction of Convection and Diffusion in the Transport of ^{131}I-Albumin within the Media of the Rabbit Thoracic Aorta," *Circ. Res.* 57, 856-863.

MIXTURE MODELS: VALIDATION AND PARAMETER ESTIMATION

C.W.Oomens[I], J.M.Huyghe[II] and J.D.Janssen[III]

1. ABSTRACT

Mixture models are used extensively for the description of the mechanical behaviour of biological tissues. Because of their increasing complexity there is a need for new methods to validate the models and to determine the material parameters in the constitutive equations of the individual components and their interaction. In the paper a mixed numerical experimental method to determine the material parameters in biphasic mixtures is summarized and a recursive estimation algorithm is derived to estimate parameters for biphasic mixtures. In a simulation it is shown that an inverse analysis to determine 4 material parameters can be done in a CPU-time of the order of one single direct analysis. Moreover, some results of experiments with a synthetic model material are discussed. The latter were used to validate a tri-phasic material model and have shown that the theory according to Snijders [7] is not complete.

2. INTRODUCTION

Since the mid-seventies mixture models, originally used to describe the mechanical behaviour of soil [1], have become popular to describe the mechanical behaviour of different types of biological materials. The first models were used to describe the behaviour of articular cartilage [2], later intervertebral disks [3], skin [4] and heart muscle [5] were described with mixture theory. The first were biphasic models describing the tissues as mixtures of a solid saturated with fluid, later models have become more complex when osmotic effects were included [6,7]. This increasing complexity causes

[I]   Senior Lecturer/ Department of Mechanical Engineering, Eindhoven University of Technology, P/O Box 513   5600 MB Eindhoven, The Netherlands

[II]   Research Fellow/ Department of Mechanical Engineering, Eindhoven University of Technology, P/O Box 513 5600 MB Eindhoven, The Netherlands

[III] Full Professor/ Department of Mechanical Engineering, Eindhoven University of Technology, P/O Box 513   5600 MB Eindhoven, The Netherlands

problems with respect to the determination of material parameters in the constitutive equations and problems with respect to model validation. Because a constitutive equation has to be found for each separate component and for the interaction terms for the components the number of parameters to be determined is rather high. Moreover, it is difficult to make test specimen from most biological materials. Add to these problems the fact that materials are usually inhomogeneous and highly anisotropic and it is clear that standard tests to determine material parameters are difficult if not impossible to perform. In   section 2 a mixed numerical/experimental method will be described extending the possibilities of standard techniques which may even open the way to *in vivo* characterization of materials.

Another problem is model validation. Experiments on biological tissues are difficult for many reasons, like those mentioned above. Probably one of the more profound problems arises from the fact that the material has to be taken as it is. It is not possible to change material parameters in an experiment in a controlled way to separate certain effects due to these changes. If there is a misfit between model prediction and experiment there are to many possible errors to identify where the discrepancy comes from. For this reason an intermediate step with synthetic model materials can be worthwhile. In section 3 some results with charged hydrated materials are discussed. The paper ends with some conclusions.

3. DETERMINATION OF MATERIAL PARAMETERS.

When using standard methods to determine constitutive equations, material specimen with a well defined shape are subjected to boundary conditions that lead to known, often uniform, stress strain fields in a defined specimen. The reason for this is that boundary conditions can be converted to local stress and strains or strain rates inside the material.   Examples are uniaxial and biaxial tests. Most of these tests are standardized and very useful in many applications. However, when applied to complex materials sooner or later the limitations of standard test methods become

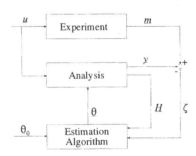

Fig. 1   Schematic representation of the mixed numerical experimental method.

clear for reasons already mentioned in the introduction. Oomens et al. [8] proposed a Mixed Numerical Experimental Technique (MNET) to characterize biological materials. It is an integral approach where measured field properties (displacements, strains, pressures) are compared to properties determined by means of a numerical model on the basis of estimated material parameters. The difference (the residual) is used to iteratively improve the parameter estimations, until the residual is minimised. Figure 1 shows the basic constituents of MNET: 1. An experiment on a more or less arbitrarily shaped specimen with typically a time-dependent (multi-axial) loading. 2. A (numerical) analysis of the experiment. Some constitutive model has to be proposed and implemented in the numerical model. The à priori unknown model parameters in the constitutive model are stored in the column $\underline{\theta}$. 3. An estimation algorithm which improves the parameters $\underline{\theta}$ by comparing the measured and calculated quantities.

Op den Camp [9] developed a recursive estimation algorithm for an efficient determination of material parameters in biphasic mixtures, which can be extended to more complicated mixtures. After the balance equations are transformed into a weak form by the weighted residual method and discretized in space they can be written as follows. The balance of mass for the mixture:

$$\underline{B}\,\dot{\underline{q}} + \underline{K}\,\underline{p} - \underline{s} = \underline{0} \tag{1}$$

The balance of momentum of the mixture:

$$\underline{A}\,\underline{q} - \underline{B}^T - \underline{t} = \underline{0} \tag{2}$$

In these equations the primary unknown variables are the nodal displacements of the solid $\underline{q}$ and the nodal fluid pressures $\underline{p}$. Matrices $\underline{A}$, $\underline{B}$ and $\underline{K}$ are all functions of the displacements. The columns $\underline{s}$ and $\underline{t}$ represent the boundary fluid flux and the boundary load respectively. At this moment already assumptions have been made for constitutive models for the components. The parameters in these equations are stored in the column $\underline{\theta}$. Let us assume that at each time step i a column $\underline{m}_i$ of measured properties comes available. The column contains displacements as well as pressures. A the same time an output $\underline{y}_i$ from the numerical model becomes available, based on previous estimates of the material parameters $\underline{\hat{\theta}}_{i-1}$ and the known output at time t_{i-1}. In the recursive algorithm at some discrete point in time a new estimate for the parameters and the output is determined by minimizing:

$$J_i = (\underline{\theta}-\underline{\hat{\theta}}_{i-1})^T\underline{W}_i\,(\underline{\theta}-\underline{\hat{\theta}}_{i-1})+(\underline{m}_i-\underline{y}_i)^T\underline{V}_i\,(\underline{m}_i-\underline{y}_i) \tag{3}$$

The second term on the right hand side minimizes the difference between the measured and calculated field properties (the residual). After successive estimation steps the estimates for the parameters become more reliable. This can be taken into account by means of the first term on the right hand side. There are different ways to chose the weighting matrices $\underline{W}_i$ and $\underline{V}_i$. Usually $\underline{V}_i$ is a constant matrix based on some error measure of the measurements. For matrix $\underline{W}_i$ updating schemes are derived [9]. Minimisation of J_i leads to:

$$\underline{W}_i\,(\underline{\theta}-\underline{\hat{\theta}}_{i-1})-\underline{H}_i\underline{V}_i\,(\underline{m}_i-\underline{y}_i) = \underline{0} \tag{4}$$

which is nonlinear in the parameters, but can be solved by standard methods. In this equation matrix $\underline{H}_i$ occurs, in literature on inverse methods known as the sensitivity matrix. This matrix can numerically be determined by means of a finite difference scheme , but it is also possible to use a method partly based on the system matrices that already are derived for the direct problem and partly on analytical derivatives of the constitutive equations to the parameters [9]. For this reason the method is called pseudo-analytical sensitivity analysis. This pseudo-analytical approach together with a recursive scheme leads to a very efficient method to determine material parameters.

Figure 2 gives a schematic view of a simulated "experiment" to show how the method works. An axi-symmetric specimen is clamped between two plates and loaded with a force that makes two cycles of an increases with a constant rate to a maximum values and then a linear decrease to zero. At points marked with D displacements are measured. At points marked with P pressures are measured. In the following it is assumed that the solid is an elastic, isotropic solid with a Young's modulus E and a Poisson's ratio υ. The interaction between solid and liquid is characterized by a strain dependent permeability. For the permeability a strain dependent equation is used:

$$K = K_0 \left(\frac{\det F - 1}{n^f_0} + 1 \right)^2 \tag{5}$$

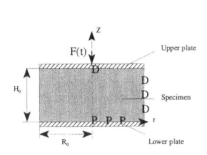

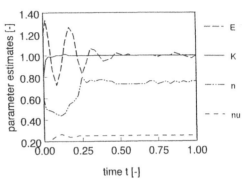

Fig.2   Schematic drawing of simulated "experiment"

Fig. 3   Parameters values as a function of iteration (=time) steps

of course these equations are too simple for most biological materials, but they can easily be extended. Parameters used to generate the "measurements" are $E = 1$, $\upsilon = 0.4$, $K_0 = 1$, $n_0^f = 0.8$. The CPU-time for a direct analysis for this problem is 900 sec on a Silicon Graphics workstation. After the observations are disturbed with a normally distributed noise with a standard deviation of 0.1 % parameters are estimated with initial estimates $\theta_0 = [0.8, 0.2, 1.2, 0.6]^T$. Figure 3 shows that all parameters converge to values very close to values used in the original analysis. Apparently in this test the permeability is difficult to estimate, but after a number of estimates also this parameter converges. The procedure to estimate all four parameters takes in this case 1200 s CPU time which is about 30 % more than one single analysis. It is clear that the recursive scheme together with the pseudo-analytical determination of the sensitivity matrix is very efficient. It is not difficult to apply this method to the more complicated tri-phasic mixtures. In this case, because the number of parameters is larger, the relative efficiency is even better.

4. VALIDATION BY MEANS OF A SYNTHETIC MODEL MATERIAL

The advantage of using model materials is evident. The material can be manufactured with reproducible and homogeneous properties, it does not degenerate as quickly as biological materials and parameters can be adjusted and controlled. In this way it is more likely that differences between numerical calculations and experimental results can be traced back either to problems in the experimental set-up or, more likely flaws and imperfections in the theoretical model and numerical solution process. Moreover, the materials can be used to develop new measuring techniques that can subsequently be applied to biological materials. We [10] developed a synthetic material to mimic intervertebral disk tissue, meaning that parameter values have comparable magnitude to those of the disk, but the model material can be used to simulate the behaviour of many charged hydrated tissues. The material is based on the concept of a charged hydrated solid, a liquid (in our case water) and a third

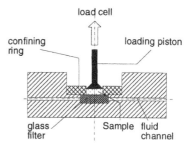

Fig. 4 Schematic view of the testing chamber for the swelling pressure measurements

component consisting of anions and cations. In a tri-phasic theory (Snijders et al [7]) de cations and anions are considered to be one single phase so electrical effects are neglected. The theory is comparable to the tri-phasic theory of Lai et al. [6] for articular cartilage. One of the major differences is that Snijders does not use the concept of a chemical expansion stress.

The material consists of an open-cell micro porous polyurethane (PU) foam containing a hydrophilic copolymer gel. The copolymer has been synthesised within the PU foam by means of *in situ* polymerization of acrylic acid (AA) and acryl amide (AAm) monomers A low concentration N,N'-methylenebisacrylamide was added to the initial solution to obtain a weakly cross linked copolymer gel. A solution of NaOH was used to neutralize the solution.

We were able to perform a number of different experiments on specimen originating from one homogenous batch allowing the direct measurement of most material parameters in the constitutive equations of the components. One example of such an experiment is illustrated in fig. 4. It is an experiment to determine the osmotic pressure of a specimen when it is in contact with salt solutions with different concentrations. The cylindrically shaped specimen were kept in a confining ring, which rested on a sintered

glass filter. The material was in contact with the NaCl bathing solution saturating the filter. By altering the ionic strength of the solution the material was loaded chemically. The sample was kept at fixed initial height by a 10 mm diameter loading piston, connected to a load cell. Depending on the external salt concentration the specimen wants to swell or shrink causing a varying load on the load cell. From a number of experiments with different salt concentrations a curve of swelling pressure against salt concentration can be constructed like in fig 5. From a curve-fit of these data with an expression for the Donnan osmotic pressure it is possible to derive a value for the fixed charge

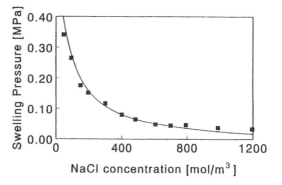

Fig. 5 Swelling pressure as a function of the concentration of the NaCl-solution [10].

density C_0^{fcd} = 0.31 $[10^3 \; mol/m^3]$ and the osmotic coefficient ϕ = 0.934 $\pm$ 0.007. By means of permeability measurements, the equilibrium values in a confined compression experiment and by measuring the wet and dry weight of the material it was possible to determine all material parameters, but two. The diffusion coefficient for anions and cations and a dimensionless parameter in the nonlinear permeability strain relationship had to be determined by fitting the transient behaviour in a confined compression experiment. This experiment is nearly the same as depicted in figure 4 except the fact that this time the piston with load cell is replaced by a cantilever loading arm which allows to apply a mechanical load and gives at the same time the tissue the opportunity to swell or shrink. The displacement of the cantilever is measured with an LVDT. The experimental protocol consisted of four stages including the conditioning stage in which the material was equilibrated against the mechanical and chemical load of 0.038 MPa and 0.6 M NaCl respectively (table I).

Table I Boundary conditions in the different phases of the confined swelling and compression experiment

phase	A	B	C	D
	conditioning	swelling	consolidation	control
mechanical load [MPa]	0.038	0.038	0.086	0.038
concentration NaCl [M]	0.6	0.2	0.2	0.2

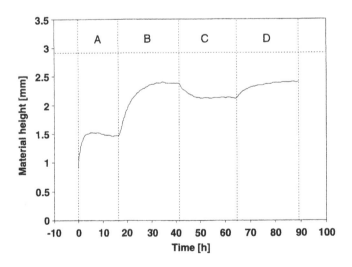

Fig. 6 Typical result of a confined swelling and compression experiment (A) Conditioning (B) swelling (C) consolidation (D) control [10].

Fig. 6 gives a typical response for such an experiment. In the conditioning phase the material swells to an equilibrium phase. Then the concentration of the bathing solution is changed so the specimen will swell again to a new equilibrium. When the mechanical load increases the material is compressed and it swells again after load is released. It was possible to fit both transients in phase B and C however at the cost of finding a diffusion coefficient of $5.95 * 10^{-9}$ m^2/s which is way to high, a phenomenon that was also reported by Snijders [7] for intervertebral disk tissue. The inconsistency is probably caused by the presence of electrical potentials due to an unequal distribution of mobile ions in the material, causing disturbances in the fluxes of liquid and ion. It is possible to account for these effects by considering anions and cations as different components in a 4-phasic model [11].

5. CONCLUSIONS AND DISCUSSION

Two techniques that extend standard methods for parameter estimation and model validation have been described. The first method is a Mixed Numerical Experimental Method, a subclass of the inverse methods. If there is one challenging area that demands the use of this method it must be the study of biological materials. On the one hand the problems to design "standard" tests for biological materials are tremendous if not impossible to solve. On the other hand biological materials can be approached with non-invasive measuring techniques that allow the experimentalist to look inside the tissue. Biological tissues consist mostly of water, proteins and sugars they can be penetrated with a large array of electromagnetic and acoustic waves. Echography, CT-scan and MRI-scan are already quite common and within reach for many research groups. Especially Magnetic Resonance Imaging offers great opportunities, from the point of view of MNET. MRI-tagging as a technique to measure displacements of discrete points inside the tissues or Magnetic Resonance Diffusion Tensor Imaging [12] to measure local fibre orientations inside tissues or transport of liquid (or other particles) while tissues are loaded. Internal pressure measurements can be performed nowadays with hardly any disturbance of the material by using very small micro-pipettes. So although application to biological materials raises several extra problems compared to technical materials the opportunity to look inside the material gives great potential to use MNET.
In the paper an efficient algorithm which may form the basis for a mixed technique is described and tested by means of a simulated experiment. Although developed for a bi-phasic mixture it can be straightforwardly extended to more complicated mixture.

The second topic of the paper is the use of a model material to validate constitutive models for mixture. In material science it is common practice to use model materials to test constitutive equations or develop measuring techniques. It is essential that the physics that is incorporated in the chosen constitutive model is working in the same way in the synthetic material. In this case the experiments with the model materials have clearly shown the lack of completeness of the tri-phasic theory when applied to charged hydrated materials. This would at least have taken much more time if only experiments on biological materials were performed.

5. ACKNOWLEDGEMENT
The research of dr.J.M.Huyghe has been made possible through a fellowship of the Royal Dutch Academy of Arts and Sciences.

6. REFERENCES

1.   Boer R. De, Highlights in the historical development of the porous media theory: towards a consistent macroscopic theory, Appl. Mech. Rev., 1996, vol. 4, 201 - 262.

2.   Mow, V., Kuei, S., Lai, W., Armstrong, C., Biphasic creep and stress relaxation of articular cartilage in compression: theory and experiments, J. Biomech. Engng., 1980, vol. 102, 73 - 84.

3.   Simon, B., Wu, J., Carlton, M., Evans, J., Kazarian, L., Structural models for human spinal motion segments based on a poroelastic view of the intervertebral disk., J.Biomech. Engng., 1985, vol. 107, 327 - 335.

4.   Oomens, C., Campen, D. van, Grootenboer, H. A Mixture Approach to the Mechanics of Skin. J.Biomechanics, 1987, vol. 20, 877 - 885.

5.   Huyghe, J., Arts, T., Campen, D. Van, Reneman, R., Porous medium finite element model of the beating left ventricle. Am. J. Physiol., 1992, vol. 262., H1256 - H1267.

6.   Lai, W.M., Hou, J., Mow, V., A triphasic theory for the swelling and deformation behviours of articular cartilage. J. Biomech. Engng., 1991, vol. 113, 245 - 258.

7.   Snijders H., Huyghe J.M., Janssen J.D., Triphasic finite element model for swelling porous media. Int.J.Num.Meth.in Fluids, 1995, vol. 20, 1039 - 1046.

8.   Oomens, C., Hendriks, M., Ratingen, M.v., Janssen, J., Kok, J. A Numerical-Experimental Method for a Mechanical Characterization of Biological Materials. Journal of Biomechanics, 1993, vol. 26, 617-621.

9.   Op den Camp, O., Identification algorithms for time-dependent materials. Ph.D.thesis Eindhoven University of Technology, Eindhoven, The Netherlands.

10. Oomens, C., Heus, H. De, Huyghe, J., Nelissen, L., Janssen, J. Validation of Triphasic Mixture Theory for a Mimic of Intervertebral Disk Tissue. Biomimetics, 1996, vol. 3, 4, 171 - 185.

11. Frijns, A.J. , Huyghe, J.M., Janssen, J.D. A validation of the quadriphasic mixture theory for intervertebral disc tissue, 1997, Int. J. Engng. Science (In Press).

12. Doorn, A. V., Bovendeeerd, P.H., Nicolay, K., Drost, H.R., Janssen, J.D. Determination of muscle fibre orientation using diffusion weighted MRI, European Journal of Morphology, 1996, vol. 34, 19 - 24.

FOUR COMPONENTS MIXTURE THEORY APPLIED TO SOFT BIOLOGICAL TISSUE

A.J.H. Frijns[1] , J.M. Huyghe[2] and J.D. Janssen[3]

1. ABSTRACT

The swelling and shrinking behaviour of soft biological tissues is described by a four components mixture theory. In this theory four components are distinguished: a charged solid, a fluid, cations and anions. By using balance equations, constitutive equations and equations of state, a set of coupled differential equations is derived. Because of the distinction between cations and anions, we are able to describe electrical phenomena like streaming potentials. A one-dimensional finite element implementation of this model is made by using a Galerkin method, an implicit time discretization and the Newton-Raphson iteration procedure. This implementation is used to simulate confined swelling and compression experiments. It appears that physically realistic values for the stiffness, permeability and diffusion coefficients are adequate to fit the experiments. We do not need a chemical expansion stress.

Keywords: porous media, swelling, streaming potential

[1]Ph.D. Student, Department of Mathematics and Computing Science, Eindhoven University of Technology, P.O. Box 513, 5600 MB Eindhoven, The Netherlands & Department of Movement Sciences, University of Maastricht, P.O. Box 616, 6200 MB Maastricht, The Netherlands

[2]Senior Researcher, Department of Mechanical Engineering, Eindhoven University of Technology, P.O. Box 513, 5600 MB Eindhoven, The Netherlands

[3]Professor, Department of Mechanical Engineering, Eindhoven University of Technology, P.O. Box 513, 5600 MB Eindhoven, The Netherlands & Department of Movement Sciences, University of Maastricht, P.O. Box 616, 6200 MB Maastricht, The Netherlands

2. INTRODUCTION

Soft biological tissues exhibit swelling and shrinking behaviour due to mechanical load-ings, and osmotic and electrical effects. Since biological tissues not only consist of water and solid, but also of small particles like ions, this swelling and shrinking behaviour can not be described by a biphasic mixture theory, because the biphasic theories are not equipped to model phenomena like Donnan osmose.

An example of the mechanical behaviour of the tissues is shown by confined swelling and compression experiments, performed at the University of Maastricht [1, 2]. In these ex-periments a cylindric sample is enclosed in an impermeable confining ring (figure 1). At

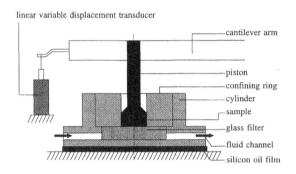

Figure 1: Schematic representation of the testing chamber.

the bottom of the sample is a glass filter through which a salt solution flows. The molarity of this salt solution can be changed during the experiments. At the top is an impermeable loading piston, by which a mechanical load can be applied to the sample. This piston can move in vertical direction. By measuring the piston displacement, the sample height dur-ing the experiments is known. Figure 2 shows a typical result of such an experiment. We use a sample made of the intervertebral disc of a large dog. In this experiment four dif-ferent periods are distinguished. In the conditioning period (0–5 [h]), from an unknown starting condition an equilibrium is reached. In this period, the mechanical load is 0.078 [MPa] and the external salt concentration is about 0.45 [mol l^{-1}]. In the next period, the swelling period (5–10 [h]), the external salt concentration is changed into 0.16 [mol l^{-1}]; the mechanical load is still the same. The tissue swells due to osmotic suction. In the con-solidation period (10–15 [h]), the mechanical load is increased to 0.195 [MPa], but the salt concentration remains the same. The tissue is compressed. In the last period, the control period (15–20 [h]), the same boundary conditions as in the conditioning period are pre-scribed.

We use a macrocontinuum model to describe the mechanical behaviour of the soft biolog-ical tissues: the mixture model. Lai et al. [3] model the behaviour by a triphasic mixture theory. Their theory incorporates the Donnan ion distribution and osmotic pressure theory for polyelectrolyte solutions. They distinguish three phases: the solid phase, the interstitial

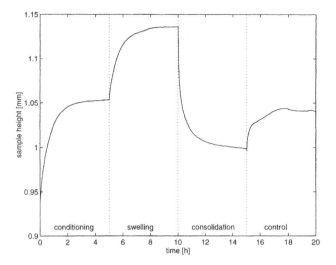

Figure 2: Result of a confined swelling and compression experiment.

fluid phase and the ionic phase. This last phase is miscible in the fluid phase. This triphasic theory unites the continuum biphasic theory and the macrocontinuum physicochemical theories for the mechanics of cartilaginous tissues. According to Lai et al. [3], the Donnan osmotic pressure alone is not sufficient to account for the tissue swelling, and they introduce the chemical expansion stress to contribute to the swelling behaviour of cartilage and to its mechanical stiffness in compression.

Snijders et al. [4] developed a similar triphasic mixture model. The difference between the two theories is that Snijders et al. do not define a chemical expansion stress, but only use Donnan theory to account for the swelling and that they neglect electrical potential gradients. Snijders et al. [4] developed a finite element model of this theory. Their model is confronted with one-dimensional confined swelling and compression experiments performed on intervertebral disc tissue. It appears that the change in sample height in the consolidation period (in which a mechanical load is applied to the sample), is well described by their triphasic mixture model. However, the change in the sample height due to chemical loadings (by changing the salt concentrations in the swelling period), can only be fitted by using diffusion coefficients, which are physically too large [1, 5, 6]. This phenomenon is caused by neglecting some electrical effects inside the sample such as streaming potentials, diffusion potentials, electro-osmosis and electro-phoresis. These effects are important when the fixed charge density is large as demonstrated by Gu et al. [7].

In order to describe these phenomena a four components mixture model is developed by Huyghe and Janssen [8]. This model can describe large deformations. In this model four components are distinguished: a charged solid (s), a fluid (f), cations $(+)$ and anions $(-)$. By distinguishing between cations and anions, these electrical phenomena can be modeled.

3. MODEL

In the four components mixture model, the material behaviour is described by a set of coupled equations: balance equations, constitutive equations and equations of state. First we
consider the balance equations. The material has to fulfill mass balance for each component:

$$\frac{\partial (n^\alpha \rho^\alpha)}{\partial t} + \nabla \cdot (n^\alpha \rho^\alpha \mathbf{v}^\alpha) = 0, \qquad \alpha = s, f, +, -. \tag{1}$$

In this equation n^α, ρ^α and $\mathbf{v}^\alpha$ are respectively the volume fraction, the density and the
velocity of component α. Next, the material has also to fulfill the momentum equation:

$$\nabla \cdot \boldsymbol{\sigma}^{eff} - \nabla p = \mathbf{0}. \tag{2}$$

In this equation $\boldsymbol{\sigma}^{eff}$ is the effective stress tensor and p is the hydrodynamic fluid pressure.
In this momentum equation the inertial terms and the body forces are neglected.

We also need some constitutive equations. The first equation is a Hooke's law:

$$\boldsymbol{\sigma}^{eff} = J^{-1}\mathbf{F} \cdot (2\mu\mathbf{E} + \lambda tr(\mathbf{E})\mathbf{I}) \cdot \mathbf{F}^c. \tag{3}$$

In this equation J is the relative volume change, $\mathbf{F}$ is the deformation tensor, μ and λ are
Lamé 's constants and $\mathbf{E}$ is the Green-Lagrange strain.

The second constitutive equation is a combination of Fick's law and the ion migration (the
Nernst-Planck equations)

$$c^\alpha n^f (\mathbf{v}^\alpha - \mathbf{v}^f) = -\frac{c^\alpha}{RT}\mathbf{F} \cdot \mathbf{D}^\alpha \cdot \mathbf{F}^c \cdot \nabla \mu^\alpha, \qquad \alpha = +, -, \tag{4}$$

where c^α, $\mathbf{D}^\alpha$ and μ^α are respectively the concentration per fluid volume, the diffusion tensor and the electrochemical potential of component α, R is the universal gas constant and
T is the absolute temperature. The objectivity is ensured by the deformation tensor $\mathbf{F}$.

The last constitutive equation we need, is an extended Darcy's equation

$$n^f (\mathbf{v}^f - \mathbf{v}^s) = -\mathbf{F} \cdot \mathbf{K} \cdot \mathbf{F}^c \cdot \left(\nabla(p - \pi) + c^+ \nabla \mu^+ + c^- \nabla \mu^- \right), \tag{5}$$

where $\mathbf{K}$ is the permeability tensor, π is the osmotic pressure. Here also, the objectivity is
ensured by the deformation tensor.

We define the electrochemical potentials for the ions by

$$\mu^\alpha = \mu_0^\alpha(T) + RT ln \left(\overline{V}^f \gamma^\alpha c^\alpha \right) + z^\alpha F\xi, \qquad \alpha = +, -, \tag{6}$$

where $\overline{V}^f$ is the molar fluid volume, γ^α is an activity coefficient of ion α, z^α is the valence
of ion α, F is Faraday's constant and ξ is the electrical potential. The subscript 0 denotes
a reference value.

We define the osmotic pressure by

$$\pi = \pi_0(T) + RT\phi(c^+ + c^-). \tag{7}$$

There is also electro-neutrality inside the material:

$$z^+ c^+ + z^- c^- + z^{fc} c^{fc} = 0. \tag{8}$$

The superscript fc denotes fixed charges.

Further, we assume complete saturation:

$$n^s + n^f + n^+ + n^- = 1. \tag{9}$$

When we neglect the cations and anions volume fractions with respect to the solid and fluid volume fractions, the volume fraction for the fluid n^f and the fixed charge density depend on the deformation of the tissue according to

$$n^f = 1 - \left(\frac{1 - n_0^f}{J} \right), \tag{10}$$

and

$$c^{fc} = \left(\frac{n_0^f}{n_0^f - 1 + J} \right) c_0^{fc}. \tag{11}$$

The subscript 0 denotes a reference state and J is the relative volume change.

4. FINITE ELEMENT MODEL

We assume that the volume fractions of the ions can be neglected with respect to the relative volume fractions of the solid and the fluid ($n^s + n^f = 1$). Thus, we can replace the mass balances for the fluid and the solid by one equation

$$\nabla \cdot \mathbf{v}^s + \nabla \cdot (n^f (\mathbf{v}^f - \mathbf{v}^s)) = 0. \tag{12}$$

Next we substitute the extended Darcy's law and the Nernst-Planck equations into the mass balances and we substitute Hooke's law into the momentum equation. Then we write the resulting coupled differential equations in a form that is suitable for a finite element solution procedure. A simple one-dimensional finite element implementation is made by using a Galerkin method. We use an implicit time discretization. The resulting equations are solved by the Newton-Raphson iteration procedure. This implementation can be improved with respect to the stability and efficiency. It also can be extended to more dimensions, which is a goal of our future research.

5. RESULTS

We perform some simulations with a one-dimensional finite element implementation. We use the swelling and the consolidation periods to validate the four components mixture model. For simplicity, we assume the diffusion coefficients D^+ and D^-, the permeability K and the stiffness H to be deformation-independent.

We need some boundary conditions. The boundary conditions on top of the sample are

$$(\sigma^{eff} - p\mathbf{I}) \cdot \mathbf{n} = p_l\mathbf{n},$$
$$(\mathbf{v}^f - \mathbf{v}^s) \cdot \mathbf{n} = 0, \tag{13}$$
$$(\mathbf{v}^\alpha - \mathbf{v}^s) \cdot \mathbf{n} = 0, \qquad \alpha = +, -,$$

where p_l is the external mechanical load.

The boundary conditions at the side of the glass filter are

$$\mathbf{u} = \mathbf{0},$$
$$p - \pi = -2\overline{\phi}RT\overline{c}, \tag{14}$$
$$\mu^\alpha = RT \, ln(\overline{V}^f \overline{\gamma}^\alpha \overline{c}), \qquad \alpha = +, -,$$

where $\mathbf{u}$ are the displacements. The overlined symbols denote the values outside the sample.

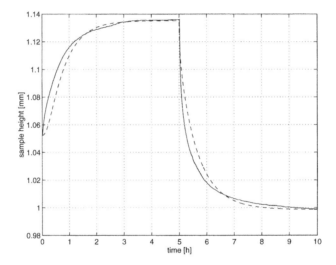

Figure 3: Computed data (dashed line) and experimental data of the swelling and consolidation periods of the experiment of figure 2, in which the stiffness $H = 1.03 \cdot 10^6$ [N m^{-2}], the permeability $K = 2.20 \cdot 10^{-16}$ [m^4 N^{-1} s^{-1}] and the diffusion coefficients $D^+ = 5.40 \cdot 10^{-10}$ [m^2 s^{-1}] and $D^- = 8.30 \cdot 10^{-10}$ [m^2 s^{-1}]. At t = 0 [h], the porosity $n_0^f = 0.786$ and the fixed charged density $c_0^{fc} = 0.12$ [mol l^{-1}].

In figure 3 a typical example of a fitted experiment is given. Although we use constant parameters, the fitting of this experiment is reasonable. From the fittings of four other experiments, it appears that physically realistic diffusion coefficients, permeabilities and stiffnesses can be used to get reasonable results (table 1) [2]. The range of the fitting values is caused by the different properties of the samples.

Table 1: Values of the parameters obtained by fitting of the experimental data.

parameter		fitting values		literature values		references
H	$[\text{N m}^{-2}]$	0.75-1.08	$\cdot 10^6$	0.36-0.96	$\cdot 10^6$	[9]
K	$[\text{m}^4\,\text{N}^{-1}\,\text{s}^{-1}]$	0.15-0.40	$\cdot 10^{-15}$	0.22-0.40	$\cdot 10^{-15}$	[10, 11]
D^+	$[\text{m}^2\,\text{s}^{-1}]$	0.45-0.54	$\cdot 10^{-9}$	0.48	$\cdot 10^{-9}$	[12]
D^-	$[\text{m}^2\,\text{s}^{-1}]$	0.70-0.85	$\cdot 10^{-9}$	0.78	$\cdot 10^{-9}$	[12]

Since we distinguish between cations and anions, we are able to simulate electrical potential differences (figure 4). In this figure we see a Donnan potential at the bottom of the
sample (x = 0) and a streaming potential inside the sample (x > 0). This streaming potential will go to zero as an equilibrium is reached.

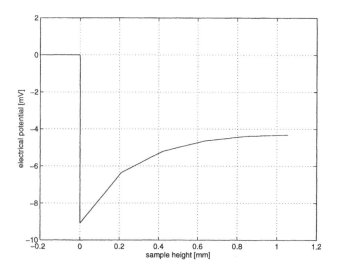

Figure 4: Electrical potential at the beginning of the swelling period (t = 180 [s]) relatively
to the potential of the electrolyte solution in the glass filter.

6. CONCLUSIONS

By using a four components mixture model, we do not need a chemical expansion stress in
order to describe the tissue swelling. We fit the one-dimensional swelling and compression
experiments using physically realistic values of the material parameters and we are able to
describe electrical phenomena, like streaming potentials.

7. ACKNOWLEDGEMENT

The research of Dr. J.M. Huyghe has been made possible through a fellowship of the Royal
Netherlands Academy of Arts and Sciences.

8. REFERENCES

[1] G. B. Houben. *Swelling and compression of intervertebral disc tissue: model and experiment.* PhD thesis, University of Maastricht, Maastricht, 1996.

[2] A. J. H. Frijns, J. M. Huyghe, and J. D. Janssen. A validation of the quadriphasic mixture theory for intervertebral disc tissue. *International Journal of Engineering Science.* in press.

[3] W. M. Lai, J. S. Hou, and V. C. Mow. A triphasic theory for the swelling and deformation behaviors of articular cartilage. *ASME Journal of Biomechanical Engineering,* 113:245–258, 1991.

[4] H. Snijders, J. M. Huyghe, and J. D. Janssen. Triphasic finite element model for swelling porous media. *International Journal for Numerical Methods in Fluids,* 20:1039–1046, 1995.

[5] H. J. de Heus. *Verification of mathematical models describing soft charged hydrated tissue behaviour.* PhD thesis, Eindhoven University of Technology, Eindhoven, 1994.

[6] J. M. A. Snijders. *The triphasic mechanics of the intervertebral disc: a theoretical, numerical and experimental analysis.* PhD thesis, University of Limburg, Maastricht, 1994.

[7] W. Y. Gu, W. M. Lai, and V. C. Mow. Transport of fluid and ions through a porous-permeable charged-hydrated tissue, and streaming potential data on normal bovine articular cartilage. *Journal of Biomechanics,* 26:709–723, 1993.

[8] J. M. Huyghe and J. D. Janssen. Quadriphasic mechanics of swelling incompressible porous media. *International Journal of Engineering Science.* in press.

[9] M. R. Drost, P. Willems, H. Snijders, J. M. Huyghe, and J. D. Janssen. Confined compression of canine annulus fibrosus under chemical and mechanical loading. *ASME Journal of Biomechanical Engineering,* 117:390–396, 1995.

[10] B. A. Best, F. Guilak, M. Weidenbaum, and V. C. Mow. Compressive stiffness and permeability of intervertebral disc tissues: Variation with radial position, region and level. In *Proceedings Winter Annual Meeting ASME,* pages 73–74, San Fransisco, 1989.

[11] F. Guilak, W. B. Zhu, M. Weidenbaum, B. A. Best, and V. C. Mow. Compressive material properties of human anulus fibrosus. In *Abstracts First World Congres Biomechanics, Vol. II,* page 41, San Diego, 1990.

[12] A. Maroudas. Nutrition and metabolism of the intervertebral disc. In P. Gosh, editor, *The biology of the intervertebral disc,* pages 1–38, Florida, 1988. CRC Press Inc.

EXPERIMENTAL VALIDATION OF A POROHYPERELASTIC
FINITE ELEMENT MODEL OF THE ANNULUS FIBROSUS

N.A. Duncan[1] and J.C. Lotz[2]

1. ABSTRACT

The material properties of the annulus used in current poroelastic finite element models of the motion segment differ considerably from recent tissue level experiments. We have developed unique experiments to determine the axial tension-compression response of the annulus during finite deformation at varied strain rates, and to measure both the fluid pressure and solid stress in the center of the annulus. The axial tension-compression response, as well as the fluid and solid stresses, were highly nonlinear and significantly affected by strain rate in compression. A porohyperelastic finite element model of the annulus was then developed and validated using this experimental data.

2. INTRODUCTION

Poroelastic finite element models (Poro-FEMs) of the lumbar motion segment have been used to predict annular stresses and strains in attempts to elucidate the pathomechanics of disc degeneration and herniation [1-4]. However, these models have primarily been validated at the motion segment level using load-displacement tests, and remain poorly validated at the tissue level. Since the gross deformation of a motion segment is dependent on several tissues within this complex composite structure, models developed solely using this technique may not accurately predict stress and strain within specific regions of the disc. Recent measurements of the material properties of the annulus fibrosus at the tissue level [5-6] suggests that the properties assumed for the annular matrix in current Poro-FEMs may differ considerably from experiment. Further, a key feature of these models is the prediction of fluid pressure

Keywords: Spine, Porohyperelasticity, Finite Element Models, Validation, Experiment

---
[1] Post-Doctoral Fellow,   [2] Assistant Professor & Director, Bioengineering Laboratory, Department of Orthopaedic Surgery, University of California at San Francisco, USA   94143-0514

within the hydrated intervertebral disc. While the total effective stress in the nucleus and annulus has been measured [7], distinction between the fluid pressure and the solid stress was not possible. Without direct measurement of the fluid pressure or the solid stress, the accuracy of these model predictions for tissue stress remains unknown. Our overall goal was to develop an experimentally validated porohyperelastic FEM (Porohyper-FEM) of the annulus fibrosus. Specifically, our aims were to: 1) measure the axial tension and compression stress-strain response of the annulus during finite deformation and at varied strain rates; 2) measure the pore pressure and solid stress in the annulus; 3) evaluate if current Poro-FEMs can accurately predict this measured behaviour, and 4) validate the predictions of a new Porohyper-FEM of the annulus.

3. MATERIALS AND METHODS

3.1 Experiment: Axial Tension-Compression

Nine fresh-frozen (-20 °C) human lumbar spines were harvested at autopsy. Only grade I or II discs were selected [8]. Motion segments were sectioned on a diamond saw (Exact) into rectangular segments of bone - annulus - bone (b-a-b) (Fig. 1). In total, thirty-five specimens (10 anterior; 25 lateral) were prepared from 20 different discs (10 L_{2-3}; 13 L_{3-4}; 12 L_{4-5}). The b-a-b segments were embedded with PMMA into fixation cups, mounted on a hydraulic materials test system (MTS Bionox) in line with a load cell (Sensotec), submerged in 0.15M NaCl and allowed to equilibrate. Next, these specimens were cycled at 0.5 Hz to ±10 % axial strain while adjusting the height of the disc till the position of zero axial stress was found and remained stable for 45 minutes. At this position, zero axial strain was also defined and the initial height (H) of the specimen measured. The dimensions of the annulus at the bone-annulus interface were also measured and used to determine the cross-sectional area (A). Axial stress and strain were calculated from the load-displacement data based on the initial dimensions (A and H). The specimens were divided into two groups and each specimen was tested sequentially at varied strain rates in axial tension and compression up to ± 25 % axial strain. Group I specimens were tested at a physiologic strain rate of 12.8 %/s and three slower rates of 3.2, 0.8 and 0.2 %/s, where each rate differed by a factor of four. Group II specimens were tested at a very slow rate of 0.003125 %/s and three faster rates of 0.0125, 0.05 and 0.2 %/s. Specimens in each group were tested at an overlap rate of 0.2 %/s to ensure that intergroup comparisons could be made. The specimens were allowed to stress relax at zero strain for 25 minutes between loading rates and loading direction. The order of strain rates was randomized to ensure that there was no cummulative effects due to the repeated experimental design.

3.2 Experiment: Pore Pressure and Solid Stress

Eleven specimens of bone - annulus - bone were prepared from the anterior and lateral regions of eight spines using the technique described above. A pressure gauge tipped catheter was inserted into the center of the annulus (Galetec; 1 mm dia.[7]). Because the annular specimens were unconfined, they developed tensile stresses and strains in the transverse plane during axial compression. Thus, if the pressure gauge was oriented transversely it only measured fluid pressure, as the gauge by design could not measure the tensile solid stresses (Fig. 2). In contrast, when the gauge was oriented axially, both the fluid pressure and axial compressive stress in the solid phase were measured. Consequently, the difference between the axial and transverse measurements (ΔP) should represent the axial stress in the solid phase. The specimens were mounted on the test system (MTS Bionix) as explained above and each specimen was compressed to -25% axial strain, with the gauge oriented first axially and then transversely, at a very slow strain rate of 0.003125%/s.

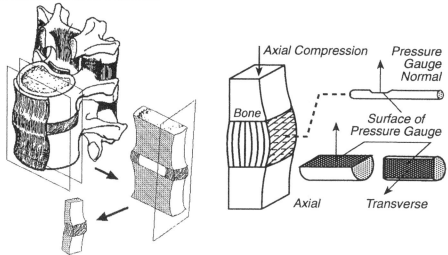

Fig. 1 Sectioning of the bone - annulus - bone specimens from the motion segment.

Fig. 2 Insertion of the pressure gauge into the bone - annulus - bone specimens.

3.3 Model

A Porohyper-FEM (ABAQUS V5.6) was developed with the average dimensions of the rectangular annular specimens (H = 10.5 mm; A = 8 x 11 mm²). Invoking symmetry, only 1/4 of this geometry was actually modeled with 204 poro brick elements (20 node) for the interlamellar matrix. The collagen fibers of the lamellae were represented by 30 layers of REBARS oriented at ±30° to the transverse plane with a collagen content of 16%. (Fig. 3), and established the anisotropic behavior of the annulus. The collagen properties were nonlinear with no compressive stiffness [4]. The interlamellar matrix was modeled as an isotropic, compressible Mooney-Rivlin hyperelastic material (N=2), with a strain energy function given by

$$U = \sum_{i+j=1}^{N} C_{ij}(\bar{I}_1 - 3)^i (\bar{I}_2 - 3)^j + \sum_{i=1}^{N} \frac{(J-1)^{2i}}{D_i} \quad , \quad (1)$$

where $\bar{I}_1, \bar{I}_2$ are the deviatoric strain invariants and J is the volume strain. The coefficients (C_{ij} , D_i) of the strain energy function were determined through finite element simulation of our radial tension experiments [5]. Comparisons were also made to data from confined compression experiments found in the literature [6]. The permeability was exponentially nonlinear [6] and implemented as a function of the porosity. The initial permeability was 2.5E-15 m⁴/N-s. No flow was permitted through the bone-annulus interface, whereas free fluid flow was permitted through the exposed annular surfaces. The model was loaded at the two strain rates (12.8 & 0.003125 %/s) using finite deformation.

4. RESULTS

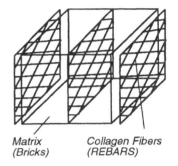

Matrix   Collagen Fibers
(Bricks)   (REBARS)

The Porohyper-FEM stress-strain predictions were matched exactly to the radial tension experiments in order to determine the coefficients ($C_{01}=C_{10}=0.01$, $C_{02}=C_{11}=C_{20}=0.1$, $D_1=D_2=35$) of the strain energy function (Fig. 4). Reasonable

Fig. 3 Basic composite unit of the annular Porohyper-FEM.

agreement was also found in comparison to the confined compression experiment using the same coefficients, however, the stress predictions of the model were lower (Fig. 5).

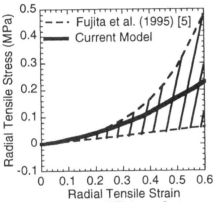

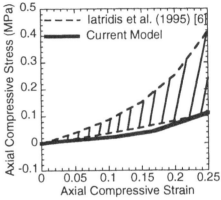

Fig. 4 Porohyper-FEM predictions vs radial tension experiments [5] (± 1 std. dev.) at equilibrium.

Fig. 5 Porohyper-FEM predictions vs confined compression experiments [6] (± 1 std. dev.) at equilibrium.

The axial tension-compression response was highly nonlinear and the stresses increased nearly eight fold at -25% axial compression when the strain rate was varied from very slow to physiological levels (Fig. 6). Excellant agreement was found between the current predictions of the Porohyper-FEM and the axial tension-compression experiment for both the physiological and very slow strain rates (Fig. 6). The predictions were smaller than experiment in tension, but still within one standard deviation. All current Poro-FEMs [1-4] were unable to accurately predict the axial tension-compression experiment. For example, the model of Argoubi and Shirazi-Adl (1996) [4] overpredicted the stresses in tension and compression, and did not predict the significant stiffening response in compression with increased strain rate (Fig. 7).

Both the fluid pressure and solid stress increased nonlinearly during axial compression and were of similar magnitude even at the very slow strain rate (Figs. 8-9). The current Porohyper-FEM successfully predicted both the fluid pressure (Fig. 8) and the solid stress (Fig. 9) in the annulus.

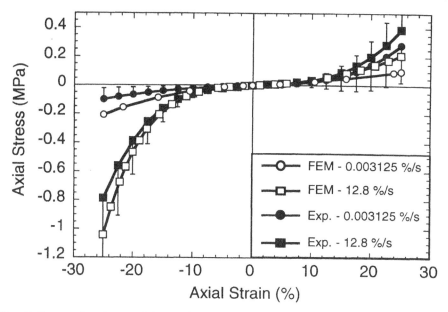

Fig. 6 Current Porohyper-FEM predictions vs experiment for the stress-strain response during axial tension and compression at very slow (0.003125 %/s) and physiological strain rates (12.8 %/s).

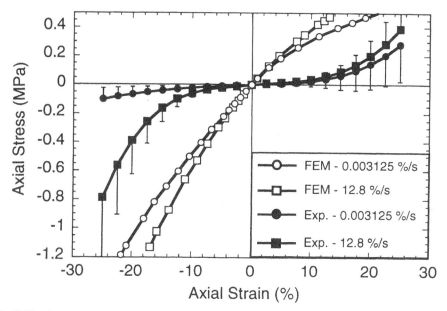

Fig. 7 The Poro-FEM predictions of Argoubi and Shirazi-Adl (1996) [4] vs experiment for the stress-strain response during axial tension and compression at very slow (0.003125 %/s) and physiological strain rates (12.8 %/s).

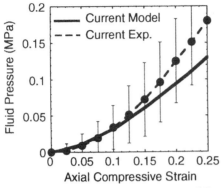
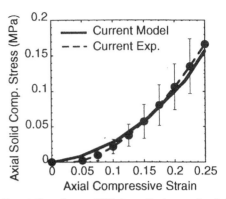

Fig. 8 Porohyper-FEM predictions of fluid
pressure vs experimental measurement in
the center of the annular tissue during
ramp loading at 0.003125%/s.

Fig. 9 Porohyper-FEM predictions of solid
stress vs experimental measurement in the
center of the annular tissue during ramp
loading at 0.003125%/s.

5. DISCUSSION

We sought to develop novel experimental protocols in order to validate a new
porohyperelastic FEM of the annulus fibrosus. Our model successfully predicted the
axial tension-compression response of the annulus in finite deformation over a large
range of strain rates, including the dramatic nonlinear stiffening in compression at a
physiologic rate. Further, our model predicted both the nonlinear increase in fluid
pressure and solid stress within the annulus. To our knowledge, this represents the
first direct measurement of both the fluid pressure and solid stress within a hydrated
soft tissue, and represents a unique validation of a Poro-FEM.

We established the material coefficients of a porohyperelastic model of the annulus
through FEM simulation of our radial tension experiments [5]. However, the use of
these coefficients in a separate FEM of confined compression experiments [6] resulted
in low stress predictions. The primary reason for this difference was the stiffer
behavior exhibited by the annulus in compression, compared with tension, in these
uniaxial tests. Though in both these experiments the load is carried predominantly by
the interlamellar matrix (with minimal tensile straining of the collagen), the behavior did
not appear to be isotropic. However, the interlamellar matrix used in our model was
isotropic, and hence, not capable of capturing this asymmetric behavior. In the
compressive range of our axial tension-compression experiment, the annulus is
deformed similar to an unconfined compression experiment, and the middle of the
annulus is in radial tension. Therefore, this may explain why our model successfully
predicted the axial tension-compression experiment using coefficents established from
the radial tension experiment. On the other hand, if the material coefficients were
matched to the confined compression experiment, then the model predictions were too
stiff for both the radial tension and the axial tension-compression experiment.

The axial response of the annulus differed greatly in tension and compression, especially at physiological strain rates. At larger values of axial tensile strain, the model predictions for stress tended to be smaller, even though a nonlinear stiffening material with finite deformation was used for the collagen. This discrepancy may be due to matrix-fiber interactions which were not modelled, but which have been reported [9]. In our bone-annulus-bone specimens, the tensile role of the collagen is reduced due to the absence of a circumferential hoop strain, as in an intact disc. Consequently, in our experiment the collagen is essentially unstrained and the nonlinear axial tensile stiffness is likely related to both the interlamellar matrix and fibrous interactions. It is not clear whether the current small underprediction of the axial tensile response will be significant once this model is extended to an intact disc and the collagen fibers constrained.

The magnitudes of the fluid pressure and the solid stress were very similar even though the strain rate was very slow. This demonstrates the long time constant of the annulus and indicates that an impractically slow strain rate would be required in order to only measure the elastic response of the annulus. Consequently, these results suggest that at any reasonable loading rate the fluid pressures will be significant, and a poroelastic or biphasic representation of the tissue is essential in order to correctly understand the observed behavior, and accurately predict the solid stresses.

The current axial tension-compression experiment also demonstrated the considerable deficiencies of current Poro-FEM's. The inability of these models to predict this behavior was due primarily to the use of incorrect material properties for the interlamellar matrix, as well as only using an poroelastic material representation, as opposed to porohyperelastic. For instance, in our model we used the same representation for the collagen fibers as reported by Argoubi and Shirazi-Adl (1996) [4]. Therefore, the differences in the model predictions must be due to our porohyperelastic material model. That they needed material properties of the annulus which are inconsistent with experimental data is likely due to adjusting the properties to match validation experiments for the whole motion segment. Due to the complexity of the motion segment, validation of the gross response does not guarantee validation of the response in the respective constituents. Consequently, given these demonstrated discrepancies in the predicted stress-strain response of the isolated annulus, it may be necessary to reevaluate the predictions of tissue failure from these earlier models.

We have successfuly demonstrated that the axial tension-compression response of the annulus at varied strain rates, as well as the fluid pressure and solid stress within the annulus, can be predicted with a Porohyper-FEM. However, under other loading conditions, such as shear or biaxial, it is possible that the current model may not be as accurate. Further, the coefficients used in this model may not be unique and other porohyperelastic models could be equally accurate. In the future we plan to eliminate the REBARs by combining the continuum composite approach of Spencer (1984) [10] with an exponential strain energy function similar to that used by Almeida et al (1995) [11] to model the annulus as a continuum reinforced with a family of two extensible fibers with asymmetric behavior along preferred material directions oriented with the collagen fibers. It is hoped that this theoretical material approach combined with our validation experiments will lead to a versatile and better validated model of the annulus.

6. REFERENCES

1. Simon, B.R., Wu, J.S.S., Carlton, M.W., Evans, J.H. and Kazarian, L.E., Structural models for human motion segments based on a poroelastic view of the intervertebral disc, J. Biomech. Eng., 1985, Vol. 107, 327-335.
2. Laible, J.P., Pflaster, D.S., Krag, M.H., Simon, B.R. and Haugh, L.D., A poroelastic-swelling finite element model with application to the intervertebral disc, Spine, 1993, Vol. 18, 659-670.
3. Meroi, E.A., Natali, A.N. and Schrefler, B.A., A poroelastic approach to the analysis of spinal motion segment, Comp. Meth. Biomech. Biomed. Eng., 1994, Eds.: Middleton, J., Jones, M.L. and Pande, G.N, 325-338.
4. Argoubi, M. and Shirazi-Adl, A., Poroelastic creep response analysis of a lumbar motion segment in compression, J. Biomech., 1996, Vol. 29, 1331-1339.
5. Fujita, Y., Lotz, J.C. and Duncan, N.A., Site and grade specific nonlinear tensile properties of the lumbar annulus fibrosus, 2nd Combined ORS, 1995, 132.
6. Iatridis, J.C., Setton, L.A., Foster, R.J., Rawlins, B.A., Weidenbaum, M. and Mow, V.C., Human annulus fibrosus behaves nonlinearly and isotropically in finite deformation confined compression, Summer ASME BED-29, 1995, 251-252.
7. McNally, D.S. and Adams, M.A., Internal intervertebral stress mechanics as revealed by stress profilometry, Spine, 1992, Vol. 17, 66-73.
8. Thompson, J.P., Pearce, R.H., Schechter, M.T., Adams, M.E., Tsang, I.K.Y. and Bishop, P.B., Preliminary evaluation of a scheme for grading the gross morphology of the human intervertebral disc, Spine, 1990, Vol. 15, 411-415.
9. Adams, M.A. and Green, T.P., Tensile properties of the annulus fibrosus: I. The contribution of fibre-matrix interactions to tensile stiffness and strength, Eur. Spine J., 1993, Vol. 2, 203-208.
10. Spencer, A.J.M., Continuum Theory of the Mechanics of Fibre-Reinforced Composites, 1984, Springer-Verlag, New York, 1-22.
11. Almeida, E.S. Spilker, R.L. and Holmes M.K., A transversely isotropic constitutive law for the solid matrix of articular cartilage, Summer ASME BED-29, 1995, 161-162.

7. ACKNOWLEDGEMENTS

We would like to acknowledge the extensive help and technical expertise provided by F.A. Ashford in conducting the experiments, as well as the financial assistance of the Department of Orthopaedic Surgery, UCSF.

MODELLING OF BRAIN TISSUE MECHANICAL PROPERTIES: BI–PHASIC VERSUS SINGLE–PHASE APPROACH

K. Miller[1] and K. Chinzei[2]

1. ABSTRACT

Recent developments in Robot–Aided Surgery — in particular, the emergence of automatic surgical tools and robots — as well as advances in Virtual Reality techniques, call for closer examination of the mechanical properties of brain tissue. The ultimate goal of our research is development of corresponding, realistic mathematical models. The paper discusses two candidates for tissue models: standard, non–linear, bi–phasic and single–phase, non–linear, viscoelastic.

The mechanical behaviour of brain tissue is highly non–linear. The stress–strain curves are concave upward containing no linear portion from which a meaningful elastic modulus might be determined. The tissue response stiffens as the loading speed increases, indicating a strong stress–strain rate dependence.

The standard methods of modeling tissue as a bi–phasic continuum face serious problems: strong stress–strain rate dependence can not be easily explained. According to our experiments, for brain tissue the stresses under fast loading can be six times higher than those under slow loading. Therefore, the use of the single–phase model is recommended. The non–linear, viscoelastic model, based on strain energy function with time dependent coefficients has been developed. The material constants for the brain tissue have been evaluated. Agreement between the proposed theoretical model and experiment is good for compression levels reaching 30% and for loading velocities varying over five orders of magnitude. One advantage of the proposed constitutive model is that it is not difficult to be employed in larger scale finite element computations.

Keywords: Brain tissue, Mechanical Properties, Mathematical modelling

[1] Lecturer, Department of Mechanical and Materials Engineering, University of Western Australia, Nedlands/Perth, WA 6907, Australia, email: kmiller@mech.uwa.edu.au
[2] Researcher, Biomechanics Division, Mechanical Engineering Laboratory, MITI, Namiki 1-2, Tsukuba, Ibaraki 305, Japan, email: chin@mel.go.jp

2. INTRODUCTION

Mechanical properties of living tissues form a central subject in Biomechanics. The properties of the musculo-skeletal system, skin, lungs, blood and blood vessels have attracted much attention. The properties of "very" soft tissues, which do not bear mechanical loads, such as brain, liver, kidney, etc., have not been so thoroughly investigated.

However, recent developments in robotics technology, in particular the emergence of automatic surgical tools and robots [1] as well as advances in virtual reality techniques [2], call for closer examination of the mechanical properties of these tissues. This research was initially motivated by the need to design a Nuclear Magnetic Resonance Image guided surgical robot. NMRI can provide rich information of tissue deformation but currently tens of seconds are required to produce a set of images. The plausible method of dealing with these delays is the prediction of the deformation based on the model [3, 4].

The appropriate "very" soft tissue models are required to equip the virtual reality surgeon training and operation planing systems with force and tactile feedback capabilities.

Knowledge of the mechanical properties of soft tissues and ultimately of their mathematical models is also required for *registration*: matching images of different modality, such as MRI and Single Photon Emission Computed Tomography (SPECT), defining relations between coordinate systems (eg., between a coordinate system associated with imaging equipment and those of robotic tools in an operating room), segmentation of reference features and defining disparity or similarity functions between extracted features [5]. Registration is a key technique for the computer–integrated surgery. Rigid tissue registration is now well–established. Registration of soft tissues is much more difficult because it requires the knowledge about local deformations. Here comes the place for accurate models of tissue deformation behaviour.

The reported experimental data on the mechanical properties of brain tissue, especially concerning stress–strain rate dependence at low strain–rates, are limited. Ommaya described mammalian brain as a "soft, yielding structure, not as stiff as a gel nor as plastic as a paste" [6]. In [7, 8] attempts to find elastic parameters of brain tissue by measuring induced changes in intra–cranial pressure are described. The experimental data, which might be used to determine constitutive relations for brain tissue at higher strain rates, was reported in [9, 10]. Based on these experimental results, non–linear constitutive relations for human brain tissue were proposed [11, 12]. However both the experimental results and theoretical investigations concentrated on rapid loading conditions resulting in high strain rates, typical for accident and injury modelling, and can not be applied in surgical procedure simulations.

2. UNCONFINED COMPRESSION EXPERIMENT OF SWINE BRAIN TISSUE

We decided to perform unconfined uniaxial compression experiments. An alternative was a confined compression experiment, used for example by Mow and coworkers [13] to validate bi–phasic models of soft cartilage tissues.

The more detailed description of the *in-vitro* unconfined compression of fresh swine brain can be found in [14, 15]. Here we would like to summarise the results only.

Cylindrical samples of diameter ~30 mm and height ~13 mm were cut. Four samples were taken from the frontal and posterial portions of the Sylvian fissure of each hemisphere for each swine brain. The ventricle surface and the arachnoid membrane formed the top and bottom faces of the sample cylinder. Thus the arachnoid membrane and the structure of the sulci remained as parts of each specimen.

Uniaxial unconfined compression of swine brain tissue was performed in a testing stand, shown in Fig. 1.

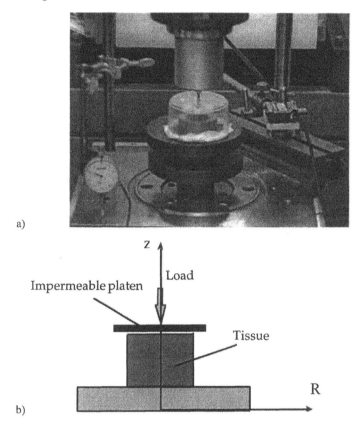

Fig. 1: Experiment set-up: a) general view, b) layout with coordinate axes

The main testing apparatus was a UTM-10T (Orientec Co.) tensile stress machine. The vertical displacement (along z–axis in Fig. 1b) was measured by a micrometer with analog output. The experiment was documented by automatically taking CCD camera images.

In the paper we will discuss the results obtained during the loading phase, for three loading velocities:

- fast: 500 mm min^{-1} — the fastest loading speed possible with our equipment, corresponding to the strain rate of about 0.64 s^{-1},
- medium: 5 mm min^{-1}, corresponding to the strain rate of about 0.64×10^{-2} s^{-1}, and
- slow: 0.005 mm min^{-1}, corresponding to the strain rate of about 0.64×10^{-5} s^{-1}.

We performed 12 fast, 13 medium speed and 6 slow tests. The number of slow tests was limited because after each tissue delivery (usually 2 brains) we could perform only one overnight test. Figure 2 shows the Lagrange stress[3] versus true strain

---

[3] To calculate Cauchy (true) stresses, the precise measurement of the cross–section area during loading or the assumption of tissue incompressibility is needed.

($\varepsilon=ln\lambda_z$, where λ_z is a stretch in vertical direction, Fig. 1b) curves for three loading velocities. The standard deviation of the measurements and the theoretical predictions are indicated. The stress–strain curves are concave upward for all compression rates containing no linear portion from which a meaningful elastic modulus might be determined. The tissue response stiffened when the loading speed increased, indicating a strong stress–strain rate dependence. The results shown in Fig. 2 are in general agreement with those published in [9].

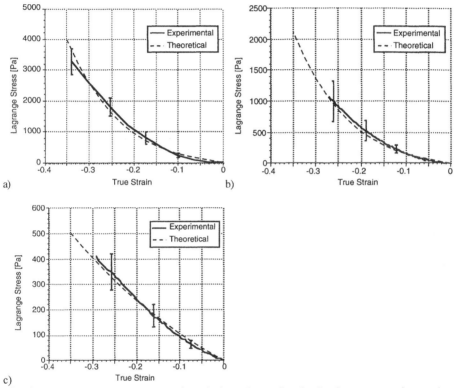

Fig. 2. Lagrange stress - true strain relations for swine brain tissue, experimental and theoretical results. Error bars indicate standard deviation. Loading speeds:
a) 500 mm min^{-1}, corresponding to the stretch rate of about 0.64 s^{-1},
b) 5 mm min^{-1}, corresponding to the stretch rate of about 0.64×10^{-2} s^{-1},
c) 0.005 mm min^{-1}, corresponding to the stretch rate of about 0.64×10^{-5} s^{-1}.

3. CANDIDATES FOR BRAIN TISSUE CONSTITUTIVE MODELS

3.1. Brain as a Bi–phasic Continuum

The concept that the soft tissues can be treated as bi–phasic continua consisting of solid deformable porous matrix and penetrating fluid is widespread. It has been particularly useful in cartilage biomechanics, see for example [13]. The linear bi-phasic model of brain was proposed, for example, in [16, 17].
The governing equations of bi-phasic continuum are:

Continuity: $\qquad \nabla(\phi^S \mathbf{v^S} + \phi^f \mathbf{v^f}) = 0$ $\qquad\qquad$ (1)

Equilibrium: $\qquad \nabla\sigma^\alpha + \Pi^\alpha = 0$ $\qquad\qquad\qquad$ (2)

where:

ϕ^α – α phase content (f - fluid phase, s - solid phase),

$\mathbf{v}^\alpha$ - velocity of α phase.

σ^α – α phase Cauchy stress tensor,

Π^α - diffusive momentum exchange between phases.

When writing the equilibrium equation we neglected inertial body forces.

If we make the additional assumption that the fluid is inviscid and that the diffusive momentum exchange is proportional to the relative velocity between phases, the constitutive equations are as follows:

$$\sigma^s = -\phi^s p\mathbf{I} + \sigma_E{}^s \qquad\qquad (3)$$

$$\sigma^f = -\phi^f p\mathbf{I} \qquad\qquad (4)$$

$$\Pi^s = -\Pi^f = -p\nabla\phi^f + K(\mathbf{v}^f - \mathbf{v}^s) \qquad\qquad (5)$$

where:

p - apparent pressure,

K - diffusive drag coefficient function,

$\sigma_E{}^s$ - Cauchy stress tensor of the solid phase.

I - identity tensor (rank three),

In general, the diffusive drag coefficient K is not constant. It is usually considered to be dependent (exponentially) on strain.

The accepted way to relate the stresses to the deformation (in a solid matrix) is by means of the Helmholtz free energy:

$$\mathbf{S} = \frac{\partial W}{\partial \mathbf{E}} \qquad\qquad (6)$$

where:

S - Second Piola–Kirchoff stress of the solid phase (measured with respect to the original configuration)

W - the Helmholtz free energy (per unit volume) function of the solid phase. This function depends on the current deformation only.

E - Green's strain tensor (relative to original configuration) of the solid phase.

This approach is based on the assumption that the solid phase stress depends only on the current deformation. Therefore, there is no energy dissipation in the solid, but the dissipation comes from interactions between phases only.

It was shown analytically [18] for the linear bi–phasic model and confirmed numerically for the non–linear case [3, 4] that the ratio of the instantaneous stress (after sudden movement of the upper platen) to the equilibrium stress (after sufficiently long time following load application), as predicted by the bi-phasic theory, *cannot be larger than* $\dfrac{3}{2(1+v)} \in \langle 1, 1.5\rangle$, where v is the Poisson's ratio of the solid phase. This poses a severe limit on the stress dependence on loading velocity. The bi-phasic theory in its present form cannot be accepted for very soft tissues (eg. brain), for which the stresses for the largest loading velocity in our experiments are about six times higher than for the smallest one (Fig. 2).

3.2 Brain as an Inelastic Single-Phase Continuum

The reason for standard bi–phasic theory's inability to describe strong stress–strain rate dependence is the underlying assumption of solid matrix hyperelasticity. In the case of the confined compression experiment the relative velocity of the phases is almost equal to the velocity of the solid phase. Therefore the dissipation may be accounted for by choosing suitably large darg coefficient K (eq. 5). In the unconfined experiment the velocity of solid phase is much larger than the relative velocity between phases. The results show that the solid phase is inherently dissipative. The dissipation in the system cannot be accounted for by adjusting the drag coefficient.

The simple, phenomenological, modelling method introducing suitable dissipation into the system at the expense of single–phase description, is discussed below.
Let's start with the modeling of non–linear stress–strain dependence using the strain energy function in the polynomial form:

$$W = \sum_{i+j=1}^{N} C_{ij}(J_1 - 3)^i (J_2 - 3)^j \tag{7}$$

Where the strain invariants are:

$$J_1 = Trace[\mathbf{B}]; \quad J_2 = \frac{J_1 - Trace[\mathbf{B}^2]}{2J_3}; \quad J_3 = \sqrt{\det \mathbf{B}} = 1 , \tag{8}$$

$\mathbf{B}$ is a left Cauchy-Green strain tensor. For infinitesimal strain conditions, the sum of constants C_{10} and C_{01} have a physical meaning of one half of the shear modulus:

$$\frac{\mu_0}{2} = C_{10} + C_{01} \tag{9}$$

The energy dependence on strain invariants only comes from the assumption that brain tissue is initially *isotropic*. The assumption of tissue incompressibility results in setting the third strain invariant equal to one. The first two terms in (7) form a well known Mooney-Rivlin energy function, originally developed for incompressible rubbers (for discussion see [19]).
In our experiment the deformation was orthogonal, and the left Cauchy-Green strain tensor had only diagonal components:

$$\mathbf{B} = \begin{bmatrix} \lambda_z^2 & 0 & 0 \\ 0 & \lambda_z^{-1} & 0 \\ 0 & 0 & \lambda_z^{-1} \end{bmatrix} , \text{ where } \lambda_z \text{ is a stretch in vertical direction (Fig. 1b)}$$

$$\tag{10}$$

In such a situation, taking $J_1 = \lambda_z^2 + 2\lambda_z^{-1}$ and $J_2 = \lambda_z^{-2} + 2\lambda_z$, the only non-zero Lagrange stress components can be computed from the simple formula:

$$T_{zz} = \frac{\partial W}{\partial \lambda_z} \tag{11}$$

To model the time dependent behaviour of the tissue we write the coefficients in the formula for energy function (7) in the form of a exponential series

$$C_{ij} = C_{ij\infty} + \sum_{k=1}^{n} C_{ijk} e^{-\frac{t}{\tau_k}} , \tag{12}$$

and the energy function in the form of convolution integral

$$W = \int_0^t \{ \sum_{i+j=1}^{N} C_{ij}(t-\tau) \frac{d}{d\tau}[(J_1-3)^i (J_2-3)^j] \} d\tau \tag{13}$$

From (11) we obtain:

$$T_{zz} = \int_0^t \{ \sum_{i+j=1}^{N} C_{ij}(t-\tau) \frac{d}{d\tau}[\frac{\partial}{\partial \lambda_z}((J_1-3)^i (J_2-3)^j)] \} d\tau \tag{14}$$

Equation (14) served as a basis for the comparison of the theory and experiment. In the case of the compression with constant velocity, the integral (14) can be evaluated analytically[4]. The result obtained with *Mathematica* [20] is long and not presented here, however it is important to note that the expression for Lagrange stress is linear in material parameters $C_{ij\infty}$ and C_{ijk} (eq. 12). Therefore it was easy to find them using least square method. To model accurately the tissue behaviour for a wide range of loading velocities, we found it necessary to use two time-dependent terms in the C_{ij} expansion (eq. 12) and to include second order terms in energy function ($N=2$ in eq. 13). We assumed also $C_{ij} = C_{ji}$ (the equality of energies of reciprocal and original deformation), $C_{11}=0$, $C_{20\infty}=C_{02\infty}=0$ and characteristic times $t_1=50$ [s]; $t_2=.5$[s].
A good agreement with experiment for all three loading velocities (Fig. 2) has been obtained for the following values of the parameters:
equilibrium parameters: $C_{10\infty}= C_{01\infty}=81$ [Pa];
parameters multiplying exponential with characteristic time τ_1: $C_{101} = C_{011} =26$ [Pa]; $C_{201} = C_{021} =395$ [Pa];
parameters multiplying exponential with characteristic time τ_2: $C_{102} = C_{012} = 163$[Pa]; $C_{202} = C_{022} =84$ [Pa]. Under above assumptions the estimated parameters are unique.

4. DISCUSSION AND CONCLUSIONS

This study discusses two distinct mathematical models of brain tissue mechanical properties. The strong stress–strain rate dependence prohibits the use of standard bi–phasic models for brain tissue modelling. Therefore, the use of the single–phase, non–linear, viscoelastic model based on the concept of the strain energy function, written in the form of convolution integral with coefficient expressed in the form of exponential series, is advocated.
One advantage of the proposed model is that the constitutive equation developed here is already available in ABAQUS [21] and can be used immediately for larger scale FEM computations.
How to use the *in vitro* experimental results in the more realistic *in vivo* environment remains an open question. Further research is needed to determine brain tissue constitutive models, which would incorporate the influence of the blood and cerebrospinal fluid pressure and flow.

ACKNOWLEDGMENTS

The authors would like to thank most sincerely all the members of Biomechanics Division of Mechanical EngineeringLaboratory, MITI and Dr. Tetsuya Tateishi not forgetting Dr. Yoshio Shirasaki's and Mr. Kazuya Machida, for their indispensable

---

[4] The result contains exponential integrals.

help. The financial support of The New Energy and Industrial Technology Development Organisation (NEDO) and Department of Mechanical and Materials Engineering of The University of Western Australia is gratefully acknowledged.

REFERENCES

1. Brett, P.N., Fraser, C.A., Henningan, M., Griffiths, M.V. and Kamel Y., Automatic Surgical Tools for Penetrating Flexible Tissues, *IEEE Eng. Med. Biol.*, 1995, pp.264-270.
2. Burdea, G., Force and Touch Feedback for Virtual Reality. Wiley. New York, 1996.
3. Miller, K. and Chinzei, K., Modeling of Soft Tissues, *Mechanical Engineering Laboratory News*, 1995, **12**, 5-7 (in Japanese).
4. Miller, K. and Chinzei, K., Modeling of Soft Tissues Deformation, *Journal Computer Aided Surgery*, 1995, **1**, Supl., *Proc. of Second International Symposium on Computer Aided Surgery*, Tokyo Women's Medical College, Shinjuku, Tokyo, 62-63.
5. Lavallée, S., Registration for Computer Integrated Surgery: Methodology, State of the Art. *Computer–Integrated Surgery*, 1995, MIT Press, Cambridge Massachusetts, pp. 77-97.
6. Ommaya, A.K., Mechanical Properties of Tissues of the Nervous System, *J. Biomech.*, 1968, **1**, 127-138.
7. Walsh, E.K. and Schettini, A., Calculation of brain elastic parameters in vivo. *Am. J. Physiol.* 1984, **247**, R637-R700.
8. Sahay, K.B., Mehrotra, R., Sachdeva, U. and Banerji, A.K., Elastomechanical Characterization of Brain Tissues, *J. Biomech.*, 1992, **25**, 319-326.
9. Estes, M.S. and McElhaney J.H., Response of Brain Tissue of Compressive Loading, *ASME Paper No. 70-BHF-13*, 1970.
10. Galford, J.E. and McElhaney, J.H., A Viscoelastic Study of Scalp, Brain and Dura, *J. Biomech.*, 1970, **3**, 211-221.
11. Pamidi, M.R. and Advani, S.H., Nonlinear Constitutive Relations for Human Brain Tissue, *Trans. ASME, J. Biomech. Eng.*, 1978, **100**, 44-48.
12. Mendis, K.K., Stalnaker, R.L. and Advani S.H., A Constitutive Relationship for Large Deformation Finite Element Modeling of Brain Tissue, *Trans. ASME, J. Biomech. Eng.*, 1995, **117**, 279-285.
13. Mow, V.C., Ateshian, G.A. and Spilker R.L., Biomechanics of Diarthrodial Joints: A Review of Twenty Years of Progress. *Trans. ASME, J. Biomech. Eng.*, 1993, **115**, 460-467.
14. Chinzei K., Miller K., "Experimental evaluation of compressive behavior of very soft tissues," *Mechanical Engineering Laboratory News (ISSN 0286-2271), No. 11, 1996*, pp. 1-3 (in Japanese).
15. Miller, K. and Chinzei, K., Constitutive Modeling of Brain Tissue; Experiment and Theory, *J. Biomech.* (submitted).
16. Nagashima, T., Tamaki, N., Matsumoto, S., Horwitz, B. and Seguchi, Y., Biomechanics of Hydrocephalus: A New Theoretical Model, *Neurosurgery*, 1987, **21**, No. 6, 898-903.
17. Basser, P.J, Interstitial Pressure, Volume, and Flow during Infusion into Brain Tissue, *Microvasc. Res.*, 1992, **44**, 143-165.
18. Armstrong, C.G., Lai, W.M. and Mow, V.C., An Analysis of the Unconfined Compression of Articular Cartilage, *Trans. ASME, J. Biomech. Eng.*, 1984, **106**, 165-173.
19. Rivlin, R.S., Forty Years of Nonlinear Continuum Mechanics, In *Proceedings of the IX Int. Congress on Rheology*, 1984, Mexico, pp. 1-29.
20. Wolfram Res., Inc., *Mathematica*, Version 2.2, Wolfram Research Inc., Champaign, Illinois, USA, 1994.
21. *ABAQUS* Theory Manual Version 5.2, Hibbit, Karlsson & Sorensen, Inc. , 1992.

A THREE DIMENSIONAL CONTINUUM MODEL OF SKELETAL MUSCLE

A.W.J. Gielen[1], P.H.M. Bovendeerd[2] and J.D. Janssen[3]

1   ABSTRACT

Skeletal muscle consists of a nonlinear, anisotropic, fibrous contractile material. Besides, these properties are distributed non-uniformly across the muscle, which itself can have a complex geometry. Traditional models can not predict the actual local behaviour of the muscle, because uniformity and/or simple geometries are assumed. We present a model, which takes into account the active contractile properties using a Distributed Moment approximated Huxley model and the passive tissue with a three dimensional nonlinear anisotropic elastic model. The model is approximated numerically with the finite element method. The main features of the model are illustrated with simulations of an isometric contraction of a geometrically simple muscle for a plane stress and a plane strain situation. Large differences between both situations demonstrate the importance of this type of modelling.

2   INTRODUCTION

The force generated by one sarcomere depends on its stimulation, the current sarcomere length and sarcomere shortening velocity. In other words, the local stress depends on the local stimulation, strain, and strain rate. Also, physiological important processes like damage and adaptation are responses of muscle cells and connective tissue to local stress and/or strain and/or strain rate [1]. Therefore, fundamental insight into the functioning of muscle tissue can only be obtained if the local loading of the tissue is known, i.e. the local state of stress and strain. Unfortunately, the local loading of muscles cannot be determined from measurements of macroscopic parameters like length and force of the muscle,

Keywords: skeletal muscle, contraction, micro mechanics.

[1] PhD student, Biomechanics / Computational and Experimental Mechanics / Department Mechanical Engineering, Eindhoven University of Technology, P.O.box 513, 5600MB Eindhoven, The Netherlands
[2] Lecturer, same address. [3] Professor, same address.

because tissue properties and strains and stresses are inhomogeneously distributed across the muscle [2].

Many of the traditional mathematical models, like parallelogram models [3] or fibre-fluid models [4], were designed to predict the macroscopic behaviour of muscles. They assume homogeneity of stress, strain and material properties, oversimplify geometry, or neglect the contribution of the three dimensional connective tissue network. While valuable in many studies, as for example gait-analysis, these models fail in predicting the actual local stess and strain.

In this paper we present a mechanical model of skeletal muscle, that enables the computation of the local state of stress and strain throughout the muscle, taking into account the complex muscle geometry, including the three-dimensional fibre field, and the inhomogeneous, anisotropic material properties. This is accomplished by using a continuum approach, as applied also in cardiac mechanics [5]. Here the constitutive equations of the contractile properties are described with a micro mechanical Huxley-type model in contrast to the phenomenological Hill model generally employed. Besides, the excitation-contraction dynamics are described in more detail. Moreover, in skeletal muscles the tendons have to be modelled, which are absent in the heart.

The main features of the model are illustrated with simulations with a simplified flat three dimensional muscle geometry as a realistic three-dimensional muscle geometry including fibre field is not yet available, and no data could be found in literature on the spatial distribution of material parameters. However, with this flat geometry two boundary condition cases, plane stress and plane strain, can be easily studied. Comparison of those cases is interesting because most existing muscle models assume a plane strain situation [3, 4] to represent the 3D muscle behaviour, while plane stress is the other extreme.

3   METHOD

In the description of the mechanical behaviour of the muscle tissue the discrete microscopic cross bridge behaviour and the connective tissue network mechanics are averaged. Now, the muscle tissue is represented by a composite of contractile fibres and a connective tissue network. Both components will undergo the same deformation F

$$F = (\vec{\nabla}_0 \vec{x})^c = (\vec{\nabla}_0 \vec{u})^c + I \tag{1}$$

where $\vec{x}$ are the current material positions, $\vec{\nabla}_0$ is the gradient operator with respect to the reference coordinates $\vec{x}_0$. The tensor I is the unity tensor. The vector $\vec{u} = \vec{x} - \vec{x}_0$ is the displacement of the material. The relative change of volume is determined with $J = \det(F)$. The stress in the tissue is the superposition of stresses of both components

$$\sigma = \sigma_a \vec{e}_f \vec{e}_f + \sigma_p \tag{2}$$

where σ_a is the one dimensional active stress that acts along the fibre direction $\vec{e}_f$. The three dimensional stress σ_p arises from the deformation of the connective tissue network.

The response of the tissue to external loading (e.g., subjection to force or lengthening) or internal loading (e.g., generation of active fibre forces) is described by the local conservation laws of mass, momentum and moment of momentum,

$$\rho J = \rho_0, \qquad \vec{\nabla} \cdot \sigma = \vec{0}, \qquad \sigma = \sigma^c \tag{3}$$

respectively, where ρ and ρ_0 are the current and initial tissue density and ^{c} is a tensor conjugation. The operator $\vec{\nabla}$ is the gradient operator.

If for both stresses σ_a and σ_p constitutive equations are specified and the boundary conditions are given, then the conservation laws (3), can be solved with the finite element method to find the displacement $\vec{u}$ of the tissue. In the next two subsections the constitutive equations are described. The last subsection shortly describes the implementation in the FEM package SEPRAN.

3.1   Active tissue stress

The contraction of muscle cells results from the interaction between two proteins, actin and myosin. The movement involves the relative sliding of two sets of filaments: the thick, myosin filaments; and the thin, actin filaments (see Figure 1 left). The forces that drive the filament motion are generated by cross bridges that extend radially from the myosin filaments and interact cyclically with the actin filaments. This process, that can only occur in the region where both filaments overlap, starts when attachment sites on the actin filaments are activated by calcium ions. Depending on the spatial configuration of the cross bridge, also called attachment length, the cross bridge is likely to attach to or detach from the actin sites (see Figure 1 right). The cross bridge process can be described

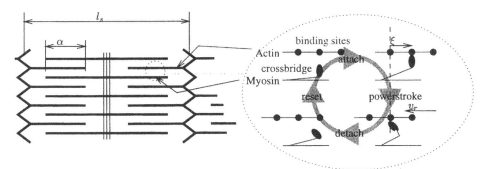

Figure 1: Left: The sarcomere in the muscle fibre is a repeating unit of lenght l_s. The actin and myosin filaments overlap with a factor α. Right: Cross bridges, which are ridgidly connected to myosin, can attach to the binding sites on the actin filament. When connected, they produce a powerstroke and move the filaments relatively to one-another with velocity v_r. At the end of the powerstroke they detach and the process can start again.

with a Huxley cross bridge model. To include the filament overlap and activation, the original equation is modified to [6, 7]

$$\frac{\partial n(\xi,t)}{\partial t} - v_r \frac{\partial n(\xi,t)}{\partial \xi} = r(t)f(\xi)[\alpha(l_s) - n(\xi,t)] - g(\xi)n(\xi,t) \qquad (4)$$

where $n(\xi,t)$ is the attachment distribution function, which represents the fraction of attached cross bridges with attachment length ξ at time t. v_r is the contraction velocity (i.e., velocity of sliding of the filaments). $f(\xi)$ is the attachment rate function, which gives the attachment probability for cross bridges with attachment length ξ, and $g(\xi)$ is the detachment rate function. They are modelled as

$$f(\xi) = \begin{cases} 0 \\ f_1\xi \\ 0 \end{cases} \quad ; \quad g(\xi) = \begin{cases} g_2 & \xi < 0 \\ g_1\xi & 0 \le \xi \le 1 \\ g_1\xi + g_3(\xi - 1) & \xi > 1 \end{cases} \tag{5}$$

where f_1, g_1, g_2 and g_3 are constants. The overlap α from equation (4) is the fraction of all actin-sites reachable by myosin-heads (i.e., the overlap factor of the filaments), and is modelled as

$$\alpha(\ell) = \begin{cases} 1 - p_1(\ell - 1)^2 & \ell \le 1 \\ 1 - p_2(\ell - 1) & \ell > 1 \end{cases} \quad \ell = l_s/l_s^0 \tag{6}$$

with l_s and l_s^0 the current and initial sarcomere length, respectively.

The factor r represents the fraction of sites on the actin filament that are activated. It is expressed as [6, 7]

$$r = C^2/(C^2 + \mu C + \mu^2) \tag{7}$$

with C the calcium concentration in the muscle fibre and μ the troponin-calcium reaction constant. The most simple model to compute C is given by [8]

$$\partial C/\partial t = \gamma(c\nu - C) \tag{8}$$

where γ represents a fibre-type dependent rate parameter, c a calcium release parameter and ν the dimension-less stimulation frequency. For constant ν and $C(t = 0) = 0$ equation (8) can be solved analytically, yielding:

$$C(t) = c\nu(1 - \exp(-\gamma t)) \tag{9}$$

The actual stress developed in the contraction is derived from the attachment distribution function n. It is generally assumed that the cross-bridges force is linear with constant κ: The force of one cross bridge is $F = \kappa\xi$. The Cauchy stress in a slice of half-sarcomeres with thickness $l_s/2$, is computed by averaging the cross-bridge forces, i.e., summing the forces and dividing by the slice area [9]

$$\sigma_a(t) = \frac{ml_s\kappa}{2l_a} \int_{-\infty}^{\infty} \xi n(\xi, t)d\xi = c_a Q_1(t) \tag{10}$$

with m the cross bridge density, l_a the actin binding site spacing, and $c_a = ml_s\kappa/2l_a$ represents the maximum active contraction stress. The term Q_1 is the first moment of function $n(\xi, t)$. In general, the λ-th moment of a distribution function n is defined as:

$$Q_\lambda = \int_{-\infty}^{\infty} \xi^\lambda n(\xi, t)d\xi \tag{11}$$

As can be seen from (10) it is only necessary to know the first moment of the function n. By choosing beforehand a reasonable shape for the function n and rewriting the Huxley equations in terms of moments, the active stress can be computed using the DM method developed by Zahalak [9, 6]. A proper choice for n is a Gaussian distribution function [9]:

$$n(\xi, t) = \frac{\Omega}{\sqrt{2\pi}\chi} \exp\left(-\frac{(\xi - \psi)^2}{2\chi^2}\right) \tag{12}$$

The variables Ω, ψ and χ of the function n can be expressed in terms of exactly the first three moments:

$$\Omega = Q_0, \qquad \psi = Q_1/Q_0, \qquad \chi = \sqrt{(Q_2/Q_0) - (Q_1/Q_0)^2} \tag{13}$$

Therefore, the Huxley equation (4), which transforms into

$$\dot{Q}_\lambda = \alpha r \beta_\lambda - r\phi_{1\lambda} - \phi_{2\lambda} - \lambda v(t) Q_{\lambda-1} \quad \lambda = 0, 1, 2 \tag{14}$$

with $\beta_\lambda = \int_{-\infty}^{\infty} \xi^\lambda f(\xi) d\xi$, $\phi_{1\lambda} = \int_{-\infty}^{\infty} \xi^\lambda f(\xi) n(\xi, t) d\xi$, and $\phi_{2\lambda} = \int_{-\infty}^{\infty} \xi^\lambda g(\xi) n(\xi, t) d\xi$, can be solved as a closed set of three nonlinear ordinary differential equations if r, α and v_r are given.

3.2 Passive tissue stress

The passive stress for both the passive muscle tissue and tendons is determined using a strain-energy function $W_p = W_s + W_v$. The functions W_s and W_v describe the strain-energy from shape and volume changes, respectively. The strain energy functions are specified in an orthonormal vector basis $[\vec{e}_1 \vec{e}_2 \vec{e}_3]$, with $\vec{e}_1$ parallel to the fibre direction. The strain tensor $\boldsymbol{E}$ is then written as the strain matrix $\underline{E}$ with components E_{ij}. W_s was chosen [5] as

$$W_s(\underline{E}) = a_0[\exp(a_1 I_E^2 + a_2 II_E + a_3 E_{11}^2 + a_4(E_{12}^2 + E_{13}^2) - 1] \tag{15}$$

where I_E and II_E represent the first and second invariant of the Green-Lagrange strain tensor, and $a_0 \ldots a_4$ are material parameters. W_v was chosen as:

$$W_v(\underline{E}) = a_5(\det(2\underline{E} + \underline{I}) - 1)^2 \tag{16}$$

The term $\det(2\underline{E} + \underline{I})$ is equal to J^2. The material behaves increasingly incompressible with an increasing bulk-modulus a_5.

3.3 FEM implementation

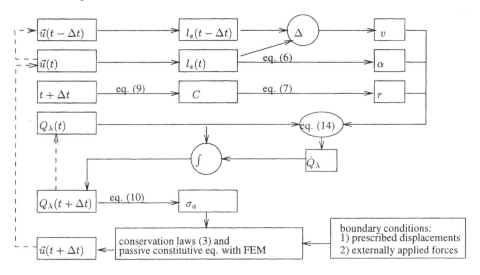

Figure 2: Solution scheme: Starting from known $\vec{u}$ and Q_λ at t and $t - \Delta t$, the values at $t + \Delta t$ must be computed. $\vec{u}$ at t and $t - \Delta t$ are used to compute v_r and α for input in the DM-equation, while r is determined analytically. Integrating $\dot{Q}_\lambda$ the values of Q_λ at $t + \Delta t$ are found and the active stress is determined. The FEM computation determines the corresponding displacements u at $t + \Delta t$ according for specified boundary conditions. After this step the variables are shifted and the scheme starts again. (The dashed arrows). For the tendons only the FEM computation is performed.

For the formulation of the finite element method a standard solid 8 nodes brick element was used. We implemented a Total Lagrange formulation which takes into account the geometrical (i.e., large deformations) and material non-linearities. Linearisation was accomplished with the Newton-Raphson method. Integration at the element level was performed using 8 point Gauss integration. The fibre orientation was allowed to vary in the element. Time integration (to solve the differential equation in Q_λ) and differentiation (to compute strain rates) were performed using explicit first order Euler equations. The following scheme (see Figure 2) was used to perform the computations.

4 SIMULATIONS

Figure 3: Left: Simplified geometry with a muscle belly (gray) and two tendons (white). Right: The fibre direction in both muscle belly and tendons

To illustrate the possibilities and capabilities of our approach we use a simplified 3D muscle geometry with constant thickness and a fibre field as shown in Figure 3. The values of the material parameters are given in table 1 and were based on data in literature. Initially, the overlap α is 1, and the moments Q_0, Q_1 and Q_2 are set nearly zero at $0.3084 \cdot 10^{-6}, 0.0809 \cdot 10^{-6}, 0.0227 \cdot 10^{-6}$, respectively. All muscle fibres were activated homogeneously and activation started at $t = 0$. For the FEM computation the geometry was divided into brick shaped elements as shown in Figure 4.

	passive tissue				active tissue					
	tendon	muscle	units							
a_0	100.0 $_a$	0.6 $_b$	kPa	f_1	35.0 $_e$	s^{-1}	c_a	200.0 $_g$	kPa
a_1	10.0 $_a$	15.0 $_b$	-	g_1	7.0 $_e$	s^{-1}	p_1	6.25 $_h$	-
a_2	20.0 $_a$	30.0 $_b$	-	g_2	200.0 $_e$	s^{-1}	p_2	1.25 $_h$	-
a_3	10.0 $_a$	-12.0 $_b$	-	g_3	30.0 $_e$	s^{-1}			
a_4	-10.0 $_c$	-15.0 $_c$	-	γ	11.25 $_f$	s^{-1}	μ	0.2 $_i$	-
a_5	2.5 $_d$	0.2 $_d$	MPa						

Table 1: Material parameter settings. $_a$ fitted to data from Hatze [8]; $_b$ fitted to data from Strumpf [10]; $_c$ set to facilitate shearing; $_d$ set to get volume changes less than 8%; $_e$ taken from Zahalak [9]; $_f$ take from Hatze [8]; $_g$ set to 200 kPa to get active stresses in the physiological range (50–100 kPa); $_h$ fitted to data from Herzog [11]; $_i$ taken from Zahalak [6].

Two extreme loading cases are studied. In the first case we simulate an isometric contraction of an isolated muscle in plane stress situation, where the muscle can expand freely in the third dimension. In the second case the situation is plane strain, where the muscle is restrained in third dimension and keeps same thickness, which is the common way to describe muscle mechanics in 2D [4, 3].

Figure 4: Division of the geometry into 3D elements

$\partial E_{11}/\partial t$ *at t=30ms center value -4.0, -6.0*

Deformed mesh at full tetanus

E_{11}, *center value -0.18, -0.23*

E_{22}, *center value 0.27, 0.16*

E_{33}, *center value 0.0, 0.21*

σ_a, *center value 62, 48 kPa*

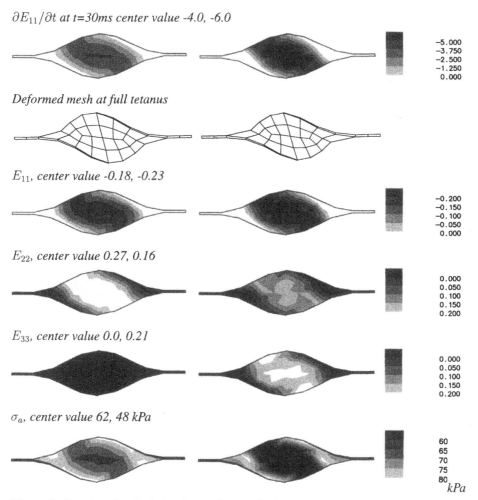

```
-5.000
-3.750
-2.500
-1.250
 0.000
```

```
-0.200
-0.150
-0.100
-0.050
 0.000
```

```
0.000
0.050
0.100
0.150
0.200
```

```
0.000
0.050
0.100
0.150
0.200
```

```
60
65
70
75
80
kPa
```

Figure 5: Results: On the left side are the results from the plane strain situation, on the right the plane stress. The values at the center of the geometry are given for both cases. The $_{11}$ component corresponds to the fibre direction, $_{22}$ to the cross fibre direction, and $_{33}$ to the thickness direction.

The results presented in figure 5 are illustrative for the mechanical behaviour of the muscle. The contour plots of the local results were made on the reference geometry and are based on the values in the mid-plane of the geometry. Around $t = 30$ms the maximum fibre strain rate was reached. The simulation stopped at full tetanus at $t = 300$ms, when $C = 0.97$. We compare the plane stress results with the plane strain results. The deformation plot shows that the muscle expands less in cross fibre direction, and shortens more in the fibre direction. This result can be read better from the strain plots. The fibre strain E_{11} in the center is -0.23 versus -0.18, and cross fibre strain E_{22} is 0.16 versus 0.27. These differences can easily be explained from E_{33}, the strain in the thickness direction. When the muscle fibres shorten, the muscle has to expand in the cross fibre direction because of the low compressibility. In the plane stress situation the volume can be displaced in two cross fibre directions $\vec{e}_2$ and $\vec{e}_3$, while in the plane strain situation the $\vec{e}_3$ direction is restricted. Since, at full tetanus the fibre shorting velocity v_r is zero and active stress only depends on the overlap factor α. The larger negative value of the fibre strain gives a lower α value, which gives a lower active strain.

5   DISCUSSION

The model presented in this paper allows us to compute fully three dimensional stresses, strains, and strain rates throughout a muscle-tendon geometry. The use of a Huxley type contraction model has two advantages. Firstly, since most parameters of the model have a well defined physical meaning, they can be determined from independent experimental data in literature [12]. Secondly, the model can be extended with additional relations for chemical energy and metabolism as proposed by [12].

Because of the simplified geometry and uncertainties in the values of the material parameters, the presented results must be regarded as an illustration of the possibilities of the model, rather than a quantitatively realistic prediction of the local tissue loading. Still, the model shows that qualitatively large differences can be found between plain stress and plain strain modelling of the muscle. These differences emphasise the need for realistic three dimensional modelling of muscles. A problem in the step towards a realistic model is the lack of reliable data on the spatial distribution of the values of the model parameters with respect to muscle geometry, fibre orientation and material properties. The theoretical setup and the numerical implementation of the model have no limitations in this respect.

6   REFERENCES

[1] C. B. Ebbeling and P. M. Clarkson. Exercise-induced muscle damage and adaptation. *Sports Medicine*, pages 207–234, 1989.

[2] C. J. Zuurbier and P. A. Huijing. Changes in geometry of actively shortening unipennate rat gastrocnemius muscle. *J. Morphology*, pages 167–180, 1993.

[3] P. A. Huijing and R. D. Woittiez. The effect of architecture on skeletal muscle performance: A simple planimetric model. *Neth. J. Zoology*, 34:21–32, 1984.

[4] J. L. van Leeuwen and C. W. Spoor. Modelling mechanically stable muscle architectures. *Phil. Trans. R. Soc. Lond. B*, 336:275–292, 1992.

[5] P. H. M. Bovendeerd, T. Arts, J. M. Huyghe, D. H. van Campen, and R. S. Reneman. Dependence of local left ventricular wall mechanics on myocardial fiber orientation: a model study. *J. Biomechanics*, 25:1129–1140, 1992.

[6] G. I. Zahalak and S.-P. Ma. Muscle activation and contraction: constitutive relations based directly on cross-bridge kinetics. *J. Biomech. Eng.*, 112:52–62, 1990.

[7] G. I. Zahalak and I. Motabarzadeh. A re-examination of calcium activation in the huxley cross-bridge model. *J. Biomech. Eng.*, 119:20–29, 1997.

[8] H. Hatze. *Myocybernetic control models of skeletal muscle: Characteristics and applications*. University of South Africa, Muckleneuk, Pretoria, 1981.

[9] G. I. Zahalak. A distribution-moment approximation for kinetic theories of muscular contraction. *Math. Biosciences*, 55:89–114, 1981.

[10] R. K. Strumpf, J. D. Humphrey, and F. C. P. Yin. Biaxial mechanical properties of passive and tetanized canine diaphragm. *Am. J. Phys.*, 265:H469–H475, 1993.

[11] W. Herzog, S. Kamal, and H. D. Clarke. Myofilament lengths of cat skeletal muscle: theoretical considerations and functional implications. *J. Biomechanics*, 25:945–948, 1992.

[12] S.-P. Ma and G. I. Zahalak. A distribution-moment model of energetics in skeletal muscle. *J. Biomechanics*, Vol. 24:21–35, 1991.

A 2D FINITE ELEMENT MODEL OF A MEDIAL COLLATERAL LIGAMENT RECONSTRUCTION

[1]R.M. Grassmann, [2]N.G.Shrive and [3]C.B. Frank

1. SUMMARY

A 2-D finite element (FE) model was developed of a lapine medial collateral ligament (MCL) autograft-bone complex with the semitendinosus tendon (ST) as graft tissue, to determine the new stress states at two insertional geometries: 90° femoral insertion and 45° tibial insertion. Cancelleous bone, cortical bone, marrow, meniscus, the overlying skin, suture material and the tendon with its crimp and its poroelastic properties have all been incorporated into this model in which the geometry was derived from measurements taken of the autograft complex immediately after surgery. Histological studies done in parallel with the finite element modelling will assist in correlating the biological changes of the autograft with the mechancial environment predicted by the FE model.

2. INTRODUCTION

Ligament injuries are prevalent in North America, with an estimated 13.5% of the population between the ages of 18 to 35 being affected (Kelsey et al[1]). These injuries are typically handled in one of two ways: by conservative treatment, e.g. physiotherapy in the case of a MCL injury; or by reconstructive surgery which may be performed in cases of an anterior cruciate ligament (ACL) injury or triad injury (MCL, ACL and posterior cruciate ligament). For reconstructive surgery, graft tissue is often taken to replace the former ligamentous tissue. However, even with reconstructive surgery, the graft material never attains the mechanical properties of the former tissue.

Keywords: Finite Element Model, MCL, Semitendinosus Tendon, Histology

---

[1] Graduate Student, Department of Civil Engineering, University of Calgary, 2500 University Dr. NW, Calgary, Alberta T2N 1N4, Canada

[2] Professor and Head of Dept., Department of Civil Engineering, University of Calgary, 2500 University Dr. NW, Calgary, Alberta T2N 1N4, Canada

[3] Professor of Surgery; Chief, Division of Orthopaedic Surgery, Faculty of Medicine, University of Calgary, 3330 Hospital Dr. N.W., Calgary, Alberta, Canada

We have examined how placement of the graft tissue affects the incorporation process, since it appears that the graft tissue adapts to some degree according to its new mechanical environment. In previous studies, the structural differences between various types of osteotendinosus and ligamentous insertions have been investigated (Benjamin et al[2], Cooper et al[3], Matyas[4]). The biological adaptation of these insertions to a changing mechanical environment (Ploetz et al[5], Gillard et al[6], Matyas et al[7]) has also been analyzed.

For this study, we adopted an ST autograft transplant for the MCL in a lapine animal model. The MCL model was selected for several reasons: its relative homogeneity, non-spiraled structure, simple transplantation surgery with its minimal impact on joint mechanics, its well understood insertional structure and geometry and its distinctly different insertions (Matyas[8]). Much is known about the MCL through mechanical, histological and biochemical studies of normal ligament, of injury models, of immobolization models and of graft transplantation models. Finite element modelling has been employed in attempts to provide some insight concerning the stress states in the tissue (e.g.: Wilson et al[9]). The MCL inserts into the femur with the fibres turning through a nearly ninety degree angle, passing through uncalcified and calcified cartilage zones prior to inserting into the bone. At the tibial insertion, on the other hand, the fibres insert directly into the bone at an oblique angle.

Using finite element analysis to model the complex environment of the ST autograft, the objective was to determine what is significantly different in terms of stress states immediately after surgical implantation of the graft at the different insertional angles into the tibia and femur. In conjunction with developing the finite element model, histological graft transplantation studies were performed with New Zealand White (NZW) rabbits. With histology and FE analysis, the intent was to correlate the initial imposed mechanical environment with the changing biological structure of the graft tissue as the two insertions experience the same loading history *in vivo*.

3. METHODS

3.1 Surgery

Twenty-four skeletally mature NZW rabbits underwent ST autograft transplant surgery. The MCL was incised from its insertions and the insertions were drilled out at 90° to the femur and 45° to the tibia. The ST of an appropriate length (incised at the same two anatomical locations) was placed in the former position of the MCL. The tendon ends were secured to the base of the insertional holes by reabsorbable suture (Vicryl suture 3.0).

3.2 Histology

The histological portion of the study involved embedding the whole femur-ST-tibia complex into either polymethylmethacrylate (PMMA) to determine the *in situ* geometry or into paraffin to understand the changing organology, histology and cytology. All embedded sections were metachromatically stained with toluidine blue and safrinin O to highlight areas of high anionic proteoglycan content which have been associated with zones experiencing compressive forces and consisting of cartilaginous material (Vogel et al[10]). Specimens were examined at zero, six, twelve and twenty-four weeks post-operatively.

3.3 Finite Element Model

The finite element model was constructed to simulate the initial post-operative *in vivo* structure of the femur-ST-tibia complex. A two dimensional finite element model, using Patran 2.4 (pre-processor) and Abaqus 5.4 (post-processor), was created even though the femur-ST-tibia complex has a three dimensional structure (Figure 1). The geometry of the model was derived from measurements made at the time of the ST autograft transplant surgery as well as from PMMA sections of the femur-ST-tibia complex immediately after surgery (Figure 2). The tibia and the femur were modelled as cortical shells filled with marrow and cancelleous bone, respectively. To minimize the computational time, the bone material was modelled using eight noded isoparametric elements as isotropic and homogeneous, having no trabecular architecture and no poroelastic behaviour i.e. no pore fluid flow. The tendon ground substance was modelled with poroelastic elements involving an eight noded isoparametric formulation for the solid and a four noded formulation for fluid pressure and flow. Spring elements were used to simulate tendon fibres with two springs placed on each element edge directed along the length of the tendon. Spring elements were also used to imitate the behaviour of the suture attached to each tendon end. To model the effect of the meniscus, pressure was applied to the bone edges of the joint cavity. Similarly, a pressure boundary was applied to the top surface of the tendon to represent the effect of the overlying skin. The sliding movement between the bone and the tendon was modelled with slideline interface elements with a coefficient of friction of 0.04 (Uchiyama et al[11]) between the two surfaces. Because in reality, the drill holes for the transplant insertions are surrounded by bone, beam elements were placed   across   the   gaps   of   the   insertional   drill   holes   in the model to simulate the stiffness of the surrounding material.

The material properties of the model were taken from published literature (Keaveny and Hayes[12], Van der Voet[13]), except that the ST matrix material properties were taken from in-house mechanical test results.

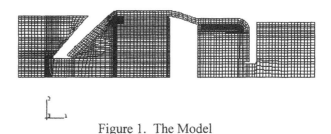

Figure 1. The Model

Figure 1. The Femur-ST Tibia Complex Embedded in Polymethalymethacrylate

Numerical instability occurred at the interface between the cortical bone and the cancelleous bone of the femoral insertion and the interface between the cortical bone and the marrow of the tibial insertion, because of a sharp change in material property. By curving the interfaces where the tendon and osseous components of the autograft meet (Figures 3), the instability was reduced but not eliminated. The insertional corners presented a challenge for computational convergence. To prevent separation between the tendon and the bone in these two areas during loading, springs were added to the tendon and the smallest smoothing factor was utilized with a Hermitian shape function to construct splines between connecting slideline elements.

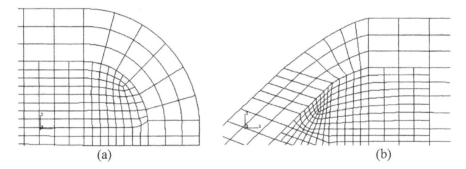

(a)   (b)

Figure 3. (a) The Femoral Insertional Corner of the FE Model
(b) The Tibial Insertional Corner of the FE Model

4. RESULTS

Results for some of the mechanical parameters at each insertional corner (Figures 4 to 7) show a distinctly different mechanical environment at each corner. The stresses (Figure 4) and strains (Figure 5) are higher at the femoral insertion; the stress and strain contour lines reveal a higher stress concentration at this location with higher gradients. The pore pressure distribution (Figure 6) as well as the strain energy density (Figure 7) are similar in terms of their contour line geometry and distribution at each insertional corner, again with the femoral insertion having higher absolute values and gradients.

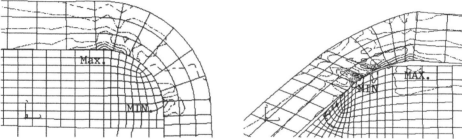

Figure 4. The Principal Stress Distribution at the (a) Femoral Insertion and
(b) Tibial Insertional Corner

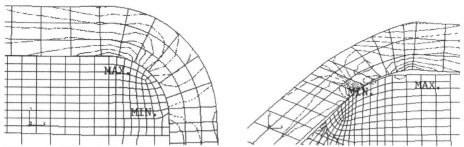

Figure 5. The Principal Strain Distribution at the (a) Femoral Insertion and
(b) Tibial Insertional Corner

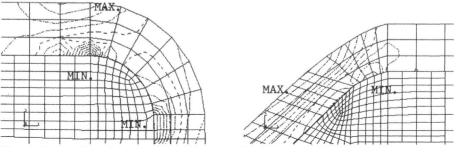

Figure 6. The Pore Pressure Distribution at the (a) Femoral Insertion and
(b) Tibial Insertional Corner

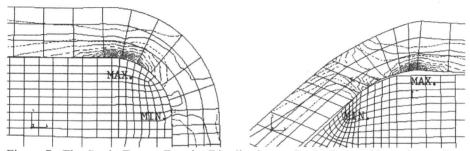

Figure 7. The Strain Energy Density Distribution at the (a) Femoral Insertion and
(b) Tibial Insertional Corner

The histological results show substantial changes in the macroscopic features (Figure 8). The net change in the geometrical shape of the tendon within the tibial insertion is minuscule compared to the changes at the tendon end within the femoral insertion. The tendon in the tibial insertion retains its rope-shape for twenty-four weeks post-operatively whereas the tendon end within the femoral insertion becomes elliptical with interdigitations of bone.

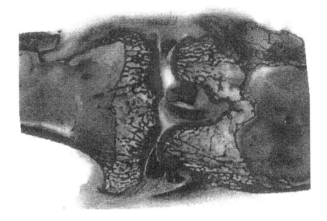

Figure 8. PMMA Section of the 24 Week ST Autograft

In terms of the microscopic features of collagen fibre orientation, vascularity and cell shape gradient, the following observations were made. After six weeks of incorporation, in the femoral insertion no clear tendon fibre orientation exists but where the tendon travels over the two bones and into the tibial insertion, the collagen fibres are aligned in a homogeneous manner. Areas showing evidence of high vascularity in the tendon are present on the distal edge of each drill hole with respect to the insertional corners and along a portion of the tendon-bone boundary of each drill hole. In these areas, the surrounding tissue has the appearance of scar (Frank et al[14] and Shrive et al[15]). The cell shape appears to respond to the initially imposed mechanical environment. The cell shape gradient is steeper at the femoral insertional corner (Figure 9a) than at the tibial insertional corner (Figure 9b). (The seperation existing between the tendon surface and the bone surface in Figure 9a as well as the gaps between collagen fibres in Figure 9b are cutting artifacts.) The cell shape gradient at either insertional corner is clear at six weeks but becomes less distinct with time. Whereas, the tendon end within the femoral insertion is completely consumed by chondrocyte-like cells at six weeks, the tendon end at the tibial insertion is not.

Figure 9. Cell Shape Gradient (a) at the Femoral Insertional Corner and (b) Tibial Insertional Corner

The chrondrocytic-like cells then decrease at the femoral tendon end. Similarly, the staining pattern of the histological sections indicates cartilaginous material at each insertional corner at six weeks which diminishes after this time period. Cartilaginous material subsequently develops in the tendon material within the insertions.

5. DISCUSSION

The mechanical parameters of stress, strain, pore pressure and strain energy density and the gradients thereof will be assessed in relation to the histologic changes occurring *in vivo*. Histological parameters of fibre orientation, cell shape, tendon-bone structure as well as vascularity will be quantified. A cell shape gradient was present at each insertional corner after six weeks of incorporation which may well correlate with the compressive stresses and strains, pore pressures and strain energy density displayed by the finite element model. However, by twelve and twenty-four weeks, any correlation between the mechanical parameters as shown by the FE model and the histological parameters may cease to exist. The biological processes of healing, particularly the increasing formation of scar tissue and bone callus appear to begin to dominate the overall appearance of the ST autograft after the first few weeks. The finite element model of the initial state of the transplant, therefore, appears to provide useful information for initial changes in the *in vivo* structure. With detailed analysis it should be possible to determine which mechanical factor correlates best with which biological change. Biological processes certainly dominate incorporation of the graft after a few weeks. Any algorithm developed to model soft tissue healing would, therefore, have to incorporate the effects of both mechanical and biological influences.

REFERENCES

1. Kelsey J.L., Epidemiology of musculoskeletal disorders. New York: Oxford University Press, 1982, 3.

2. Benjamin, M., Evans E. J. and Copp L., The histology of tendon attachments to bone in man, J. Anat., 1986, Vol. 149, 89-100.

3. Cooper, R. R. and Misol S., Tendon and ligament insertion, J. Bone Jt. Surg., 1970, Vol 52A, 1-20.

4. Matyas, J. R., The structure and function of tendon and ligament insertions into bone. Master's thesis, 1985, Cornell University Graduate School of Medical Sciences.

5. Ploetz E., Funktioneller bau and funktionelle anpassung der gleitsehnen, Z. Orthopadie, 1938, Vol 67, 212-234.

6. Gilliard, G. C., Reilly H.C., Bell-Booth P.G. and Flint M.H., The influence of mechanical forces on the glycosaminoglycan content of the rabbit flexor digitorum profundus tendon, Conn. Tissue Res., 1979, Vol 7, 37-46.

7. Matyas, J., Edwards P., Miniaci A., Shrive N., Wilson J., Bray R. and Frank C., Ligament tension affects nuclear shape *in situ*: an *in vitro* study, Conn. Tissue Res., 1994, Vol 31, 45-53.

8. Matyas, J. R., The structure and function of the insertions of the rabbit medial collateral ligament: an experimental morphological and biomechanical study of the insertions of the medial collateral ligament in the skeletally mature rabbit. Doctoral thesis 1990, University of Calgary Graduate School of Medical Sciences.

9. Wilson, A. N., Shrive N. G. and Frank C. B., Three dimensional model of the rabbit medial collateral ligament using the finite element method, Proceedings of the 1994 Engineering Systems and Design and Analysis Conference, 1994, 17-26.

10. Vogel, K. G., Ordog A., Pogany G. and Olah J., Proteoglycans in the compressed region of human tibialis posterior tendon and in ligaments, 1993, J. Orthop. Res., 1993, Vol 11, 68-77.

11. Uchiyama, S., Coert J.H., Berglund L, Amadio P.C. and An K.-N., Method for the measurement of friction between tendon and pulley, J. Orthop. Res., 1995, 13, 83-89.

12. Keaveny, T. M. and Hayes W. C., A 20-year perspective on the mechanical properties of trabecular bone, J. Biomech. Eng., 1993, 115, 534-542.

13. Van der Voet, A.F., Finite element modelling of load transfer through articular cartilage. Doctoral thesis 1992, University of Calgary Graduate School of Civil Engineering.

14. Frank, C.B., Loitz B.J. and Shrive N.G., Injury location affects ligament healing, Acta. Orthop. Scand., 1995, 66, 455-462.

15. Shrive, N., Chimich D., Marchuk L., Wilson J., Brant R. and Frank C., Soft-tissue "flaws" are associated with the material properties of the healing rabbit medial collateral ligament, J. Orthop. Res., 1995, 13, 923-929.

Acknowledgements

We would like to thank: Medical Research Council of Canada, the Alberta Heritage Foundation for Medical Research, Ted Zjiaka of University Computer Services and Dr. John Matyas of McCaig Research Center at Foothills Hospital for their support.

FINITE ELEMENT MODEL OF THE HUMAN ANTERIOR CRUCIATE LIGAMENT.

Pioletti DP[1,2], Rakotomanana L[1,2], Benvenuti JF[1,2] and Leyvraz PF[1]

1. ABSTRACT

In the present study, we developed a finite element model of the human ACL taking into account the anatomical insertion zones of the ligament, a knee passive 3D kinematics and a realistic constitutive law. This study was performed in three steps. The first step was to determine the three dimensional kinematics of the knee during a passive flexion. The second step was to quantify the mechanical properties of the human ACL. Identification process allowed to determine an elastic potential which describes the non linear elastic behavior of the ligament. This potential formulation was suitable for large strain situations. Finally the third step was to incorporate the measured kinematics, the ligament insertion zones and the identified elastic law into a three dimensional finite element model.

Different situations were then tested. Stress within the ligament was calculated for knee flexion till 70° under neutral, internal and external flexion. Anterior tibial drawer tests at 20° of flexion were also performed with the knee in neutral, internal and external rotation. As illustration, for the anterior tibial drawer tests, the hydrostatic stress field was almost comparable for the knee in neutral and external position. It was found that the hydrostatic and von Mises stresses during an anterior tibial drawer test were more important when the knee was in internal rotation.

2. INTRODUCTION

The ligaments play a central role in the stability of the knee. Due to the increase in sport activities of the young population, rupture of the anterior cruciate ligament (ACL) has become a frequent clinical problem. However, the success of surgical treatments is

Keywords: ACL, stress, finite element model

---

1. Orthopaedic Hospital, Av. Pierre-Decker 4, 1005 Lausanne, Switzerland.
2. Biomedical Engineering Laboratory, Swiss Federal Institute of Technology, 1015 Lausanne, Switzerland.

inconsistent. Therefore, biomechanical studies were performed in order to understand the stabilizing role of the ligaments inside the knee. These biomechanical studies are concerned on either theoretical, or experimental or numerical aspects.

Numerical models of the knee joint often consider ligaments as linear or non-linear spring elements [1, 2]. With this type of model, it is not possible to calculate the stress distribution in the ligaments. The stress distribution can be calculate with a finite element model. A few works in this area can be found in the literature. In a recent study [3], the stress was calculated in the patellar tendon graft. Elastic spring was used for the constitutive law of the patellar tendon and a simplified kinematics was applied. A transverse isotropic constitutive law was numerically implemented to calculated the stress in soft tissues under simulated traction tests [4]. Numerical stress calculations in the cruciate ligaments made using realistic constitutive laws and kinematics seem to be lacking. Hence the objective of this study is to perform numerical simulations in which the stress distribution in the ACL is calculated. A realistic identified constitutive laws is used for the material description and the three dimensional kinematics of the knee flexion are used as boundary conditions.

3. METHODS

Numerical simulations can be performed under the following conditions: 1) the geometry of the specimens is provided; 2) specimens' mechanical behaviors are given; 3) the existence of an adequate spatial discretization (meshing); 4) the existence of appropriate boundary conditions. These conditions are described in this section and tests for their validity are given. A geometrical model and mesh design (pre-processing) were created using the software package program PATRAN 1.4.3 (The MacNeal-Schwender Corporation, Los Angeles-USA). The data analysis (post-processing) was performed with the same software. The numerical resolutions (solver) are achieved with the software package program ABAQUS 5.4 (Hibbitt, Karlsson, & Sorensen Inc., Newpark-USA).

3.1. Geometrical model

The insertion zones of the ACL, obtained from the kinematical study (see 3.4. for details), were delimited by 5 pellets in both the femur and tibia. An interpolation method based on an isoparametrisation was developed to reconstruct the insertion zones [5]. The external contours of the ACL at the tibial and femoral insertion sites were determined. These external contours took into account the fibers orientation of the ligament. Since, the exact three dimensional shape of the ligament was not available, a solid that joins the two reconstructed insertion zones was created (Fig. 1.). The bone insertions around the ligament was constructed. These bone insertions were separated into two distinct zones corresponding to the cortical (3mm of depth) and spongious bones (12mm of depth).

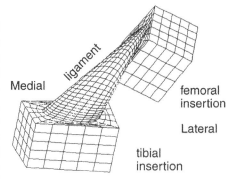

Fig. 1. Antero-posterior view of the geometrical model and corresponding meshing of the ligament

3.2. Material properties

Mechanical tests were performed on human ACL (5 male specimens, mean age 74.8 ± 2.1 years) which were collected with their bone insertions. These specimens were fixed in a custom made traction machine [6]. Tractions performed with an elongation rate of 0.3mm/s were applied under constant 37°C and 100% humidity. The obtained stress-strain curves were identified to the following elastic potential W_e proposed by Veronda (1970) with the incompressibility assumption:

$$W_e = \alpha \exp[\beta(I_1 - 3)] - \frac{\alpha\beta}{2}(I_2 - 1)$$

where α and β are two parameters and I_1 and I_2 are the first and second strain invariants. Determination of the two parameters α and β were obtained with a least-square fit of the constitutive law resulting from the potential W_e and the experimental stress-strain curve e.g. [7]. An isotropic linear constitutive law was used to model the cortical and spongious bony insertions. The values of the Young's modulus and Poisson's ratio for cortical bone are 17000 [MPa] and 0.36, respectively and for the spongious bone they are 14000 [MPa] and 0.40, respectively e.g. [8].

3.3. Meshing

The ligament was discretized with 600 isoparametric quadratic elements (3350 nodes) which support large deformations (element name C3D20H). The mesh was refined near the insertion zones to take into account high stress gradients. Both cortical and spongious bony insertions were discretized with 130 isoparametric quadratic elements (727 nodes) and 240 elements (1996 nodes), respectively which support large deformations (element name C3D20).

3.4. Boundary conditions

Initial stress. The ACL has no stress free state. At all the knee flexions, a stress is present in the ligament [9]. The precise experimental distribution of this stress field is not known. However, some experimental works have furnished the resulting force of the ACL as a function of the knee flexion. In this study, a resulting force of 100N along the ligament axis at full extension was adopted. This value represents a mean value between different studies e.g. [10, 11].

Three dimensional kinematics. Two intact fresh frozen postmortem knees (mean age: 69) were mounted in a device which allowed knee flexion angles [12]. Three flexions were performed for each knee: a neutral one with no torque applied about the tibial axis, an external one with a -3Nm torque applied about the tibial axis and an internal one with a +3Nm torque. Six radio opaque markers were inserted in both the femur and the tibia. A Roentgen Stereophotogrammetric Analysis system was used to reconstruct the 3D positions of these artificial landmarks with an accuracy of 0.01 mm. The knees were rotated from full extension to 150 degrees of knee flexion in 15 degrees increments. The knees were then dissected in order to reach the insertion sites of the ACL. Five fibers were identified and then marked at their femoral and tibial insertions with additional radio opaque markers. An additional Roentgenogram provided the relative position of these new markers with respect to the previous markers in the femur and the tibia. In the present work, an algorithm was developed to get the rotational matrix and translation vector between a set of points at 2 different positions. Hence, the kinematics of the ACL insertions zones were determined e.g. [5]. These kinematics were used as boundary conditions for the numerical model. Three kinds of kinematics can be applied: neutral, internal or external corresponding to a normal flexion, a flexion with an internal rotational torque or a flexion with an external rotational torque, respectively.

Tibial drawer test. The tibial drawer test allows the clinician to determine the antero-posterior laxity of a knee when the patient is supine and his knee is at 20° of flexion. The clinician applies an anterior load perpendicular to tibial axis and evaluates the corresponding laxity. This test was numerically simulated in the following way. Next the initial stress and the kinematics corresponding to the knee flexion of 20° were applied, an anterior 4mm displacement was imposed on the tibial bony insertion. Four millimeter tibial drawer tests represent a mean value of what can be found during clinical tests with an intact ACL [13].

3.5. Validations

The influence of the ligament geometry on the stress distribution was tested. Different ACL geometries were generated using identical insertion sites. Minor stress differences were found between the geometries. In the limit of an "acceptable" geometry (i.e. with some ACL resemblances), the general stress trends were similar regardless of the geometry.

Experimental traction tests were simulated with two different boundary conditions. First, the ligament was represented by a cylinder. This cylinder was free to deform in the transverse plane, even at its extremities. This particular problem has an analytical solution which is compared to the numerical one at different strains (Fig. 2. A). Verification of the computed constitutive law was achieved in this way. Secondly, the reconstructed ligament geometry was used. Loads similar to the experimental values were applied and the ligament extremities were not allowed to deform in the transverse plane. This situation corresponds to the experimental case. Here, no analytical solutions exist and therefore numerical solutions (in the middle of the reconstructed ligament) were compared to the present experimental results to check the meshing refinement. Obviously, the precision of the numerical solution increases with the mesh refinement. A compromise must be found between the number of elements used (computational time consuming) and the desired precision of the solution. This precision was judged acceptable (Fig. 2. B) when cubic elements were used and higher element density was applied near the insertion zones.

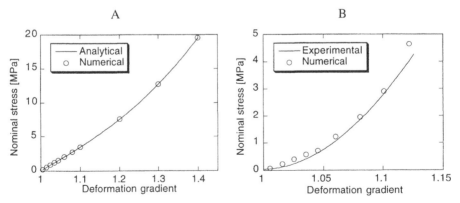

Fig. 2. A: comparison between analytical and numerical solutions in case of traction tests performed on a cylinder; B: comparison between experimental and numerical stress in case of traction tests.

4. RESULTS

4.1. Results variables

Two stress invariants results (hydrostatic stress and Von Mises stress) are presented. The stress invariants fields in the ACL are presented under an antero-posterior view of the knee and are reported in the initial position of the knee (full extension) in order to compare the results. The results are displayed in the form of fringe plots where the black and white represents the largest and the smallest values, respectively of hydrostatic and Von Mises stresses.

4.2. Flexion of the knee

We present the results at 20° for the neutral case (Fig. 3.). During knee flexion, an

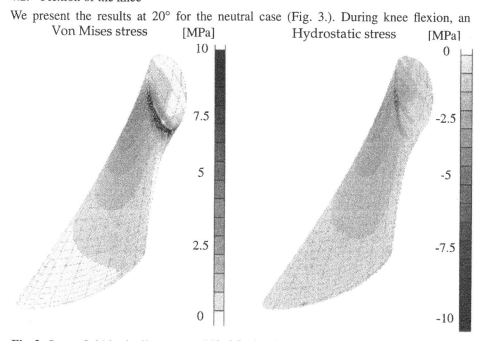

Fig. 3. Stress field in the ligament at 20° of flexion for the neutral case.

increase in both the Von Mises and hydrostatic stress are found near the anterior femoral insertion of the ligament. The region around the tibial insertion is not affected much by the knee flexion. The stress in the ligament is increased with the knee in internal rotation.

4.3. Tibial drawer test

Numerical tibial drawer tests are presented for the neutral case with a 4 mm anterior displacement (Fig. 4.).

The tibial drawer tests clearly increase the stress in the ligament. This test allows one to verify that the ACL is a major restraint to antero-posterior motions. This increase of the ligament stress was found for all three different cases. However, external case was found to slightly decrease the stress in regards to the neutral case, while the internal case increases the ligament stress.

Von Mises stress   [MPa]   Hydrostatic stress   [MPa]

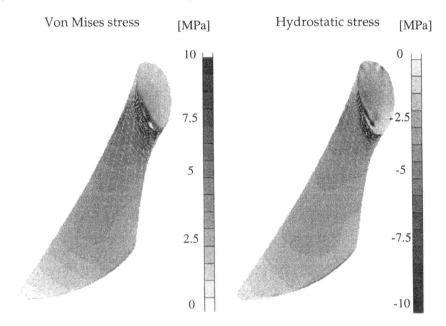

Fig. 4. Stress filed in the ligament during a tibial drawer test for the neutral case.

5. DISCUSSION

The numerical results were compared to experimental studies of knee flexion and tibial drawer tests. In the present study, the calculated stress field was inhomogeneous in the ACL during flexion and tibial drawer tests. The highest calculated stress was located in the anteriormedial portion of the ligament near the femoral insertion. Clinical study has shown that ruptures of the ACL mainly occur in the intrasubstance of the ligament [14]. Moreover, partial ruptures of the anteriormedial ACL bundle have been described as the consequence of stress increase during knee flexion [15]. It was concluded that the highest stress is located in the anteriormedial intrasubstance of the ligament which confirms the present numerical results.

The simulated tibial drawer test clearly highlighted that the anteriormedial fibers of the ACL are an important restraint to anterior displacement of the tibia. This result is also confirmed by *in vitro* studies [9, 16]. Internal rotation of the knee increased the calculated stress in the ACL for the flexion and tibial drawer test. Effects of this internal rotation on the ACL was experimentally investigated with either knee flexion or tibial drawer tests [9]. Results showed that internal rotation increases the stress in the ACL and they are, therefore, in accordance with the present numerical study.

Despite good correlation were found between the present numerical study and experimental studies, several limitations of the numerical model should be noted. The present study considers the ligament as a single bundle. The ACL is in fact composed of two principal bundles: an anteriormedial and a posteriorlateral bundles. *In vitro* studies demonstrated that the anteriormedial bundle of the ACL is mainly stressed during knee flexions or antero-posterior motions e.g. [16]. Addition of the posteriorlateral bundle would certainly increase the reliability of the numerical model when the knee is at full extension as it was shown that the posteriorlateral bundle contributes to the stability of the knee at that position e.g. [15]. During knee flexion, the posteriorlateral bundle plays a secondary role for the knee stability. The cruciate ligaments wrap around each other during knee flexion. The consequence of this wrapping could be an increase of the stress value in the ACL. Despite the above limitations, the present numerical study gives results that are in accordance with experimental studies. Therefore, the numerical simulations furnished reliable results for comparing the effects of different tests.

A comparison with other numerical studies is difficult because, to the best of our knowledge, no finite model elements for cruciate ligaments are available in the literature. Until now, numerical studies involving flexion of the knee considered the cruciate ligaments as linear or non-linear springs e.g. [1, 2]. The results of these studies can at best give the resultant force of the ligament but not the stress field.

The present numerical study gives complementary results to the experimental studies. It was shown that the resultant force decreases in the ligament as the knee flexion increases e.g. [10, 11]. Similar results are obtained in the present numerical study if only the posterior part of the ligament near the tibial insertion is considered. This is not the case if anteriormedial part of the ligament is considered. In the present numerical study, anteriormedial fibers clearly showed a stress increase when the flexion increased, a situation also observed in a study concerned with only anteriormedial fibers e.g. [17]. Hence, differences in stress distribution can be important and may be missed with experimental studies. *In vitro* studies, which furnish the resultant force in the ACL, can only describe partially the mechanical behavior of the ligament during a flexion or a tibial drawer test. The present numerical simulations, which gives the stress field, provides a model that is complementary to experimental results.

6. REFERENCES

1. Andriacchi T. P., Mikosz R. P., Hampton S. J., and Galante J. O., Model studies of the stiffness characteristics of the human knee joint, J Biomech, 1983, Vol. 16, 23-29.

2. Mommersteeg T. J. A., Blankevoort L., Huiskes R., Kooloos J. G. M., and Kauer J. M. G., Characterization of the mechanical behavior of human knee ligaments: a numerical-experimental approach, J Biomech, 1996, Vol. 29, 151-160.

3. Harper K. A., and Grood E. S., A three dimensional finite element model of an ACL reconstruction, *In* ASME Bioengineering Conference, 1993, 592-595.

4. Weiss J. A., A constitutive model and finite element representation for transversely isotropic soft tissues., PhD Thesis, University of Utah, 1994

5. Pioletti D. P., Heegaard J. H., Rakotomanana R. L., Leyvraz P. F., and Blankevoort L., Experimental and mathematical methods for representing relative surface elongation of the ACL, J Biomech, 1995, Vol. 28, 1123-1126.

6. Pioletti D. P., Rakotomanana L., Gilliéron C., Leyvraz P. F., and Benvenuti J. F. Non-linear viscoelasticity of the ACL: Experiments and theory. *In* Computer Methods in Biomechanics and Biomedical Engineering. Gordon & Breach. 1996, 271-280.

7. Pioletti D. P., Viscoelastic properties of soft tissues: application to knee ligaments and tendons, PhD Thesis, EPFL, 1997

8. Cowin S. C., Bone mechanics, CRC Press, 1989.

9. Dürselen L., Claes L., and Kiefer H., The influence of muscle forces and external loads on cruciate ligament strain, Am J Sports Med, 1996, Vol. 23, 129-136.

10. Roberts C. S., Cummings J. F., Grood E. S., and Noyes F. R., In-vivo measurement of human anterior cruciate ligament forces during knee extension exercises, *In* 40th ORS, New Orleans, 1994, 84-15.

11. Wascher D. C., Markolf K. T., Shapiro M. S., and Finerman G. A., Direct in vitro measurement of forces in the cruciate ligaments. Part I: the effect of multiplane loading in the intact knee, J Bone Joint Surg, 1993, Vol. 75-A, 377-386.

12. Heegaard J. H., Leyvraz P.-F, van Kampen A., Rakotomanana L., Rubin P. J., Blankevoort L., Influence of soft structures on patellar three-dimensional tracking, Clin Orthop, 1994, Vol. 299, 235-243.

13. Benvenuti J. F., Vallotton J. A., Meystre J. L., and Leyvraz P. F., Objective assessment of the anterior tibial translation in Lachman test position: comparison between three types of measurement, in revision for Knee Surg Sports Trauma Arthro, 1997.

14. Duncan J. B., Hunter R., Purnell M., and Freeman J., Meniscal injuries associated with acute anterior cruciate ligament tears in alpine skiers, Am J Sports Med, 1995, Vol. 23, 170-172.

15. Fruensgaard S., and Johannsen H. V., Incomplete ruptures of the anterior cruciate ligament, J Bone Joint Surg, 1989, Vol. 71-B, 526-530.

16. Blomstrom G. L., Livesay G. A., Fujle H., Smith B. A., Kashiwaguchi S., and Woo S. L. Y., Distribution of in-situ forces within the human anterior cruciate ligament, *In* ASME Bioengineering Conference, 1993, 359-362.

17. Takai S., Woo S. L. Y., Livesay G. A., Adams D. J., and Fu F. H., Determination of the in situ loads on the human anterior cruciate ligament, J Ortho Res, 1993, Vol. 11, 686-695.

AN AXISYMMETRIC MODEL FOR JOINT STATIC CONTACT STRESS ANALYSIS

M. Sakamoto[1] and T. Hara[2]

1. ABSTRACT

The static stress distribution on the surface of contact between an anisotropic layered elastic sphere and an anisotropic layered elastic cavity is estimated utilizing an analytical axisymmetric model to simulate contact of diarthrodial joints. The problem is equivalent to a mixed boundary-value problem of the theory of elasticity. In stead of using the Fredholm integral equations, an exact and complete analytical solution is obtained through a system of simultaneous equations. Numerical results are obtained with indication of the effect of material anisotropy of the cartilage layer, layer thickness and joint conformity on the contact stress distributions.

2. INTRODUCTION

It has been proposed that abnormal mechanical stress applied to an anatomical joint has an effect on the initiation or progression of degenerative joint disease. Common biomechanical articular joint models treated articular cartilage as isotropic and elastic layer [1,2]. However the ultrastructure of articular cartilage has been described to have a superficial tangential zone, a layer tightly-woven fibrils, thus providing a high tangential stiffness [3].

The purpose of this study was to define the mechanical environment of the joint, specifically, the contact stress distribution, through the development and analysis of an axisymmetric joint model. The axisymmetric contact stress distribution in an elastic, homogeneous and transversely isotropic cartilage layer was solved by a formulated mixed-boundary value problem satisfying the field equations of the theory of elasticity.

Keywords: Joint contact, Elasticity, Anisotropy, Stress analysis, Articular cartilage

[1]Associate Professor, Biomechanics Laboratory, Department of Mechanical Engineering, Niigata College of Technology, 5-13-7, Kamishin'ei-cho. Niigata, 950-21, Japan
[2]Professor, Biomechanics Laboratory, Department of Mechanical Engineering, Niigata University, 8050, Ikarashi 2, Niigata, 950-21, Japan

The problem was reduced to solutions of two sets of unknown coefficients from a system of simultaneous equation by means of a technique [4,5] of expressing the normal contact stress as an appropriate series. Significant effects of the joint conformity and the cartilage layer thickness on the contact stress distributions were demonstrated with numerical results.

3. MATHEMATICAL FORMULATION AND SOLUTION

We simulate an axisymmetric diarthrodial joint model of which a subchondral bone is a rigid substrate and an articular cartilage is a transversely isotropic layer (z-axis: principal direction of anisotropy) as shown in Fig. 1. Using assumptions of Hertz theory, the boundary conditions on the layer as follows:

$$(w_z)_{z=0} = (\alpha - r^2 / 2R)/2, \quad (0 \le r \le a), \tag{1a}$$
$$(\sigma_z)_{z=0} = 0, \quad (a < r < \infty), \tag{1b}$$
$$(\tau_{rz})_{z=0} = 0, \quad (0 \le r < \infty), \tag{1c}$$
$$(u_r)_{z=h} = (w_z)_{z=h} = 0, \quad (0 \le r < \infty), \tag{1d}$$

where α is a relative displacement of two solids and

$$R^{-1} = R_1^{-1} - R_2^{-1}. \tag{2}$$

Following Elliott [6], to satisfy the field equations of the linear theory of elasticity, displacements and stresses for transversely isotropic medium without torsion are expressed in terms of two quasi-harmonic potential functions ψ_1 and ψ_2 through the relations

$$u_r = \partial \psi_i / \partial r, \quad w_z = k_i \partial \psi_i / \partial z,$$
$$\sigma_z / c_{44} = \mu_i (1 + k_i) \partial^2 \psi_i / \partial z^2, \quad \tau_{rz} / c_{44} = (1 + k_i) \partial^2 \psi_i / (\partial r \partial z), \tag{3}$$

where μ_i and k_i, $(i = 1, 2)$ are dependent on the five elastic constants c_{11}, c_{13}, c_{33} and c_{44}. The parameters μ_i and k_i are defined by

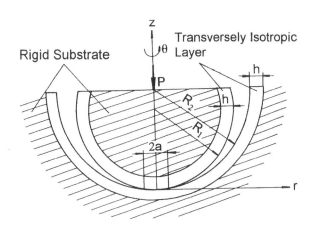

Fig. 1. Geometry of the axisymmetric joint model.

$$c_{11}c_{44}\mu^2 + \{c_{13}(c_{13} + 2c_{44}) - c_{11}c_{33}\}\mu + c_{33}c_{44} = 0, \qquad (4)$$

$$k_i = (c_{11}\mu_i - c_{44})/(c_{13} + c_{44}), \quad (i = 1,2). \qquad (5)$$

The potential functions ψ_i can be taken as

$$\psi_i = \int_0^\infty \{A_i(\lambda)\cosh\lambda z_i + B_i(\lambda)\sinh\lambda z_i\}J_0(\lambda r)d\lambda, \quad (i = 1,2), \qquad (6)$$

where $J_n(\lambda r)$ is the Bessel function of the first kind of order n, $A_i(\lambda)$ and $B_i(\lambda)$ are functions which can be obtained by matching appropriate boundary conditions, and

$$z_i = z/\mu_i^{1/2}, \quad (i = 1,2). \qquad (7)$$

By substituting Eq. (6) into Eq. (3) and using boundary conditions (1c) and (1d), we obtain

$$A_1(\lambda) = g_1(\lambda)B_2(\lambda)/g_0(\lambda),$$
$$A_2(\lambda) = g_2(\lambda)B_2(\lambda)/g_0(\lambda), \qquad (8)$$
$$B_1(\lambda) = -\gamma(1 + k_2)B_2(\lambda)/(1 + k_1),$$

where

$$g_0(\lambda) = k_2\cosh\lambda h_1 \cdot \sinh\lambda h_2 - k_1\sinh\lambda h_1 \cdot \cosh\lambda h_2/\gamma,$$
$$g_1(\lambda) = k_2(1 + \gamma k_2\sinh\lambda h_1 \cdot \sinh\lambda h_2) - \cosh\lambda h_1 \cdot \cosh\lambda h_2, \qquad (9)$$
$$g_2(\lambda) = 1 + k_1\sinh\lambda h_1 \cdot \sinh\lambda h_2/\gamma - k_2\cosh\lambda h_1 \cdot \cosh\lambda h_2,$$
$$h_i = h/\mu_i^{1/2}, \quad (i = 1,2), \qquad (10)$$
$$\gamma = (\mu_1/\mu_2)^{1/2}. \qquad (11)$$

The normal displacement w_z and stress σ_z of Eq. (3) can be derived using functions ψ_i of Eq. (6). The unknown function $B_2(\lambda)$ can be determined by substitute w_z and σ_z into boundary conditions (1a) and (1b), these two boundary conditions can be written by the following dual integral equations:

$$(w_z)_{z=0} = -\frac{1 - k_2}{\mu_2^{1/2}}\int_0^\infty B_2(\lambda)\lambda J_0(\lambda r)d\lambda = \frac{1}{2}\left(\alpha - \frac{r^2}{2R}\right), \quad (0 \le r \le a), \qquad (12)$$

$$(\sigma_z)_{z=0} = c_{44}(1 + k_2)\int_0^\infty s(\lambda)B_2(\lambda)\lambda^2 J_0(\lambda r)d\lambda = 0, \quad (a < r < \infty), \qquad (13)$$

where

$$s(\lambda) = \{2 + (\gamma k_2 + k_1/\gamma)\sinh\lambda h_1 \cdot \sinh\lambda h_2 - (k_1 + k_2)\cosh\lambda h_1 \cdot \cosh\lambda h_2\}/g_0(\lambda). \qquad (14)$$

The dual integral equations (12) and (13) are usually transformed into a Fredholm integral equation [2]. In this study, a different technique is utilized, where the normal contact stress $(\sigma_z)_{z=0}$ is expressed as an appropriate series function as follow [4,5]:

$$(\sigma_z)_{z=0} = \frac{2c_{44}}{\pi r(a^2 - r^2)^{1/2}}\sum_{n=0}^\infty a_n T_{2n+1}(r/a), \quad (0 \le r \le a), \qquad (15)$$

where a_n $(n = 0, 1, 2, ...)$ are unknown coefficients and $T_{2n+1}(r/a)$ represent Tchebycheff polynomials. Since the contact stress $(\sigma_z)_{z=0}$ is finite as $r \to a_{-0}$, we obtain

$$\sum_{n=0}^{\infty} a_n = 0. \tag{16}$$

Using the entity

$$\int_0^\infty \lambda Z_n(\lambda) J_0(\lambda r) d\lambda = \begin{cases} 0, & (a < r < \infty), \\ \dfrac{2T_{2n+1}(r/a)}{\pi r (a^2 - r^2)^{1/2}}, & (0 < r < a), \end{cases} \tag{17}$$

where

$$Z_n(\lambda) = J_{n+1/2}(\lambda a/2) \cdot J_{-n-(1/2)}(\lambda a/2), \quad (n = 0, 1, 2, ...), \tag{18}$$

the Hankel inversion of Eqs. (13) and (15) yields

$$\lambda B_2(\lambda) = \frac{t(\lambda)}{1 + k_2} \sum_{n=0}^{\infty} a_n Z_n(\lambda), \tag{19}$$

where

$$t(\lambda) = 1 / s(\lambda). \tag{20}$$

Substituting Eq. (19) into Eq. (12) and using Gegenbauer's formula

$$J_0(\lambda r) = \sum_{n=0}^{\infty} (2 - \delta_{0m}) X_m(\lambda) \cos m\phi, \quad (r = a \sin(\phi/2)), \tag{21}$$

we obtain

$$\sum_{n=0}^{\infty} a_n \int_0^\infty t(\lambda) Z_n(\lambda) \sum_{m=0}^{\infty} (2 - \delta_{0m}) X_m(\lambda) \cos m\phi \, d\lambda$$
$$= -\frac{\mu_2^{1/2}(1 + k_2)}{2(1 - k_2)} \left\{ \alpha - \frac{a^2}{4R}(1 - \cos\phi) \right\}, \quad (0 \le r \le a), \tag{22}$$

where δ_{0m} is Kronecker's delta function and

$$X_m(\lambda) = J_m^2(\lambda a/2), \quad (m = 0, 1, 2, ...). \tag{23}$$

The unknown coefficients a_n can be written as

$$a_n = -\frac{\mu_2^{1/2}(1 + k_2)}{2(1 - k_2)} \left\{ \alpha b_n - \frac{a^2}{4R} c_n \right\}, \quad (n = 0, 1, 2, ...), \tag{24}$$

where b_n and c_n are also unknown coefficients. Substituting Eq. (24) into Eq. (22) and matching the coefficients of $\cos m\phi$ on both sides of Eq. (22), we obtain the following infinite system of simultaneous equations:

$$\sum_{n=0}^{\infty}(b_n,c_n)I_{mn} = (\delta_{0m},\delta_{0m}-\delta_{1m}/2), \quad (m = 0,1,2,...),\tag{25}$$

where

$$I_{mn} = \int_0^{\infty} t(\lambda)X_m(\lambda)Z_n(\lambda)d\lambda.\tag{26}$$

By means of Eqs. (16) and (24), we obtain

$$\frac{a^2}{4R\alpha} = \sum_{n=0}^{\infty}b_n / \sum_{n=0}^{\infty}c_n = \xi.\tag{27}$$

The above analysis shows that by expressing the normal contact stress in the series function (15), the dual integral equations (12) and (13) can be simplified in to a set of simultaneous equations expressed in Eq. (25). The normal contact stress $(\sigma_z)_{z=0}$ in Eq. (15) can be rewritten in the following form:

$$(\sigma_z)_{z=0} = -\frac{\kappa}{\pi r(a^2-r^2)^{1/2}} \sum_{n=0}^{\infty}(b_n - \xi c_n)T_{2n+1}(r/a), \quad (0 \le r \le a),\tag{28}$$

where

$$\kappa = c_{44}\alpha\mu_2^{1/2}(1+k_2)/(1-k_2).\tag{29}$$

The relationship between the relative approach α, and the applied normal load P, is obtained by equating load P with the integration of $(\sigma_z)_{z=0}$ over the contact area, i.e.

$$P = -2\pi\int_0^a r(\sigma_z)_{z=0}dr = 2\kappa \sum_{n=0}^{\infty}(-1)^n(b_n - \xi c_n)/(2n+1).\tag{30}$$

The anisotropic parameters μ_i and k_i obey the following relations:

$$\mu_i = \{\chi \pm (\chi^2 - 4\omega)^{1/2}\}/2, \quad (i = 1,2),$$
$$k_i = \frac{(\zeta - v'^2)\mu_i - \rho\{(1-v)\zeta - 2v'^2\}/2}{(1+v)v'\zeta + \rho\{(1-v)\zeta - 2v'^2\}/2}, \quad (i = 1,2),\tag{31}$$

$$\zeta = E'/E, \quad \rho = G'/G,$$
$$\chi = 2\zeta(1+v)(\zeta - v'\rho)/\{\rho(\zeta - v'^2)\},\tag{32}$$
$$\omega = \zeta^2(1-v^2)/(\zeta - v'^2),$$

where E, G and v are Young's modulus, shear modulus and Poisson's ratio in the isotropic plane, respectively. E', G' and v' are engineering constants along the principal direction of anisotropy. The thermodynamic constraint is also imposed on these constants [7], i.e.

$$E, E', G, G' > 0, \quad |v'| < \zeta^{1/2}, \quad -1 < v < 1 - 2v'^2 / \zeta. \quad (33)$$

4. NUMERICAL RESULTS AND DISCUSSION

The material properties of both isotropic and transversely isotropic cartilage layer are listed in Table 1. In Table 2, we show the convergence of the coefficients $b_n - \xi c_n$ for the transversely isotropic layer. Even though the convergence of $b_n - \xi c_n$ becomes slow as the ratio h/a decreases, 10 terms of the series are sufficient to calculate the stress distribution in the case of $h/a = 0.3$.

Table 1. Material properties of the cartilage layer.

	E (MPa)	E' (MPa)	G (MPa)	G' (MPa)	v	v'
Trans. Iso.	50	10	17.9	5.4	0.4	0.2
Iso.	15	15	5.4	5.4	0.4	0.4

Table 2. Values of coefficients $b_n - \xi c_n$.

n	h/a=0.3	h/a=0.5	h/a=1.0	h/a=4.0
0	5.535473824	3.382040977	2.025694132	1.289977312
1	-6.306169510	-3.619312763	-2.053349972	-1.290034294
2	0.799835265	0.252764106	0.029167790	0.000057219
3	-0.020161286	-0.014854137	-0.001562245	-0.000000405
4	-0.008699657	-0.000780000	0.000050583	0.000000003
5	-0.000517010	0.000137569	-0.000000181	
6	0.000206735	0.000006375	-0.000000090	
7	0.000037358	-0.000001858	0.000000005	
8	-0.000003961			
9	-0.000001725			

To compare the difference of contact stress in the joint due to different layer thickness of articular cartilage, three layer thickness, h=2mm, 3mm and 4mm are chosen to represent a thin, a mid-thickness and a thick cartilage, respectively. Figure 2 shows the radial distribution of the normal contact stress $(\sigma_z)_{z=0}$ for the transversely isotropic layer as a function of layer thickness h. The maximum normal contact stress increases as the cartilage layer thickness becomes smaller. With increasing thickness, the contact area increases. The distribution of $(\sigma_z)_{z=0}$ for the transversely isotropic layer is shown in Fig. 3 for various radii R_1 of sphere. It is noted that the decrease in the indenting radius results in decreased contact area, with the increase in maximum stress. Figure 4 shows the distribution of $(\sigma_z)_{z=0}$ for both isotropic and transversely isotropic cartilage layer. It is noted that the maximum stress for transversely isotropic layer is lower than the corresponding value for isotropic layer.

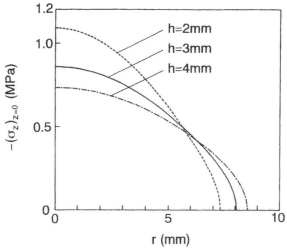

Fig. 2. Radial distribution of normal contact stress (P=100N, R_1=30mm, R_2=40mm).

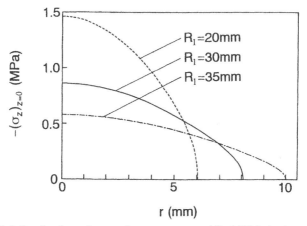

Fig. 3. Radial distribution of normal contact stress (P=100N, h=3mm, R_2=40mm).

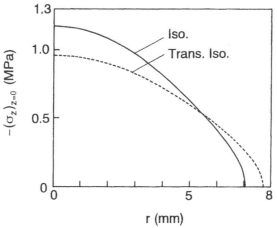

Fig. 4. Radial distribution of normal contact stress
(P=100N, h=2.5mm, R_1=30mm, R_2=40mm).

The radial distribution of the normal contact stress $(\sigma_z)_{z=0}$ is normalized as the dimensionless stress variable $(\overline{\sigma}_z)_{z=0} = \pi a^2 (\sigma_z)_{z=0} / P$ in Fig. 5. The normalized stress $(\overline{\sigma}_z)_{z=0}$ is independent of the radii R_1 and R_2. It should be noted that $(\overline{\sigma}_z)_{z=0}$ for isotropic and transversely isotropic layers in the case of $4 \le h/a$ agree with the following Hertz solution [8]:

$$(\overline{\sigma}_z)_{z=0} = \pi a^2 (\sigma_z)_{z=0} / P = -3(1 - r^2/a^2)^{1/2} / 2. \tag{34}$$

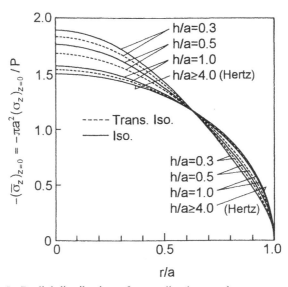

Fig. 5. Radial distribution of normalized normal contact stress.

5. REFERENCES

1. An, K. N., Himeno, S. Tsumura, H., Kawai, T. and Chao, E. Y., Pressure distribution on articular surfaces: application to joint stability evaluation, J. Biomech., 1990, Vol. 23, 1013-1020.

2. Eberhardt, A. W., Keer, L. M., Lewis, J. L. and Vithoontien, V., An analytical model of joint contact, J. Biomech. Eng., 1990, Vol. 112, 407-413.

3. Mow, V. C., Lai, W. M. and Redler, I., Some surface characteristics of articular cartilage, part I: a scanning electron microscopy study and a theoretical model for the dynamic interaction of synovial fluid and articular cartilage, J. Biomech., 1974, Vol. 7, 449-456.

4. Sakamoto, M., Hara, T., Shibuya, T. and Koizumi, T., Axisymmetric contact problem for two rigid spheres coated with elastic layers, Trans. JSME, 1993, Vol. 59A, 963-969.

5. Sakamoto, M., Li, G., Hara, T. and Chao, E. Y. S., A new method for theoretical analysis of static indentation test, J. Biomech., 1996, Vol. 29, 679-686.

6. Elliott, H. A., Three-dimensional stress distributions in hexagonal aeolotropic crystals, Proc. Camb. Phil. Soc., 1948, Vol. 44, 522-533.

7. Lempriere, B. M., Poisson's ratio in orthotropic materials, AIAA J., 1968, Vol. 6, 2226-2227.

8. Johnson, K. L., Contact Mechanics (Cambridge University Press, 1985), 84-106.

TWO DIMENSIONAL FINITE ELEMENT MODEL OF A TRANSVERSE SECTION OF THE TRANS-FEMORAL AMPUTEE'S STUMP

V.S.P Lee[1], P. Gross[2], W.D. Spence[3], S.E. Solomonidis[4], and J.P. Paul[5]

1. ABSTRACT

A two dimensional finite element (FE) model of a transverse section of a trans-femoral stump at a level 30 mm distal to the ischial tuberosity was generated. The model was used to investigate the three major aspects that affect the stress distribution in the stump : non-homogeneity of soft tissue material properties, intermuscular sliding and various stump / prosthetic socket boundary conditions. The results indicated that high stresses exist within the stump at the interface between the muscles and intermuscular tissues. The stress magnitude within the stump could be increased by as much as 60% when material non-homogeneity was considered in the model. In general the stresses were found to be reduced by approximately 20% when muscles were allowed to slide within the stump tissues.

2. INTRODUCTION

The main aims of a lower limb prosthesis is to replace some of the lost limb's functions and cosmesis due to amputation. Its construction consists of a socket enclosing the amputee's stump, and other components like the knee joint, shank and foot. The socket is complex in shape. It has to match the shape of the stump closely, but at the same time provide adequate support during standing and walking. In recent years, the biomechanics of the prosthetic socket has been investigated using the finite element (FE) methods (1). The goal is to apply a desirable surface stress distribution to FE

Keywords: Finite element analysis, biomechanics, amputee, stump, socket, artificial limb, soft tissue

---

[1]   Research Fellow, Bioengineering Unit, University of Strathclyde, Wolfson Centre, Glasgow, G4 0NW. UK.
[2]   Student, Bioengineering Unit, University of Strathclyde, Wolfson Centre, Glasgow, G4 0NW. UK.
[3]   Prosthetist/Orthotist, Bioengineering Unit, University of Strathclyde, Wolfson Centre, Glasgow, G4 0NW. UK.
[4]   Senior Lecturer, Bioengineering Unit, University of Strathclyde, Wolfson Centre, Glasgow, G4 0NW. UK.
[5]   Professor, Bioengineering Unit, University of Strathclyde, Wolfson Centre, Glasgow, G4 0NW. UK.

models of stump, using FE analysis to solve for the resultant surface displacements, and output this shape to create a suitable socket. Previous FE models **(1)** of the stump have demonstrated their potential to predict the interface pressures at the patient / prosthesis interface. These models were attempted with limited geometrical, material, loading and boundary considerations. Some of these considerations have been shown to be highly sensitive to the predicted stress distribution in a two dimensional model of a transverse slice of the trans-tibial stump **(2)**. In this study, it is envisaged that the future aim is to create a three dimenional (3-D) model of the stump with sufficient detail to predict its mechanical responses accurately. The approach is to model a transverse section of the stump with accurate geometrical data in 2-D. The 2-D model will concentrate on studying the three biomechanical aspects that affect the stress distribution in the stump, material non - homogeneity, muscle slippage and changes in loading and boundary conditions.

3. METHOD

3.1 Geometry

A transverse section of the stump was selected for modelling in 2-D. The transverse slice selected described the stump of a trans-femoral amputee located at a level 30 mm distal to the ischial tuberosity. Its geometry was obtained from magnetic resonance imaging (MRI) scans (Figure 1). The scanning procedures have been previously reported **(3)**. The selected

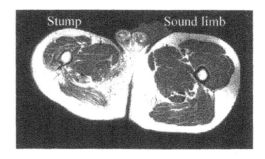

Fig. 1 MRI image of the stump without the socket at a level immediately distal to the ischial tuberosity.

level was located where the stump tissue is subjected to large deformations caused by the socket brim, thus modelling is attempted at a severe loading situation. The muscles at this level are still actively used by the amputee and are geometrically well defined. The soft tissues in the model are generally divided into three main types, fascia, intermuscular tissues and muscles. The intermuscular tissues are basically the deep fascia layers separating the different muscle groups. The contours of the different tissues are manually digitised and represented as x, y co-ordinates in the FE software ANSYS 5.0 (Swanson Analysis Inc., USA).

3.2 Finite element mesh

The general element used was the quadrilateral 4-noded element (ANSYS PLANE42). In the model for studying muscle slippage, link and interface elements were also included to describe the biomechanics at the interface. Figure 2 shows the final mesh of the transverse slice of the stump in its natural unloaded shape. The basic mesh consisted of 2826 elements and 8441 nodes.

3.3 Loading and boundary conditions

The 2-D model of the stump was loaded by displacing its external contour to the shape of the donned socket. A MRI image of the stump donned with a socket was therefore required The image was of the same subject and at the same transverse level as the natural unloaded stump; its geometry could therefore be compared directly. The bone in the model was assumed to be infinitely rigid compared to the soft tissues and fixed in position relative to the soft tissue (Figure 3).

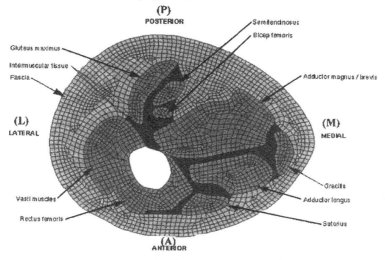

Fig. 2 Finite element mesh with the different types of soft tissue

4. ANALYSIS

4.1 FE models studying tissue non-homogeneity

The first model (model 1a, see Table 1) assumed all soft tissue to be of uniform material properties, linear elastic and isotropic. This was considered a suitable first step to understand tissue non-homogeneity. The next model 1b assigned different material properties for the different tissue types.

4.2 FE models studying musculature movement

The aim of this analysis was to incorporate the intermuscular septum which surrounds the muscles. The septum is a tough connective tissue which binds muscle or muscle groups together and provides the means for the muscles to slide over each other. The approach was to model

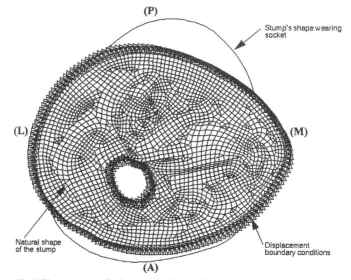

Fig. 3 The stump was displaced to the shape of the socket based on surface nodal displacements.

the septum (model 2a) using link elements (LINK10). LINK10 is a 2-noded spar element active only in uniaxial tension, very much like that of a cable. This assumption seems feasible since anatomically the septum only resists tension forces. The Young's modulus for the septum was assumed to be 400 kPa and Poisson's ratio of 0.4.

Model no.	Elastic modulus (kPa) / Poisson's ratio				Loading and boundary conditions
	Fascia	Inter-muscular tissues	Muscles	Femoral bone	
1a	36.6 / 0.49	36.6 / 0.49	36.6 / 0.49	Rigid	Nodal displacements at stump surface and fixed femoral bone.
1b	36.6 / 0.49	36.6 / 0.49	95.6 / 0.49	Rigid	Nodal displacements at stump surface and fixed femoral bone.
2a	36.6 / 0.49	36.6 / 0.49	95.6 / 0.49	Rigid	Nodal displacements at stump surface and fixed femoral bone. Link elements were used to model the intermuscular septum.
2b	36.6 / 0.49	36.6 / 0.49	95.6 / 0.49	Rigid	Nodal displacements at stump surface and fixed femoral bone. Interface elements were used to model muscle slippage.
3a	36.6 / 0.49	36.6 / 0.49	95.6 / 0.49	15.8GPa / 0.3	Nodal displacements at stump surface and femoral bone was **not** fixed.
3b	36.6 / 0.49	36.6 / 0.49	95.6 / 0.49	Rigid	Sliding allowed at stump surface and fixed femoral bone.

Table 1 Material properties, loading and boundary conditions applied to the 2-D model of the stump

In place of the LINK10 elements, interface elements were introduced between ·the muscles in order to allow muscle sliding to take place (model 2b). In some areas around the muscle tissues, a clear separation from the intermuscular tissues could be observed. Interface elements in the model were thought to be most suitable since they allow muscle sliding and separation to take place. The type of interface element used is CONTACT48. A fully symmetrical contact condition was assumed in which interface elements were specified on both the intermuscular tissues and the muscles, allowing them to act as either contact or target surfaces. The model incorporated a total of 3279 interface elements. Three values of coefficient of friction at the interface between the muscles and the intermuscular tissues were tested, zero i.e. frictionless, 0.1 and 0.3.

4.3 FE models studying different loading and boundary conditions

In addition to the first set of conditions discussed in section 3.3, a further two different sets of conditions were investigated. In Model 3a the bone was assumed not to be fixed in position but allowed to move among the soft tissues. The bone was not modelled as a rigid surface but assigned a Young's modulus of 15.8 GPa and a Poisson's ratio of 0.3.

It is thought that a more realistic loading and boundary condition can be achieved by introducing only compressive forces at the stump / socket interface. The stump in model 1a-3a was loaded by displacing the nodes at the stump surface therefore surface nodes that were displaced towards the femur would generate compressive forces while nodes that were displaced away from the femur would generate tension forces. In order to introduce only compression forces, the set of conditions applied to model 3b allowed the external contour of the stump to slide within a specified boundary, which was equivalent to the shape of the socket. The bone was assumed as fixed and rigid. To model this boundary effect, the nodes defining the external contours of the stump were firstly duplicated. Interface elements (CONTACT 48) were set up between the original nodes and the new replica set of nodes. Finally, to bring the model to the shape of the socket i.e. to load the model, displacements were only introduced to the new duplicated set of nodes. The interface elements let the stump slide along the new duplicated set of nodes acting as the socket.

All the models in this study were attempted under plane stress conditions assuming unit thickness. Large deflection analysis, i.e. geometrical non - linearity was also assumed.

5. RESULTS AND DISCUSSIONS

5.1 Tissue non-homogeneity

High stress was found at the medial side of model 1a where large nodal displacements were imposed (max. displacement 32 mm). The maximum stress (21.9 kPa) was located at the stump / socket interface at the medial side. However in model 1b, the maximum stress (35.3 kPa) was not predicted at the stump / socket interface, instead it was located at the interface of the adductor magnus and the intermuscular tissues separating the adductor magnus/brevis and adductor longus (Figure 4). The stress

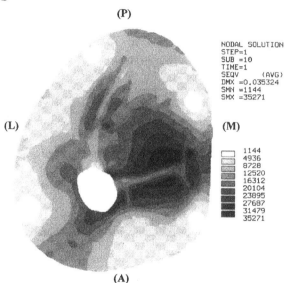

Fig. 4 Von Mises stress distribution of model 1b (non-uniform material properties. (N/m²)

pattern could be explained by the displacements of the muscular tissue as shown in Figure 5. The medial side of the stump was highly compressed which caused the gracilis which lies in a perpendicular fashion with respect to the adductor magnus and longus to push the adductor magnus posteriorly and longus anteriorly. Model 1a assuming uniform material properties and uniform geometry did not detect regions of maximum stress, which occurred near the muscle/intermuscular tissue interface in model 1b. The authors reckon that the internal stress distribution might be a more important factor in deciding socket fit than stump / socket interface stresses. This is because the former stresses are of higher magnitude which can give rise to tissue breakdown. This hypothesis is also supported by investigators studying pressure sores who predicted that pressure sores are initially generated from the inside near bony prominences and extend outward to the skin surface in a pyramidal fashion.

5.2 Soft tissue interface characteristics

The link elements in model 2a gave rise to numerical problems in the analysis at 95% of the applied displacements. This was likely due to the stiffening effect of the link elements which restricted the movements in the quadrilateral elements. Evaluating the predicted von Mises stress distribution at 80 % of the applied displacements, the maximum stress recorded was 27.4 kPa at the medial side of the adductor magnus/brevis. As seen in Figure 3, the nodal displacements in the stump model introduced compression forces in the medio-lateral plane and tension forces in the anterio-posterior plane. The link elements at the posterior region surrounding the gluteus maximus were particularly effective in restraining the applied tension forces. This lead to an overall decrease in the stresses in

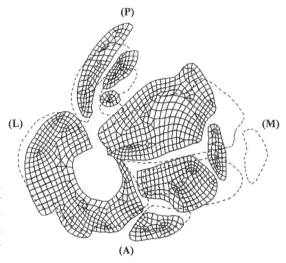

Fig. 5 Predicted displacement plot of the different muscles in the model. (Dotted lines denote the original shape)

the posterior region where the gluteus maximus was located. Similarly, in the anterior region, the link elements which were connected to the femur were under high axial stresses (Figure 6). The link element has the effect of protecting the muscles which it surrounds by absorbing tension forces leaving the muscles unstressed. However, the axial stresses on the link elements at the medio-lateral regions of the stump were almost zero as they appeared as slack cables not resisting compression. This explains the stress distribution in model 2a at the medial side which was exactly the same as that of model 1b. The latter model did not have any link elements.

The model incorporating interface elements encountered numerical convergence problems at 47% of the applied displacement. Increasing the coefficient of friction also increased the computation time and the risk of convergence difficulties. Computation time was also excessive, more than 30 hours with a standalone Silicon Graphics Indigo 2 workstation. Due to numerical problems, the model was examined at 20% of the applied displacement. A deformed

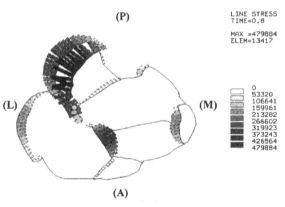

Fig. 6 Predicted axial stresses at the link elements of model 2a at 80% of the applied displacement. (N/m²)

geometry plot of model 2b which assumed frictionless interface elements is as shown in Figure 7. Several muscles were separated from the intermuscular tissue. The largest separation was observed at the anterior region of the stump between the intermuscular tissue and the rector femoris and vasti muscles. The rectus femoris and the vasti muscles were practically rigid compared to the intermuscular tissue, since they were connected firmly to the bone. The forces applied were thus taken up by the anterior intermuscular tissue leaving the muscles unstressed. Due to the compression at the medial side of the stump, the softer intermuscular tissue surrounding the adductor longus was forced to buckle, leaving a gap. There was a clear indication that a lower stress state was maintained when the muscles were allowed to slip. Near zero tensile stresses were predicted at the regions where the muscles were allowed to slip i.e. regions with interface elements. At the medial side of the stump where high stress existed in the gracilis and the adductors of model 1b, the von Mises stresses in model 2b were 20% lower.

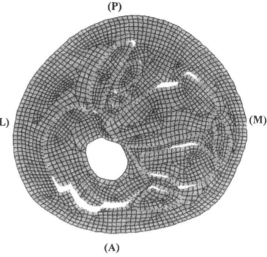

(P)

(L)

(M)

(A)

Fig. 7 Deform geometry of model 2a at 20% of the applied displacement. The deform geometry was enlarged five times for clarity.

An increase in maximum stress was observed when the coefficient of friction was increased. Maximum von Mises stress at 10% of the applied displacement for frictionless, $\mu = 0.1$ and $\mu = 0.3$ models were 3.8 kPa, 4.0 kPa and 4.1 kPa respectively. As friction at the interface between muscles and intermuscular tissue was increased, localised stresses began to surface along the interface. However, the general stress pattern were quite similar regardless of the different coefficient of friction.

5.3 Loading and boundary conditions

The femur was permitted to move among the soft tissue in model 3a by not restraining its boundary in the x and y directions. As expected, such a modification reduced the stress around the bone. At 100% of the applied load, the bone was found to move laterally 2.9 mm and anteriorly 8 mm. Comparing it to model 1b where the bone was fixed, the stress just anterior to the femur was 23 kPa. At the same location in model 3a, the stress was 12 kPa. The stresses at the medial region of the stump were lower in model 3a than in 1b since the bone had moved laterally providing stress relieve.

The aim of model 3b was to eradicate the tension at the anterio-posterior plane. The predicted stress pattern remained unchanged at the medio-lateral plane when compared to model 1b. However at the anterio-posterior plane, compressive stresses began to emerge with model 3b, whereas in model 1b stresses at this region were mainly tensile. The boundary conditions applied in model 3b have successfully reduced the tensile stress at the anterio-posterior plane and maintained the loading at the medio-lateral plane similar to the previous models. However, the model was not able to preserve the shape of the socket in the predicted solution. The stump was in contact at the medial and lateral walls of the socket. However, at the anterior and posterior walls, the stump tissue was not in contact with the socket wall. The compression at the medial and lateral wall was not able to deform the stump so that it filled the anterior and posterior space of the socket.

5.4 Model's assumptions

The tissue components in the model were grouped into four types, fat, intermuscular tissue, muscles and bone. Skin was omitted because it was impossible to geometrically define it in the MRI scans. Furthermore, the mechanical behaviour of skin was predominantly to resist tension, while in this case loading in the stump was mainly compressive. It was therefore assumed that the omission of the skin in the model would not affect the internal stress distribution of the stump severely. The model did not include any of the major vessels and nerves. These types of soft tissues were modelled as either fat or intermuscular tissue instead. Some muscles were difficult to define clearly based on the MRI images. These were the vastus intermedius, lateralis and medialis, and the adductor magnus, brevis and lateralis. In the model, these muscles were grouped together forming the vasti muscles and the adductor magnus/brevis muscle respectively. Restricting the model to only linear elastic material enabled an accurate evaluation of other forms of non linearity which arise due to geometry and interface characteristics of the soft tissues.

The 2-D approach in modelling the transverse section of the stump could only provide limited information. Anatomical joint loading or loading transmitted due to ground reaction forces cannot be modelled. The 2-D model was only feasible in predicting the stress and deformation generated on the stump when deforming to the shape of the socket. The 2-D analysis also assumed that the geometry of the stump was uniform along its length which was not realistic. Therefore the results of the analysis were only applicable to the modelled transverse slice.

6. CONCLUSION

The 2-D model generated was able to predict stresses and deformations within the stump. The internal stress distribution was found to be sensitive to the different tissues' material properties (tissue non-homogeneity), geometry, interface characteristics and loading and boundary conditions.

The complexity of the 2-D model and the non convergence solution in some models have shown that the best possible initial approach to study the internal stress distribution of the stump is a two, instead of a three dimensional model. The 2-D model was able to incorporate geometrical musculature details, material non - homogeneity, soft tissue interface characteristics and complex loading and boundary conditions. It would be an impossible task to incorporate similar specifications in a 3-D model with the present state of the art. Though limited, useful information was provided by the 2-D model in further understanding the biomechanical behaviour at the internal structure of the stump.

REFERENCES
1. Zachariah S.G. and Sanders J.E. Interface mechanics in lower-limb external prosthetics : A review of finite element models. IEEE Trans. Rehab. Eng., 1996, Vol.4, No.4, 288-302.

2. Huang D. and Mak A.F.T. The effects of sliding between muscle groups on stress distribution within below-knee stump. 8th International Conference on Biomedical Engineering, Singapore, 1994, 348-350.

3. V.S.P. Lee, S.E. Solomonidis, W.D Spence. Geometrical study of the transfemoral residual limb (MRI technique). 8th World Congress of the International Society for Prosthetics and Orthotics. Melbourne, Australia, 1995, 127.

NET REINFORCED AORTA: A NON LINEAR ELASTIC MODEL

F. Carli[1], M. Martelli[2]

1. ABSTRACT

In this paper a simplified model for the description of the mechanical behaviour of a reinforced blood vessel is presented. The main aim is to analyse the possible benefit, in terms of mechanical strength, gained in patients affected by pathologies inducing sensible modifications to the physical characters of the artery wall by the insertion and complete inclusion of a prosthesis net with wide openings.

2. INTRODUCTION

In the pathogenesis of aortic aneurysms, acquired or congenital diseases induce modifications of the vascular wall. Those alterations are often characterised by the degeneration process of the elastic fibres causing a lower resistance of the vessel structure and abnormal dilatation of the aorta evolving in time until the eventual rupture or dissection. Prosthetic substitution of thoracic and abdominal aortic aneurysms has recently been successfully accomplished by endovascular techniques, avoiding thoracotomy or laparotomy, however there are some limitations on their use. At first these procedures can not be used in those tracts of the aorta where important collateral branches originate, since these would be occluded by the prosthesis, and on the other hand the prosthetic substitution is associated with a very high level of complication risk for the patient.

An alternative possible way to obtain a wall strengthening is supposed to be the insertion into the aortic lumen of a prosthesis formed out of prolene threads. The rationale is that the net prosthesis maintained in contact with the intimal wall is gradually covered by new intima, invaded by fibroblasts and therefore in stable configuration with the aortic wall [Fig.1]. The expected behaviour of the final aortic structure is expected to give increased mechanical performances. Moreover that approach can be applied by suitable surgical techniques, such as endoscopic percutaneous insertion, allowing a very low level of risk and complications.

[1] Professor, Department of Structural Mechanics, University of Pavia, I-27100 Pavia, Italy
[2] Ph. D. Student, Department of Structural Mechanics, University of Trento, I-38100 Trento, Italy

The analysis, performed using numerical tools, is based on some simplifying hypothesis: in particular, from the geometrical point of view, the blood vase is modelled as a cylindrical tube without collateral branches and reinforced by a prosthesis net with a rectangular mesh directed along the main axis; on the other hand, from the mechanical point of view, the arterial wall is supposed to be an isotropic, homogeneous and incompressible material with a constitutive law derived by the mathematical model recently proposed in Ref. [3]. Viscous phenomena, dynamic and cyclic loads, creep and relaxation phenomena are not considered at the current stage of the research.

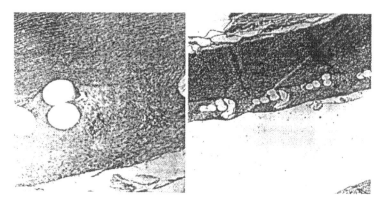

Fig. 1 Histology shows a very thick neo-intima layer completely including the net prosthesis which lies in contact with the media. The square area shows the net pattern

3. METHODS

3.1 The Hart-Smith model

In the first numerical analysis the Hart-Smith model is employed to model the mechanical behaviour of the arterial wall: it postulate an exponential dependence of W with respect to the first invariant of the metric tensor of the deformed configuration I_1 and a logarithmic and a polynomial dependence with respect to respectively the second and the third invariant I_2 and I_3, as follow:

$$W = \int_{I_1} \exp[C(I_1 - 3)^2]dI_1 + 3BlnI_2 + D(I_3^2 - 1) + E(I_3 - 1)^2 \tag{1}$$

where the coefficient C assumes very low values, next to zero, A, B are evaluated by uniaxial traction tests and D is calculated so as to reach the equilibrium in the zero stress state.

The uniaxial tension is:

$$\frac{N}{A_0} = 2\left(\lambda - \frac{1}{\lambda^2}\right)\left(\frac{\partial W}{\partial I_1} + \frac{1}{\lambda}\frac{\partial W}{\partial I_2}\right) \tag{2}$$

where λ is the stretch and

$$I_1 = \lambda^2 + \frac{2}{\lambda} \tag{3}$$

$$I_2 = 2\lambda + \frac{1}{\lambda^2}$$

By dividing equation (3) for $2\left(\lambda - \dfrac{1}{\lambda^2}\right)$, it is possible to obtain the correspondent

$\left(\dfrac{\partial W}{\partial I_1} + \dfrac{1}{\lambda}\dfrac{\partial W}{\partial I_2}\right)$ values.

With respect to equation (1) it is possible to calculate:

$$\frac{\partial W}{\partial I_1} = A\ \exp[C(I_1 - 3)^2]$$

$$\frac{\partial W}{\partial I_2} = \frac{3B}{I_2} \tag{4}$$

If λ is small enough the influence of C coefficient is negligible, then we can assume

$$\frac{\partial W}{\partial I_1} = A \tag{5}$$

In particular the plot of $\left(\dfrac{\partial W}{\partial I_1} + \dfrac{1}{\lambda}\dfrac{\partial W}{\partial I_2}\right)$ with respect to $\dfrac{1}{\lambda I_2}$ is represented by a straight

line with slope equal to 3B and the intersection with the ordinate axis rappresent the A coefficient.

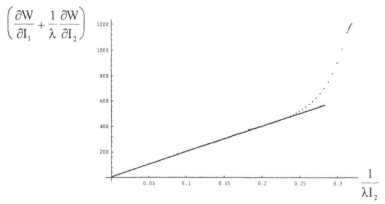

Fig. 2 Energy density function: experimental fit and interpolation function

For example, the interpolating function for the experimental fit shown in Fig. 2 is:
$$y = 6.60044 + 1989.58\,x \tag{6}$$
then

$$\begin{aligned} A &= 6.60044\ \text{kPa} \\ 3B &= 1989.58\ \text{kPa} \end{aligned} \tag{7}$$

The parameter C controls the stiffness increment for large strains and, for the present problem, it can be set equal to zero.

Finally, forcing the general stress component to be equal to zero for $\varepsilon_{ij} = 0$ the value of the D parameter is found as:

$$D = -\frac{1}{2}(A + 2B) \tag{8}$$

3.2 Biomechanical behaviour of abdominal aortic aneurysm

In order to analyse the biomechanical behaviour of aortic aneurysm the model recently proposed in [3] is followed. This kind of approach is based on the observation that soft biological tissue contains two primary load-bearing fibres: elastin and collagen.
The first is characterised as compliant whereas the second is much stiffer and tortuous as soon as the tissue is load-free or, at least, not much strained.
As the tissue is stretched, the collagen recruitment starts to became relevant: in particular the collagen fibres gradually become taut and start contributing to the load bearing.
The mathematical relationship proposed to describe such behaviour is

$$\varepsilon = \left(K + \frac{A}{B + \sigma} \right) \sigma \tag{9}$$

where K, A and B are model parameters and they can be simply derived by tensile tests observing that, on the basis of experimental evidence, the aortic aneurysm tissue can be considered isotropic since there are not significant differences between longitudinal and circumferential measurement on K, A and B parameters.
In particular in the equation (9) the relation $K + \dfrac{A}{B} = \lim\limits_{\sigma \to 0} \dfrac{d\sigma}{d\varepsilon} = E_E$ represents the total stiffness of the elastin fibres alone at the beginning of the load process when the contribution of the collagen fibres is not relevant, whereas $K = \lim\limits_{\sigma \to \infty} \dfrac{d\sigma}{d\varepsilon}$ represents the total stiffness of the elastin and fully recruited collagen fibres at the and of load process.

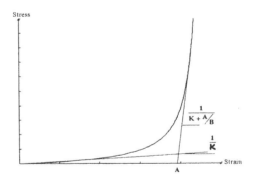

Fig. 3 Typical elastic response

4. NUMERICAL ANALYSIS

A model to simulate the biomechanical behaviour of an aortic segment is presented in the numerical example.
The part of blood vessel here analysed is considered without collateral branches and is studied as a cylindrical tube subjected to a constant blood pressure.
Since the main aim of the numerical example is to evaluate the possible benefits derived from the insertion of the net prosthesis, in terms of mechanical behaviour, a simplified geometry is considered [Fig.3]. In particular the aortic segment studied is 30 mm in length and with the external radius of 15 mm and total transversal thickness of 2.5 mm.

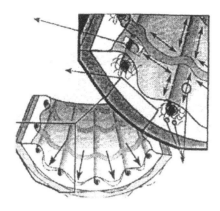

Fig. 4 Geometry

The load to be applied is chosen as the maximum physiological pressure of a medium individual (150 mmHg=0.02 MPa). In the present work we assume to fix the net prosthesis to the vessel at an ideal zero-stress state, increasing then the load at the assumed value.

Four numerical analysis are proposed. In the first three only the normal vessel is studied and so, due to the symmetry of the problem, only a quarter of the tube is considered and the different behaviour of the three layers (intima, media and adventitia) is taken into account. In particular, in the first the cylindrical vessel is studied so that it is possible to know stress and strain fields in the natural state (Fig. 6). In the second one the vessel is reinforced by transversal threads uniformly spaced at 6 mm (Fig. 7). In the third in order to prove the sensitivity of the analysis with respect to the refinement of the geometrical F.E. mesh, a vessel detail is studied (Fig. 9). In the last one the values of all the parameters are chosen accordingly to the table shown in Fig. 5 and a geometrically simplified zone affected by aneurysm is chosen.

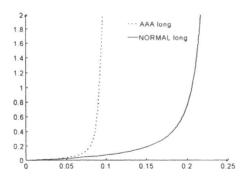

GROUP	K [MPa]$^{-1}$	A	B [MPa]	σ_u [MPa]
AAA	0.39	0.091	0.039	0.864
NORMAL	0.30	0.223	0.893	2.014

Fig. 5 Stress-Strain relationship for normal aorta and for abdominal aortic aneurysm (AAA).

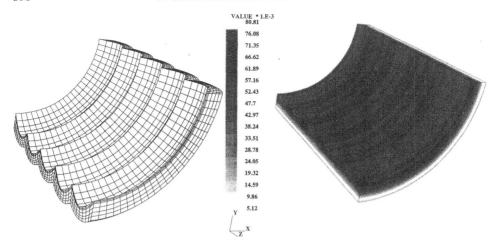

VALUE * 1.E-3

Fig 6 Aortic wall with transversal reinforcement only: deformed configuration and equivalent stress map

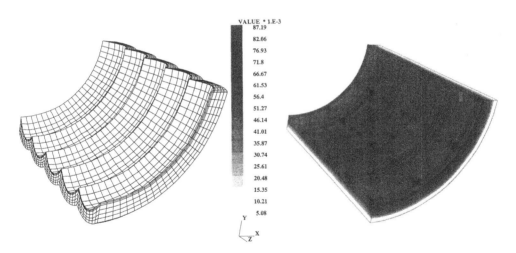

VALUE * 1.E-3

Fig. 7 Aortic wall with complete net reinforcement: deformed configuration and equivalent stress map

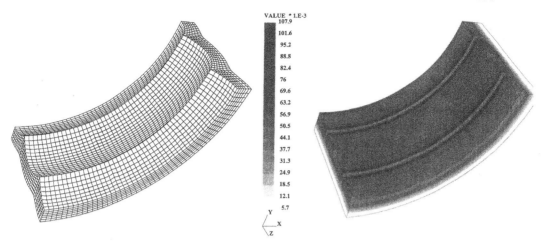

Fig 8 Aortic wall with complete net reinforcement: deformed configuration and equivalent stress map in a particular

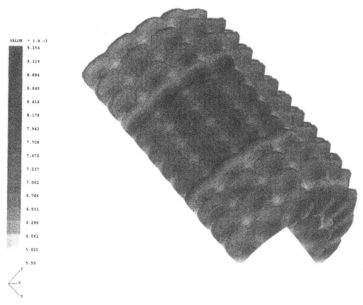

Fig. 9 Aortic wall with complete net reinforcement and with the aneurysm zone: equivalent stress map on the deformed configuration

5. CONCLUSIONS

In order to analyse the possible benefits gained in patients affected by pathologies, inducing sensible modifications to the physical characters of the artery wall, a model for the description of the mechanical behaviour of a blood vessel, reinforced by a prosthetic net with rectangular mesh, is presented.
The preliminary results obtained seem to encourage deeper investigations both for surgical and the mechanical aspects. In particular a better mechanical characterisation of the arterial wall with experimental validation ex-vivo tests is planned in order to obtain a more refined analysis of the problem. Under these circumstance a F.E. analysis with a prescribed load history can be implemented and estimation of the behaviour of the neo-intima layer surrounding the net roads can be evaluated in order to characterise the stress-strain level needed to support the load transfer from the artery to the net.

6. ACKNOWLEDGEMENTS

The present work was partially supported by grant from the Italian Ministry of University and of Scientific and Technological Research (MURST).

7. REFERENCES

1. Fung Y.C., Biomechanics, Springer-Verlag, New York, 1993.
2. Humphrey J.D., Mechanics of the arterial wall, Critical reviews in Biomedical Engineering, 1995, Vol. 23.
3. Raghavan M.L., Webster M.W. and Vorp D.A., Ex vivo biomechanical behaviour of abdominal aortic aneurysm: assessment using a new mathematical model, Ann. Of Biomedical Engineering, 1996, Vol. 24, 573-582.
4. Robicsek F., Thubrikar M.J., Hemodynamic considerations regarding the mechanism and prevention of aortic dissection, Ann. Thoracic Surgery, 1994, Vol. 58, 1247-1253.
5. Stringfellow M.M., Lawrence P.F., Stringfellow R.G., The influence of the aorta-aneurysm geometry uponstress in aneurysm wall, Journal of Surgery Res., 1987, Vol. 42, 425-433.
6. Parodi J.C., Palmaz J.C.and Barone H.D., Transfemoral intraluminal graft implantation for aortic aneurysm, Ann. Vascular Surgery, 1991, Vol. 5, 491-499.
7. Laborde J.C., Parodi J.C., Clem M.F., Tio F.O., Barone H.D., Rivera F.J.,. Encarnacion C.E, Palmaz J.C., Intraluminar bypass of abdominal aortic aneurysm, Radiology, 1991, Vol 184, 185-190.
8. Moon M.R., Dake M.D., Walker P.J., Fann J.I., Semba C.P., Mitchell R.S., Miller D.C., Endovascular stent-grafts for descending thoracic aortic pathology, Proceedings IV Aortic Surgery Symposium, New York, 1994.
9. S. Matteo, Rinforzo della parete aortica con una protesi a rete per arrestare la progressione dell'aneurisma e prevenirne la rottura"Pavia, Italy, 1995.
10. Carli C., Cinquini C., Martelli M., Modello meccanico di vaso sanguigno con rete di rinforzo, XII AIMETA Congress, Napoli, Italy, 1995.
11. SAMCEF reference manual, Samtech, Liege, 1993.

TRANSMITRAL FLOW ANALYSIS BY MEANS OF COMPUTATIONAL FLUID DYNAMICS ON UNSTRUCTURED GRIDS

C. Capozzolo[1], F. M. Denaro[2], F. Sarghini[3]

1. ABSTRACT

This paper presents a numerical methodology to investigate the flow developing through the mitral valve with the aim of realising a simple model able to support clinicians in assessing non-invasive methodologies. An unstructured grid generation and a high order control volume method, developed in the framework of the immersed boundary method, are the main computational tools adopted to simulate the problem. Results are presented to demonstrate the potentiality of the approach.

2. INTRODUCTION

Diastolic dysfunction is a primary mechanism in the pathogenesis of congestive heart failure, it is an important cause of cardiac morbidity and it seems to be one of the most premature alteration in many pathological conditions that affect the heart. Several indexes obtained from the Doppler derived velocity/time profiles are currently employed to characterise the diastolic function of the heart. Nevertheless, what we evaluate represent always a measurement of the filling and not of the actual function and some researches have demonstrated inconclusive, if not contradictory, relationships with the invasive parameters. Many times, clinicians adopt the Bernoulli equation to correlate velocity and pressure. Unfortunately, the required hypotheses for the validity of the

Keywords: Numerical simulation of mitral flows , Unstructured grids, Immersed boundary method, M-mode color Doppler

[1] PhD student, Dipartimento di Cardiologia - II Policlinico, Università di Napoli "Federico II", Italy
[2] Researcher, Dipartimento di Ingegneria Aerospaziale, Seconda Università di Napoli, Aversa, Italy
[3] PhD, Department of Mechanical Engineering, University of Maryland at College Park, USA

Bernoulli equation are not fully verified in the heart. Furthermore, the lack of multidimensional information can add some misleading. Instead, the blood flow motion is governed by the Navier-Stokes equations for incompressible flows that represent the correct equations for unsteady, viscous flows providing the velocity as function of time and space. At present, the potentialities that Computational Fluid Dynamics (CFD) tools could provide to clinical investigations, are not fully exploited. In the past years, CFD methodologies have been applied in an invaluable series of papers of Peskin *et al.* [e.g., 1-5] where both 2-D and 3-D geometry were adopted and examples of using such methodologies for clinical observations are shown by Yellin *et al.* in [6]. Helpful, but much more simple mathematical models, are also adopted by Thomas *et al.* [7] with the goal to analyse pathological relevance in modifications of the physiological variables that are measured by Doppler tools.

The general aim of the present research is to develop some CFD methodologies in order to investigate transmitral flows developing under several conditions of the physiological variables and, in such a way, to support clinical diagnostic. The analysis we present in this paper, mainly deals with the physical hypotheses that can be adopted to simplify the study and with the mathematical and numerical tools to simulate the problem. To this goal, we adopt a 2-D geometry to solve the unsteady incompressible Navier-Stokes equations on unstructured grids. A Newtonian fluid model is considered, as generally proposed in [8] for the blood. Owing to the rapid time variation of the flow, a high accuracy order Control Volume (CV) method is used [9-11]. In 2-D case, the dynamic of the mitral valve can be simulated by considering each leaflet as a chain of elastic links as approached in [1, 3]. The material line, immersed in the fluid, acts as source of additional stresses thus leading to global elastic-fluid dynamics interactions. Based on such a model, we will present the methodology and some results for a normal case showing that, even though the model is limited in its reality by the adopted hypotheses, the main characteristics of diastole are well represented.

3. THE DIASTOLIC PHASE OF THE CARDIAC CYCLE: A PHYSICAL DESCRIPTION AND THE ADOPTED COMPUTATIONAL HYPOTHESES

The diastolic function is the part of the cardiac cycle characterised by the fact that blood flows through the mitral valve orifice causing left ventricular filling. To quantify the diastolic left ventricular performance, it has been generally useful the analysis of the pattern of left ventricular filling obtained, as example, with the aid of Doppler measurements (Fig.1). Myocardial relaxation begins in the latter part of systole and causes a steep exponential fall in infra-ventricular pressure. This fall produces a pressure gradient that accelerates the blood from left atrium into the left ventricle, resulting in a rapid early filling (E wave). During the filling, left ventricular pressure increases, with general properties depending on ventricular compliance, and atrio-ventricle gradient pressure decreases and transiently reverses. Thus, little left ventricular filling occurs during the midportion of diastole. Finally, atrial contraction increases atrial pressure late in diastole, producing blood to flow again in the left ventricle (A wave) until to the closure of mitral valve. It appears evident that, to qualify the diastolic performance, the only velocity patterns result insufficient being them related only to pressure difference and therefore, in some cases, subject to misleading [e.g., 6, 7]. This description of the diastole results useful to understand the hypotheses adopted in the mathematical model. In particular, the 2-D geometry clearly represents only a rough assumption and fully 3-D case should be considered.

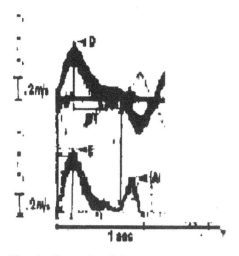

Fig. 1: Example of Doppler measurement during left ventricular filling. The top diagram represents the recorded pulmonary flow the bottom one the recorded transmitral flow.

However, as reported in [1, 3], the plane of calculation can be physically located as that bisecting the leaflets of the mitral valve, the root of the aorta and passing through the apex of the heart as sketched in Fig.2. Only two pulmonary inflow are assumed on opposite faces of the atrium and the atrial contraction is modelled by assuming a suitable normal velocity distribution along the wall. At present, another simplification is that no deformation of the ventricular boundary is modelled. In order to respect mass balance, blood will flow through the ventricular wall just as if the surfaces were permeable. For what it concerns the valve, the mitral leaflets will be massless and the presence of the chordae tendineae will have no influence on the flow apart the effects of the forces they apply to the tips of the mitral leaflets.

Furthermore, the leaflets will be able to produce only normal stresses (i.e., directed along the tangent direction of the line).

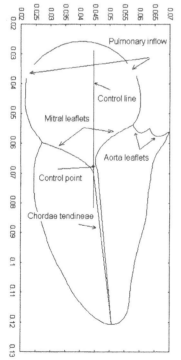

Fig. 2: Sketch of the adopted 2-D configuration. Dimensions are expressed in meters.

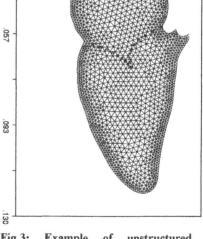

Fig.3: Example of unstructured grid generation with refinement near the walls. The number of points is 1627 allowing to obtain the smallest grid size of $\approx 4.4 \ 10^{-4}$ m. The post-optimisation procedure makes the nodes have only 5, 6, or 7 edge connections.

4. THE MATHEMATICAL MODEL

The global viscous-elastic phenomenology is governed by coupled elastic and fluid-dynamics system of equations. As the present problem deals with the interaction of an elastic material immersed in an incompressible fluid, we adopt some guidelines of the immersed boundary method presented by Peskin *et al.* in [1-5]. In particular, in the present paper, we focus on the interaction of the mitral valve with the fluid in order to analyse some influences of the physiological variables on the transmitral flow pattern. In doing so, we make some assumptions (that in the future should be removed) that mainly deal with the 2-D geometry and the fixed ventricular boundary.

The blood flow motion is governed by the Navier-Stokes equations for incompressible flows that can be written in integral form over a domain Ω as:

$$\int_{\Omega} \frac{\partial \mathbf{v}}{\partial t} \, dV = \int_{B\Omega} \mathbf{n} \cdot \mathbf{F} \, dS \tag{1}$$

associated to the divergence-free constraint. In Eq.(1) $\mathbf{v}$ is the velocity field and the tensor flux function $\mathbf{F}$ is expressed by:

$$\mathbf{F} = -\mathbf{v}\mathbf{v} - \frac{\mathbf{I}p}{\rho} + \nu \underline{\nabla} \mathbf{v} + \frac{\mathbf{T}}{\rho} \tag{2}$$

being ρ the constant blood density ($1.05 \cdot 10^3$ Kg/m^3), $\nu = \mu/\rho$ the constant kinematics viscosity ($6 \cdot 10^{-6}$ m^2/s), p the pressure[1], $\mathbf{I}$ the unity matrix, and $\mathbf{T}$ is the stress tensor taking into account for the elastic contribution of the mitral leaflets. In the immersed boundary method, the elastic material is considered as a part of the fluid where additional stresses are applied. In this way, the equation governing the flow motion can be solved on a fixed domain (Eulerian description) while the motion of the leaflets can be described by a suitable set of discrete points (Lagrangian description) that occupy part of the domain moving over it and producing local stresses depending on the fibre strain. If we formally call σ the tension acting on the elastic material, ε the strain rate, s the displacement of the material point, than we require a relation that link the density forces acting on the fluid contained in Ω, to the leaflet stresses $\sigma[\varepsilon(s)]$ caused by the motion. If we consider the elastic leaflet capable of exercising only tension or compression stresses, then in a local curvilinear reference system (ξ, η) the tensor $\mathbf{T}$ has only the element $t_{11} = \sigma_\xi \neq 0$. The fibre strain will be $\varepsilon_\xi = \partial s_\xi / \partial \xi$ so that, for a given stiffness E, the tension will be $\sigma_\xi = E \, \varepsilon_\xi$. Thus, the vector stress associated to the ξ direction, will generally produce both x- and y-components according to: $\sigma = \sigma_\varepsilon(\xi)\left(\xi_x \mathbf{i} + \xi_y \mathbf{j}\right)$. Finally, the fibre strain will be determined by the fact that any material point, belonging to the leaflet, is subject to the governing equation:

$$\frac{\partial \mathbf{x}(\xi,t)}{\partial t} = \mathbf{v}\left[\mathbf{x}(\xi,t),t\right] \tag{3}$$

namely the leaflet moves at the local fluid velocity. All the steps concerning the determination of the leaflet stresses are done on a Lagrangian system and, therefore, to compute the contribution in the Eulerian reference, it is necessary to express the functional relation $\sigma(\mathbf{x}) = M\sigma\left[\xi(\mathbf{x})\right], \forall \mathbf{x} \in \Omega$ where M is some suitable interpolation operator that links the Lagrangian to the Eulerian system (see [4]).

---

[1] It is worthwhile to remark that for isothermal incompressible fluids no primitive pressure equation exists, being the pressure p derived only from Eq. (1) with the divergence-free constraint.

5. THE NUMERICAL PROCEDURE

The type of computational grid that is generally suitable for not simple geometry, being possible to refine it in a flexible way, is that based on a triangulation of the domain. In our study, this is accomplished based on a frontal-Delaunay unstructured grid generation with a post-optimisation [11]. In principle, with the aid of such grid generation, it is also possible to consider the mitral leaflets as part of the boundaries of the domain, to adopt a moving adaptive grid and to use a fully Eulerian formulation. However, this procedure is very expensive resulting both in a step-by-step grid adaptation and mesh-to-mesh interpolation. Such a procedure is planned to be implemented in future but only to take into account the changes of ventricular chamber. In Fig. 3 it is shown an example of triangulation of the heart with a grid refinement near the walls obtained with 1627 points number. Even with such a few points, this grid allows to obtain the smallest grid size of about $4.4 \ 10^{-4}$ (m) i.e., about 80 times smaller than the adopted diameter of the mitral annulus. A greater points number can be easily computed with present workstations. Markers show an example of discretization for the mitral leaflets in the initial condition.

The numerical procedure is based on a Control Volume based projection method developed for unstructured grids [9-11]. Initially, we define a partition of Ω in N CV's Ω_i (see Fig. 4), over which we will solve the equation:

$$\int_{\Omega_i} \left(\mathbf{v}^{n+1} - \mathbf{v}^n \right) \, \mathrm{d}V = \int_{t^n}^{t^{n+1}} \mathrm{d}t \int_{B\Omega_i} \mathbf{n} \cdot \mathbf{F} \, \mathrm{d}S \tag{4}$$

From Eq.(4), we first compute a field $\mathbf{v}^{*\,n+1}$ based on the the flux $\mathbf{F}^* = \mathbf{F} - I p/\rho$ i.e., disregarding the pressure contribution. In order to compute $\mathbf{F}^*$ we solve the equation for the leaflets stresses. Each leaflet is discretized in p elements where we assume linear behaviour of the nodal displacement s, each element provides a stress matrix depending on the elastic properties and finally, by assembling the global stress matrix, one obtains the stresses on the $2(p+1)$ points of the leaflets depending on the fibre strain. The set of stresses on the Lagrangian system is then mapped in the Eulerian system by means of the interpolation operator that, in our framework, is a cosine hill centred in $\mathbf{x}_i$ and having an influence radius corresponding to $2\sqrt{|\Omega_i|/\pi}$. In this way, each sub-domain will be subject to an influence region related to its characteristic area. Now we can compute:

$$\int_{t^n}^{t^{n+1}} \mathrm{d}t \int_{B\Omega_i} \mathbf{n} \cdot \mathbf{F}^* \, \mathrm{d}S \cong \sum_{l=1}^{b_i} \int_{B\Omega_{il}} \mathbf{n}_l \cdot \mathbf{F}_l^{*(k,\,m)} \, \mathrm{d}S \tag{5}$$

where b_i is the number of boundaries of Ω_i, $\mathbf{n}_l$ is the normal to the l-th section $B\Omega_{il}$ in outward direction (see Fig.4) and $\mathbf{F}_l^{*(k,\,m)}$ is the high order numerical flux function computed according to a general procedure presented in [10, 11]. In the following, we fix the order of the accuracy to k=3 and m=2. In Eq.(5), for each l-th section, one defines for $\mathbf{F}_l^*$ a fully second degree polynomial having a support region individuated by the vector velocity according to an upwind criterion as illustrated in Fig.5. Based on the polynomial reconstruction, Eq. (5) can be computed and one evaluates the velocity $\mathbf{v}^{*\,n+1}$. In order to ensure the final divergence-free velocity field, one must take into account the pressure contribution. An equation for the time average pressure $\langle p \rangle$ is herein obtained by taking the divergence on both sides of Eq.(4) where the velocity field $\mathbf{v}(\mathbf{x}, t^{n+1})$ is imposed to be divergence-free and by replacing Eq.(5). In the present approach, the pressure equation is associated to Neumann boundary conditions and it is solved by means of an optimized SOR procedure. The intermediate velocity field $\mathbf{v}^*$ results so projected onto the space of divergence-free vector fields. Eventually, the single step time marching formula can be written as:

$$v_j(\mathbf{x}_i, t^{n+1}) = v_j^*(\mathbf{x}_i, t^{n+1}) - \frac{\Delta t}{|\Omega_i|} \int_{B\Omega_i} \mathbf{n} \cdot \mathbf{e}_j \langle p \rangle \, dS \qquad (6)$$

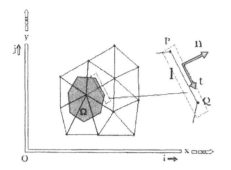 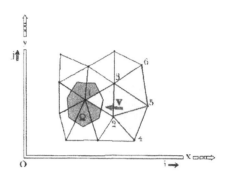

Fig. 4: Control Volume definition over 2-D unstructured grids and details of the l-th flux section.

Fig. 5: Definition of the "upwind" criterion on unstructured grid for (v·n) < 0 and (v·t) < 0. The four-triangle region is adopted as support for a fully second degree polynomial.

6. RESULTS

The model we have presented has several simplifications we want to assess have as less implications as possible on the results. In particular, the absence of a real deformation of the ventricular chamber could modify in unrealistic way the behaviour of some physiological variables. As validation, we analyse a normal case in order to examine the main characteristic of a filling pattern during diastole. To this goal, a pulmonary flow is assigned as Dirichlet inflow condition on the veins according to:

$$V_p(t) = V_{max} \exp(1/2) \exp(-t^2/2T^2_{max}) \, t/T_{max} \qquad \text{for } t < T_{ac} \qquad (7)$$

where V_{max} is the peak velocity reached at the time T_{max} and T_{ac} is the time corresponding to the starting of the atrial contraction during which the pulmonary flow reverses according to a polynomial 3-th degree law. A time dependent pressure is imposed over the walls of the ventricle chamber. As main geometrical parameters we fix the diameter of the mitral ring to 3.2 cm and the atrium-to-ventricle apex length of 10 cm (see Fig.2). The dynamic parameters for the normal case are: V_{max}= 0.55 m/s , T_{max} =0.08 s, T_{ac} = 0.375 s. Each mitral leaflet was discretized by 40 elements . The time integration is performed over a total time interval of 0.53 s. As control parameters, we register in time the flow along a 5 cm control line posed between the tips of the mitral leaflets, starting in the atrium at x=3 cm, crossing the mitral ring up to the early part (x=8 cm) of the ventricle (see: Fig.2). In this way, a contour map of the fluid dynamics quantities in the (t, x) plane can be computed and compared with clinical measurement obtained with the M-mode color Doppler. In Fig.6 we show the E-A waves along some locations on this line and in Fig.7b the corresponding velocity map in the (t, x) plane. As a comparison, we slightly modify only the value of T_{max} by decreasing it to 0.07 s (Fig.7 a) and by increasing it to 0.09 s (Fig.7c) in order to examine the relevance in changing the physiological variables. It appears that this modifies the area characterising the rapid filling in its extension and in terms of local peak velocities. This qualitative picture can already provide some discriminations in the analysis of pathologies.

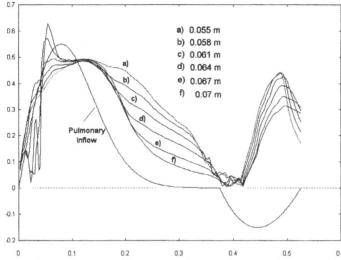

Fig. 6: Patterns of E-A waves at different locations along the atrium-ventricle axis for the normal case. For sake of completeness it is also reported the curve representing the pulmonary velocity.

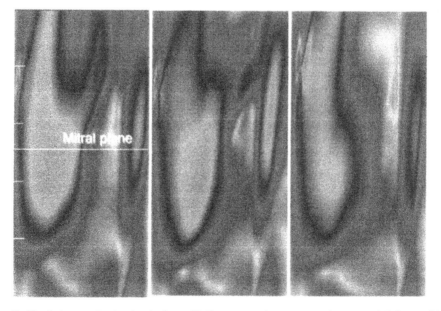

Fig.s 7: Shaded maps in the (t, x) plane. b) Base case. As a comparison we slightly modify only the value of T_{max} by decreasing it to 0.07 s a) and by increasing it to 0.09 s b) in order to examine the relevance in changing the physiological variables. It appears that the area characterising the rapid filling modifies its extension and local maxima.

7. CONCLUSIONS AND PERSPECTIVES

The main difficulty in analysing diastole by means of non-invasive measures, appears to be the lack of information on the pressure field because, the necessary direct pressure measurements can be obtained only resorting to intra-cardiac catheterization. Such information can be in part recovered if fully multidimensional Doppler measurements were available. At present, it is possible to have, at acceptable accuracy, a scansion along one dimension by means of the M-mode color Doppler. In this way, in principle it is possible to compute along this line, the pressure gradient and the consequent map in the (t, x) plane can be used to add information. A balance of the effective forces can drive to some interesting indexes of diastolic function. The perspective of our work is to develop a numerical model that could support the validation of this clinical experimentation. In the future, depending on some physiological variable changes boundary conditions can be imposed in order to allow a parametric analysis. Thus, numerical results should provide information about the time varying atrio-ventricular pressure gradient and its modification in pathological cases.

REFERENCES

1. C. S. Peskin, Numerical Analysis of Blood Flow in the Heart, J. Comp. Physics, 25, 1977.
2. C. S. Peskin, D. M. McQueen, A Three-Dimensional Computational Method for Blood Flow in the Heart I. Immersed Elastic Fibers in a Viscous Incompressible Fluid, J. Comp. Physics, 81, 1989.
3. C. S. Peskin, B. F. Printz, Improved Volume Conservation in the Computation of Flows with Immersed Elastic Boundaries, J. Comp. Physics, 105, 1993.
4. C. S. Peskin, D. M. McQueen, Computational Biofluid Dynamics, Contemporary Mathematics, 141, 1993.
5. C. S. Peskin, D. M. McQueen, A General Method for the Computer Simulation of Biological Systems Interacting with Fluids, The Society for Experimental Biology, pp.265, 1995.
6. E. L. Yellin, S. Nikolic, R. W. M. Frater, Left Ventricular Filling Dynamics and Diastolic Function, Progress in Cardiovascular Diseases, Vol. XXXII, 4, 1990.
7. J. D. Thomas, C. Y. P. Choong, F. A. Flachskampf, A. E. Weyman, Analysis of the Early Transmitral Doppler Velocity Curve: Effect of Primary Physiologic Changes and Compensatory Preload Adjustment, J. Am. Coll. Card., Vol. 16, 3, 1990.
8. T. J. Pedley, The Fluid Mechanics of Large Blood Vessels, Cambridge University Press, 1980.
9. G. De Felice, F.M. Denaro, C. Meola, F. Sarghini, Model Free Numerical Simulation of High Reynolds Compressible Flows on Adaptive Unstructured Grid, Conference of Numerical Method for Fluid Dynamics, Oxford, 3-6 April 1995.
10. F. M. Denaro, Toward a New Model-Free Simulation of High Reynolds Flows: Local Average Direct Numerical Simulation, Int. J. Num. Methods in Fluids, 23, 125-142, 1996.
11. F. Sarghini, Simulazioni ai Volumi di Controllo per Campi Compressibili Subsonici su Griglie Non Strutturate, Basate su Tecniche di Ricostruzione delle Variabili Dirette e Filtrate, Tesi di dottorato, Facoltà di Ingegneria, Università "Federico II" di Napoli, Italy, 1996.

EFFECTS OF CAVAL VELOCITY PROFILES ON PULMONARY FLOW IN THE TOTAL CAVOPULMONARY CONNECTION: CFD 3-D MODEL AND MAGNETIC RESONANCE STUDIES

F. Migliavacca[1], G. Pennati[1], G. Dubini[2], R. Pietrabissa[1], R. Fumero[1],
P. J. Kilner[3] and M. R. de Leval[4]

1. ABSTRACT

In this paper results from a study based on the application of computational fluid dynamics techniques are compared to those from *in vivo* measurements performed by means of magnetic resonance with regard to a case of the total cavopulmonary connection. This is a surgical procedure adopted to treat complex congenital malformations of the right heart and consists basically in a connection of both venae cavae directly to the right pulmonary artery. The distribution of the blood flow in the pulmonary arteries, which depends on the geometry of the surgical reconstruction, is analysed quantitatively with both techniques on a single patient.

2. INTRODUCTION

Total cavopulmonary connection (TCPC) is an operation to treat hearts with essentially a single ventricular chamber. It is well described in the literature[1-3] and it consists basically (Fig.1) of an anastomosis of the superior vena cava (SVC) to the superior aspect of the right pulmonary artery (RPA) and the construction of a right atrial lateral tunnel that connects the inferior vena cava (IVC) to the RPA. A window is sometimes made in the tunnel wall in patients who are not ideal candidates for this operation. This fenestration, that can be subsequently closed,[4] allows a communication with the atria,

Keywords: Paediatric surgery, Computational fluid dynamics, Magnetic resonance

[1]Dipartimento di Bioingegneria, Politecnico di Milano, Piazza Leonardo da Vinci 32, 20133 Milano, Italy
[2]Dipartimento di Energetica, Politecnico di Milano, Piazza Leonardo da Vinci 32, 20133 Milano, Italy
[3]Magnetic Resonance Unit, Royal Brompton Hospital, Sydney Street, London SW3 6NP, UK
[4]Cardiothoracic Unit, Great Ormond Street Hospital for Children NHS Trust, Great Ormond Street, London WC1N 3JH, UK

limiting the level of pressure inside the tunnel. The aim of the TCPC is the total bypass of the right part of the heart. Since this operation creates a circulation driven by a single ventricular pump, it is of the utmost importance to minimise any flow disturbance as well as to guarantee a balanced blood flow to the lungs.

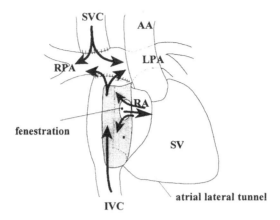

Fig.1. Anatomical sketch of the total cavopulmonary connection with the atrial fenestration (AA: ascending aorta, IVC: inferior vena cava, LPA: left pulmonary artery, SV: single (left) ventricle, RA: right atrium, RPA: right pulmonary artery, SVC: superior vena cava).

In vitro studies[1,5,6] based on simplified models of the connection demonstrated the presence of complex flow patterns, strongly dependent on geometrical changes of the connections. Recently, computational fluid dynamics (CFD) techniques[7-9] have been applied, as well. These studies aimed at the hydraulic optimisation of the caval connections with particular regard to the shape and function of the anastomoses. In order to complement our previous CFD studies,[8,9] magnetic resonance (MR) imaging techniques were applied to a patient previously submitted to TCPC. *In vivo* data were obtained on both the anatomy (*i.e.*, dimensions of the vessels and locations of the caval anastomoses, necessary to create the CFD three-dimensional model) as well as the blood flow distribution in the pulmonary arteries and the venae cavae.

Purpose of this study is to evaluate the effects of different inlet (caval) velocity profiles on the flow in the pulmonary arteries in a realistic three-dimensional model of the TCPC.

3. MATERIALS AND METHODS

A 13-year-old boy was studied after TCPC. Univentricular heart syndrome, with atrial septal defect and severe pulmonary stenosis, had been diagnosed at birth. He had a right modified Blalock-Taussig shunt (*i.e.* a palliative connection between the subclavian artery and the pulmonary artery) performed five days after birth. He was submitted to TCPC (and shunt closure) at the age of 8.

The MR scan was performed with the patient at rest, lying supine in the magnet, without need of any sedation. The mean heart rate was 1.18 Hz (71 bpm) during the scan. A 0.5 T Picker magnet (Picker MR Division, Highland Heights, OH, USA) was used to acquire cardiac gated spin echo images in multiple transaxial, coronal (Fig.2) and sagittal planes. Gradient echo cine imaging, echo time 14 ms, was then performed, with phase contrast velocity mapping[10] used to measure flow through each of four planes. Voxel size was $8 \times 1.5 \times 1.5$ mm. The four planes were located from spin echo images so as to transect the superior vena cava, the lower part of the right atrial tunnel and right and left pulmonary arteries. Twenty cine frames were acquired, recording flow throughout the 855 ms average cardiac cycle, which allowed axial flow profiles and volume flow changes to be plotted.

Figure 3 shows the instantaneous volume flow rate in the SVC and the IVC, and the RPA and LPA derived from MR measurements during the cardiac cycle. The mean volume flow rates were: $\overline{Q}_{SVC} = 18.17$ cm^3 s^{-1} (1.09 l min^{-1}), $\overline{Q}_{IVC} = 20.17$ cm^3 s^{-1} (1.21 l min^{-1}), $\overline{Q}_{LPA} = 25.51$ cm^3 s^{-1} (1.53 l min^{-1}) and $\overline{Q}_{RPA} = 15.50$ cm^3 s^{-1} (0.93 l min^{-1}).

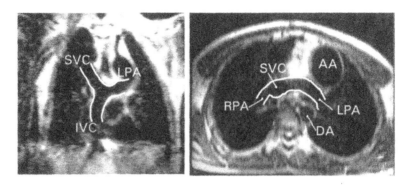

Fig.2. Coronal spin echo image (left) and transaxial spin echo image (right) at the level of the pulmonary bifurcation, showing the venae cavae as well as the left and right pulmonary arteries. AA: ascending aorta, DA: descending aorta, IVC: inferior vena cava, LPA: left pulmonary artery, RPA: right pulmonary artery, SVC: superior vena cava. Solid white lines show the TCPC shape.

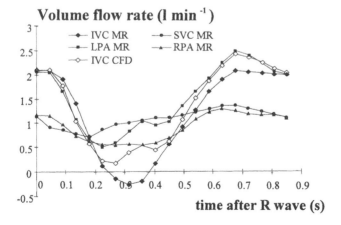

Fig.3. Instantaneous volume flow rate in the SVC, IVC, RPA and LPA derived from MR measurements during the cardiac cycle (1 l min^{-1} = 16.67 cm^3 s^{-1}). Adopted time function for the IVC inlet flow in the CFD study is depicted as well.

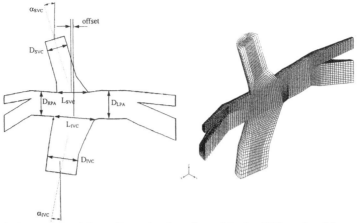

Fig.4. Geometry and three-dimensional mesh used in the CFD study of the total cavopulmonary connection. The geometrical data are as follows: diameters: D_{IVC} = 13 mm, D_{SVC} = 20 mm, D_{LPA} = 17 mm, D_{RPA} = 13 mm, length of anastomoses: L_{IVC} = 26 mm, L_{SVC} = 20 mm, angles: α_{IVC} = 8°, α_{SVC} = 16°, *offset* = 1 mm. The total number of nodes is 37,847.

The pulsatility of the IVC flow was presumably due to the atrial pressure changes. This effect was perceptible in the IVC only, because of the presence of the atrial fenestration. The CFD model was developed according to the method described by Dubini *et al.*[8]

The geometry of the three-dimensional model of the TCPC was built on the basis of anatomic data taken from the MR scan. FIDAP, a commercial CFD code (FDI - Fluid Dynamics International, Evanston, IL, USA), was used to generate the mesh as well as to impose the boundary conditions. The main geometric data of the model are reported in Fig.4. The mesh consists of 37,847 nodes and 41,829 elements. Eight-node isoparametric brick elements for the fluid domain and four-node isoparametric quadrilateral elements for the inlet and outlet sections and the vessel walls were used.

For a discussion on the main assumptions of the CFD model - rigid and impermeable vessel walls, homogeneous and Newtonian fluid (viscosity and density equal to 0.003 kg m^{-1} s^{-1} and 1,060 kg m^{-3}, respectively), co-planar vessel axes - we refer to our previous works.[8,9]

Unlike our previous studies, a pulsatile flow was imposed at the SVC and IVC inlet sections, while zero uniform pressure was used at the outlet sections.

MR-acquired caval flows were used as inlet boundary conditions for the CFD simulations. The time function adopted for the SVC inlet was equal to that acquired during the MR scan, while the IVC one was slightly different from that acquired. Actually, we observed that the measured mean volume flow rates in the SVC, IVC, LPA and RPA did not agree with the mass conservation law, with a difference (($\overline{Q}_{SVC} + \overline{Q}_{IVC}$) - ($\overline{Q}_{LPA} + \overline{Q}_{RPA}$)) equal to -2.67 cm^3 s^{-1} (-0.16 l min^{-1}) throughout a cardiac cycle. Indeed, the presence of the atrial fenestration allowed an additional blood inflow into the right atrial lateral tunnel through the fenestration. As the CFD model did not include the atrial fenestration, the CFD time function for the IVC inlet was modified in order to guarantee mass conservation (Fig.3). The quantity:

$$d(t) = Q_{in}(t) - Q_{out}(t) = (Q_{SVC}(t) + Q_{IVC}(t)) - (Q_{LPA}(t) + Q_{RPA}(t)) \qquad (1)$$

was computed at each instant from the MR data and then subtracted from the instantaneous volume flow rate at the IVC inlet.

Peak Reynolds numbers in the SVC and IVC were 1,295 and 2,636, respectively, while Womersley numbers were 10.6 and 16.0, respectively.

Two different simulations were performed: a flat velocity profile was adopted for the two caval inlets in the former, while a parabolic velocity profile was used in the latter. In each simulation four cardiac cycles were computed, in order to obtain a solution corresponding to the developed pulsatile flow.

The full Navier-Stokes equations were solved by FIDAP, which uses the Galerkin's weighted residual approach in conjunction with finite element approximation. A segregated solver was chosen for the solution method and conjugate gradient based iterative solvers were adopted to solve the symmetric and non-symmetric linear systems. The time integration technique was the implicit backward Euler, implemented with a variable time increment determined by control of the local time truncation error at each time step. The adopted value for the maximum relative local truncation error was 0.001, combined with a 'tolerance window' equal to 0.16 (*i.e.*, the time step n+1 was repeated only if $(dt_{n+1}/dt_n) < 1-0.16 = 0.84$).

The computations were carried out on a HP 9000/735 workstation with 128 Mb RAM.

4. RESULTS

The volume flow curves at LPA and RPA outlet sections *vs.* time, obtained in the CFD study, are plotted in Fig.5. The corresponding curves from MR measurements are plotted, as well. The CFD-calculated LPA flow was higher than that from the MR study for both types of inlet (caval) velocity profiles, while the RPA flow showed an opposite behaviour. With regard to the whole cardiac cycle, the CFD study gave a mean volume flow rate to the left lung equal to 28.51 cm^3 s^{-1} (1.71 l min^{-1}) for the model with flat velocity profiles and 28.01 cm^3 s^{-1} (1.68 l min^{-1}) for that with parabolic velocity profiles, *vs.* 25.51 cm^3 s^{-1} (1.53 l min^{-1}) obtained with MR.

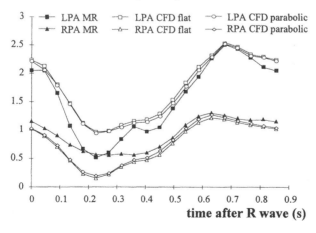

Fig.5. Instantaneous volume flow curves at LPA and RPA outlet sections obtained in the CFD study compared to the corresponding MR curves (1 l min^{-1} = 16.67 cm^3 s^{-1}).

Similarly, the mean blood flow to the right lung predicted by CFD was 12.34 cm^3 s^{-1} (0.74 1 min^{-1}) and 13.00 cm^3 s^{-1} (0.78 1 min^{-1}), respectively, *vs.* 15.50 cm^3 s^{-1} (0.93 1 min^{-1}) from the MR study.

The percentage relative errors (ε) on the mean volume flow rate in the two pulmonary arteries were calculated as follows:

$$\varepsilon_i = \frac{\overline{Q}_{i,CFD} - \overline{Q}_{i,MR}}{\overline{Q}_{i,MR}} \times 100 \qquad\qquad (i = LPA, RPA). \qquad (2)$$

ε_{LPA} was +11.6 % for the model with flat velocity profiles and +9.8 % for that with the parabolic ones. ε_{RPA} was -20.5 % and -17.2 %, respectively.

Figure 6 illustrates an example of the velocity profiles in the LPA, 1.5 cm distally from the SVC anastomosis, 630 ms after R wave, obtained with both techniques. Profiles were plotted along two perpendicular diameters. CFD and MR velocity profiles show a similar shape. The MR velocity profile appears skewed towards the upper body, which is well reproduced by the CFD model, both qualitatively and quantitatively. MR highest velocities are 30.5 cm s^{-1} and 31.7 cm s^{-1} *vs.* 28.0 cm s^{-1} and 27.5 cm s^{-1} for the model with flat inlet velocity profiles, and 24.2 cm s^{-1} and 28.4 cm s^{-1} for that with parabolic velocity profiles. The consequences of the assumption of the existence of a symmetry plane for the CFD model can be clearly observed, as well.

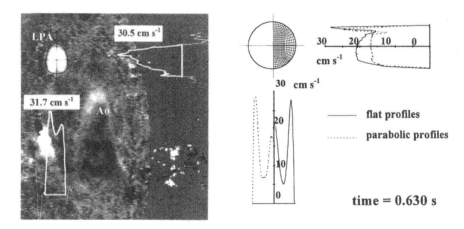

Fig.6. Velocity profiles along two perpendicular diameters in the LPA, 1.5 cm distally from the SVC anastomosis, 630 ms after R wave. Left: MR profiles in a sagittal spin echo image; right: CFD profiles, represented in the same set up. Ao, aorta.

5. DISCUSSION

The presented results were obtained in a preliminary study aiming at the evaluation of CFD techniques as a tool to design (as well as to optimise) surgical repairs of complex cardiac congenital malformations and predict the postoperative local haemodynamics created by the repairs themselves. Predictive capabilities of such models rely on both an accurate description of the anatomy of the patient and proper boundary conditions.

The postoperative anatomy clearly depends upon the preoperative anatomy of the patient as well as on the adopted surgical procedure. Hence, building a model of the postoperative anatomy is fairly an easy task. On the contrary, it appears rather difficult to estimate preoperatively the postoperative flows in the TCPC. Indeed, postoperative flow assessment is usually performed only in case of failure of the surgical treatment (where failure has to be intended as need for reoperation). In such circumstances, cardiac catheterisation measurements are performed, which are heavily invasive. Moreover, non-invasive quantitative Doppler flow measurements are often unreliable due to the complex haemodynamics and vessels' curvature in the anastomoses region.

With regards to the CFD models, it is well known that boundary conditions affect the results in the fluid domain.

This study was devised in order to assess the effects of different caval velocity profiles on flows in the pulmonary arteries in a realistic three-dimensional model of a patient submitted to the TCPC. MR investigation was used to obtain reliable flow data. In this regard, one should note that MR scans are not used to evaluate flows in the routine paediatric cardiology, due to their high cost and long time that the young patient has to spend lying in the magnet.

Our study's results show that the different velocity profiles adopted at the caval inlets of a CFD model of the TCPC have only minor effects on blood flow distribution into the lungs. Since we assumed the same pressure at the pulmonary outlets, the fact that the blood flow to the left lung is greater than that to the right one has to be attributed to the role of the local geometry of the anastomoses. This means that the surgeon can effectively achieve the desired pulmonary flow balance acting on the caval anastomoses, provided their size is compatible with the anatomic constraints.

Percentage relative error on mean volume flow rate in the pulmonary arteries look rather high. Actually, the presented CFD model study suffers from some limitations, such as the absence of compliance in the vessel walls, the presence of a symmetry plane, the discarded effects of the respiratory rhythm as well as the correction to the IVC flow rate. On the other hand, MR techniques also may suffer from errors arising in the definition of the region of interest for the velocity measurements, due to image resolution in the currently available MR devices.

In spite of the above remarks, we believe that CFD models of the TCPC based on the preoperative anatomic data would be extremely important and powerful tools for predicting postoperative haemodynamics as well as optimising the hydraulic design of the caval connections.

6. ACKNOWLEDGEMENTS

Francesco Migliavacca was supported by the Italian 'Consiglio Nazionale delle Ricerche' and Philip Kilner by the British Heart Foundation.

7. REFERENCES

1. de Leval, M. R., Kilner, P., Gewillig, M. and Bull, C., Total cavopulmonary connection: a logical alternative to atriopulmonary connection for complex Fontan operations, J. Thorac. Cardiovasc. Surg., 1988, Vol. 96, 682-695.

2. Jonas, R. A. and Castaneda, A. R., Modified Fontan procedure: atrial baffle and systemic venous to pulmonary artery anastomotic technique, J. Cardiac. Surg., 1988, Vol. 3, 91-96.

3. Puga, F. J., Chiavarelli, M. and Hagler, D. J., Modifications of the Fontan operation applicable to patients with the left atrioventricular valve atresia or single atrioventricular valve, Circulation, 1987, Vol. 76, III-53-III-60.

4. Bridges, N., Lock, J. E. and Castaneda, A. R., Baffle fenestration with subsequent transcatheter closure: modification of the Fontan operation for patients at increased risk, Circulation, 1990, Vol. 82, 1681-1689.

5. Low, H. T., Chew, Y. T. and Lee, C. N., Flow studies on atriopulmonary and cavopulmonary connections of the Fontan operations for congenital heart defects, J. Biomed. Eng., 1993, Vol. 15, 303-307.

6. Sharma, S., Goudy, S., Walker, P., Panchal, S., Ensley, A., Kanter, K., Tam, V., Fyfe, D. and Yoganathan, A., In vitro flow experiments for determination of optimal geometry of total cavopulmonary connection for surgical repair of children with functional single ventricle, J. Am. Coll. Cardiol., 1996, Vol. 27, 1264-1269.

7. Van Haesdonck, J.-M., Mertens, L., Sizaire, R., Montas, G., Purnode, B., Daenen, W., Crochet, M. and Gewillig, M., Comparison by computerized numeric modeling of energy losses in different Fontan connections, Circulation, 1995, Vol. 92, II-322-II-326.

8. Dubini, G., de Leval, M. R., Pietrabissa, R., Montevecchi, F. M. and Fumero, R., A numerical fluid mechanical study of repaired congenital heart defects. Application to the total cavopulmonary connection, J. Biomechanics, 1996, Vol. 29, 111-121. [Erratum: J. Biomechanics, 1996, Vol. 29, 839].

9. de Leval, M. R., Dubini, G., Migliavacca, F., Jalali, H., Camporini, G., Redington, A. and Pietrabissa, R., Use of computational fluid dynamics in the design of surgical procedures: application to the study of competitive flows in cavopulmonary connections, J. Thorac. Cardiovasc. Surg., 1996, Vol. 111, 502-513.

10. Mohiaddin, R. H. and Longmore, D. B., Functional aspects of cardiovascular nuclear magnetic resonance imaging: techniques and application, Circulation, 1993, Vol. 88, 264-281.

G. Riccardi[1], A. Iafrati[2], G. Tonti[3], F. Fedele[4], R.M. Manfredi[4]

1   Abstract

The flow in the left atrium-ventricle system has been numerically investigated by assuming the blood as an *inviscid* incompressible fluid. The geometry is taken as axisymmetrical around the ventricle axis and the motion of the heart walls is given. The massive separation from the mitral valve is described in terms of a standard Kutta condition, by producing filaments at the separation line. The motion of the valves, forced by pressure and elastic forces, is integrated during the cardiac cycle.

Preliminary results have been discussed in the diastole, with the aim to understand the influence of the motion of the heart walls on the time history of pressure and velocity fields and the resulting dynamics of the mitral valve.

2   Introduction

In the clinical practice, the transmitral flow field is currently observed during a cardiac cycle through Eco-Doppler. In this way all the flow details become available for diagnostic aims and they can be used to investigate relevant cardiac patologies. As a consequence, the deloptment of non-invasive tools able to use this great quantity of data is needed. In the past, many simplified analytical models have been devoloped [1], with the aim to motivate some particular aspects of the cardiac cycle.

Keywords: Boundary Integral Methods, Mitral Flow, Blood Dynamics in the Heart

(1)   Dip. di Ingegneria Aerospaziale, SUN - via Roma, 29; 81031 Aversa
(2)   CIRA - via Maiorise; 81043 Capua
(3)   Servizio di Cardiologia con UTIC - Sez. di Penne; ASL Pescara
(4)   *III* Cattedra di Cardiologia, Dip. Scienze Cardiovascolari e Respiratorie,
      Università *"La Sapienza"* - viale del Policlinico, 155; 00161 Roma

At the present time, the numerical simulation of the flow in the heart appears to be one of the most attractive approach, even if the analysis of the numerical results is rather difficult. As an example, the flow into the left part of the heart has been numerically investigated in Peskin *et al.* [2], where the coupling of the blood dynamics with the elastic fibers motion is considered, accurately describing the structure of the ventricle muscle and of the valves.

In the present work, particular attention is devoted to fluid dynamical aspects of the problem and strong simplifications are posed on the cardiac cycle and on the heart geometry, whose motion is assigned at the present stage, by providing a sequence of geometries with a *constant* time step. The cardiac cycle is modelled by assuming an *ideal* behaviour characterized by four separate phases. Starting at the time when the aortic valve closes, the first phase is the *iso-volumetric relaxation* of the ventricle, in which it elongates preserving its volume. During this stage the pressure on the ventricle face of the mitral valve drops, up to the corresponding value on the atrium face. When this occurs, the valve opens and a new phase of the cycle begins, the diastole, along which the ventricle fills up. In the first stage of the filling the blood moves into the ventricle according to the pressure gradient (*E*-wave of the transmitral Doppler pattern velocity). When the ventricle is filled, the blood flow from the atrium to the ventricle becomes very small, up to the atrium systole (*A*-wave). Then the mitral valve closes and the *iso-volumetric compression* of the ventricle begins: in this phase the ventricle decreases in length keeping its volume constant. At the same time the pressure on the ventricle face of the aortic valve grows, up to reach the aortic pressure. Then, the *aortic valve* opens and the ventricle empties during the *systole*. The time history of the ventricular and atrial volumes during the cycle is shown in Fig. 1.

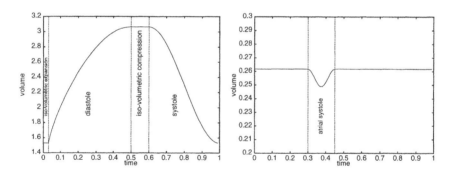

Figure 1: Ventricle (*left*) and atrium (*right*) volumes during the ideal cardiac cycle.

The motion of the valves (mitral and aortic) is integrated by considering the dynamics of a material surface forced by the external pressure field and by elastic forces acting only in the tangential direction. The numerical integration of the resulting system is briefly described in Section 4.

In Section 6, preliminary results are discussed for the diastolic phase, in terms of the velocity and pressure fields and of the resulting motion of the mitral valve. Finally, a plane of the future work is given in Section 7.

3 Integral formulation for the velocity

In the present Section, the flow into the bounded domain Ω_h (see Fig. 2), composed by the left ventricle and atrium separated by the mitral valve, is considered.

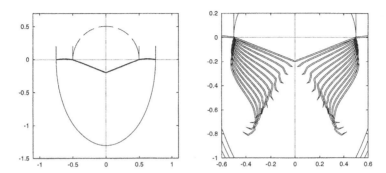

Figure 2: The geometry of the system is depicted on the left, while preliminary results about the motion of the mitral valve are shown on the right. Here, starting from time $t = 0.03$, the position of the mitral valve is represented with a time step 0.02.

The velocity field $\mathbf{u}$ in any point $\mathbf{x}$ inside Ω_h is written in terms of its limit values on $\partial\Omega_h$. For the sake of simplicity, $\mathbf{u}$ is defined on the whole plane, by assuming $\mathbf{u} \equiv \mathbf{0}$ outside Ω_h. The classical Helmholtz decomposition of $\mathbf{u}$ in terms of a *scalar potential* Φ and a *vector* potential $\mathbf{A}$ is considered:

$$\mathbf{u} = \nabla\Phi + \nabla \times \mathbf{A} \,, \tag{1}$$

where $\mathbf{A}$ is chosen divergence free. Using a simple layer representation for both potentials, it follows:

$$
\begin{aligned}
\Phi(\mathbf{x}) &= \int_{\partial\Omega_h} dS(\mathbf{y})\phi(\mathbf{y})G(\mathbf{x} - \mathbf{y}) \\
\mathbf{A}(\mathbf{x}) &= \int_{\partial\Omega_h} dS(\mathbf{y})\mathbf{a}(\mathbf{y})G(\mathbf{x} - \mathbf{y}) - \int_{\Omega_h} dV(\mathbf{y})\, G(\mathbf{x} - \mathbf{y})\omega(\mathbf{y}) \,,
\end{aligned}
\tag{2}
$$

where $G(\mathbf{x}) = -1/(4\pi|\mathbf{x}|)$ while ϕ and $\mathbf{a}$ are the layer densities.

The density ϕ of the scalar potential is related to the normal velocity on the boundary $u_\nu = \mathbf{u} \cdot \boldsymbol{\nu}$. Actually, being the tangential derivatives of $\mathbf{A}$ continuous across $\partial\Omega_h$, the jump of the normal component of $\mathbf{u}$ in passing from outside to inside of Ω_h results:

$$[\mathbf{u} \cdot \boldsymbol{\nu}] = 0 - \mathbf{u} \cdot \boldsymbol{\nu} = \phi \,. \tag{3}$$

Let us now evaluate the density $\mathbf{a}$: since $\mathbf{A}$ is divergence free, it can be shown that the density $\mathbf{a}$ is tangential to $\partial\Omega_h$ and, by evaluating the jump of $\mathbf{u} \times \boldsymbol{\nu}$ across $\partial\Omega_h$, it follows:

$$[\mathbf{u} \times \boldsymbol{\nu}] = 0 - \mathbf{u} \times \boldsymbol{\nu} = \mathbf{a} \,. \tag{4}$$

Using the densities given in equations (3, 4) into equation (2) it follows that the velocity in a point $\mathbf{x}$ inside the domain Ω_h is:

$$\mathbf{u}(\mathbf{x}) = -\nabla \int_{\partial\Omega_h} dS(\mathbf{y})\mathbf{u}(\mathbf{y}) \cdot \boldsymbol{\nu}(\mathbf{y})G(\mathbf{x}-\mathbf{y}) - \nabla \times \int_{\partial\Omega_h} dS(\mathbf{y})\mathbf{u}(\mathbf{y}) \times \boldsymbol{\nu}(\mathbf{y})G(\mathbf{x}-\mathbf{y}) +$$

$$- \nabla \times \int_{\Omega_h} dV(\mathbf{y})\boldsymbol{\omega}(\mathbf{y})G(\mathbf{x}-\mathbf{y}) . \tag{5}$$

For $\mathbf{x} \in \partial\Omega_h$, a vector integral equation is obtained by evaluating the limit form of equation (5) from the inside of Ω_h:

$$\frac{1}{2}\mathbf{u}(\mathbf{x}) + \nabla \int_{\partial\Omega_h} dS(\mathbf{y})\mathbf{u}(\mathbf{y}) \cdot \boldsymbol{\nu}(\mathbf{y})G(\mathbf{x}-\mathbf{y}) + \nabla \times \int_{\partial\Omega_h} dS(\mathbf{y})\mathbf{u}(\mathbf{y}) \times \boldsymbol{\nu}(\mathbf{y})G(\mathbf{x}-\mathbf{y}) =$$

$$= -\nabla \times \int_{\Omega_h} dV(\mathbf{y})\boldsymbol{\omega}(\mathbf{y})G(\mathbf{x}-\mathbf{y}) , \tag{6}$$

which results in a Fredholm equation of second kind in the limit velocity vector on the boundary. If one of the two velocity components (tangential or normal) is given on the heart wall $\partial\Omega_h$, equation (6) yields the other one. For example, on the ventricle wall the normal velocity of the fluid is assigned and the tangential one is calculated, while at the inlet of the pulmunary flow the tangential velocity is assumed to be zero and the inflow normal velocity is calculated.

In the hypotesis of axisymmetric no-swirl ($\omega_x \equiv 0$) flow around the x axis, the azimuthal integrations can be performed analitically. Actually, in the cylindrical frame of reference it results $\mathbf{x} = (x, r\cos\vartheta, r\sin\vartheta)$ and the following equality holds:

$$|\mathbf{x} - \mathbf{y}| = \Delta \left(1 - k^2 \cos\frac{\theta - \vartheta}{2}\right)^{1/2} , \tag{7}$$

with $\Delta = \left[(x-\xi)^2 + (r+\rho)^2\right]^{1/2}$ and $k^2 = 4r\rho/\Delta^2$. It can be shown that $k \in [0,1]$, with $k = 0$ only if r or ρ vanishes and $k = 1$ if $x = \xi$ and $r = \rho$, i.e. $\mathbf{x}$ and $\mathbf{y}$ belong to the same circle around the x axis. In terms of the *complete elliptic integrals*

$$F(k) = \int_0^{\pi/2} d\beta \left(1 - k^2\cos\beta\right)^{-1/2} , \quad E(k) = \int_0^{\pi/2} d\beta \left(1 - k^2\cos\beta\right)^{1/2} \tag{8}$$

and introducing the function $H = 2F/k^2 + (k^2 - 2)E/[k^2(1-k^2)]$, it follows for the first integral in equation (5):

$$-\nabla \int_{\partial\Omega_h} dS\, \mathbf{u} \cdot \boldsymbol{\nu}G = -\frac{1}{\pi}\int_C ds u_\nu \frac{\rho}{\Delta^3} \left[\mathbf{e}_x (x-\xi)\frac{E}{1-k^2} + \mathbf{e}_r \left(r\frac{E}{1-k^2} + \rho H\right)\right] . \tag{9}$$

The second contribution in equation (5) represents the velocity field induced by a vortex sheet on the boundary of Ω_h, according to the effect of a boundary layer of vanishing thickness for growing Reynolds number.

In absence of swirl, the tangential velocity lies in the symmetry plane Π, hence $\mathbf{u} \times \boldsymbol{\nu} = -u_\tau(0, -\sin\theta, \cos\theta)$, u_τ being the tangential velocity. Note that the density of circulation on the boundary of Ω_h is opposite to the tangential velocity, due to the counter-clockwise orientation of $\partial\Omega_h$. Performing the azimuthal integration one obtains:

$$-\nabla \times \int_{\partial\Omega_h} dS\mathbf{u} \times \boldsymbol{\nu}G = \frac{1}{\pi}\int_C ds u_\tau \frac{\rho}{\Delta^3}\left[-\mathbf{e}_x\left(rH + \rho\frac{E}{1-k^2}\right) + \mathbf{e}_r (x-\xi)H\right] . \tag{10}$$

The third term on the right-hand side of equation (5) represents the Biot-Savart contribution related to the vorticity field $\boldsymbol{\omega}$. To account for the massive separation at the mitral rims a standard Kutta condition for panel methods [3] is used. The shed vorticity is accumulated during every time step in which the mitral valve is open and then it is relased in the form of a vortex filament, the motion of which is followed using a Lagrangian method [4]. Having a set of N vortex filaments placed at the points (x_i, r_i) of a symmetry plane, the term becomes:

$$-\nabla \times \int_{\Omega_h} dV \, \omega G = \frac{1}{\pi} \sum_{i=1}^{N} \Gamma_i \frac{r_i}{\Delta_i^3} \left\{ \mathbf{e}_x \left[r H(k_i) + r_i \frac{E(k_i)}{1 - k_i^2} \right] - \mathbf{e}_r \left(x - x_i \right) H(k_i) \right\} , \quad (11)$$

where Γ_i is the circulation of the i-th filament, $\Delta_i^2 = (x - x_i)^2 + (r + r_i)^2 + \alpha \varepsilon_i^2$ and $k_i^2 = 2 r r_i / \Delta_i^2$. The cut-off parameter ε_i is updated according to the actual length of the i-th filament.

4   Valve dynamics

During the systole, the blood flow from the ventricle to the atrium is prevented by the presence of the mitral valve, which is kept closed by the action of the chordae tendineae. On the contrary, when the diastole starts the pressure drop in the ventricle enables the opening of the valve, leading to a quite fast filling of the ventricle (E-wave). A similar behavior enforce the opening of the aortic valve at the beginning of the systole. However, in the present paper only preliminary results concerning the initial stage of the diastole are discussed.

Due to its geometrical configuration, the mitral motion cannot be considered as axisymmetric being absent elastic forces in the azimuthal direction due to valve deformations. For this reason the valve motion is considered as a two-dimensional one, in the symmetry plane Π. In absence of viscosity, the motion of the valves (mitral and aortic) is induced by the pressure field and by elastic forces which contrast any variation of the valve size. By discretizing the valve section with a set of N straight panels with vertices at $\{\mathbf{x}_i\}_{i=1,\dots,N+1}$, where $\mathbf{x}_1$ is on the heart wall, the dynamics of the k-th node on the valve is

$$m_k \ddot{\mathbf{x}}_k = \mathbf{P}_k + \chi_k \left(\frac{l_k - l_k^0}{l_k^0} \boldsymbol{\tau}_k - \frac{l_{k-1} - l_{k-1}^0}{l_{k-1}^0} \boldsymbol{\tau}_{k-1} \right) , \quad (12)$$

where m_k is the mass associated with the k-th node, $\mathbf{P}_k$ is the resulting pressure force on the node, l_k is the length of the k-th panel (having edges in $\mathbf{x}_k$ and $\mathbf{x}_{k+1}$) at the current time, while l_k^0 is at the initial time. Moreover, $\boldsymbol{\tau}_k$ is the unit tangent vector on the k-th panel and χ_k is an elastic constant. The pressure force $\mathbf{P}_k$ on the k-th node is approximated by summing half the force acting on the two adiacent panels. Finally, a three points first order approximation of the node acceleration $\ddot{\mathbf{x}}_k$ is used, togheter with Crank-Nicolson treatment of the elastic terms.

It is worth to notice that the pressure cannot be employed to drive the opening of the mitral valve, being its calculation not accurately feasible with the valve orifice smaller than the mean length of a panel. For this reason, at the beginning of the diastole, a vanishing tangential velocity is assigned on the ventricle side of the mitral valve: in this way the normal velocity is computed from equation (6) and it is employed to move the mean line of the valve.

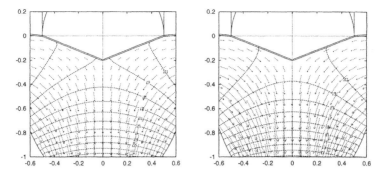

Figure 3: Iso-volumetric relaxation: pressure and velocity fields at time $t = 0.010$ (*left*) and $t = 0.025$ (*right*).

5   Integral formulation for the pressure

For the dynamical pressure $P = p + |\mathbf{u}|^2/2$, a simple layer representation is employed. To this aim, let $R(\mathbf{x}) = \int_{\Omega_h} dV(\mathbf{y})G(\mathbf{x}-\mathbf{y})\nabla_y \cdot [\boldsymbol{\omega}(\mathbf{y}) \times \mathbf{u}(\mathbf{y})]$, continuous across $\partial\Omega_h$, it follows

$$P(\mathbf{x}) = -\int_{\partial\Omega_h} dS(\mathbf{y})G(\mathbf{x} - \mathbf{y})\sigma(\mathbf{y}) - R(\mathbf{x}) \ . \tag{13}$$

To evaluate σ, the normal derivative of P is written on the heart boundary and it is requested to satisfy the Euler equation in the limit on the boundary of Ω_h. Actually, by considering that from the Euler equation $\partial_\nu P = -\boldsymbol{\nu} \cdot \partial_t \mathbf{u}$ on $\partial\Omega_h$, the following Fredholm integral equation of the second kind is obtained:

$$\frac{1}{2}\sigma(\mathbf{x}) - \int_{\partial\Omega_h} dS(\mathbf{y})\partial_{\nu(x)}G(\mathbf{x} - \mathbf{y})\sigma(\mathbf{y}) = -\boldsymbol{\nu} \cdot \partial_t \mathbf{u}(\mathbf{x}) + \partial_\nu R(\mathbf{x}) \ , \tag{14}$$

which enables to calculate the density σ.

The evaluation of the Eulerian acceleration normal to the heart boundary in equation (14) requires some considerations, in order to ensure the absence of convective contributions. First of all, we recall that the following boundary condition is enforced on the velocity field $\mathbf{u}$ *for all times* t:

$$\mathbf{u}\left[\mathbf{x}(\mathbf{a}, t), t\right] \cdot \boldsymbol{\nu}(\mathbf{a}, t) = \mathbf{u}_h(\mathbf{a}, t) \cdot \boldsymbol{\nu}(\mathbf{a}, t) \ . \tag{15}$$

In the left-hand-side of equation (15) the limit process for $\mathbf{a}$ going on the initial heart boundary is understood. Equation (15) states the equality between the fluid velocity normal to the boundary and the normal velocity of the boundary $\mathbf{u}_h$ itself. Here the Lagrangian coordinate $\mathbf{a}$ has been introduced.

Performing the time derivative of both sides of equation (15) and taking into account that the vorticity field has vanishing limit on the heart boundary, one obtains:

$$\boldsymbol{\nu} \cdot \partial_t \mathbf{u} = \mathbf{a}_h \cdot \boldsymbol{\nu} + (\mathbf{u}_h - \mathbf{u}) \cdot \partial_t \boldsymbol{\nu} - \partial_\nu \frac{|\mathbf{u}|^2}{2} \ , \tag{16}$$

where $\mathbf{a}_h$ is the acceleration of the heart boundary. While the first and second terms of the right-hand side of equation (16) are easily evaluated, the last term,

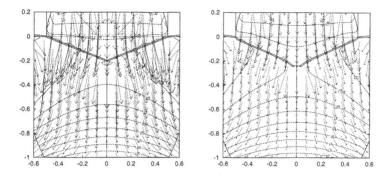

Figure 4: Pressure and velocity fields soon after the mitral valve opening: times $t = 0.040$ (*left*) and $t = 0.050$ (*right*).

involving the normal derivative of the fluid velocity on the heart boundary, cannot be accurately computed by a numerical derivative. The normal derivatives of **u** should be replaced by the tangent ones, using the continuity equation and the condition of vanishing vorticity on the heart boundary.

6   Preliminary Results

The motion of the mitral valve during the initial phase of the diastole is shown on the right of Fig. 2. A crucial role in the integration of the valve motion is played by the ratio $\sqrt{m/\chi}$, which results as a characteristic time for the propagation of the elastic perturbations along the leaflet of the valve. To perfom the numerical integration

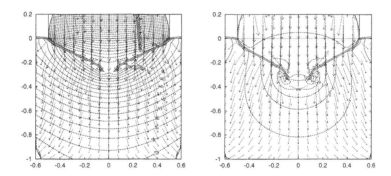

Figure 5: Pressure and velocity fields at the beginnig of the diastole: times $t = 0.060$ (*left*) and $t = 0.080$ (*right*).

of the valve motion a time step of the same order of this one is needed. For this reason, with the coarse Δt used ($\Delta t = 1/200$) the valve has a mass larger than the real case and an elastic constant quite small, which leads to a slower dynamics and to an elongation of the valve leaflet. To overcome this difficulty, an integration with a variable Δt is needed.

In terms of fields of pressure and velocity, at the present two phases of the cardiac cycle are analyzed: the iso-volumetric relaxation and the beginning of the diastole. The behavior of the pressure and velocity fields in the ventricle during the iso-volumetric relaxation is depicted in Fig. 3. The results are in a non-dimensional form and the jump between two consecutive iso-lines is 10, while the scale of the velocity is 1/10. The figure shows as the ventricle walls motion produces a favorable pressure gradient, allowing for the mitral valve opening.

Preliminary results obtained in absence of vorticity generation, show that for some times after the mitral valve opening the pressure gradient remains favorable leading to the establishment of a intense transmitral flow (see Fig. 4). After this phase, the pressure gradient becomes adverse as depicted in Fig. 5. Due to this adverse pressure gradient the trasmitral flow is reduced up to reverse its direction. At the moment the causes of this strange behaviour are not clear: probably it depends on the time history assumed for the ventricular geometry. The generation of geometries with a variable time step, according to the acceleration level on the boundary, is presently under development.

7   Perspectives

In the present work some features of the blood flow in the left atrium-ventricle system have been briefly discussed. A simplified model usefull to investigate about the unsteady velocity and pressure fields during one cardiac cycle has been developed. At present, strong simplifications on the heart geometry have been made and the motion of the heart wall has been assigned in a simple way, with reference to an *ideal* cardiac cycle. These lacks of the present model should be overcome in the future, by coupling the numerical description of the blood flow with a suitable model for the myocardium motion and assuming the elastic tension in the muscle fibers as input data [5]. Moreover, an extension to three-dimensional flow seems strongly necessary both for fluid dynamics and elasticity.

References

[1] Pedley, T.J. *The fluid mechanics of large blood vessels*, Cambridge Univ. Press (1980)

[2] Peskin, C.S., McQueen, D.M. (1989) *A Three-Dimensional Computational Method for Blood Flow in the Heart. I. Immersed Elastic Fibers in a Viscous Incompressible Fluid*, J. Comput. Physics 81, 372

[3] Riccardi, G., Iafrati, A., Piva, R., (1994) *Vorticity Shedding from a Lentil-Shaped Body at Large Incidence in Uniform Flow*, Meccanica 29, 159

[4] Leonard, A., (1985) *Computing Three-Dimensional Incompressible flows with vortex elements*, Ann. Rev. Fluid Mech. 17, 523

[5] Peskin, C.S. (1989) *Fiber Architecture of the Left Ventricular Wall: An Asymptotic Analysis*, Comm. Pure and Applied Math. XLII, 79

7. DENTAL MATERIALS, BEHAVIOUR AND BIOMECHANICS

NUMERICAL MODELLING OF ORTHODONTIC BRACKETS

B. Kralj[1], J. Knox[2], J. Middleton[3], M. L. Jones[4]

1. ABSTRACT

This paper describes the use of computational modelling in evaluating the mechanical response of fixed orthodontic brackets. The method of homogenisation is applied to model the complex geometry of the bracket base where both single and double mesh layers are investigated. Here results are given for the equivalent material properties and it is shown that double layer impregnated wire meshes offer reduce stress regimes. Material modelling is extended to include the enamel/adhesive interface and here a novel technique has been employed to predict adhesive penetration. This method allows the composite nature of the enamel surface to be determined and material constants and validation procedures are presented

2. INTRODUCTION

The objective of this analysis is to evaluate the mechanical response of fixed orthodontic brackets with the final aim of improving design. Here, the emphasis will be placed on mechanics assuming that the clinical requirements will be satisfied by the proposed design. In order to achieve good bracket design, from the mechanical point of view, the following requirements should be considered:

- safe and predictable transfer of treatment load from arch wire to the tooth,
- good, reliable and robust bond attachment of the bracket during the treatment period to ensure resistance to mastication and other oral forces,

Keywords: FEM, orthodontic bracket, homogenisation

[1] Honorary Lecturer, Dental School, University of Wales College of Medicine/ University of Wales Swansea, Wales, UK

[2] Senior Registrar, Dental School, University of Wales College of Medicine Cardiff, Heath Park, Cardiff CF4 4XN, Wales, UK

[3] Reader, Department of Civil Engineering, University of Wales Swansea, Swansea SA2 8PP, Wales, UK

[4] Professor, Dental School, University of Wales College of Medicine Cardiff, Heath Park, Cardiff CF4 4XN, Wales, UK

- predictable and easy removal of the bracket at the end of the treatment life, with easy 'clean-up' of adhesive remaining on the tooth.

After taking into account these requirements and accompanying loading conditions the system is influenced by:

- design of the bracket itself,
- design of the bracket base and its bond to the adhesive,
- adhesive properties, especially its adhesion to the bracket and enamel surfaces.

The finite element method is a particularly suitable tool for analysing the mechanical aspects described above since:

- arbitrary geometry can be modelled,
- material properties can be readily varied,
- various loading and boundary conditions can be taken into account,
- changes in the design can easily be introduced in the model and quantitatively evaluated,
- compared to experimental tests FE is both a rapid and cost efficient method of design improvement.

This study is concerned with the application of FE modelling to the fixed appliance orthodontic bracket, with a special emphasis being placed on the bracket base design, adhesive properties and adhesive - tooth interface. Illustrative results of parametric studies are given.

3. THE FINITE ELEMENT MODEL

A maxillary first premolar together with a stainless steel bracket (Master Series BI PAD II) manufactured by American Orthodontics were used in this study. The bracket was fixed to the tooth using an adhesive whose mechanical properties will be the subject of parametric studies. A view of the complete finite element mesh which consists of 15324 nodes and 2971 twenty noded isoparametric finite elements is shown in Fig. 1.a.

The system to be investigated can be divided into five distinct parts, listed in Table 1, together with the basic material and modelling characteristics.

Part	Material	Characteristics	
bracket	stainless steel	homogeneous	isotropic
bracket base	wire mesh + adhesive	composite	anisotropic
adhesive lute	adhesive	homogeneous[*]	isotropic
tooth - adhesive interface	enamel + adhesive	composite	anisotropic
tooth	enamel	homogeneous	isotropic

[*] can contain inclusions and/or air voids

Table 1. Description of orthodontic fixed bracket system

The FE mesh of the bracket was based on original drawings supplied by the manufacturer, while the mesh for the tooth was based on the computer imaging of the cross sections of a typical adult premolar. While the modelling of these components was straightforward, special attention had to be given to modelling the bracket base and adhesive-tooth interface which are highly complex composite structures.

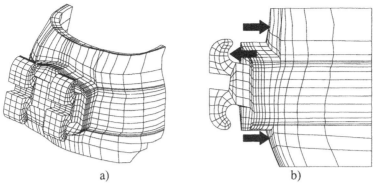

a)   b)

Figure 1. - a) FE mesh of the tooth and bracket, b) loading conditions for removal of the bracket using LODI

In order to evaluate the mechanical behaviour of the bracket, several loading conditions (treatment force, pulling off, mastication) were imposed on models with variable design parameters: adhesive mechanical properties, thickness of adhesive lute, wire diameters and wire spacings in the bracket base and single-double layer wire meshes. These analyses created a large amount of data and only a representative selection will be presented here. The load case considered here is that of removal of the bracket at the end of treatment using a lift-off debracketing instrument (LODI) where removal is achieved by pulling one wing of the bracket with the LODI. As mentioned earlier this is one of the most critical phases since it is important to remove the bracket without causing damage to the enamel whilst ideally leaving as little adhesive as possible on the tooth.

3. BRACKET BASE

The base of the bracket under consideration consists of a very fine steel wire mesh welded to a thin steel foil, see Fig. 2. During the bonding procedure the spaces in the mesh are penetrated by the adhesive creating a nonhomogeneous material. One way to model this is to follow a spatial distribution of different materials with finite elements ensuring that there is only one material within each element. Taking into account the number of mesh openings/wires in the bracket base this approach would clearly lead to a huge number of finite elements. Another approach, adopted here, is to homogenise this nonhomogeneous material into an equivalent homogeneous one (on the basis of equal deformation energies) which could be discretised with fewer finite elements while still giving equal overall behaviour.

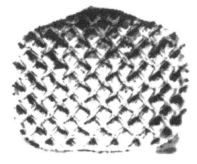

Figure 2. - base of the bracket - fine steel wire mesh

The homogenisation process involves subjecting a unit cell of the this material (which is its smallest representative volume) to a number of independent loading conditions and finding its response. By employing the averaging rules and kinematic and equilibrium conditions (see ref [1,2]):

$$\sigma_h = \frac{1}{V} \int \sigma_m dV$$

$$\varepsilon_h = \frac{1}{V} \int \varepsilon_m dV \tag{1}$$

$$div(\sigma) = 0$$

$$\varepsilon_{ij} = a_{ijkl} \sigma_{kl}$$

equivalent material properties can be found. In the above equations subscripts h stands for homogenised and m for real (or micro) values, a_{ijkl} is a constitutive (compliance) tensor and V is the volume of the unit cell.

Due to the complex geometry of the mesh base the only viable way of determining the behaviour of the unit cell, i.e. solving the problem (1), is to use the finite element method. As a consequence of spatial distribution of the wires the equivalent material was found to be transversely orthotropic, i.e. the plane of the mesh is a plane of isotropy with the different material properties being in the perpendicular direction.

As a variation of this design there is also a bracket base with a 'double mesh' in which two layers of the mesh, with different geometrical properties, are welded to the bracket base. In this case the mesh closer to the bracket has smaller openings and smaller diameter of wires, see Fig. 3 for an example. Homogenising this material in a rigorous manner can pose certain problems since:

- it can be difficult to define a unit cell, or,
- if the ratio of wire spacings is not an integer, that could lead to a large unit cell,
- for large unit cells the number of the cells per bracket base might be too small to justify homogenisation.

In the model used here this was taken into account by homogenising each layer separately and introducing them into the overall FE model as two different (homogeneous) materials.

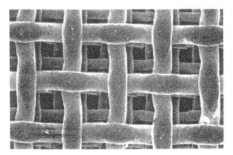

Figure 3 - example of a double layered mesh

Figure 4 shows the variation of the mechanical properties of the homogenised single mesh material with the change in properties of its constituents, i.e. wires and adhesive.

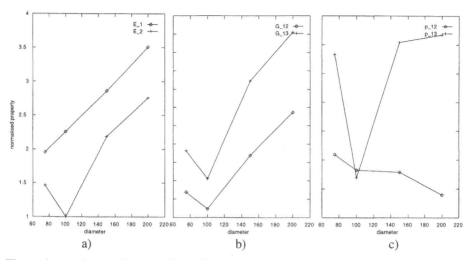

a)   b)   c)

Figure 4. - variation of the mechanical properties of the homogenised material with wire diameter, a) Young's moduli, b) shear coefficient and c) Poissons ratios, normalised to adhesive properties

Parametric studies were undertaken in order to evaluate the effects of different properties of the single layered wire mesh on the final stress distribution together with variations of cement lute thickness. These results are shown in Fig. 5 for each part of the fixed bracket. The effects of the introduction of the double layered mesh are shown in Fig. 6 and are compared with the results for the single layered mesh. It is shown from these figures that:

- variation of the mesh geometry has most influence on stresses in the bracket and the adhesive,
- thickness of the adhesive lute has the greatest influence on the stresses in the enamel,
- double layered mesh design reduces stresses in almost all parts of the appliance, the only exception being the mesh layer close to the bracket.

4. ADHESIVE TOOTH INTERFACE

In order to enhance the bond between enamel and the adhesive it is usual that the surface of the tooth is prepared by etching with an acid (phosphoric, maleic acid). During the etching a demineralisation process takes place creating a rough, porous surface. The depth of the etching and the pattern of the roughened surface depend on the type of acid used as well as on the length of time it is applied, ref [3]. This roughened surface in conjunction with modern hydrophilic adhesives enhances the bond between the adhesive and the enamel considerably. This is due to the adhesive penetrating the pores of the etched surface thus increasing its contact area with the enamel and at the same time ensuring that there is increased mechanical locking between these two materials.

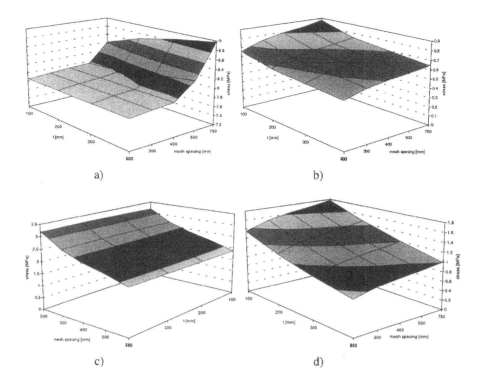

Figure 5 - variation of major principal stresses with adhesive lute thickness t and wire mesh spacing, a) bracket, b) adhesive, c) wire mesh layer and d) enamel

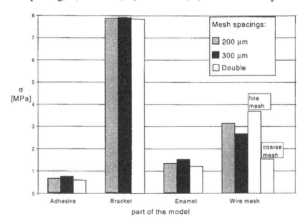

Figure 6 - variation of major principal stresses with the design of the bracket base during the bracket debonding (removal) procedure

In the numerical model this interface phase, consisting of the enamel and the adhesive, also has to be treated as a non-homogeneous material. However this nonhomogeneous material cannot be treated as a regular periodic material meaning that the 'unit cell' cannot be treated in the same way as for the wire mesh. Here other variables have to be taken into account such as pore size, their direction when reaching the free surface and the pattern/distribution. Following from this it is obvious that this phase cannot be homogenised using the same FE approach as the wire mesh and that analytical models for homogenisation of random composite materials are more appropriate.

A model suitable for this material will be briefly described here, see ref. [4]. The model is based on the Eshelby-Mori-Tanaka theory of orthotropic or transversely isotropic composite materials. The homogeneous matrix is assumed to be 'reinforced' by a set of monotonically aligned elliptic cylinders, which in this case correspond to tags of the adhesive penetrating the pores of the enamel. The basic idea is again an equivalence of responses between a homogeneous (subscript h) and nonhomogeneous material consisting of a matrix (subscript m, in this case enamel) and inclusions (subscript i, adhesive tags). Stresses in the matrix can be described as:

$$\sigma_m = \sigma_h + \tilde{\sigma} = L_m(\varepsilon_m + \tilde{\varepsilon})$$

with L being the compliance of the material. At the same time stresses in the inclusions can be expressed as:

$$\sigma_I = \sigma_h + \tilde{\sigma} + \sigma_p = L_I(\varepsilon_h + \tilde{\varepsilon} + \varepsilon_p)$$
$$= L_0(\varepsilon_h + \tilde{\varepsilon} + \varepsilon_p - \varepsilon^*)$$

where subscript p stands for perturbation values and ε^* is Eshelby's equivalence transformation strain and is related to perturbed strains through the relationship:

$$\varepsilon_p = S\varepsilon^*$$

where S is Eshelby's tensor. By applying a volume weighted averaging rule for stresses in the components the following can be established:

$$\tilde{\sigma} = c_i \sigma_p$$

with c_i being the concentration of inclusions. Similarly, it can be shown

$$\varepsilon_h = \varepsilon_m + c_i \varepsilon^*$$

which can gives the elastic moduli tensor of the homogenised material.

Using this approach the relationship between the homogenised properties of this layer and the concentration of penetrated adhesive can be found and this is given diagramatically in Fig. 7. In this example the material properties used are given in table 2 (note that E for enamel stands for an apparent Young's modulus):

	Young's modulus [MPa]	Poisson's ratio
Enamel	47000	0.3
Adhesive	12000	0.21

Table 2. - material properties used for homogenisation given in Fig. 7

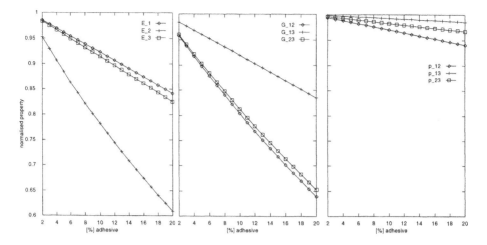

Figure 7 - variation of homogenised material properties of an enamel-adhesive layer
        normalised to enamel properties

4.1. Penetration coefficient

In order to estimate the extent and depth of penetration of the adhesive into the pores of
the enamel, measurements to determine a penetration coefficient were undertaken. The
method used is based on theory described in [5] and [6] and is briefly described here with
the apparatus used for these measurements being shown schematically in Fig. 8.

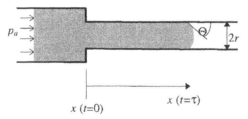

Figure 8 - measurement of a penetration coefficient

From the theory described in [4] and [5] it follows that the relationship between time and
depth of penetration is given by

$$x^2 = \frac{p_a + \dfrac{2\gamma}{r}\cos\Theta}{8\eta}r^2 t$$

$$= \left(\frac{p_a r}{4\eta} + P_c\right) rt$$

(2)

where γ is a surface tension, Θ is contact angle and η is viscosity of the adhesive and P_c is
a penetration coefficient given by:

$$P_c = \frac{\gamma \cos \Theta}{2\eta}$$

Equation (2) contains two unknowns, namely η and P_c meaning that at least two measurements for different applied pressures have to be made. It should be noted that Eq. (2) actually represents a second order (square) polynomial which can be rewritten as:

$$x^2 = at \qquad (3)$$

with

$$a = \left(\frac{p_a r}{4\eta} + P_c \right) r$$

During the experiments, measurements were made of the adhesive penetration into a tube under a constant applied pressure and contact angle. One of the typical resulting diagrams obtained in this way is given in Fig. 9. The curve shown here was found by fitting a polynomial of the type given by Eq. (3) to the set of measured results using the least square method and gives directly the values of the parameter a. Knowing two values of the parameter a for two different applied pressures p_a gives a set of two equations with two unknowns (η and P_c) which can be uniquely solved.

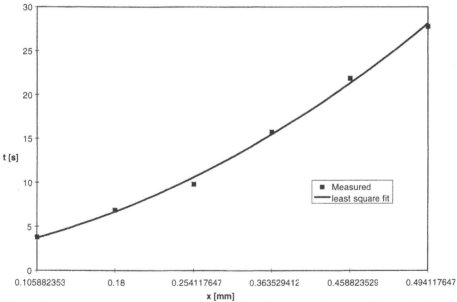

Figure 9 - measurement and least square fit for the penetration coefficient for Concise (least square fit using 2nd order polynomial)

5. CONCLUSIONS

It has been shown that the finite element method can be successfully used to quantify the mechanical response of fixed orthodontic brackets. Here it is demonstrated that the homogenisation method can be applied to determine the orthotropic and anisotropic material parameters that describe the bracket/adhesive/enamel composite form. This method has been extended to the modelling of adhesive penetration of the enamel forming the tooth surface and hence provides a method for predicting the mechanical bonding of the bracket to the tooth surface. The numerical models described also demonstrate the successful prediction of material response, at both micro and macro level, which are intractable by other means and hence provide a most powerful design and assessment method.

The authors wish to thank the Welsh Scheme for the Development of Health and Social Research for providing funds to undertake this research.

6. REFERENCES

1. Hubsch P.F., Knox J, Jones M.L., Middleton J, 'Global and local behaviour of the stress field at the base of an orthodontic bracket'.2nd International Symposium on Biomechanics and Biomechanical Engineering, Swansea, UK, 1996, pp 361-369.

2. Hollister S. J., Fyhrie D. P., Jepsen K. J. and Goldstein S. A., 'Application of homogenization theory to the study of trabecular bone mechanics', *J. Biomechanics*, 1991, pp. 825-839.

3. Legler L. R., Retief D. H., Bradley E. L., 'Effects of phosphoric acid concentration and etch duration on enamel depth of etch: An in vitro study', *Am. J. Orthodont. Dentofac. Orthop.*, 1990, pp. 154-160.

4. Zhao Y. H., Weng G. J., 'Effective elastic moduli of ribbon reinforced composites', *Transactions of ASME - Journal of Allied Mechanics*, 1990, pp. 57:158-167.

5. Washburn E. D., 'The dynamics of capillary flow', *The Physical Review*, 1921., pp. 273-283.

6. Fan P. L., Seluk L. W. and O'Brien W. J., 'Penetrativity of sealants', *J. Dent. Res.*, 1975., pp. 262-264.

FINITE ELEMENT MODEL FOR TRANSMISSION AND ABSORPTION OF OCCLUSAL IMPACT FORCE

T. Ishida[1], N.Haibara[2] and K.Soma[3]

1. ABSTRACT

The purpose of this study was to investigate the dynamic responses of periodontal ligament and particularly the absorption of stress derived from occlusal and orthodontic force. A 3-dimensional finite element model of a mandibular central incisor, together with its periodontal ligament and alveolar bone, was constructed. A simulated occlusal force was applied to the tooth in the model. Under these conditions, the stress and the displacement at various points of the model were computed by a finite element non-linear dynamic analysis. The duration of simulated loading was varied to represent occlusal force and orthodontic force. The results showed that occlusal force and orthodontic force have different effects on periodontal ligament and alveolar bone.

2. INTRODUCTON

In the field of dentistry, increasing importance has been placed on understanding the mechanism associated with the transmission and absorption of occlusal force by the teeth, periodontium, maxilla and mandible.

In order to accomplish this, dynamic analysis should be conducted at the time of occlusal contact. If the occlusal force at the time of occlusal contact is considered to be

Keywords: Occlusal impact force, Dynamic simulation, Periodontal ligament, Finite element non-linear dynamic analysis, Viscosity

[1]Lecturer. The First Department of Orthodontics. Faculty of Dentistry, Tokyo Medical and Dental University. 1-5-45 Yushima. Bunkyo-ku. Tokyo, Japan
[2]Special Student. The First Department of Orthodontics, Faculty of Dentistry, Tokyo Medical and Dental University. 1-5-45 Yushima. Bunkyo-ku. Tokyo, Japan
[3]Professor. The First Department of Orthodontics, Faculty of Dentistry, Tokyo Medical and Dental University. 1-5-45 Yushima, Bunkyo-ku. Tokyo. Japan

an impact force, then movement of the periodontal ligament and alveolar bone may change over time. In such a case, these movements may influence many biological reactions.

Finite element non-linear dynamic analysis is considered the best method for examining the transmission and absorption of occlusal forces. This method make it easy to adjust the morphology and material properties of the samples. This allows for the assessment of not only temporal elastic deformations but also of chronological changes in movement 1,2).

A general assumption in clinical orthodontics is that teeth do not move as a result of occlusal force, but rather due to orthodontic force. Therefore, treatment has been focused on using the latter. The duration of force applied to teeth is believed to affect their movement.

In the present study, we investigated the chronological changes in the forces applied to teeth. In particular, the correlation between loading duration and the movement of teeth was examined by a computer-assisted dynamic analysis. This involved an analysis of the differences in the fluctuations of the thickness of the periodontal space and the stress generated in the periodontal ligament between two experimental conditions. In one condition, occlusal force was simulated (dynamic load with a short load duration) and in the other condition, orthodontic force was simulated (static load with a long load duration).

2. METHODS

2.1. Establishing a finite element model

2.1.1. Development of the configuration model

First, based on the cross-sectional configuration of the lower anterior teeth region in the alveolar bone shown in textbooks 3), a three-dimensional finite element model of the lower central incisors, together with the periodontal ligament and alveolar bone, was constructed (Fig. 1).

The region of interest was the lower central incisors, as well as the alveolar bone and periodontal ligament located between the two cross-sectional planes perpendicular to the mesiodistal direction that pass through the interdental contact point. This planar symmetrical model was sectioned into 1275 eight-node solid elements consisting of 1722 nodes with respect to the plane passing through the tooth axis and perpendicular to the mesiodistal direction.

2.1.2. Establishing constraining and loading conditions

The base region of the alveolar bone furthest from the loading site was completely constrained. For the cross-sectional region, only the components perpendicular to the cross-sectional plane were constrained.

The loading sites consisted of nodes on the plane corresponding to the incisal edge, and the simulated occlusal load was equally distributed.

2.1.3. Establishing the physical properties of the models

The elasticity, inertia and viscosity of the model were set. Material constants such as the Young's modulus, Poisson's ratio, weight density 4,5) and coefficient of viscosity of the teeth, periodontal ligament and alveolar bone are shown in Tab. 1. The coefficient of viscosity of the periodontal ligament was determined as described in the next section.

	Young's modulus (MPa)	Poisson's ratio	Density (Kg/mm³)	Viscosity (MPa·s)
Enamel	8.1×10^4	0.33	1.8×10^{-6}	
Dentin	1.8×10^4	0.31	1.8×10^{-6}	
Pulp	2.0	0.45	1.8×10^{-6}	
PDL	4.8×10	0.45	1.2×10^{-6}	8.3×10^{-2}
Cancellous bone	1.3×10^4	0.38	1.5×10^{-6}	
Cortical bone	3.3×10^4	0.26	1.5×10^{-6}	

Tab. 1 Properties of FEM model

The enamel, dentine, pulp, cancellous bone and cortical bone were homogenous isotropic elastic bodies and the periodontal ligament was an isotropic visco-elastic body.

In the present study, two factors were incorporated into the 4-element model representing the periodontal ligament to provide the characteristics of an isotropic visco-elastic body (Fig.2). First, two different types of elements that make up the periodontal ligament were set: one type possessed the characteristics of a homogeneous isotropic elastic body and the other type possessed the characteristics of a homogeneous isotropic viscous body. Next, the two types of elements were alternately arranged in three dimensions so as to ensure an equal quantity and volume. In addition, the elastic and viscous bodies were arranged in series as well as in parallel in all directions (Fig. 3).

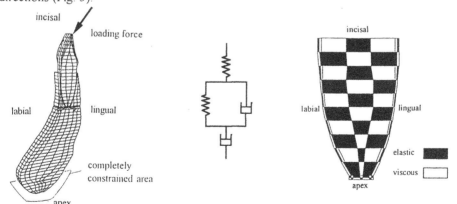

Fig.1 View of 3D FEM model Fig.2 4-element model Fig.3 Longitudinal cross-sections of PDL surface

2.1.4. Determining the coefficient of viscosity

When conducting a dynamic analysis, the mechanical characteristics of the periodontium should be considered. Ever since Kurashima 6) first reported the importance of this point, many studies have examined the displacement of teeth (as visco-elastic bodies). However, since movement of the whole model was considered, the material constants could not be compared.

Thus, to conduct a dynamic analysis of the transmission and absorption of occlusal force and the movement of the teeth and periodontium, the coefficient of viscosity of the periodontium was determined.

After establishing the temporary coefficient of viscosity for the finite element model, various finite element non-linear dynamic analyses were conducted by applying the occlusal load specified by Kurashima 6). The coefficient of viscosity that was obtained from the diagram of load displacement that most closely resembled that of Kurashima 6) was used in the primary analysis of the present study (Fig 4a).

In addition, the present study examined whether the model that was developed for analyzing static loads was also valid for analyzing impact loads observed in the body. We measured the displacement of anterior teeth by applying an impact load using an optical fiber displacement sensor (ST-3713, Iwatsu) 7) without actually touching the teeth. In Fig. 4b, the same impact load was applied to the present finite element model. As shown in this figure, waveforms with similar attenuation were obtained. Thus, the present model appears to be valid for analyzing impact loads.

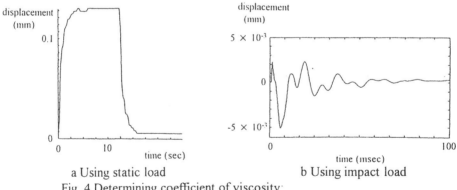

a Using static load   b Using impact load
Fig. 4 Determining coefficient of viscosity:
Diagramatic representation of tooth displacement

2.2. Primary analysis

For the simulated load, force (0.5 N) was applied from the lingual side at a 70 degree angle with respect to the tooth axis (Fig. 1).

An occlusal force with a short loading time (0.01 second) was applied in one case, and an orthodontic force with a long loading time (10.0 seconds) was applied in the other.

Finite element non-linear dynamic analysis was performed using DYNA3D (Lawrence

Livermore National Laboratory, U.S.A.) under the above conditions at different times to measure the thickness of the labiocervical periodontal space and the stress generated in the periodontal ligament, which both fluctuated when force was applied to the model.

3. RESULTS AND DISCUSSION

First, the thickness of the periodontal space is discussed. As shown in Fig. 5a, when the loading time was short, the thickness of the periodontal space was hardly affected. On the other hand, as shown in Fig. 5b, when the loading time was long, the periodontal space gradually flattened out and eventually stabilized as loading continued.

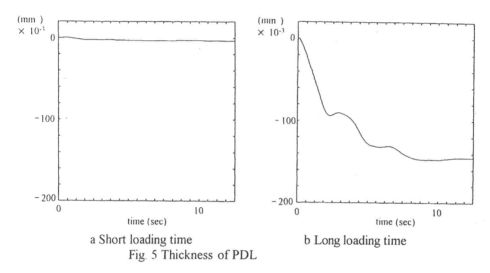

a Short loading time   b Long loading time
Fig. 5 Thickness of PDL

Next, the role of the viscosity of the periodontal ligament was investigated in relation to the occlusal and orthodontic forces.

In general, when a load is continuously applied to a structure, oscillation is attenuated and becomes stabilized by viscosity. This condition is the same as that observed when conducting a static analysis using a static load. During maximum biting or in the application of orthodontic force, although the teeth oscillate at first, the oscillations attenuate immediately and are believed to eventually stabilize.

The following changes in stress in the periodontal ligament were observed:

Fig. 6a shows the reactions of a viscous body element in response to a dynamic load, and Fig. 6b shows the reactions of an elastic body element in response to a dynamic load.

When the dynamic load was applied, changes in the stress generated in the viscous body element were detected immediately and attenuated over time. On the other hand, the stress generated in the elastic body element did not tend to fluctuate.

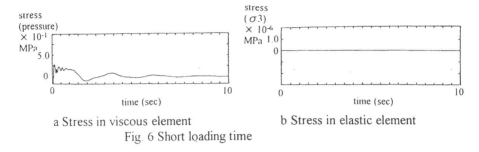

a Stress in viscous element   b Stress in elastic element
Fig. 6 Short loading time

Based on these findings as well as the lack of chronological changes in the thickness of the periodontal ligament and the elastic body element, we thought that when occlusal force, which is an impact load with a short loading time, was applied, elastic body elements (various delicate structures of the periodontal ligament) were protected by viscosity.

When the static load was applied, a large stress value was detected immediately after loading in the viscous body element. This stress became constant over time. In addition, after removing the load, the stress value rapidly returned to the original zero level and eventually stabilized (Fig. 7a).

On the other hand, although very little stress was detected in the elastic body element soon after loading, stress was gradually generated by oscillation over time (Fig. 7b).

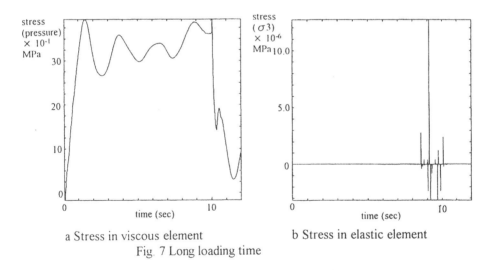

a Stress in viscous element   b Stress in elastic element
Fig. 7 Long loading time

Therefore, when a static load with a long loading time is applied, although the viscous body element initially absorbs the load, the elastic body element eventually takes on the load.

Consequently, the various delicate structures of the periodontal ligament are assumed to be affected mechanically as the loading time increases. Thus, even though the periodontal ligament is not flattened out by an occlusal force with a short loading time, it mechanically reacts to orthodontic forces with a longer loading time 8), which in turn leads to additional absorption by the alveolar bone, thus causing teeth to move.

Furthermore, even when force with a short loading time was repeatedly applied, the stress generated in the elastic body element did not tend to fluctuate.

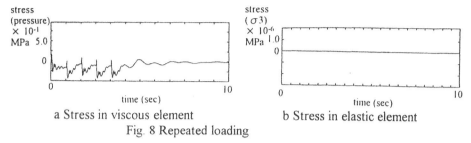

a Stress in viscous element   b Stress in elastic element

Fig. 8 Repeated loading

4. CONCLUSIONS

The results of the present study suggest that varying the loading time causes different reactions in the periodontal ligament. Therefore, we were able to obtain a better understanding of the mechanical induction of various reactions in the periodontal ligament in response to occlusal and orthodontic forces.

5. REFERENCES

1. Satou, F., Simulation on the research & development of sports goods, Journal of the Japan Society for Simulation Technology Vol. 11(3): 174-179, 1992.
2. Hallquist, J.O., Theoretical manual for DYNA3D, Lawrence Livermore National Lab. UCID-19401. March, 1983.
3. Williams, P.L. and Warwick, R., Gray`s anatomy, Churchill Livingstone, 315-319, 1980.
4. Williams, K.R., Edmundson,J.T., Morgan,G., Jones,M,L., and Richmond,S.: Orthodontic tooth movement analysed by the finite element method. J. Biomaterials, Vol. 5: 357-361, 1984.
5. Japanese Society of ME, Handbook of bio-medical engineering, Tokyo, Japan, 1984, Corona publishing Co., LTD., 1-960.
6. Kurashima, K., The viscoelastic properties of the periodontal membrane and alveolar bone, J Stomatol Soc. Vol. 30:361-385, 1963.(in Japanese).
7. Shinozaki, N., Fujita, Y., Oka, M., Soma, K., The basic study of measurement generated by impacted occlusal force, 55th annual meeting of Japan Orthodontic Society, 198,1996.(in Japanese).
8. Proffit, W.R., Contemporary orthodontics 2nd ed., 270-275, Mosby Year Book , St Louis, 1989.

AN INTEGRATED FEM MODEL SYSTEM
OF THE CRANIOFACIAL COMPLEX APPLIED
TO BIOMECHANICAL RESEARCH IN ORTHODONTICS

K. Tanne[1], S. Matsubara[2], A. Sasaki[3] and M. Ishida[4]

1. ABSTRACT

This study was designed to describe modelling procedures of the human craniofacial complex for finite element analysis (FEA) and to investigate biomechanical responses of the complex to artificial and functional loadings during orthopaedic therapy and biting functions. Stresses in the nasomaxillary sutures were varied by the direction of orthopaedic force and directing the line of force closer to the CRe produced the most optimal sutural modification for controlling natural maxillary growth. Differences in bite force direction due to posterior cross-bite with bone defect significantly affected mechanical loadings on bony structures and the subsequent maxillofacial growth, indicating an important role of bone grafting in maintaining the normal bony architecture and the subsequent growth of the maxillary complex.

2. INTRODUCTION

In the field of clinical orthodontics, various therapeutic forces are used to induce tooth and jaw movements as well as growth control of the craniofacial skeleton[1-5]. Biomechanical responses of living bones to these external forces, therefore, are of great importance to achieve optimal outcome of treatment. Craniofacial complex complies the cranial and facial bones connected together at the sutural interfaces. Therefore, maxillary growth is generated by sutural modification at the interfaces between various bones in the nasomaxillary complex. However, the nature of biomechanical responses of the

Keywords: Finite element analysis(FEM), craniofacial growth, orthopaedic therapy, biting function, bone remodelling

1. Professor and Chairman, 3. Graduate Student and 4. Assistant Professor, Department of Orthodontics, Hiroshima University School of Dentistry

2. Former Assistant Professor, Department of Oral and Maxillofacial Surgery, Nara Medical University, and currently in private clinic.

craniofacial bones and the sutures to external forces has not been studied in detail, because most previous studies were limited in the quantification of internal changes of living structures. The purpose of this study was to describe modelling procedures for thecraniofacial bones, and to investigate biomechanical responses of the nasomaxillary bones to orthopaedic therapeutic and functional forces by use of FEM.

3. MATERIALS AND METHODS

A three-dimensional finite element model of the craniofacial skeleton was developed from the dried skull of a young human. First, the skull was cut into transverse sections parallel to the Frankfort horizontal plane. Photographs of both the dorsal and ventral aspects were taken of the sections and the anatomic structures observed on the photographs were traced on acetate paper. These two-dimensional drawings were divided into a finite number of elements, ensuring that the geometric shape of the model to the anatomic structures was well maintained. Mesh refinement was also repeated by means of distortion parameter, which allows a reasonable distortion of element and convergence or accurate analysis. Then, all the meshed sections were stacked in the perpendicular direction to the Frankfort horizontal plane to develop a three-dimensional configuration of the model. The model consists of 2918 nodes and 1776 solid element[6], comprising the teeth and compact and cancellous bones (Fig. 1). This model, hereafter referred to as the normal model, was modified to CLP and bone graft models by excluding and including elements in the alveolar and palatal regions, simulating patients with CLP and bone graft.

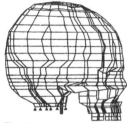

Fig. 1
A three-dimentional model of the craniofacial complex for FEM.
Solid triangles denote restraints for the model during loading.

Table 1

Mechanical properties of the tooth and compact and cancellous bones

Material	Young's modulus (Kgf/mm²)	Poisson's ratio
Tooth	2.07×10^3	0.30
Compact bone	1.37×10^3	0.30
Cancellous bone	8.00×10^2	0.30

This model includes 18 cranial and facial sutures. At the sutural interface, two node with different nodal numbers and same coordinates were established. This allows for the independent displacement of the bones adjacent to the sutural systems. The mechanical properties of the components in the model were defined based on previous data[7,8], as shown in Table 1. The model was fixed at the area around the foramen magnum to avoid a sliding movement of the entire model during loading (Fig. 1).

Before stress analysis, the nature of sutural displacement during natural growth of the maxillary complex was studied by artificially displacing the complex in the anteroinferior direction. A posteriorly directed 1.0 Kg-force was then applied at the upper first molars in five different directions to the functional occlusal plane[9], including a 52.4 degrees superior direction demonstrated in advance to be that connecting the centre of resistance (CRe) of the nasomaxillary complex and the upper first molars[10]. For the CLP and bone graft models, a biting force of 10Kg was loaded on the upper first molars or canines in varying directions according to clinical findings on biting or occlusal forces in CLP patients[11].

Stress analysis was executed by use of an analysis programme FEM5 provided by Kyoto

University Computation Centre. For orthopaedic forces, mean principal and shear stresses were evaluated for four sutures and lamina cribrosa. For CLP and bone graft models, equivalent stresses were analysed in the nasomaxillary sutures, nasal septum, and alveolar and palatal bones.

4. RESULTS

Fig. 2 shows a schematic illustration of sutural separation induced by natural displacement of the maxillary complex[9]. This result indicates several sutures with substantial reaction to the displacement of the complex. Therefore, the following   stress analyses will be focused on these anatomical structures.

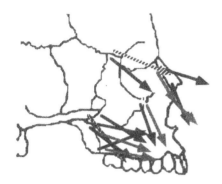

Fig. 2
A schematic illustration of sutural separation resulted from natural growth of the maxillary complex

Fig. 3 shows stresses in three anatomical structures resisting posterior displacement of the complex. In the superior and medial regions of the sphenozygomatic suture, compressive stresses were induced by all forces. Meanwhile, stresses in the inferior area of the suture were tensile in nature and exhibited a substantial decrease when the force direction became upward. In the inferior region of the temporozygomatic suture, slight tensile stresses were induced in loading with all   forces, whereas stresses in the superior area were negligible. In the superior and inferior areas of the sphenomaxillary suture, compressive stresses exhibited substantial decrease in loading with more superiorly directed   forces. These changes were more obvious in the inferior than in the superior region. On the other hand, stresses in the medial region were tensile in nature, however, the magnitude was decreased by more superiorly directed forces approaching 52.4 and 60 degrees.

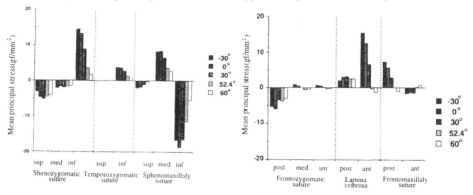

Fig.3
Mean principal stresses in the anatomical areas resisting backward displacement of the complex

Fig.4
Mean principal stresses in the anatomical areas resisting upward displacement of the complex

Fig. 4 shows stresses in three anatomical structures resisting superior displacement of the complex. In the posterior area of the frontozygomatic suture, almost uniform compressive stresses were generated irrespective of force direction. Meanwhile, stresses in the remaining points in this suture were tensile in nature and changed to compressive when more superiorly directed forces were applied. In the frontomaxillary suture, the forces applied in 52.4 and 60 degrees superior directions produced almost uniform tensile and compressive stresses in the anterior and posterior regions and the nature of stresses became opposite in loading with forces applied in the remaining directions. In the lamina cribrosa, stresses were tensile in most loading cases, however, two superiorly directed forces produced slight compressive stresses in the anterior region.

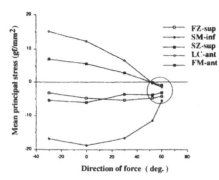

Fig. 5
Association between changes in mean principal stress at five different points and force directions to the functional occlusal plane determined by the maxillary dention
FM-frontomaxillary suture ; SM-sphenomaxillary suture ; SZ-sphenozygomatic suture ; LC-lamina cribrosa ; and FM-frontomaxillary suture.

Fig. 5 shows changes in stresses at five points in different anatomical areas, which exhibited relatively large stresses associated with varying force directions. The stresses exhibited gradual increase or decrease approaching a uniform level of compressive stress as the direction was varied from -30 to 60 degrees. Thus, the nature of stress distributions became more uniform when directing the line of force closer to the CRe of the complex.

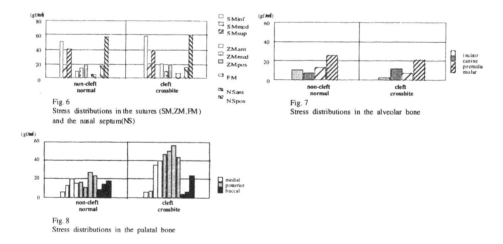

Fig. 6
Stress distributions in the sutures (SM,ZM,FM) and the nasal septum(NS)

Fig. 7
Stress distributions in the alveolar bone

Fig. 8
Stress distributions in the palatal bone

Figs. 6 to 8 show stress distributions for the sutures and the alveolar and palatal bones in the normal non-cleft and CLP[11]. In the sphenomaxillary and zygomaticomaxillary sutures, the differences in stresses were slightest between the non-cleft and CLP models, if posterior occlusal relation and the relevant force direction are just the same for both models (Fig.6). Meanwhile, the stresses exhibited substantial differences between two models when bone defect and posterior cross-bite are co-existed in the CLP model. For the frontomaxillary suture and nasal septum, similar findings were found in association with varying occlusal force directions due to posterior cross-bite and an existence of bone defect in the CLP model. Differences in stress distributions in the alveolar and palatal bones were substantial between posterior normal-bite and cross-bite models, and normal non-cleft and CLP models (Figs. 7 and 8). In particular, the differences were prominent at the canine region adjacent to the bone defect. Thus, it is suggested that differences in stress distribution produced by posterior cross-bite with bone defect, common to CLP patients, significantly affect mechanical loadings on bony structures and the subsequent maxillofacial and dentoalveolar growth.

For the nasomaxillary sutures and nasal septum, pattern of stress distribution in the non-cleft, CLP and bone graft models was essentially very similar to (Fig. 9). For stress distribution at the alveolar and palatal bones, however, the differences between normal non-cleft and bone graft models were less than those between the normal non-cleft and CLP models (Figs. 10 and 11), indicating an important role of bone grafting in maintaing the normal bony architecture and enhancing the subsequent growth potential of the maxillary complex[11].

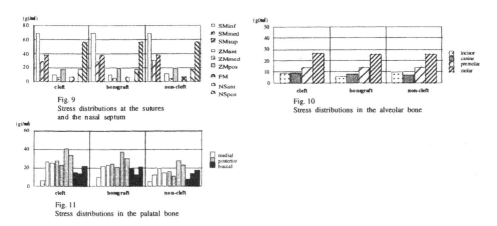

Fig. 9
Stress distributions at the sutures
and the nasal septum

Fig. 10
Stress distributions in the alveolar bone

Fig. 11
Stress distributions in the palatal bone

5. DISCUSSION

In a biomechanical aspect, many variables on force system are pertinent to the nature of displacement and stress in the periodontium and bony structures from orthodontic and orthopaedic treatments[12-15]. Among these variables, the direction of forces is suggested to be of great importance to determine the pattern of displacement of the nasomaxillary complex produced by orthopaedic forces[9]. In this study, biomechanical responses of the craniofacial bones and its sutural systems to therapeutic and functional forces were thus investigated.

With respect to the nature of maxillary growth with or without therapeutic forces, various approaches have been conducted. Cephalometric studies demonstrated that maxillary retraction forces produced significant skeletal changes by decreasing forward and/or downward maxillary growth[1-4]. Further, animal experiments have been conducted to investigate the histologic changes, indicating that orthopaedic forces for maxillary retraction produce primary displacement of the bones in the complex and subsequently initiate bone remodeling at the sutural interfaces[3,5]. Based upon these preliminary findings, it is hopefully anticipated to analyse or measure biomechanical changes in the sutures, however, the nature of stress distributions in the sutural interfaces have not been elucidated because the direct measurement without any tissue damage is not available for living structures.

For orthopaedic maxillary retraction forces, normal stresses approached gradually a certain level of uniform compressive stress in association with a gradual decrease in shear stresses, as the force direction passed more closely to the CRe of the nasomaxillary complex[9]. If the nasomaxillary complex presents pure displacement or translation in the posterior or superior direction, normal stresses at the sutural interfaces should exhibit almost uniform compressive stresses with slightest shear stresses. More uniform stress distribution with smaller shear stresses produced by a 52.4 degrees superiorly directed force may indicate almost translatory displacement without substantial counter-clockwise rotation of the complex. With these considerations, it is indicated that directing the line of therapeutic orthopaedic force closer to the CRe produces more optimal sutural modification for controlling maxillary growth. Further, it is strongly emphasised that different maxillary retraction forces may produce different morphologic changes of the nasomaxillary complex during maxillary retraction therapy[1-4], as was demonstrated in morphological studies with cephalograms and experimental animals.

For functional biting forces, stress distributions in the nasomaxillary bones and the surrounding sutures were substantially affected by differences in biting force direction due to posterior cross-bite and bone defect, which are common anatomical structure to CLP subjects[16]. Therefore, it may be assumed that maxillary growth in CLP patients is influenced by these differences in mechanical loadings on bony structures. Such assumption are verified by a morphological study for CLP patients, which has demonstrated retarded growth of the maxillary complex in both the anterior and inferior directions[17].

Morphological changes of the maxillaly complex, experienced in clinical orthodontics, would be due to the differences in sutural responses and mechanical loading to various external forces applied therapeutically and during orofacial function. Therefore, it may be confirmed that sutural response, the stresses at the sutural interfaces and the bony structures in particular, is a key determinant to morphological changes of the craniofacial complex[18-19]. Based upon the present findings, histological verification is anticipated to integrate analytical and biological findings at the sutural interfaces. Since the model used in this study is too complicated to integrate both quantities, a more simplified model system, comprising a single suture with two bones, would be more useful for understanding the interactions between biomechanical components and tissues reactions.

6.ACKNOWLEDGEMENT

This study was supported in part by Grant-in-aid (B-2, No.09470471) from the Ministry of Education, Science, Sports and Culture in Japan.

7.REFERENCES

1. Boecler, P. R., Riolo, M. L., Keeling S. D. and TenHave T. R., Skeletal changes associated with extraoral appliance therapy: an evaluation of 200 consecutively treated cases. Angle Orthod., 1989, Vol. 59, 263-270.
2. Poulton, D. R., Changes in Class II malocclusion with and without occipital headgear therapy. Angle Orthod., 1959, Vol. 34, 181-193.
3. Elder, J. R. and Tuenge R. H., Cephalometric and histologic changes produced by extraoral high-pull traction to the maxilla in macaca mulatta. Am. J. Orthod., 1974, Vol. 66, 599-617.
4. Barton, J. J., High-pull headgear versus cervical traction: a cephalometric comparison. Am. J. Orthod., 1972, Vol. 62, 517-529.
5. Droschl, H., The effects of heavy orthodontic forces on the sutures of the facial

bones. Angle Orthod., 1975, Vol.45, 26-33.

6. Tanne, K., Miyasaka J., Yamagata T., Sachdeva R., Tsutsumi S. and Sakuda M., Three-dimensional model of the human craniofacial skeleton: method and preliminary results using finite element analysis. J. Biomed. Eng., 1988, Vol. 10, 246-252.

7. Carter, D. R. and Hayes W. C., The compressive behavior of bone as a two-phase porous structure. J. Bone Joint Surg., 1977, Vol. 59A, 954-962.

8. Orr, T. E. and Carter D. R., Stress analysis of joint arthroplasty in the proximal humerus. J. Orthop. Res., 1985, Vol. 3, 360-371.

9. Tanne, K. Matsubara S. and Sakuda M., Location of the centre of resistance for the nasomaxillary complex studied in a three-dimensional finite element model. Brit. J. Orthod., 1995, Vol. 22, 227-232.

10. Tanne, K. and Matsubara S., Association between the direction of orthopedic headgear force and sutural responses in the nasomaxillary complex, Angle Orthod., 1996, Vol. 66, 125-130.

11. Ishida, M., Biomechanical responses of the maxillofacial complex to occlusal forces in patients with cleft jaw and palate: three-dimensional analysis by finite element method, J. Hiroshima Univ. Dent. Soc., 1997, Vol. 29, in press.

12. Tanne, K., Sakuda M. and Burstone C. J., Three-dimensional finite element analysis for stresses in the periodontal tissue by orthodontic forces. Am. J. Orthod. Dentofac. Orthop., 1987, Vol. 92, 499-505.

13. Tanne, K., Koenig H. A. and Burstone C. J., Moment to force ratios and the center of rotation. Am. J. Orthod. Dentofac. Orthop., 1988, Vol. 94, 426-431.

14. Armstrong, M. M., Controlling the magnitude, direction and duration of extraoral force. Am. J. Orthod., 1971, Vol. 59, 217-243.

15. Kragt, G., Duterloo H.S. and Algra A. M., Initial displacement and variation of eight human skulls owing with laser holography. Am. J. Orthod., 1986, Vol. 89, 399-406.

16. Graber, T. M., A cephalometric analysis of the developmental pattern and facial morphology in cleft palate. Am.J.Orthod., 1949, Vol. 19, 91-100.

17. Brader, A. C., A cephalometric X-ray appraisal of morphological variations in cranial bace and associated pharyngeal structures implicationa in cleft palate therapy, Angle Orthod.,1957, Vol. 27, 179-195

18. Miyawaki, S. and Forbes D., The morphologic and biochemical effects of tensile force application to the interparietal suture of the Sprague-Dawley rat. Am. J. Orthod. Dentofac. Orthop., 1987, Vol. 92, 123-133.

19. Steenvoorden, G. P., Velde J. P. and Prahl-Andersen B., The effect of duration and magnitude of tensile mechanical force on sutural tissue in vivo. Eur. J. Orthod. 1990., Vol. 12, 330-339.

A NEW SYSTEM FOR 3D PLANNING OF ORTHODONTIC TREATMENT AND 3D TOOTH MOVEMENT SIMULATION

M. Alcañiz, [1] F. Chinesta, [2] S.Albalat[1], V.Grau[1], C.Monserrat[1]

1. ABSTRACT

This paper presents a new system for three-dimensional orthodontics treatment planning and movement of teeth. We describe a computer vision technique for the acquisition and processing of three-dimensional images of the profile of hydro-colloids dental imprints. The profile measurement is based on the triangulation method which detects deformation of the projection of a laser line on the dental imprints. The 3-D image of the imprint is segmented in order to identify different teeth presents. We develop an original simplified model of arch-wire behaviour and a viscoplastic behaviour law for the alveolar bone, in order to simulate teeth displacements during orthodontic treatments. The proposed algorithms enable to quantify the effect of orthodontic appliance on tooth movement. The developed techniques have been integrated in a system named MAGALLANES. This original system presents several tools for 3D simulation and planning of orthodontic treatments. The prototype system has been tested in several orthodontic clinics with very good results.

2. INTRODUCTION

Orthodontics is the branch of dentistry that is concerned with the study of growth of the craniofacial complex. The detection and correction of malocclusions and other dental abnormalities is a significant area of work in orthodontic diagnosis, thus information concerning the shape and position change of a tooth is a fundamental reference for diagnostic and treatment evaluation. Teeth position is based on the observation, by the orthodontist, of a plaster model of the mouth (see Figure 1 (a)).

Keywords: orthodontics, 3D laser capturing, numerical simulation, simplified models, non-linear behaviour law

[1]Professor, Design and Image Developments Laboratory, Universidad Politecnica Valencia, Camino vera, 46022, Spain

[2]Professor, Departamento Mecanica Medios Contínuos, Universidad Politecnica Valencia, Camino vera, 46022, Spain

Another source of information is obtained by tracing teeth on X-ray films (Ricketts et al, 1972) taken at different stages of the treatment.The gathering of information required to lay down the diagnosis is a time consuming and costly operation. When taken too frequently, recordings of X-rays may become harmful to the patient. Distance measurement on plaster cast is carried out by using simple instruments such as a caliper and a ruler, such as one shown in figure 1 (b). The problem with classical cast analysis is that manual measurements are both inaccurate and one-dimensional so the estimation of occlusion is biased and yields poor results. Hence, the development of computer-aided techniques for the orthodontic field has long been an important goal. These techniques must enable automatic calculation of diagnostic parameters and visual checking of different treatments by moving teeth on the digital model. And on the other hand, once decided necessary teeth movements and appliances, to simulate the effects that selected appliances will have on teeth movements, that is, teeth movements.

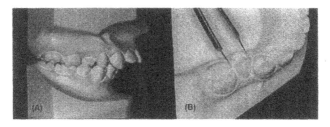

Figure 1: Plaster cast (a) and manual measurement (b)

For biomechanic modeling of tooth movement, we must distinguish between the mechanical behaviors on the one hand of the arch and on the other hand of the alveolar bone. Although the behavior of the steel arch is well known (linear elastic behavior), and there are powerful strategies for the simulation of its deformation (Finite Elements Method and other simplified methods based on the Materials of Strenght); the arising of a plastic behavior in large deformations and the consequent incapability for returning to the original configuration (residual deformations) have led to a progressive replacement of these materials by other ones which exhibit a pseudoelastic behavior (shape memory alloys), as for instance NITI alloys. These materials show a strong non-linearity and an hysteresis in the loading-unloading cycle. Such behaviours have been described during the last ten years from the experimental observation in uniaxial loadings. On the other hand, there are several models which allow the simulation of the load-deformation-temperature behaviour of shape memory alloys (Muller,1986) or another more general models based on the Generalized Plasticity (Lubliner, 1990; Lubliner et al., 1996). It is very recently that we find attempts to raise the simulation of orthodontic appliances made of pseudoplastic materials (Auricchio, 1996; Hempowitz et al. 1997) by means of three-dimensional non linear finite elements.

In this paper we present a computer vision technique for the acquisition and processing of three-dimensional images of the profile of hydrocolloids dental imprints. The profile measurement is based on the triangulation method which detects deformation of the projection of a laser line on the dental imprints. We develop a one-dimensional simplified model, which allows an efficient simulation of the non-linear behaviour of the arch, and which proves to be easily generalizable to the arising of hysteresis in the loading-unloading cycle, or to the case of angular corrections by means of the arch pre-torsion. Finally, we present a 3D treatment editor that allows to visualize the effect of teeth movements in malocclusion corrections.

3. METHODS

3.1 Dental anatomy 3D reconstruction

In last years, several 3D dental anatomy capturing techniques have been developed, not for orthodontics use but as data acquisition ways for CAD/CAM systems for design and fabrication of dental restorations (Rekow et al., 1995). Rekow (Rekow, 1991) has applied stereophotogrammetry techniques to the morphological study of teeth. Duret (Duret et al., 1988) makes use of a hand-held optical probe whose data are processed with Moire fringe techniques and a commercial CAD/CAM software package that creates a 3D model. The CEREC system (Moermann et al., 1986) also uses an optical topographic scanning procedure. An original system developed by Laurendeau (Laurendeau et al,. 1991) makes use of wax wafer for acquisition of 3D images of both maxillaries. The acquisition of 3-D images is based on the absorption of light by a dispersive medium and uses standard CCD cameras. More recently, Yamamoto (Yamamoto et al., 1991) has described a system for measuring three-dimensional profiles of dental casts from laser triangulation techniques. Main problem with these systems consist in that they obtain only external surface of patient's mouth, use high cost and cumbersome capturing devices, and not avoid the use of high cost and cumbersome plaster casts.

Our 3D capturing method is based in the use of hydrocolloids moulds used by orthodontists, for stone model obtention. These moulds of the inside of patient's mouth are composed of very fluent materials like alginate or hydrocolloids that reveal fine details of dental anatomy. Moulds are very easy to obtain and have very low cost. Once captured patient's dental geometry by means of hydrocolloids moulds, we digitize moulds by means of a 3D digitizing system. There are several 3D laser based commercial systems that enable 3D capturing of small pieces with great accuracy and speed. However, these systems are not well suited for dental anatomy capturing due to the high geometrical complexity of these structures. For this reason, we have designed and manufactured an optimized optical measuring system based on laser structured light. The developed scanner is based in the optical triangulation method. This method consists in the projection of a laser line on the alginate mould surface. The deformation of the linegives uncalibrated information of the shape of alginate mould. With a relative linear movement of the mould with respect to the sensor head, more sections are digitized obtaining a full 3D uncalibrated model of the dentition, (see Fig. 2). This device, makes use of redundant CCDs in the sensor head and servocontrolled linear axis to move the mould. The last step is calibration to get a real and precise X,Y,Z image. All the process is done automatically. The scanner has been specially adapted for this application in order to fullfill some specific requirements such as: scanning time, accuracy, security and correct acquisition of 'hidden points' in the alginate mould.

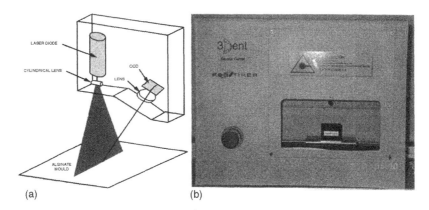

Figure 2: Laser scanner diagram (a) and photograph (b)

For our purposes, it is absolutely necessary that the diferent teeth are segmented from the gum, so that they can be individually moved, rotated, etc.A number of measures, relevant for treatment planning, are directly extracted from the segmented teeth. Also the visual aspect of the reconstruction strongly depends on the segmentation. These facts make the segmentation step extremely important for the correct behaviour of the system. To make the manual part of the segmentation, we work on a 2D image which corresponds to a height map of the arcade (Fig. 3). The systems asks the clinician to locate, more or less exactly, the centers of the teeth, starting from the left side and moving to the right side. For molars and premolars, an automatic process is carried out to adjust the contour of the tooth to the image. This process is based in the search of an approximately circular contour, with the given center. The search of the points of this contour is performed following radial lines starting from the center in order to locate local maxima of directional gradient.For incissors and canines, since they have approximately the same shape in all patients, with differences in size, position and orientation, they are reconstructed using dental models contained in a database, which are placed, scaled and rotated according to the tooth scanned. To extract these parameters from the tooth, three points are detected: the two limits of the mesio-distal line of the crown and the center of the base line of one of the vestibular or lingual faces , depending on the angle of the tooth.Figure 3 shows some results of the segmentation process.

3.2 Biomechanic modelling

Although several techniques has been used successfully to measure three-dimensional dental anatomy from casts, at the moment, no techniques has been described for teeth movement simulation due to orthodontic appliances. It can be noted that the above cited methods are only measuring ones and none of them performs tooth movement simulation.

Non-linear bending of the arch-wire
For each material we know the local stress-strain relationship:

$$\sigma \ = \ g(\epsilon) \tag{1}$$

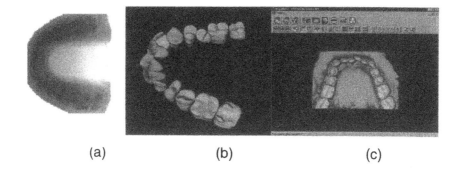

(a)   (b)   (c)

Figure 3: Range image (a,) Teeth segmentation results and 3D treatment editor screen (c)

The bending moment - curvature relationship is obtained from the equation 1, Bernoulli Kinematic hypothesis and the cross-section geometry

$$\frac{-y}{\Delta dx} = \frac{\rho}{dx} \rightarrow \epsilon = -\frac{y}{\rho}$$

From this, the stress distribution results

$$\sigma = g\left(-\frac{y}{\rho}\right)$$

and integrating in the cross-section it results

$$M_z = -\int_A y \, g\left(-\frac{y}{\rho}\right) \, dA \qquad (2)$$

We can write the bending moment - curvature relationship as:

$$\left. \begin{array}{l} M_z = f\left(\frac{1}{\rho}\right) \\ \frac{1}{\rho} = f^{-1}(M_z) \end{array} \right\}$$

Simplified model

We start from the complementary work variational formulation of the elastic problem

$$\int_\Omega Tr(\underline{\epsilon} \, \underline{\delta\sigma}) \, d\Omega = \int_{\partial\Omega} \underline{\delta F^T} \, \underline{u} \, dS$$

being $\underline{\sigma}$ and $\underline{\epsilon}$ the stress and strain tensors.

If we denote the complementary work by R, and a system of N concentrated loads $\underline{P_1}, \cdots, \underline{P_N}$ is applied on the elastic body, from the above equation results

$$\delta\tilde{R} = \int_\Omega \delta R \, d\Omega = \sum_{i=1}^{i=N} \underline{\delta P_i} \, \underline{u_i} = \sum_{i=1}^{i=N} \left(\frac{\partial\tilde{R}}{\partial\underline{P_i}}\right)^T \underline{\delta P_i}$$

From this results Castigliano's theorem

$$\frac{\partial\tilde{R}}{\partial\underline{P_i}} = \underline{u_i} \qquad \forall i \in [1, \cdots, N]$$

Since $\underline{P_i}$ and $\underline{u_i}$ have the same direction, the relationship above may be written in a scalar form, although with the direction fixed by the vector load $\underline{P_i}$.

In the bending of the arch-wire, stresses are in the axes direction, thus $\delta R = \epsilon_x \delta \sigma_x = \epsilon \delta \sigma$ and from the integration in the elastic body it results

$$\delta \tilde{R} = \int_L \int_S \delta R \, dS \, ds = \int_L \delta R_S \, ds$$

being L the arch-length of the arch-wire and S the cross-section. For δR_S we have the following relationship

$$\delta R_S = \int_S \delta R \, dS = \int_S \epsilon \, \delta \sigma dS = \int_S \frac{-y}{\rho} \delta \sigma \, dS =$$

$$= \frac{1}{\rho} \int_S -y \, \delta \sigma \, dS = \frac{\delta M_z}{\rho}$$

Thus, it results

$$\tilde{R} = \int_L \left\{ \int_0^{M_z} \delta R_S \right\} ds = \int_L \left\{ \int_0^{M_z} \frac{\delta M_z}{\rho} \right\} ds$$

Applying Castigliano's theorem, we obtain

$$u_k = \frac{\partial \tilde{R}}{\partial P_k} = \int_L \left\{ \frac{\partial}{\partial P_k} \int_0^{M_z} \frac{\delta M_z}{\rho} \right\} ds$$

taking into account $M_z(s) = P_1 \overline{M}_1(s) + \cdots + P_N \overline{M}_N(s)$ being $\overline{M}_i(s)$ the unitary bending diagram associated with a unitary load applied in the point i in the direction of the real load $\underline{P_i}$, results

$$\frac{\partial M_z}{\partial P_k} = \overline{M}_k$$

and using the Leibnitz integration rule

$$u_k = \int_L \frac{\partial M_z}{\partial P_k} \frac{1}{\rho} (M_z(s)) \, ds$$

in conclusion

$$\left. \begin{array}{l} u_k = \int_L \frac{1}{\rho} \overline{M}_k \, ds \\ \frac{1}{\rho} = f^{-1}(M_z) \end{array} \right\} \tag{3}$$

Linearization

Equation 3 depends on the loads P_i, $i \in [1, \cdots, N]$, and this relationship is non-linear. In this case a linearization of equation 3 is required. Using a Newton-Raphson method result

$$R_k^{(n)} = u_k - \int_L \left(\frac{1}{\rho} \right)^{(n)} \overline{M}_k \, ds \quad \forall \, k \in [1, \cdots, N]$$

where (n) es the iteration number and R_k is the residual vector. We obtain the tangent matrix from

$$R_k^{(n+1)} = 0 \approx R_k^{(n)} + \sum_{i=1}^{i=N} \left(\frac{\partial R_k}{\partial P_i} \right)^{(n)} \Delta P_i^{(n)} \quad \forall \, k \in [1, \cdots, N]$$

where

$$\left(\frac{\partial R_k}{\partial P_i}\right)^{(n)} = -\int_L \left(\frac{\partial f^{-1}(M_z)}{\partial M_z}\right)^{(n)} \overline{M}_i \, \overline{M}_k \, ds$$

Solving $\forall \, k$ we obtain $\Delta P_i^{(n)}$, and $P_i^{(n+1)} = P_i^{(n)} + \Delta P_i^{(n)}$

Non-linear torsion of the arch-wire

Been introducing the stress function Ψ, so that

$$\left.\begin{array}{rcl} \tau_{xy} &=& \frac{\partial \Psi}{\partial z} \\ \tau_{xz} &=& -\frac{\partial \Psi}{\partial y} \end{array}\right\}$$

and taking into account the non-linear behaviour law obtained from :

$$\tau = F(\gamma)\gamma$$

the tension function verifies the equation

$$\Delta \Psi - \frac{1}{F} \, Grad(F) \, Grad(\Psi) = -2\Theta \, F$$

in the section S, with the boundary condition

$$\Psi|_{\partial S} = 0$$

The non-linearity is solved by a continuation method

$$\tau = \lambda \, G \, \gamma + (1 - \lambda) \, F(\gamma) \, \gamma$$

starting from the linear elastic solution $\lambda = 1$, and iterating towards the desired solution $\lambda = 0$

In this way, Ψ having been stablished, for $\lambda = 0$, it will be obtained the torque to wich is submitted the section depending on the angle Θ

$$M_T = M_T(\Theta)$$

or its invers

$$\Theta = \Theta(M_T)$$

For the resolution if the inverse problem of torsion it is again taken into account the generalized Castigliano's theorem, according to wich the rotation angle in the point k

$$\Theta_k = \int_L \Theta(M_T(s))\overline{M}_k(s) \, ds$$

where $\overline{M}_k$ is the internal torque diagram originated by an unitary torque located in the point k.

The rotation angles system $\Theta_1, \dots, \Theta_N$ being known, the asociated torques $M_1, \dots, M_N$ are obtained by the Newton-Rapson method.

3.3 Bone rehological behaviour

The alveolar bone, where the teeth are located, has a behaviour characterized by an elastic component and other viscous component, with irreversible deformation. Due to this, we can consider its behaviour to be equal to an elastoviscoplastic solid.

We evaluate in an uncoupled form the displacement of each tooth, with respect to the rest of teeth because they only interact in a direct form through the contact zone or through the arch wire. In this way, the analysis is reduced to an onedimensional simplified model. Asuming the initial hypothesis, we can say that the global deformation is given by the expresion:

$$\epsilon = \epsilon^e + \epsilon^v$$

The elastic component is:

$$\epsilon^e = \frac{\sigma}{E}$$

and the viscous component is, according to law:

$$\frac{d\epsilon^v}{dt} = \left(\frac{\sigma}{\lambda}\right)^N$$

where N is the viscous law exponent and λ is the viscous law coeficient. Then:

$$\frac{d\epsilon}{dt} = \frac{1}{E}\frac{d\sigma}{dt} + \left(\frac{\sigma}{\lambda}\right)^N$$

This differential equation is solved with Finite Differences :

$$\frac{\epsilon_{i+1} - \epsilon_i}{\Delta t} = \frac{1}{E}\frac{\sigma_{i+1} - \sigma_i}{\Delta t} + \left(\frac{\sigma_{i+1}}{\lambda}\right)^N \tag{4}$$

On the other hand, the stress that tooth makes on the bone is determined by the arch wire deformation; and that is in relation to teeth position, so we obtain a coupled sistem.

For each time step we need to solve the bending moment of the arch wire, that is a consequence of teeth determinated position; and the deformation that we can obtain in a following time step, is a consequence of this deformation.

The expression 4 is an implicit model that results to be a non linear and coupled one. We can write this equation as an explicit model:

$$\frac{\epsilon_i - \epsilon_{i-1}}{\Delta t} = \frac{1}{E}\frac{\sigma_i - \sigma_{i-1}}{\Delta t} + \left(\frac{\sigma_i}{\lambda}\right)^N$$

expression that we can solve in an uncoupled form.

4. RESULTS

4.1 3D anatomy capturing

For measurement resolution and accuracy, we have scanned phantoms of known geometry and performed a point by point comparison. Obtained results are discussed in next section.

In the 3D treatment simulation editor, the user can simulate different orthodontic treatments and visualise the effects on teeth movements. For this task, the system presents a 3D realistic dental arcade model of the patient and the orthodontic appliances used. Based on these models, the editor facilitates treatment simulation by providing realistic visualisation of dental arcades, and 3D navigation tools (zoom, rotations, translations, etc.) and several teeth movement functions. A screen of the system is shown in Fig. 3.

4.2 Biomechanical results

We have been working with a β-Ti arch-wire, with a 0'018 inch of diameter. With our simplified model, we have obtained the displacements and forces along the treatment simulation. The displacements and forces of one canine are represented in fig. 4.

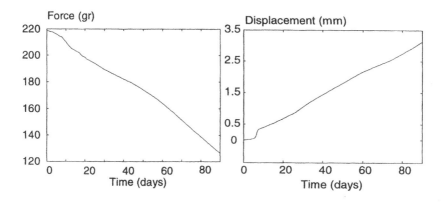

Figure 4: Force results (a) displacement results (b)

5. DISCUSSION

For 3D dental anatomy capturing, proposed technique present a great accuracy in geometry capturing (0,2 mm in X,Y and 0,1 mm in Z), with a scanning time for each arch of 45 seconds.These values are totally acceptable for orthodontic treatments simulation. For biomechanical modelling, we have in perspective to introduce the shape memory alloys (Nitinol,etc) and to design the experimental test to obtain the reologhical characterization.

6. ACKNOWLEDGEMENTS

MAGALLANES is a joint academic-industrial project supported by IBM Corp (Healthcare Industry Europe, Advanced Health Applications Division, Proj. contract num. 20067). The authors thank Dr. Jose Antonio Canut and his team from the Department of Orthodontics of the Dentistry faculty of Valencia for discussions in regard to the clinical needs. Javier García Tejedor (ROBOTIKER, Zamudio) is kindly acknowledged for laser capturing device designing and developing.

7. REFERENCES

1. Ricketts, R. W. Bench, J. J. Hilgers, and R. Schulhof. "An overview of computerized cephalometrics.". *Amer. J. Orthod.* Vol. 61. Pp 1-28. 1972.

2. Muller, I. Psudoelasticity and Shape Memory. In Large deformation of Solids. Ed. Elsevier Applied Science. 1986.

3. Lubliner, J. and Auricchio, F. Generalized Plasticity and Shape-Memory Alloys. *Int. J. Solids Structures.* Vol. 33. Pp. 991-1003. 1996.

4. Lubliner, J. "Plasticity Theory". Ed. Macmillan. New York. 1990.

5. Auricchio, F. Joint Conference of Italian Group of Computational Mechanics and Ibero-Latin American Association of Computational Methods in Engineering. Padova (Italy). 1996.

6. Hempowitz, H., Sander, F.G., Wichelhaus, A., Heigele, C., Edrich, J. and Franke, R-P. Nonlinear Finite Element Calculations of Orthodontic Appliances Made of Pseudoelastic Niti-alloys. Third International Symposium on Computer Methods in Biomechanics and Biomedical Engineering. Barcelona (Spain). 1997.

7. Rekow, D., and Nappi B. "Automation and expert systems for design and fabrication of dental restoration". In Taylor R., Lavallée S. and Burdea G. (ed.), Computer-Integrated Surgery, MIT Press, 1995.

8. Rekow. "CAD/CAM for dental restorations-some of the curious challenges" *IEEE Trans. Biomedical Engineering.* Vol. 38. Pp. 314-318.1991

9. Duret, J.L. Blouin and B. Duret. "CAD/CAM in dentistry". *Journal of American Dental Association.* vol. 117. Pp. 715-720. 1988.

10. Moermann, H. Jans, M. Brandestini, A Ferru and F. Lutz. "Computer machined adhesive porcelain inlays: margin adaptation after fatigue stress Abstract(*J. Dent. Res.* Vol. 65. Pp 762. 1986

11. Laurendeau, D., Guimond L., and Poussart D. (1991) A computer-vision technique for the acquisition and processing of 3-D profiles of dental imprints: an application in Orthodontics. *IEEE Trans. Med. Imag.*, 10, 453-461.

12. Yamamoto, S. Hayashi, H. Nishikawa, S. Nakamura and T. Mikami. "Measurement of Dental Cast Profile and Three-Dimensional Tooth Movement During Orthodontic Treatment". *IEEE Trans. Biomed. Eng.* Vol. 38-4 pp 360-365. 1991.

BONE REMODELLING IN ORTHODONTICS

Ch.Provatidis[1]

1. ABSTRACT

This paper presents a computational model for the prediction of orthodontic tooth movement with bone remodelling. A theoretical explanation is provided to sustain the application of the finite element method in successive linear elastic steps using a large Poisson's ratio. Physiological considerations lead to a simple algorithm capable of simulating the alveolar bone remodelling. The proposed algorithm can be integrated in most of the commercial finite element codes.

2. INTRODUCTION

The application of the finite element method (FEM) in dental investigation began in the seventies by using two-dimensional [1-6] and three-dimensional models [7,8]. In orthodontics, finite element studies have focussed on the analysis of orthodontic appliances [9,10] as well as dental movement in relation with the connective tissues [11-15]. Patterns of instantaneous displacements associated with various root lengths and alveolar bone heights have been studied for an upper incisor [16] as well as for a mandibular canine [17]. In the previous studies linear-elastic material behaviour has been considered, although visco-elastic phenomena may occur [18].

The problem to be dealt in this paper consists of the development of a model to accurately predict the primary and secondary tooth movement with bone remodelling. Presently, no theoretical or experimental model is available to predict orthodontic tooth movements accurately. Little is known about the processes leading to the movement of the teeth other than that it involves a process of 'bone remodeling' whereby bone is resorbed on one side and added on the other. In orthopaedic research these 'bone remodeling' processes are generally explained as a result of 'Wolff's Law' [19],

Keywords: Orthodontic tooth movement, Bone remodelling, Finite element method

[1] Assistant Professor, Biomechanics/ Mechanical Engineering Department, National Technical University of Athens, P.O.BOX 64078, 157 10 Athens, Greece

referring to observations that bone is added at overloaded and resorbed at underloaded locations. This process has been simulated and validated using computer models, in which mathematical rules to relate bone mass to bone loading are incorporated with FEM computer models [20]. The application of these models to orthodontic bone remodelling however, is not trivial, as 'Wolff's Law' can not explain bone resorption at the overloaded side during repositioning of teeth, as it is the case during orthodontic tooth movements. It is hypothesised that the periodontal ligament distributes the forces to the bone in an unexpected way, resulting in the observed bone resorption patterns.

3. A THEORETICAL APPROACH

In this section it will be shown that linear elastic analysis can be applied in successive steps in order to simulate the orthodontic movement. An equivalent alternative would be the hydrodynamic modelling of tooth movement.

Let us consider an arbitrary point $P(x, y, z)$ lying into the biological structure that consists of (a) the tooth, (b) the periodontal ligament (PDL) and (c) the alveolar bone. After the application of the orthodontic force system (force and/or moment), the new position of the point P' will be determined by the displacement vector $u(x,y,z) = PP' = [u, v, w]^T$. The former displacement field results in strains ε and stresses σ that are given by the following equations

$$\varepsilon = \begin{bmatrix} \varepsilon_x & \varepsilon_y & \varepsilon_z & \gamma_{xy} & \gamma_{yz} & \gamma_{zx} \end{bmatrix}^T \quad \sigma = \begin{bmatrix} \sigma_x & \sigma_y & \sigma_z & \tau_{xy} & \tau_{yz} & \tau_{zx} \end{bmatrix}^T \quad (1)$$

In linear elasticity, stresses and strains are related with Hooke's law (v = Poisson's ratio):

$$E \cdot \varepsilon_x = \sigma_x - v \cdot (\sigma_y + \sigma_z)$$
$$E \cdot \varepsilon_y = \sigma_y - v \cdot (\sigma_z + \sigma_x)$$
$$E \cdot \varepsilon_z = \sigma_z - v \cdot (\sigma_x + \sigma_y)$$
$$G \cdot \gamma_{xy} = \tau_{xy} \quad\quad\quad\quad\quad\quad (2)$$
$$G \cdot \gamma_{yz} = \tau_{yz}$$
$$G \cdot \gamma_{zx} = \tau_{zx}$$

where E is the Young's modulus and G the shear modulus which are related as

$$E = 2 \cdot G \cdot (1 + v) \quad\quad\quad\quad\quad (3)$$

By adding the three first equations of (2) one obtains that the first invariants of the stress tensor (σ_v) and the strain tensor (ε_v) are related as

$$E \cdot \varepsilon_v = (1 - 2v) \cdot \sigma_v \quad\quad\quad\quad (4)$$

Therefore, from eq.(4) it becomes evident that the compressibility $\left(\varepsilon_v\right)$ of a material vanishes when Poisson's ratio $v = 1 / 2$ and the material obtains the properties of an incompressible fluid. This case holds approximately for the periodontal ligament (PDL), for which literature [11-14,17] suggests the value of $v = 0 \cdot 49$.

It is worth mentioning that the elasticity matrix E in Hooke's law can be split in two submatrices depended on Lamé's constants $\left(\lambda, \mu\right)$ as follows:

$$\sigma = \left(E_\mu + E_\lambda\right) \cdot \varepsilon \tag{5}$$

where

$$E_\mu = \mu \cdot \begin{bmatrix} 2 & & & & & \\ & 2 & & & 0 & \\ & & 2 & & & \\ & & & 1 & & \\ & 0 & & & 1 & \\ & & & & & 1 \end{bmatrix} \quad , \quad E_\lambda = \lambda \cdot m \cdot m^T \quad , \quad m = \begin{bmatrix} 1 \\ 1 \\ 1 \\ 0 \\ 0 \\ 0 \end{bmatrix} \tag{6}$$

and also

$$\mu = G \quad , \quad \lambda = 2\mu \frac{v}{1 - 2v} \tag{7}$$

After proper manipulation of equations (2,4), equation (5) can be written as

$$\sigma - \frac{v}{1 + v} m \cdot \sigma_v = E_\mu \varepsilon \tag{8}$$

For $v \to 1 / 2$ eq.(7) involves that $\lambda \to \infty$ and, consequently, eqs.(2) do not lead to well defined solutions $\left(\sigma, \varepsilon\right)$. In this case the PDL behaves like an incompressible fluid possessing shear deformations and the RHS of eq.(8) is replaced by the material viscosity μ'. Therefore, the usual elastic solution obeying Hooke's law can be also applied to the Newtonian fluid (PDL) and vice versa. In other words, when Poisson's ratio $v \to 1 / 2$ the constitutive equations become singular leading to strain rate deformations $\dot{\varepsilon}$ instead of definite strains ε; this means that for a certain prescribed extremal force system, the resulting displacement vector u does not have a definite value but it varies in time! So, when applying a procedure of successive linear-elastic displacements, in fact the motion of a tooth that is *slowly* moving into a fluid of extremely *high viscocity* is simulated. In conclusion, from the computational point of view, both hydrodynamic and linear-elastic approaches become identical.

4. THE PROPOSED MODEL

When the orthodontic force system is induced, then, under the assumption that PDL behaves as a continuum, the stresses involved become proportional to the variation of its thickness and obtain both compressive and tensile values. Bone resorption/apposition is assumed to be proportional to these stresses and, consequently, the alveolar bone moves towards a parallel position to the tooth. From this new hypothetic position the force system at the tooth bracket is applied for a second time, and so on.

5. NUMERICAL IMPLEMENTATION AND RESULTS

The proposed linear-elastic algorithm has been implemented into the general purpose FEM-program COSMOS/M. At each iteration step the tooth is discretized at its current deformed position and the PDL as well as the alveolar bone are put around it.

The algorithm was applied on a maxillary canine on a plane parallel to its buccal plane, using the following geometrical data:
- Length of tooth root = 18.0 mm
- Distance between the bracket and the alveolar crest = 2.5mm
- Root diameter at the level of the alveolar crest = 7.0mm
- Root shape: Parabolic
- Tooth thickness = 4.5 mm (constant)
- PDL thichness = 0.3mm (constant)

Material properties have been selected as follows:

Material	Young's modulus (N/mm^2)	Poisson's ratio
Tooth	20390	0.15
PDL	1.8	0.49
Alveolar bone	14270	0.15

Two-dimensional analysis (plane stress triangular finite elements on the x-y cartesian coordinate system, y=long tooth axis) has been applied for the following two different force systems:
- A constant pure force F=1N towards the horizontal x-axis.
- A constant pure moment M=5Nmm

In the first case the alveolar bone has been discretized and restrained at a certain distance as it is shown in the upper side of Figure 1, while in the second case it has been ignored (Figure 2).

The long tooth axis (y) was assumed to be vertical at the starting position (α=0 deg.). In both force systems, the calculated displacement 'u' (parallel to x-axis) at the bracket has been plotted versus the rotational angle 'α' that varies between 0 and 20 degrees.

The computational results have been compared with OMSS [21-22], an experimental system that is based on averaged parameters gathered by clinical and experimental observations. The comparison is as follows (Figures 1,2):
1) In the case of a pure force, the relative error of FEM with respect to OMSS is about 5%.
2) In the case of a pure moment, the relative error of FEM with respect to OMSS is about 25%. The main reason for this difference lies on the fact that the two-dimensional

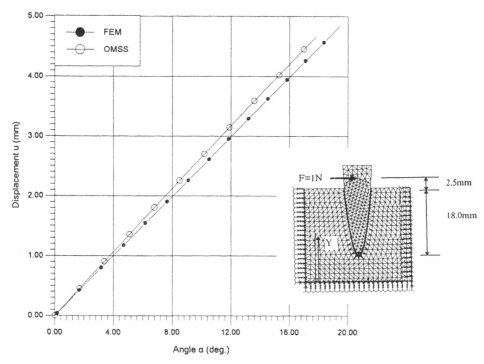

Figure 1. Tooth displacement at the bracket versus rotational angle
(Force system: pure force F=1N)

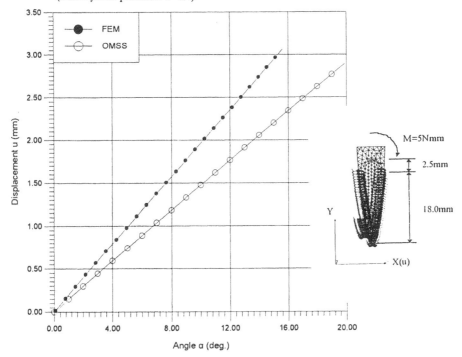

Figure 2. Tooth displacement at the bracket versus rotational angle
(Force system: pure moment M=5Nmm)

FEM analysis defines a centre of resistance at 53% down the alveolar crest while OMSS assumes this at 33%. Axisymmetric and full three-dimensional FEM models are currently under investigation and are expected to diminish this difference.

6. CONCLUSIONS

A new computational model for the prediction of orthodontic tooth movement with bone remodelling has been developed. Although it is generally applicable to arbitrary tooth shapes and force systems, in this work results are presented for a two-dimensional parabolic tooth of constant thickness under constant pure force or moment.

The algorithm was implemented in the general purpose finite element program COSMOS/M and it was applied on a plane parallel to the buccal plane of a maxillary canine for two different constant force systems (a pure force of 1N and a pure moment of 5Nmm); it required 1300 to 2400 iterations that correspond to 2.5 and 5 days, respectively, of continuous run on a PC486/66MHz/16MB RAM under DOS. The computational results were found to be in close cooperation; beneath 5-25%, with experimental data provided by the OMSS-device.

7. ACKNOWLEDGEMENTS

This work was supported by the European Comission (BIOMED1: Contract No. BMH1-CT93-1712 CAO). I acknowledge Prof.A.Kanarachos for his overall assistance, Dipl-Ing.N.Photeas for support in implementation, as well as the Poliklinik für Kieferorthopaedie, University of Bonn, Germany (PD.Dr.D.Drescher and Dr.rer.nat. C.Bourauel) for providing the OMSS curves in Figures 1 and 2.

8. REFERENCES

1. Tesk, J.A., Widera O., Stress distribution in bone arising from loading on endosteal dental implants, J.Biomed Mater Res Symp., 1973, Vol. 4(7), 251-261.
2. Takahashi, N., Dedsuya K., Komori T., Analysis of stress on a fixed partial denture with a blade-vent implant abutment, J. Prosth. Dent., 1978, Vol. 40, 186-191.
3. Craig R.G., Farah J.W., Stresses from loading distal-extension removable-partial dentures, J.Prosth.Dent., 1978, Vol. 39, 274-277.
4. Wright, K.W.J., Yettram AL., Reactive force distributions for teeth when loaded singly and when used as fixed partial denture abutments, J.Prosth.Dent., 1979, Vol.42, 411-416.
5. Anusavice, KJ., Dehoff P.H., Faihurst C.W., Comparative evaluation of ceramic-metal bound tests using finite element stress analysis, J. Dent.Res., 1980, Vol.59, 608-613.
6. Atmaram, GH., Mohammed H., Estimation of physiological stresses with a natural tooth considering fibrous PDL structure, J. Dent. Res., 1981, Vol. 60, 873-877.
7. Cook, S.D., Weinstein A.M., Klaweiter J.J., A three-dimensional finite element analysis of a porous rooted CO-Cr-Mo alloy dental implant, Materials Science, 1982, Vol. 61, 25-29.

8. Cook S.D., Klawitter J.J., Weinstein A.M., A model for the implant-bone interface characteristics of porous dental implants", J. Dent. Res., 1982, Vol. 61, 1006-1009.

9. DeFranco JC., Koenig HA., Burstone C.J., Three-dimensional large displacement analysis of orthodontic appliances, J. Biomech., 1976, Vol. 9, 793-801.

10. Fotos, P.G., Spyrakos C.C., Bernard D.O., Orthodontic forces generated by stimulated appliance evaluated by the Finite Element Method, Angle Othod., 1987, Vol. 60(4), 277-282.

11. Tanne, K., Sakuda M., A dynamic analysis of stresses in the tooth and its supporting structures: the use of the finite element method as numerical analysis, J. Jpn. Orthod. Soc., 1979, Vol. 38, 372-382.

12. Tanne, K., Stress induced in the periodontal tissue at the initial phase of the application of various types of orthodontic force: three-dimensional analysis by means of the finite element method, J.Osaka Univ.Dent.Soc., 1983, Vol. 28, 209-261.

13. Tanne, K., Sakuda M., Burstone C.J., Three-dimensional finite element analysis for stress in the periodontal tissue by orthodontic forces, Am. J. Orthod. Dentofac. Orthopedics, 1987, Vol. 94, 499-505.

14. McGuinness N.J.P., Wilson A.N., Jones M.L., Middleton J., A stress analysis of the periodontal ligament under various orthodontic loadings, Eu J Orthod., 1991, Vol. 13, 231-242.

15. Andersen, KL., Pedersen E.H., Melsen B., Material parameters and stress profiles within the periodontal ligament, Am.J.Orthod.Dentofac.Orthop., 1991, Vol. 99, 427-440.

16. Tanne, K., Nagataki T., Inoue Y., Sakuda M., Burstone C.J., Patterns of initial tooth displacements associated with various root lengths and alveolar bone heigths, Am. J. Orthod. Dentofac. Orthop., 1991, Vol. 100, 66-71.

17. Cobo, J., Sicilia A., Argtelles J., Suarez D., Vijande M., Initial stress induced in periodontal tissue with diverse degrees of bone loss by an orthodontic force: Tridimensional analysis by means of the finite element method, Am. J. Orthod. Dentofac. Orthop., 1993, Vol. 104, 448-454.

18. Middleton, J., Jones M. and Wilson A., The role of the periodontal ligament in bone modelling: The initial development of a time-dependent finite element model, Amer.J.of Ortho.& Dentofac. Orthoped., 1996, Vol.109(2), 155-162.

19. Wolff, J., Das Gesetz Transformation der Knochen, Hirschwald Verlag, Berlin, 1892.

20. Huiskes,R., Weinans H., Grootenboer H.J., Dalstra M., Fudala B. and Slooff T.J., Adaptive bone-remodeling theory applied to prosthetic-design analysis, J.Biomechanics, 1987, Vol. 20(11/12), 1135-1150.

21. Drescher, D., Bourauel C. and Thier M., Application of the Orthodontic Measurement and Simulation System (OMSS) to orthodontics, Eur.J.Orthod., 1991, Vol. 13, 169-178.

22. Bourauel, C., Dieter D. and Thier M, An experimental apparatus for the simulation of three-dimensional movements in orthodontics, J.Biomed.Eng., 1992, Vol. 14, 371-378.

Biomechanical Modelling in Orthodontics - Following a Theme? The Development of a Validation System for Modelling Orthodontic Tooth Movement.

M.L.Jones[1], J. Middleton[2], C. Volp[3], J. Hickman[4], & C. Danias[5].

Abstract

A progressive approach to the modelling of orthodontic tooth movement is described. 3 dimensional finite element models of maxillary canine tooth movement are presented and the distinction is drawn between instant and time dependent analysis. A more recent model of a human maxillary central incisor is described in detail together with a novel method of experimental validation. The Elastic Modulus and Poisson's Ratio for the periodontal ligament were determined to be 40N/mm2 and 0.45 respectively.

Key Words: Tooth Movement, FEM modelling, Experimental validation, Periodontal Ligament Properties.

Introduction

Orthodontics is a dental speciality concerned with the movement and alignment of teeth in reaction to the placement of a load, usually applied by some form of intra-oral appliance. The health gain of such measures relates to an improvement in occlusal function and dental health together with better aesthetics of the teeth and facial profile.

Although teeth are moved routinely in orthodontic practice little is known as to the precise mechanism of tissue response following the loading of a tooth crown. The remodelling of the load-bearing tissues within the human body has been considered for many years to be influenced by the loads they carry. This reaction is well known to orthodontists however the association between loading and structure has proved difficult to quantify. A number of investigators have attempted to relate tooth

---

[1] Professor, Division Of Dental Health & Development. University of Wales Coll. Of Medicine. Heath Park. Cardiff.CF64 3NY. Wales. UK.
[2] Reader, Centre for Research in Biomedical Engineering, University of Wales Swansea, Singleton Park, Swansea SA2 8PP. Wales. UK.
[3] Lecturer in Dental Technology. UWCM.
[4] Postgraduate Student in Orthodontics. UWCM.
[5] Research Assistant. UWCS.

movement to the applied force, developing theories based on very simple and imprecise experimental techniques on human subjects[1-2].

Most of the experimental work performed in the area since that time has been based on animal experimentation[3-12] which can only give a crude indication of the likely mechanical consequences in the human since animal tissues in this instance are poor morphological reflections of the matching human tissues. More recently, tissue culture systems have been have been developed to examine the effects of stress on osteoblast cells [13]. Although of interest, this approach cannot reflect the complex stress patterns involved and it will be some time before the results of such studies influence clinical practice.

Other methods used to predict tissue response to load have included theoretical mathematical techniques[14], photoelastic systems[15], and laser holographic interferometry[16]. However, such techniques have the disadvantage of only examining surface stress whilst being supported by poor validation systems. In the last decade the application of a well proven predictive technique, the finite element method has revolutionised dental biomechanical research.

Early work focused on the development of crude 2D models using existing information on the properties of dry/wet bone and other tissues, the validation systems were very limited in scope[17]. Since that time 3D FEM models of the tooth, periodontal ligament and bone continuum have been produced, also introducing simple time dependency and viscoelastic properties [18,19]. Such predictive models were found to follow existing experimental data[20] on tooth displacement following loading although such information is sparse and the methodology often questionable. Some workers have developed computational models of tooth movement in animals using animal experimental techniques for validation [21]. However, the current paper describes a new technique of direct measurement of tooth movement on humans that may then be used to validate finite element models.

Aim of Investigation

- To develop an accurate and validated 3D computer model of initial time dependent tooth movement in a typical human subject.

- To thus have a model to study the behaviour of teeth and surrounding tissues under load.

Objectives

- Establish the response of teeth to loading and the material properties of the periodontal ligament through measurement on human volunteers.

- Develop a 3D computer simulation of maxillary incisor tooth movement applying the finite element method.

- Examine, in detail, the stresses and strains in the surrounding tissues associated with tooth movement and, in particular, the nature of the displacement of the periodontal ligament (PDL).

Method and Materials

- An extensive review of the literature was undertaken and values for physical properties of many of the tissues were confirmed against historical data.

- Literature on behaviour of teeth and tissues under load was reviewed, this has largely involved experimental work on animals.

- Laser measurement equipment that has been previously described in detail [22] was further developed and used in six human volunteers to test the *in vivo* tooth response to load over time. This was also to be used to determine the properties of the periodontal ligament (PDL) and was therefore calibrated against a known material (a perspex rod).

- In the experimental phase the loading system was adjusted to achieve a low , well defined and precise force on the tooth. This was of a continuous nature. Figure 1a shows the equipment in operation with one of the volunteers, the tooth displacement calculated from the plots is shown graphically in Figure 1b. Such an approach facilitates the detailed examination of time dependent change in tooth displacement, there is the capacity in the apparatus to sample every $1/100^{th}$ of a second. The accuracy has been previously confirmed[22] at 0.001mm. Measurements in this current study were taken over a 2 minute cycle.

- Detailed load displacement plots were obtained for 6 patients who were judged to be typical of young healthy adults (a number of subjects had been previously examined, and it was found that age and periodontal health were important factors in the PDL response). This data was to be used to establish PDL behaviour under orthodontic load (Figure 2).

- The experimental results were used to assist in the development of a 3 dimensional finite element method model. This was designed to reflect the anterior maxillary teeth and jaws of a typical human subject (Figure 3). In parts of the model modules were included for automatic adaptive mesh refinement and since it was anticipated from previous work by the authors [23] that the PDL was particularly important in tooth movement, considerable effort was committed to modelling this area in detail. The PDL was designed to be initially elastic then basically visco-elastic and non-linear based on the experimental data.

- A mesh of 15,000 elements was constructed, the element used being of the four-noded linear tetrahedral type. This element was chosen since it is good at meshing arbitrary geometries - a prerequisite in this project. The tooth was divided into two basic materials: dentine and enamel. The surrounding alveolar bone (with compact and cancellous layers) and periodontal ligament was also included in the model. The basic material properties used are shown in Table 1 and a graphical sample from the model is shown in Figure 2.

MATERIAL	YOUNGS MODULUS (N/mm^2)	POISSON'S RATIO
Enamel	84100	0.2
Dentine	18600	0.31
Cancellous bone	345	0.31
Cortical bone	13800	0.26
Periodontal ligament	50	0.45

Table 1: Material Properties used in FEM model

- Using the FEM model the predicted response to load was initially validated against previous data form earlier models, historical data and then the experimental results in the parallel experimental study.

Parallel Validation of FE Model

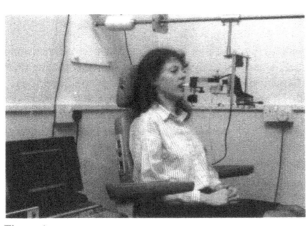
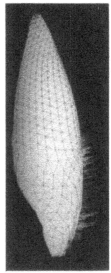

Figure 1:
a) Volunteer in apparatus for measuring tooth displacement following placement of a load on a maxillary central incisor.
b) Calculated movement for maxillary incisor from generated displacement plots.

Results: Meaning & Significance

- To achieve a steady displacement recording volunteers chewed a standard material for 5 minutes prior to a load of 0.39 N being placed on the incisor tooth to be measured. This eliminated a variable 'fatigue' response that had been noted in the pilot studies and gave more consistent readings. Initially 10 volunteers were measured 6 times but this was reduced to 6 volunteers that were able to give a consistent reproducible reading. One difficulty with measurement was maintaining the point of force application and a method was developed to address this difficulty, which involved attaching a glass ionomer marker to the surface of the incisor. The overall mean value for total displacement of the PL over a two minute cycle was 0.012 mm, with a range of 0.021 - 0.002 mm. The elastic modulus of the ligament was calculated for the PL and applied to the FE model. A typical plot of an incisor tooth under load is shown in Figure 2.

Measured Tooth Displacements

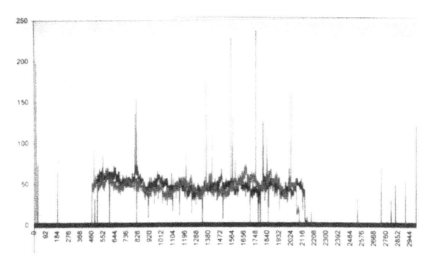

Accuracy of 0.001 mm - recording displacement every 0.01 seconds over 2 minutes in first instance

Figure 2: Example of a typical plot of time dependent tooth movement

- The detail of the model is discussed elsewhere, but it was comprised of a basic 15,000 3D trihedral elements. The tooth movement analysed on the model was found to follow rigid body motion with an initial instantaneous centre of rotation (Figure 1b). The movement of the root of the tooth was comfortably within the PDL space and stresses were found to be within Lee's maximum for physiological movement. The stress patterns within the modelled PDL were found to be very complex.

Time Dependent Behaviour of Incisor

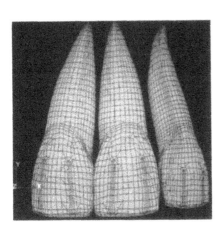

Figure 3: Components constituting the FEM model

- The maximum displacement occurred at the cervical margin (neck of tooth). This is where one might expect greater, sometimes excessive, stresses, leading to a potential for cell hyalinisation/under-mining resorption and a more pathological response. Such a localised reaction can have a significant effect on the predictability of the tooth movement. In this area the strains noted were largely shear across the PDL.

- Interestingly, large strains were found to be localised to the periodontal ligament but only negligible strains were found at the surface of the tooth root and bony socket. This data, together with that obtained from a largely theoretical previous model, suggests that orthodontic tooth movement must be largely mediated via the periodontal ligament rather than by any cellular remodelling response originating in the local bone.

- Finding an absolute value for the elastic value of the periodontal ligament based on the experimental data proved difficult and is in any case probably inappropriate since there are large variations between individuals. A good working assumption is that in a young adult with a healthy PDL the elastic modulus is likely to be under 0.18N/mm2, most probably usually in the region of 1N/mm2.

Results of Incisor FE Model
Contour Plots of Principle Strains

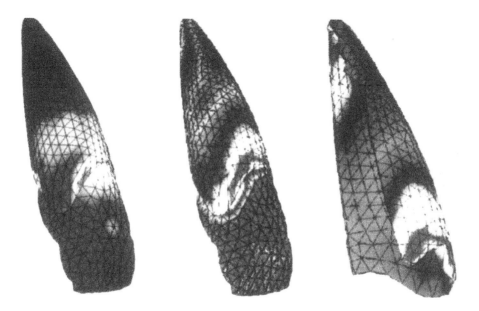

Figure 4: Plots of strain fields in model. Ligament is shown alone on the right. Note areas of strain towards neck and apex of PDL.

- The experimental work on human volunteers examining tooth displacements by the application of laser technology has been established as a valid approach and the apparatus has provided a number of interesting findings. Firstly, the initial elastic and then basic visco-elastic behaviour of the ligament has been confirmed. Secondly, it is apparent that in order to determine the physical properties of the ligament by back calculation the apparatus needed significant modification to be able to place accurate light loads over time. This gives a slower initial displacement and gives a clearer picture of early response ligament behaviour under load.

Results of Incisor FE Model
Contour Plots of Principle Strains

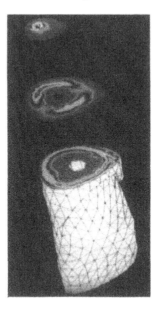

Figure 5: Detail of strains in PDL

- Since a whole series of loading and recovery has been performed on the teeth of volunteers, it has been very interesting to accurately chart fatigue behaviour in the periodontal ligament. This can be seen from the figure attached to the back of this report. We are in the process of making contact with physiologists working in this area to determine exactly the mechanism of this fatigue behaviour. The findings have great relevance to the development of the computer model and indeed, are of significance in the variable behaviour of teeth to continuous and intermittent loading in the clinic. This will be investigated in more detail in the future.

Conclusion and Summary of Findings

- A computer based three dimensional finite element model of a maxillary left incisor tooth together with its neighbouring teeth and surrounding tissues was created. A novel laser displacement apparatus to measure the movement of teeth under a constant load on human volunteers was constructed and underwent extensive modification to support this study. Detailed accuracy trials at low loading were performed.

- The experimental results acquired from this apparatus were fed into the finite element model and thus the time dependent computer simulation of tooth movement was validated.

- The material properties of the periodontal ligament, a notoriously difficult material to quantify, was calculated from the experimental data. An early finding, not previously reported in humans was a "long-term fatigue response" of the periodontal ligament on repetitive loading. A method of overcoming this response and achieving a stable reading was investigated and parameters established for the behaviour of the periodontal ligament under load.

- An important incidental finding was a variability of response which seems likely to be related to early disease of the ligament and perhaps the apparatus could form the basis of a non-invasive detection mechanism for this common disease process.
- This initial computer model has demonstrated that such an approach can be valid in the detailed study of orthodontic biomechanics. The movement of a tooth was charted, validated and the centre of rotation was described in detail.

References

1. Storey E., Smith R., Force in orthodontics and its relation to tooth movement., Aust. J. Dent. 1952, 56, 11 -18.
2. Lee B., Relationship between tooth movement rate and estimated pressure applied., J. Dent. Res. 1965, 44, 1053.
3. Sandstedt C., Nagra bidrag til tandregleringens teori., 1901. Pub. P.A. Nordstedt & Stoner. Stockholm.
4. Reitan K., Some factors determining the evaluation of force in orthodontics, Am.J.Orthod. 1957, 43, 32 - 45.
5. Picton DCA., Davies WIR., Dimensional changes in the periodontal membrane of monkeys due to horizontal thrusts applied to teeth. Arch. Oral. Biol. 1967, 12, 1635 - 1643.
6. Rygh P., Elimination of hylanized periodontal tissues associated with orthodontic tooth movement., Scan. J. Dent. Res. 1974, 82, 57 - 73.
7. Davidovitch Z., Shanfield JL., Cyclic AMP levels in alveolar bone of orthodontically treated cats., Arch. Oral. Biol. 1975, 20, 567 - 574.
8. Wills DJ., Picton DCA., Davies WTR., An investigation of the visco-elastic properties of the periodontium in monkeys., J.Perio.Research. 1972, 7, 42 - 51.
9. Rygh P., Orthodontic root resorption studied by electron microscopy. Angle Orthod. 1977, 47, 1 - 16.
10. Chiba M., Ohkawa S., The response of the tensile strength of the rat periodontium in the rat mandibular first molar. Arch. Oral Biol. 1980, 25, 569 - 572.
11. Yamasaki K., The role of cyclic AMP, calcium and prostaglandins in the induction of osteoclastic bone resorption associated with experimental tooth movement. J. Dent. Res. 1983, 62, 877 - 881.
12. Hong RK., Effect of experimental tooth movement on the mechanical properties of the periodontal ligament in the rat maxillary molars. Am.J.Orthod.& Dentofac.Orthopaed. 1990, 98, 553 - 543.
13. Sandy J., DNA changes in mechanically deformed osteoblasts: a new hypothesis. Brit.J.Orthod. 1993, 20, 1 - 11.
14. Steyn CL., Verwoed WS., Van der Merwe D., Fourie DL. Calculations of the position of the axis of rotation when single rooted teeth are orthodontically tipped. Brit.J.Orthod. 1978, 5, 153 - 156.
15. Caputo A., Chaconis SJ., Hayashi RK. Photoelastic visualisation of orthodontic forces during canine retraction. Am.J.Orthod. 1974, 65, 250 - 259.
16. Burstone CJ, Pryputniewicz RJ., Holographic determination of centres of rotation produced by orthodontic forces. Am.J.Orthod. 1980, 77, 396 - 409.
17. Williams KR., Edmondson JT., Morgan G., Jones ML., Richmond S. Orthodontic movement of a canine into an adjoining extraction site. J. Biomed. Eng. 1986, 8, 115 - 120.
18. McGuinness NJP., Wilson AN., Jones ML., Middleton J. A stress analysis of the periodontal ligament under various orthodontic loadings. Eur.J.Orthod. 1992, 13, 115 - 120.
19. Wilson AN., Jones ML., Middleton J. The effect of the periodontal ligament on bone remodelling. Proc.Methods in Biomech. & Biomech.Eng. 1992, Pub, BJI. Oxford.150 - 158.
20. Ross CG., Lear CS., De Lour R. Modelling the lateral movement of teeth. J.Biomech. 1976, 92, 723 - 734.
21. Tanne K., Sakuda M., Burstone CJ., Three dimensional finite element analysis for stress in the periodontal ligament by orthodontic forces. Am.J.Ortho.&DentoFac.Orthopaed. 1987, 92, 499 - 505.

22. Volp CR., Weston BJ., Knox J., Williams K., Jones ML., A method of evaluating dynamic tooth movement, Computer Methods in Biomechanics and Biomedical Engineering. 1996., Pub. Gordon & Breach., 461 - 470.

23. Middleton J., Jones ML., Wilson AN., The role of the periodontal ligament in bone remodelling - the initial development of a time dependent finite element model. 1996., Am.J.Orthod. & Dentofac.Orthopaed., 109, 155 - 162.

Acknowledgements

The authors are grateful to the Dr. Hadwen Trust for Humanity in Research who kindly sponsored this research project.

STRESS ANALYSIS AFTER YIELDING
AT THE RESIN COMPOSITE/DENTINE INTERFACE

Kunio Wakasa[1], Y. Yoshida[2], Y. Nakayama[3], A. Nakatsuka[2] and M. Yamaki[4]

1. ABSTRACT

Bond strengths in dentine bonding systems (DBSs) were measured during shear or tensile bond testing, and their characteristics were estimated by the following methods. The effect of the resin composite/dentine interface (bonding area) for copolymerization with resin composite on the distribution of interfacial stress during tensile bond testing was determined using finite element method (FEM). In addition, average stresses during shear bond testing were calculated along the interface and the strength values were normalized by elastic moduli of bonding area which were obtained by the same nano-indentation test as an earlier report. The interfacial stress had greater values along dentine site than resin composite site in FEM model and the normalized stress values ranged from 0.1×10^{-3} to 1.7×10^{-3}.

2. INTRODUCTION

In recent studies, dentinal adhesion between resin composite and dentine was evaluated in terms of adhesive bond strength value[1-3], and the wide scatter of bond strengths was explained by the hypothesis that failure was due to cohesive fracture of dentine because the bond between adhesive resin and dentine exceeded the cohesive strength of the dentine itself. The 4-META/MMA-TBB adhesive system formed tenacious bonds to enamel or dentine, and the formation of hybrid layer on decalcified dentine which was etched by dentine conditioners was proposed as the cause of dentine adhesion [2]. Higher bond strength values were achieved because of the hybridization in the resin-impregnated zone. By etching and priming with dental DBSs, the bonding site for copolymerization with the resin composite was obtained as the bonding area by an adhesive resin. The bonding area has been represented schematically in terms of its morphological aspects in DBSs [1, 3, 4]. Analytical models were thus needed to

[1] Associate Professor, Department of Dental Material, Hiroshima University School of Dentistry, Kasumi 1-chome, Minamiku, Hiroshima City, 734 Japan

[2] Instructor, Department of Dental Material, Hiroshima University School of Dentistry, Kasumi 1-chome, Minamiku, Hiroshima City, 734 Japan

[3] Research Associate, Toray Research Center, 3-7 Sonoyama 3-chome, Shiga, 520 Japan

[4] Professor and Chairman, Department of Dental Material, Hiroshima University School of Dentistry, Kasumi 1-chome, Minamiku, Hiroshima City, 734 Japan

calculate mean values of interfacial stresses in relation to the quality of the resin composite/dentine interface. Conditioning with an acidic agent demineralized the dentine surface to a certain depth and left behind a collagen-rich mesh-work [5]. Adhesive resin monomers altered as predicted the collagen-fibre arrangement in a way that facilitated the penetration of an adhesive monomer. FEM analysis exhibited that nonuniform stress acted upon the bonded interface and also maximum principal stress occurred along the dentine/adhesive resin interface in the bonding area during uniform tensile testing [6]. Interfacial failure was promoted when critical stress reached to the stress level of initiation and propagation of cracks [6, 7].

This study examined shear bond or tensile bond strength in dental DBSs, and measured the thickness of the interface (resin composite/dentine) and elastic moduli. Also, the strength value (mean) along the interface (dentine site) in the bonding area was determined theoretically, and the normalized strength value by interfacial elastic modulus was analyzed using an analytical calculation model.

3. MATERIALS AND METHODS

1. Shear or tensile bond strength in dental DBSs

The DBSs used in this study are coded A1 (Imperva Bond, Shofu Inc, Kyoto, Japan), A2 (Light Bond, Tokuyama Co, Tokuyama City, Yamaguchi, Japan), A3 (Clearfil Liner Bond, Kuraray Co, Okayama, Japan), A4(Clearfil Liner Bond II, Kuraray Co), A5 (Superbond D-liner Plus, Sun Medical Co, Kyoto, Japan), A6 (Scotchbond Multipurpose, 3M Co, St. Paul, MN, USA), A7 (Gluma Bonding System, Bayer Dental, Germany), A8 (Tenure, Dent-Mat Co, Santa Maria, CA, USA). for the shear or tensile bonding test. The bonding steps of the experimental treatments are listed according to Pashley et al [3] and Van Meerbeek et al [4]. The notation was partly modified for the materials used in this study ; a = apply conditioner, b = rinse conditioner, c = air dry, d = apply primer, e = cure primer, f = apply resin, g = thin resin with air, h = visible light cure, i = chemical cure, j = apply low viscosity resin, cure, k = apply composite. The dentine bonding sequences were dcfhk (5 steps) for code A1, abcdcfghk (9 steps) for A2, abcdcfgijhk (11 steps) for A3, dcfhk (5 steps) for A4, abcdefick (9 steps) for A5, abcdcfhk (8 steps) for A6, abcdfgck (8 steps) for A7 and abcdefghk (9 steps) for A8. Bonding was performed on bovine dentine whose sites were polished with #600 emery paper, which corresponded to cervical dentine sectioned vertically. The bonding area was limited by the use of double-sided adhesive tape containing a 6 mm diameter hole. A 6 x 6 mm cylindrical nylon matrix was placed on the treated surface. Visible light-cured resins (C1 for code A1, C2 for code A2, C3 for code A3 and A4, C4 for code A6, and C1 for code A5, A7 and A8) were placed in two 1.5 mm increments by photocuring for 60 s (C1 ; Occulsin, ICI Co, UK, C2 ; Graft LCII, Shofu Inc, C3 ; Clearfil Photo Posterior, Kuraray Co, C4 ; Z-100, 3M Co).

Between 45 and 60 min after bond, shear or tensile bond strength was tested to failure at a crosshead speed of 0.5 mm/min. Ten specimens were prepared for the shear bond strength test (kilograms were divided by the surface area of the bond and then converted to MPa;. DCS-500, Shimadzu Co, Kyoto, Japan). The thickness of the bonding area was measured using a scanning electron microscope (EPMA-8705, Shimadzu Co), and elastic moduli were estimated using the results measured by nano-indentation testuing (DUH-200, Shimadzu Co) [4, 6].

2.Calculation models

Fig.1 shows FEM model for tensile bond testing which calculates the stress distribution along the interfaces of bonding area/dentine (dentine site). Fig.2 shows a schematic diagram of a composite which includes smear layer, hybrid layer, adhesive resin (bonding area) (bonding area) and resin composite on the dentine, when the operation in DBSs was applied to decalcified dentine.

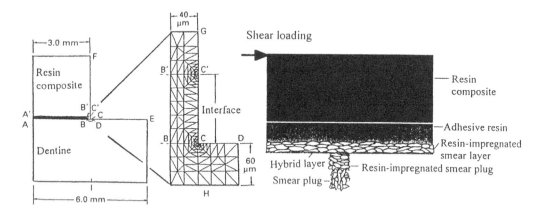

Figure 1
FEM calculation model(elastic modulus =3
(Resin composite),0.3(Interface),30GPa
(Dentine))

Figure 2
Schematic diagram for a composite
after an operation of DBSs

The procedure to calculate average stress during shear bond testing is described by an earlier report [7], showing that a calculation model employing thickness of the interface (bonding area) (d), diameter of bonding area (*l*) and height of samples including composite resin (D). Shear stress direction along the top of the surface of a sample was schematically indicated as a test arrangement, where a two-dimensional plane strain section was assumed.   Shear stress direction along the top of the surface of a sample was considered as a test arrangement, where a two-dimensional plane strain section was assumed [6], and the stress value is calculated according to the following equation,

$$< \sigma_i > = E \, e \, [\, 1 + \, (B - 1) \, tanh(\, Al)/ \, Al]$$

where E = elastic modulus of the interface, e = applied strain, A = $(2/Dd)^{1/2}(\mu /E)^{1/2}(1/ln(D/d))^{1/2}$, B= (interfacial bond strength) / $E \, e$, and the other notations as described in an earlier report [7].

4.RESULTS

1.Bond strength evaluation

Table 1 indicates shear or tensile bond strengths and thicknesses of resin composite/dentine interface (bonding area) in DBSs, showing widely varied strength values of 4.0 to 21.0 MPa (shear), 4.5 to 20.1 MPa (tensile) and bonding area thickness of 20 to 120 μ m. Based on these observations, bonding area thickness values of 1, 10, and 100 μ m were assumed in this calculation model.

Table 1
Shear bond or tensile bond strength (MPa) and thickness of bonding area (μ m)
of bovine dentine in dental DBSs. See text for key.

Code	Shear bond strength (MPa)	Tensile bond strength (MPa)	Thickness of bonding area (μ m)
A1	6.8(3.4)	10.2(4.2)	20(5)
A2	6.1(2.6)	4.5(1.9)	115(10)
A3	4.5(1.4)	7.4(1.0)	90(7)
A4	19.1(6.9)	18.2(5.2)	50(12)
A5	21.0(5.2)	20.1(4.5)	40(15)
A6	7.5(4.0)	11.4(3.8)	26(4)
A7	4.0(2.1)	6.9(2.2)	120(25)
A8	6.0(1.5)	4.8(1.8)	84(22)

Mean (standard deviation)

2. Analytical results

Fig.3 shows principal stress along the X-Y interface, i.e. from resin composite site to
dentine site in FEM calculation model, which was 10 MPa as an applied stress. The
thickness of the bonding area (interface) was supposed to be 25, 50, 100 and 200 μ m.
The stress values had more at dentine site than resin composite site in a case that elasic
modulus of resin composite/dentine interface (bonding area) was 0.3 GPa.

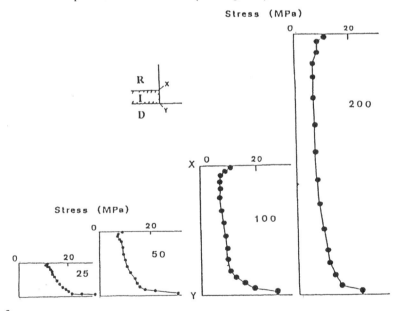

Figure 3
Principal stress along the X-Y interface, i.e. from resin composite site to dentine site in
FEM calculation model (25, 50, 100 and 200 μ m = thicknesses of the interface)

5.DISCUSSION

In dental DBSs, low-viscosity adhesive resin with about 10 GPa was effective in preventing a separation of the dentine/resin interface[4, 9], and resin-based adhesives shrank toward the point where they begin curing. On the contrary, 4-META/MMA-TBB system shrank toward the tissue, and BPO-amines and visible light-cured adhesives tend to shrink away from the dentine surface[2]. Based on these experiments, we studied theoretically the behavior of polymerized adhesives (bonding area) on the hybrid layer as the resin composite/dentine interface.

Fig.4 shows a change of maximum value of principal stress with respect to interfacial elasticity of the bonding area from 0.3 to 12 GPa at applied stress = 1, 5 and 10 MPa, and maximum stress increased with increasing their elasticity values. Also, the maximum values were larger at greater applied stress at respective interfacial elastic moduli. Thus, the magnitude of interfacial stress was affected by interfacial elastic moduli of adhesive resin /dentine inteface as a bonding area. The mean value along the interafce, i.e.interfacial stress, $< \sigma_p >$, which was described as average stress in this study, at applied stress = 10 MPa was calculated by equation proposed as shown in Fig 5. Average stress within the dentine/adhesive resin interface increased with greater interafcial bond stress (constant) between adhesive resin and dentine, which was supposed in this study, at each interface thickness of 1,10 and 100 μ m.

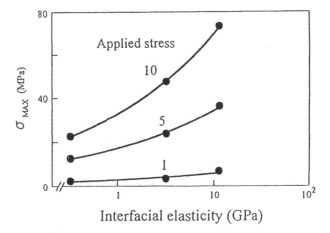

Figure 4
Maximum value of principal stress with respect to
interfacial elastic moduli of the interface (0.3 to
12 GPa) at applied stress = 1, 5 and 10 MPa

To clarify the effect of elasticity on the average stress value, average stress-to-Ee ratios with respect to elasticity and thickness of bonding area were calculated (Fig 6). The average stress/ Ee ratio decreased with increasing elasticity values from 10^2 to 10^4 MPa at each value of interfacial stress. Examining current DBSs using respective resin composites, based on shear or tensile bond testing, calculation models proposed to estimate the stress distribution within resin composite/dentine interface (Figs 3, 4), average stress along the interface with different interfacial bond strength (Fig 5), and normalized value of average stress by elastic modulus (Fig 6).

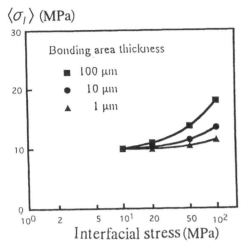

Figure 5
Average stress at the interface with different interfacial bond strength

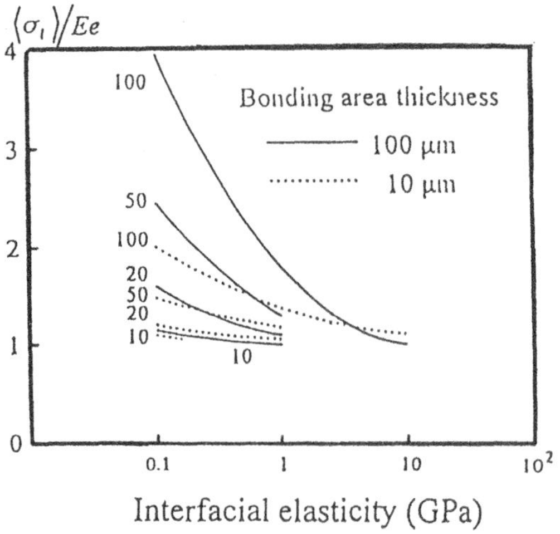

Figure 6
Normalized value of average interfacial stress

The interfacial stress at the resin composite/dentine interface is very important in analysis of fracture processes during shear or tensile bond testing [4,7,9]. It is difficult to measure the magnitude of interfacial stress. In nano-indentation test [7], the plastic deformation zone size of the interface (bonding area), as examined, depended upon the interfacial elastic moduli (E) as well as the E (elasticity)/H (hardness) ratio for each dentine bonding system. It is thus necessary to estimate average stress along the interface during bond testing and to clarify the effect of estimated interfacial bond stress, as suggested [7,8], on average interfacial stress values. The macroscopic mode of failure appears to be adhesive in nature which indicates failure at the dentine/composite interface, because the strength values range from 1 to 23 MPa [10] : a three-level classification scheme based on shear bond strength ; 5 to 7 MPa , 8 to14 MPa, and up to 20 MPa in recent DBSs [1-10]. Bond strength varied inversely with the dentine's proximity to the pulp. To clarify the quality of the resin composite/dentine interface as the bonding area as observed by repeated experiments [1-3], it is of importance in calculating theoretically average stress and strain behaviour of the interface formed during bonding. As described in their results showing that the interface thickness is measured to relate bond strength and that elastic moduli are dependent upon the quality of the interface, i.e. bonding area or interdiffusion zone between resin composite and dentine[4,6,8], theoretical attempts are needed to clarify their estimates.

The elasticity value of the bonding area depended strongly upon the zone size after elastic/plastic deformation [6,7]. At the bonding area analyzed by nano-indentation testing, average strengths estimated using these results were lower than values obtained during a bond test as indicated in Table 2. The change of average stress in the bonding area with

Table 2

Calculated stresses (mean) and their normalized values by elastic moduli of bovine dentine during shear bond testing of dental DBSs

Code	Calculated stress (MPa)	Normalized value
A1	7.5	1.3×10^{-3}
A2	9.5	1.3×10^{-3}
A3	9.0	0.1×10^{-3}
A4	10.5	0.3×10^{-3}
A5	8.0	1.7×10^{-3}
A6	5.0	0.3×10^{-3}
A7	10.0	0.3×10^{-3}
A8	6.5	0.2×10^{-3}

interfacial stress was related to elastic moduli of the bonding area in dental DBSs. Average interfacial stresses ranged from 5.0 to 10.5 MPa, showing that their values were small for each magnitude of interfacial bond stress, 10, 20 50 and 100 MPa. Also, normalized values of average stress by elastic moduli were 0.1×10^{-3} to 1.7×10^{-3}. These results suggest that the interfacial stress along the interface in DBSs is very important in estimating the interfacial stress at the failure of the resin composite/dentine interface as a future work. In order to the interfacial stress along the interface, the average interfacial stress with respect to elastic modulus at constant applied strain gave the trend of adhesive behaviour in dental DBSs (Fig 6).

To examine the stress behaviour of the interface during shear bond testing, the average stress in bonding area was calculated when applied to the composite including resin composite/dentine interface according to the calculation model. First, shear or tensile bond strengths were measured in conventional bond test for dental DBSs. Secondly, principal stresses as interfacial stress were analyzed within the interface with 25, 50, 100 and 200 μ m using FEM calculation model, and average interfacial stress values changed with interfacial bond strength, interfacial elasticity of the interface (bonding area), and thickness of the interface. Thus it was clear that the quality of the interface (resin composite/dentine) as a polymerized site determined the dentinal adhesion in DBSs.

The shear or tensile bond strengths had a wide scatter and also the thicknesses of the interface were different among dental DBSs tested. From these results, FEM model exhibited the stress distribution along the stress direction within the interface and average interfacial stress was analyzed with respect to the interfacial bond stress for each of the interface thickness. The inhomogeneous deformation model clarified that the elasticity, interfacial (dentine/adhesive resin) bond stress, and interface thickness (bonding area) were used to estimate the average interfacial stress. Based on these calculations, theoretical normalized values were determined by elastic moduli through nano-indentation analysis of the resin composite/dentine interface.

ACKNOWLEDGMENTS

The authors express deep thanks and appreciation to the Central Research Laboratory at the Hiroshima University School of Dentistry for granting preliminary use of the Biomaterial Combined Analysis System. This study was supported in part by a Grant-in-Aid from the Ministry of Education, Science, Sports, and Culture, Japan, (C)07672112 and (A)08771789.

REFERENCES

1. Tagami, J., Tao, L. and Pashley, D.H., Correlation among dentin depth, permeability and bond strength of adhesive resins, Dent. Mater.,1990, Vol 6, 45-50.

2. Nakabayashi, N., Ashizawa, M. and Nakamura, M., Identification of a resin dentin hybrid layer in vital human dentin created in vivo: durable bonding to vital dentin, Quint. Int., 1992, Vol.23,135-141.

3. Pashley, E.L., Tao, L., Matthews, W.G. and Pashley, D.H., Bond strengths to superficial, intermediate and deep dentin in vivo four dentin bonding systems, Dent. Mater., 1993, Vol.9, 19-22.

4. Van Meerbeek, B., Inokoshi, S., Braem, M., Lambrechts, P. and Vanherle, G., Morphological aspects of resin-dentin interdiffusion zone with different dentin adhesive systems, J. Dent. Res., 1992, Vol.71,1530-1540.

5. Pashley, D.H., The effects of acid etching on the pulpodentin complex, Oper. Dent., 1992, Vol.17,229-242.

6. Wakasa,K. and Yamaki,M., Bond strength between dentine and restorative resins - Calculation model, Dentistry in Japan , 1994, Vol.31, 81-84.

7. Wakasa,K., Yamaki, M. and Matsui, A., Calculation models for average stress and plastic deformation zone size of bonding area in dentine bonding systems, Dent. Mater.J.,1995, Vol.14 ,152-165.

8. Van Noort, R., Cardew, G.E., Howard, I.C., and Noroozi, S.,The effect of local interfacial geometry on the measurement of the tensile bond strength to dentin, J. Dent. Res., 1991, Vol.70, 61-67.

9. Van Meerbeek, B., Braem,M., Lambrechts, P. and Vanherle,G., Assessment by nano-indentation of the hardness and elasticity of the resin - dentin bonding area, J .Dent. Res., 1994,Vol.72, 1434-1442.

10. Eick, J.D., Cobb, C.M., Chappel, R.P., Spencer, P. and Robinson, S.J.,The dentinal surface: Its influence on dentinal adhesion. Part I, Quint. Int., Vol.22, 967-977.

AN EX VIVO EVALUATION OF THE PHYSICAL AND GEOMETRIC PROPERTIES OF THE ORTHODONTIC BRACKET CEMENT INTERFACE.

J. Knox[1], M. L. Jones[2], B. Kralj[3], J. Middleton[4]

1. ABSTRACT

The contributors to the physical and geometric properties of the bracket cement interface were determined by ex vivo evaluation allowing the validation of a three dimensional FE model of the orthodontic bracket-cement-tooth system.

2. INTRODUCTION.

In the search for a method of orthodontic bracket attachment, which avoids the use of unaesthetic and clinically cumbersome bands, the acid etch technique has, to date, proved to be the only reliable method of achieving an adhesive bond with enamel. However, one of the difficulties experienced in the evaluation of orthodontic attachment has been the method of bond strength testing. Evaluation of standard techniques of shear testing by means of numerical analysis suggests that the stress generated at the bracket tooth interface may differ by a factor of 2 to 2.5 times when an identical load is applied at different points on the bracket stem. This is probably one of the reasons for the large standard deviations observed in such in vitro studies and emphasises the need to examine the range of such measurements over a much larger sample than is currently employed. In addition, current methods of bond strength evaluation test the cohesive strength of the cement and the strength of bracket-cement and cement enamel interfaces recording only the weakest element of this system.

Keywords: Orthodontic brackets, cement, penetration coefficient

[1] Senior Registrar, Dental School, University of Wales College of Medicine, Heath Park, Cardiff CF4 4XN, Wales, UK
[2] Professor, Dental School, University of Wales College of Medicine Cardiff, Heath Park, Cardiff CF4 4XN, Wales, UK
[3] Honorary Lecturer/Research Officer, Dental School, University of Wales College of Medicine/ University of Wales Swansea, Wales, UK
[4] Reader, Department of Civil Engineering, University of Wales Swansea, Swansea, Wales, UK

The Finite Element Method of stress analysis is suggested to provide an alternative method of evaluating orthodontic attachment. However, the validity of such models is dependant on the accurate determination of the geometric and physical properties of the system. It is, therefore, the objective of this study to determine the influence of bracket base morphology and cement type on the physical and geometric properties of the bracket-cement interface by ex-vivo evaluation.

3. THE PHYSICAL PROPERTIES OF THE CEMENT-BRACKET INTERFACE

3.1. Materials and method

Eight bracket base designs[5] and seven cements[6] were evaluated in this study. The strength of the bracket-cement interface was determined by aligning and opposing identical bracket bases and 'sandwiching' a given cement between the two at a prescribed lute dimension. In most instances lower incisor brackets were opposed due to the relative flatness of their bases (Figure 1).

To avoid the moments induced during shear testing (Fox et al 1994, [1]) all samples were tested in tension (Figure 2). The plane of failure was recorded for each sample ensuring that failure was at the bracket cement interface rather than cohesively within the cement.

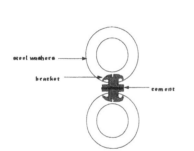

Figure 1 - Model for bracket cement   Figure 2 - Tensile evaluation of the
interface evaluation   bracket cement interface.

3.2. Results

The strength of the interface was recorded with each of the bracket bases and orthodontic cements studied is summarised graphically in figures 3-4.

Of the orthodontic cements evaluated (Figure 3), Concise provided a uniformly strong interface with all bracket bases. The strength of attachment achieved with Right On was

[5] *Master Series,* 60, 80 and 100 mesh disc samples- (American Orthodontics, Sheboygan, Wi.)
Mini Twin, Victory, Dynalock - (3M Unitek, St Paul, Minnesota)
Omni - (GAC International, New York)
[6]*Concise, Transbond* (3M Unitek, St Paul, Minnesota)
Right On (TP Orthodontics Inc., La Porte, Indiana)
Fugi LC (GC Corporation, Japan)
G552, OA1, OA2 (Associated Dental Products, Swindon, UK)

comparable to that achieved with Concise except when used with Omni and Dynalock bases. Transbond provided a comparatively weak attachment with Master Series, 60, 80 and 100 mesh bases. With the Mini Twin and Omni bases, Transbond provided an interface that was comparable in strength to Right On and Concise. With Dynalock and Victory bases, the strength of attachment was comparable to that achieved with Right On and Fugi LC but weaker than that resulting from the use of Concise.

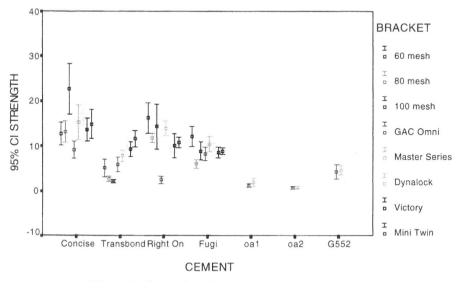

Figure 3 - Strength of the bracket cement interface

The Fugi LC hybrid glass ionomer cement provided a strength of attachment that was comparable to Right On and Concise with the 60 mesh base, weaker than Concise and Right On with the 80 mesh and Omni bases, and similar to Concise and Transbond with the 100 mesh base. With the Master Series, Dynalock and Victory bases, the strength of attachment achieved with Fugi LC was similar to that achieved with Right On and Transbond but weaker than that achieved with Concise. When combined with the Mini Twin base, Fugi LC provided a weaker attachment than Concise, Right On and Transbond.

The prototype glass ionomer cements (G552, OA1, OA2) provided a strength of attachment that was weaker than that achieved with the commercially available cements, with all base designs tested.

When comparing bracket bases (Figure 4), it would appear that an inferior attachment results from the use of Omni bases if Concise, Transbond or Right On cements are used. In addition, if Transbond is to be used, plain 60, 80 and 100 mesh bases can be expected to provide inferior attachment. However, all base designs appear to perform similarly with Fugi LC.

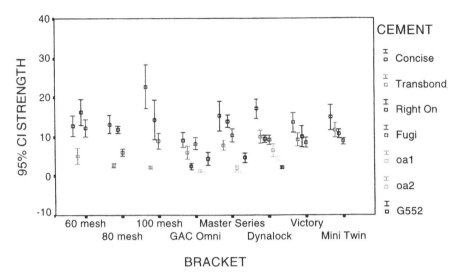

BRACKET

Figure 4 - Strength of the bracket cement interface.

3.3. Discussion

This study attempted to reduced the number of variables associated with conventional methods of bond strength evaluation. The bracket cement interface was tested in isolation with the influences of shear and peel minimised. It would appear that the physical properties of the bracket cement interface are significantly influenced by the bracket base design and the orthodontic cement chosen.

4. THE GEOMETRY OF THE CEMENT-TOOTH INTERFACE

The geometry of the bracket cement interface is determined primarily by the penetration of the orthodontic cement into the undercuts of the bracket base design. Representation of this complex layer demands the use of homogenisation techniques (Hubsch 1996, [2]) and the assumption that cement penetration is complete.

To determine the nature of cement flow into the bracket base undercut a theory developed by Washburn [3] and later used by Fan et al [4] can be utilised. This theory deals with a penetration of fluids into a capillary tubes. In its simplified form, which can be applied to this problem, a relationship between an applied pressure p_a and the time dependent penetration depth x can be expressed as:

$$x(t)^2 = \frac{p_a + \dfrac{2\gamma}{r}\cos\Theta}{8\eta} r^2 t$$

$$= \left(\frac{p_a r}{4\eta} + P_c\right) rt$$

where γ stands for surface tension, Θ is contact angle, r is the tube radius, η is viscosity and P_c is penetration coefficient.

The apparatus, schematically shown in Figure 5, was used to measure x-t for the orthodontic cements in this study, using a range of different pressures.

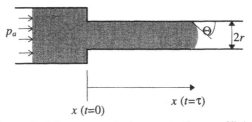

Figure 5 - Measurement of a penetration coefficient

Figure 6 shows the typical relationship between measured and ideal results. Table 1 presents the calculated values for a typical cement. A good agreement among results for '$\gamma \cos(\Theta)$' can be observed with a lower consistency of results being recorded for η .

Using the results of this investigation in combination with Scanning Electron Microscope evaluation of the bracket cement interface has confirmed the uniformly good penetration of the orthodontic cements studied. This has allowed a more clinically valid model of the bracket cement interface (Figure 7) to be developed (Kralj et al. [5]).

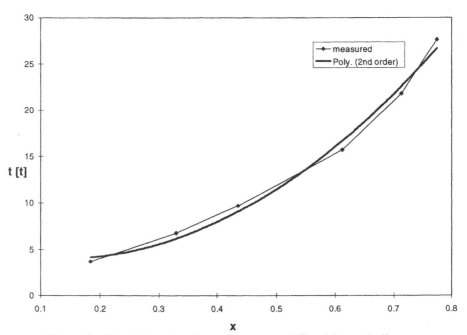

Figure 6. - Typical results of measurement and 'fitted theoretical' curve

	η	$\gamma \cos(\Theta)$
	48.1E+3	-518.6E+0
	35.2E+3	-561.2E+0
	26.7E+3	-589.1E+0
	31.0E+3	-575.1E+0
	132.9E+3	-615.2E+0
	60.5E+3	-649.0E+0
	87.9E+3	-636.2E+0
	170.2E+3	-635.8E+0
	67.2E+3	-660.9E+0
	102.8E+3	-652.2E+0
	145.1E+3	-622.0E+0
	62.9E+3	-653.3E+0
	93.1E+3	-641.8E+0
mean	81.8E+3	-616.2E+0
st. dev.	45.6E+3	43.0E+0

Table 1 - Penetration coefficient results (Concise)

Figure 7. - FE mesh of the bracket - tooth interface

5. CONCLUSION.

It would appear that although many of the variables that influence ex vivo bond strength testing can be controlled, many more remain. As the quality of orthodontic attachment is primarily determined by the stresses generated in the cement and impregnated wire mesh areas of the bracket-cement tooth system, FE techniques are best suited for evaluation.

By studying the influence of bracket base design and cement physical property on the stresses generated, it is possible to design a system which is sufficiently robust for clinical service. In addition, the system should fail at a prescribed interface when brackets are removed reducing the risk of enamel damage. The current presentation represents a small step towards such a system.

6. REFERENCES.

1. Fox N A, McCabe J F, Buckley J G., 'A critique of bond strength testing in orthodontics. *Br J Orthod.* 1994, pp. 33-43.

2. Hubsch P.F., Knox J, Jones M.L., Middleton J, 'Global and local behaviour of the stress field at the base of an orthodontic bracket'.2nd International Symposium on Biomechanics and Biomechanical Engineering, Swansea, UK, 1996, pp 361-369.

3. Washburn E. D., 'The dynamics of capillary flow' , *The Physical Review*, 1921., pp. 273-283.

4. Fan P. L., Seluk L. W. and O'Brien W. J., 'Penetrativity of sealants' , *J. Dent. Res.*, 1975., pp. 262-264.

5. Kralj B., Knox J, Jones M.L., Middleton J, 'Numerical modelling of orthodontic brackets.' 3rd International Symposium on Biomechanics and Biomechanical Engineering, Barcelona, Spain, 1997 (In press)

THREE DIMENSIONAL MODEL OF AN EDENTULOUS MANDIBLE WITH DENTAL IMPLANTS

B.Merz[1], R. Mericske-Stern[2] M. Lengsfeld[3], J. Schmitt[4]

1. ABSTRACT

Dental implants are of considerable importance for the treatment of partially or completely edentulous patients. In the case of an edentulous mandible, the use of a overdenture supported by only two implants can improve the masticatory function and help prevent the further atrophy of the proximal alveolar ridge.

In this study a 3D inhomogeneous voxel model of an edentulous mandible with two ITI dental implants was created based on CT scans.

The loading was determined by in vivo measurements of the 3D forces on the implants in five patients with implants successfully osseointegrated for more than five years. For this purpose the patients were fitted with special suprastructures on their implants with integrated three dimensional load measuring piezoelectric transducers. Loads were registered for maximum bilateral and unilateral biting as well as for several functional activities such as chewing of bread, grinding or tapping.

The measured loads were then imposed on the implants in the model together with loads directly imparted from the prosthesis onto the alveolar ridge.

The results indicate that the non-vertical components of the loads on implants must not be neglected, which reach values of 30-50 % of the vertical components. The highest loads found on the implants were located in the medio-frontal area where the implants penetrate the cortical bone. With respect to the overall loading of the mandible a relatively even distribution of stress and strain was found over the whole bony structure, indicating the functional adaptation for chewing and biting.

Keywords : Force Measurement, FEM, Occlusal Load, Mandible

---

[1] Dr sc. techn., Medical Development Group, Institut Straumann AG, CH-4437 Waldenburg, Switzerland
[2] PD Dr med. dent., Associate Professor, Dept. of Removable Prosthodontics, School of Dental Medicine, University of Berne, Switzerland
[3] PD Dr med., Associate Professor, Orthopaedic Surgeon, Dept. of Orthopaedic Surgery, Philipps-University Marburg, Germany
[4] Dr med., Orthopaedic Surgeon, Dept. of Orthopaedic Surgery, Philipps-University Marburg, Germany

2. INTRODUCTION

Dental implants are subject to functional and parafunctional loads. Since the protective function of sensible receptors as found in the periodontal ligament is missing in the case of osseointegrated implants, overload of implant based restorations might be a problem. While there is considerable knowledge about microbial pathologic processes in the context of implants, it is not known, to what degree functional and parafunctional loads could be responsible for resorption and angular recession in the marginal bone.

In most papers published today, the loads directed onto implants were simply estimated or measured only in the vertical axis of the implant [1,2,3]. The assumption of the forces acting only in the implant axis, however, is an oversimplification of the clinical situation. This could be shown by first results from the three dimensional load measurements [4].

It was therefore the goal of this study, to measure the real three dimensional loads transmitted to dental implants in a first group of completely edentulous patients with two mandibular implants supporting an overdenture. The measured loads were applied to a finite element (FE) model of an edentulous mandible with implants, in order to study the stress imposed on the implants and the surrounding bone.

3. METHODS

In the course of a clinical study, 5 edentulous test subjects were selected, each having two ITI implants (Institut Straumann AG, Waldenburg/Switzerland) in the interforaminal region of the mandible [4]. All implants had been in function for over 5 years. In order to measure the three-dimensional loads on the implants when biting with the overdenture, special suprastructures with integrated piezoelectric transducers measuring in three dimensions (Kistler AG, Winterthur/Switzerland) were used in duplications of the original overdentures. The transducers and the whole measuring equipment were subjected to extensive validation and calibration [5]. Figure 1 gives an overview of the situation in the mouth.

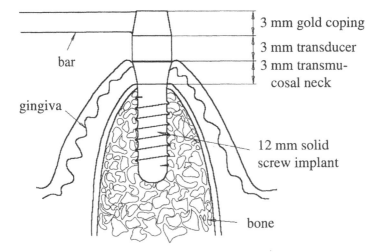

Figure 1 Measuring set-up in the alveolar ridge, example with bar support.

The test subjects were measured in different test modalities, e.g. maximum occlusal force in symmetric and unilateral occlusion, grinding and chewing test food.

For unilateral maximum occlusion the patients bit on a bite-plate allowing for simultaneous registration and comparison between the inter-occlusal and the implant forces. For control purposes two bite-plates were also used in bilateral maximal occlusion. Figure 2 shows a patient biting on the bite-plate.

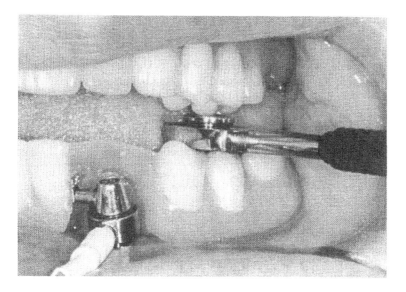

Fig. 2. Biting on a bite-plate with simultaneous registration of inter-occlusal force and forces measured on the implant. In the lower left an implant with mounted transducer and connecting bar is visible. The overdenture is custom made for the measurement.

In order to create a Finite Element (FE) model of an edentulous human mandible a macerated specimen was scanned with a Somatom Plus-S CT scanner (Siemens, Erlangen/Germany). The slice thickness was 1 mm and the scans were taken in high-resolution mode. By means of a special, validated interface program [6] a three-dimensional voxel model was created and transferred to the ANSYS 5.2 FE code (ANSYS Inc., Houston PA/USA). The dimensions of a single voxel, an hexahedron were $1.15 \times 1.15 \times 1.0$ mm^3. The Young's modulus was calculated for each element by averaging its Hounsfield value (HU), by comparing the range of values to those found in radiological tables (corrected for maceration) and by finally mapping it linearly to a range of Young's moduli representing the whole bandwidth of bone values as found in the literature. The Poisson number was set to 0.33.

Within ANSYS the model was edited in order to modify or add elements representing the geometry of the implants. Implant positions were selected according to the positioning in the patient with the most similar mandibular geometry with respect to the specimen (Figure 3). The finished model was comprised of approximately 28'000 elements with 117'000 degrees of freedom.

The loading was derived from the in vivo measurements of the corresponding patient. In addition to the implant loads, the forces directly transferred from the overdenture to the gum on the alveolar ridge were estimated and imposed in the vertical direction in the area of the first molar. Simultaneous measurements with bite-plates and transducers in symmetric maximal occlusion showed that the absolute occlusal force was about 25 % higher than the values measured in the implants.

The restraints consisted of a fixation of all translational degrees of freedom in one node of each condyle thus modelling the joint as hinge joint. In addition the Masseter and Temporalis muscles were modelled as four times 2 rods connecting the areas of muscle insertion on the mandible to fixed nodes in the approximate position of the muscular insertions on the mastoid or temporal bone of the skull. Figure 3 shows the model with restraints.

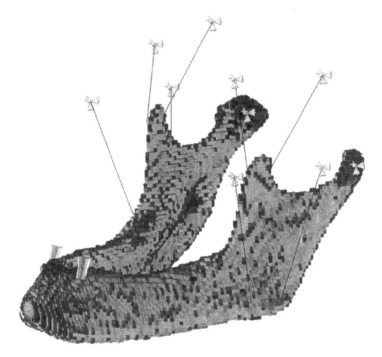

Figure 3 Model of an edentulous mandible with 2 ITI implants in the interforaminal region. The rods represent the restraints imposed by the Masseter and Temporalis muscles.

For the present calculation average load values were used, obtained in bilateral maximal occlusion as measured in the selected patient with a telescoping overdenture. The imposed load amounted to 17.1 N directed from lateral towards medial, 32.4 N in dorso-ventral direction towards ventral and 114 N vertically towards caudal.

4. RESULTS

The three-dimensional force measurements in different patients during different activities and with different suprastructures revealed that substantial non-vertical components of loading occur. They can reach values of 30-50 % of the axial loading. Figure 4 shows a chewing cycle for the selected patient while chewing bread. In this case the patient was chewing predominantly on the right side, where the vertical loads are positive (towards caudal).

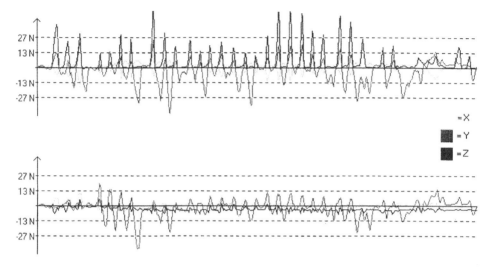

Fig. 4 Chewing cycle when eating bread. X = medial-lateral axis (top: positive towards medial, bottom: negative towards medial), Y = anterior-posterior axis (positive towards anterior), Z = vertical axis (positive towards caudal). The patient chewed predominantly on the right side (upper diagram) where the vertical loads are positive while on the left there were small negative axial loads measured as the prosthesis seems to have a slight tendency to lift on the telescope.

Fig. 5. Strain Energy Density in the loaded mandible. The distribution is relatively even, higher values appear in the ramus ascendens.

Figure 5 shows the distribution of Strain Energy Density which is in discussion as one of the triggers of bone remodelling. While the ramus shows higher values due to the thin cortical structure and the restraints by the joint and the muscles, the distribution is generally quite even.

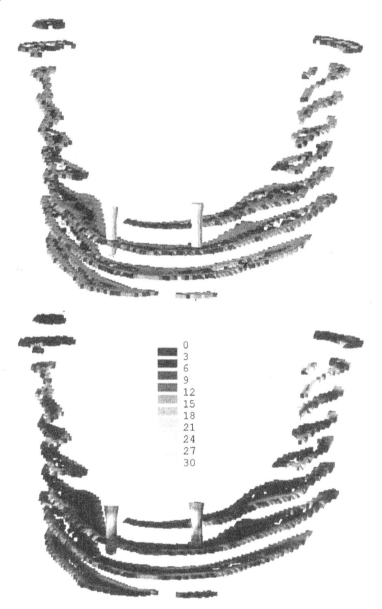

Fig. 6. Stiffness distribution (top, stiffer elements appear brighter) and distribution of equivalent stress (bottom) in several slices of the model.

Figure 6 gives an insight into the interior stress distribution and compares it to the distribution of denser and therefore stiffer elements. The loads imposed on the implants are transferred to the bone in the area of the medio-frontal passage through the cortical bone. The maximal equivalent stress found in the implants is about 40 MPa. In the bone the loads are transferred by the buccal and the lingual cortical plate to the ramus where the highest stress values in the bone are found with around 30 MPa.

5. DISCUSSION

The role of loading for long-term success of dental implants is not well known. Usually the literature recommends the axial loading of implants. While evidence exists, that small static transversal loads as used in orthodontics have no adverse effects [7,8], it is not clear to what extent the dynamic transversal loads have an influence on the remodelling of the bone adjacent to the implant.

The three-dimensional measurements show that the non-vertical components of implant loading reach substantial values such that they must not be neglected when studying an implant-bone system. Although this study investigated only completely edentulous patients with implants retaining an overdenture, first results of measurements with bridges and single crowns confirm the presence of these transversal loads. In order to obtain a comprehensive overview of the three-dimensional loads that act on dental implants in the whole range of applications, i.e. completely edentulous patients, bridges and single tooth replacement, more patients will have to be measured with implants in both maxilla and mandible.

The present FE model is based on the voxel technique. This technique allows for an easy and straight-forward creation of bone models on the basis of CT scans. The technique may therefore serve to create models of arbitrary bones and is applicable also in clinical settings. The program used in the present study was developed and validated on orthopaedic applications in long bones [6].

While the human mandible shows clearly anisotropic behaviour [9,10], the present study was restricted to isotropic, non-homogeneous material properties. This being clearly a simplification, evidence exists that accounting for inhomogeneity represents a viable approximation for the anisotropic behaviour [6,11,12] of both, spongy and cortical bone.

The results of the present simulation show that higher stress intensities apply in the anterior rami as described by Hart et al. [10]. Apart from the rami, stress is relatively evenly distributed, indicating a physiologically compatible loading which fits the natural functional adaptation when employing dental implants to support the overdenture. While the model is restricted to an overview, due to the rough resolution of the voxel elements, a more refined model is needed for obtaining detailed information on the stress in the implant and the surrounding bone. Nevertheless, the present model allows the study of different load cases and different suprastructures with respect to their overall effects.

6. REFERENCES

1. Siegele, D., Soltész, U., Numerical investigations of the influence of implant shape on stress distribution in the jaw bone, , *Int J Oral Maxillofac Implants* 1989; 4: pp. 333-340.

2. Mejer H.J.A., Kuper, J.H., Starmans, F.J.M., Bosman, F., Stress distribution around dental implants: influence of superstructure, length of implants, and height of mandible, *J Prosthet Dent* 1992; 68: pp. 96-101.

3. Van Zyl, P.P., Grundling, N.L., Jooste, C.H., Terblanche, E., Three-dimensional finite element model of a human mandible incorporating six osseointegrated implants for stress analysis of mandibular cantilever prostheses, , *Int J Oral Maxillofac Implants* 1995; 10: 51-57.

4. Mericske-Stern, R., Geering, A.H., Bürgin, W.B., Graf, H., Three-dimensional force measurements on mandibular implants supporting overdentures, *Int J Oral Maxillofac Implants* 1992; 7: pp. 185-194.

5. Mericske-Stern, R., Assal, P., Bürgin, W.B., Simultaneous force measurements in 3 dimensions on oral endosseous implants in vitro and in vivo - a methodological study, *Clin Oral Impl Res* 1996; 7: pp. 378-386.

6. Schmitt, J., Lengsfeld, M., Alter, P., Leppek, R., The use of voxel-oriented femur models in stress analysis - preprocessing, calculation and validation of CT-based finite element models, *Biomed Technik* 1995; 40: pp. 175-181.

7. Roberts, W.E., Smith, R.K., Zilberman, Y., Mozsary, P.G., Smith, R.S., Osseous adaptation to continuous loading of rigid endosseous implants, *Am J Orthod* 1984; 81: pp. 95-111.

8. Wehrbein, H., Glatzmaier, J., Yildirim M., Orthodontic anchorage capacity of short titanium screw implants in the maxilla. An experimental study in the dog, *Clin Oral Impl Res* 1997; 8: pp. 131-141.

9. Arendts, F.J., Sigolotto, C., Standard dimensions, Young's modulus and strength of the human mandible. A contribution to the description of the biomechanics of the mandible - Part I, *Biomed Technik* 1989; 34: pp. 248-255.

10. Hart, R.T., Hennebel, V.V., Thongpreda, N., Van Buskirk, W.C., Anderson, R.C., Modelling the biomechanics of the mandible: a three-dimensional finite element study, *J Biomechanics* 1992; 25: pp. 261-286.

11. Lotz, J.C., Gerhart, T.N., Hayes, W.C., Mechanical properties of trabecular bone from the proximal femur: a quantitative CT study, *J Comput Assist Tomogr* 1990; 14: pp. 107-114.

12. Harp, J.H., Aronson, J., Hollis, J.M., Measured mechanical stiffness of the canine tibia predicted by a QCT based finite element analysis, *Trans Orthop Res Soc* 1991; 16: p. 487.

THE INFLUENCE OF BONE MECHANICAL PROPERTIES AND IMPLANT FIXATION UPON BONE LOADING AROUND ORAL IMPLANTS

H. Van Oosterwyck[1], J. Vander Sloten[1], M. De Cooman[2], S. Lievens[2], R. Puers[2], J. Duyck[3], I. Naert[3]

1. ABSTRACT

Two-dimensional finite element models were created to study the stress and strain distribution around a solitary Brånemark implant. The influence of a number of clinically relevant parameters was examined : the nature of the interface between the implant and the bone, bone elastic properties, unicortical versus bicortical implant fixation and the presence of a lamina dura. Bone loading patterns in the vicinity of the implant are very sensitive to these parameters. Hence they should be integrated correctly in numerical models of in vivo behaviour of oral implants. This necessitates the creation of patient-dependent finite element models.

2. INTRODUCTION

The use of implant-supported dental prostheses for the treatment of edentulous patients has nowadays grown to a viable -although still expensive- alternative to removable dentures, certainly for patients who cannot be treated satisfactorily with removable dentures. The Brånemark implant has proven to be a very successful implant design and is certainly the best documented implant system available on the market, revealing success rates of more than 90 percent.[1-4]
Osseointegration is a necessary condition for the long-term success of an oral implant. Good quality bone must be formed at the interface with the implant during the healing phase to achieve a rigid fixation and must be maintained during functional loading.

Keywords : Oral implant, Finite element analysis, Bone loading.

[1]Division of Biomechanics and Engineering Design, KU Leuven, Celestijnenlaan 200A, B-3001 Heverlee, Belgium
[2]Division ESAT-MICAS, KU Leuven
[3]Department of Prosthetic Dentistry, KU Leuven

Undue bone loading can lead to a loss of osseointegration and implant failure. In this study the finite element method is used to study bone loading around a free-standing solitary Brånemark implant for a number of clinically relevant situations. In this way the influence of different bone, interface and loading characteristics can be determined.

First the effect of different interface conditions between the bone and the implant is examined. Although direct bone apposition around titanium oral implants can be observed [5,6] it remains uncertain whether some physico-chemical interaction will take place between living bone tissue and the titanium oxide that covers the implant surface. One of the only experimental studies that suggests the existence of such a bond was published by Steinemann et al [7] who found evidence of resistance to pure tensile stresses (perpendicular to the interface) of the magnitude of 4 MPa.

In a clinical situation the surgeon will sometimes try to place the implant tip in cortical bone. This is called bicortical fixation of the implant in contrast to unicortical fixation where only the implant neck contacts the (crestal) cortical bone. It is assumed that bicortical fixation would enhance primary stability of the implant by reducing micromotion. The question is whether bicortication has also a beneficial effect on the stress and strain distribution in the bone surrounding the implant. This aspect is also examined. As described by Ulm et al [8] the trabecular bone density in the human jaw can vary enormously. Accordingly the elastic modulus can strongly vary. Therefore different Young's moduli are considered for trabecular bone.

Another effect that is examined is the presence of a lamina dura : for implants which have become successfully osseointegrated bone densification at the interface can often be observed. The thin dense layer of bone surrounding the implant is called the lamina dura.

3. MATERIALS AND METHODS

Two-dimensional finite element models of a solitary oral implant surrounded by a cylindrical bone volume were created, as shown in fig. 1. The finite element software MARC k6.2 was used for all analyses.

Axisymmetric and plane strain models were used to study the bone loading patterns for a 100 N axial (vertical) and a 20 N lateral (horizontal) force respectively. All nodes on the lateral edges of the bone mesh are fully constrained.. While for the axisymmetric model the implant is completely surrounded by bone tissue this is not the case for the plane strain model, resulting in a lower resistance against deformation under lateral (i.e. in a direction perpendicular to the implant axis) tensile or compressive forces This is compensated for by the use of a sideplate.[9]

Linear elastic isotropic material properties are assigned to all materials involved in the analyses. Three different values for the Young's modulus were chosen for trabecular bone : 200 MPa, 700 MPa and 1370 MPa. Cortical bone was assigned an elastic modulus of 13700 MPa. The Young's modulus of titanium (the implant material) is 110000 MPa. For all materials a Poisson coefficient of 0.3 was applied.

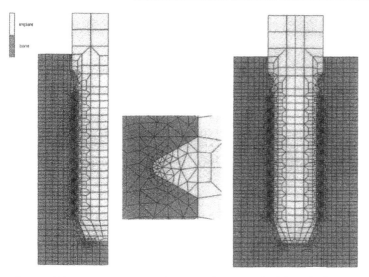

Fig. 1. Finite element mesh for the axisymmetric (a) and plane strain (c) model; a fine mesh is created at the tips of the screw-thread.

Fig. 2 displays the different bone arrangements that were used in this study. Only the arrangements for the axisymmetric model are depicted, but the same bone distributions are applied to the plane strain model. Two different values of elastic modulus were chosen for the lamina : 13700 MPa (cortical bone) and 2727 MPa (trabecular bone). The latter follows from the relationship between trabecular bone density and elastic modulus, as established by Rice et al [10], extrapolated for a volume fraction of 100 %.

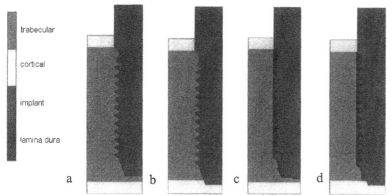

Fig. 2. Different cases of bone arrangement for the axisymmetric model : unicortical fixation without (a) and with (c) a lamina dura, bicortical fixation without (b) and with (d) a lamina dura.

Two extreme situations are modelled for the interface between bone and implant : a fixed bond with an infinite strength in tension and shear and a free contact, where relative displacement is allowed. Both conditions are implemented by the contact option that is provided by the MARC software. To describe the contact between bone and implant non-penetration constraints are automatically imposed by the programme. Implant and bone are both defined as (deformable) bodies. All boundary nodes from one body (e.g. the implant) will be prevented from crossing the boundary surface of an other body (e.g. the bone).

4. RESULTS

A free contact interface results in much higher stress and strain concentrations in trabecular bone at the interface, for an axial as well as a lateral force : e.g. for an axial force in case of a trabecular bone modulus of 1370 MPa, the maximum Von Mises stress (equivalent strain) occurring at the tips of the screw-thread amounts to 2.8 MPa (1764 $\mu\varepsilon$) for a fixed bond, while for a free contact this is 5.6 MPa (3608 $\mu\varepsilon$). The Von Mises stress distribution in bone at 0.2 mm of the interface is illustrated in fig. 3 for an axial force and in fig. 4 for a lateral force.

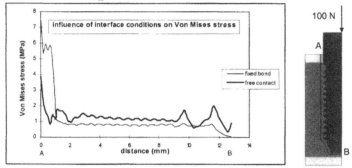

Fig. 3. Von Mises stress distribution for an axial force at 0.2 mm of the interface : values are taken along the path A-B, as shown on the right (the first 0.9 mm of the path run through cortical bone, the rest through trabecular bone).

For an axial force a much higher portion of the external force is carried by the cervical cortex in case of a fixed bond, resulting in much higher von Mises stresses for a fixed bond than for a free contact. For a lateral force a huge difference in stress values between the cervical cortex and the trabecular bone can be noticed. For a free contact interface only the cortex on the left side of the implant (the side to which the lateral force is directed) is highly loaded in compression : at the opposite side no contact between the cortical bone and the implant is maintained during loading because there is no resistance to tension, leading to low stresses. The only stresses encountered at this side originate from the presence of the sideplate which holds the bone together. For a fixed bond both sides are equally loaded : the left side is loaded in compression while the right side is equally loaded in tension.

For a unicortical fixation a high strain peak appears in trabecular bone at the implant apex (marked with 'I' in fig. 5), especially for a free contact, as shown in fig. 5a for an axial force and in fig. 5b for a lateral force. Its value becomes very high for the low modulus (200 MPa) trabecular bone. For a bicortical fixation this deformation peak is completely eliminated.

Bicortication results in a stress and strain decrease along the entire implant length in the cervical as well as the trabecular bone. The effect of bicortication on the average equivalent strain at 0.2 mm of the interface is summarised in table 1 (for a free contact). The decrease is only substantial in the case of the low modulus trabecular bone. The influence of a bicortical fixation on the strain values in the cervical cortex is negligible for the lateral force. When a fixed bond is considered the effect of implant fixation is much smaller : e.g. for an axial force and a trabecular bone elastic modulus of 700 MPa the average strain is lowered by 4.6 % in the cervical cortex and by 5.4 % in the trabecular bone; for a free contact the corresponding decreases are 9.8 % and 9.5 % respectively.

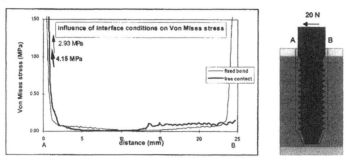

Fig 4. Von Mises stress distribution for a lateral force at 0.2 mm of the interface : values are taken along the path A-B, as shown on the right (the first and the last 0.9 mm of the path run through cortical bone, the rest through trabecular bone).

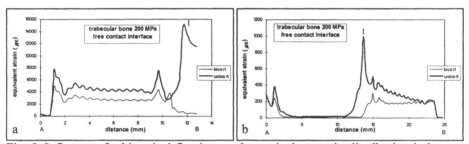

Fig. 5. Influence of a bicortical fixation on the equivalent strain distribution in bone at 0.2 mm of the interface for an axial (a) and a lateral force (b) in case of a free contact interface. The value of the trabecular bone elastic modulus is 200 MPa. Notice the different scales for the vertical axis.

modulus trabecular bone (MPa)	relative decrease (%) cervical cortex		relative decrease (%) trabecular bone	
	axial force	lateral force	axial force	lateral force
1370	3.5	0.4	2.4	4.2
700	9.8	1.5	9.5	13.3
200	33.6	5.1	35.7	35.0

Table 1. Decrease (relative to the value for a unicortical fixation) in average equivalent strain in the cervical cortex and trabecular bone as a result of bicortical fixation for a free contact interface.

When a lamina dura is present the strain peaks at the interface are almost entirely eliminated for an axial as well as a lateral force. It has also a stress- and strain-reducing effect on the trabecular bone adjacent to the lamina in case of an axial force (fig. 6). The latter effect is much less pronounced for a lateral force.

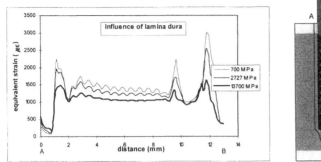

Fig. 6. Equivalent elastic strain in bone at 0.3 mm of the interface in case of a free contact and a trabecular bone elastic modulus of 700 MPa : no lamina dura ('700 MPa' in the legend), a lamina dura with an elastic modulus of 2727 MPa and a lamina dura with an elastic modulus of 13700 MPa.

5. DISCUSSION

Because the bone anatomy for the solitary implant model is simplified to a cylindrical volume the absolute value of calculated stresses and strains will probably differ from the actual values. However because this study is merely concentrated on the comparison of different cases the relative values can still lead to valuable conclusions.

Bone-implant interface conditions have a very strong effect on the bone loading patterns around the implant : not only the level of stresses and strains, but also the stress and strain distribution in bone are highly affected by the interface conditions It is likely that the high strain peaks, found at the tips of the screw can lead to bone resorption and soft tissue formation. This is confirmed by histological research of retrieved implants.[11] Considering the much higher strain peaks in case of a free contact this could suggest that the actual nature of the bone-titanium interface for a smooth implant surface tends to be more like a free contact, without the development of a firm physico-chemical bond.

Bicortical fixation eliminates apical peak strains and reduces the trabecular bone stresses and strains. The decrease only becomes substantial when the Young's modulus of trabecular bone is low. In general the influence of bicortication on the stress and strain values is smaller than one would expect. This may be explained by the fact that the load-bearing capabilities of trabecular bone are higher than first assumed. The results suggest that it is unnecessary to place the implant tip in the lower cortex when trabecular bone of sufficient quality is present. The problem is of course to determine what would be 'sufficient quality'. In the analysed models an elastic modulus of 700 MPa can be termed as sufficient, in a sense that the bicortication only results in a moderate stress and strain decrease by 10 %. However as already discussed the model geometry is too simple to determine some 'safety limit' for the elastic modulus of trabecular bone. Only three-dimensional patient-dependent finite element models can possibly give an answer to that kind of questions.

For the cervical cortex the effect of bicortication is dependent on the applied load : for an axial force the decrease is qualitatively and quantitatively similar to trabecular bone, but for a lateral force bicortication has almost no effect on the stress and strain values. This suggests that if marginal bone loss would be caused by excessive lateral forces or bending moments bicortication cannot reduce the risk of bone resorption.

Few detailed information can be found in literature describing the bone density changes that take place at the implant interface. Therefore the lamina dura was modelled here as a thin dense layer with a uniform elastic modulus, although this will not be the real situation. When a lamina dura is present strain concentrations in the bone at the interface are almost completely eliminated. In this way the presence of the lamina dura is a possible mechanism to prevent bone resorption. For an axial force the presence of a lamina also results in a decrease of trabecular bone stresses and strains at a distance of the interface. For a lateral force the latter effect does not appear.

All results suggest that the bone loading patterns are highly sensitive to the considered characteristics. This stresses the importance of patient-dependent finite element models of the human jaw, in which the individualised bone properties, bone anatomy, prosthetic design and implant loading are implemented. To determine the stress and strain patterns in the patient's jawbone during functional loading of the prosthesis the Division of Biomechanics at the KU Leuven is currently involved in a research project together with the Department of Prosthetic Dentistry and ESAT-MICAS (Medical and Integrated Circuits and Sensors). The individualised bone properties and bone anatomy and the correct position of the oral implants are derived from (post-operative) CT-images of the patient's jaw. The in vivo loading conditions are determined from measurements with strain gauges, mounted on the abutment cylinder of each implant. To obtain the correct bone geometry, contours (polylines) that border the segmented bone volume are determined for each CT-slice. In a next step a NURBS surface is calculated that passes through the contours. The obtained surface forms the boundary of the bone volume that is subsequently meshed with tetrahedral elements by means of a three-dimensional automatic mesh generator. A first three-dimensional CT-based finite element model has already been created for an excised human mandible, in which two Brånemark implants were installed in the anterior region post mortem. The correct (i.e. CT-based) bone elastic properties are not yet implemented. Further research will be concentrated on the optimisation and validation of these models.

6. REFERENCES

1. Cox J.F. and Zarb G., The longitudinal clinical efficacy of osseointegrated dental implants : a 3-year report, Int. J. Oral Maxillofac. Implants, 1987, Vol. 2, 91-100.
2. Albrektsson T. et al, Osseointegrated oral implants : a Swedish multicenter study of 8139 consecutively inserted Nobelpharma implants, J. Periodontol., 1988, Vol. 59, 287-296.
3. Adell R. et al, A long-term follow-up study of osseointegrated implants in the treatment of totally edentulous jaws, Int. J. Oral Maxillofac. Implants, 1990, Vol. 5, 347-359.

4. Naert I. et al, A study of 589 consecutive implants supporting complete fixed prostheses. Part II : prosthetic aspects, J. Prosthet. Dent., 1992, Vol. 68, 949-956.

5. Sennerby L., Ericson L.E., Thomsen P., Lekholm U. and Åstrand P., Structure of the bone-titanium interface in retrieved clinical oral implants, Clin. Oral. Impl. Res., 1991, Vol. 2, 103-111.

6. Steflik D.E., Parr G.R., Sisk A.L., Hanes P.J. and Lake F.T., Electron microscopy of bone response to titanium cylindrical screw-type endosseous dental implants, Int. J. Oral Maxillofac. Implants, 1992, Vol. 7, 497-507.

7. Steinemann S.G., Eulenberger J., Maeusli P.A. and Schroeder A., Adhesion of bone to titanium, in: Christel P., Meunier A. and Lee A.J. (editors), Biological and biomechanical performance of biomaterials, Elsevier Publishing Company, Amsterdam, 1986: 409-14.

8. Ulm C.W., Kneissel M., Hahn M., Solar P., Matejka M. and Donath K., Characteristics of the cancellous bone of edentulous mandibles, Clin. Oral Impl. Res., 1997, Vol. 8, 125-130.

9. Vander Sloten J., The functional adaptation of bones in vivo and consequences for prosthetic design, PhD thesis, University of Leuven, Belgium, 1990, 96-100.

10. Rice J.C., Cowin S.C. and Bowman J.A., On the dependence of the elasticity and strength of cancellous bone on apparent density, J. Biomechanics, 1988, Vol. 21, 155-168.

11. Albrektsson T. and Sennerby L., State of the art in oral implants, J. Clin. Periodontol., 1991, Vol. 18, 474-481.

AN IMPLANT-ORIENTED METHOD FOR
DENTAL DIGITAL SUBTRACTION RADIOGRAPHY

Niclas Börlin[1], Tomas Lindh[2]

1. ABSTRACT

The purpose of this study was to develop a method for navigation in digital dental radiographs. The proposed navigation method uses edge detection algorithms to locate the threads of the dental implant and construct an intra-image coordinate system. The coordinate system facilitates navigation in physical coordinates and calculations independent of differences in magnification. Our method can be combined with any contrast correction method for digital subtraction radiography. In this study, our method was combined with the Rüttimann et al. normalization method. The results show that simulated peri-implant bone density changes, invisible on clinically acceptable radiographs, could be detected. Furthermore, the method was sufficiently robust to allow for deviation in projection geometry of more than ten degrees. This study shows that the method can be used to analyze radiographic data from clinical studies on implants.

2. INTRODUCTION

When lost teeth are replaced with dental implants, the status of the peri-implant bone is crucial for a predictable prognosis. In this context, a healthy bone/implant interface is characterized by a maintained marginal bone level and often an increasing density over time of the bone close to the implant[1].

Radiography is considered to be a non-invasive method for control and follow-up of implant treatment[2]. In recent years, digital subtraction radiography has been used experimentally to investigate the time-related changes in bone density that occur in the vicinity of dental implants[3].

Keywords: Subtraction radiography, dental implants, photogrammetry, peri-implant bone change, medical imaging.

---

[1]Dept. of Comp. Sci., Umeå Univ., 90187 UMEÅ, Sweden, Niclas.Borlin@cs.umu.se.

[2]Dept. of Prosth. Dent., Umeå Univ., 90187 UMEÅ, Sweden, Tomas.Lindh@protetik.umu.se.

Typically, the radiographic subtraction technique employs a video camera for digitization of the radiographs which allows for adjustments of the alignment of the
radiographic pair prior to subtracting the digitized images[4].

With the development of intra-oral direct digital radiography several advantages
over film based technique have been achieved[5]. However, the possibility to view
and make analog adjustments of the image alignment is lost.

The objective of this study was to develop an algorithm for intra-image navigation
on digitized intra-oral dental radiographs and to verify its robustness and sensitivity
in a experimental model. The method bypasses the alignment problems arising with
direct digital radiography, and simplifies detection and preliminary quantification
of relative changes in bone mineral adjacent to dental implants.

3. MATERIALS

3.1 Experimental model

The experimental model (Fig. 1) consisted of a Brånemark implant placed in a
bovine rib using the standard procedure and instrumentation for placing dental
implants in edentulous jaws. On the focus side of the rib, a 1.6 mm thick Plexi-glass
plate was fixed to the bone. An L-shaped metal reference structure was attached
to the plate. Also, small markings were made on the Plexi-glass plate to indicate
where thin wedges of cortical bone, 2×8 mm and 0.08 mg, 0.2 mg, or 0.4 mg in
weight, were placed. Thus, radiographs of a positive bone mineral change adjacent
to the threaded implant surface could be simulated with the superimposed bone
wedges. The wedge-shape made it possible to indicate an increasing mineralization
toward the implant.

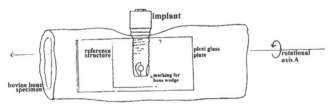

Figure 1:   Illustration of the experimental model. Beside the indicated rotational axis,
rotational axis B extends out from the paper.

The rib was fixed in a device allowing rotation of the implant about its hypomoklion
(rotational axis A). The X-ray detector was placed in a holder that could be rotated
in the film-plane (rotational axis B). The holder was mounted at a fixed distance
from the rib opposite to the collimator of the Anatom X-ica dental X-ray machine
operating at 65 kV/7.5 mA with a 2 mm Al filtration. The film-focus and film-object
distances were maintained throughout the experiments.

We adopt the term *angulation* to indicate rotation about axis A and the term
rotation to indicate rotation about axis B.

3.2 X-ray detector

The X-ray detector used in this study was the Sens-A-Ray system, which is based
on a charge coupled device (CCD) detector designed for direct exposure to X-ray
radiation[5].

3.3 Images

Several image sets were exposed, each varying a different parameter. Unless otherwise stated, the image sets were exposed for 0.32s with no rotation, no angulation, and no simulated bone mineral change.

Exposure time. Six images exposed for 0.13s, 0.20s, 0.25s, 0.32s, 0.40s, and 0.50s, respectively. Of these, the 0.20–0.32s were considered clinically acceptable.

Rotation. Sixteen images taken with a rotation of $0, 1, \ldots, 15$ degrees.

Angulation. Twenty-five images taken with an angulation of $0, 0.5, \ldots, 13$ degrees.

Reference. Three reference images taken at the clinically acceptable exposure times.

Simulated change. Three sets of images were produced using the 0.08 mg, 0.2 mg, and 0.4 mg bone wedges. Each set was exposed at the three clinically acceptable exposure times. The density changes were classified as clinically *invisible, difficult,* and *easy* to detect, respectively, both in the Sens-A-Ray images and in clinically acceptable Ectaspeed radiographs.

4. METHODS

The methods presented here are described in more detail in [6].

4.1 Navigation

Define the *upper* and *lower* thread regions, $\mathcal{R}_u$ and $\mathcal{R}_l$, respectively, according to Fig. 2 (left). Also, denote the thread tip edges as *rising* and *falling*, with corresponding angles α_r and α_f, respectively (Fig. 2, right).

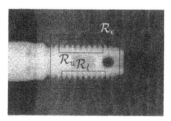

Figure 2:   Left: The upper thread region $\mathcal{R}_u$, the lower thread region $\mathcal{R}_l$, and the enclosing region $\mathcal{R}_e$. Right: Illustration of α_r and α_f.

To calculate the rising and falling edge directions α_r and α_f, the thread regions $\mathcal{R}_u$ and $\mathcal{R}_l$ were filtered with four quadrature edge detectors and the output was combined into *coxels* (complex-valued pixels) according to the *double angle representation*[7]. The double angle representation (Fig. 3, left) produces a separation between the edge orientation *value*, indicated by the coxel argument, and the edge orientation *certainty*, indicated by the coxel magnitude.

A set $\mathcal{S}$ was constructed from the coxel values from $\mathcal{R}_u$ and $\mathcal{R}_l$ and the matrix

$$A = \begin{bmatrix} x_1 & x_2 & \ldots & x_n \\ y_1 & y_2 & \ldots & y_n \end{bmatrix}^T$$

was constructed, where (x_i, y_i) are the real and imaginary parts, respectively, of each coxel value in $\mathcal{S}$. The primary axis of $\mathcal{S}$ was taken as the dominant eigenvector of $A^T A$ [8] (Fig. 3, center).

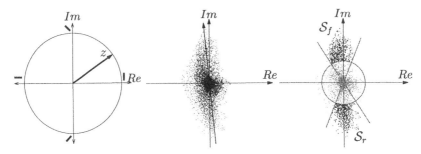

Figure 3: Left: Illustration of the direction of the output coxel z as a function of the edge orientation. Center: The set $\mathcal{S}$ with the primary axis indicated. Right: The subsets $\mathcal{S}_r$ and $\mathcal{S}_f$ of $\mathcal{S}$ with subset delimiting lines indicated.

Two subsets $\mathcal{S}_r$ and $\mathcal{S}_f$ of $\mathcal{S}$ were constructed by thresholding the coxel values in $\mathcal{S}$ on magnitude (to remove directions with low certainty) and on argument (to remove directions deviating too much from the primary axis) (Fig. 3, right).

Finally, the thread directions α_r and α_f were calculated from the mean argument γ_r and γ_f of the coxel values in $\mathcal{S}_r$ and $\mathcal{S}_f$, respectively.

To calculate the location of the rising and falling edges, the enclosing region $\mathcal{R}_e$ (Fig. 2, left), was filtered as described in the previous section and segmented into rising, falling, and background regions. The segmentation was done by thresholding on the projection of each coxel onto $e^{i\gamma_r}$ and $e^{i\gamma_f}$ (Fig. 4, left).

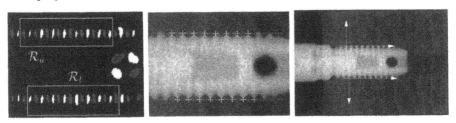

Figure 4: Left: The segmented edge image with rising regions (gray), falling regions (white), and background (black). Center: The implant with the thread edge lines (solid black), thread tips (white crosses), and thread outline (dashed, gray). Right: The implant related coordinate axes.

Considering the upper thread (the lower thread was handled analogously with the meaning of "rising" and "falling" exchanged), the rising and falling regions with center of gravity outside $\mathcal{R}_u$ were discarded. Parallel, equidistant lines $\ell_{ri}(x)$ with slope $\tan^{-1}\alpha_r$ were fit to the remaining rising regions by least squares fitting[9]. Likewise, parallel, equidistant lines $\ell_{fi}(x)$ with slope $\tan^{-1}\alpha_f$ were fit to the remaining falling regions (Fig. 4, center). The position of each thread tip was approximated to where the corresponding edge lines $\ell_{ri}(x)$ and $\ell_{fi}(x)$ intersect and the thread outline was approximated by fitting two straight, parallel lines to the thread tips (Fig. 4, center).

Two coordinate systems were established, one for each thread. The origins were fixed in selected tips, and the scale was established from the physical size of the implant (Fig. 4, right).

4.2 Normalization

As a contrast correction, or normalization, method, the method of Rüttimann et al.[10] was chosen. This method is based on modifying the histogram of a normalization reference area and was the best normalization methods in a comparative study by Fourmousis et al.[11]

Two different reference areas were tried, excluding and including, respectively, the area of simulated bone density change (Fig. 5).

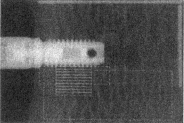

Figure 5:   Measurement area partitioning (solid grid), normalization reference areas (dotted box), and region of simulated bone density change (dashed box).

The sample area was selected as 0–4 mm out from the implant and -0.5–4.5 mm along the implant, partitioned into 0.2 mm wide "stripes" along the implant (Fig. 5). This includeded the region of the simulated positive bone mineral change. The measurements were taken as mean gray values within each "stripe".

5. RESULTS

5.1 Navigation results

The navigation algorithm worked for all images except at angulation above 12 degrees. In the following results, the two images with higher angulation have been removed from the angulation image set.

To measure the sensitivity of the navigation method to different variations of exposure conditions, the following parameters were defined: The diameter d of the implant, the thread tip distance u and l of the upper and lower threads, respectively, and the angle β between the implant and the x axis (Fig. 6).

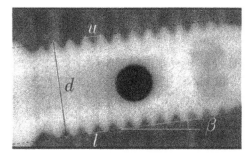

Figure 6:   Illustration of the different test parameters.

The parameters were calculated for all images, and the results are shown in Table 1.

The deviation from zero for the mean in-plane angle β is explained by a physical mis-alignment in the experimental model.

Since the angulation affects the projected geometry in a non-trivial way, the angulation image set was excluded from the in-plane angle β results. The thread distance results for the angulation image set are also shown separately in Fig. 7. Both the decreasing thread distance with increasing angulation and the separation between the upper and lower thread distances at low angulation are clearly seen.

In-plane angle β			Implant diameter d		
Image set	Mean	Std. dev.	Image set	Mean	Std. dev.
Rotation	0.77	0.12	Rotation	86.66	0.22
Exposure time	0.77	0.08	Exposure time	86.79	0.26
			Angulation	86.63	0.18

Upper thread distance u			Lower thread distance l		
Image set	Mean	Std. dev.	Image set	Mean	Std. dev.
Rotation	14.61	0.046	Rotation	14.58	0.037
Exposure time	14.59	0.031	Exposure time	14.62	0.044

Table 1: Navigation results. The in-plane angle table units are degrees, all other pixels. The in-plane values for the rotation image set are taken as the difference between the setup and the measured value.

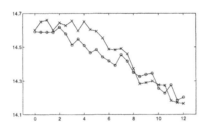

Figure 7: Measured upper (x) and lower (o) thread distances as a function of angulation.

5.2 Normalization results

The normalization results are shown in Fig. 8. For both normalization reference areas, the density changes are clearly distinguishable, with approximately the same gradient. However, for the normalization reference area affected by the density change, the positive change is underestimated and the negative change false.

6. DISCUSSION

The introduction of direct digital radiography for obtaining digital intra-oral radiographs presented an image alignment problem in subtraction radiography compared to the earlier electronic subtraction technique utilizing video cameras. To overcome this problem, a method for navigation in the digital image to allow identification of matching areas in a consistent manner was developed. The method is by construction insensitive to differences in magnification, and the results presented show that with this algorithm, the implant position, size, and alignment in the image can be calculated with high accuracy. Notable is also that the projection differences between the upper and lower thread due to the thread inclination as described by Sewerin[12] are measurable.

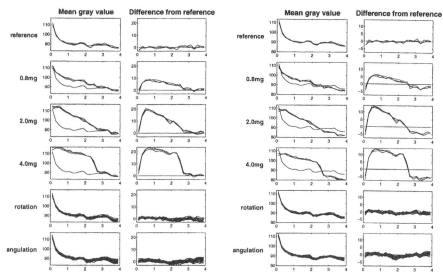

Figure 8: Measurements after normalization with unaffected normalization area (left block column) and affected normalization area (right block column). The left columns within each block show measured gray values, while the right columns show difference from a reference. From top to bottom, the rows show the results from the reference, increasing simulated change (3 rows), rotation, and angulation image sets.

The method automates the placement of all implant-related areas, e.g. normalization and measurement areas, and facilitates navigation in physical coordinates. To our knowledge, this is the first method to use the implant as a geometrical reference object.

In this study, the navigation algorithm was combined with the histogram normalization method for contrast correction described by Rüttimann et al.[10] This method has recently been shown to be superior to methods based on linear mapping of gray levels between invariant regions of the two images to be subtracted.[11]

However, this study shows that for the *values* in the subtracted image to be reliable, the histogram normalization method depends on that the selected reference areas are unaffected by the density change. When the simulated positive density change was included in the reference area, the positive change was underestimated when compared to the reference image. In addition, the normalization method introduced a false negative change in the area where the experimental model remained unaltered. This can present a serious problem in clinical studies attempting to exact quantify a change in bone mineral based on the density shift since it may be impossible to locate reference areas where no change in density have taken place.

Nevertheless, this study also shows that when using the *gradient* in the subtracted image as an diagnostic indicator of relative bone mineral changes in the peri-implant bone, these methods are adequate.

7. ACKNOWLEDGEMENTS

This work was partially funded by the Swedish Institute for Applied Mathematics (ITM), the Swedish National Board of Industrial and Technical Development (NUTEK), Umeå University, the Kempe Foundation and Västerbottens läns landsting.

8. REFERENCES

[1] P.-I. Brånemark, B. O. Hansson, R. Adell, U. Breine, J. Lindström, O. Hallén, and A. Öhman. *Osseointegrated implants in the treatment of the edentulous jaw. Experience from a 10-year period.* Almqvist & Wiksell, Sweden, 1977.

[2] K.-G. Stridh. Radiographic results. In P.-I. Brånemark, G. A. Zarb, and T. Albrektsson, editors, *Tissue-Intregrated Prostheses. Osseointegration in clinical dentistry.*, chapter 11, pages 187–198. Quintessence Publ., Chicago, 1985.

[3] U. Brägger, W. Bürgin, N. P. Lang, and D. Buser. Digital subtraction radiography for the assessment of changes in peri-implant bone density. *Int J Oral & Maxillofac Implants*, 6:160–166, 1991.

[4] H.-G. Gröndahl, K. Gröndahl, and R. L. Webber. A digital subtraction technique for dental radiography. *Oral Surg*, 55(1):96–102, 1983.

[5] P. Nelvig, K. Wing, and U. Welander. Sens-A-Ray: A new system for direct digital intraoral radiography. *Oral Surg Oral Med Oral Pathol*, 74:818–823, 1992.

[6] N. Börlin and T. Lindh. An implant-oriented method for dental digital subtraction radiography. Technical Report UMINF-97.09, Department of Computing Science, Umeå University, Sweden, 1997.

[7] G. H. Granlund and H. Knutsson. *Signal Processing for Computer Vision.* Kluwer, 1995.

[8] G. H. Golub and C. F. Van Loan. *Matrix Computation.* John Hopkins Press, 3rd edition, 1996.

[9] Å. Björck. *Numerical methods for least squares problems.* SIAM, 1996.

[10] U. E. Ruttimann, R. L. Webber, and E. Schmidt. A robust digital method for film contrast correction in subtraction radiography. *J Periodon Res*, 21:486–495, 1986.

[11] I. Fourmousis, U. Brägger, W. Bürgin, M. Tonetti, and N. P. Lang. Digital image processing: I. Evaluation of gray level correction methods *in vitro. Clin Oral Impl Res*, 5:37–47, 1994.

[12] I. P. Sewerin. Radiographic image characteristics of Brånemark titanium fixtures. *Swed Dent J*, 16:7–12, 1992.

THREE DIMENSIONAL FINITE ELEMENT ANALYSIS FOR DIRECT FIBRE - REINFORCED COMPOSITE DENTAL BRIDGE

W. Li [1], G. P. Steven [2] and C. P. Doube [3]

1. ABSTRACT

In this study, three-dimensional finite element (FE) models were developed for determining the stress and displacement distributions in a direct composite dental bridge reinforced with high density etched polyethylene fibres. The accuracy of the FE model was established via a convergence study and an appropriately fine FE mesh with around 21,000 degrees-of-freedom (*dof*) was adopted for the investigations. Model variations were made by changing the number of the fibre spans and the bond conditions on the proximal surface between the artificial and the abutment teeth. The significance of modelling with and without an adhesive layer was also investigated. The results, based on the finite element analysis, are expected to provide dental clinicians with some structural implications, which will aid the implementation of such composite cantilever dental bridges.

2. INTRODUCTION

Composite and fibre reinforced resins are being adopted in many application areas that were once the domain of metals. A direct fibre bridge technique for constructing light-cured composite dental bridge reinforced with high density etched polyethylene fibres has been developed in dentistry as an alternative to metal framework Maryland type bridges. A direct fibre reinforced dental bridge system consists of human teeth as abutments, light-cured composite made into single or multiple artificial teeth as pontics, light-cured adhesive as bond and the fibre span(s) as a connector. The fibre connector may be a single fibre or two or more fibres splinted together and bonded to

Keywords: Finite Element Analysis, Direct Fibre Dental Bridge, Stress Analysis

[1] PhD student, Department of Aeronautical Engineering, The University of Sydney, NSW 2006, Australia
[2] Professor, Department of Aeronautical Engineering, The University of Sydney, NSW 2006, Australia
[3] Director, Nulite System International Pty Ltd, PO Box 388, Hornsby, NSW 2077, Australia

both abutments and pontics by adhesive. Usually those direct fibre bridges are made as cantilevers for reducing the torque stress associated with two fixed abutments and for facilitating flossing. As there is little or no tissue removals, no metal de-bonds, and very low laboratory costs as well as minimal tooth preparation, the direct fibre technique is increasingly being adopted in modern aesthetic dentistry.

As a numerical method, finite element analysis (FEA) makes it possible to analyse structures such as dental bridges with complicated geometry, boundary and load conditions. It can also accommodate non-linear and orthotropic materials. This method was first introduced for the stress analysis of dental structure in 1970s [1]. Recently, there have been several papers devoted to the 3D finite element stress analysis of dental restorations and tooth structures [2-8]. However, such finite element-based stress analysis, has not yet involved any direct fibre-reinforced dental bridge structures.

In this paper, a two-unit anterior cantilever bridge was used as the basis for all three-dimensional FE models. It includes a human maxillary right incisor as an abutment, a maxillary left incisor as a cantilever pontic made of composite assembly, and two fibre spans as a connector (Fig. 1(a)). The main aims of this research were (1) to develop a fundamental 3D finite element model for composite dental bridges by establishing accurate FE models via convergence tests; (2) to investigate the stress and displacement distribution patterns, particularly the values and the locations of the maximum principal stresses and the maximum displacements for the bridge structures; (3) to evaluate the structural influence of the number of fibres; and (4) to examine the effect of the adhesive layer on the FE models.

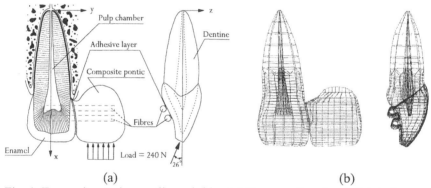

Fig. 1 Two-unit anterior cantilever bridge (a) Cross-sectional and distal views;
(b) Three-dimensional finite element mesh

3. MATERIALS AND METHODS

As shown in Fig. 1(b), a three-dimensional FE model representing the two-unit anterior cantilever bridge with two fibre spans was established. This model was based on the real bridge geometry and the average dimensions of human teeth [9]. Two basic cases were taken into account in the present study. In Case 1, the proximal surface between pontic and abutment tooth was assumed to be perfectly bonded, while in Case 2, the interface was allowed to debond. For the latter case, compression only gap elements

[10] were used to model the contact or non-contact situation on the interfaces. Each of these two cases was analysed with the following variations : (1) pontic and abutment were lingually bonded by a single fibre; (2) pontic and abutment were bonded by two fibres; (3) a 0.2 mm simplified adhesive layer on fibre-pontic, fibre-abutment, and pontic-abutment interfaces were incorporated into the finite element models; (4) there was no adhesive layer on any interface. Thus, eight finite element models were set up in this study.

In all these models, the bone structure which supports the root of the abutment was assumed to be rigid. Thus all nodes, which located on the external surface of the root portion of dentine, were taken to be fixed. As illustrated in Fig. 1(a), a uniformly distributed bite force of total 240N [11] was applied at the incisal margin of the pontic. It was further assumed that the bite force was 26° to the vertical axis (X axis) which represents the angle of first contact during biting [4, 12].

The materials adopted in the study include: the light-cured composite for the pontic, the adhesive for the bond, the fibres for the connector and a human central incisor for the abutment. All materials were assumed to be homogeneous, isotropic, and linear elastic. The Young's modulus [13] and Poisson's ratio [13,14] of the materials are given in Table 1.

Table 1 Material properties used in the analysis

Material	Young's Modulus (MPa)	Poisson's Ratio
Enamel	60,000	0.33 [14]
Dentine	15,000	0.31 [14]
Fibre	50,000	0.30 [13]
Adhesive	2,000	0.30 [13]
Composite	18,000	0.30 [13]

The Young's modulus of pulp is negligibly small comparing with enamel and dentine. Therefore, the effect of pulp on the structural analysis can be neglected and the pulp chamber was considered to be empty.

The FE mesh (Fig.1(b)) consisted mostly of twenty-node quadratic hexahedra brick elements and a few of fifteen-node wedge elements to characterise the complex curved regions of the abutment, the composite pontic and the fibres. Gap elements were applied to model the debonded adhesive layer and to simulate the contact or non-contact situation on the proximal surface between abutment and pontic.

To verify the accuracy of the FE models, the convergence tests were carried out. In the tests, four FE meshes with increasing numbers of nodes and elements were constructed. All models had the same load and boundary conditions.

4. RESULTS AND DISCUSSION

The finite element analyses were carried out using G+D STRAND6 FEA package [10] running on a Silicon Graphics workstation to determine the stress distribution in the pontic, fibres and supporting structure as well as the displacement of the structure. Note

that the current research, in its preliminary stage, only focused on conducting a
parametric study and providing an overview of the finite element analysis of the bridge.

4.1. Convergence Test

A series of four FE models with increasing mesh refinement was analysed in the
convergence test. The mesh size varied from Mesh 1 with 257 elements (4,359 dof),
Mesh 2 with 702 elements (10,698 dof), Mesh 3 with 1,418 elements (21,354 dof) to
Mesh 4 with 2,451 elements (33,996 dof). Node A at point (17.5, 7.2, -2.13) in the
adhesive and Node B at point (19.2, -2.16, -2.58) in the second fibre were chosen as the
points where the displacements were compared between different meshes. The
computing time which varied from only 4 minutes for Mesh 1 to 43 hours for Mesh 4
was also recorded for the efficiency analysis.

Fig. 2 shows the variation of displacement at the two points (Nodes A and B) and
computing time against the number of degrees-of-freedom (*dof*). The convergence of
the FE model is well indicated in the displacement curves. Differences in displacements
were found about only 2% between Mash 3 and Mesh 4 whereas the differences in the
number of *dof* and computer time were as large as 59% and 200%, respectively. From
the viewpoints of both accuracy and efficiency, Mesh 3 with around 21,000 degrees-of-
freedom was recommended as a fundamental mesh size for an accurate model.

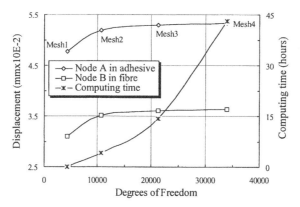

Fig. 2 Nodal displacements and computing time against the
        number of degrees-of-freedom for the four FE meshes

4.2 Principal Stress Peak and Location

The maximum principal stresses predicted by FEA for all models are summarised in
Tables 2 and 3. It is obvious to note that the peak stress values occurred in the high
stress concentration areas (which are circled), as illustrated in Fig. 3. The magnitudes
and the locations of the stress peaks varied with the model variations.

In Case 1 (*Bonded*), the highest values of tensile stress in the entire model were
occurred on the crown distal surface near cervical line of the abutment (Fig. 3(a)) with
about 179 MPa in all model variations. While the maximum compressive stresses were
occurred on the crown mesial surface near cervical line with a value of 249 MPa and
267 MPa, respectively, for the models with and without adhesive layer.

Table 2 Principal stress peaks (MPa) in Case 1 (pontic-abutment interface bonded)

| | With adhesive layer | | Without adhesive layer | |
	Single fibre	Double fibres	Single fibre	Double fibres
Max. tensile stress σ_1	179.3	179.7	178.3	179.1
Max. comp. stress σ_3	248.9	248.6	267.2	266.3

In Case 2 (*Debonded*), all stress peaks were found located on the fibre span(s) and with unacceptably higher values than those in Case 1. It is also worth noticing that the principal stress peaks distinctly differ with the changes of fibre number and the inclusion of adhesive layer. With the simulated debond of pontic-abutment interface, the fibre(s) not only forms a connection but also fully supports the pontics. For this case, the mechanical behaviour of the fibre(s) with each model variations is particularly important for investigating the failure patterns of the dental structure.

Table 3 Principal stress peaks (MPa) in Case 2 (pontic-abutment interface debonded)

| | With adhesive layer | | Without adhesive layer | |
	Single fibre	Double fibres	Single fibre	Double fibres
Max. tensile stress σ_1	3772.6	1207.8	6410.0	1468.5
Max. comp. stress σ_3	3295.0	806.0	3068.0	698.2

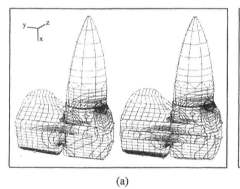 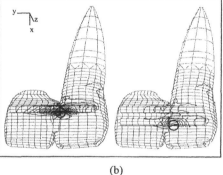

(a)   (b)

Fig. 3   Contours of principal stress σ_1 (a) Case 1 (pontic-abutment interface perfectly bonded) with single and double fibre models; (b) Case 2 (pontic-abutment interface debonded) with single and double fibre models
Note: The locations of the peak stress are circled

4.3 Structural Stiffness

To compare the stiffness of various structural designs, the maximum upward displacements (x direction) are tabulated in Table 4. It is noticed that when pontic-abutment interface was debonded (Case 2), the maximum upward displacement of the structure increased comparing with the corresponding models in Case 1 with the perfectly bonded interfaces. This means that the bond condition on the pontic- abutment interface will significantly affect the structural stiffness. Taking the models with adhesive layer as examples, the maximum upward displacements in Case 1 are only 2% and 13% of those in Case 2 with single and double fibres respectively. It is also interesting to note that the connection quality of pontic-abutment interface and the number of fibres can significantly improve the clinic effect when debonded.

Table 4   Comparison of maximum upward displacement (mm)

	Case 1 (*Bonded*)		Case 2 (*Debonded*)	
	Single fibre	Double fibres	Single fibre	Double fibres
With adhesive layer	0.125	0.122	6.59	0.96
Without adhesive layer	0.103	0.0999	2.731	0.396

4.4 The Effects of Adhesive Layer

In order to investigate the effects of adhesive layer, a 0.2 mm simplified adhesive layer with specific material property was incorporated into the finite element models. The results were compared against those without adhesive layer. In Case 1 (*Bonded*), it can be found that the presence of the adhesive layer had relatively little effects on peak stresses and structural stiffness. In single fibre design, for instance, the differences were only 0.5% in maximum tensile stress and 21% in maximum upward displacement. In Case 2, however, the adhesive layer considerably changed the structural responses. The differences in the above mentioned examples were increased by 70% and 141% respectively.

From the FEA it is clear that there are stress concentrations within the adhesive layer on the interfaces between abutment-pontic, fibre-abutment and fibre-pontic. The values of peak stresses in the adhesive layer are summarised in Table 5; and the contours of the corresponding principal stresses are presented in Fig. 4.

Table 5   The stress peaks (MPa) within adhesive layer between pontic and abutment

	With single fibre	With double fibres
Max. tensile stress σ_1	113.3	103.2
Max. comp. stress σ_3	185.3	184.4
Max. shear stress τ	75.5	75.4

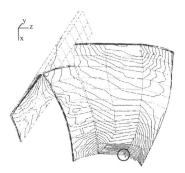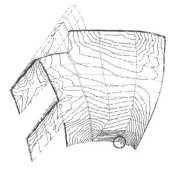

Fig. 4   Contours of principal stress σ_1 in adhesive layer on the pontic-abutment interface with single fibre model (left) and double fibre model (right)
     Note: The locations of the peak stress are circled

If the tensile and shear bond strengths between etched enamel and composite are considered to be 20 MPa [15, 16], then Table 5 shows that the predicted tensile and shear stress peaks in the adhesive layer exceed their corresponding bond strengths by many times. This implies a more detailed study of mitigating effects due to load reduction with pontic movement as adjacent teeth take more of the redistributed load.

4.5 The effect of fibre number

As summarised in Tables 2 to 4, the number of fibres has little effects on both principal stresses and the structural stiffness in Case 1 (*Bonded*). This means that the pontic is mostly supported by the bond on the pontic-abutment interface. In this case, the fibre(s) could only play a constraining role on the bridge designs. Whereas in Case 2, there is no bond on the pontic-abutment interface to support the pontic. All loading has to be fully supported by the fibre(s). In this situation the number of fibres will significantly affect the structural response. Taking the models with adhesive layer as examples, the tensile stress peak, compressive stress peak and the maximum upward displacement with double fibre design are respectively 32%, 24% and 14.6% of those with single fibre design. In other words, the introduction of one more fibre can improve around three times the tensile strength, four times the compressive strength and six times the structural stiffness in debonded case. Hence not only can the increase in strength of the fibre but the increase in fibre number improve the structural reliability of direct dental bridges.

5. CONCLUSION

The FEA modelling of direct composite cantilever bridge reveals that high stress concentrations occur in some areas at the crown on the cervix of the abutment, in the fibre span(s), as well as on the interfaces of pontic-abutment, fibre-pontic and fibre-abutment. The magnitudes of these concentrations and their variations with the arrangement of the bridge have been presented. Due to the lower tensile and shear bond strengths in the adhesive layer, FEA results suggested that the adhesive layer on pontic-abutment interface may be the most likely failure location.

It is found that the number of fibres in the bridge had negligible influence when a perfectly bonded pontic-abutment interface was adopted. But its effect became significant when the pontic-abutment interface was debonded. This debonding also caused considerable stress concentration in the fibre(s). Thus careful consideration of the design arrangement and the proper material properties of the fibre span(s) are suggested in the clinical situation since there exist significant differences of strength and stiffness in the various design cases investigated in the present study.

Comparing the results from the models with and without adhesive layer, it is concluded that the FE analysis ignoring the adhesive layer is not capable of predicting the actual mechanical behaviour of the bridge under given loading and design variations.

Finally, as the first attempt to conduct FEA for direct composite cantilever bridge, the accuracy of the three-dimensional model has been assessed by the convergence tests. However, before the results from a FEA can be accepted, the validity of the analysis must be objectively established. Therefore a further investigation on the comparison of the FEA results with clinical observations and laboratory tests of the bridge is proposed.

6. ACKNOWLEDGMENT

This study has been supported by funds from Nulite System International Pty Ltd. We also appreciate the educational discount made available by G+D Computing Pty Ltd, Australia, for the use of the STRAND6 FEA program.

REFERENCES

1. Farah, J. W., Craig R. G. and Sikarski D. L., Photoelastic and finite element stress analysis of a restored axisymmetric first molar, J. Biomech., 1973, Vol. 6, 511-520.
2. Awadalla, H. A., Azarbal M., Ismail Y. H. and El-Ibiari W., Three-dimensional finite element stress analysis of a cantilever fixed partial denture, J. Prosthet. Dent., 1992, Vol. 68, 243-247.
3. Rubin, C., Krishnamurthy N., Capilouto E. and Yi H., Stress analysis of the human tooth using a three-dimensional finite element model, J. Dent. Res., 1983, Vol. 62, 82-86.
4. Darendeliler, S., Darendeliler H. and Kinoglu T., Analysis of a central maxillary incisor by using a three-dimensional finite element method, J. Oral. Rehabil., 1992, Vol. 19, 371-383.
5. Hojjatie, B. and Anusavice K. J., Three-dimensional finite element analysis of glass-ceramic dental crowns, J. Biomechanics, 1990, Vol. 23, 1157-1166.
6. Huysmans, M. C. D. N. J. M. and Van deer Varst P. G. T., Finite element analysis of quasistatic and fatigue failure of post and cores, J. Dent., 1993, Vol. 21, 57-64.
7. Ho, Ming-Hsun, Lee Shyh-yuan, Chen Hsiang-Ho and Lee Maw-Chang, Three-dimensional finite element analysis of the effects of posts on stress distribution in dentine, J. Prosthet. Dent., 1994, Vol. 72, 367-372.
8. Hart, R. T., Heennebel V. V. et al, Modeling the biomechanics of the mandible: a three-dimensional finite element study, J. Biomechanics, 1992, Vol. 25, 261-286.
9. Ash, M. M., Wheeler's Atlas of Tooth Form, 5th ed. W.B.Saunders Co., 1984.
10. G+D Computing Pty Ltd, Australia, STRAND6 finite element analysis system reference manual and user guide, 1993.
11. Waltimo, A. and Kononen, M., Bite force on single as opposed to all maxillary front teeth, Scand. J. Dent. Res., 1994, Vol. 102, 372-375.
12. Tylman, S. D. and Malone, W. P. F., Tylman's theory and practice of fixed prosthodontics, 7th Edition, CV Mosby Co., St. Louis, 1978.
13. Information from Nulite System International Pty Ltd.
14. Farah, J. W., Craig R. G. and Meroueh K. A., Finite element analysis of three- and four-unit bridges, J. Oral. Rehabil., 1989, Vol. 16, 603-611.
15. Craig, R. G. (ed.), Restorative dental materials. 9th Edition, CV Mosby Co., St. Louis, 1993.
16. Sturdevant, C. M., Roberson T. M., Heyman H. O. and Sturdevant, J. R. (eds.), The art and science of operative dentistry, CV Mosby Co., St. Louis, 1995.

COMPARATIVE FEM ANALYSIS OF THE STRESSES TRANSMITTED BY INTRAMOBILE ELEMENTS OF THE IMZ IMPLANT

Ulbrich, N. L.[1], Hecke, M. B.[2], Possobom, A. L.[3] and Bassanta, A. D.[4]

1. ABSTRACT

Natural teeth possess fibbers, which are the periodontal ligaments, and whose function is to neutralise the masticatory forces transmitted to the bone. The majority of dental implants is not able to reduce this force. One implant design of the IMZ system (Interpore International) incorporates an intramobile resilient element (IMC) of polioximethylene resin between the implant and the prosthesis. According to its proponent, this IMC element imitates the function of the periodontal ligament and the natural movement inherent to the tooth. IMZ also produces a titanium IME identical to the first, and a third type made of titanium and polioximethylene resin with a different geometry. The Finite Element Method (FEM) is an approximate solution technique for mathematical models which seek to represent the mechanical behaviour of bodies, here the prosthetic structure, subjected to loads and their support. In the present work, axisymmetric elements were employed, since they exhibit geometrical symmetry about the central axis, with non-axisymmetric loads. The objective of the present work was to simulate the mechanical behaviour of a system composed of prosthesis-implant-bone support under masticatory forces. Studies have shown such forces to be in the range 113-190 N, reaching 100 N in the incisors region and 500 N in the molar region. Based on this information, it was decided to test and compare the three IME of the IMZ system applying 500 N loads in the vertical direction, which is considered to be the principal

Keywords: finite element, stress analysis, IMZ, intramobile element.

[1] Professor, Department of Dentistry, Federal University of Paraná, Brazil
[2] Professor, Numerical Methods for Engineering Graduate Course, Federal University of Paraná, Caixa Postal 19011, 81531-970, BRAZIL email:mildred@cesec.ufpr.br
[3] Undergraduate Mechanical Engineering student, Federal University of Paraná, Brazil
[4] Professor, Department of Prosthesis Dental, S. Paulo University, Brazil

direction of the forces of the posterior teeth. In another study, a 173 N load was applied at a 30° angle, as an average load for the anterior teeth. For vertical and inclined loads, there was not a significant difference in the stress transmission to the bone although some differences appear on the stress concentration in the retaining screw.

2. INTRODUCTION

There is no disagreement concerning the role of the PDL in providing support for the teeth as it and the natural body fluids contained in the bone dampen these forces so that the underlying bone is not damaged by the high stress. In essence, the tooth and surrounding structures provide shock absorption. The material property that enables this to occur is know as viscoelasticity. This characteristic of the periodontal ligament and bone provides a dampening of the transmission of peak stress in natural dentition[1].

In an osseointegrated implant, however, occlusal loads are transmitted directly to supporting bone with no relative motion. Thus, the close apposition of an osseointegrated implant to bone and the absence of connective tissue at the bone-implant interface, which is considered to be the key to the long-term stability of implant supported prostheses, results in low shock resistance.

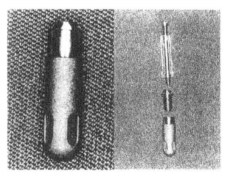

The goal of the IMZ researchers was to create an implant design that ensured an even transfer of occlusal forces into the bone similar to that which occurs with natural dentition. The system incorporates a resilient polyoxymethylene (POM)[2] resin intramobile element (IME) between the fixture and prosthesis. According to its proponents, the IME mimics the function of the periodontal ligament and the natural movement inherent to the tooth. (Fig 2).

Fig 1 - Titanium IME

The IMZ system also produces a titanium IME identical to the first (Fig 1), and a third type made of titanium and polyoxymethylene resin with a different geometry (Fig 3).

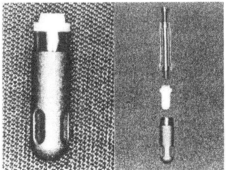

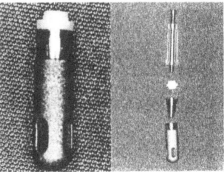

Fig 2 - Polyoxymethylene (POM) IME   Fig 3 - (POM) and titanium IMC

Using a finite element analysis, Papavasiliou[3] evaluated one model with IME (titanium). It appeared that although the IME (POM) may result in prosthesis mobility similar to the one provided by the human PDL it did not seem to act as a shock absorber.

Through photoelasticity, McGlumphy et al[4] compared the differences in the stress patterns generated from an IMZ implant with a resilient (polyoxymethylene) or from a rigid (titanium) IME. Under standardised, static loading, there was no difference in the stress pattern generated in the (modelled) bone with either a resilient or a rigid IME. An interesting finding was that the same load (5 pounds) caused a cantilever deflection of 1 mm regardless of the IME material used. According to the authors, this suggested that it was the retaining screw that allowed bending in the system.

In a first work, Holmes[5] using an axisymmetric model IMZ implant placed in the trabecular bone and restored using a complete gold crown compared stresses generated when using a resilient or a rigid IME. Load inclination was either vertical or at a 30° angle to the long axis of the implant. Results indicated that when the IME was modelled of polyoxymethylene under oblique loading conditions, significant deflection of the superstructure occurred with a corresponding flexure of the retaining screw. The results of this finite element method study demonstrated that the titanium retaining screw bear[5] the highest stresses of any component of the IMZ system.

In his second work, Holmes[6] used the same analysis he did in his first one to evaluate the connection between the deflexion of the super structure and the strain concentration in the retaining screw of the IMZ system when submitted to different magnitudes of stresses and angles. Holmes concluded that the use of IME(POM) allows a higher deflexion of a prosthetic super structure and higher concentration of strain in the retaining screw.

Chapman[7] studied 15 cases of persons who had the IMZ implant system restoration to investigate the forces of occlusal level when comparing the effects of a IME (POM) versus a titanium internal element. The forces were measured with the T - Scan computerised occlusal sensing system. Chapman concluded that occlusal forces were reduced when an internal shock absorber such as the IME (POM) were used.

3. MATERIALS AND METHODS

The present study used a Finite Element Technique to predict the stress and strain distributions in the bone surrounding a dental implant. The Finite Element Method (FEM) is a numerical technique to solve differential partial equations and is used in the solution of engineering problems. Typically, the FEM software is used to help designing better products from jumbo jets to cars and electronic equipment. This method has been widely used in the literature for biomechanical analysis to evaluate implant design and function[3,5,6].

For FEM, continuous bodies of any shape are modelled by subdividing them into a finite number of elements with simpler geometric shapes which behaviour is specified by a finite number of parameters. The problem is transformed into one of matrix algebra and a system of simultaneous equations is generated and solved by computers.

Stress and strain distributions were calculated using the ANSYS Finite Element Program, Revision 5.0 (Swanson Analysis Systems, Inc. , Houston , Pa) running on Pentium 166Mz processor.

The assumption made, following similar analysis of technical literature, is that all of the materials are isotropic, homogeneous and linearly elastic. A fixed bond in all interfaces, which means that the interfaces are continuous and under loading, relative motion between interfaces will not occur. The value of material properties assumed in the model are listed in Table 1. There is wide rigidity variation among part making that difficult modelling and generates distortion in the results. Tooth modelled are assumed symmetric in cast gold and we adopted static loading with small deformation.

Table 1: Elastic Properties of Materials Modelled[5,6]

Material	Elasticity Modulus in GPa	Poisson's/ratio
Titanium	115	0.35
Cast gold	96	0.35
Cortical Bone	13.7	0.30
Trabecular Bone	1.37	0.30
Composite Resin	9.7	0.35
Polyoxymethylene	3.45	0.35

Previously to modelling we performed measurement check in the Federal University of Paraná Laboratory. This problem has axial symmetry in cylindrical co-ordinates, that is geometric symmetry about a central axis we could use an axisymmetric model. This model increases greatly the efficiency and accuracy of the analysis over that that of an equivalent three-dimensional model. The software ANSYS has a harmonic element, that (if applied) makes possible nonaxisymmetric loading in an axisymmetric model through a series of harmonic functions (Fourier series). Therefore solution results of an axisymmetric analysis may be calculated in three dimensions at any circumferential location usually yielding more accurate results than those obtained from an equivalent three dimensional model.

In the analyses we used as a comparative criterion among the results the equivalent Von Mises stress that is calculated from the principal stresses according to the function

$$\sigma_{eqv} = \frac{1}{\sqrt{2}}\left[\left(\sigma_1 - \sigma_2\right)^2 + \left(\sigma_2 - \sigma_3\right)^2 + \left(\sigma_3 - \sigma_1\right)^2\right]^{\frac{1}{2}} \qquad (1)$$

In this work, we were interested in comparative stresses, so it is not so important which yielding criteria is adopted. Anyway, this was the criterion used by all the authors we have researched on this subject.

We generated three models adopting only a part of the maxillary where there is stress distribution. The first model, is an implant with the IME in titanium, fixed to the bone. For mesh generation, we tried higher homogeneity and lower distorting among the elements. Thus we used around 2740 with approximately 8000 nodes (Fig 4).

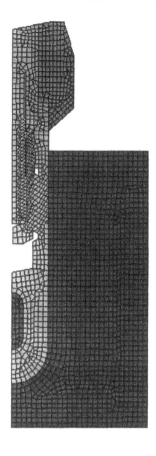

Fig 4 - Finite Element Mesh

Previous studies[5,6] have reported average occlusal forces with implant-support prostheses ranging from 110 N to 200 N in the incisor's region that can amount to 500 N in the molar region. As a comparative strategy and after a bibliographical research[3-4] we adopted a vertical force of 500 N, in first analysis. Afterwards, we adopted a 173 N force with a 30° skewed. For the displacement restrictions, we used the nodes on the external lateral edge of the bone. For the second model, we changed only the material of IME from titanium to polyoxymethylene (POM) and the model is similar to the first. Finally, for model three, there was a change in the IMC model for another of titanium retainers in addition a ring in composed.

So, we studied six different situations for models:
1. Titanium IME with vertical loads;
2. Polyoxymethylene IME with vertical loads;
3. Polyoxymethylene and titanium IMC with vertical loads;
4. Titanium IME with inclined (30°)
5. Polyoxymethylene IME with inclined (30°) loads;
6. Polyoxymethylene and titanium IMC with inclined (30°) loads

4. RESULTS

The major differences in the stresses occurred in the fastening screw. The stress distribution in the implant and the bone for model one with a 500 N vertical loading is shown in figure 4. In the model two, appear a discontinuity IME stress distribution and a concentration on the retrains screw without significant changes in the bone as we can see in figure 5. In the model three the bone stress continues similar although the stresses on the screw have been reduced as are shown in figure 6. In the analysis with 30° oblique load of 173 N there were an elevation in the stress distributed in the bone and mainly in the screw in the three models, when comparing with the vertical load. The figures 7, 8 and 9 show the stress distribution of three models with the load inclined. Concluding, we noticed that there was not significant changes with IME change, however in what concerns to the retaining screw, the third model has had a better performance regarding to the screw integrity against fatigue.

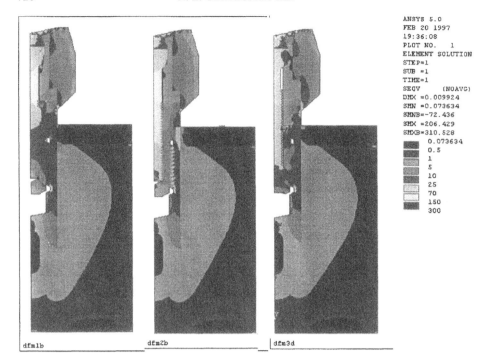

ANSYS 5.0
FEB 20 1997
19:36:08
PLOT NO.   1
ELEMENT SOLUTION
STEP=1
SUB =1
TIME=1
SEQV   (NOAVG)
DMX =0.009924
SMN =0.073634
SMNB=-72.436
SMX =206.429
SMXB=310.528
0.073634
0.5
1
5
10
25
70
150
300

dfm1b   dfm2b   dfm3d

Fig 4 - Mises stresses   Fig 5 - Mises stresses   Fig 6 - Mises stresses
for Titanium IME with   Polyoxymethylene IME   Polyoxymethylene and
vertical loads;   with vertical loads;   titanium IMC with
                                                   vertical loads;

5. DISCUSSION AND CONCLUSION

Computer modelling offers many advantages over other methods in considering the complexities that characterise actual clinical situations. However, no model can be better than its assumptions and input data. The inherent limitations of the finite element method must therefore be acknowledged[2]. The method and model used here implied several assumptions regarding the simulated structures.

The oblique forces have higher stress concentration in the anatomic structures and the system of implant than the axial forces. The deflexion of superstructure and the stress concentration in the retaining screw were much higher with polioximethilene IME than of the others two models. This happens in reason of the mechanic materials properties (POM + titanium) submitted to pressure and fatigue, once the resistance of titanium is between 160 and 550 MPa[5-6] and the one of POM is nearly 70 MPa[5-6]. So, the third model had a smaller stress in the retaining screw and a better load distribution over the whole model group. Probably, this is the reason of a larger life time of IMC(POM + titanium) compared to that one of IME(POM).

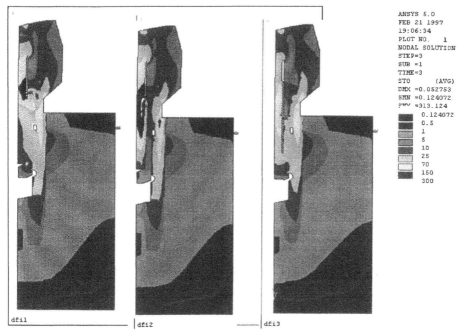

```
ANSYS 5.0
FEB 21 1997
19:06:34
PLOT NO.    1
NODAL SOLUTION
STEP=3
SUB =1
TIME=3
STO        (AVG)
DMX =0.052753
SMN =0.124072
SMX =313.124
     0.124072
     0.5
     1
     5
     10
     25
     70
     150
     300
```

dfil   dfi2   dfi3

Fig 7 - Mises stresses   Fig 8 - Mises stresses   Fig 9 - Mises stresses
for Titanium IME with   Polyoxymethylene IME   Polyoxymethylene and
inclined (30°) loads;   with inclined (30°) loads;   titanium IMC with
                                                          inclined (30°) loads;

In all model cases, there was not a significant difference in the stress transmission to the bone in any of the three intramobile components. All implant systems share a common target which is the stable anchorage to the bone. If the adding of resilient components to the implant design makes easier and keeps this anchorage, still there are controversies. The properties of the periodontal ligament to absorb and share stresses are extremely complex and hard to copy. Despite of many arguments in favour of the inter mobile resilient components look to be logical, there are few scientific evidences which can support it.

6. REFERENCES

1. Caputo, A. and Jon P. S., Biomechanics in clinical dentistry, Quintessence Publishing Com Inc. 1997, 55-84 and 205-219.
2. Kirsch, A. and Karl L. A., The IMZ osseointegrated implant system, Dent. Clinical N. America, 1989, Vol. 33, Num. 4, 733-791.
3. Papavasiliou, G., Phophi K., Stephen C. B. and David A. F., Three-dimensional finite element analysis of stress-distribution around single tooth implants as a function of bony support, prosthesis type and loading during function, J. Prosth. Dent., Vol. 76, Num. 6, 633-640.

4. McGlumphy, E. A., Campagni. W. V. and Petersen L.. J.., A comparison of the stress transfer characteristics of a dental implant with a rigid or a resilient internal element. J. Prosthet. Dent., 1989,Vol. 62, 586-593.

5. Holmes, D. C., William R. G., Vijay K. G. and John C. K., Comparison of stress transmission in the IMZ implant system with polyoxymethylene or titanium intramobile element: a finite element stress analysis, Int. J. Oral & Maxillofac. Implants, 1992, Vol.7, Num. 4, 450-458.

6. Holmes, D. C., Chris R. H.and Steven A. A., Deflection of superestructure and stress concentrations in the IMZ implants system, The Int. Journal of. Prosthodontics., 1994, Vol. 7, Num. 3, 239-246.

7. Chapman, R.J. and Axel K., Variations in internal implants shocks absorber, Int. J. Oral & Maxillofac. Implants, 1990, Vol. 5, 369-374

CUSPAL FLEXURE AND THE FAILURE OF CLASS V RESTORATIONS

J. S. Rees[1]

1. ABSTRACT

Class V restorations is a clinical occurrence which is often blamed on inadequate moisture control. However, the effects of occlusal forces and cuspal movement may also have an effect. The aim of this study was to examine the effects that cuspal movement had on the shear forces around a buccal Class V composite restoration in a lower second premolar. A lower first premolar restored with a buccal Class V composite restoration and a class I composite or amalgam restoration were modelled using a 2-D plane strain mesh. The width of the occlusal restoration was varied from 2.1-3.7 mm and the cavity depth was varied from 1.7-3.7 mm. Compared to the unrestored premolar, the presence of an occlusal cavity restored with composite increased the interfacial forces around the buccal Class V restoration by 1-67%. Similarly, the presence of an amalgam occlusal restoration increased the interfacial forces by 9-228%. It was found that the presence of an occlusal restoration increased cuspal movement which in turn increased the shear forces around the buccal Class V cavity. This effect was more pronounced with increases in cavity depth compared to cavity width, and when amalgam was the occlusal restorative material.

2. INTRODUCTION

The premature loss of restorations from Class V cavities is well documented with some clinical trials reporting failure rates of up to 58 % over a two year period[1]. Gabel[2] was the first to consider the possibility of occlusal forces as an aetiological factor in this loss. He suggested that excursive mandibular movements placed the buccal cusp in tension which in turn opened up the occlusal margins. More recent *in vitro* work has tended to confirm this suspicion[3,4].

The failure rates of Class V composite restorations have also been reported to be higher with high modulus macrofilled materials[5], higher in mandibular teeth[6,7] and higher in patients with large occlusal loads identified by the presence of wear facets[5].

1. Senior Lecturer in Restorative Dentistry, Dental School, UWCM, Cardiff, Wales, UK.
Keywords: Dental restorations, Finite element analysis, Class V restorations

A tooth flexure theory of retention has been suggested to explain these findings by Heymann et al[5]. They suggest that two mechanisms operate to cause failure; lateral excursive movements resulting in lateral cuspal movement generate tensile stresses along the tooth-restoration interface. In addition, heavy forces in centric occlusion cause vertical deformation of the tooth (the 'barrelling effect') leading to compressive and shear stresses at the tooth-restoration interface.

Cuspal flexure has also been shown *in vitro* to increase as the extent of a coronal cavity preparation increases[8,9]. If the tooth flexure hypothesis is correct, then the presence of an occlusal restoration may have a deleterious effect on the retention of a Class V restoration. To date the effect of an occlusal restoration does not seem to have been considered in any clinical trial involving the retention rates of Class V restorations.

The aim of this study was to use the finite element method to examine the effect that an existing occlusal amalgam or composite restoration might have on the stress distribution around the periphery of a buccal Class V cavity. The hypothesis used was that the presence of a Class I restoration would increase cuspal flexure under occlusal loading and thereby increase the stresses around the Class V restoration.

3. MATERIALS AND METHODS

A caries free lower first premolar had a buccal Class V cavity 1.6 mm deep x 1.5 mm long prepared partly in enamel and partly in dentine. The cervical cavity was restored with composite resin and the tooth embedded in a low exotherm epoxy resin. The premolar was sectioned bucco-lingually through the centre of the tooth and the outline was used to develop a two dimensional plane strain mesh. The mesh contained 5023 eight noded quadrilateral elements (Fig. 1). The enamel cap was modelled as an anisotropic material[10] with a principal elastic modulus $E_x = 80$ GPa and $E_y = E_z = 20$ GPa. The principal elastic modulus direction E_x was rotated through 180° in 10° increments to model the radial distribution of the enamel prisms that is thought to give rise to its anisotropic properties[11]. The physical properties used in this analysis are given in Table 1 and a load of 100 N was applied 0.4 mm inside the buccal cusp tip to simulate a lateral excursive load. The inferior border of the alveolar bone was rigidly fixed in the x and y directions.

Table 1: Physical properties used in the analysis

Material	Elastic modulus (MPa)	Poisson's ratio
Enamel	80 000*	0.30
Dentine	15 000	0.31
Compact bone	13 800	0.26
Periodontal ligament	50	0.49
Cervical composite	5 000	0.30
Occlusal composite	20 000	0.30
Occlusal amalgam	21 200	0.35

* Principal elastic modulus E_x

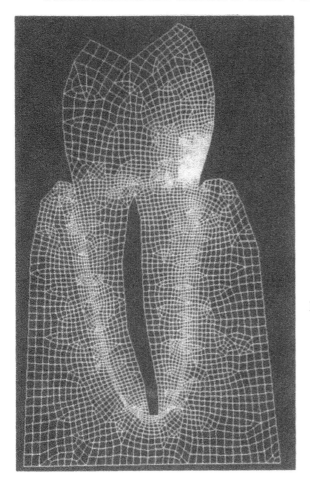

Fig. 1 The finite element
mesh

The initial load case run was with no occlusal restoration present and this formed the baseline measurement by which to judge all further analyses. The first variable investigated was the width of the occlusal cavity. The cavity depth was held constant at 2.6 mm and three bucco-lingual widths were used; 2.1 mm, 2.95 mm and 3.7 mm measured at the base of the occlusal cavity.

The second variable investigated was the cavity depth. The cavity width (measured at the base of the cavity) was held constant at 2.4 mm and three cavity depths were used; 1.74 mm, 2.6 mm and 3.7 mm.

The third variable investigated was the effect of amalgam or composite as the coronal restorative material. It is well known that composite bonds well to enamel and is capable of restoring some of the strength of the tooth lost during cavity preparation[12]. However, the bond of composite resin to dentine is more tenuous, while amalgam forms no bond to enamel or dentine and merely obturates the cavity space.

To model the discontinuity between amalgam and the tooth, a layer of gap elements 1 μm wide was introduced around the coronal tooth-restoration interface. For the occlusal composite restoration the 'worst case' was modelled and it was assumed that there was failure along the entire composite-dentine interface. Therefore, with each change in cavity depth or width, the mesh was run with either amalgam or composite as the occlusal restorative material. To overcome problems of interpreting interfacial nodal stresses at the Class V tooth-restoration interface, joint elements were placed at each node which was common to the tooth and the Class V restoration. Each joint element gave a value for the x and y forces in both tension and compression together with a shear value. However, to reproduce all these values would be impossible and as the shear forces were numerically greater it was decided to concentrate on the shear forces.

The easiest way to compare different load cases was to summate all the separate shear nodal forces along each interface and divide these by the length of the interface to express their output as a force per unit length for each interface. This was carried out for each of the four interfaces forming the margins of the class V cavity (Fig. 2).

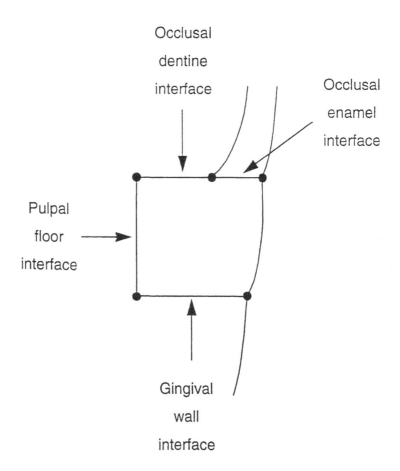

Fig. 2 Interfaces formed between the composite restoration and the Class V cavity

4. RESULTS

The results are shown graphically in Figs. 3-6. The joint element forces around the Class V cavity for variations in cavity width when the occlusal cavity was restored with composite are shown in Fig. 3. The figures given in parenthesis are the percentage increase or decrease relative to the premolar with no occlusal restoration. The joint element forces for variation in cavity width when the occlusal cavity was restored with amalgam are shown in Fig. 4.

Fig. 3

CHANGES IN INTERFACIAL SHEAR FORCES WITH INCREASES IN
OCCLUSAL COMPOSITE CAVITY WIDTH (Percentage change in parentheses)

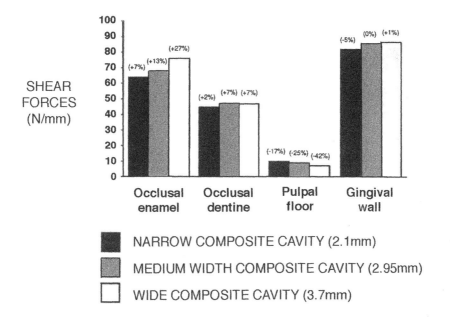

Fig. 4

CHANGES IN INTERFACIAL SHEAR FORCES WITH INCREASES IN
OCCLUSAL AMALGAM CAVITY WIDTH (Percentage change in parentheses)

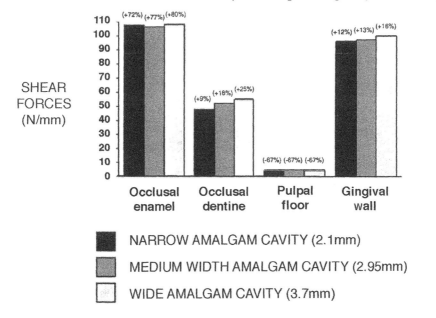

Fig. 5

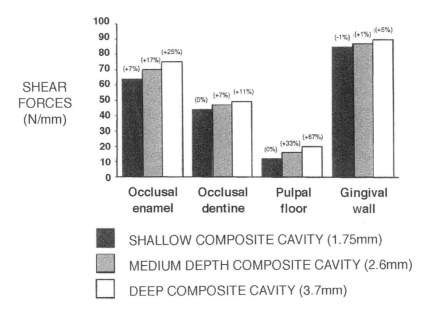

CHANGES IN INTERFACIAL SHEAR FORCES WITH INCREASES IN OCCLUSAL COMPOSITE CAVITY DEPTH (Percentage change in parentheses)

SHEAR FORCES (N/mm)

SHALLOW COMPOSITE CAVITY (1.75mm)

MEDIUM DEPTH COMPOSITE CAVITY (2.6mm)

DEEP COMPOSITE CAVITY (3.7mm)

Fig.6

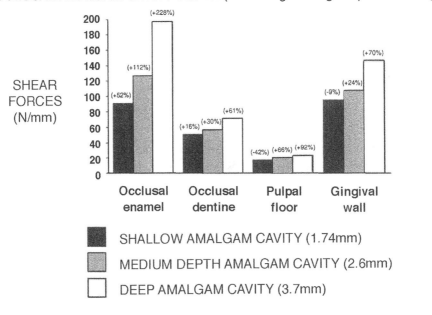

CHANGES IN INTERFACIAL SHEAR FORCES WITH INCREASES IN OCCLUSAL AMALGAM CAVITY DEPTH (Percentage change in parentheses)

SHEAR FORCES (N/mm)

SHALLOW AMALGAM CAVITY (1.74mm)

MEDIUM DEPTH AMALGAM CAVITY (2.6mm)

DEEP AMALGAM CAVITY (3.7mm)

The joint element forces around the Class V cavity for variations in cavity depth when the occlusal cavity was restored with composite are shown in Fig. 5 and the joint element forces for variations in cavity depth when the occlusal cavity was restored with amalgam are shown in Fig. 6.

5. DISCUSSION

For the variation in the width of the occlusal cavity the overall trend was an increase in interfacial forces around the Class V cavity. The only exception to this was the pulpal wall interface that showed a decrease although the reasons for this are unclear. The percentage increases in the interfacial forces were much larger for the amalgam occlusal cavity compared to the composite occlusal cavity. Presumably this was due to the composite restoration bonding the buccal and lingual cusps together, thereby reducing cusp movement compared to the occlusal amalgam cavity. For the variation in the depth of the occlusal cavity the overall trend was an increase in interfacial forces as the cavity depth increased. It is also interesting to note that for the shallow occlusal composite cavity, the occlusal dentine interface and the pulpal wall interface of the Class V cavity showed no increase in the interfacial forces. The magnitude of the increases in the interfacial forces around the Class V cavity were much larger for the occlusal cavity restored with amalgam compared to composite. The changes were also much larger compared to the changes produced by varying the occlusal cavity width.

These effects may be explained by a cuspal flexure theory, since the phenomena of cuspal flexure may be likened to the displacement of a cantilever beam[13]. The displacement of a cantilever beam is directly proportional to the cube of its length and inversely proportional to its width. Therefore, it is not too surprising that the forces around the cervical cavity, that is in the region of the fulcrum of the cantilever beam, increase more dramatically with increases in cavity depth than increases with cavity width, since cuspal flexure will be more pronounced with increases in L.

In conclusion, it appears that the presence of an occlusal cavity does influence the force distribution around a buccal Class V restoration. These forces are influenced to a smaller degree by the width of the occlusal cavity and to a greater degree by its depth. The type of restorative material also had an influence, with a composite coronal restoration reducing the force distribution around the cervical cavity for comparable coronal cavity dimensions.

In spite of these findings, the results must be interpreted with a certain amount of caution. This analysis was a two dimensional plane strain analysis but teeth are three dimensional objects. Therefore this study could only model a bucco-lingual slice of a Class I cavity and effects such as the presence of an approximal box could not be modelled. However, the presence of an approximal box has been demonstrated to increase cuspal flexure by approximately 60% compared to a Class I amalgam cavity[13]. Therefore, the presence of an MOD cavity in a tooth may well lead to even greater force concentrations around a buccal Class V restoration.

Clinically, the findings of this study suggest that the morphology of an occlusal cavity and the material chosen to restore the cavity may influence the retention of a cervical restoration. It also suggests that reducing cuspal flexure by using a composite restoration or a gold onlay may reduce the force distribution around a cervical restoration which may then result in reduced failure rates for the cervical restoration. The presence of an occlusal restoration does not seem to have been considered as being important in clinical trials of Class V restorations, but the influence of cuspal flexure could explain, at least in part, the widely different failure rates reported by various workers.

6. REFERENCES

1. Hansen E.K. Five year study of cervical erosions restored with resin and dentine bonding agent. *Scand. J. Dent. Res.*1992; **100**:244-247.
2. Gabel A.B. American Textbook of Operative Dentistry. 4th Edition. McGraw-Hill, London, 1956.
3. Hood J.A.A. Experimental studies on tooth deformation: Stress distribution in class V cavities. *N.Z. Dent. J.* 1972; **68**:116-131.
4. Jørgensen K.D., Matono R. and Shimokobe H. Deformation of cavities and resin fillings in loaded teeth. *Scand. J. Dent. Res.*1976; **84**:46-50.
5. Heymann H.O., Studevant J.R., Bayne S. et al Examining tooth flexure effects on cervical restorations: A two year clinical study. *J. Am. Dent. Assoc.*1991; **122**:41-47.
6. Van Meerbeek B., Inokoshi S., Davidson C.L. et al Clinical status of ten dentine adhesive systems. *J. Dent. Res.* 1994; **73**:1690-1702.
7. Ziemicki T.L., Dennison J.B. and Charbeneau G.T. Clinical evaluation of cervical composite resin restorations placed without retention. *Oper. Dent.*1987; **12**:27-33.
8. Grimaldi J.R. Measurement of the lateral deformation of the tooth crown under axial compressive loading. M.D.S. Thesis, University of Otago, 1971.
9. Hood J.A.A. Discussion paper: Methods to improve fracture resistance of teeth *In*: Posterior composite resin dental restorative materials. Vanherle G. and Smith D.C. (Eds.). Peter Szulc Publishing Co., The Netherlands, 1985; pp.443-450.
10. Rees J.S. and Jacobsen P.H. Modelling the effects of enamel anisotropy with the finite element method. *J. Oral Rehabil.*,1995; **22**:451-454.
11. Spears I.R., van Noort R., Crompton R.H. et al The effects of enamel anisotropy on the distribution of stress in a tooth. *J. Dent. Res.* 1993; **72**:1526-1531.
12. Burke F.J.T. Tooth fracture *in vivo* and *in vitro*. *J. Dent.* 1992;**20**:131-139.
13. Douglas W.H. Methods to improve fracture resistance of teeth *In*: Posterior composite resin dental restorative materials. Vanherle G. and Smith D.C. (Eds.). Peter Szulc Publishing Co., The Netherlands, 1985; pp.433-442.

THE INFLUENCE OF BOUNDARY CONSTRAINTS ON NUMERICAL SIMULATIONS OF A MANDIBULAR CANINE TOOTH.

C. J. Richardson[1], R. van Noort[1], I. C. Howard[2]

1. ABSTRACT.

This numerical study investigates the effect on four Finite Element models of a lower canine tooth of different boundary constraints, similar to those used previously by other researchers. The results demonstrate the necessity of incorporating the Periodontal Ligament (PDL) into numerical simulations to produce a valid stress distribution, and that the supporting bone should also be included to give realistic displacement data. Hence, it is concluded that boundary conditions must be appropriate for the particular situation under consideration.

2. INTRODUCTION.

As finite element analysis (FEA) has become more user-friendly, it has found, and will continue to find, an increased use in dentistry. It is an ideal tool for improving our understanding of the often subtle interactions between material properties and the design of natural and restored teeth under load (Spears *et al.*, 1993). This is particularly important, since teeth are complex structures with small dimensions, and hence are not amenable to experimental investigation to determine the surface and internal stresses and strains arising from the multitude of possible loading conditions.

The finite element method is increasingly being used in dentistry to assess the effects of occlusal loading on the movement of teeth and the resulting stress distribution. Previous researchers have analysed a variety of finite element models of the whole or parts of the teeth, using widely different boundary conditions. For example, the crown alone of a molar tooth was considered in an analysis by Rees and Jacobsen (1992) and van Noort *et al.* (1988), while Yettram *et al.* (1976) analysed a normal and a restored mandibular premolar which were cut off approximately 3mm below the crown.

Keywords: Finite Element, Teeth, Dentistry, Boundary Constraints

---

[1] Department of Restorative Dentistry, University of Sheffield, Claremont Crescent, Sheffield, S10 2TA.

[2] Department of Mechanical Engineering, University of Sheffield, Mappin Street, Sheffield, S1 3JD.

A 3-D model of an entire canine, including the periodontal ligament(PDL) and jaw bone, has been modelled to study the PDL under orthodontic loading (McGuinness *et al.*, 1994; Wilson *et al.*, 1994). In another case, a model of the tooth set in the jaw bone has been analysed, but the tooth was regarded as composed of a single, isotropic substance (Cobo *et al.*, 1993).

This numerical study examines the effects of different boundary conditions on the stress distribution within, and the movement of, a mandibular canine tooth. The influences of the PDL (often excluded from FE models because of its small size), the supporting bone, and anisotropic as opposed to isotropic models of the enamel on the mechanical behaviour of the natural tooth are explored.

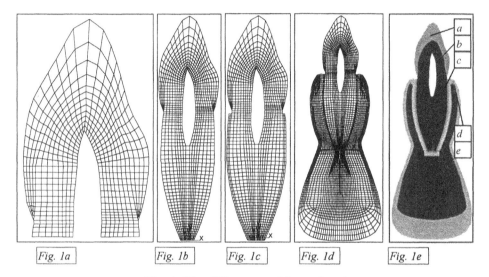

| Fig. 1a | Fig. 1b | Fig. 1c | Fig. 1d | Fig. 1e |

Fig. 1: The different models analysed.

a = Enamel
b = Dentine
c = PDL
d = Cortical Bone
e = Cancellous Bone

3. MATERIALS AND METHODS.

3.1. Method.

For the purposes of this study, the models of other researchers were classified into four categories. These ranged from the crown alone (model A) through to a model encompassing the entire tooth (B), PDL (C) and supporting bone (D). (*See Fig. 1.*)

Model A examines the case where the tooth is cut off and constrained just below the crown (Fig. 1a). The tooth alone is considered in model B, constrained along the root (Fig. 1b). Model C comprises the tooth and the PDL, with constraints applied to the outer surface of the PDL. (Fig. 1c.) The entire tooth, set in the jaw bone, and incorporating the PDL is examined in Model D. Constraints have been applied to the base of the jaw bone. (Fig. 1d.) Fig. 1e indicates the different materials used in the analyses.

Four-noded, isoparametric, plane strain elements were used to construct the models. Models A, B, and C are reduced versions of model D, made by unselecting sets of nodes and elements to ensure the consistency of the mesh. All four models are constrained against translation in the global X and Y-directions. All analyses were performed using the commercial *ANSYS5.0a* software.

Both isotropic and anisotropic material properties were ascribed to the enamel (Spears *et al.*, 1993). The analyses of the four models were undertaken with each subject to an occlusal load of 100N, as shown in Fig. 1.

3.2 Materials.

The elastic moduli (E_x and E_y) of the materials used in this study are shown in Table 1, with appropriate references cited there. For anisotropic enamel, E_x denotes the Young's Modulus (or elastic modulus) of the material in the local *x*-direction, which is parallel to the EDJ (Enamel-Dentine Junction) in this study. Similarly, E_y denotes the Young's modulus of the enamel in the local *y*-direction, which is perpendicular to the EDJ. All of the other materials are assumed to be isotropic; i.e. to have the same material properties in all directions.

Finding an appropriate elastic modulus for the PDL presented a particularly difficult problem, since it is essentially a soft tissue with non-linear elastic properties. An appropriate value of Young's Modulus for the PDL was determined by a quasi-static analysis using the commercial software *LS-DYNA (DYNA)*. The non-linear, stress-strain behaviour of natural tissue can be represented by an exponential strain energy density function, as described by Huang *et al.* (1990). The resulting curve has been defined as a material type in the *DYNA* package (Chew *et al.*, 1994), and is shown in Fig. 2.

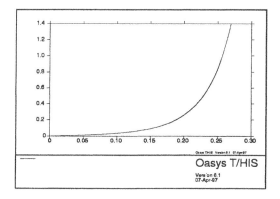

Fig. 2: The Stress-Strain Curve Describing the Behaviour of the PDL.

Furthermore, an interface, called *DYNASYS* (Thornton, Chew and Yoxall, 1994) between the two software packages used, *ANSYS5.0a* and *DYNA*, has been created by the Heart Valve Research Group at the University of Sheffield. Using *DYNASYS*, it was possible to transfer Model C from *ANSYS* to *DYNA*, using the material properties given in Table 1, except that the Young's Modulus of the PDL was described by the strain energy function in DYNA. A vertical load of 100N was then applied to the cusp tip.

When the analysis was complete, the values of stress found at various points on the PDL were noted and a curve fitting program (Chew *et al.*, 1994) was used to determine the Young's Modulus for the stresses predicted by *DYNA*. As a result of this analysis an average value of 30MPa was selected and used in the subsequent analyses in *ANSYS*. Further work on the modelling of the PDL under different loading conditions is in progress.

Table 1: Elastic moduli of materials used in the FEA models.

MATERIAL	E (GPa)	v_{xy}	DENSITY kg/m^3
Isotropic Enamel	80[c]	0.3[c]	2.8[c]
Anisotropic Enamel	$E_x = 10$[b] $E_y = 80$[b]	0.3[b]	2.8[c]
Dentine	15[c]	0.3[c]	1.96[c]
PDL	0.030	0.3	1000
Cancellous Bone	0.689[a]	0.3[a]	/
Cortical Bone	13.7[a]	0.3[a]	/

[a]Cook *et al.*,1982: [b]Spirings *et al.*, 1984: [c]Spears *et al.*, 1993.

4. RESULTS.

Fig. 3 shows an example of the distribution of stresses at each node along the EDJ (Enamel - Dentine Junction) for all four models subjected to an occlusal load of 100N and analysed with anisotropic enamel. The EDJ was chosen as a convenient, arbitrary line along which to compare the stresses, with the first point on the graph being the stress at the first node along the EDJ, working from buccal to lingual. The stresses in the global vertical direction (the Y-stresses) only are presented because these are largest in magnitude; other stresses were analysed, but are not presented here.

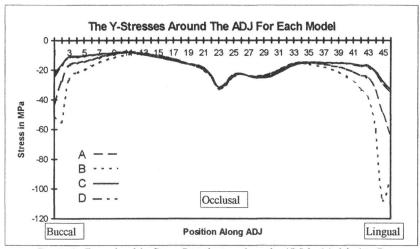

Fig. 3: An Example of the Stress Distribution along the ADJ for Models A to D.

Fig. 3 demonstrates that models C and D have very similar stress distributions, and that these are distinctly different from those of models A and B. The regions of greatest difference are the cervical margins; a significant finding for the field of restoration design, as this is where Class V lesions develop.

Fig. 4 shows the deformations of the four models; Fig. 4a showing the resultant displacements for model A, Fig. 4b representing that of model B, *etc*. The original shape is indicated by a dotted outline, and the resultant displacement is given (in mm)on the right hand side of the figure.

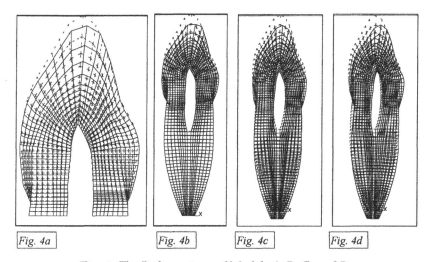

Fig. 4: The Deformations of Models A, B, C, and D.

It can be seen from Fig. 4 that for models A and B, only the crown deforms, whereas in model C, the whole tooth displaces almost vertically, and in model D, both vertical displacement and tipping of the tooth occur. Movement of the type predicted by model D was described by Parfitt (1965) and has been recorded more recently by other researchers. The constraints applied to models A and B prevent them from deforming in this manner, and, since the PDL is assumed to be only 0.33mm thick, model C cannot displace sufficiently for this type of deformation to be observed.

Fig. 5 shows a comparison of the Y-stresses along the EDJ for model D analysed with isotropic and with anisotropic enamel. It is evident from Fig. 5 that much greater compressive stresses are predicted in the enamel when it is regarded as being isotropic rather than anisotropy. This is again particularly significant in the cervical regions of the crown. Furthermore, if the region directly under the point of application of the load is considered, the load channelling effect of the anisotropy can be clearly observed.

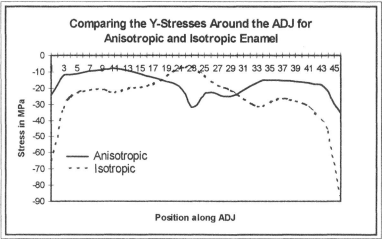

Fig. 5: A Comparison of the Y-Stresses Along the ADJ for Model D Analysed with Anisotropic and Isotropic Enamel.

5. DISCUSSION.

It is standard engineering practice in FEA to ensure that the constraints are placed far enough from the area of interest not to influence the results. This study shows that the way in which constraints are applied affects the stiffness of the tooth in models A, B, and C.

The stresses, their distribution and the displacements of models A and B differ significantly from those of model D. It is also interesting to note that the largest stresses occur in models A and B. The stresses in model C do not differ significantly from those obtained from analysis of the tooth in model D. However, the displacements do differ appreciably between these two models.

Hence, if only the stresses in the tooth are of interest, then a simplified model such as model C would suffice. Conversely, if information about the displacements was required, then it would be necessary to incorporate the mandible into the model too.

A fifth model was analysed to explore whether or not the constraints imposed upon model D influenced the stresses in the tooth. This comprised model D with the boundary constraints attached by beams. There were no changes in the stress or displacement plots for the tooth resulting from these two analyses. Thus it can be inferred that the constraints applied to model D do not affect the behaviour of the tooth.

Limitations of the models used in this study include the facts that they are only 2-D, and that the PDL is modelled as an isotropic, linear elastic ligament of uniform thickness (0.33mm). Assumptions concerning the PDL could introduce errors in the estimation of the magnitudes of the stresses and displacements.

Concerns regarding the application of boundary constraints to FE studies of teeth are similarly applicable to 2-D and 3-D simulations. Since the root of a canine tooth can be regarded as being roughly conical in shape, it will remain approximately symmetrical when sectioned, regardless of the plane of section. Therefore, the stress distribution in the root will also remain roughly the same in 2-D and in 3-D simulations, and hence, the effect of the constraints applied to the root in 2-D can be extrapolated to apply similarly to 3-D models.

The most important finding of this work is that the boundary constraints imposed upon such models can significantly affect the results. If the softer, supporting structures of the PDL and the jaw bone are neglected, then the root experiences very little stress. This would suggest that modelling much of the root was superfluous, and hence that models such as model A are adequate. However, analyses of models C and D show that, as a proportion, considerably more stress is directed into the root when the supporting tissues are incorporated. This implies that the method of constraint of the tooth cannot be ignored in numerical simulations.

The application of a boundary constraint to a node in a numerical simulation introduces a point of infinite stiffness at that node. Hence, the root of the tooth will be much stiffer when constraints are applied directly to it than if they are applied to the supporting tissue. This could therefore lead to erroneous interpretation of the results of such an analysis.

The comparison between anisotropic and isotropic enamel shows clearly that it is necessary to incorporate anisotropy in numerical simulations in order to predict realistic load transfer patterns. Incorporation of the mandible further alters the displacements of the tooth and the PDL, but has no significant effect upon the stress distribution.

6. CONCLUSIONS

It is essential to model the root and the supporting structure of the tooth when investigating the behaviour of either a restored or an unrestored tooth.

To analyse the stresses in the crown of the tooth, it is sufficient to constrain the PDL only, but the mandible must be incorporated to produce displacements of the right order of magnitude.

An anisotropic model of the enamel should be used to produce a realistic load transfer mechanism in such analyses.

ACKNOWLEDGEMENT.
We are indebted to Ove Arup and Partners, and Dr. J. Miles for access to *LS-DYNA*.

7. REFERENCES.

Chew, G. G., Howard, I. C., Patterson, E. A., *Non-Linear Modelling of Porcine Bioprosthetic Valves,* Engineering Failure Analysis, 1994; **1**: 231 - 242.

Cobo, J., Sicilia, A., Argüelles, J., Suavez, D., Vijande, M. *Initial Stresses Induced in Periodontal Tissue with Diverse Degrees of Bone Loss by Orthodontic Force; A 3-D Analysis by FEA,* Am. J. Ortho. Dentofacial Orthop.1993; **104**: 448 - 454.

Cook, S. D., Weinstein, A. M., Klawitter, J. J. *A 3-D Finite Element Analysis of a Porous-Rooted Co-Cr-Mo Alloy Implant,* J. Dent. Res. 1982; **61**:25 - 29.

McGuinness, N. J. P., Wilson, A. N., Jones, M. L., Middleton, J. *A Stress Analysis of the Periodontal Ligament Under Various Orthodontic Loads,* Euro. J. Orthodont. 1994; **13**: 231 - 242.

Rees, J. S., Jacobsen, P. H. *Modelling the Effects of Enamel Anisotropy With FEM,* J. Oral Rehab.1995; **22**: 451-454.

Spears, I. R., Van Noort, R., Crompton, R. H., Cardew, G. E., Howard, I. C. *The Effects of Enamel Anisotropy on the Distribution of Stress in a Tooth,* J. Dent. Res.1993; **72**: 1526 - 31.

Spirings, A. M., de Vree, J. H. P., Peters, M. C. R. B., Plasschaert, A. J. M. *The Influence of Restorative Dental Materials on Heat Transmission in Human Teeth,* J. Dent. Res.1984; **63**: 1096 - 1100.

Thornton, M.A., Chew, G.G., Yoxall, A. *DYNASYS. An Interface Between ANSYS and OASYS LS-DYNA3D,* Proprietary Software, Department of Mechanical Engineering, The University of Sheffield, 1994.

Van Noort, R., Cardew, G., Howard, I. C. *A Study of the Interfacial Shear and Tensile Stresses in a Restored Molar Tooth,* J. Dent.1988; **16**: 286 - 293.

Wilson, A. N., Middleton, J., Jones, M. L., McGuinness, N. J. P. *The Finite Element Analysis of the Stress in the Periodontal Ligament When Subject to Vertical Orthodontic Forces,* Brit. J. Orthodont. 1994; **211**: 161 - 167.

Yettram, A. L., Wright, K. W. J., Pickard, H. M. *Finite Element Stress Analysis of the Crowns of Normal & Restored Teeth,* J.Dent. Res. 1976; **55**: 1004 - 11.

8. CRANIO-FACIAL MECHANICS AND DIAGNOSTIC METHODS

3-DIMENSIONAL MEASUREMENT OF HEAD INJURIES
FOR THEORETICAL MODEL VALIDATION

J. Subke[1] , H.-D. Wehner[2] and C. Götz[3]

1. ABSTRACT

Requirements for the reconstruction of traumatic effects are:
1. The generation of an individual geometrically correct and mechanically adequate body model by which the effect (e.g. the injury pattern) of the special injury dynamics, which has to be examined, can be validated by simulation.
2. The integrative and time to scale representation of traumatic lesions documented by different data systems.

For a head injury the first requirement was realized using a Finite-element-model (FEM) to reconstruct an individual skull-model from computer tomographic data. To each FEM-cell the realistic material property was added to obtain an object which could be exposed by simulation to the dynamics of the destroying forces. The second requirement is fulfilled by getting a precise superposition of computer tomographic and photogrammetric data by the kind of matching presented in the paper.

2. INTRODUCTION

For the estimation of prognosis and (perhaps) for an adequous therapeutic procedure of severe head trauma it is necessary to get more information about the causal connections between impacts acting on the head and the damages caused by these. This can be done by impact simulations acting on a general finite element model of the skull [2, 5, 6].

Keywords: Matching, FEM-skull-model, Skull fracture pattern, Computer tomography, Photogrammetry

[1]) Reader, Institut für Gerichtliche Medizin, Universität Tübingen, Nägelestr. 5, 72074 Tübingen
[2]) Professor, Institut für Gerichtliche Medizin, Universität Tübingen, Nägelestr. 5, 72074 Tübingen
[3]) Co-worker, Institut für Astrophysik, Universität Tübingen, Auf der Morgenstelle 10 c, 72076 Tübingen.

In order to examine the validity of such calculations and to improve their reliability it is of advantage

1. to develop an individual finite element model skull with material properties as close as possible to their real properties
2. to find a possibility which allows the comparison of simulated damages with real damages (e.g. fractures and contusions).

The presented paper shows such a development of an individual finite element model of the skull and the possibility to match real injury patterns to such a model.

3. METHOD

In order to install an appropriated and skull related co-ördinate system (called marker system) at first titan screws were brought into the bone structures of the skull (fig. 1) which had the following positions: M1 region of the right temple M2 centre of the forehead, M3 region of the left temple (hidden point in fig. 1), M4 occipital region, M5 top of the head. Subsequent to this about 200 computertomographical layers (fig. 2) were made containing not only the bone and brain structures but also the marker system (M1, M2, M3, M4, M5). After selecting the bone structures by a density filter it was possible to re-compose the layers again giving a precise representation of the skull geometry including the marker system.

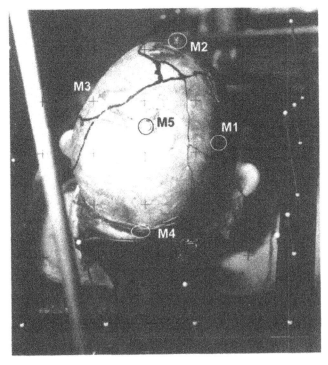

Fig. 1 The real injury pattern including the marker system of five screws (M1, M2, M3, M4, M5)

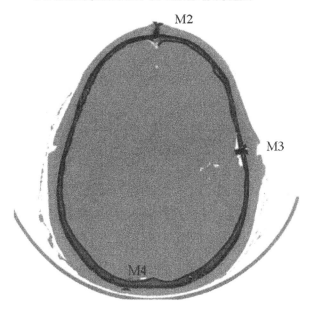

Fig. 2 A computer tomographical layer containing three of the five marker screws

Using this geometry by a Medical-Imaging-System [3] a finite element model (fig. 3) was generated consisting of 732 triangular elements. This model was expanded into a biomechanical one by adding the corresponding mechanical properties to each triangle. This model allows the realisation of strain and stress simulation calculations leading for instance to a virtual fracture pattern shown in fig. 3.

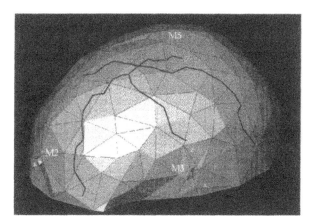

Fig. 3 Generation of a finite element model of the skull with a virtual injury pattern (fracture system)

The injury patterns and their space co-ördinates are measured by photogrammetry [1, 4]. For this purpose the subject [here the injury pattern (e.g. a fracture system) and the marker system (M1 - M5)] is photographed by a calibrated camera in relation to a reference system consisting of an u-shaped frame wearing points of fixed distances each to another (fig. 4). The photographs must be taken from two different positions at least.

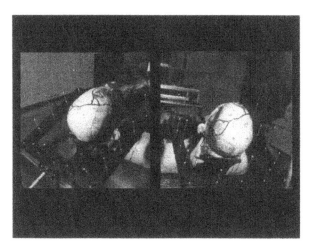

Fig. 4 Photogram measurement of the injury pattern from at least two different camera positions

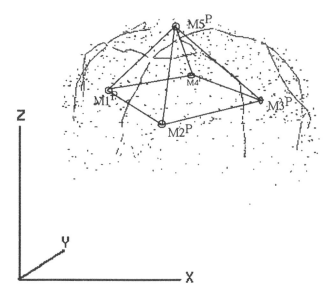

Fig. 5 The representation of the isolated fracture pattern in relation to the marker system

After this procedure the co-ördinates of each point of the injury pattern and of the marker system in relation to the reference frame are known and stored. They are available for visualisation from different points of view (fig. 5).

After the generation of the individual finite element model of the skull and the preparation of the photogrammetrically measured data the last task exists in matching the measured injury dates to the surface of the finite element model of the skull. For this purpose it has to be taken into consideration that there is a difference between the origin of the co-ördinate system representing the finite element model and representing the structures obtained by photogram. The matching task must be solved therefore by a transformation bringing each point of the marker system (M1 - M5) of the two representations into one another (fig. 6).

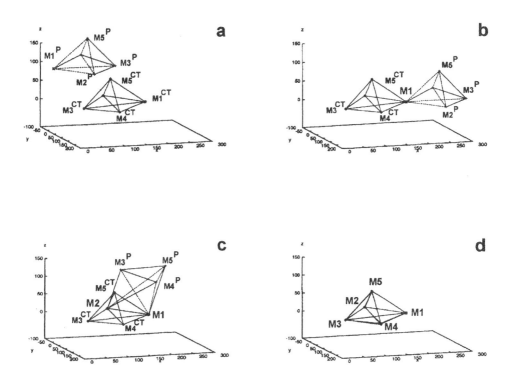

Fig. 6 The transformation process necessary for the matching operation

With the co-ördinates $M1^{CT}$, $M2^{CT}$, $M3^{CT}$ ... $M5^{CT}$ of the marker systems of the skull model and with the co-ördinates $M1^{P}$, $M2^{P}$ ... of the photogram the first translation has to bring in congruence the points $M1^{P}$ and $M1^{CT}$ (fig. 6a $M1^{P} \longrightarrow M1^{CT} = M1$). Subsequent to this procedure a rotation around M1 brings $M2^{P}$ to $M2^{CT}$ (fig. 6b; $M2^{P} \longrightarrow M2^{CT} = M2$). Finally a last rotation around the $\overline{M1M2}$-axis brings markers and therefore all points of the different representation into congruence.

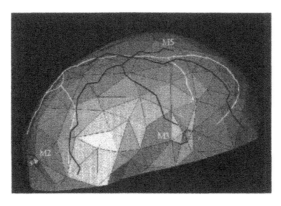

Fig. 7 The superposition of the real injury pattern and the individual finite element model of the skull (black line: a virtual fracture pattern; white line: real fracture pattern)

4. RESULT

The result is shown in fig. 7 indicating that the matching process leads to a precise superposition and therefore to an integrated representation of the individual finite element skull model and the injury patterns, which is the basic requirement for examination of quality of simulation calculations. Using the finite element model of the skull described above only such impact calculations are of validity predicting injury patterns coming most close to the real ones.

5. REFERENCES

1. Brüschweiler, W., Braun, M., Fuchser, H.J., Dirnhofer, R. (1997) Photogrammetrische Auswertung von Haut- und Weichteilwunden sowie Knochenverletzungen zur Bestimmung des Tatwerkzeuges - grundlegende Aspekte. Z Rechtsmedizin 7: 76 - 83
2. Chu, C.-S., Lin, M.-S., Huang, H.-M., Lee, M.-C. (1994) Finite element analysis of cerebral contusion. J. Biomechanics 27 Nr. 2: 187 - 194
3. Grunert, Th, Ehricke, H.-H., Straßer, W (1994) Ein Medical Imaging Entwicklungssystem. In Arnold (B. u. Müller, H. (Hrsg.) Digitale Bildverarbeitung in der Medizin. Gesellschaft für Informatik
4. Kraus, K. (1994) Photogrammetrie-Grundlagen und Standardverfahren. Dümmler-Verlag, Bonn
5. Ruan, J.S., Khalil, T., King, A.I. (1991) Dynamic response of human head to impact by three. dimensional finite-elemente-analysis. J. Biomech.Eng. 113: 276 - 283
6. Willinger, R., Ryan, G.A., Mc Lean A.J., Kopp, C.M. (1994) Mechanismus of Brain Injury Related to Mathematical Modelling And Epidemiological Data Accid. Anal. and Prev. 26 Nr. 6: 767 - 779

MODELLING OF THE HUMAN SKULL INCLUDING LOADING, DETERMINATION OF THE LIMIT LOAD AND FRACTURE PREDICTION.

Maria del Pilar A M Rodriguez C Gomes.[1]

1.ABSTRACT

The purpose of this work is the determination of the limit load of the human skull skeleton as a structure, using a computer program based on finite elements method with degenerate quadratic elements for shells. This program also provides the possibility of determining of the damaged elements and making a prediction of the bone fracture of the skull skeleton as a shell structure, since the basic feature of this program is the analysis of any shell, regardless of its shape and its material. The structure was analysed with several loading cases. With the results of the analysis it was also possible to see what is the best kind of application of loads in the skull in case, for example when the use forceps is needed in the birth of babies or the best kind of removing the head loads in case of accidents. The main results of these analyses are included in this work.

2.INTRODUCTION

The computer program was developed in the Civil Engineering Department of COPPE/UFRJ and it was included in the D. Sc. author's thesis. The main objective of the program is the shells limit load determination. This program also provides the possibility of determining of the damaged elements and making a prediction of the structure fracture in the analysis of any shell, with any shape or material. It is also possible the use of anisotropic materials. The analysis may be physically or geometrically nonlinear. The elements options are Serendipity, Heterosis or Lagrangian type. The integration methods may be reduced or selective.

3.KEYWORDS: Skull modelling, Skull fracture, Bone limit load, Skull limit load.

----

[1]Professor, Civil Engineering Department, FAU/COPPE, Federal University of Rio de Janeiro, RJ, Brazil

4. METHODS

The bone structure of the present work was prepared using the shape obtained from an human skull X-ray photograph. This structure was analysed by the computer program described in ref.(1) with several types of loading cases. Only the most important ones were included in this work. The properties of the material were partially obtained in ref.(2). The S.I. units are used in this work.

5. THE HUMAN SKULL BONE STRUCTURE

The structure was prepared using the joint co-ordinates obtained from the X-ray photograph. The thickness of the skull is variable and it is possible to see it in the same X-ray photograph.

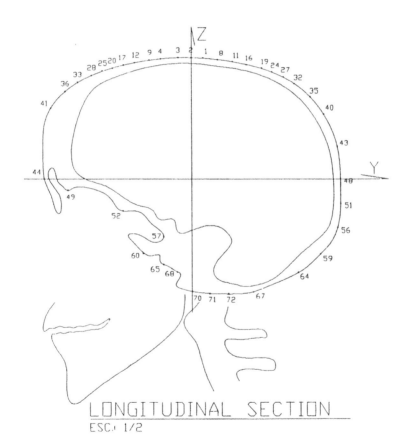

6.FINITE ELEMENTS MODEL

The finite element model of the present work was prepared using the joint co-ordinates mentioned. The elements were divided in layers according to their thickness. The thickness of the skull is variable. The model contains seventeen elements of Serendipity type, the integration is reduced, and the analysis is physically non linear.

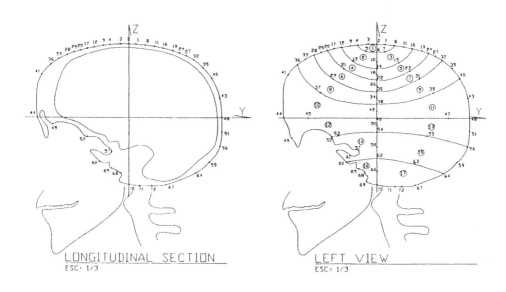

LONGITUDINAL SECTION
ESC: 1/3

LEFT VIEW
ESC: 1/3

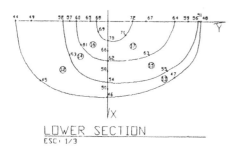

LOWER SECTION
ESC: 1/3

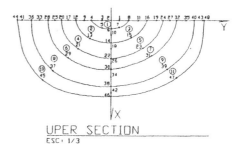

UPER SECTION
ESC: 1/3

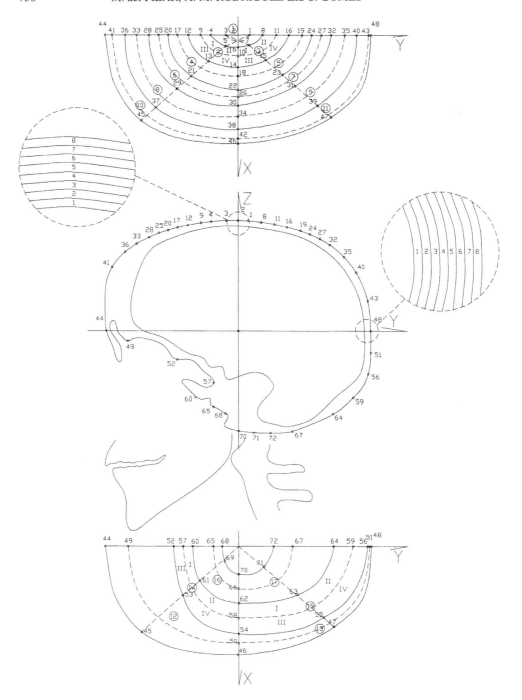

7. LOADING CASES

The structure was analysed with several loading cases and the following were select to this work.

Loading case (1)

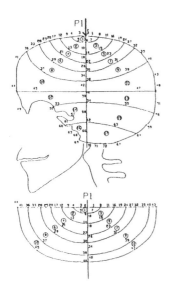

Loading case (2)

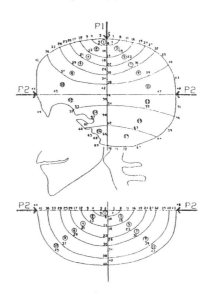

Loading case (3)

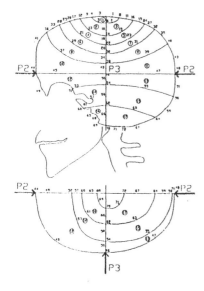

Loading case (4)

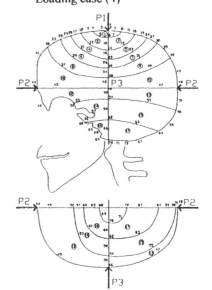

8. LIMIT LOAD

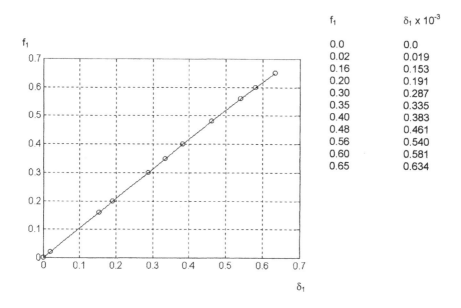

f_1	$\delta_1 \times 10^{-3}$
0.0	0.0
0.02	0.019
0.16	0.153
0.20	0.191
0.30	0.287
0.35	0.335
0.40	0.383
0.48	0.461
0.56	0.540
0.60	0.581
0.65	0.634

δ_1 = vertical deflection

Increasing the load for a factor that varies from 0.01 to 0.65 the plastification starts with the factor 0.02 in the Gauss point I , of the layer # 8 of the # 1 element. When the factor reaches 0.48 it occurs plastification in the layers # 8, # 7,# 6, # 2 and # 1. The table indicates the increasing factor and the correspondent plastified layers of the elements # 1.

With the factor 0.65 the plastification extends to elements # 2 and # 3 and later to elements # 14. These are the areas where with the load P1 occurs the damaged elements and where it is also possible to make a prediction of the skull fracture.

load	Plastification with loading case (1)			
factor	element	1	2 - 3	14
0.02		8		
	plastified			
0.48		8 - 7 - 6 2 - 1		
	layers			
0.56		8 - 7 - 6 - 5 - 4 - 3 - 2 - 1		
0.65		8 - 7 - 6 - 5 - 4 - 3 - 2 - 1	8	8 - 7 - 6 - 5 - 4 - 3 - 2 - 1

9.PLASTIFICATION OF THE ELEMENTS WITH THE LOADING CASE (1)

Regarding this analysis it was possible to determinate not only the limit load but also to predict the bone fracture in the skull damaged area.

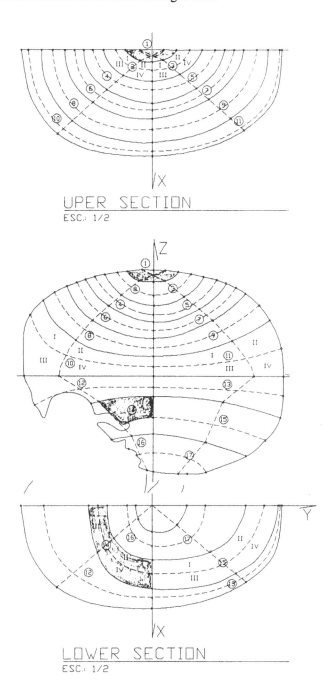

UPER SECTION
ESC.: 1/2

LOWER SECTION
ESC.: 1/2

10.CONCLUSION

According to this method it is possible to determinate with great precision the limit load of the human skull. It is possible to make the same analysis using a specific biomaterial based on their mechanical characteristics, it is   also possible the selection   of appropriate biomaterial   to replace any part of the human skull and make computer tests with this material.

The results with loading case (1) and the results with loading case (2), (3) and (4) were compared according to tables about limit loads. The loads P2 and P3 used neutralize the stresses from the   load P1 as due to the use of the loads P2 and P3 at the same time no element plastifies. This technique therefore provides a way to confirm that the design of the usual forceps is not right. It also shows the importance of the employment of the loads P2 and P3 at the same time in the skull.

11.ACKNOWLEDGMENTS

The author was incentivated by prof. Fernando Luiz Lobo Barboza Carneiro from Civil Engineering Department,   and prof. Antonio Fernando Catelli Infantosi   from Biomechanics Engineering Department of COPPE/UFRJ.

12.REFERENCES

1. Gomes, M.P.A M.R.C., Elasto-plastic and Geometrically Non-linear Analysis of Shells
   ' by Finite Element Method, COPPE/UFRJ, 1993, pp 131
2. Wise, D.L., Trantolo, J.T., Altobelli,D.E., Yaszemski, M.J., Gresser,J.D., Schwartz,E.R., Encyclopedic Handboock of Biomaterials and Bioengineering, Marcel Dekker,Inc., 1995, U.S.A .
3. Habal,M.B., Implantable Biomaterals in Plastic, Reconstrutive, and Esthetic Surgery, Encyclopedic Handboock of Biomaterials and Bioengineering, Marcel Dekker,Inc., 1995, U.S.A ., Vol.1, 127-131.

THE ACTION OF GRAVITY ON FRONTAL FACIAL AGING

D. Pamplona[1], H. I. Weber[2], M. Giuntini[3]

1. ABSTRACT

From the photos of the American astronaut Shannon Lucid, it could be observed that she looked much younger without the gravitational field acting on the muscles of her face. Facial aging is a biological phenomenon but the facial movement of the skin during aging is also due to mechanics. The present research is a first step in identifying the mechanical forces responsible for facial deformation owing to time. With the use of a Nonlinear Finite Elements formulation in a facial mask with the mechanical properties of the skin, it was possible to discuss the effects of gravity in the mechanics of facial aging. A strong correlation between the effects of gravity and the behavior of the aging parameters of our recent work was found.

2. INTRODUCTION

The human face changes with age and some factors that are decisive to this process have been identified in the literature (Pitanguy, I. *et al,* 1977). One main factor is atrophy in osteocartilages, and another is that the skin loses elasticity and thins, causing the wrinkling process and decay of the skin. The decay of the palpebral pouches, the lateral pouches of the face and the eyelids, the narrowing of the lips, the formation of the Nasogenian fold and the enlargement of the ear are some factors that appear on the aging face. The research to obtain an adequate model to study the process of aging is fundamental not only for the improvement of the techniques used in face lifting, but also possibly for finding a way to control the aging process mechanically. The literature on

Aging, Biomembrane, Finite-element.

[1] Associate Professor, Civil Engineering Department, Pontifícia Universidade Católica do Rio de Janeiro, Marquês de São Vicente, 225, Rio de Janeiro, 22453-090 - Brazil
[2] Full Professor - Visiting, Mechanical Engineering Department, Pontifícia Universidade Católica do Rio de Janeiro, Marquês de São Vicente, 225, Rio de Janeiro, 22453-090 - Brazil
[3] Graduate student, Civil Engineering Department, Pontifícia Universidade Católica do Rio de Janeiro, Marquês de São Vicente, 225, Rio de Janeiro, 22453-090 - Brazil

aging lacks an identification of the mechanical forces related to the aging process.

The present work is concerned with a qualitative comparison of the effects of gravity acting on the face and the pattern of change of the aging parameters that were defined, measured and discussed in our previous work (Pamplona *et al*, 1996), whose partial results can be seen in Fig.1.

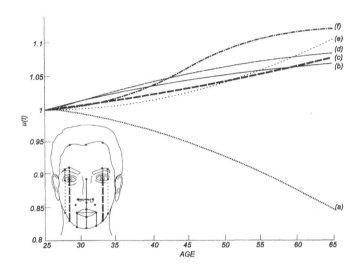

Fig. 1. General curves u(t)for: (a) the eyelids; (c), (d), (e) and (b) the central and lateral pouches of the face; (f) width of the nose.

3. MATERIAL AND METHOD

The skin is an anisotropic and inelastic material (Fung Y.C., 1980), i.e., there is no single-valued relationship between stress and strain, where relaxation, creep and hysteresis can be observed. The material presents different rates for loading and unloading. To test the skin, *preconditioning* is used; several cycles of loading are done so that afterwards the stress-strain relationship is repeatable. The pseudo potential energy function of deformation $\underline{W}$, which is defined for these preconditioned tissues, has the same thermodynamic meaning as the strain energy function, W, and the following expression (Fung Y. C., 1980) was used in the present study to represent the human skin,

$$\underline{W} = C_1 \exp[C_2(\gamma_{11}^2 + \gamma_{22}^2 + \gamma_{33}^2) + (C_3 + 2C_2)(\gamma_{11}\gamma_{22} + \gamma_{33}\gamma_{22} + \gamma_{11}\gamma_{33}) - C_3(\gamma_{12}^2 + \gamma_{23}^2 + \gamma_{31}^2) \quad (1)$$

where: γ_{ij} are deformations and the Cs are constants. Those constants were unknown to us but their values could be obtained through the solution of a very well known problem where the displacement field is prescribed in a strip of the experimental work done by Daly (1982),

$$C_1 = 4.2\text{MPa}, \quad C_2 = 0.56 \quad \text{and} \quad C_3 = -0.67. \tag{2}$$

In the simplex element, the displacement field is approximated by linear functions of the coordinates in the undeformed configuration. Knowing the coordinates of the nodal points of the finite element is possible to obtain the interpolation functions, Ψ_N, and so the displacement field can be written as a function of the nodal displacements,

$$u(x) = \psi_N u^N. \tag{3}$$

The Equilibrium Equations that rule the finite deformation of membranes in Pane Stress (Oden J.T., 1972) are as follows:

$$\int_{v_0} t^{ij} \psi_{N,i} (\delta_{ij} + u_k^M \psi_{M,j}) dv_0 = p_N \tag{4}$$

i, j=1,2 and M, N=1,2,3 and

$$t^{ij} = \frac{\partial W}{\partial \gamma_{ij}} \tag{5}$$

where: γ_{ij} is the Green-Lagrange strain vector,

$$\gamma_{ij} = \frac{1}{2}\left(G_{ij} - \delta_{ij}\right), \tag{6}$$

and G_{ij} is the metric tensor in the intrinsic coordinate system.

Where t^{ij}; $\psi_{N,i}$; δ_{ij}; u_k^M; v_0 and p_N are the second Piolla-Kirchhoff stress tensors,

the derivatives of the interpolation functions with respect to the coordinates, the Kronecker-Delta tensor, the displacement field, the initial volume of the element and the nodal body force, respectively.

The element used is a simplex element, that eases the integration of the equilibrium equation of the element. In this way the problem is reduced to the solution of a system of nonlinear equations, which could be solved using the Modified Newton-Raphson Method.

The numerical simulation was conducted by discretizing a plane mask of human facial proportions with a triangular mesh as shown in Fig. 2. The body force acting in the element was considered proportional to the initial volume of the element. The boundary conditions were chosen with no displacements in both directions on the upper boundary and on three nodal points.

4. RESULTS

The movement of the contour of the mask and Finite Element mesh can be seen in Fig. 2, in which there can be noted the decay of the eyelids, the formation of the lateral pouches of the face, and the enlargement of the upper lip. The qualitative comparison of the observed movement with the measured parameters of aging (Pamplona *et al*, 1996), Fig. 1, points out the importance of gravity in the aging process.

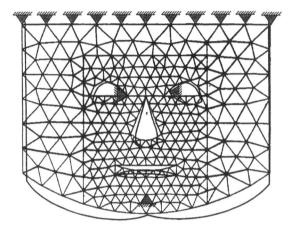

Fig. 2. Finite Element mesh in a mask that represents the facial skin and the movement of the boundary.

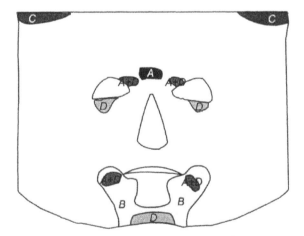

Fig. 3. Regions of extreme stresses (A and B) and extreme thickness (C and D).

The extreme negative principal stresses can be observed in regions A and B and the change in thickness can be observed in regions C and D (Fig. 3). Region A refers to compression in the first principal stress explaining the formation of the horizontal fold in the forehead and in the extremes of the mouth. In regions B there is compression in the second principal stress, explaining the Nasogenian fold. The extreme changes in thickness are shown in region C, where the skin becomes thinner and in region D, where the skin becomes thicker, explaining the formation of the palpebral pouches and the increase in the formation of the Nasogenian fold.

5. DISCUSSION

This paper takes the first step in describing the action of gravity on the facial aging process, using a plane stress model. The study has shown a strong correlation between the action of gravity and the deformation of the face during a life time. At the present an experimental research using a visco-elastic material is being carried out.

The main point of this paper is to introduce the need of identification of the mechanical forces in the mechanics of facial aging. As far as we know, there lacks a complete study of the mechanics of aging. We understand that the knowledge of this process can lead to improvement in ways of reverting the effects of aging or even controlling it mechanically.

6. ACKNOWLEDGEMENTS

We are grateful to CNPq and CAPES for their financial support for this study.

6. REFERENCES

1. Pitanguy, I. *et al.,* Anatomia do envelhecimento da face, Rev. Bras. Cir., 1977, Vol. 67, 79-84.

2. Pamplona, D.*et al.,* Defining and measuring aging parameters, In: Appl. Math. Comp., 1996, Vol. 78/2,3, 217-227.

3. Fung, Y.C., On the Pseudo-elasticity of Living Tissues, Mech T., 1980, Vol. 5, 49-66.

4. Daly, H.C. Biomechanics Properties of Dernis, J. Invest. Derm., 1982, Vol. 79, 17-19.

5. Ogden, J.T., Finite Elements of Nonlinear Continua, MacGraw-Hill, 1972.

ASSESSING RABBIT MOVEMENT OVER LONG PERIODS USING A MOTION ANALYSIS SYSTEM

E.L.N. Spelier[1], N.G.Shrive[2] and C.B Frank[3]

1. ABSTRACT

A method was developed to use a Motion Analysis System to assess a rabbit's movements over long periods of time. Real-time data acquisition, analysis and filtering were implemented in three programs running in parallel on a Sun Sparc Station and two terminals. The method proved to perform satisfactorily, detecting $74\pm3\%$ of all movements of the rabbit over four tests.

2.INTRODUCTION

As part of a project to determine the in vivo loading on a rabbit Medial Collateral Ligament, the movement of a rabbit in its cage had to be assessed over several hours. However, the Motion Analysis System (MAS) VP320 available acquires data for only 15 s. Intial attempts with sequential batch programming to acquire data for longer periods led to data collection for about 75% of the test time. Vast quantities of data were collected and all "motion" files had to be checked manually for rabbit movement. Real-time data acquisition, analysis and filtering therefore had to be implemented. Programming these tasks was complicated by the fact that the software (EV) associated with the MAS has no capacity for looping or conditioning. Short, sequential motion files were therefore collected and processed immediately in two parallel processes. Each motion file contained about 15 s of movement data. The programs and performance of this new method are described.

Keywords: movement, long periods, data acquisition, data analysis, real-time

[1]Consultant, CMG Den Haag B.V., Division Advanced Technology, Postbus 187, 2501 CD Den Haag, The Netherlands

[2]Professor and Head, Department of Civil Engineering, University of Calgary, 2500 University Drive N.W., Calgary, AB, T2N 1N4

[3]Professor of Surgery, Chief, Division of Orthopaedic Surgery, Faculty of Medicine, University of Calgary, 3330 Hospital Drive N.W., Calgary, AB, T2N 4N1, Canada

3. SYMBOLS

Several flow charts will be used to illustrate the programs. The following symbols are used.

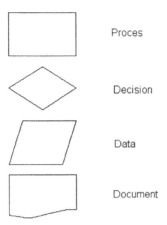

Proces

Decision

Data

Document

4. COMPUTER METHOD

Three programs were written to control the movement assessment experiments: "Motion", "Exp.ev" and "Dataselect". These programs run in parallel on a Sun Sparc station and two terminals. "Motion" is a "UNIX C-shell" program that runs on one of the terminals. In motion files, data of animal movement are acquired. Data can be acquired for about 15 s before buffers are full and data acquisition ceases. To extend the time of data acquisition, "Motion" collects video data by using a loop which triggers the collection of the motion files. "Motion" also saves the start time of each data file. Time is required between each sequential acquisition of motion data for the VP320 to reset. This takes about 5 s, which results in data being acquired for roughly 15 of

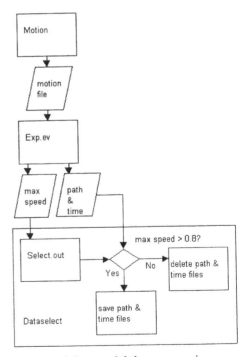

Figure 1 Sequential data processing

every 20 s. After the first few samples have been collected, "exp.ev" is started on the Sparc station in EV mode. "Exp.ev" is a batch file which calls the custom written EV user program "dir.ev" to process each motion file. In "dir.ev" the centroid, path and speed files are calculated. From the speed file the maximum speed is derived and saved in an ASCII file. After the first motion file has been processed by "exp.ev", "dataselect" begins on the second terminal. This latter program extracts the maximum speed from the ASCII file and calls the "FORTRAN" routine "select.out" to check the maximum speed against the threshold value stipulated by the user. Speed was measured in pixels/s in these experiments and the threshold value was set at 0.8 pixels/s. If the maximum speed of a motion file exceeded 0.8 pixels/s, the path and time files were saved while the other files for that time period were deleted to save disk space. If the maximum speed was below the stipulated minimum, all files were deleted. The sequence of data processing is illustrated in Figure 1. All program listings and logical diagrams are provided at the end of the paper. Figure 2 shows how the programs pass down data and run parallel in time.

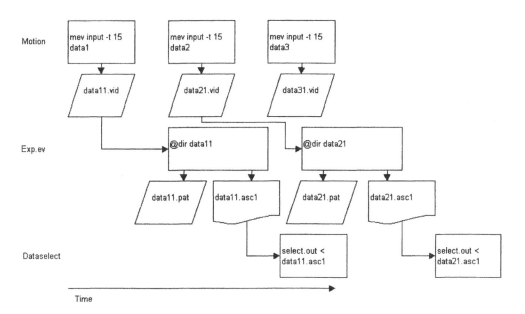

Figure 2: Dataflow in parallel processing by the control programs

5. TEST METHOD

To test the performance of the system with the new control programs, the method was tested four times on the same animal over periods ranging from one and a half to four hours. The experimental setup is shown in Figure 3. The rabbit was placed in a plastic crate filled with wood shavings. A plastic crate was used as opposed to a metal crate to prevent glare in the video image. A reflective marker was attached with tissue glue to a patch of bare skin in the middle of the lower back of the rabbit. One motion analysis video camera was placed almost vertically above the crate. The camera has infrared LEDs placed around the lens to illuminate the marker. The camera is connected to the MAS and to a video recorder. Motion files were collected and processed as described above. During the experiment, the rabbit's movements were also recorded on videotape. After each experiment, a small "C-shell" program, named "post-anal", checked which path files had been saved and printed their names and start

Figure 3: Experimental setup

times. An EV user program was created to plot all the path files on screen. By comparing these results with movements observed on the videotape, the percentage of the actual movements detected was obtained.

6.RESULTS AND DISCUSSION

The results of the movement detection tests are displayed in Figure 4. An overall average of 74±3% of all movements was detected with the new method. Results from the 4 individual tests are shown in Figure 4. This result reflects very closely the roughly 75% of the test time in which data can actually be collected. In the previous attempt of long term movement assessment with batch programming, data were also only collected for 75% of the test time. The performance of the new method is therefore satisfactory, particularly as the method is more stable, more efficient with memory space and automatic. Manual analysis of all the collected path files is no longer required.

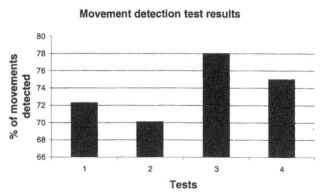

Movement detection test results

7.ACKNOWLEDGEMENTS

This work was supported by the Medical Research Council of Canada, The Alberta Heritage Foundation for Medical Research and the Arthritis Society of Canada. We are grateful for this support.

8.PROGRAM LISTINGS AND LOGICAL DIAGRAMS

"Motion"

```
#
@A=1
@B=11
date | cat > exp/start.time
while ($A < "641")
        date | cat > data$B.time
        mev input -t15 data$A
        @A= 'expr $A +1
        @B= 'expr $B +10
end
```

"exp.ev"

```
                        "dir.ev"
@dir data11             cent$1
@dir data21
@dir data31             path$1
@dir data41             path$1
@dir data51
@dir data61             path$1
@dir data71
@dir data 81            smoo $1.pat$1.pat7
```

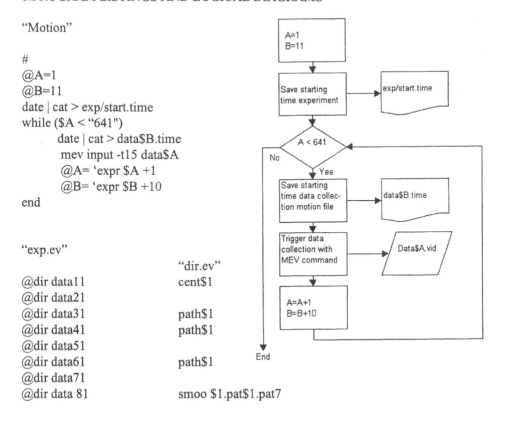

```
@dir data91
@dir data 101              join $1 1 2
and so on                  speed $1.pat
                           list/ma/uf $1.spd
                           $1.asc 1
```

"dataselect" (logical diagram on next page)

```
#
@ A = 11
@ B = 21
while ($A < "6411")
        while ((! -r data$A.asc1)&&(! -r data$B.cen))
                echo waiting
        end
        if(-r data$B.cen) then
                echo data$A.asc1 does not exist
                rm data$A.vid
        else
                vi data $A.asc1 < edit
                select.out < data$A/asc1
                set D = `line` < work.doc
                if(D = = y) then
                        cp data$A.pat exp
                        cp data$A.time exp
                endif
                rm data$A.*
                rm work*
        endif
        if(-r data$A.tmp) then
                rm data$A.tmp
                rm evlog.tmp
        endif
        @ A = `expr $A + 10`
        @ B = `expr $B + 10`
end
```

"select.f"

```
        INTEGER A,B
        REAL C,X
        OPEN (UNIT=10, FILE='work.doc', FORM='FORMATTED',
        +      ACCESS='DIRECT', RECL=11, STATUS='NEW')
1000    FORMAT (A,T1, A)
        READ *, A, B, C, X
        IF (X .GT. 0.80) THEN
            WRITE (UNIT=10, FMT=1000, REC=1) 'y'
        ENDIF
        END
```

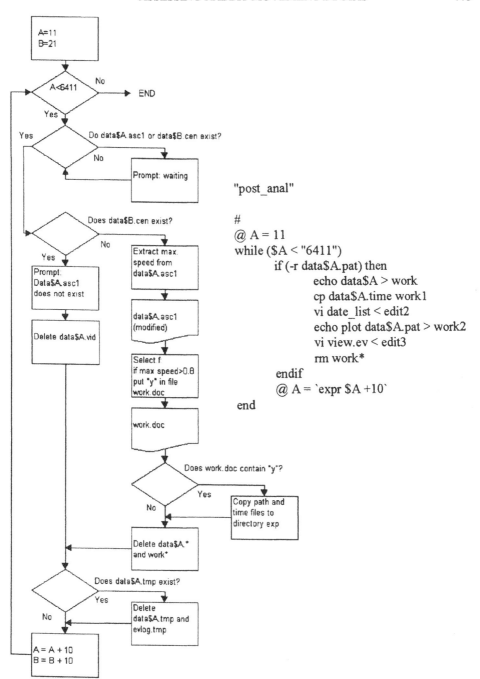

```
"post_anal"

#
@ A = 11
while ($A < "6411")
        if (-r data$A.pat) then
                echo data$A > work
                cp data$A.time work1
                vi date_list < edit2
                echo plot data$A.pat > work2
                vi view.ev < edit3
                rm work*
        endif
        @ A = `expr $A +10`
end
```

BIOMECHANICAL MODEL AND ARTIFICIAL CONTROL DESIGN OF UPRIGHT STANDING IN HUMANS

Luciano Luporini Menegaldo[1], Hans Ingo Weber[2]

1. ABSTRACT

Neural prostheses and motor control studies may find in computational simulation studies helpful aids. This work, focusing on human postural dynamics and control, developed a model that includes: rigid body mechanics, muscular contraction and neural excitation dynamics, and also an associated geometric musculoskeletal model. A methodology for controller design was established, using the LQR approach, and the pseudo-inverse matrix for distribution of control torques among redundant musculotendon actuators, employing also an inverse model of muscular contraction dynamics. Numerical results show some effects on simulations for certain initial conditions, with LQR weight matrix variations. Muscular coordination pattern is addressed.

2. INTRODUCTION

Some of the most powerful tools for understanding human locomotor apparatus has came from mathematical modelling techniques. It's amazing complexity, despite the very interesting problems that emerge from it, raised plenty difficulties in orthopaedic medicine (traumatic injuries, congenital deficiencies, articular affections, etc), not yet overcomed. In the field of neurological clinics, important interactions between the central nervous and musculoskeletal systems are the scenario of several diseases with very strong social and personal impacts: basal ganglia, cerebelar and vestibular diseases, sclerosis, medullar and peripheral nerve lesions, etc.

Two main paradigms in mathematical modelling and simulation of the neuro-musculo-skeletal system are gait [1,2] and single joint tasks [3]. Both have been very useful to understand motor control strategies, like firing-rate and recruitment duality in muscle

Keywords: Posture, Biomechanical Model, Simulation, Control, Static Optimization, Functional Electrical Stimulation.

[1] PhD. Candidate, Escola Politécnica da Universidade de São Paulo, Departamento de Engenharia Mecânica, Av. Prof. Mello Moraes 2231, CEP 05508-900, São Paulo-SP, Brasil, e-mail:lmeneg@usp.br.
[2] Visiting Professor, PUC-RIO University, e-mail: hans@mec.puc-rio.br.

control of force, or the role of Central Pattern Generators in the emergency of gait. These studies have been useful also in assessment of pathological gait patterns, in design of Functional Electrical Stimulation (FES) neural prostheses for paraplegics, besides ergonomics and sport-science applications. An intermediate approach for integrating central nervous control of movement and musculo-skeletal biomechanics is modelling the upright posture. Posture models, dealing with multi-joint systems, allow for considering dynamic and anatomical couplings among limbs and muscles. For example, Muscle Induced Accelerations [4], redundant and bi-articular musculotendon actuators control problems and also motor-unit level force sharing can be studied. Moreover, complexity is situated in more tractable levels than most of the gait models.

In the field of posture simulation models, Jaeger [5] formulated two joint static and a single joint dynamic inverted pendulum model, controlled by a PID regulator. Barin [6] tested four dynamic models of posture, based on previous work of Hemani and Jaswa [7] and Stockwell et al. [8], with one, two, three and four degrees of freedom, in the saggital plane; by formulating an inverse dynamical model to calculate joint torques and with knowledge of the input variables (angular displacements, velocities and accelerations), he identified by multiple regression analysis a K matrix of feedback gains, introducing perturbations in the support platform. Comparisons between angular trajectories obtained from numerical simulation and real data showed good agreement.

Khang e Zajac [9,10] introduced the concept of musculotendon actuator in postural dynamics, dealing with muscular mechanics and activation dynamics of FES systems. The control system was designed as a linear optimal regulator and static optimisation of muscular activation, with a energy-based cost function, was used.

This paper, using a model resembling Khang and Zajacs's, minimises the sum of the squared forces, estimating from control torques the muscle forces apt to bring the system to an equilibrium position from a given initial state. This static optimisation was obtained by a pseudo-inverse matrix of muscle moment arms, and the time series from the numerical integration was analysed for three frequencies of stimulation.

3. METHODS

3.1. Muscle Dynamics

A version of Zajac [11] Hill-type model was implemented. Taking into account the equation for the tendon force rate

$$\dot{\widetilde{F}}^T = \widetilde{k}^T \left(\widetilde{v}^{MT} - \widetilde{v}^M \cos\alpha \right) \tag{1}$$

and balance of forces in the muscle model *aponeurosis*

$$\cos\alpha \left(\widetilde{F}^{PE} + \widetilde{F}^{DE} + \widetilde{F}^{CE} \right) = \widetilde{F}^T \tag{2}$$

where $\widetilde{F}^T$: tendon force

$\widetilde{k}^T$: tendon stiffness

$\widetilde{v}^{MT}$: musculotendon actuator velocity

$\widetilde{v}^M$: contractile element velocity

α :   peenation angle

$\widetilde{F}^{PE}$:   force in the parallel elastic element

$\widetilde{F}^{DE}$:   force in the damping element

$\widetilde{F}^{CE}$:   force in the contractile element

Introducing appropriate expressions for $\widetilde{F}^{PE}, \widetilde{F}^{DE}$ and $\widetilde{F}^{CE}$ the velocity of shortening may be obtained by solving a simple 2nd degree algebraic equation [12]. Once the solution for isometric contraction is computed, a 1st order linear model of muscle contraction dynamics is found. The mean values of the musculotendon parameters were those presented by [13]. Elasticity and damping coefficients were taken from [14].

The model of activation dynamics was taken from the work of Khang e Zajac [9], considering instantaneous raise time.

3.2. Rigid body and geometrical model

Equations of motion for a three-link (legs, thighs and head, arms and trunk - HAT) inverted pendulum were derived and linked to muscle forces by means of a moment arm matrix:

$$
\begin{pmatrix} U_1 \\ U_2 \\ \vdots \\ U_j \end{pmatrix} = \begin{pmatrix} r_{11} & r_{12} & \cdots & r_{1m} \\ r_{21} & r_{22} & \cdots & r_{2m} \\ \vdots & \vdots & \cdots & \vdots \\ r_{j1} & r_{j2} & \cdots & r_{jm} \end{pmatrix} \begin{pmatrix} F_1 \\ F_2 \\ \vdots \\ F_m \end{pmatrix} \tag{3}
$$

where $U_{1,\dots j}$ are control torques at each jth joint, r_{jm} is the mth muscle moment arm with respect to the jth joint and F_m is the mth muscle force. Masses, centres of mass, and moments of inertia were calculated by regression curves [15,16] using antropometric data measured from an adult male subject. The equations of motion with muscle mechanics resulted in a non-linear state-space model (4)

$$
\begin{pmatrix} \dot{x}_1 \\ \dot{x}_2 \\ \dot{x}_3 \\ \dot{x}_4 \\ \dot{x}_5 \\ \dot{x}_6 \\ \dot{x}_7 \\ \dot{x}_8 \\ \vdots \\ \dot{x}_{n+6} \end{pmatrix} = \begin{pmatrix} x_4 \\ x_5 \\ x_6 \\ \left[M(x_1,x_2,x_3)^{-1} \right] \left([D] \begin{pmatrix} r_{11} & r_{12} & \cdots & r_{1n} \\ r_{21} & r_{22} & \cdots & r_{2n} \\ r_{31} & r_{32} & \cdots & r_{3n} \end{pmatrix} \begin{pmatrix} x_7 \\ x_8 \\ \vdots \\ x_{n+6} \end{pmatrix} - [C(x_1,x_2,x_3)] \begin{pmatrix} x_4^2 \\ x_5^2 \\ x_6^2 \end{pmatrix} - g(x_1,x_2,x_3) \right) \\ \dot{F}_1^T = f_1(a, \widetilde{L}^{MT}, \widetilde{F}^T, \widetilde{k}^T, \widetilde{L}_s^T, F_0^M, \dots) \\ \dot{F}_2^T = f_2(a, \widetilde{L}^{MT}, \widetilde{F}^T, \widetilde{k}^T, \widetilde{L}_s^T, F_0^M, \dots) \\ \vdots \\ \dot{F}_n^T = f_n(a, \widetilde{L}^{MT}, \widetilde{F}^T, \widetilde{k}^T, \widetilde{L}_s^T, F_0^M, \dots) \end{pmatrix} \tag{4}
$$

where [D] is a matrix relating joint to rigid-body torques, [M] is the mass matrix, [C] is the centripetal matrix, $\mathbf{g}$ is the gravity vector and $\dot{F}_1^T = f_1(a, \widetilde{L}^{MT}, \widetilde{F}^T, \widetilde{k}^T, \widetilde{L}_s^T, F_0^M, \dots)$ represents muscle contraction dynamics.

Moment arms for each muscle about the joints it spans were calculated as:

$$(\mathbf{r} \times \mathbf{Fu}) \cdot \mathbf{n} = \tau \qquad\qquad (4)$$

where $\mathbf{r}$ is a vector from the centre of the joint to some point between origin and insertion, $\mathbf{Fu}$ is an unitary force vector (representing the muscle line of action) and $\mathbf{n}$ is an unitary vector orthogonal to the plane of movement [12]. Muscle origins, insertions, scale factors, etc. were taken from [17]. Nine muscles were considered: gluteus maximus, gluteus medius, isquiotibialis, ilipsoas, rectus femoris, group of vasti, gastrocnemius, soleus and group of dorsiflexors.

3.3 Control systems design

The problem of bringing to an equilibrium position such a three link inverted pendulum, with redundant and multi-joint musculotendon actuators, by means of an electrical pulse signal may be divided in three sub-problems:

3.3.1. Torque control

A simple state-space feedback control system was designed to calculate the torque signals able to control a linearized, close-to-equilibrium equivalent system, multiplying angular displacements and velocities observations by a gain matrix K; LQR approach was used, in which Q and R were chosen by trial-and-error, simulating a simpler model with no muscles, but only static torque actuators.

3.3.2. Distribution of forces

A next step is the distribution of the calculated control torques between the pertinent musculotendon actuators, by means of a static optimisation. Considering equation (3), the number of muscles m is greater than the number of joints j. A solution that minimises a performance index J equal to the square-root of the sum of the squared muscle forces, subject to the linear constraints $\mathbf{u}=\mathbf{Rf}$, can be computed by a pseudo-inverse matrix $\mathbf{R}^+$, such that $\mathbf{f}=\mathbf{R}^+\mathbf{u}$.

However, an inequality constraint must be satisfied, sice muscles can produce only tensile forces ($f_m \geq 0$). A method for eliminating negative forces from the minimum solution given by the pseudo-inverse matrix was used, based on [18].

3.3.3. Calculating muscular activation a(t)

When the desired muscle force is known, the control variable a(t) may be calculated from an inverse model of muscle contraction dynamics [12]. The results shown below were determined with the inverse linear equivalent model. Prescribing an interpulse time-interval (a frequency of stimulation), the excitation signal u(t) was formed with a pulse in the beginning of each interval. This approach is a model of pulse-width modulation of FES. In the reality, pulse width is related to the muscle's degree of activation by a sigmoidal recruitment curve. Given the amount of activation desired, a pulse-width between the limits of saturation is prescribed. Usual pulse-width is about 10^{-6} s and interpulse time-interval 10^{-2}, therefore an amplitude-modulated impulse of activation a(t) may be a good approximation of the pulse-width-modulated excitation u(t) at the same frequency without using a very small time-step of integration.

4. RESULTS

Integrating the model numerically with *Matlab* (MathWorks Inc.) 4/5th order Runge-Kutta function, results of angular displacement (Degress) and velocity (Degress/s) in time (seconds) are obtained, for 200 Hz and 100 Hz (Figures 1 to 4). For 50 Hz of stimulation the system loses stability. Figures 5 to 10 shows muscular forces (N) in time: 200 Hz left, 100 Hz right.

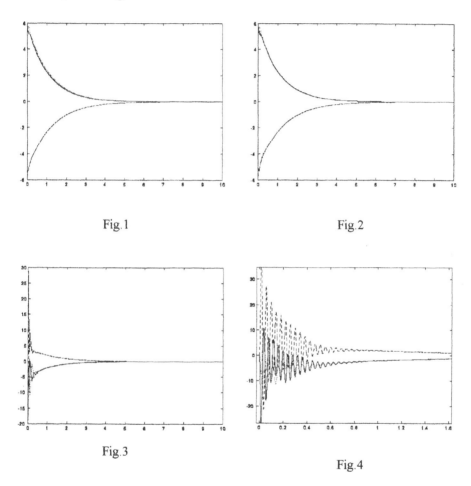

Fig.1   Fig.2

Fig.3

Fig.4

Figures 1 to 4: angular displacements (degrees) and velocities (degrees/s) with time for 200 Hz of stimulation (1 and 3) and 100 Hz (2 and 4).

| —— leg | ----- thigh | ········ HAT |

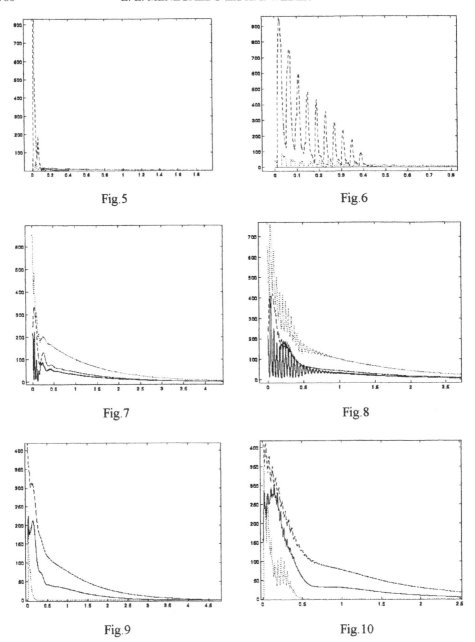

Figures 5 to 10: Muscular activation patterns for 200 Hz (Figs. 5, 7, 9) and 100 Hz (Figs. 6, 8, 10).

Figs. 5-6	—— rectus femoris	----- ilipsoas	········· gluteus maximus
Figs. 7-8	—— gluteus medius	----- isquiotibialis	········· vasti
Figs. 9-10	—— gastrocnemius	----- soleus	········· dorsiflexors

5. DISCUSSION

Results from the performed simulations show a reasonably soft curve for angular displacements, for 200 Hz (Fig. 1) and 100 Hz (Fig. 2), leading the angular displacements from 0.1 rad of initial flexion for the three limbs to 0 rad in about 4 seconds. The velocity curves (Figs. 3 and 4), however, reveal an oscillatory behaviour for both cases. For a frequency of 50 Hz (not shown), the system looses stability and the subject falls. Each numerical simulation in Matlab environment spent about 8 hours in a Sun Spark 1000 workstation.

The muscular activity pattern obtained reveals at first sight a triphasic pattern of antagonists recruitment, resembling those usually obtained in free arm movements (Ghez, 1991): the activity of an agonist muscle is followed by a silent period of the same muscle when the antagonist is activated. It can be observed comparing the three HAT extensors activity (gluteus maximus, gluteus minimus and isquiotibialis) with it's flexor ilipsoas (Figs 5 and 7 for 200 Hz; 6 and 8 for 100 Hz). Confronting the relative force production between these extensors, analogous levels were found in gluteus medius and isquiotibialis; the fact of being greater than gluteus maximus is easily explained, since the moment arm considered for gluteus maximus (0.2 cm) was smaller than gluteus medius (~1 cm). Nevertheless, the moment arm calculated for isquiotibialis was about 5 cm: it means that the control system decides for saving this muscle, because it is also knee flexor, an undesirable ability for the task performed.

The knee flexors activity show almost no force being exerted by rectus femoris, but expressive recruitment of the vasti group; it can be explained by the simultaneous HAT flexor and knee extensor properties of the vasti group.

Observing the plantarflexors gastrocnemius and soleus (Fig. 9: 200 Hz; Fig 10: 100 Hz) , in spite of having nearly the same moment arm (~3.5 cm), the last presented levels of force about twice the former. The gastrocnemius besides extending the ankle also flexes the knee joint. Coherently, the optimisation procedure minimises the use of gastrocnemius. A kind of triphasic pattern of antagonists recruitment is observed in the ankle muscles too, with activation of the dorsiflexors while the plantarflexors were presenting some reduction of force.

Comparing the responses obtained with 200 Hz and 100 Hz for HAT muscles, thiphasic pattern of antagonists recruitment repeats about 10 times in the latter condition (Figs. 5 and 6), until the stabilisation after 0.4 s; for 200 Hz, it was obtained in 0.1 s. The results of Fig. 8 shows an interesting coincidence of isquiotibialis and gluteus medius responses, but the oscillatory pattern is found only in gluteus medius. In this case, the time constant of the gluteus medius muscle is about 8 times smaller than isquiotibialis, giving enough time for the force to decay during the interpulse interval.

The proposed control system was able to control the biomechanical model formulated with linear musculotendon actuators; despite the possibility of improving the inverse model of muscular contraction dynamics for controlling the non-linear case, the optimisation cost function $\left(\sqrt{\sum F_i^2}\right)$ seems not to be the most appropriate for the postural task proposed, as pointed out by [19] in the optimal control problem, due to the rapid raising of force in the beginning of the movement. More realistic cost functions should be tested, for instance the sums of the time-derivatives of muscle forces. Artificial control of paralysed limbs must employ a low computational cost strategy. Combined state-feedback and static optimisation by the pseudo-inverse approach may indicate a feasible solution for small deviations from equilibrium.

7. REFERENCES

1. Yamaguchi, G.T., Zajac, F.E., Restoring unassisted natural gait to paraplegics via Functional Electrical Stimulation: a Computer simulation study, IEEE Transactions Biomed.Eng., 1990, Vol. 37, 886-902.
2. Taga, G., A Model of the neuro-musculo-skeletal system for human locomotion, Biological Cybernetics, 1995, Vol. 73, 97-111.
3. Enoka, R.M., Neuromechanical basis of kinesiology, Human Kinetics, 2nd ed., 1993.
4. Zajac, F. E., Gorndon, M.E., Determinig muscle's force and action in multi-articular movement, Exercise and Sport Scic.Revs. Pandolf, K. ed., 1989, Vol.17, 187-230.
5. Jaeger, R. J., Design and simulation of close-loop electrical stimulation orthoses for restoration of quiet standing in paraplegia, J. Biomech., 1986 Vol. 19, No. 10, 825-835.
6. Barin, K., Evaluation of a generalized model of human postural dynamics and control in the saggital plane, Biol. Cybernetics, 1989, Vol. 61, 37-50.
7. Hemani, H., Jaswa, V. C., On a three-link model of the dynamics of standing up and sitting down, IEEE Transactions on Syst. Man and Cybernetics, 1978,Vol. 8,115-120.
8. Stockwell, C. W., Koozekani, S. H., Barin, K., A Physical model of human postural dynamics, Ann. NY Academy of Science, 1981, Vol. 374, 722-730.
9. Khang, G., Zajac, F. E., Paraplegic standing controlled by functional neuromuscular stimulation: Part I- Computer model and control-system design, IEEE Transactions on Biomed.Eng. 1989, Vol. 36, No.9, 873-884.
10. Khang, G., Zajac, F. E., standing controlled by functional neuromuscular stimulation: Part II- Computer simulation studies, IEEE Transactions on Biomed.Eng. 1989, Vol. 36, No.9 885-894.
11. Zajac, F. E., Muscle and tendon: properties, models, scaling and application to biomechanics and motor control, CRC Critical Revs. Biomed. Eng., 1989, Vol.17, No.4, 359-411.
12. Menegaldo, L. L., Mathematical model, simulation and and control of upright posture in humans (in portughese), Master Thesis, Mechanical Eng. Institute, State University of Campinas, Brasil, 1997.
13. Hoy, M. G., Zajac, F. E., Gordon, M. E., A Musculoskeletal model of the human lower extremity: the effect of muscle, tendon and moment arm on the moment-Angle relationship of musculotendon actuators ant the hip, knee and ankle, J.Biomechanics, 1990, Vol.23, No. 2, 157-169.
14. Imbar, G. F., Adam, D., Estimation of Muscle Active State, Biol. Cybernetics, 1976, Vol. 63, 61-72.
15. Yeadon, M. R., Morlock, M., The appropriate use of regression equations for the estimation of segmental inertial parameters, J. Biomechanics,1989, Vol. 22, 683-689.
16. Vaughan,C.L., Davis,B.L., O'Connor,J.C., Dynamics of human gait, 1st. ed., Human Kinetics Publishers, Chicago IL, 1992.
17. Brand, R. A., Crowninshield, R. D., Wittstock, C. E., Pederson, D. R., Clark, C. R., van Krieken, F. M, A model of lower extremity muscular anatomy, J. Biomech. Eng., 1982, Vol.104, 304-310.
18. Yamaguchi, G. T., Moran, D. W., Si, J., A Computationally efficient method for solving the redundant problem in biomechanics, J. Biomech., 1995, Vol.28, 999-1005.
19. Pandy, M. G., Garner, B. A., Anderson, F. C., Optimal Control of non-ballistic muscular movements: a constraint-based performance criterion for rising from a chair, J. Biomech. Eng., 1995, Vol. 117, 15-26.

MOMENT-BASED PARAMETERIZATION OF EVOLVING CYCLOGRAMS ON GRADUALLY CHANGING SLOPES

A. Goswami[a], E. Cordier[b]

1 ABSTRACT

The problem under study is the systematic characterization of a series of gradually evolving cyclograms with a small number of features. The mathematical quantities derived from the geometric properties of the hip-knee cyclograms are the main features considered in this study. Our thesis is that the gradual evolution of gait on inclined planes, which is manifested by progressive shape change of these closed-contour curves can be tracked by observing the evolution of their geometric moments. Experimental slope-walking data obtained for each 1° interval within the range of −13° to +13° (±23.1%) on a variable-inclination treadmill was used in this study.

The parameterization procedure presented here is fairly general in nature and maybe employed without restriction to any closed curve such as the phase diagram and the moment-angle diagram of human gait. The technique may be utilized for the quantitative characterization of normal gait, global comparison of two different gaits, clinical identification of pathological conditions and for the tracking of progress of patients under rehabilitation program.

2 INTRODUCTION

Characterization of human gait in a quantitative and objective manner has many potential benefits in clinical diagnosis and rehabilitation as well as in the enhancement of our basic understanding of the complicated gait mechanism. In spite of the impressive sophistication of our present day data collection systems, communicable descriptions of certain gait conditions remain surprisingly difficult. Consider an ideal situation of slope-walking where we have a compact mathematical description which assigns a gait pattern number, say $\mathcal{G}(\alpha)$, for the gait on slope α. As the ground slope changes, this number also changes reflecting the adaptation of the gait. Such a tool would have several practical utilities such

Keywords: gait parameterization, slope walking, cyclogram, moments, gait kinematics, invariants

[a]Research Staff, INRIA Rhône-Alpes, 655 avenue de l'Europe, ZIRST, 38330 Montbonnot St. Martin, France

[b]Ph.D. Student, INRIA Rhône-Alpes, 655 avenue de l'Europe, ZIRST, 38330 Montbonnot St. Martin, France

as in the objective characterization of normal gait, global comparison of two different gaits, clinical identification of pathological conditions and in the tracking of the progress of patients under rehabilitation program.

The objective of the paper is to find such a gait representation technique, although, in reality, given the complexity of human gait, we would need several quantities to completely describe a single gait. In a multidimensional space these quantities can be represented as a point which would characterize the gait. Different gaits will be represented by different points and the evolution of a gait can be characterized by the locus of the point in that space. Emphasis in this paper is on the method of gait parameterization with a general applicability.

By *parameterization* we refer to the systematic objective description of the evolution of the gait descriptors (the cyclogram moments) with respect to a parameter (the ground slope). Although most of the measurable variables of the human gait respond to a parameter change, and parameterization may be performed, in principle, with any of these variables, the analysis based on closed trajectories such as the cyclograms bring in special advantages. Cyclograms have forms or shapes that reflect the gait kinematics during the total gait cycle which is different from having other discrete measures such as the step length, or walking speed, which are more common in the literature[12][20]. Our choice of moment-based characterization of the cyclogram features is justified by the fact that moments (of different orders) can be viewed as a generalization of most of the commonly used geometric features.

3   CYCLOGRAMS AND THEIR MOMENTS

Although it has not been seen very frequently in the literature of the recent past, the concept of cyclograms is known to the biomechanics community. Interested readers are directed to [6][14][7][3][2][1]. A hip-knee cyclogram is formed from the time angle curves of hip and knee joint by ignoring the time axis of each curve and directly plotting the knee angle VS hip angle. Readers interested in the slope-walking literature are directed to [4][20][12][16][17] [15][19].

Slope-walking experiments were performed with two healthy male subjects without any history of lower extremity injury. A motor-driven variable inclination treadmill was used for all the trials. The subjects chose the "most comfortable" treadmill speed for each slope. A video camera registered the marker positions. The marker position data were filtered with a 4th-order Butterworth filter with a cut-off frequency of 12 Hz. The inclination of the treadmill was varied from $-13°$ to $+13°$ with data collected at each $1°$ interval. Out of a complete 8 minutes of walk on each slope, we have carefully selected one cycle which is fairly repetitive for the particular slope and which does not show any transients. This paper uses data from only one of the subjects.

Several techniques for quantifying planar shapes are available in the digital image processing literature[10]. Some of the important works on the use of area-based moments in identifying 2D shapes can be found in [9][18]. Efficient algorithms to compute moments were presented in [13][11]. We use perimeter-based moments in this work.

The perimeter-based moments are calculated by making a physical analogy of the cyclogram with a thin wire loop with uniform mass distribution along its length. The two-dimensional moments of order $(p + q)$ of a polygonal curve of length L is expressed

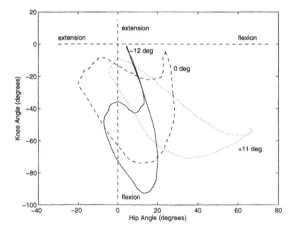

Figure 1: Superposition of three cyclograms at different slopes showing their evolution.

as,

$$M_{pq} = \sum_{i=1}^{n} \int_{0}^{L_i} x^p y^q dl = \sum_{i=1}^{n} \mu^i(p,q), \tag{1}$$

where L_i is the length of the ith segment of the curve which is in the xy-plane. For hip-knee cyclograms $x = \theta_h$, hip angle and $y = \theta_k$, knee angle. $\mu^i(p,q)$ is expressed as

$$\mu^i(p,q) = \int_{0}^{L_i} x^p y^q dl = \int_{min(x_i,x_{i+1})}^{max(x_i,x_{i+1})} x^p y^q \sqrt{(1+s_i^2)} dx, \tag{2}$$

where s_i is the slope of the ith line segment, a constant.

The details of the calculations and the recursive equations for the higher order moments are available in [5], adapted from [11].

4   PARAMETERIZATION OF CYCLOGRAM FEATURES

We now consider the parameterization of the moment-based descriptors of the cyclograms as they evolve as the ground slope changes from $-13°$ to $+13°$. For each case, we define the gait descriptor, plot its evolution as a function of the ground slope and discuss its trend.

Perimeter   The perimeter of a cyclogram is simply its zeroth moment M_{00}. As we show in Fig. 2(a) the cyclogram perimeter (normalized against the perimeter at $0°$ slope) is a linearly decreasing function of the ground slope.

Although the "jerkiness" of the joint motion was given as a possible reason of an increase in the perimeter of similar curves in [7], it is unlikely to be reason here. In order to interpret the cyclogram perimeter, we express the length of the straight line segment L_i connecting two successive data points $\theta_{h_i}, \theta_{k_i}$ and $\theta_{h_{i+1}}, \theta_{k_{i+1}}$ as,

$$\begin{aligned} L_i &= \sqrt{(\theta_{h_{i+1}} - \theta_{h_i})^2 + (\theta_{k_{i+1}} - \theta_{k_i})^2} \\ &= \Delta t_i \sqrt{(\omega_{h_i})^2 + (\omega_{k_i})^2} \end{aligned} \tag{3}$$

where ω_{h_i} and ω_{k_i} are the average angular velocities of the hip and the knee joints, respectively during the interval, and Δt_i is the corresponding time interval. The above equations can be extended to the entire cyclogram.

According to the first of the Eqns. 3 the cyclogram perimeter is the "total distance traveled" by the two joints in their respective joint spaces. The second of the Eqns. 3 relates the cyclogram perimeter with the average velocity of the two joints during a complete gait cycle. For cycles of equal duration, the perimeter is proportional to the average joint velocity. The duration of gait cycle is a linearly increasing function of slope as shown in Fig. 2(b). From the plot of the perimeter/cycle duration ratio we can infer that the average joint velocity linearly decreases while the slope changes from $-13°$ to $+13°$.

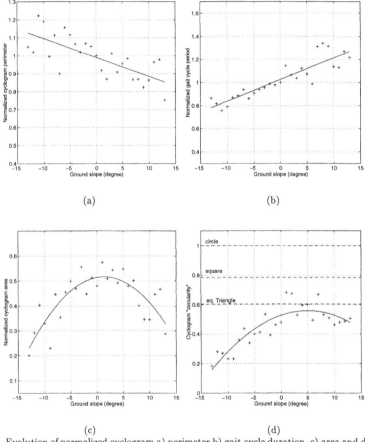

(a)   (b)

(c)   (d)

Figure 2: Evolution of normalized cyclogram a) perimeter b) gait cycle duration, c) area and d) circularity with change in ground slope. The superposed lines show linear (a,b) and quadratic (c,d) fit of the data. In d) the circularity of well-known geometric entities, the circle, the square and the equilateral triangle are shown for comparison.

Area   The area of a polyline contour can be computed as $A = \sum_{i=1}^{n} x_i y_{i+1} - x_{i+1} y_i$ which assigns a positive value to the area inside a counterclockwise contour and a negative value to one inside a clockwise contour. This introduces errors in our calculations for the cyclograms with intersecting loops.

Fig. 2(c) shows the normalized cyclogram area as a function of the ground slope. The normalization is done with respect to the area of a circle whose perimeter is equal to that of the cyclogram corresponding to the $0°$ slope. The parabolic line represents a quadratic fit of the data. Relating the cyclogram area with the conjoint range of joint movements [7] we can say that the range is maximum for level and feeble uphill slopes.

Circularity or compactness or roundedness A few versions of the dimensionless criterion γ characterizing its circularity, compactness or roundedness are in use in the literature [10][7]. We use $\gamma = \frac{4\pi A}{P^2}$ since it is unity for a circle (which is the maximum possible value). For comparison we note that the values of γ for a square and for an equilateral triangle are $\frac{4}{\pi}$ and $\frac{3\sqrt{3}}{\pi}$, respectively. Generally oblong objects have smaller circularity values (meaning they are less circular). Fig. 2(d) presents the evolution of the circularity criterion $\frac{P^2}{A}$ as a function of the ground slope. The figure indicates that the cyclogram circularity is the maximum and stays constant within the range $1° - 7°$. The low circularity of the negative slope cyclograms is due to the long stretch of impact-cushioning around the heel-strike that adds to the perimeter but encloses little area. Circularity is related to the eccentricity of the cyclogram[5].

Location Location, in this context, means the position of the center of mass (CM) of a wire of uniform mass in the shape of the cyclogram. The position of the CM $(\theta_{h_{CM}}, \theta_{k_{CM}})$ is given by

$$\theta_{h_{CM}} = \frac{M_{10}}{M_{00}} \text{ and } \theta_{k_{CM}} = \frac{M_{01}}{M_{00}} \tag{4}$$

where M_{10} and M_{01} are the two first-order moments of the polygonal contour, and M_{00}, as seen before, is its perimeter. In Fig. 3(a) we plot the distance of the cyclogram CM from the coordinate origin as a function of the ground slope. The distance is normalized with that corresponding to the cyclogram for $0°$ slope. Fig. 3(b) shows the locus of the CM on the θ_h/θ_k plane. It is clear from the plots that the CM is the nearest to the origin for downhill walk on feeble slopes and moves away for positive as well as negative slopes.

For the joint angle assignment convention used in the paper the coordinate origin of the cyclogram plane corresponds to a straight leg configuration aligned with the trunk. Thus the distance of a point from the origin to the cyclogram represents the deviation of the current configuration from this configuration. If the cyclogram CM is viewed as the "average" leg configuration during a complete walk cycle, its distance from the origin will be a quantification of the deviation of this average configuration from the passive leg configuration which is vertical[1].

Orientation The angle (bounded between $\pm 90°$) between the positive abscissa and the line of least second-order moment of the contour is traditionally called the *orientation* of a contour. Since the line of the least second moment passes through the CM of the contour $(\theta_{h_{CM}}, \theta_{k_{CM}})$, its equation will be simpler if we displace the coordinate frames such that the new origin coincides with the center[8]. The moments calculated with respect to a coordinate frame situated at the CM of an object are called its *central moments* and are denoted by $\overline{M}_{pq}$. In the new coordinates $\overline{M}_{10} = \overline{M}_{01} = 0$.

The orientation of the line of minimum second-order moment, ϕ can be computed from the eigenvectors of the so-called 2×2 matrix of second moments, the eigenvalues of the matrix being the magnitudes of the moments. The most characteristic feature in the evolution of orientation with ground slope (Fig. 4(a)) is the two constant orientation regions connected by a discrete jump which occurs around $+5°$ slope. The cyclograms for higher uphill slopes are decidedly inclined. The orientation analysis quantifies this geometric feature. Fig 4(b)

---

[1]We note that this "average" is a purely geometric average and time is not involved in this definition.

shows the ratio of $\frac{\overline{M}_{20}}{\overline{M}_{02}}$.

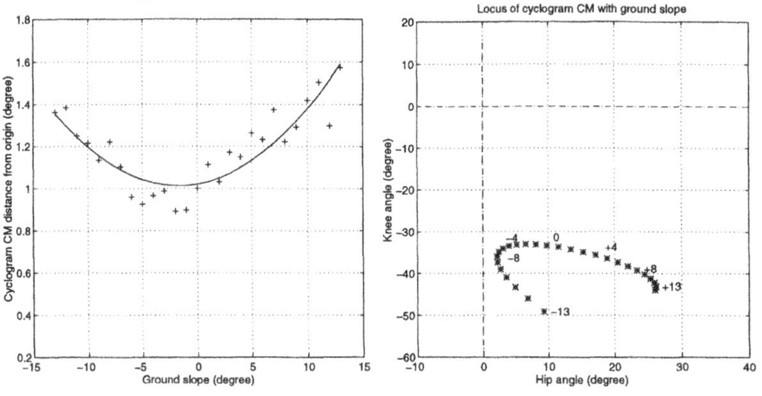

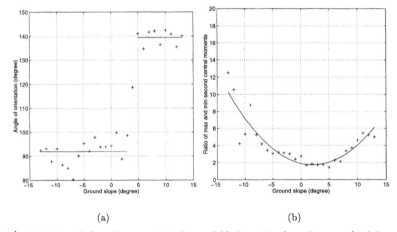

(a)   (b)

Figure 3: Evolution of normalized cyclogram CM distance with change in ground slope. A quadratic approximation is superposed on the data points. In the second figure we see the locus of the CM position with slope.

(a)   (b)

Figure 4: Evolution of a) cyclogram orientation and b) the ratio of maximum and minimum second central moments with change in ground slope. The first curve is fitted with two constants. The line in the second figure shows cubic fit of the data.

Quantification of cyclogram evolution   In order to demonstrate how we can provide a global picture of the evolution of cyclograms with the help of the gait descriptors we plot in a 3D space the triplet – the normalized perimeter, the ratio of maximum and minimum second moments and the distance of the cyclogram CM from the origin for each slope (Fig. 5(a)). The figure represents a signature of normal walking by compactly capturing a multitude of information.

Higher order moment – an invariant of slope walking?   The higher-order central moments of the contour are calculated with respect to the coordinate axes fixed to the CM and aligned with the maximum and minimum second moments. In general, unless normalized, the higher order moments have high numerical values (orders of 10^4 to 10^5) as we can anticipate from Eq. 1. However, if we consider the ratio of $\frac{\overline{M}_{03}}{\overline{M}_{30}}$ as shown in Fig. 5(b) we notice that these quantities remain remarkably constant for the entire range of positive and negative slopes. Even without having a sound physical interpretation of the third-order cyclogram moments, these quantities have potentials to play the role of *invariants* of slope-walking.

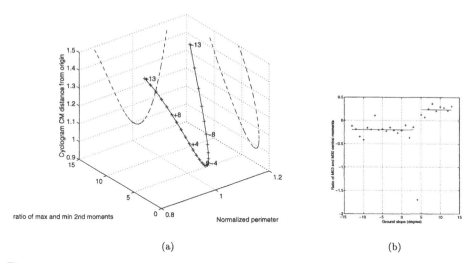

(a)   (b)

Figure 5: a) 3D plot of normalized perimeter, the ratio of maximum and minimum second moments and the normalized cyclogram CM distance from the origin. The projection of the curve on the two vertical planes are shown in dashed lines. b) Evolution of the ratio of $\overline{M_{03}}$ and $\overline{M_{30}}$ with change in ground slope.

5   DISCUSSION

Gait parameterization by means of cyclogram moments can be of potential use in a number of fields such as the quantitative characterization of normal gait, global comparison of two different gaits, clinical identification of pathological conditions and in the tracking of the progress of patients under rehabilitation program.

Recognizing the fact that the cyclograms continuously deform in a predictable manner as the ground slope changes, we have assigned a set of numbers to each cyclogram to quantify its moment-based geometric characteristics. The standard qualitative features of the hip-knee cyclograms for uphill and downhill walk are acknowledged in the literature. Based on this we can expect that the general trends of the slope walking gait descriptors will be similar for all individuals. What is to be seen is whether the precise moment-values reported in this work are also subject-independent. In order for the cyclograms to become an effective tool for the clinical identification of pathological conditions we need the analytical skill of correlating a pathology with a corresponding feature of the cyclogram. If we represent a cyclogram as a point in a multidimensional space capturing all its moments, a pathology might be suspected or identified by noting the position of this point relative to other points representing normal gaits. Thus more work in concert with the clinicians is necessary.

Acknowledgments

Helpful discussions with Bernard Espiau of INRIA Rhône-Alpes at all stages of this work are gratefully acknowledged. Prof. Jean-Pierre Blanchi of the laboratory UFR STAPS at Joseph Fourier University, Grenoble lent us his data registration system and Prof. Alain Belli of the Laboratory of Exercise Physiology, GIP Exercise, St. Etienne made his laboratory equipped with a variable inclination treadmill available to us. A preliminary version of the slope-walking experiment was performed on a fixed 10° slope in the UFR STAPS Movement Analysis Group laboratory of Prof. Thierry Pozzo in the University of Dijon.

References

[1] P. Cavanagh. *Biomechanics of Distance Running.* Human Kinetics Books, Champaign, IL, 1990.

[2] J. Charteris. Human gait cyclograms: Conventions, speed relationships and clinical applications. *International Journal of Rehabilitation Research,* 5(4):507–518, 1982.

[3] J. Charteris, D. Leach, and C. Taves. Comparative kinematic analysis of bipedal and quadrupedal locomotion: A cyclographic technique. *Journal of Anatomy,* 128(4):803–819, 1979.

[4] L. Erickson, E. Simonson, H. L. Taylor, H. Alexander, and A. Keys. The energy cost of horizontal and grade walking on the motor-driven treadmill. *American Journal of Physiology,* (145):391–401, 1946.

[5] A. Goswami. A new gait parameterization technique by means of cyclogram moments: Application to human slope walking. *Gait & Posture,* 1997 (submitted).

[6] D.W. Grieve. Gait patterns and the speed of walking. *Biomedical Engineering,* 3:119–122, 1968.

[7] C. Hershler and M. Milner. Angle-angle diagrams in the assessment of locomotion. *American Journal of Physical Medicine,* 59(3):109–125, 1980.

[8] B.K.P. Horn. *Robot Vision.* The MIT Press, Cambridge, MA, 1986.

[9] M-K. Hu. Visual pattern recognition by moments invariants. *IRE Transactions on Information Theory,* IT-8(8):179–187, 1962.

[10] A.K. Jain. *Fundamentals of Digital Image Processing.* Prentice-Hall International, Inc., Engelwood Cliffs, N.J., 1989.

[11] X.Y. Jiang and H. Bunke. Simple and fast computation of moments. *Pattern Recognition,* 24(8):801–806, 1991.

[12] K. Kawamura, A. Tokuhiro, and H. Takechi. Gait analysis of slope walking: a study on step length, stride width, time factors, and deviation in the center of pressure. *Acta Medica Okayama,* 45(3):179–184, 1991.

[13] B-C. Li and J. Shen. Fast computation of moments invariants. *Pattern Recognition,* 24(8):807–813, 1991.

[14] M. Milner, D. Dall, V. A. McConnel, P. K. Brennan, and C. Hershler. Angle diagrams in the assessment of locomotor function. *S.A. Medical Journal,* 47:951–957, 1973.

[15] M. S. Redfern and J. J. DiPasquale. Biomechanics of descending ramps in young and elderly. In *Fifth Injury Prevention Through Biomechanics Symposium,* 1995.

[16] K.J. Simpson, P. Jiang, P. Shewokis, S. Odum, and K. Reeves. Kinematic and plantar pressure adjustments to downhill gradients during gait. *Gait & Posture,* 1:172–179, 1993.

[17] J. Sun, M. Walters, N. Svensson, and D. Lloyd. The influence of surface slope on human gait characteristics: a study of urban pedestrians walking on an inclined plane. *Ergonomics,* 39(4):677–692, 1996.

[18] M.R. Teague. Image analysis via the general theory of moments. *J. of Optical Soc. of America,* 70(8):920–930, 1980.

[19] J.C. Wall, J.W. Nottrodt, and J. Charteris. The effects of uphill and downhill walking on pelvic oscillations in the transverse plane. *Ergonomics,* 24(5):807–816, 1981.

[20] M. Yamasaki, T. Sasaki, S. Tsuzuki, and M. Torii. Stereotyped pattern of lower limb movement during level and grade walking on treadmill. *Annals of Physiological Anthropology,* 3(4):291–296, 1984.

CORRELATION AND PREDICTION OF JOINT TORQUES FROM JOINT MOTION USING NEURAL NETWORKS

Z. Taha[1], R. Brown[2], D. Wright[3], M. Parnianpour[4], K.Khalaf[5], P.Sparto[6]

1. ABSTRACT

The development of a dynamic strength model for humans would be extremely useful in predicting the forces and torques to be applied when performing a physical task. This will enhance the design of equipment, workstation and workplace in order to maximise efficiency and at the same time minimising energy and stress. In the past biomechanic strength models tended to be based on static controlled exertions . This is partly due to the mathematical complexity of dynamic biomechanical analysis. One approach in resolving this problem is to develop a model based on empirical data rather than attempting to model it as a mechanical system with rigid links, rotational springs and dampers. There have been attempts to develop polynomial equations determining joint torque as a function of joint position and velocity using the least squares regression technique. In this paper we present an alternative approach using neural networks.

Keywords: Dynamic strength, Models, Correlation, Prediction, Neural networks

---

[1]Research Fellow, Design Engineering Research Center, University of Wales Institute Cardiff, Western Avenue, Llandaff, Cardiff CF5 2YB, Wales, UK
[2]Professor and Director, Design Engineering Research Center, University of Wales Institute Cardiff, Western Avenue, Llandaff, Cardiff CF5 2YB, Wales, UK
[3]Visiting Professor, Design Engineering Research Center, University of Wales Institute Cardiff, Western Avenue, Llandaff, Cardiff CF5 2YB, Wales, UK
[4]Associate Director, Biodynamics Lab, Department of Industrial, Welding and Systems Engineering, The Ohio State University, 210 Baker Systems, 1971 Neil Avenue, Columbus, OH 43210-1271, USA
[5]Research Associate, Biodynamics Lab, Department of Industrial, Welding and Systems Engineering, The Ohio State University, 210 Baker Systems, 1971 Neil Avenue, Columbus, OH 43210-1271, USA
[6]Research Associate, Biodynamics Lab, Department of Industrial, Welding and Systems Engineering, The Ohio State University, 210 Baker Systems, 1971 Neil Avenue, Columbus, OH 43210-1271, USA

2. INTRODUCTION

There is an increasing concern for taking into consideration forces and torques to be applied when performing a task, into the design of equipment, workstation and workplace. The task may involve the movement of load by lifting, lowering, pushing, or pulling. In addition, the task may have to be performed in restricted space, at a high frequency and with extreme or ackward postures. The resulting forces and torques may give rise to stress and fatigue which will affect the performance and health of the user.

In the past, static models have been used to determine the forces and torques arising from slow and controlled exertions. Static strength models only takes into consideration the resultant moments acting on the joints and the consequent reactive moments and forces to maintain stability and posture. The reactive moments and forces are dependent on the strength producing capability of the joints. These are usually based on a database of population strength data derived from static strength tests[1,2]. Obviously, these models are not appropriate in tasks where there are significant limb movements.

The difficulty of developing dynamic strength models for a human lies in its mathematical complexity. Although the musculoskeletal system can be simplified as a mechanical system with rigid links, nonlinear springs and dampers [3], it will still need to be verified with empirical data. However if empirical data already exists or is easily measurable then developing a model based on this data would be more attractive and accurate. Pandya et. al [4,5] have attempted to develop polynomial equations determining joint torque as a function of joint displacement and velocity using the least quares regression technique. In this paper we attempt to develop a model correlating joint displacement and joint torque based on neural networks.

3. NEURAL NETWORK MODELS

Studies on cats have shown that there exists an internal representation of motion [6]. This internal representation can be modelled by a neural circuit functioning as a pattern generator stimulating muscle activity in a predetermined fashion. Based on these observations, we propose a simple biological model for the generation of human motion (figure 1). The next step is to determine the input and output. In this study we are concerned with the posture sequence and strength prediction of each joint whilst performing a task. Posture or joint displacement sequence will facilitate animation of the task whilst strength prediction will allow anticipation of joint torques. Thus joint displacements and torques will form the output of this model. The difficulty lies in determining the appropriate inputs.

Figure 1. Simple biological model of human motion generation.

A human consciously decides on the task to be performed and the speed it is performed. The speed is fuzzy in nature ranging from slow, medium and fast. These parameters will have an effect on the joint displacement and torques. Any other parameters which can affect the performance of the task will also have to be included. For example in the task of lifting, the load to be lifted will certainly affect the strength capability of each joint. Performance may also vary because of differences in anthropometric parameters. Prediction equations have been developed from lean body mass for static strength [5]. Using a easily measurable parameter would be of significant value in predicting dynamic strength. In this study, in addition to mass, height is also a parameter for prediction. The body internal clock provides a time frame for the execution of the joint displacements in time. Thus timing sequence has to be an input into the neural network model. We call these parameters 'behaviour parameters' as the performance of a task is sensitive to variations of these parameters.

The mathematical model of the neural network can only accept numeric values therefore qualitive input values are referreed to by an index[7]. Thus if there are 10 tasks to be performed, each task is referred to by an index from 1 to 10. However in this study we are studying only one task, that is lifting, so the index is defaulted to 1. The speed of lifting varies from fast, medium and slow which are also assigned an index of 1,2 and 3 respectively. The same is applied to the load (light, medium, heavy) and style of lifting (straight, bend and normal). Table 1 summarizes the input and output of the proposed model. The time frame is taken as a percentage of the task cycle.

INPUT	OUTPUT
Speed: slow(1),medium(2),fast(3) Style:preferred (1), straight(2), bent(3) Load: light(1), medium(2), heavy(3) Time Frame: % of cycle time	Joint displacement Joint Torques

Table 1. Input and output parameters for neural network model of lifting.

3.1 Lifting data

Joint positions, velocities, accelerations and torques were measured from ten subjects of different mass and heights (Table 2). In this study only data for fast lifting (index = 3), preferred style (index = 1) and heavy load (index = 3) was used. The variables for prediction are mass and height.

Subject	A	B	C	D	E	F	G	H	J	K
Mass (kg)	94.7	84.5	77.1	92.0	66.9	97.7	79.2	68.8	79.9	111.9
Height(cm)	185	174	178	178	161	196	194	171	171	179

Table 2. Mass and height of ten subjects measured for load lifting.

3.2 Correlation between behaviour parameters and joint displacements.

In this section we investigate how well the behaviour parameters correlate with joint displacements. The network of figure 2 represents the central pattern generator. The neural network used is of the fully connected type whilst learning is by the method of back propagation [7].

Figure 2. Correlating behaviour parameters and joint displacements.

Initially data from subject A was used to establish correlativity. Figures 3 shows the comparison between actual and neural network values after 50,000 iterations. The graphs suggest that the behaviour   parameters and joint displacements correlates very well .

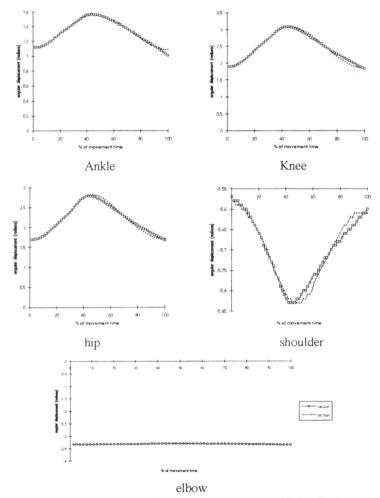

elbow

Figure 3. Correlation between behaviour parameters and joint displacements

3.3 Correlation between joint displacements and joint torque.

In this section we study the correlation between joint displacement and joint torque. The neural network in this case represents the muscle in figure 1.

Figure 4. Correlating joint displacements and joint torques.

Again with data from the same subject and the same number of iterations , the results are shown in figure 5.

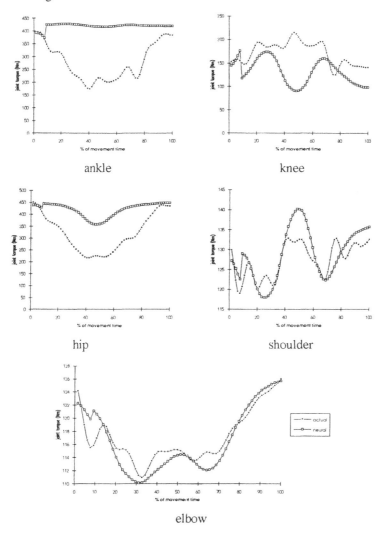

Figure 5. Correlation between joint displacement and joint torques

It is observed that there is a significant amount of error with the ankle, knee and hip. However there is a closer match between actual and neural network values for the shoulder and elbow.

3.4 Correlation between behaviour parameters, joint displacements and joint torques.

In this section we propose that the input parameters into the neural network representing the muscle consist of the joint displacements and behaviour parameters as well. Again data from the same subject was used and with the same number of iterations a closer matching between actual and neural network values are observed for all joints (figure 6).

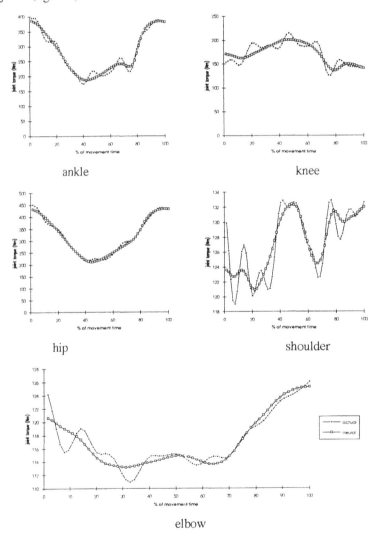

Figure 6. Correlation between joint displacement, behaviour parameters and joint torques

4. PREDICTION OF JOINT TORQUE.

To test the prediction capability of the neural network model, data from nine subjects was used for training. Subject H was randomly chosen and left out from the training data. Figure 7 shows the predicted torque for subject H.

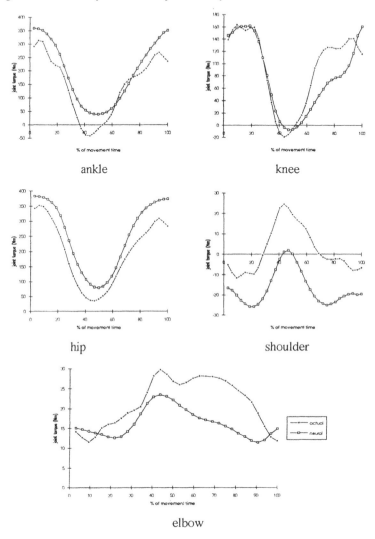

Figure 7. Prediction of joint torques for subject H.

It is observed that there is a close match for the ankle and knee. However there are constant errors for the hip and shoulder throughout the whole cycle and significant errors in the second half of the cycle for the elbow. But in general the neural network predicted the shape of the graphs relatively accurately. This is significant as the slopes of these graphs represents the rate of change of torque.

5. DISCUSSION AND CONCLUSION

The neural network model proposed in this study have performed very well in terms of correlating the input parameters that specifies the human behaviour, joint displacement and joint torques. It has predicted accurately the rate of change of joint torque. Although there are discrepancies in the predicted joint torques particularly for the shoulder and elbow, this may be attributed to the number of training data used. The greater the number of subjects, the more accurate would be the prediction. However a major drawback in this case would be the longer time it takes to learn. But neural network offers the flexibility that a database approach lacks that is the ability to accommodate additional data from a wider range of population as they become available and increasing the accuracy of its prediction at the same time.

We conclude from this study that a dynamic strength model for humans can be built by correlating behaviour parameters and empirical data of joint displacements and joint torques. This model is extremely useful in predicting joint forces and torques to be applied when performing a task. The model is able to accommodate a wider range of population samples and with increasing accuracy.

7. REFERENCES

1. Chaffin D.B. and Baker W., A Computerized Biomechanical Model for the Study of Manual Materials Handling, AIIE Transactions, 1970, 2(1).

2. Stobbe T., The Development of a Practical Strength Testing Program in Industry, Ph.D. Dissertation, Department of Industrial and Operations Engineering, The University of Michigan, Ann Arbor, 1982.

3. Buck B. and Wolfel H.P, A Dynamic Model for Human Whole-Body Vibration with Detailed Representation of the Lumbar Spine, 10th Conference of The European Society of Biomechanics, Leuven, August 28-31, 1996, 338.

4. Pandya A.K, Maida J.C, Aldrige A.M, Hanson S.M and Woodford B.J, The Validation of a Human Force Model To Predict Dynamic Forces Resulting From Multi-Joint motions, NASA Technical Paper 3206, 1992.

5. Pandya A.K, Maida J.C, Aldrige A.M and Hanson S.M, Correlation and Prediction of Dynamic Human Isolated Joint strength From Lean Body Mass", NASA Technical Paper 3207, 1992.

6. Wetzel M.C and Stuart D.G., Ensemble Characteristics of Cat Locomotion and its Neural Control, Prog Neurobiol 7, 1-98

7. Taha Z., Brown R. and Wright D., Realistic Animation of Human Figures Using Artificial Neural Networks, Medical Engineering and Physics, 1996, Vol. 18, No.8, 662-669.

OPTIMAL REJECTION OF ARTIFACTS IN THE PROCESSING OF SURFACE EMG SIGNALS FOR MOVEMENT ANALYSIS

S. Conforto[1] and T. D'Alessio[2]

1. ABSTRACT

In this work, the problem of movement artifacts, affecting dynamic myoelectric recordings, is studied. Dynamic myoelectric recordings allow the study of muscle activation intervals which provide a fundamental information for both the movement analysis and the clinical assessment of motion disorders. The detection of muscle activation patterns can be detrimentally affected by the presence of artifacts due to the movement of the surface electrodes on the skin. In order to overcome this difficulty, three artifacts filtering procedures have been tested: a high-pass filtering procedure, a moving average procedure and a moving median procedure. For the monitoring of the correct detection of muscle activation intervals, before and after the artifacts removal, a new original double-threshold statistical detector, recently developed by one of the authors, is utilised. This detector shows good performance and it is not affected by subjective settings, such as the classical activation intervals detector based on a not automatical thresholding of the integrated myolectric signal. Preliminary results illustrate the effectivness of the artifacts removal. The moving median filtering procedure seems to perform better than the other two proposed methods.

2. INTRODUCTION

The information obtained from the surface electromyographic signal (SEMG) recorded

Keywords: Myoelectric signals, Artifacts removal, Filtering.

---

[1]   PhD Student, INFOCOM Departement, University of Rome 'La Sapienza', Via Eudossiana 18, 00187, Rome, Italy; Mechanics ans Industrial Engineering Departement, University of Rome III, Via Segre 60, 00146, Rome, Italy.
[2]   Professor, Mechanics and Industrial Engineering Departement, University of Rome III, Via Segre 60, 00146, Rome, Italy.

in dynamic conditions represent a valuable tool for both the clinical assessment of motion disorders and the movement analysis. Dynamic recordings allow the study of muscle activation intervals and the analysis of the relationship between signal amplitude and muscular force. The last one is a well studied problem for isometric contractions but much is unknown in dynamic conditions. The information related to the muscle activation intervals is considered by several authors [1-6], as a fundamental support into orthopedics, neurology and several other clinical fields. Moreover, the muscular activity timing represents a valid integration of the biomechanic parameters in the movement analysis.

The methods proposed in the literature for the detection of activation patterns [7-8] are based on a thresholding of the integrated SEMG (the signal is rectified and then low-pass filtered in order to extract its envelope). The drawbacks of this technique consist of the subjective choices of both the time constant of the filter which, affecting the envelope shape (high values guarantee the envelope reproducibility, low values allow the detection of sudden amplitude variations), needs a trade-off value and of the thresholding procedure, based on an operator choice, which affects the estimate by subjective errors. These features make the standard methods not reliable for the detection of activation patterns [7-9], because they are affected by noise, by subjective settings of the operators and by the presence of artifacts.

A recent work, developed by one of the authors, presents an original double-threshold statistical detector of muscle activation, which shows good performance and it is not affected by subjective settings [10]. The method is operator free because the thresholds are automatically calculated on the basis of the signal to noise ratio (SNR) of the signals. An iterative procedure step-by-step allows the improvement of the activation patterns estimate, by means of an updating of the SNR estimate of the signal. The convergence is obtained when the estimated SNR reaches a plateau value. The effects of spurious transitions, due to low SNR values, are controlled by the algorithm rejecting transitions shorter than 30 ms, which have no sense from a biomechanic point of view. The obtained results present a 5% percentage of activation patterns misdetection; the estimate is characterized by a bias of 10 ms and a standard deviation of 15 ms, for SNR values higher than 8 dB.

Anyhow, the method performance can degrade if the recordings are affected by movement artifacts. These are basically due to the movement of the electrodes over the skin, stressed during dynamic measurements, which gives rise to a high power drift superimposed on the useful signal. This drift can detrimentally affect the timing detection, by widening the muscular burst. For this reason a filtering procedure able to remove these artifacts from the signal trace is highly recommended. The usefulness of such a filtering procedure is amplified in the case of multielectrode measurements, where the artifacts presence even on only one channel can compromise an entire recording session. This implies a great loss of time, especially for long measurements as those realised in order to monitor the muscular fatigue.

3. MATERIALS AND METHODS

In this work, different filtering procedure, aiming at the artifacts removal, have been tested.

A standard time-frequency analysis of signals affected by movement artifacts reveals that these ones are located in the low part of the frequency axis well separated by the signal frequency components (see Fig.1. for an example). A high-pass filtering procedure is the most immediate solution to the problem, but a particular attention has to be paid to the phase distortion induced by the filter, which can completely distort the timing detection. In order to save the performance of the detector, the temporal information of the signals has to be preserved by the rejection of the artifacts through the use of a suitable filtering.

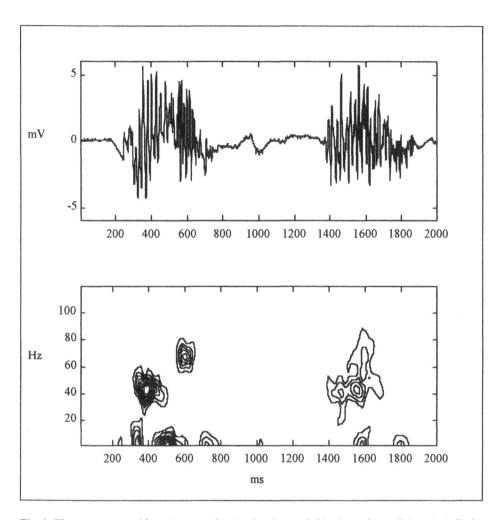

Fig. 1. The movement artifacts in a myoelectric signal recorded in dynamic conditions (up), lie in the low part of the frequency axis, as it is shown in the bottom figure where an isolevel contour plot representation of a Short Time Fourier analysis of the signal is given.

3.1. Processing

Aiming at this goal, three different solutions have been tested:

1. a moving average procedure realised by moving a window, composed of 11 samples, on the signal trace, substituting the average value calculated on the window to the central sample and then subtracting the so obtained trace from the original one;

2. a moving median procedure which differs from the previous one in the value assigned to the central sample of the window which is the median instead of the average value;
3. a high-pass anticausal filter of the eighth order and with a cutting frequency of 50 Hz.

The efficiency and the performance of the proposed methods have been evaluated on the basis of the correct detection of muscle activation intervals.

3.2. Experimental protocol

The study procedure is the following:

- simulation of a non stationary EMG signal, realised by modulating a gaussian white noise, in order to exactly know the activation timing;
- extraction of an artifact patterns from an experimental signal, by using each of the proposed filters;
- superimposition of the real artifacts on the simulated signal;
- filtering using each one of the proposed algorithms;
- detection of muscle activation intervals by means of the double-threshold algorithm.

4. EXPERIMENTAL RESULTS

Preliminary results allow to consider the median filter as the optimal procedure in the artifacts removal (see Fig. 2). In fact, this one gives a bias in the timing estimates lower than 5 ms in both the onset and the falling edge of muscle activity. The moving average filter shows similar results in the bias related to the falling edge, but the behavior gets worse in the onset detection with bias of 15 ms (as an example of detection see Fig. 3). The high-pass filter is not able to get over the sudden artifacts lying between two muscular activity patterns. For this reason the high-pass filter is not suitable in order to remove movement artifacts before detecting the muscular activation patterns. Further statistical tests have been realised over a testing set composed of 10 artifacts patterns treated following the protocol described in the paragraph 3.2. The values of both bias and standard deviations calculated on the testing set are shown in Tab. 1.
From Tab. 1 it is evident that the moving median filter gives more reduced bias and variance in the estimation of the activation patterns.

	Bias [ms]		Standard Deviation [ms]	
	Onset	Offset	Onset	Offset
Moving Median	-2.67	3.66	2.88	0.58
Moving Average	-15.33	2.54	6.11	4.35

Tab. 1. Values of bias and standard deviation in the onset and offset of the activation patterns using the moving median and the moving average procedures.

The procedure has also been used in the analysis of experimental signals with good results.

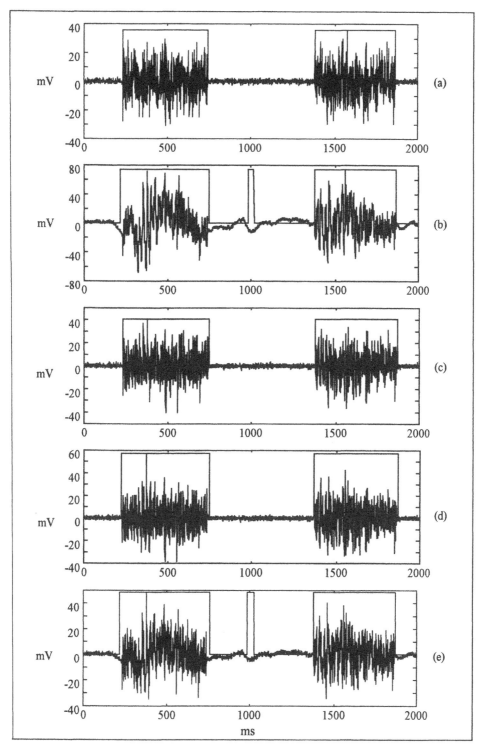

Fig.2. Performance of the different filters on a simulated signal (a), corrupted by a real artifact (b), and then filtered by: (c) moving median procedure, (d) moving average procedure, (e) high-pass anticausal filter. Solid lines represent the output of the activation patterns detector.

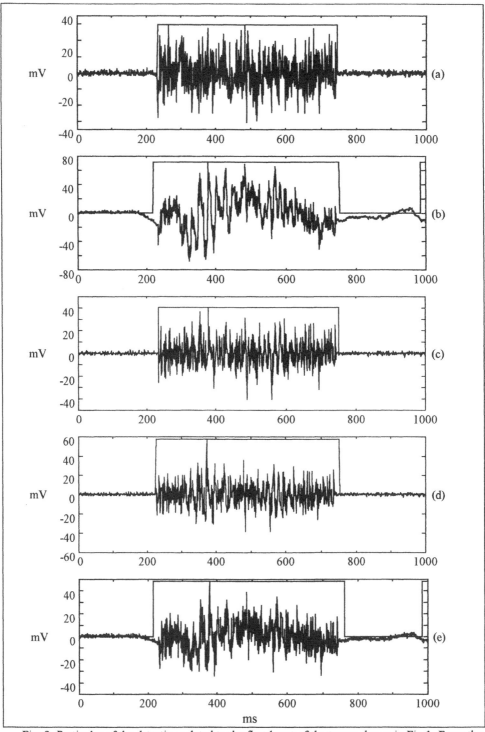

Fig. 3. Particular of the detection related to the first burst of the traces shown in Fig.1. From the figure, it is evident that: the filtering through the moving average procedure (d) gives rise to a greater bias on the detection of activation patterns, the high-pass filter does not avoid misdetection.

5. CONCLUSIONS

In this paper an analysis of different filtering procedures for artifacts rejection have been presented. The procedure which makes use of a moving median filter rejects more efficiently the artifacts while preserving tthe information on the activation patterns.
In conclusion, the proposed procedures represent an useful tool in the artifacts removal in the activation pattern detection.

6. AKNOWLEDGMENT

This work has been partially supported by MURST, Roma.

7. REFERENCES

1. Perry, J., Hoffer, M.M., Giovan, P., Antonelli, D., Greenberg, R., Gait analysis of the triceps surae in cerebral palsy, J. Bone Joint Surg., April 1974, vol. 56-A, 511-520.
2. Perry, J, Hoffer, M.M., Antonelli, D., Plut, J., Lewis, G., Greenberg, R., Electromiography before and after surgery for hip deformity in children with cerebral palsy, J. Bone Joint Surg., 1976, vol. 58-A, 201-208.
3. Perry, J., Hoffer, M.M., Preoperative and Postoperative dynamic electromyography as an aid in planning tendon transfers in children with cerebral palsy, J.Bone Joint Surg., 1977, vol.59-A, 531-537.
4. Steiner, M.E., Simon, S.R., Pisciotta, J.C., Early changes in gait and maximum knee torque following knee arthroplasty, Clinical Orthopedics and Related Research, January 1989, n.238, 174-182.
5. Demottaaz, J.D., Mazur, J.M., Thomas, W.H., Sledge, C.B., Simon, S.R., Clinical study of total ankle replacement with gait analyis, J.Bone Joint Surg., 1979, vol.61-A, 976-988.
6. Andriacchi, T.P., Galante, J.O., Fermier, R.W., The influence of total knee replacement design on walking and stair-climbing, J.Bone Joint Surg., 1982, Vol.64-A, 1328-1335.
7. Bekey, G.A., Chang, C.W., Perry, J., Hoffer, M.M., Pattern recognition of multiple EMG signals applied to the description of the human gait, Proc. of IEEE, 1985, vol. 65, 674-681.
8. Schiavi, R.G., Griffin, P., Representing and Clustering Electromyographic Gait Patterns with Multivariate Techniques, Med. & Biol. Eng. & Comput., 1981, vol. 19, 605-611.
9. Winter, D.A., Pathologic Gait Diagnosis with Computer-Averaged Electromyographic Profiles, Arch. Phys. Med. Rehabil, July 1984, vol.65, 393-398.
10. Bonato, P., Catani F., D'Alessio, T., Knaflitz, M., Evaluation of activation patterns in gait analysis from surface myoelectric signals, 8-th East Coast Clinical Gait Conference, Rochester, MN, 1993.

WAVELET NETWORKS: AN EFFICIENT TOOL FOR HANDLING GAIT KINEMATIC DATA

Espiau F.X.[1], Cordier E.[2] and Espiau B.[3]

1. ABSTRACT

When having to deal with kinematic data, most of the biomechanists use the well known Butterworth filter. But this is also the most badly used filtering algorithm, regarding the determination of the cut-off frequency. This paper explains in which way using the wavelet networks (called waveNets) is a better method to smooth kinematic data. Unlike recursive digital filters, such as Butterworth filter, wich start from a vector of values, and give another filtered one, the wavenets give an analytic expression of the data. This allows easier further computations, such as velocity, acceleration and closed-curve area.

2. INTRODUCTION

The data produced by the kinematic recording of movement are generally very noisy. In order to make reliable and accurate further computations, one of the first treatments to apply to the data is to suppress the noise. This is often done by using the popular Butterworth recursive digital filter, due to it's apparent handling facility. But the determination of the cut-off frequency is not a trivial task. Furthermore, this frequency does not adapt to signal variation. This conduct to senseless oversmoothed data with loss of local characteristics.

The purpose of this paper is to show how this can be avoided, by using the new wavelet networks (called waveNets) method. The wavenets allow to automatically suppress the noise of the data, while keeping the local signal characteristics. Among the others main features of the wavenets can be found that: they provide an analytical expression of the data, allowing easy realistic velocity, acceleration and area computations; they can be used for data compression; they are a multi-scaling approximating method.

KEYWORDS: wavelet networks, smoothing, data reconstruction, multi-scaling

1 M.Sc. student, French National Institute for Research in Computer Science and Control (INRIA)
2 Ph.D. student, INRIA, 655 avenue de l'Europe, ZIRST, 38330 Montbonnot, FRANCE
3 Research Director, INRIA - FRANCE - espiau@inrialpes.fr

In it's first part, this paper will show the path from Butterworth filtering to wavelet networks multi-scaling approach. Then, the different methods seen before will be compared when applied to an exemple of real data.

3. SMOOTHING KINEMATIC DATA

3.1 With Butterworth filter

The kinematic experiments give large amount of noisy data. The first treatment to apply to them consists to suppress the noise. Numerous methods are available for this task. The most used is the filtering one, with the Butterworth filter. But the data are frequently so noisy, that it become impossible to choose the best cut-off frequency (and corresponding coefficients) for the filter: a too high frequency do not suppress all the noise, while a too low frequency suppress some parts of the signal. Automatic methods exists to determine the optimal frequency, but a unique one is found for all the data, too low for some parts of the signal, and too high for some other parts of the signal. The wavenets solve the often encountered problem of distinguishing the noise from the signal.

The second disadvantage of filtering methods is that, starting with a vector of noisy discrete time dependent data, a filter gives another vector of discrete time dependent data (fig. 1).

Fig. 1: Butterworth filter gives a vector

This leads to use discrete differentiation techniques to compute velocities, accelerations and closed curves areas speeds, giving unreliable results.

The new approach proposed is based on a smoothing method called the "mean-squares approximation", above which are applied the wavelet networks method, developped by Zhang and Benveniste [BZ92].

3.2 From Butterworth Filter to Wavelets model

The purpose of the mean-squares approximation method is to construct a theoretical model of the signal, starting from the time-dependent discrete noisy data. A model is defined by a given set of functions, called the base of functions. This method consists in choosing a model which depends on parameters. Then these parameters are to be computed to best fit the data. Finally, the signal is a composition of the basis-functions and the corresponding parameters.

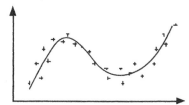

Fig. 2 : least square approximation

This is done by minimizing the following expression on the parameters:

$$\sum_{k=1}^{N} (y_k - f(t_k))^2$$

(1)

where : (y_k, t_k) are the N noisy data

$f(t)$ is the chosen model with parameters to compute

It's clear that if the model is linear onto his parameters, the minimization will be reduced to the computation of a linear system, giving the exact unique solution. This is the case with models like polynomial functions, B-splines and Fourrier-series (frequency interpretation of the data).

The drawback of finding the best, unique solution is that, the basis-functions of linear models can not be moved. For exemple, the Fourrier-series model base of functions is: *cos(t), sin(t), cos(2t), sin(2t), ... cos(nt), sin(nt)*. This base is fixed. By computing the parameters, are computed the weights of the basis-functions. If the data are very noisy, a lot of basis-functions, consequently a lot of parameters, are generally needed to best fit the data. But each basis-function used to fit a part of the data will add noise everywhere else on the signal. This is called a "global approximation" approach, because one can not act on a part of the signal without without modifying the rest of it. A fixed base of functions is called a "regular grid".

The linear models have the advantage of providing an analytic expression of the data, but generate noise in this expression, that was not present in the data in the beginning.

At the opposite of these previous models, there are the non-linear models, once more used with the mean-squares approximation.

This is an example of non-linear model:

$$f(t) = A_0 + \sum_{k=1}^{P} A_k \sin(\omega_k t + \varphi_k)$$

(2)

where the *(p+1)* parameters to compute are : A_0, A_k, ω_k and φ_k.

This model allows to adjust the basis-functions and to fit better than with the linear models (the basis is not fixed). Here are computed the best basis-functions.

Nevertheless, this kind of models requires a non-linear-system resolution method, like Levenberg-Marquart method [Lev44][Mar63], or conjuguate-gradient method. These algorithms are strong to determinate the parameters but some new problems arise. The first and the most important one, is the fact that, initial values must be given to the parameters. The solution will then depend from the initialization of the parameters. In fact, the best solution is obtained when the parameters are initialized with values given in it's neighbourhood. The second problem is the fact that, generally the system to solve has got many solutions. Like the resolution of the mean-squares approximation is to minimize an expression, in the case of a non linear model can be found many minima, each called a local minimum. Therefore, the global minimum, corresponding to the best solution, has to be found.

However, these models allow to fit the data with a small number of parameters.

The wavelets will give another aspect of these non-linear approach. It is better to be able to move and adjust the basis-functions. But the main problem is, how to choose these basis-functions and how to initialize the parameters?

The foundation of the wavelet theory may be found in [Mey90] et [Dau92]. Here is given the main characteristics of a wavelet φ :

> φ is infinitely derivable
> φ is defined on a compact set

If a wavelet, called mother-wavelet (Fig. 3a), is dilated and/or translated (Fig. 3b), a wavelet is obtained again, and all of them form an orthogonal family in $L^2(\Re)$.

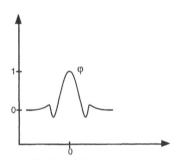
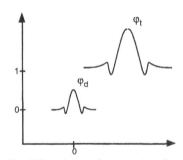

Fig. 3a : Mother wavelet   Fig. 3b : Dilated wavelet, translated wavelet

The fact that a wavelet can be dilated and translated everywhere, makes that a wavelet can be if needed set on each part of the signal, to best fit the data (Fig. 4a). And, due to it's compact support design, an added wavelet don't disturb the rest of the signal (Fig. 4b).

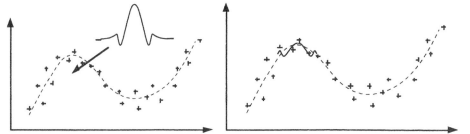

Fig. 4a : Local characteristic...   Fig. 4b : ...fitted with a dilated and translated wavelet

This approach is called "local approximation". The model can be easily be adjusted to the data: Often is found in the data a peak which seems not to be a noise; this peak, called a local characteristic, can be kept by adding a wavelet (Fig. 4a, 4b).

But, like the model presented in equation (2), finding the best base of functions is not trivial. The model which is a wavelet composition, is a linear one.

$$f(x) = \sum_{i=1}^{N} \omega_i \varphi_i (x)$$

(3)

In fact, the basis-functions are already known, and the only thing to compute is their weights. But in pratice, it is quite difficult to compute basis-functions from the original data, that is, exactly the good coefficients of dilatation and translation for each wavelet. Here again, we talk about "regular grid".

From filter to wavelet composition, some usefull properties for smoothing and signal reconstruction have been pointed out. The wavelet composition gives the guarantee of a good smoothing, but requires to compute basis-functions before, which is very difficult. And non-linear models allow to move these basis functions fitting at best the signal. The wavelet networks, also called wavenets, assemble these two advantages and offer some others, as it will be see in the next paragraph.

3.3 The wavelet networks

First of all, it must be noted that Zhang and Benveniste [BZ92] insure that global approximation property is conserved. Wavelet networks further properties are the following: the basis-functions are wavelets and are adjustable, which means the model is not linear. But it still stay the parameters initialization problem. Zhang has developed an algorithm which initialize the parameters in function of the data. The advantages are: initial values have not to be chosen, and the original data can be given by paquets (Fig. 5). It means that it's not necessary to use fixed-rate sampled data. The algorithm uses an interpolation if there is a "hole" in the data.

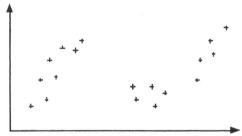

Fig. 5 : Irregular rate sampled data and holes

The theoretical form of a wavelet network is given by :

$$f(x) = \sum_{i=1}^{p} \omega_i \varphi(t_i, d_i, x) + \sum_{j=1}^{p} \omega_j \psi(t_j, d_j, x)$$

(4)

where: ψ is the scaling function

              φ is the mother wavelet (for exemple the gaussian function)

              ω_i the weights of the wavelets

              t_i the translation parameters

              d_i the dilatation parameters

Now is explained the idea of the construction of the network. At the begining, the algorithm will give a first wavelet which will try to go through the data. After, it refines by adding another wavelet, then another, etc. But if there is a characteristic to keep, we just have to add a wavelet, which will be well translated and well dilated to the right place (Fig 6a, 6b). And if a fine analysis of a part of the signal is needed, we can add again small wavelets and again smaller. This is called the multi-scaling resolution.

This model insure us that the model constructed is completely denoised, but also keeps all the little essential characteristics, even if there are some peaks which we could have seen like noise before.

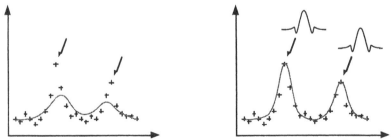

Fig. 6a, 6b : Wavenets allows local characteristics conservation

So, it clearly appears this network model is theoretically the most performing method to smooth data.

As Zhang's algorithm finds the best parameters and gives a very good reconstruction of the signal, another application can be found to wavelet networks: the data compression. Indeed, just the coefficients of the wavelets are needed to reconstruct a signal.

Moreover, the package [WN] developped by Zhang incudes a routine which gives the optimal number of wavelets (i.e. basis-functions). Finally, it can be noted that this model may be used for two dimensions signals.

4. RESULTS AND COMMENTS

We applied different models to very noisy kinematic data. Figure 7a represents raw ankle angle variation versus time.

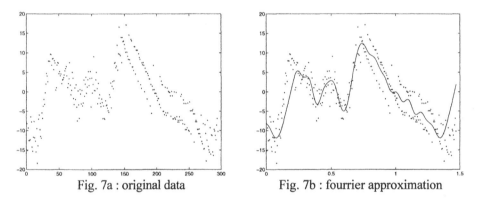

Fig. 7a : original data   Fig. 7b : fourrier approximation

The next figures 7b, 7c and 7d show the signal reconstruction by the mean of sinus-series, Butterworth filter and wavelet network.

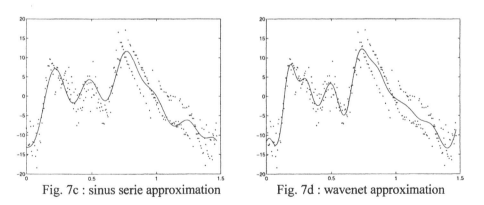

Fig. 7c : sinus serie approximation   Fig. 7d : wavenet approximation

As we can see, the Butterworth filter suppress a great part of the noise of the original data, but it also deletes some peaks which are local characteristics, and when there is a great amount of variations, it does not really fit well. And as we already point it out in the first part of this paper, we obtain a new vector of points.

The wavelet network gives a very good reconstruction of the signal, just taking off the noise and keeping on the real signal, even providing very noisy raw data. We can see that if there are great variations like peaks, the model returns them correctly, and when there are some missing points, it interpolates well. For this exemple, we just need fourteen wavelets for two hundred and twenty data. So we have a good rate of compression.

5. CONCLUSION

In conclusion, the wavelet networks offer the possibility to smooth very noisy data while keeping all the essential characteristics, this with a small number of parameters (good data compression rate). The properties of the wavelets allow a multi-scalling resolution, which means we can apply a wavelet locally (zoom effect). Moreover, this model gives an analytical expression of the signal, which allows the user to make theoretical computations, rather than using unreliable discrete differentiation techniques.

Furthermore, wavelet network model can be extended to multidimensionnal data as well.

6. REFERENCES

[BZ92]   Benveniste A., Zhang Q. : Wavelet network. IEEE transactions on neural networks 3(6), November 1992.

[Lev44]   Levenberg K. : A method for the solution of certain non-linear problems in least-squares. Quarterly of applied mathematics, 2(2) : 164-168, July 1944.

[Mar63]   Marquardt D.W. : An algorithm for the least-squares estimation of nonlinear parameters. SIAM Journal of applied mathematics, 11(2) : 431-441, June 1963.

[Mey90]   Meyer Y. : Ondelettes et opérateurs. Hermann, Paris, 1990.

[Dau92]   Daubechies I. : Ten lectures on wavalets. Society for industrial and applied mathematics, 1992.

[WN]   ftp.irisa.fr/local/wavenet/wnet2.1.tar.Z

COMPARISON OF LOWER-EXTREMITIES BIOIMPEDANCE MODELS DIAGNOSTIC EFFICIENCY

S. Tonković[1], I. Tonković[2], D. Voloder[1]

1. ABSTRACT

Compartmental syndrome, a very frequent and severe ischaemic circulation disturbance in lower extremities, is a condition in which increased tissue pressure compromises the circulation and function of the tissue within that space. Usually, it requires an immediate diagnostic finding and surgical treatment. One method of compartmental syndrome detection using multiple frequency bioimpedance measurement is described. Measurements were performed using bipolar technique with self-adhesive cutaneous electrodes. The results confirmed this possibility, especially in the patients with "one extremity syndrome" (only one leg with compartmental syndrome), showing high correlation between bioimpedance changes and the disease seriousness or compartmental pressure. Trying to achieve more quantitative diagnostic criteria, we essayed to fit measured results to some bioimpedance models using quantitative results (values) of model parameters for diagnostic purposes. One new *"bioimpedance model"* is presented and discussed.

2. INTRODUCTION

Ischaemia of the lower extremities may be the first manifestation of systemic disease or

Keywords : Ischaemia, Compartmental syndrome, Bioimpedance measurement and modelling,

[1]- Faculty of Electrical Engineering and Computing, 3 Unska, 10000 Zagreb, Croatia
[2]- Department of Vascular Surgery, School of Medicine, University Hospital Rebro, 12 Kišpatićeva, 10000 Zagreb, Croatia

a disease of a distant organ system, or it may represent the end stage of progressive atherosclerosis. The most common complaint of patients with lower extremities arterial occlusive disease is pain. The symptom of lower-extremity arterial insufficiency range from claudication to night pain, and finally to tissue necrosis and gangrene. Compartmental syndrome [1,2], a very frequent and severe circulation disturbance in lower extremities, is a condition in which increased tissue pressure compromises the circulation and function of the tissue within that space. Usually, it requires an immediate diagnostic finding and surgical treatment. Although there is no consensus regarding which procedures are most appropriate for the evaluation of lower extremity arterial insufficiency, some of them such as Doppler arterial survey, limb pressure measurements, stress testing, ultrasound angiography, pressure index, appear most appropriate. The standard clinical diagnostic procedures are either painful for patients (for example, pressure measurements), complex, time consuming, expensive or require highly skilled team. We explored the possibility of compartmental syndrome detection using multiple frequency bioimpedance measurements and bipolar technique [3] , measuring the impedance magnitude and phase angle in the frequency range from 1 kHz to 1 MHz. The electrical impedance (or "bioimpedance") of living tissues is determined by the shape, alignment, and distribution of cells, and the amount and distribution of interstitial and extracellular fluids [4,5]. The results proved this possibility, specially on the patients with "one extremity syndrome" (only one leg with compartmental syndrome) and show high correlation between bioimpedance changes and the disease seriousness or compartmental pressure, measured using Wick technique [6].

We attempted to fit measured results to some bioimpedance models using quantitative results (values) of model parameters for diagnostic purposes. First trials with widely used bioimpedance models, such as RRC and Fricke models, gave relatively deficient fitting results. One new *ameliorated bioimpedance model"* of lower extremities skeletal muscles is presented, discussing its possible diagnostic use in early non-invasive compartmental syndrome diagnostics [7].

3. MATERIALS AND METHODS

During these studies we measured the impedance magnitude and phase angle, or real and imaginary impedance part using HP LCR - Meter 4284A and two electrode technique with constant current of 0.1 mA (Fig.1). Measurements on the examined three lower-extremities groups (m. gastrocnemius - longitudinally and transversally, m. tibialis and m. fibialis) were performed by use of "one-use" self adhesive cutaneous electrodes. Position of electrodes was determined by palpation of the major muscle bulge, strictly controlling the distance of 12 cm between electrodes (Fig. 2).

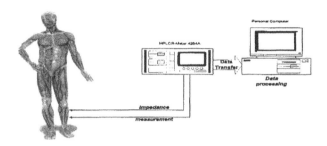

Fig 1.: Measuring set-up

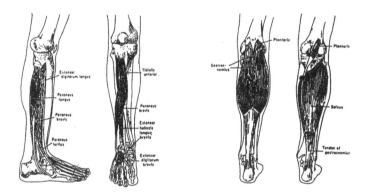

Fig 2.: Electrode's position for four channel bioimpedance measurement

Two groups of patients were examined, one suffering of "**severe**", "**medium**" or "**low**" compartmental syndrome and the other with patients suffering from acute ischaemia signs. All the patients had exaggerate " one extremity syndrome". In the first group, a comparison between results obtained by impedance measurement and direct tissue pressure measurement, using Wick technique, gave significant correlation, especially in the frequency range between 15 and 100 kHz (always greater then 0.9).

Patient: X.Y.

Leg: right

Muscle: m. gastrocnemius trans.

f (Hz)	|Z| (Ω)	φ (deg)
20	160000	--72,00
25	133000	--73,00
35	100000	-74,30
45	81200	-75,40
55	68500	-76,00
65	59200	-76,50
80	49400	-77,10
100	40600	-77,70
200	21900	-78,80
500	9590	-78,40
800	6290	-77,20
1000	5160	-76,30
1500	3610	-74,20
2000	2820	-72,20
3000	2010	-68,60
5000	1350	-62,60
8000	973	-55,80
10000	847	-52,20
15000	678	-45,50

20000	592	-40,80
50000	429	-27,70
80000	382	-22,10
100000	365	-19,60
150000	341	-15,60
200000	329	-13,10
500000	303	-6,95
800000	296	-4,40
1000000	293	-3,18

Fig. 3.

Because of difficult comparison using the results as tabled values (typical example on Fig.3.; for one patient eight tables !), we tried to model measured bioimpedance by some of well-known and widely used schemes (Fig.4).

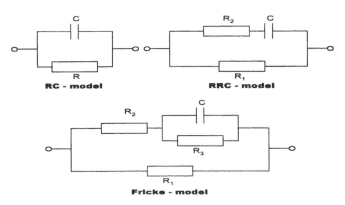

RC - model

RRC - model

Fricke - model

Fig 4.

As the results of fitting were significantly inferior than expected, we tried to apply a new model with added few components, representing the existence of nucleus and nucleolus, as well as the nuclear pore effect in the striated muscle cells, i.e. R_4, C_2 and R_{C2}, respectively (Fig.5).

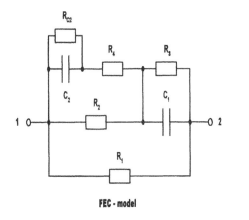

R_1	extracellular liquid conductivity
R_2	intracellular liquid conductivity
R_4	conductivity of nuclei
R_3	conductivity of cell membranes
R_{C2}	conductivity of nuclei membranes
C_1	capacity of cell membranes
C_2	capacity of nuclei membranes

Fig.5.

One example of measured and fitted curves (calculated by the MATLAB software routines), indicating values for real part of impedance magnitude is shown on Fig.6.

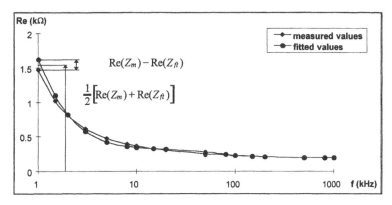

Fig.6.

The fitting successfulness was estimated by use of consideration of calculated curves and, taking into account the values of Aberrance Coefficient (**AC**), calculated using some sort of normalised expression (due to the differences in the Y-axis scale values when calculating real and imaginary parts of bioimpedances) :

$$AC = \frac{1}{N}\sum_{i=1}^{N}\left[\frac{Re(Z_m(f_i)) - Re(Z_{ft}(f_i))}{\left(\frac{Re(Z_m(f_i)) + Re(Z_{ft}(f_i))}{2}\right)}\right]^2 + \left[\frac{Im(Z_m(f_i)) - Im(Z_{ft}(f_i))}{\left(\frac{Im(Z_m(f_i)) + Im(Z_{ft}(f_i))}{2}\right)}\right]^2 \qquad (1)$$

where Z_m denotes measured impedance value, Z_{ft} fitted impedance value and, N number of measuring frequencies.

4. RESULTS

Measurements, performed on 24 patients suffering of severe, medium or low compartmental syndrome and 32 patients suffering of acute ischaemia states, showed that, if the patients were suffering of " one leg compartmental syndrome", an early and reliable diagnostics is possible.

It was shown that diagnostically most interesting frequency range, as well as due to electrode- skin interface impedance artefacts supression is between 15 kHz and 100 kHz.

Typical results of fitting for one patient, using Fricke and FEC (Faculty of Electrical Engineering and Computing) models are shown in Fig. 7.a,b, and c. It can be easily conceived that the results of fitting applying FEC model are significantly better and that the values of *"added"* components could be used for diagnostic purposes.

Fricke - model

Fitted values

Gastrocnemius

	Right leg	Left leg
R_1 (Ω)	2009.74	1409.10
R_2 (Ω)	477.22	244.92
R_3 (Ω)	221.78	195.17
C (nF)	11.79	11.59
AC	0.0454	0.076

Peroneus

	Right leg	Left leg
R_1 (Ω)	494.17	919.33
R_2 (Ω)	191.55	213.10
R_3 (Ω)	3.76	98.91
C (nF)	8.88	12.14
AC	0.087	0.0867

a)

SAME PATIENT – FEC model

FEC - model

Fitted values

Gastrocnemius

	Right leg	Left leg
R_1 (Ω)	6003.13	6047.42
R_2 (Ω)	886.02	498.02
R_4 (Ω)	772.46	384.52
R_3 (Ω)	14.31	4.89
R_{C2} (Ω)	24.39	395.86
C_1 (nF)	33.79	36.58
C_2 (nF)	4.415	2.53
AC	0.0084	0.0115

Peroneus

	Right leg	Left leg
R_1 (Ω)	6013.84	6169.33
R_2 (Ω)	278.82	396.99
R_4 (Ω)	279.90.	323.02
R_3 (Ω)	2.146	4.293
R_{C2} (Ω)	371.43	39.95
C_1 (nF)	76.67	48.33
C_2 (nF)	3.30	2.66
AC	0.0132	0.0121

b)

ABERRANCE COEFFICIENT

MODELS	GASTROCNEMIUS		PERONEUS	
	RIGHT LEG	LEFT LEG	RIGHT LEG	LEFT LEG
FEC	0.0084	0.0115	0.0132	0.0121
FRICKE	0.0454	0.0760	0.0870	0.0870

c)

Fig.7. a,b and c.

5. CONCLUSION

The performed measurements, as well as the results of fitting, showed and proved presumption and idea that the proposed method of measurement allows early non-invasive diagnostics of compartmental syndrome and that the proposed model allows reasonably better fitting to the measured results than other, widely used models, such as RRC or Fricke models. The results of calculated parameters of model can be used as quantitative diagnostic parameters for predicting or compartmental syndrome seriousness or follow-up successfulness of surgical treatment. Moreover, diagnostics of acute ischaemia states is possible, non requiring hospital surrounding or highly skilled staff.

6. REFERENCES

1. Matsen,F.A., Compartmental syndromes, 1980, *Grune&Stratton,* N.Y.

2. Grace,A.M., Ischaemia - reperfusion injury, British J. of Surgery, 1994, 81, 637 - 647

3. Woo,E.J.,Hua, P.,Webster,J.G., Thompkins,W.J., Pallas-Areny,R., Skin impedance measurements using simple and compound electrodes, Med.&Biol. Eng.&Comp., 1992, 30, 97 - 102

4. Lukaski,H.C., Methods for the assessment of human body composition: traditional and new, *Am.J.Clin.Nutr.,*1987, Vol. 46, 537-546

5. ...., Bioelectric impedance and body composition, *Dietosystem Scientific Monographs,* 1991, Milan

6. Tonković,I., Ratkaj,D., Tonković,S., Petrunić,M., A method for tissue pressure evaluation of lower-extremity muscles, *Period.Biolog.,*1993, Vol. 95., 1, 97 - 100

7. Tonković,S., Tonković, I., Petrunić,M., Modelling tissue pressure changes in lower extremities, Med.&Biol. Eng.&Comp., 1996, Vol.34., Suppl.1., Part 2, 171 - 172

CALIBRATION OF AN ULTRASOUND SCANNER BASED ON THE URTURIP TECHNIQUE USING A LEAST SQUARE METHOD

B. Migeon [1], P. Deforge [2], P. Marché [3]

1. ABSTRACT

This paper presents a method to solve the calibration problem for the URTURIP Technique. It is based on a least squares fitting and is compared to the reference method in order to discuss the results in terms of accuracy and time-consuming.

2. INTRODUCTION

As a part of our project concerned with the development of an ultrasound scanner dedicated to limb study, the URTURIP Technique (Ultrasound Reflection-mode Tomography Using Radial Image Processing) has recently been developed [1]. It consists in using classical B-scan images instead of projections [2-5] and gives qualitative images instead of quantitative images. The final goal of this project is the 2D and 3D reconstruction of anatomical structures at limb level by using echographic image processing. The developed process consists of several successive steps like : multiple reflection removing [6], 2D reconstruction [1], segmentation [7], contour association, contour interpolation [8], 3D reconstruction and visualization [9]. It has been validated by in vitro experiments on anatomical pieces of limbs of new-borns using a simple acquisition system prototype [9]. To validate the URTURIP Technique by in vivo experiments with the help of a new acquisition system, the subpixel coordinates of the rotation center must be accurately determined. The calibration consists in solving this problem of rotation center determination and is the purpose of this paper. A reference

Keywords : Calibration, Echography, 2D Reconstruction, URTURIP Technique

[1] Ph.D., Laboratoire Vision Robotique, 63 Av. de Lattre de Tassigny, 18020 Bourges, France

[2] Ph.D. Student, Laboratoire Vision Robotique, 63 Av. de Lattre de Tassigny, 18020 Bourges, France

[3] Ph.D., Head of Dept, Head of the institute of technology of Bourges, Laboratoire Vision Robotique, 63 Av. de Lattre de Tassigny, 18020 Bourges, France

calibration method has recently been developed [10] and gives the coordinates of the rotation centre with a subpixel accuracy. But, this reference method is very time-consuming and cannot be used systematically for each routine exam. This paper presents a method based on a least squares fitting and is compared to the reference method in order to discuss the results in terms of accuracy and time-consuming.

3. ACQUISITION AND 2D RECONSTRUCTION

3.1. The acquisition system

The acquisition system is composed of two parts : an electronical one and a mechanical one.

The electronical part includes the echograph device (AU 3 Partner from Esaote Biomedica) with a 3-5 MHz electronical sectorial scanning probe, a graphic card which allows to digitalize the video signal of the echograph device, and a micro-computer (166 MHz Pentium) on which the images are stored and then processed.

The mechanical part is composed of a water tank in which the probe can turn around the object to be studied, and can move with a vertical translation device in order to access various slices.

3.2. The URTURIP Technique

The Ultrasound Reflection-mode Tomography Using Radial Image Processing (URTURIP Technique) consists in using B-scan images [1] instead of projections, so that the obtained images are qualitative ones instead of quantitative ones. However, the quality of reconstructed images using very few radial directions of investigation is sufficient for our application, and it is possible to distinguish clearly various anatomical parts, and correctly extract some relevant structures, using image processing, such as the skin, the skeleton and even the muscular masses.

Thus, considering L as the image to be reconstructed, each pixel of coordinates (x,y) has an L(x,y) luminance in grey levels. Then, the goal of the 2D reconstruction methods using radial B-Scan image processing is to compute the luminance $L(x,y)$ of each pixel, from the N luminances $Li(x,y)$, i=1..N, of the N radial B-Scan images readjusted beforehand. The readjustment step consists in rotating each image of the angle relative to its corresponding radial direction of investigation so that the pixel (x,y) of each readjusted image corresponds to the same point of the investigated cross-section.

4. CALIBRATION PROBLEM

4.1. Introduction

Before applying a method of the URTURIP Technique, the radial images must be adjusted according to the rotation center of the probe. The object of the calibration stage is to determine the rotation center, the coordinates of which are identical on each radial image.

4.2. The reference method

A reference method based on an energy minimization technique has been developed [10]. It allows obtaining the subpixel coordinates of the rotation center with an accuracy as much as one could wish for by using several radial images of a very hyperechogenic solid object.

The basic idea is that the real center is such that the reconstruction of a cross section of this object is perfect i.e. all radial information of the external contour overlaps perfectly with the others. Then, if an energy is able to represent the non-overlapping, the real rotation center minimizes this energy.

Let $E(x_c, y_c)$ be the energy of a reconstructed image after an adjustment of the radial images by considering the rotation center (x_c, y_c). $E(x_c, y_c)$ is defined by the number of the occupied pixels of the image reconstructed by the *maxima* method [1]. In other words :

$$E(x_c, y_c) = \sum_{(x,y) \in L^{(x_c, y_c)}} \delta^{(x_c, y_c)}(x, y) \text{ with } \delta^{(x_c, y_c)} = \begin{cases} 0 & \text{if } L^{(x_c, y_c)}(x, y) = 0 \\ 1 & \text{if } L^{(x_c, y_c)}(x, y) \neq 0 \end{cases} \quad (1)$$

$$\text{and} \qquad L^{(x_c, y_c)}(x, y) = \max_i L_i^{(x_c, y_c)}(x, y)$$

where $L^{(x_c, y_c)}$ is the reconstructed image after an adjustment considering the (x_c, y_c) rotation center, i is the index of the radial direction of investigation, $L_i^{(x_c, y_c)}$ is the ith radial image adjusted according to the rotation center (x_c, y_c).

To accede to the desired accuracy, each pixel of the images is subdivided into an adequate number of subpixels. Then, by minimizing this energy with a solid reference object, the subpixel coordinates of the rotation center are determined.

This reference method is very time-consuming and it is difficult to apply it in a routine exam if one often wants to change the probe (for each change, the calibration problem has to be solved). It is the reason why one has to develop new methods, which are faster and comparable with the reference method in terms of accuracy.

5. THE PROPOSED METHOD BASED ON LEAST SQUARES TECHNIQUE

5.1. Principle

Let us consider a point P belonging to the exploration plane, and Pi, i=1..N, corresponding to P on the N radial images. Superimposing the N radial images (which comes to express the different rotation of P in the same referential), the N points Pi belong to a circle whose rotation center is unknown and demanded.

In practice, a vertical taut nylon thread of 0.3 mm diameter is used and N radial images taken according to N angular equidistant radial directions give the N points Pi. A segmentation step is required to determine the points Pi which appear as a noisy spot on the images [11].

The rotation center has to be determined from these extracted points Pi. A simple method consists in computing the gravity center of these points [11] because they are angular equidistant. It is very fast and gives an estimation of the rotation center of the probe. However, the accuracy is insufficient due to the uncertainty on the Pi position after segmentation.

The method presented here consists in determining the center of the circle which is best fitted by the points Pi to a least squares sense.

5.2. Theoretical description

Let us consider the gravity center O of the points Pi as a first approximation of the center of a circle with a radius R defined to be the average of the radius OPi.

The origin being chosen on the point O (center of the circle with a radius R), the points Pi can be expressed in polar coordinates, and the center $C(u,v)$ of the circle with the radius r has to be determined (Fig.1).

Let ξ_i and e_i be respectively the distances between Pi and the two circles with radii R and r (Fig. 2).

We have :
$$OP_i = OB + BA + AP_i = u.\cos\theta_i + v.\sin\theta_i + AP_i$$

and
$$OP_i = R + \xi_i = \rho_i = r + e_i$$

O being close to C, the following hypothesis can be made :
$$CP_i \approx AP_i \tag{2}$$

We can then write :
$$R + \xi_i = u.\cos\theta_i + v.\sin\theta_i + r + e_i$$

and :
$$e_i = \xi_i - u.\cos\theta_i - v.\sin\theta_i - \Delta R \qquad \text{with} \qquad \Delta R = r - R$$

Let us consider the function W defined by :
$$W = \sum_{i=1}^{n} e_i^2 = \sum_{i=1}^{n} \left(\xi_i - u.\cos\theta_i - v.\sin\theta_i - \Delta R\right)^2 \tag{4}$$

One can then minimize W to determine u, v, and ΔR.

We have :
$$\frac{\partial W}{\partial u} = -2\sum_{i=1}^{n}\left[\left(\xi_i - u.\cos\theta_i - v.\sin\theta_i - \Delta R\right)\cos\theta_i\right]$$

$$\frac{\partial W}{\partial v} = -2\sum_{i=1}^{n}\left[\left(\xi_i - u.\cos\theta_i - v.\sin\theta_i - \Delta R\right)\sin\theta_i\right] \tag{5}$$

$$\frac{\partial W}{\partial \Delta R} = -2\sum_{i=1}^{n}\left[\xi_i - u.\cos\theta_i - v.\sin\theta_i - \Delta R\right]$$

and
$$\begin{cases} \dfrac{\partial W}{\partial u} &= 0 \\[2mm] \dfrac{\partial W}{\partial v} &= 0 \\[2mm] \dfrac{\partial W}{\partial \Delta R} &= 0 \end{cases}$$
gives the linear system to be solved which can be written as

follows :

$$
\begin{pmatrix}
\sum_{i=1}^{n}\cos\theta_i & \sum_{i=1}^{n}\sin\theta_i & n \\
\sum_{i=1}^{n}\cos\theta_i\sin\theta_i & \sum_{i=1}^{n}\sin^2\theta_i & \sum_{i=1}^{n}\sin\theta_i \\
\sum_{i=1}^{n}\cos^2\theta_i & \sum_{i=1}^{n}\sin\theta_i\cos\theta_i & \sum_{i=1}^{n}\cos\theta_i
\end{pmatrix}
\begin{pmatrix} u \\ v \\ \Delta R \end{pmatrix}
=
\begin{pmatrix}
\sum_{i=1}^{n}\xi_i \\
\sum_{i=1}^{n}\xi_i\sin\theta_i \\
\sum_{i=1}^{n}\xi_i\cos\theta_i
\end{pmatrix}
\tag{6}
$$

In our case, the points Pi are angulary equidistant so that :

$$
\sum_{i=1}^{n}\cos\theta_i = \sum_{i=1}^{n}\sin\theta_i = 0 \qquad \text{and} \qquad \sum_{i=1}^{n}\sin\theta_i\cos\theta_i = 0
$$

and the system to be solved is simplified to a diagonal one, which allows computing directly the three unknowns u, v and ΔR by :

$$
u = \frac{\sum_{i=1}^{n}\xi_i\cos\theta_i}{\sum_{i=1}^{n}\cos^2\theta_i}, \qquad
v = \frac{\sum_{i=1}^{n}\xi_i\sin\theta_i}{\sum_{i=1}^{n}\sin^2\theta_i}, \qquad
\Delta R = \frac{\sum_{i=1}^{n}\xi_i}{n}
\tag{7}
$$

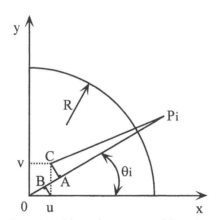

Figure 1 : the points Pi can be expressed in polar coordinates

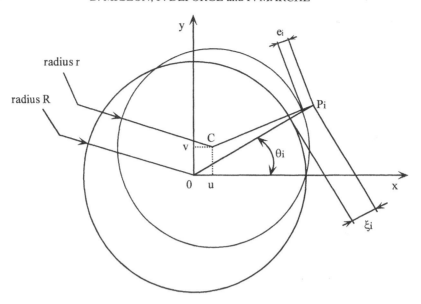

Figure 2 : ζ_i and e_i are respectively the distance between Pi and the two circles with radii R and r

6. RESULTS

The method has been compared to the reference method in terms of accuracy and the results are presented in Table I. The results are different but nevertheless interesting.

It has to be remarked that a great number of points Pi are required to have a reasonable accuracy. The method seems to be quite sensitive to the uncertainty of the extracted points Pi due to the noise on the images. In fact, the first estimation of the rotation center, given by the computation of the gravity center (of the segmented points Pi) is noisy, and the hypothesis (2) (verified for a small gap inferior to 0.1R) may be incorrect in certain cases because of the noise present on the images. A choice of a greater R seems to be a solution to this problem but it generates a greater luminance disparity between the spots corresponding to the points Pi on the images ; thus, it does not solve the problem and it is difficult to have a good compromise.

Moreover, compared to the reference method which requires about 30 minutes for a significant accuracy of 1 mm, the time-consuming is now reasonable, less than one second.

Number of points Pi	4	8	16	32	64
Distance between the result and the center given by the reference method	+/- 1.6mm	+/- 0.7 mm	+/- 0.5mm	+/- 0.4 mm	+/- 0.4 mm

Table 1: Comparison with the reference method for a required accuracy of 1mm.

7. CONCLUSION AND PERSPECTIVES

A calibration method of an ultrasound scanner using the URTURIP Technique has been developed. It is based on a least squares technique and gives interesting results with a reasonable time-consuming.

However, the method is sensitive to the noise present on the echographic images and it requires in return very numerous radial acquisitions for a compensation. Thus, a perspective is to develop a new method which is less sensitive to noise.

REFERENCES

[1] Migeon, B., and Marché P., Ultrasound tomography by radial image processing, Innov. Tech. Biol. Med., vol. 13, n°3, pp. 292-304, 1992.

[2] Sehgal, C. M., et al., Ultrasound transmission and reflection computerized tomography for imaging bones and adjoining soft tissues, IEEE Ultrasonic Symp. Chicago, IL, vol. 2, pp. 849-852, 1988.

[3] Hiller, H., Hermert H., System analysis of ultrasound reflection mode computerized tomography, IEEE Trans. Sonics Ultrason., vol SU 31, pp. 240-250, 1984.

[4] Friedrich, M., , et al., Computerized ultrasound echo tomography of the breast, Europ. J. Radiol., vol 2, pp. 78-87, 1982

[5] Ylitalo, J., et al., Ultrasonic reflection mode computed tomography through a skullbone, IEEE Trans. Biomedical Engineering, Vol. 37, N° 11, Nov. 1990.

[6] Migeon, B., Vieyres P., Marché P., A simple solution for removing echo bars for URTURIP Technique, Acoustical Imaging, Proc. of 22nd International Symposium on Acoustical Imaging, Firenze, Italie, 4-6 Sept., Vol. 22, 1995.

[7] Migeon, B., Serfaty V., Gorkani M., Marché P., An Adaptive Smoothing Filter for URTURIP Images Applying the Maximum Entropy Principle, IEEE Engineering in Medicine and Biology, pp. 762-765, Nov/Dec 1995.

[8] Migeon, B., et al., Interpolation of star-shaped contours for the creation of lists of voxels : application to 3D visualisation of long bones, Int. J. of CADCAM and Computer Graphics, Vol.9, n°4, pp. 579-587, 1994.

[9] Migeon, B., Marché P., Echographic Image Processing for Reconstructing Long Bones, Proc. of 17th Annual International Conference IEEE-EMBS, Montreal, Canada, 20-23 Sept., 1995.

[10] Migeon, B., Deforge P., Marché P., Calibration for the URTURIP Technique using an energy minimization method, Proc. of the 23rd Int. Symp. On Acoustical Imaging, Boston, April, 1997.

[11] Deforge, P., Migeon B., Marché P., Calibration du système d'acquisition d'un scanner à ultrasons médical basé sur la technique URTURIP, Proc. of the 4th French Congress on Acoustics, Marseille, April 1997.

Printed and bound by CPI Group (UK) Ltd, Croydon, CR0 4YY

23/10/2024

01777679-0010